AF776911

REDOX BIOCATALYSIS

REDOX BIOCATALYSIS

FUNDAMENTALS AND APPLICATIONS

Daniela Gamenara
Gustavo A. Seoane
Patricia Saenz-Méndez
Pablo Domínguez de María

WILEY

A JOHN WILEY & SONS, INC., PUBLICATION

Published by John Wiley & Sons, Inc., Hoboken, New Jersey.
Published simultaneously in Canada.

Library of Congress Cataloging-in-Publication Data:

Redox biocatalysis : fundamentals and applications / Daniela Gamenara ... [et al.].
p. cm.
Includes bibliographical references and index.
ISBN 978-0-470-87420-2 (cloth)
1. Environmental chemistry. 2. Environmental chemistry–Industrial applications.
3. Oxidation-reduction reaction. 4. Enzymes. I. Gamenara, Daniela, 1964–
TD193.R43 2012
660.6′34–dc23

2012025734

ISBN: 9780470874202

10 9 8 7 6 5 4 3 2 1

CONTENTS

PREFACE

The use of enzymes for redox processes has gained an increasing interest in the last decades, becoming in many cases "the first choice" for scouting novel industrial synthetic routes. This has been realized by solving issues related to cofactor regeneration, often needed for these enzymes, together with the developments in molecular biology areas that have enabled the provision of enzymes in large and reproducible scale in a fermentative sustainable manner. The development of environmentally sound synthetic protocols is mandatory in this century and, in this regard, oxidoreductases are ideally suited to the task, providing efficient and green alternatives to conventional synthetic procedures. This is particularly remarkable in oxidative processes, where oxidases and oxygenases perform clean and selective oxidations by activation of molecular oxygen with no need of heavy metals or expensive chemocatalysts. On the reductive side, these enzymes find ample application in the industry and academia for the generation of enantioenriched compounds.

This book provides a comprehensive and updated overview on the use of redox enzymes and enzyme-mediated oxidative processes. Chapters 1 and 2 provide an introduction on biochemical features of redox enzymes, together with aspects related to cofactors, and cofactor regeneration methods. Chapters 3–5 describe in detail the biocatalytic applications of different redox enzymes, namely, dehydrogenases (Chapter 3), oxygenases (Chapter 4), and oxidases and peroxidases (Chapter 5). Enzyme-mediated oxidative processes based on biocatalytic promiscuity (e.g., of hydrolases) are covered in Chapter 6. Chapter 7 focuses on the necessary steps starting from the discovery of a certain enzyme with a catalytic activity to a robust industrial process (e.g., directed evolution, high-throughput-screening methods, and medium engineering). Last but not least, Chapter 8 provides an overview on industrial cases using oxidoreductases, already commercialized or close to, showing that academic research is ending up with successful cases at the industrial arena. Overall, we believe that our contribution may well serve as a complete and first approach to academic and industrial research groups in the field of redox biocatalysis. It is our hope that readers will find this book an attractive and useful tool.

Finally, we would like to acknowledge Ms. Anita Lekhwani, Senior Acquisitions Editor at John Wiley & Sons, as well as the whole editorial team for the trust, hard work, interest, and patience that they have put into this project.

DANIELA GAMENARA
GUSTAVO SEOANE
PATRICIA SAENZ-MÉNDEZ
PABLO DOMÍNGUEZ DE MARÍA

Montevideo, Uruguay, and Aachen, Germany,
December 2011

CHAPTER 1

Enzymes Involved in Redox Reactions: Natural Sources and Mechanistic Overview

1.1 MOTIVATION: GREEN CHEMISTRY AND BIOCATALYSIS

Current environmental concerns are pressuring Chemical and Pharmaceutical industries to develop novel synthetic approaches that may operate under more benign conditions. This trend has paramounted the appearance of the "Green Chemistry" as a core discipline, with an increasing importance both in academia and industry. In a nutshell, Green Chemistry—as well as Green Engineering—has been compiled under several principles, as stated by Anastas and Zimmerman, and Tang and coworkers [1,2]. From the Green Chemistry approach, these principles are gathered in the acronym "PRODUCTIVELY":

- Prevent wastes
- Renewable materials
- Omit derivatization steps
- Degradable chemical products
- Use safe synthetic methods
- Catalytic reagents
- Temperature, ambient pressure
- In-process monitoring
- Very few auxiliary substances
- E-factor, maximize feed in product
- Low toxicity of chemical products
- Yet, it's safe

Redox Biocatalysis: Fundamentals and Applications, First Edition.
Daniela Gamenara, Gustavo A. Seoane, Patricia Saenz-Méndez, and Pablo Domínguez de María.

On the other hand, from the Green Engineering perspective, those principles are grouped in the acronym "IMPROVEMENTS":

- Inherently nonhazardous and safe
- Minimize material diversity
- Prevention instead of treatment
- Renewable material and energy inputs
- Output-led design
- Very simple
- Efficient use of mass, energy, space, and time
- Meet the need
- Easy to separate by design
- Networks for exchange of local mass and energy
- Test the life cycle of the design
- Sustainability throughout product life cycle

In this respect, the use of enzymes and whole-cells as biocatalysts for synthetic purposes (White Biotechnology) is an increasingly important field that may fit, in many cases, with all or some of these Green Chemistry principles. In fact, enzymatic living processes are often conducted under extremely mild reaction conditions, for example, neutral pH, or no need of high pressures or temperatures, which may provide energy savings for the overall process. Albeit biocatalytic processes are not always in line with all Green Chemistry principles (e.g., wastewater generation after downstream processing), they often provide advantages when compared to other chemical approaches. These assets have triggered the development of biocatalysis, reaching today the status of established technology, and occupying a prominent role as "synthetic organic chemistry tool" [3–16].

1.2 SOURCES OF BIOCATALYSTS

In the early stages of biocatalysis, plant tissues, and animal organs were the most important sources of enzymes, representing in the 1960s about 70% of the biocatalysts used for synthetic purposes [17,18]. The trend rapidly changed, and 20 years later most of industrially used enzymes were already being obtained from microbial sources. There are still some commercially available enzymes from animal origin, mostly hydrolases, accounting for approximately 10% of total of enzymes used at industrial level [19]. In this group, catalase from liver (EC 1.11.1.6), triacylglycerol lipase (EC 3.1.1.3), and trypsin from pancreas (EC 3.4.21.4) are the most relevant animal enzymes currently used, mainly in food industries [20,21]. Enzymes from vegetable origins, such as papain and cysteine proteases from papaya latex (*Carica papaya*, *Carica candamarcensis*), have industrial relevance as well, representing almost 5% of the market [17,22,23]. Other enzymes, such as invertase [24], peptidases [25], and acid phosphatase [26] are produced *in vivo* in plant cell cultures, but their production involves highly complex and expensive processes, thus showing a limited use at industrial scale [27]. However, some glycoenzymes, such as glutamine synthetase, which are not easily produced as recombinant proteins in microbial hosts, are suitable candidates for the *in vitro* production with adequate cell lines [28].

While the origin of the biocatalysts can obviously be highly diverse, microorganisms are a rich source of enzymes, and thus their use as whole-cells, or the use of isolated microbial enzymes as biocatalysts has been vastly reported in the literature. Since the 1960s microbial enzymes have been replacing the biocatalysts from other origins, and to date, represent over 90% of the total market [17]. Currently, the exploitation of microbial diversity in the quest for new enzymes with novel activities is one of the major research goals in biocatalysis. This is complemented by the rational design of enzymes, and their production and overexpression in adequate microbial hosts through genetic engineering techniques [29]. Remarkably, the use of recombinant microorganisms was originally envisaged for the production of proteins of therapeutic interest. However, its real advantage is the reduction of production costs for a wide variety of proteins, especially when compared with the fermentation of wild-type microorganisms [30].

1.2.1 Plants and Animals as Sources of Redox Biocatalysts

As stated above, animal and plant tissues are classic sources of biocatalysts. Enzymes from higher eukaryotes have been traditionally used in food industry as food additives, in fruit processing or in wine production, as well as pharmaceutical additives. Some examples are the use of papain, lipoxygenase (LOX), or pepsin in processes already developed in the 1980s, which are still widely used [31,32]. Many hydrolases such as porcine pancreas lipase (PPL), pig liver esterase (PLE), or chymotrypsin and trypsin have been isolated from animal tissues, and have been widely used as biocatalysts [33]. Oxidoreductases (EC 1.-), hydrolases (EC 3.-), lyases (EC 4.-), and isomerases (EC 5.-) provide the vast majority of examples of higher eukaryotic enzymes for industrial applications, and no commercial processes using plant or animal enzymes from other enzyme classes have been reported [32].

Some of these animal or plant tissues can provide high amounts of enzymes (up to 1% of wet weight), for example, pancreatic enzymes or others involved in specific metabolisms (from liver or heart), or enzymes located in plant reserve organs such as seeds. However, recovering these enzymes from tissues is often cumbersome, and thus alternative sources must be found for their production at large scale for synthetic purposes. In addition, in case of pancreatic enzymes, after the discovery of pancreas as insulin-producing organ in 1921, the tissue became very expensive as a source of enzymes for biocatalysis. The enzymes that are still obtained from pancreas—trypsin and chymotrypsin—are actually by-products in insulin metabolism. Furthermore, nowadays insulin is mostly produced by recombinant microorganisms (*Escherichia coli* or yeast cells), as other enzymes originally obtained from pancreas do as well.

Higher eukaryotic enzymes usually need to be isolated and purified, or cloned and overexpressed in suitable hosts in order to obtain sufficient amounts for biocatalytic applications. The use of synthetic host-adopted genes and codon-optimized *E. coli* strains, and the development of highly successful eukaryotic expression systems such as the yeasts *Pichia pastoris* and *Hansenula polymorpha*, have enabled the production of large quantities of eukaryotic enzymes within a short time [34].

Among redox enzymes from animal origin, horse liver alcohol dehydrogenase (HLADH, EC 1.1.1.1), or cytochrome P450 monooxygenases (CYP450, EC 1.14.-.-) are the most extensively used, both in academic and industrial settings [32]. CYP450s belong to a superfamily of hemeproteins with high catalytic versatility, which generally perform the monooxygenation of aliphatic compounds as a key step in the production of fine chemicals, and catalyze the metabolism of a wide variety of endogenous and exogenous compounds (see Section "Heme iron monooxygenases cytochromes P450"). They are involved in reactions as diverse as hydroxylations, *N*-, *O*-, and *S*-dealkylation, sulfoxidation, epoxidation, deamination, desulfuration, dehalogenation, peroxidation, and *N*-oxide reduction (see Chapter 4) [35–43]. Their substrates include fatty acids, steroids, prostaglandins, and a number of exogenous compounds such as drugs, anesthetics, pesticides, and carcinogens [35]. This diverse catalytic potential attracted researchers from different fields to study cytochrome P450 systems. Industrial applications of human P450s involve two different objectives: the production of active pharmaceutical intermediates and the simple and fast production (in mg-scale) of metabolites for drug development. Human CYP450 has been expressed in different systems such as mammalian cells, yeasts, and bacteria such as *E. coli* [32]. For this application, *E. coli* expression systems are the easiest and less expensive to operate, yielding high quantities of recombinant proteins. At present, complete sets of ready-to-use liver enzymes expressed in *E. coli* are commercially available, providing kits with all six major human liver cytochromes for simple application. With regard to HLADH, three groups of isoenzymes with different substrate specificities are known, each one containing one main form [44–47]. The enzyme is a dimer, and these three forms correspond to the possible dimeric combinations of two protein chains [48–50] that are not interconvertible, therefore having different primary structures. The three isoenzymes were crystallized [51,52], and are similar in their amino acid analysis [51], but distinguishable by immunological methods [53]. Regarding synthetic applications of HLADH, its ability for the stereoselective reduction of a broad range of carbonyl compounds is outstanding, including aromatic, open-chain, or cyclic ketones, and α- or β-ketoesters (see Chapter 3).

Apart from the above-described two main enzymes, bovine liver glutamate dehydrogenase (GluDH, EC 1.4.1.2) is another example of dehydrogenases from higher eukaryotes [32]. The enzyme reversibly catalyzes the reductive amination of α-ketoglutarate to L-glutamate using NADH as cofactor. The crystal structure of this homohexameric mitochondrial enzyme has been solved at 2.8 Å resolution and its catalytic mechanism has also been demonstrated [54].

Plant cells exhibit an ample potential for the biosynthesis of secondary metabolites [55]. Although in cell cultures the formation and accumulation of these metabolites do not occur, such cultures retain the ability to transform exogenous substrates into products of interest. Thus, plant cells have been widely used in biocatalysis, as isolated enzymes as well as whole-cells (cell cultures or crude parts such as leaves, roots, seeds, etc.), either free or immobilized, displaying hydrolase and oxidoreductase activities [55]. When performing biotransformations mediated by "native" plant organs or tissues, the main drawback is the lack of reproducibility of the experiments. Some biochemical features of these biocatalysts can vary easily depending on several factors, such as the origin of the plant, the season of the year, or particular climatic conditions [56]. Likewise, other drawbacks are the localization and concentration of enzyme in the cell, and the presence of side reactions leading to undesired by-products, or enzymes that could degrade the desired products. In addition, another issue is the mass transfer limitations involved in the transport between the bulk medium and the enzyme. Overall, these problems contribute to the often observed low efficiency of these processes [55]. Strategies such as elicitation, permeation of the cells by the addition of dimethyl sulfoxide (DMSO) or organic alcohols to promote the substrate uptake and product release, or the variation of the pH, have been developed to improve efficiencies. Cyclodextrins were also used as additives in cases of water-insoluble or poorly soluble substrates because they are able to form inclusion complexes with a variety of apolar ligands [57]. Another major drawback of cultured cells as biocatalysts is the somaclonal variation, which may lead to unstable biochemical behavior [58]. This problem can be circumvented either by the continuous screening to maintain productive lines, or by the use of organized tissues, such as roots or root cultures, which are able to provide biotransformations with potential applicability for the production of added-value products, and even for scaling up purposes. In this context hairy roots, which are obtained by the integration of a region (T-DNA) of the Ri plasmid of the bacterial soil pathogen *Agrobacterium rhizogenes* to the plant genome [59], have been used as biocatalysts.

The use of dehydrogenases from vegetal origin has been widely exploited for the reduction of carbonyl groups to the corresponding alcohols. A broad range of structurally diverse carbonyl compounds including aromatic and aliphatic ketones, diketones, ketoesters, aldehydes, steroids, alkaloids, terpenoids, coumarins, and lignans, among others, can undergo redox reactions catalyzed by plant dehydrogenases (see Chapter 3) [55,56,60–77]. Among crude plant cells or plant cell cultures used as sources of dehydrogenases, *Daucus carota* has been one of the most widely used [60,62–71,75,76,78]. Cells of *Raphanus sativus* [56], *Passiflora edulis* [77], *Cocos nucifera* [61], celeriac (*Apium graveolens*) [62,63,70], horseradish (*Armoracia lapathifolia*) [62], legumes such as *Pisum Sativum* [73], *Phaseolus angularis* [72], and *Phaseolus aureus* [74], among others [79–82], have been also used as biocatalysts in dehydrogenase-catalyzed reductions or oxidations. Likewise, *R. sativus* hairy roots were used in the stereoselective reduction of a series of prochiral alkylaryl ketones. Most of the reactions proceeded with high yields and excellent enantioselectivities [56]. Hairy roots of *Brassica napus* [83] and *D. carota* [84] were also employed as biocatalysts for the stereoselective reduction of aromatic and aliphatic ketones, diketones, and ketoesters.

In addition to the vast group of dehydrogenases, other enzyme types of vegetal origin are of synthetic use, such as oxidases and peroxidases. Glycolate oxidase (EC 1.1.3.15) is a peroxisomal oxidase isolated either from mammals including pig liver, rats and humans, or from green plants such as spinach leaves, pea, cucumber, and pumpkin [32]. Of all of them, spinach redox enzymes displayed the highest yields also having the best specific activity for the oxidation of glycolic acid to glyoxylic acid [85]. The biocatalytic oxidation was performed in high yields (>99%) and resulted in a few undesirable by-products, in the presence of oxygen and ethylenediamine, using both glycolate oxidase and catalase. Importantly for practical purposes, spinach glycolate oxidase is also active for the oxidation of other α-hydroxy carboxylic acids [86]. Yet, purified enzymes from spinach leaves are not suitable for stable enzyme preparations for industrial applications [32]. Therefore, its gene was cloned and overexpressed in *E. coli* [87,88], *Saccharomyces cerevisiae* [89], and *P. pastoris* [90]. This was the first example of expression of a plant gene in *P. pastoris*, and also the first engineered *P. pastoris* as a whole-cell catalyst developed for a commercial bioprocess. The protein crystal structure was determined at 2.0 Å resolution, showing a catalytically active tetramer or octamer made up of identical 40 kDa subunits, which form an eight-stranded α/β barrel [91]. Optically pure (*R*)-2-hydroxyacids were also obtained on semipreparative scale with >99% *ee* and good-to-excellent conversions by α-hydroxylation of long-chain carboxylic acids with molecular oxygen, through a reaction catalyzed by an α-oxidase from peas (*P. sativum*) [92].

Finally, the importance of peroxidases is emphasized by their wide distribution among living organisms and by their multiple physiological roles. They have been divided into two main superfamilies according to their source and mode of action: plant (nonanimal) peroxidases and animal peroxidases (see Section 1.3.4) [93]. The plant peroxidases superfamily, which contains enzymes from both prokaryotic and eukaryotic origin, can be in turn divided in three classes, based on structural similarities and in a suspected common evolutionary origin [94]: peroxidases from prokaryotic origin (Class I), fungal peroxidases (Class II), and plant peroxidases (Class III). Horseradish peroxidase (HRP, EC 1.11.1.7), peanut peroxidase (PNP) [95], soybean peroxidase (SoP) [96,97], tobacco peroxidase (TobP) [98], tomato peroxidase (TomP) [99], and barley peroxidase (BaP) [100] are examples of Class III peroxidases. They contain an *N*-terminal signal peptide for secretion, two conserved calcium ions, four conserved disulfide bridges, an extra helical region that plays a role in access to the heme edge, and a carbohydrate content between 0% and 25% [93]. Of all of them, HRP is the most intensively studied peroxidase from plant origin, catalyzing a variety of reactions such as reduction of hydroperoxides, epoxidation, sulfoxidation, halogenation, and oxidation of phenols and aromatic amines (see Chapter 5).

A general strategy to improve the biocatalytic efficiency is the biocatalyst immobilization [101]. Whole-cells as well as isolated enzymes can be immobilized in order to overcome stability problems and to enable the biocatalyst reuse for cost reduction. Immobilized whole-cells have additional advantages over freely suspended cells. General methods for immobilization of plant cells are gel entrapment by ion exchange, precipitation, polymerization, and by fixing them into preformed structures [102]. Enzymes can be adsorbed into insoluble supports by hydrogen bonding, dipole–dipole interactions, or hydrophobic interactions. Most commonly used supports are polypropene (e.g., Accurel TM), and diatomaceous earth (Celite). Immobilization of enzymes by ion exchange is possible when the optimum pH of the enzyme is not close to its isoelectric point. Polyacrylamides are common matrixes used for covalent linking of the enzymes. High degrees of cross-linking prevent leakage and loss of the biocatalyst, but are not suitable for bulky substrates. Microencapsulation, forming a microsphere of polymeric membranes around the enzyme in solution, is another frequently used method [55].

1.2.2 Wild-Type Microorganisms

Microorganisms have been widely exploited as biocatalysts in the area of medicine, agriculture, and food industry, and their industrial applications have an increasing impact. Actually, as mentioned before, most of the enzymes currently used in industry have microbial origin. Since the beginning of biotechnology, microbial wild-type strains have been used for food and beverages production. In turn, native isolated enzymes from microbial origin, as well as recombinant proteins and microorganisms, are having an increasingly widespread use in pharmaceutical, chemical, or biofuels industries, being mainly designed through genetic engineering [103].

1.2.2.1 Yeasts The large-scale use of yeasts in enzyme-catalyzed processes dates back several centuries ago, with the production of ethanol from glucose in alcoholic beverage manufactures. *S. cerevisiae* (baker's yeast, BY) was often the microorganism of choice, mainly due to its wide availability and low cost. Moreover, BY does not need sterile growth media, and remains viable and easy to work in a nonmicrobiology-specialized laboratory. In addition to these desirable features, as biocatalyst, baker's yeasts (as well as other yeasts) can be used in chirality generation, in racemic resolutions, or in the regioselective conversion of functional groups. For such purposes, besides redox-related enzymes (mainly dehydrogenases), other biocatalytic yeast enzymatic systems include hydrolytic enzymes (lipases, epoxide hydrolases) and lyases for C–C bond formation. Clearly, *S. cerevisiae* is still to date the most widely used yeast biocatalyst [104].

Among redox enzymes, dehydrogenases are versatile and powerful biocatalysts for synthetic organic chemistry, mainly involved in the reduction of carbonyl groups—generating chiral alcohols—or in the asymmetric reduction of enones or imines. On the other hand, yeast dehydrogenase-mediated oxidations generally involve the destruction of chiral centers, and thus their practical use has been mainly guided by environmental considerations, a major advantage over conventional chemical oxidations (see Chapter 3). In the reduction of carbonyl groups—usually associated to the synthesis of (*S*)-secondary alcohols—most of the yeast dehydrogenases follow the so-called Prelog's rule [105]. For these purposes, yeast dehydrogenases can be used either as whole-cells (wild type or recombinant), or isolated enzymes with the adequate cofactor supply and a suitable recycling method [106]. As mentioned above, *S. cerevisiae* is the most widely used yeast as biocatalyst, being efficient for selectively reducing monocarbonylic compounds (aldehydes and ketones) with alkyl or aryl substituents and dicarbonylic compounds (cyclic and acyclic α- and β-diketones, α- and β-ketoesters), thus obtaining (*S*)-secondary alcohols (see Chapters 3 and 7). Sterically hindered ketones are not usually substrates for yeast dehydrogenases except with the exception of methyl ketones (see Chapter 3). Apart from baker's yeast, a wide number of oxidoreductases from *Candida* sp. have also been characterized and used as biocatalysts [107]. These enzymes are rather diverse, enabling not only highly enantioselective reductions of carbonyl groups or deracemizations, but also some promiscuous catalytic imine reductions. In addition, some *Candida* sp. dehydrogenases have been extensively used for cofactor regeneration purposes, with the aid of sacrificial substrates.

1.2.2.2 Fungi Fungi have traditionally been one of the most studied whole-cell enzymatic systems as biocatalysts [108]. They can be identified in nature through the screening of a wide variety of environments and habitats. Bioprospecting for new biocatalysts even in extreme environments—thermal, cold, or hypersaline ecosystems—can lead to the discovery of new fungal enzymes able to catalyze a wide variety of synthetically interesting reactions. Among the most extensively explored fungal enzyme systems for biocatalytic purposes, redox enzymes can be highlighted, catalyzing oxygenase- as well as peroxidase-mediated hydroxylations, sulfoxidations, epoxidations and Baeyer–Villiger oxidations, dehydrogenase-catalyzed stereoselective reductions, oxidations or deracemizations [104,108].

Fungal dehydrogenase-catalyzed reactions, in particular of carbonyl compounds to furnish alcohols regio- and enantioselectively, are the most widely described in the literature. The broad substrate specificity of these enzymes allows for the preparation of structurally different compounds, including aromatic, cyclic, open chain aliphatic (*R*)- or (*S*)-secondary alcohols, or α- or β-hydroxyesters and hydroxyketones [108]. Dehydrogenase-catalyzed racemization or deracemizations are also efficient strategies for interconversion of enantiomers, with a number of fungal enzymes reported for these types of reactions [108].

Oxygenase-, oxidase-, and peroxidase-mediated oxidations, introducing oxygen atoms into nonactivated hydrocarbon chains of organic compounds, are useful for organic synthesis as well. These fungal enzymes catalyze Baeyer–Villiger reactions [109], epoxidations [110], sulfoxidations [111], and hydroxylations [112] using growing or resting whole-cells, as well as isolated enzymes. Likewise, fungal laccases catalyze oxidation and coupling reactions [113,114]. Although these enzymes are widely distributed in fungi, higher plants, bacteria and insects, the majority of laccases characterized so far are of fungal origin, spread in more than sixty strains belonging to various classes such as *Ascomycetes*, *Basidiomycetes*, and *Deuteromycetes* [115]. Especially, white rot *Basidiomycetes* are efficient lignin degraders, being a valuable source of laccases. Moreover, fungi are also an important source of peroxidases. Due to the catalytic versatility and stability of these enzymes, fungal peroxidases are, together with fungal laccases [114,116], of particular interest as potential biocatalysts for redox processes at industrial level [93,117–119].

1.2.2.3 Bacteria The variety of bacterial genera and species, and the different enzyme types according to their metabolism, role and environment, make these microorganisms a rich source of biocatalysts. Bacteria used in biocatalysis belong to a wide variety of genera, including *Escherichia*, *Rhodococcus*, *Bacillus*, *Lactobacillus*, *Nocardia*, *Pseudomonas*, *Acinetobacter*, *Alcaligenes*, *Corynebacterium*, and so on [30,33]. Similar to enzymes from eukaryotic cells, bacterial enzymes can be used both as isolated and as whole-cells, and either free or immobilized. Due to management requirements in the laboratory, the use of isolated enzymes is mainly restricted to extracellular and cofactor-independent-enzymes, such as hydrolases, whereas the whole-cells of microorganisms are preferred in synthetic procedures involving cofactor-dependent or intracellular enzymes. However, at industrial level examples of redox enzymes comprise both isolated enzymes and whole-cells (see Chapters 7 and 8).

Bacterial enzymes or whole-cell-catalyzed biotransformations are used in the industrial preparation of a variety of compounds [33]. The use of redox enzymes from bacterial origin for the preparation of chiral intermediates for fine chemicals is highly widespread. Dehydrogenases, involved in the reduction of carbonyl groups or the enantioselective oxidation of alcohols, as well as mono- and dioxygenases for hydroxylations of arenes and unfunctionalized alkanes, Baeyer–Villiger reactions and alkene monooxygenation, are outstanding examples [104,120]. Most of the monooxygenases used in biocatalysis are from bacterial origin, and aromatic dioxygenases have only been found in bacteria, mainly in genus *Pseudomonas*, *Sphingomonas*, and *Rhodococcus*, being intracellular enzymes. Although to date many oxygenases are known and thoroughly described, their use for synthetic purposes is still limited due to their low availability in large quantities, instability, and high cost of the required cofactors. Moreover, many of them are membrane-associated proteins, reducing their synthetic potential and hampering their use as isolated enzymes. However, efforts have been done to overcome these drawbacks, for example, using recombinant whole-cells overexpressing the desired enzymes (see Chapter 4).

1.2.3 Metagenomic Assessments

An ample range of microbial diversity, not yet accessed or explored, might be a valuable source for possible novel biotechnological applications. In this respect, only 0.1–10% of natural biodiversity can be cultured under conventional laboratory conditions [121,122]. Metagenomics, that is, the genomic reconstruction of uncultivable microorganisms that emerged in the late 1990s, refers to the extraction of the entire genetic material from all organisms present in an environmental sample (metagenome) [123,124]. It involves the culture-independent analysis of the collective microbial genomes contained in the metagenome by using two approaches: (i) the function-driven analysis and (ii) the sequence-driven analysis, to extract biological information from metagenomic libraries [125–127]. The function-driven analysis identifies the clones that express a desired feature, and then characterizes the active ones by sequencing and biochemical analysis, rapidly finding clones with potential applications. The limitation of the method is that it requires the expression of the function of interest in the host cell and the clustering of all of the genes required for the function. It also depends on the availability of an adequate assay for the function of interest that can be performed on vast libraries, because the frequency of active clones is quite low. To overcome these limitations, improved systems for heterologous gene expression have been developed with shuttle vectors that facilitate screening of the metagenomic DNA in diverse host species and with modifications of *E. coli* cells to expand the range of gene expression. Conversely, sequence-driven analysis is based on the use of conserved DNA sequences to design hybridization probes or polymerase chain reaction (PCR) primers to screen metagenomic libraries for clones that contain sequences of interest. Significant discoveries have resulted from random sequencing of metagenomic clones [127]. Metagenomics relies on the efficiency of four main steps, which are: (i) the isolation of genetic material, (ii) its manipulation, (iii) the transfer of the genetic material into a surrogate organism to create a metagenome clone library, and (iv) the analysis of the genetic material in the metagenomic library [121,122,128]. The information about diversity and community structure of microbes is further obtained through the corresponding metagenome library sequency [129,130]. The specific activities within the metagenome can be screened for particular enzymes, either *via* DNA sequences or enzymatic functions [131,132].

Metagenomic libraries have been screened for enzymes [133], and a number of metagenomic biocatalysts displayed novel substrate ranges and high stability under extreme conditions, improving their potential for industrial applications [134]. Many oxidoreductases have been incorporated into the redox-biocatalytic toolbox through metagenomic assessment [134]. As some examples, in the quest of microorganisms accepting 4-hydroxybutyrate, five clones were found displaying novel 4-hydroxybutyrate dehydrogenase activity [135]. The genes involved in metabolism of poly-4-hydroxybutyrate were also successfully screened in environmental libraries [136]. Short-chain dehydrogenases/reductases were found with $<35\%$ similarity to known enzymes, and thus could not have been detected using hybridization-based techniques such as PCR. Likewise, alcohol dehydrogenases (ADHs) oxidizing short chain polyols were obtained from twenty-four positive clones and characterized [137,138]. Polyphenol oxidase (PPO; tyrosinase), catechol oxidase, as well as multicopper oxidases, oxygenases, laccases, and L-amino acid oxidases (LAAOs) were also isolated from marine metagenome [139,140]. Clones capable of indirubin and indigo production in *E. coli* were isolated from forest soil metagenome [141]. A novel ADH, $ADHE_{Meta}$, was isolated from a waste-water treatment metagenomic library [142]. Although this enzyme showed a reasonably high sequence identity to the well-characterized ADHs from *E. coli* (60%) and *Clostridium acetobutylicum* (61%)—which catalyze the ethanol and butanol formation in a two-step reductive process—it was not deactivated by oxygen [123]. This functional property suggests that the *in vivo* role of the enzyme was catabolic, rather than the typical anabolic role of ADHs.

1.3 OVERVIEW OF REDOX ENZYMES

Enzymes catalyzing oxidation–reduction reactions—oxidoreductases, EC 1.—always act on their substrates involving an electron transfer. These biocatalytic redox processes have been attracting an increasing interest for synthetic organic chemistry, especially for applications in the chemical and pharmaceutical industries [143]. Besides the inherent enantioselectivity of enzymes, other challenging chemical transformations like the regioselective introduction of oxygen from O_2 into chemically inert C–H bonds can be performed. Furthermore, oxidoreductases catalyze the formation of chiral products from prochiral substrates with a theoretical 100% yield, as well as other reactions such as the oxifunctionalization of C–C double bonds and C–N bonds, Baeyer–Villiger oxidations, and the oxidation of alcohols, aldehydes, acids, and aromatic compounds, or else the reduction of ketones, aldehydes, C–C and C–N double bonds, and reductive aminations [120,144,145]. Applications of these enzymes comprise mainly asymmetric oxyfunctionalization of nonactivated hydrocarbons and enantioselective reductions of carbonyl compounds, synthesis and modification of polymers, as well as oxidative degradation of pollutants and construction of biosensors for a variety of analytical and clinical applications [146].

Presently, more than 25% of the known enzymes are oxidoreductases [147,148]. Due to their biodiversity and different activities, oxidoreductases have been classified according to the nature of the oxidizing substrate as dehydrogenases, oxygenases (monooxygenases and dioxygenases), oxidases and peroxidases (Figure 1.1). Alternatively, they can be classified according to their coenzyme requirements [149]. Peroxidases are sometimes considered as a subset in the group of oxidases [106], but taking into account the differences in terms of their catalytic mechanism, as well as their synthetic applications, in this book peroxidases will be treated as independent enzymes.

Dehydrogenases catalyze reversible redox reactions and are thus used as oxidative or reductive biocatalysts. Conversely, oxygenases, oxidases and peroxidases catalyze irreversible oxidations due to the reaction thermodynamics. This is attributed to the highly exothermic reduction of O_2 or H_2O_2, which act as electron acceptors in the case of oxygenases and oxidases, or peroxidases respectively [144]. In contrast to oxidases, oxygenases incorporate one (monooxygenases) or both (dioxygenases) oxygen atoms of O_2 into their substrates.

Redox enzymes often require cofactors, which are direct products of the primary metabolism of the cells. These requirements impose special constraints on the development of bioprocesses involving oxidoreductases. These cofactors or coenzymes, such as NAD(P)H or $NAD(P)^+$, act as electron donating or accepting molecules, being coupled to the redox metabolism within cells [14,150–153]. Therefore an efficient cofactor recycling system is required to achieve economically efficient processes (see Chapters 2 and 7–8 for applying biocatalysis at practical level) [104,154–158]. Exceptions include peroxidases, which couple the reduction of hydrogen peroxide to water with the two electron oxidation of the substrate, and oxidases, which couple the reduction of O_2 to hydrogen peroxide or water with the two- or four-electron oxidation of an activated carbon scaffold. All reactions dealing with O_2 or H_2O_2 reduction involve reactive oxygen species such as oxygen radicals or hydrogen peroxide itself, which often cause extensive damage to biomolecules, including enzymes. These reactive oxygen species and their damage can be handled by the metabolism

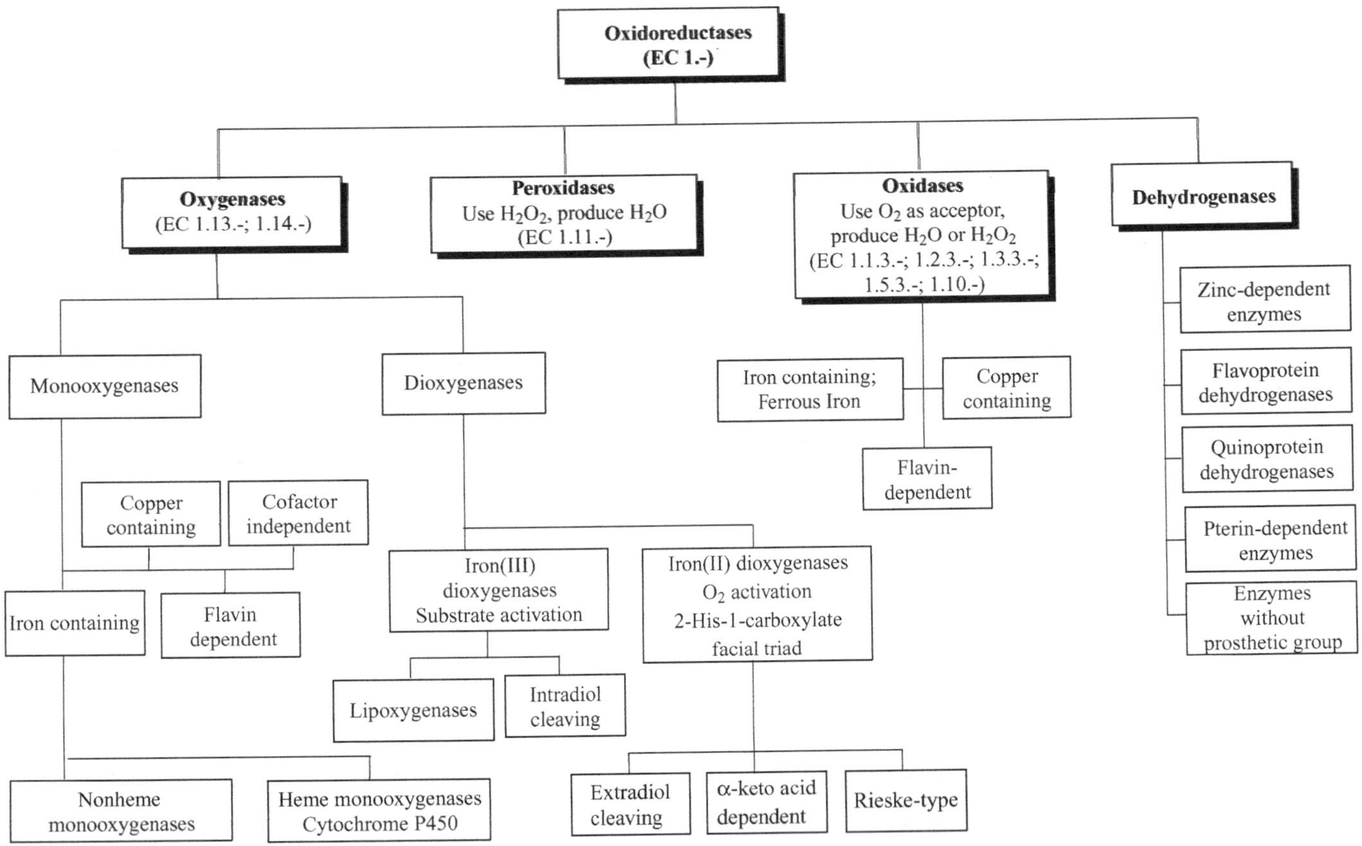

FIGURE 1.1 Classification of redox enzymes.

of resting or growing cells, an advantage that can be exploited for productive redox biocatalysis [144].

1.3.1 Dehydrogenases

Dehydrogenases (also known as reductases) are the most widely used oxidoreductases for synthetic purposes [104,143,158–160], and they are classified in groups EC 1.1.1.- (acting on the CH–OH group of donors), EC 1.2.1.- (acting on the aldehyde or oxo group of donors), or EC 1.4.1.- (acting on the CH–NH_2 group of donors). They always use NAD^+ or $NADP^+$ as cofactors [159].

Regarding substrate specificity of dehydrogenases, enantiopure alcohols, hydroxyacids, hydroxyesters, amines, and amino acids can be obtained through dehydrogenase-catalyzed reductions of the corresponding prochiral compounds [104,158,159,161]. The enzymes catalyze the reversible reaction, in which one atom of hydrogen and two electrons of the donor species are transferred to the C-4 position in the nicotinamide of the cofactor (NAD^+ or $NADP^+$). The reverse reaction involves the transfer of one atom of hydrogen and two electrons from the reduced form of the nicotinamide cofactor (NADH or NADPH) to the carbonyl (or imino) group, resulting in the reduction of the substrate, being often a highly specific reaction. The hydrogen atom transfer occurs either on the *Si* face or on the *Re* face of the carbonyl group, resulting in the corresponding (*R*)- or (*S*)-alcohol, respectively. Both, NADH and NADPH have two diastereotopic hydrogen atoms (*pro-R* and *pro-S*), which can be transferred as a hydride to an oxidized substrate (aldehyde, ketone, or imine). In turn, these substrates have two diastereotopic or enantiotopic faces (*Re*-face or *Si*-face) in the sp^2 carbon atom that will be reduced. Theoretically, any NAD(P)H-dependent redox reaction can occur stereospecifically *via* any of the four possible pathways to generate any isomer on the substrate (Figure 1.2).

In practice, most ADHs catalyze the transfer of the *pro-R* hydrogen atom at C-4 of the nicotinamide cofactor to the *Re* face of the carbonyl substrate, furnishing the corresponding (*S*)-alcohol as the main reduction product. This general behavior

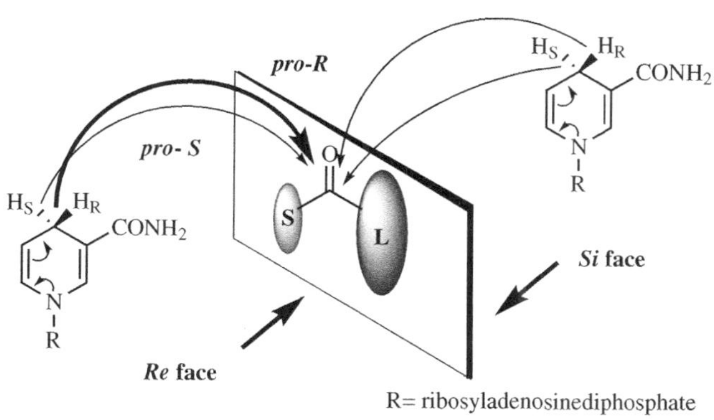

FIGURE 1.2 Stereochemical possibilities for the hydride transfer in the dehydrogenase-mediated reduction of carbonyl compounds. **L**, large substituent; **S**, small substituent.

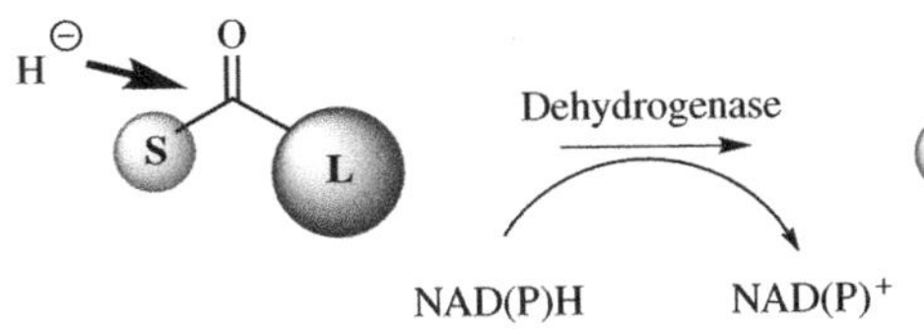

FIGURE 1.3 Stereochemical preference for the hydride transfer in the dehydrogenase-catalyzed reduction of carbonylic compounds, according to Prelog's rule. **L**, large substituent; **S**, small substituent.

is described by the Prelog's rule [105], where the stereochemical course of the reaction is predictable, and largely depends on the steric requirements of the substrate (Figure 1.3). Depending on the binding site of the cofactor in the enzyme, either the *pro-R* or *pro-S* hydrogen at C-4 of the nicotinamide is abstracted. The Prelog's rule is an empirical rule based on the stereochemistry of microbial reductions using whole-cells of *Curvularia falcata*.

Dehydrogenases can be classified into five categories depending on their prosthetic group: (i) zinc-dependent enzymes, (ii) flavoprotein dehydrogenases, (iii) pterin-dependent enzymes, (iv) quinoprotein dehydrogenases, and (v) dehydrogenases without prosthetic groups [144].

1.3.1.1 Zn-Dependent Dehydrogenases Zinc-dependent dehydrogenases comprise mainly ADHs that are grouped according to their substrate spectrum and structure into long-, medium- and short-chain ADHs [162]. These enzymes catalyze the (reversible) oxidation of primary and secondary alcohols to the corresponding aldehydes and ketones respectively, often in a chemo-, regio-, and stereoselective way. For synthetic purposes, the most relevant enzymes of this group are HLADH and ADH from *Thermoanaerobacter brockii* (TbADH).

Structure–function studies have been undertaken for the Zn-dependent medium-chain HLADH, revealing a nucleotide binding domain and a catalytic domain, with the catalytically active zinc buried between them [163]. In HLADH, the zinc is tetra-coordinated to three conserved residues (Cys46, His67 and Cys174) of the active site, and to a water molecule [164]. There are two main mechanisms proposed for HLADH-catalyzed alcohol oxidation, which differ in the coordination of the zinc atom during the redox process. One proposes that the zinc ion stays in tetrahedrical coordination during catalysis, as the bound water molecule is replaced by the hydroxyl group of the alcohol. Then, two hydrogen atoms are removed from the substrate by a coupled process of proton abstraction and hydride ion transfer. The intermediate turns to the corresponding oxidation product, with the concomitant hydride transfer to $NAD(P)^+$ to complete the reaction. The other mechanistic proposal suggests a change in the coordination symmetry of the zinc, leading to a pentacoordinated species, by the binding of the substrate in addition to the water molecule, and followed by the same steps than in the previously proposed mechanism. Recent X-ray studies on HLADH and TbADH provided evidence for the latter mechanism, involving a pentacoordinated zinc during catalysis [165,166]. Kleifeld and coworkers proposed

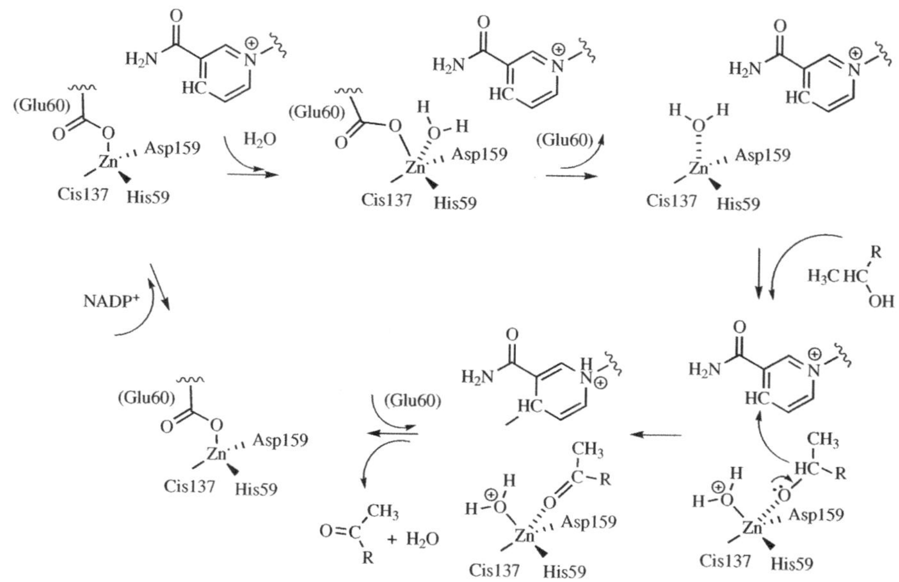

FIGURE 1.4 Proposed mechanism for the oxidation of secondary alcohols mediated by alcohol dehydrogenase from *Thermoanaerobacter brockii* (TbADH).

that TbADH differs from other ADHs, since the catalytic Zinc is bound to three conserved residues (Cys37, His59, and Asp159), and a residue of glutamic acid (Glu60) instead of water in the forth coordination site. This Glu60 is exchanged by a water molecule during the catalysis, generating a pentacoordinated intermediate (Figure 1.4).

1.3.1.2 Flavin-Dependent Dehydrogenases Enzymes containing flavin adenine dinucleotide (FAD) as prosthetic group, either covalently or noncovalently bound, known as flavoprotein dehydrogenases, are mostly involved in O_2 activation reactions. They are important enzymes in redox metabolism, yet less used for synthetic applications [144]. As in the case of other flavin-dependent enzymes such as oxidases, FAD serves as an electron acceptor during the oxidation of the substrate. Within this group, acyl-CoA dehydrogenases are a family of flavin-dependent enzymes whose members are highly conserved in sequence, structure, and function. These enzymes catalyze key steps in the mitochondrial oxidation of lysine, hydroxylysine, and tryptophan. Glutaryl-CoA dehydrogenase (GCD, EC 1.3.99.7) has a homotetrameric structure containing one noncovalently bound FAD in each monomer, and catalyzes the oxidative decarboxylation of glutaryl-CoA to crotonyl-CoA [167]. The reductive half reaction generates the enzyme-bound intermediate glutaconyl-CoA and CO_2 (Figure 1.5). Upon binding the substrate glutaryl-CoA, a glutamate residue functioning as catalytic base abstracts an α-proton from the substrate, followed by hydride transfer from the C-3 position of the substrate to the flavin, yielding the two electron reduced form of FAD and the enzyme-bound desaturated

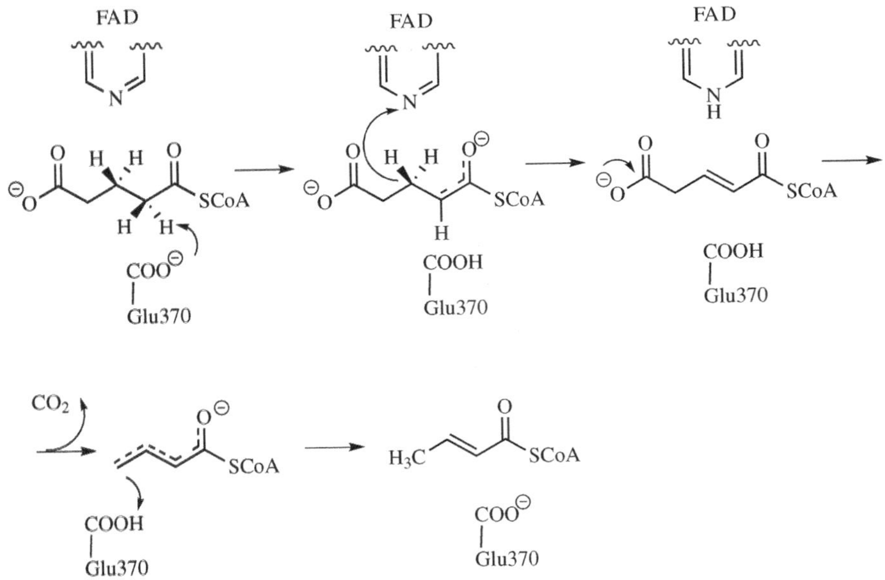

FIGURE 1.5 Proposed mechanism for the oxidative decarboxylation of glutaryl-CoA to crotonyl-CoA (human GCD numbering).

intermediate glutaconyl-CoA [168,169]. Subsequently, the latter is decarboxylated resulting in the formation of a dienolate anion and CO_2. The dienolate anion is protonated *via* proton transfer to C-4 from the catalytic base Glu370, which exchanges its proton with solvent water, giving rise to the crotonyl-CoA product (numbering of the residues is for the human GCD) [170]. The two electron reduced form of FAD is reoxidized in two consecutive steps, transferring the electrons to the respiratory chain.

Besides flavoproteins containing only the flavin in the catalytic center, there are also two-component dehydrogenases, having a covalently bound flavin in one subunit and a cytochrome in the other [171]. Examples for hydroxylating flavocytochromes include *p*-cresol dehydrogenase [172,173], and eugenol dehydrogenase [174], found in *Pseudomonas fluorescens* strains and denitrifying bacteria. After a fast electron transfer from the flavin to the cytochrome, *p*-cresol dehydrogenase transfers the electron to azurin or nitrate as physiological electron acceptors. During the reaction, a proton is removed from the hydroxyl group of the substrate *p*-cresol and a hydride ion is transferred from the methyl group to the N-5 atom of the covalently bound flavin. The formed quinone methide intermediate undergoes a nucleophilic attack of water at the methylene group, yielding *p*-hydroxybenzyl alcohol.

1.3.1.3 Pterin-Dependent Dehydrogenases Pterin-dependent dehydrogenases also belong to the group of hydroxylating enzymes, which use water as electron donor. They contain a so-called molybdopterin cofactor, where molybdenum is usually coordinated to the two thiol groups of a pyranopterin-based cofactor [144].

These enzymes can be composed of a variable number of subunits and contain different additional cofactors such as hemes, iron–sulfur clusters, and flavins, which transfer electrons from the molybdopterin cofactor to an electron acceptor. These molybdoenzymes catalyze the hydroxylation of activated carbon atoms, such as heteroaromatic ring carbons. Furthermore, selenium has been reported to participate in reactions catalyzed by xanthine dehydrogenase [175] and nicotinic acid dehydrogenases of anaerobic strains [176], for which NAD^+ and $NADP^+$ serve as electron acceptors, while nicotinic acid dehydrogenases from aerobic strains are selenium independent [177]. Molybdenum-enzymes are found in nearly all organisms, with *Saccharomyces* as a prominent eukaryotic exception [178]. The reaction mechanism has been characterized for xanthine dehydrogenase. A molybdenum center (as Mo(VI) in the resting state) with a water-derived oxo- or hydroxyl group in the ligand sphere is reduced to Mo(IV) during the substrate oxidation. Electrons derived from substrate hydroxylation are transferred through a [2Fe–2S] cluster and FAD to NAD^+ [179].

1.3.1.4 Quinoprotein Dehydrogenases Quinoproteins contain an amino acid-derived *o*-quinone as a cofactor. Different kind of quinone cofactors are known, but only pyrroloquinoline quinone (PQQ), which is not covalently bound to the enzyme, is involved in oxidoreduction reactions [180]. Quinoprotein dehydrogenases are highly interesting enzymes regarding redox metabolism, transferring reducing equivalents directly to the bacterial aerobic respiratory chain [144]. These enzymes are located in the periplasm, and are mainly involved in the dehydrogenation of primary or secondary hydroxyl groups in alcohols and sugars. These enzymes are frequently applied in biosensors [181]. Mainly two dehydrogenases of this class, in which PPQ is the only cofactor, have been extensively studied [182,183]. These are the membrane-bound quinoprotein glucose dehydrogenase (m-GDH, EC 1.1.5.2) and the soluble methanol dehydrogenase (s-MDH, EC 1.1.1.244), which catalyzes the oxidation of methanol to formaldehyde. Glucose dehydrogenase was found in a variety of bacteria including Gram-negative facultative anaerobes such as enteric bacteria and *Zymomonas*, as well as aerobic bacteria such as *Pseudomonas* and acetic acid bacteria [184]. The enzyme was originally investigated in *Acinetobacter calcoaceticus* in early 1960s by Hauge [185–187], and subsequently in late 1970s by Duine et al. [188], who reported that the enzyme was a quinoprotein. Mechanistic studies on s-MDH showed that upon substrate binding, the oxidation is initiated by an aspartate residue located in the active site, which is responsible for the abstraction of the proton from the alcohol group [189]. Then, a Ca^{2+} ion coordinated to the C-5 of the cofactor acts as a Lewis acid during substrate oxidation, stabilizing the C-5 carbonyl atom for the electrophilic attack by the hydride [144]. The proton abstraction is followed by a direct hydride transfer from the methyl group of methanol to the C-5 carbonyl of PQQ, followed by tautomerization to $PQQH_2$ [190]. The oxidized product, formaldehyde, is then released while $PQQH_2$ is reoxidized by sequential single-electron transfer to a *c*-type cytochrome in the electron transport chain.

The mechanism of the oxidation for GDH is similar to the one described for methanol oxidation by MDH. The proton in the anomeric hydroxyl group of glucose is abstracted by an aspartate residue in the active site of the enzyme, and a hydride is transferred from the C-1 of the glucose molecule to the C-5 carbon of the PQQ cofactor, yielding gluconolactonate and $PQQH_2$. Further oxidation of $PQQH_2$ involves either ubiquinone or *c*-type cytochromes as electron acceptors in the electron transport chain. The enzyme from *E. coli* is a monomeric protein of 87 kDa [191]. The *N*-terminal domain (154 residues) has five transmembrane segments, and it is likely where the ubiquinone-binding site is contained. The remaining periplasmic region (641 residues) is similar to the α-subunit of MDH. The prosthetic group PQQ in MDH is held in place in the active site by stacking interactions between a tryptophan residue and a disulfide ring structure made up of adjacent cysteine residues [192,193]. In GDH from *E. coli*, this disulfide ring is replaced by a histidine residue (His262), which is essential for binding PQQ [191]. The entrance to the active site is more open than in MDH, and His262 is located at the entrance, together with a second histidine residue (His775). GDH from *E. coli* has two ubiquinone binding sites: one of them is located near the active site and contains a tightly bound ubiquinone that performs single electron transfer steps; the second site is in the amphiphilic segment of the membrane bound GDH and reversibly binds ubiquinone from the membrane pool [194]. In the cell, the enzyme is responsible for the direct oxidation of glucose to gluconate in the periplasm, but also showed the ability to oxidize a wide range of pentoses and hexoses [191]. Disaccharides such as maltose, lactose, trehalose, and sucrose are neither substrates for GDH, nor was the trisaccharide raffinose, or D-fructose, *myo*-inositol, methanol, or ethanol. None of these compounds inhibited the oxidation of glucose, showing that they are not able to bind at the active site. L-Hexoses were not substrates, including L-glucose, L-mannose, L-allose and L-rhamnose. The only common feature of these L-hexoses is that they have a C-6 hydroxymethyl (or methyl) group below the plane of the ring, suggesting that this prevents binding by steric hindrance. As understanding of the mechanism of catalysis and electron transfer of quinoproteins progressed, the application of these enzymes as biosensors became attractive, and thus electrochemical studies aimed at the design of specific quinoprotein-based electrodes have increased [184]. Soluble glucose dehydrogenase (s-GDH) from *A. calcoaceticus* was the first quinoprotein applied to such a biosensor, and a D-glucose sensor with s-GDH in a single-use electrochemical test strip containing ferricyanide as a mediator is already in the market [184].

1.3.1.5 Dehydrogenases without Prosthetic Group Although most redox enzymes require metal or organic prosthetic groups for the hydride transfer, several dehydrogenases have neither metal ions nor any other prosthetic group bound to their active sites. The most prominent example of such enzymes is formate dehydrogenase (FDH), well known for its use in enzyme-coupled cofactor regeneration systems (see Chapters 2 and 8). The enzyme catalyzes the oxidation of formate to CO_2, cleaving a single carbon–hydrogen bond in the substrate and concurrently forming a new one in the nicotinamide cofactor, which also accepts the electrons [195]. During catalysis, a hydride anion from the substrate formate attacks the electrophilic C-4 of the nicotinamide cofactor NAD^+. Consequently, two uncharged products, CO_2 and NADH, are formed [144,196].

1.3.2 Oxygenases

Oxidative biotransformations with oxygenases are very valuable tools for synthetic organic chemistry. The most outstanding reason is that these enzymes catalyze a wide number of reactions including aliphatic and aromatic hydrocarbon hydroxylations, epoxidation of alkenes and arenes, *N*-, *S*-, and *O*-dealkylations, *N*-hydroxylations, *N*- and *S*-oxidations, deaminations, dehalogenations, and Baeyer–Villiger oxidations, with high regio- and stereoselectivity [35,149,197], while their chemical counterparts either do not exist or do not reach the required selectivity (see Chapter 4). In fact, many types of reactions that can be easily performed by oxygenases, such as asymmetric hydroxylations at saturated or unsaturated carbon atoms, or oxidative desymmetrizations, are difficult to carry out with conventional stoichiometric oxidants or by using organic or inorganic catalysts. Furthermore, when considering environmental aspects of many chemical oxidations, the replacement of hazardous procedures by more benign ones is highly desirable (see Section 1.1). The goal of reducing stoichiometric waste requires catalytic processes, and the quest for sustainable chemistry puts oxygenases in a central role for a variety of asymmetric oxidations [143,145,197–201]. Furthermore, oxygenases use molecular oxygen as a readily available and environment friendly oxygen source. Moreover, oxygenase-catalyzed reactions can be used to prepare highly valued building blocks and pharmaceutical intermediates, as well as to modify natural products with biological activities [197].

Oxygenase-catalyzed hydroxylations, namely the conversion of a C–H bond into a C–OH bond, are widespread enzymatic reactions occurring in all forms of life, from bacteria to mammals. Oxygenases are capable to insert one (monooxygenases) or two (dioxygenases) oxygen atoms from O_2 into their substrates under mild conditions [37,144]. In monooxygenase-catalyzed reactions, the second oxygen atom is reduced with electrons from NADH or NADPH, yielding water. This relatively complex reaction often requires a metal center, and electron transfer from reduced cofactors usually requires additional proteins. Dioxygenases catalyze the regio- and stereoselective insertion of both oxygen atoms from molecular oxygen into a substrate. Such highly specific oxyfunctionalizations, and especially the activation of C–H bonds, are synthetically interesting and often inaccessible by chemical means [202]. However, the use of oxygenases in organic synthesis is still limited due to their scarce availability in adequate quantities, their low stability and specific activity, and the high cost of required cofactors. Moreover, many of these enzymes are membrane-associated proteins, which reduce their potential application in synthetic procedures even more. Some of these drawbacks of oxygenases could be solved by the use of whole-cell biocatalysts, either as growing or resting cells, and preferentially recombinant microorganisms overexpressing the desired enzyme. Thus, presently the use of dioxygenases, such as arene dioxygenases, has mostly been described in whole-cell systems.

There are two major subclasses of oxygenases which incorporate dioxygen into their substrates: EC 1.13.- and EC 1.14.-. The enzymes of the former subclass do not need external hydrogen donors, such as NAD(P)H, for oxygenation and act on single substrate molecules, whereas those of the latter act on paired hydrogen donors [37]. Both, monooxygenases and dioxygenases are found in these subclasses.

$$\text{Substrate} + O_2 + H^+ + \text{NAD(P)H} \xrightarrow{\text{Monooxygenase}} \text{Substrate-O} + \text{NAD(P)}^+ + H_2O$$

FIGURE 1.6 General scheme for monooxygenase-catalyzed reactions.

1.3.2.1 Monooxygenases Monooxygenases are enzymes capable of activating molecular oxygen and oxidizing a wide diversity of almost nonreactive organic compounds such as alkanes, alkenes, aromatic compounds, or heteroatoms [203–207]. These enzymes incorporate only one oxygen atom from molecular oxygen, while the other one is reduced to yield water. A broad spectrum of useful reactions can be developed with monooxygenases as biocatalysts, for example, hydroxylations, epoxidations, Baeyer–Villiger oxidations, oxidations of heteroatoms such as N and S, and dealkylations of heteroatoms [143]. The catalytic mechanism differs significantly depending on the enzyme subtype involved, but in all cases the activation of molecular oxygen takes place in a similar way. Molecular oxygen is activated by reduction at the expense of NADH or NADPH (donor). One oxygen atom is transferred to the substrate, and the other one is reduced to form one molecule of water (Figure 1.6).

Monooxygenases can be classified according to their prosthetic group in iron-containing monooxygenases (heme and nonheme iron monooxygenases), copper containing monooxygenases, flavin-dependent monooxygenases, multicenter monooxygenases, namely Fe-flavin monooxygenases (FMOs) and Fe-pterin monooxygenases, and cofactor-free monooxygenases [118,144].

Iron-containing Monooxygenases

NONHEME IRON-CONTAINING MONOOXYGENASES Nonheme iron-containing monooxygenases are mono- or binuclear enzymes which are able to catalyze the oxidation of aliphatic or aromatic hydrocarbons to the corresponding monohydroxylated compounds. Binuclear nonheme iron monooxygenases catalyze monooxygenations of alkyl-, benzyl- and aryl-nonactivated carbon atoms, as well as fatty acid desaturations and ribonucleotide reduction. Prominent examples of these enzymes are soluble methane monooxygenase (MMO), alkane monooxygenase (AMO), xylene monooxygenase (XMO), toluene-2-monooxygenase (T2MO), and toluene-4-monooxygenase (T4MO) [144,208–210].

The soluble MMO (s-MMO) is structurally and biochemically the best characterized member of this oxygenase family [210–213]. These enzymes are unique among the di-iron monooxygenases catalyzing the NADH-dependent insertion of one oxygen atom from O_2 into the exceptionally stable C–H bond of methane (bond dissociation energy of 104 kcal/mol) to form methanol, and typically accept a broad substrate spectrum. The soluble form of the enzyme (s-MMO) contains a 251 kDa dimeric hydroxylase protein (MMOH) with three subunits in a $(\alpha\beta\gamma)_2$ stoichiometry, together with a reductase (MMOR) and a regulatory "B" component (MMOB) [210,214–216].

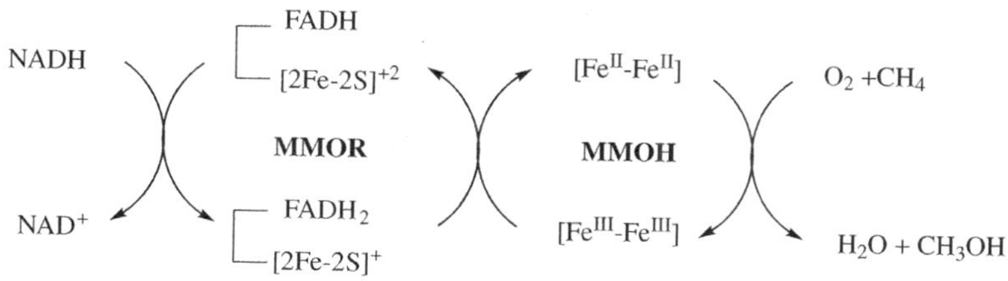

FIGURE 1.7 Catalytic mechanism of MMO.

Each subunit of MMOH contains a carboxylate- and a hydroxo-bridged dinuclear iron cluster, which is the site of catalysis, where dioxygen activation and methane hydroxylation occur. Typically, four glutamate and two histidine residues coordinate the di-iron center. The remaining of the coordination sites is occupied by solvent-derived ligands. MMOR is a 38.5 kDa iron–sulfur flavoprotein that shuttles electrons from NADH through its FAD and [2Fe–2S] cofactors to the hydroxylase active site. MMOB (15.9 kDa) does not contain prosthetic groups and modulates MMO reactivity by forming specific complexes with the hydroxylase that indirectly affect the structure and reactivity of the di-iron site (Figure 1.7) [208,215].

MMO is a classic monooxygenase in which two reducing equivalents from the cofactor are required to split the O–O bond of O_2. These reducing equivalents are used to form H_2O with one of the oxygen atoms, while the second is incorporated into methane with nearly 100% efficiency [211]. AMO and XMO show similar reaction mechanisms but differ from water-soluble proteins of the di-iron-carboxylate type in that their oxygenase component is membrane bound and the di-iron cluster is only complexed by histidines [217–219]. These enzymes contain a characteristic 8-histidine motif and a variable number of subunits. AMO consists of an integral membrane hydroxylase and two soluble proteins, rubredoxin and rubredoxin reductase, whereas XMO is a two- component enzyme, consisting of an integral membrane hydroxylase and a soluble NADH-acceptor reductase. Both enzymes accept a large substrate spectrum [144]. AMO, which catalyzes in nature the NADH-dependent hydroxylation of medium chain alkanes (C5–C12), accepts a wide range of linear, branched, and cyclic alkanes as well as alkyl benzenes [220,221]. Epoxidation of terminal alkenes, sulfoxidations and demethylations have also been reported [222–225]. XMO accepts *m*- and *p*-ethyl-, methoxy-, nitro-, and chloro-substituted toluenes, *m*-bromo-substituted toluenes and styrene, which is transformed into (*S*)-styrene oxide with high enantiomeric excesses [226]. The enzyme expressed in *E. coli* catalyzes the multistep oxidation of xylenes to the corresponding alcohols, aldehydes, and acids *via* individual monooxygenations [227,228].

Phenylalanine hydroxylase (PheOH, phenylalanine 4-monooxygenase, EC 1.14.16.1), tyrosine hydroxylase (tyrosine 3-monooxygenase, EC 1.14.16.2) and tryptophan hydroxylase (tryptophan 5-monooxygenase, EC 1.14.16.4) constitute a small family of monooxygenases that use tetrahydropterins as substrate [229]. Enzymes from eukaryotic sources are composed of a homologous catalytic domain attached to *N*-terminal regulatory domains and short *C*-terminal tetrameric domains, whereas the

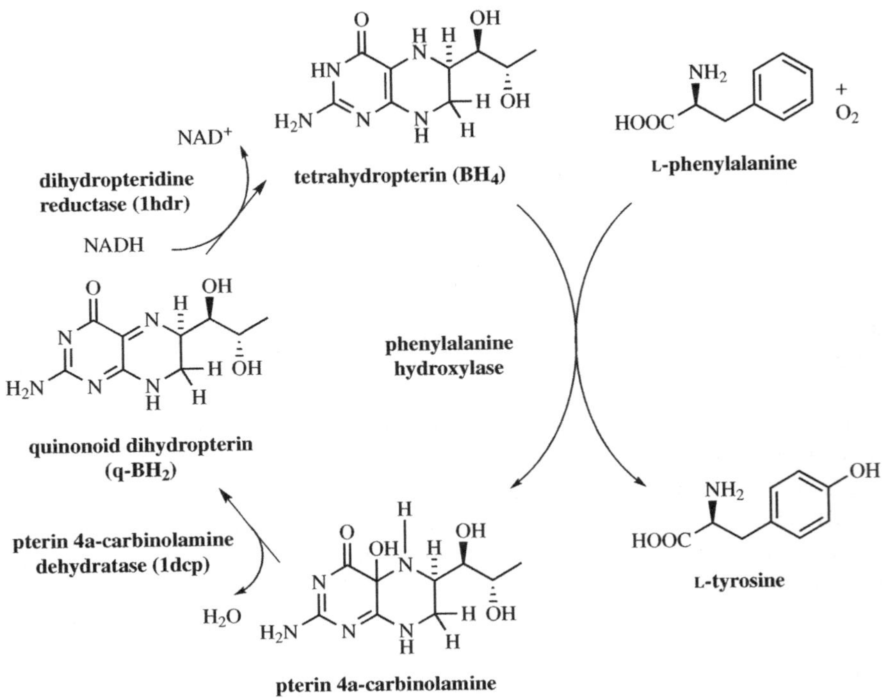

FIGURE 1.8 Reaction of phenylalanine hydroxylase (PheOH) and concomitant cofactor regeneration.

bacterial enzymes lack the *N*- and *C*-terminal domains. Each enzyme contains a single ferrous iron atom bound to two histidines and a glutamate. Although the hydroxylating intermediates in these enzymes have not been identified yet, the iron is likely to be involved. The mononuclear PheOH hydroxylates L-phenylalanine (L-Phe) into L-Tyrosine (L-Tyr) in the presence of (6*R*)-L-*erythro*-5,6,7,8-tetrahydrobiopterin (BH_4) as cofactor and dioxygen [230,231]. The enzyme is allosterically regulated by its substrates, phenylalanine and tetrahydrobiopterin. During the hydroxylation, the BH_4 cofactor undergoes a two-electron oxidation to a 4a-hydroxydihydrobiopterin (pterin 4a-carbinolamine), which is subsequently dehydrated to quinonoid dihydropterin (q-BH_2) by a carbinolamine dehydratase (EC 4.2.1.96), before being reduced by a dihydropteridine reductase (EC 1.6.99.7) (Figure 1.8) [232].

Structurally and functionally homologous to PheOH are TyrOH and TrpOH, all forming the family of the tetrahydrobiopterin-dependent aromatic aminoacid hydroxylases [232]. The three enzymes have a three-domain structure with an *N*-terminal regulatory domain (residues 1–142 in human PheOH), a central catalytic domain (residues 143–410 in human PheOH) that includes the active site-iron ligated to a 2-His-1-Glu facial triad [233,234], and a *C*-terminal tetrameric domain (residues 411–452 in human PheOH) [235–238]. The crystal structure of rat TyrOH with BH_2 has been reported, and the pterin was found to be located in the second coordination sphere of the iron center [239]. The pterin binding site has been proposed to be in a highly conserved region in all aromatic amino acid hydroxylases [231].

HEME IRON MONOOXYGENASES CYTOCHROMES P450 Cytochromes P450 (P450s) are all heme-thiolate proteins whose prosthetic group is a protoheme (iron protoporphyrin IX) with a cysteine residue as the axial ligand [240]. With the exception of soluble bacterial and fungal P450s, the majority of these enzymes are membrane bound, associated with either the inner mitochondrial or the microsomal membranes [36]. While the number of reported nonheme monooxygenases is still limited, there are more than 10,000 notified genes for these proteins, divided in over 970 families [241] and widely distributed in almost all eukaryotes, prokaryotes, and even in hyperthermophilic archaea. With more than 500 isozymes, cytochromes P450 are widespread in nature. Molecular weights are typically about 40–60 kDa, and they all show high similitude in terms of global topology, position, and orientation of their characteristic secondary structures [242]. Although enzymes with less than 40% sequence identity are placed in different gene families, the tertiary structure in the active site for proteins of different families remains unchanged even at as low as 20% sequence identity, thus pointing to a common mechanism of oxygen activation [38,243]. Regarding the primary structure, only three amino acids are absolutely conserved in all P450 enzymes, playing a key role in the maintenance of the secondary and tertiary structures, as well as in catalysis [38,244]. A cysteine residue provides the iron heme of a thiolate ligand, which is essential for the catalytic activity and common to all P450s. In contrast, variable regions are represented by the flexible substrate recognition region (SRS) [245], which enable P450s to be highly versatile biocatalysts. This explains the remarkable variety of chemical reactions catalyzed and the wide range of structurally different substrates accepted by these enzymes. Each P450 has a preferred set of substrates ranging from steroids to polycyclic aromatic hydrocarbons to fatty acids. Therefore, each must have a unique method of substrate recognition and an active site allowing regio- and stereospecific monooxygenation, yet they all appear to have similar structural cores [38]. These enzymes are essential in primary and secondary metabolism, as well as in the detoxification of xenobiotics and drugs. In mammals, P450s are involved in several processes in steroidogenesis, mainly in the gonads and adrenal glands, with six different P450s functioning in the three steroid pathways, and in the arachidonic acid cascade, producing prostaglandins and epoxy- and hydroxy eicosanoic acids. Low concentrations of these enzymes are found in the brain and skin.

This type of monooxygenases catalyzes the hydroxylation of a wide variety of substrates, involving the transfer of molecular oxygen to –C–H, –N–H or –S–H bonds in aliphatic or aromatic nonactivated compounds, from simple benzene to complex ring systems such as flavonoids or aromatic steroids (see Chapter 4) [37,246,247]. Moreover, many P450 enzymes epoxidize C–C double bonds [248]. In addition to oxygen transfer, P450s are responsible for other chemical reactions including dealkylations, *S*- and *N*- oxidations, alcohol, and aldehyde oxidations as well as dehalogenation and denitration, which are performed with remarkable selectivity [249]. P450 enzyme-catalysis requires a redox partner to transfer electrons from reduced nucleotides (NAD(P)H) to the enzyme for the binding and activation of the iron-bound molecular oxygen [37,38]. Some P450s do not require any other protein component for the reductive activation of molecular oxygen [250], but the vast majority of P450s performs the diverse range of chemical reactions after the interaction with one or more redox partners to provide their redox equivalents from electron transfer chains. Most of the electron transfer reactions begin with the transfer of electrons from NAD(P)H and end with the reductive cleavage of oxygen, with the concomitant incorporation of one oxygen atom into the substrate by the terminal P450 enzyme [251].

Although P450s are divided into gene families, they can also be classified into different groups, depending on the required redox partner. After the discovery of cytochromes P450, two main classes concerning the involved redox partners had been defined, the adrenal mitochondrial/bacterial type (Class I) [252], obtaining electrons from NADPH *via* adrenodoxin reductase and adrenodoxin, and the liver microsomal P450 (Class II) [253,254] obtaining electrons from NADPH *via* FAD and flavin mononucleotide (FMN)-containing P450 reductase [35,39,144,199,206, 255,256].

Class I: Most bacterial and mitochondrial eukaryotic enzymes belong to this class of P450 systems. The electron transfer system comprises three separated proteins: (i) a FAD-containing reductase, which transfers reduction equivalents from a pyridine nucleotide (NAD(P)H) to the second component of the system, (ii) a NAD(P)H ferredoxin reductase of the iron–sulfur [2Fe–2S] protein type, which shuttles electrons between the reductase, and (iii) the P450, reducing the cytochrome P450 itself [243]. In bacteria, all three proteins are soluble; in eukaryotes, only the ferredoxin is a soluble protein of the mitochondrial matrix, whereas the reductase and the P450 are membrane associated and membrane bound to the inner mitochondrial membrane, respectively [35,257]. Examples of Class I P450s are involved in the side-chain cleavage of cholesterol ($P450_{scc}$, CYP11A1, mammalian), the 11β-hydroxylation of 11-deoxycortisol, the production of aldosterone, in vitamin D biosynthesis, or in the 5-*exo*-hydroxylation of camphor (bacterial $P450_{cam}$, CYP101).

Class II: They occur often in eukaryotes and perform diverse reactions. In mammals they are responsible of the oxidative metabolism of endogenous compounds (fatty acid, steroids, prostaglandins) as well as exogenous compounds (therapeutic drugs, environmental toxics and carcinogens) [258]. Class II P450s from plant origin are involved in the secondary metabolism as well as in the synthesis of cutin and lignin barriers, and of defense substances [259]. In fungi, their activity includes the synthesis of membrane sterols and mycotoxins, and the detoxification of phytoalexins [243]. Microsomal cytochromes P450s are membrane-bound enzymes, and accept electrons from a microsomal NADPH-cytochrome P450 reductase (CPR), containing FAD and FMN. All drug- and xenobiotic-metabolizing cytochrome P450s, including CYP102 (P450(BM3)) isolated from *Bacillus megaterium*, as well as most of the other bacterial P450s belong to this class. These P450 systems consist on a polypeptide chain with two different domains, one containing the heme-protein and the other one containing an FAD-reductase. These soluble enzymes obtain the required electrons from an NADH-dependent FAD-containing reductase *via* an iron–sulfur protein of the [2Fe–2S] type. While FAD serves as an electron acceptor from NADPH, the FMN moiety transfers the electrons to the oxygenase protein.

After the discovery of the first bacterial P450 system, the camphor hydroxylase [260,261], it was found that this system was organized analogously to the mitochondrial ones, transferring electrons from NADH *via* an FAD-containing reductase (putidaredoxin reductase) and an iron–sulfur protein ([2Fe–2S] type, putidaredoxin) to CYP101 ($P450_{cam}$) [243]. The discovery of CYP102 (P450(BM3)) in the 1980s showed an unusual property, a fusion of the P450 reductase and reductase components into one polypeptide chain [262,263]. Since then, several different electron transfer chains for diverse P450s have been described.

Class III: In 1998, Peterson et al. described Class III P450s as enzymes that do not require an exogenous source of electrons, as they employ either endoperoxide or hydroperoxide substrates, which already contain oxygen. The mammalian enzyme thromboxane synthase (CYP5A1) and the plant enzyme allene oxide synthase (CYP74) were included in this classification. However, more recently Bernhardt and coworkers described Class III P450s as enzymes with a novel redox partner system, which were described in 2002 [264], having high similarity to those of the classical bacterial system. These enzymes are also three-component P450 systems, in which electrons are transferred from the primary electron donor (NAD(P)H) *via* a NAD(P)H-dependent FAD-containing ferredoxin reductase, and a second auxiliary redox protein. An example is the novel cytochrome CYP176A1 (P450(cin)) from *Citrobacter braakii*, in which the intermediate electron donor of the cytochrome P450 is suggested to be a flavodoxin, instead of the iron–sulfur protein characteristic of the Class I systems. Thus, the electrons are delivered *via* the redox centers FAD and FMN, and not *via* FAD and iron–sulfur cluster, resembling the eukaryotic microsomal system, though the two redox centers FAD and FMN belong to separate proteins [264,265].

Class IV: A ferredoxin and its corresponding partner, 2-oxoacid:ferredoxin oxidoreductase constitute an efficient redox chain in this type of P450s. The soluble CYP119 (EC 1.14.14.-) [266,267] from the extremely acidothermophilic archaeon *Sulfolobus solfataricus* was the first example of a P450 enzyme that does not obtain its reducing equivalents from a NAD(P)H-dependent flavoprotein. More recently, another thermophilic cytochrome P450 has been found in *Sulfolobus tokodaii* [268], but apart from the enzyme structure, named P450(st), no other information is available so far [267].

Class V: This class of P450 enzymes consists of two separated protein components: a putative NAD(P)H-dependent reductase, and a cytochrome P450-ferredoxin fusion protein [243]. There is only one known example belonging to this novel class, and it is the sterol 14α-demethylase CYP51 (EC 1.14.13.70) from *Methylococcus capsulatus*, which shows a unique primary structural organization [269]. In this enzyme, a P450 heme-monooxygenase domain is fused at the *C*-terminus to a [3Fe–4S] type ferredoxin domain *via* an alanine-rich linker region, which is thought to act as a flexible hinge allowing interactions between the two domains [243].

Class VI: This class is composed of a putative NAD(P)H-dependent flavoprotein reductase and a flavodoxin-P450-fusion protein, thus standing somewhere in between the P450(BM3) and P450(cin) systems, which principally use the same redox centers (FAD, FMN and heme) but differ in the number and characteristics of separate proteins comprising the system [243]. The first example found of this class VI P450s is the cytochrome P450-like gene from *Rhodococcus rhodochrous* strain 11Y (xp1A), in which the cytochrome P450 is fused to a flavodoxin domain at its *N*-terminus [270,271].

Class VII: In a unique type of structural organization, a cytochrome P450 is *C*-terminally fused to a reductase domain, a phthalate dioxygenase reductase domain, which is usually not associated with P450 systems [243]. The first reported enzyme of this class was the cytochrome CYP116B2 (P450(RhF)) from *Rhodococcus* sp. strain NCIMB 9784 [272], whose cytochrome P450 domain displays 55% homology to the class I CYP116 (P450(ThcB)) from *Rhodococcus erythropolis* NI86/21 [273]. In this protein, the *N*-terminal cytochrome P450 domain and the *C*-terminal reductase domain are separated only by a short linker region of 16 amino acids. In the case of P450(RhF) enzyme, three distinct functional parts can be identified: (i) a FMN-binding domain, (ii) a NADH-binding domain, and (iii) a [2Fe–2S] ferredoxin domain. The flavin cofactor of the reductase receives the electrons from the pyridine nucleotide, and in this system is FMN rather than FAD as in the other P450 systems [273].

Class VIII: This class contains P450 proteins fused to their eukaryotic-like diflavin reductase partner (CPR) in a single polypeptide chain and therefore are catalytically self-sufficient as monooxygenases. These enzymes have been found in various prokaryotes and lower eukaryotes [274]. The most deeply studied enzyme of this class is the cytosolic fatty acid hydroxylase flavocytochrome CYP102A1 (P450(BM3), EC 1.14.14.1) from *B. megaterium* [243,275]. This was the first P450 discovered as a fusion (*N*-terminal) to its NADPH-CPR (EC 1.6.2.4). The enzyme is a soluble single polypeptide with a molecular weight of 119 kDa [262,276,277], composed of a heme-containing P450 oxygenase domain connected *via* a short protein linker to a diflavin reductase domain, containing one equivalent each of the cofactors FAD and FMN [262]. The fusion of the heme domain in CYP102A1 to its diflavin reductase in a single polypeptide is an arrangement similar to that of the nitric oxide synthases [263]. CYP102A1 catalyzes the NADPH-dependent hydroxylation of medium and long-chain saturated fatty acids at ω-1, ω-2, and ω-3 positions [263,278] displaying high monooxygenase activity [279,280]. NADPH reduces the protein with electrons transferred to the FAD cofactor in the CPR domain, and subsequently they are transferred one by one from the FAD to FMN and further on to the heme iron in the substrate-bound P450 domain [278,281]. The self-sufficient CYP505A1 (P450(foxy), EC 1.14.14.1) from *Fusarium oxysporum* displays an eukaryotic counterpart of the bacterial CYP102 [282]. The protein is a fusion of a P450 and a CPR domain on one gene, and is produced as a single polypeptide [283].

Class IX (so-called Class IV according to Peterson's classification) [38]: There is only one known example belonging to this group of P450s. Nitric oxide reductase, $P450_{nor}$ (CYP55A) receives its electrons directly from reduced pyridine nucleotides without the need of an electron carrier. Among heme-thiolate proteins, $P450_{nor}$ of the filamentous fungus *F. oxysporum* was identified as a P450 with particular features [243]. The enzyme is located in the mitochondrial and cytosolic fractions of the cells, and represents the first and so far only soluble eukaryote P450 protein [284,285]. Its physiological role is part of the denitrification reaction, protecting the fungus from NO inhibition of mitochondria, especially when oxygen becomes limiting [286]. $P450_{nor}$ uses NADH as electron donor, and catalyzes, independent of other electron transfer proteins, the conversion of two molecules of nitric oxide to nitrous oxide as a P450s unique reductive process [287].

Class X: Most of P450 enzymes behave as monooxygenases, requiring a consecutive delivery of two electrons *via* different types of redox proteins. However, Class X P450 systems catalyze substrate conversion using an independent intramolecular transfer system. These enzymes include the independent cytochrome P450 allene oxide synthase (CYP74A, AOS), fatty acid hydroperoxide lyase (CYP74B/C, HPL), divinyl ether synthase (CYP74D, DES) (EC 4.2.1.92), prostacyclin synthase (EC 5.3.99.4), and thromboxane synthase (EC 5.3.99.5) [243]. AOS, HPL, and DES, located in membranes of chloroplasts, are key enzymes of plant LOX pathway, and are relatively atypical members of the cytochrome P450 family CYP74, which have not been detected in mammals [288]. These enzymes do not require O_2 or an NADPH-dependent CPR, or even the electron source NAD(P)H, for the rearrangements of fatty acid hydroperoxides [289–291]. They use the acyl hydroperoxide of the substrate as oxygen donor, thus forming new carbon–oxygen bonds in the products.

Regarding structural features of cytochromes P450, in the past years significant progresses have been made in the determination of the active-site structure of these enzymes [39]. The information has been obtained through the crystal structures of mammalian P450s [292–298], and complemented with the information available on bacterial and fungal cytochromes. The structural core of P450s is conserved, and formed by a four-helix bundle composed of three parallel helices labeled D, L and I and one antiparallel helix E [299]. The prosthetic heme group is located between the distal I helix and the proximal L helix and bound to an adjacent Cys-heme-ligand loop. This cysteine residue is absolutely conserved and is the proximal of "fifth" ligand to the heme iron, forming two hydrogen bonds with neighboring backbone amides. A further interaction is observed in some P450s with a Gln residue, and an example is CYP152A1. Mutations of these and similarly placed residues, for example, Phe393 in CYP102 (P450(BM3)), have pronounced effects on the reduction potential of the catalytic activity and on the stability of the bond between the heme iron and its ligands [39]. A highly conserved Thr residue located in the long I helix in the heme pocket preceded by an acidic residue is positioned in the active site and believed to be involved in catalysis [300,301].

The basic concepts of the catalytic cycle of P450s were developed in the early 1970s from the discovery of the role of these monooxygenases in steroid biosynthesis, and outstanding progress in the detailed understanding of the catalytic mechanism has been made in the past decade [39]. The use of a wide set of methodologies including systematic directed mutagenesis, high-resolution X-ray crystal structure determination, spectroscopic characterization of intermediates, isolation of critical steps using cryogenic or fast kinetic techniques, and computational studies gave

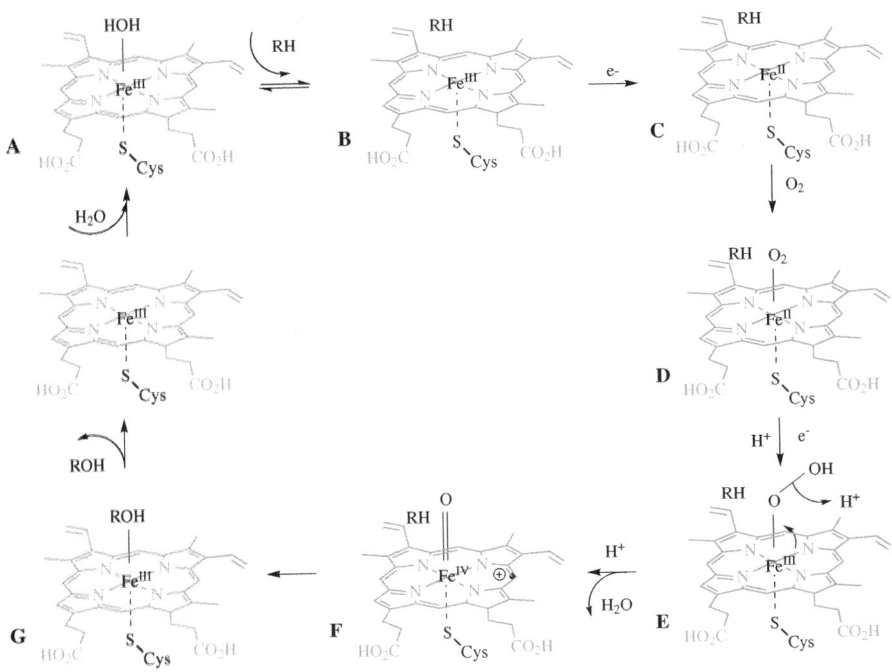

FIGURE 1.9 Cytochrome P450 catalytic cycle. The cysteine thiolate provided by the protein is represented as an S. RH, Substrate hydrocarbon; ROH, Resulting hydroxylated product. The ⊕• over the heme indicates that the radical cation is located on the porphyrins [240,302].

valuable information on the mechanism and species involved in the catalysis. The catalytic mechanism of P450 may be studied focusing on the states of heme iron and oxygen (see Figure 1.9) [240,302]. The main steps are (i) oxygen binding to the reduced heme iron and formation of an oxygenated heme Fe^{2+}–OO or Fe^{3+} –OO^-; (ii) reduction of this complex to a ferric peroxo state Fe^{3+}–OO^{-2}, which is easily protonated to form the hydroperoxo Fe^{3+}–OOH^-; (iii) a second protonation of the latter Fe^{3+}–OOH^- complex at the distal oxygen atom to form an unstable transient Fe-OOH_2, followed by the heterolytic scission of the O–O bond and release of a water molecule; and (iv) the reactions of the remaining higher valent porphyrin metal-oxo complex, named as a ferryl-oxo π-cation porphyrin radical [39]. Thus, the catalytic cycle of P450s starts with the resting enzyme in the ferric state, having a thiolate proximal ligand, which is usually a water molecule (**A**, Figure 1.9) [303], despite a few P450 enzymes do not have a ligand coordinated to the iron ion on the distal side (e.g., the case of human CP2D6) [304]. Substrate binding causes the displacement of the water ligand, resulting in a shift in the redox potential of the heme iron atom by up to 300 mV, enabling the electron transfer from the redox partner (**B**) [240]. The ferric cytochrome P450 is reduced by an associated reductase with an NADPH-derived electron to the ferrous cytochrome P450 (**C**). The formed ferrous substrate-bound protein then binds to molecular oxygen to yield the ferrous cytochrome P450-dioxygen complex (**D**). A second electron transfer generates the ferric peroxy anion, which is protonated to give the ferric hydroperoxo complex (**E**). This second electron transfer is usually (with exceptions) [305] the rate-limiting step of the catalytic cycle. The ferric hydroperoxo intermediate is unstable and upon protonation, gives a porphyrin radical cation Fe(IV) as intermediate (**F**) so-called "compound I," or, as it was alternatively proposed, an Fe(V)=O species [35]. This ferryl intermediate is two oxidation states above the resting ferric state. It attacks the substrate to produce the hydroxylated metabolite (**G**) and, after product release and equilibration with water, the resting ferric state of the enzyme is achieved to start the cycle again. The ferric peroxy anion and the ferric hydroperoxo complexes could be observed by spectroscopic and electron paramagnetic resonance (EPR) methods [306–310], showing that the oxygen of the hydroxylated compound is coordinated to the heme iron atom [311].

The interaction of the P450s with their corresponding electron donor is necessary for the catalysis. Its specificity assures a sufficient catalytic rate and discrimination among different potential donors and acceptors of electrons to protect the system from shunt reactions [35]. Since in liver microsomes many different isoenzymes interact with only one reductase type, it is expected that the binding site for reductase is similar in various cytochrome P450s. The electrons for the reduction of the heme iron atom to the ferrous state in the P450s catalytic cycle are provided—depending on the specific enzyme—by (i) CPR, which is a membrane-bound protein containing a FAD and a FMN as prosthetic groups in enzymes such as CYP3A4 and CYP2C9; (ii) an iron–sulfur protein which shuttles electrons from a flavoprotein with a single FMN prosthetic group, and this is the case of CYP101 and CYP11A1; or (iii) a P450 reductase-like domain fused to the P450 heme domain within a single polypeptide, such as in CYP102 [243,244,312]. Isolated examples of other electron donor partners such as flavodoxin [264], a 2-oxoacid ferredoxin oxidoreductase [313], and a fusion protein which incorporates both, a ferredoxin and a ferredoxin reductase into a single polypeptide with the P450 enzyme have also been reported [314]. Salt bridges are responsible for the recognition of the reductase by the P450, and for the correct orientation of both proteins to each other [257].

In mitochondrial steroid hydroxylases such as the camphor hydroxylating bacterial P450$_{cam}$ (CYP101) systems, a charge-pair interaction mechanism has been demonstrated by chemical modification, site-directed mutagenesis, and structural data of electron-transfer complexes [35,315]. Similar to microsomal reductases, the mitochondrial ferredoxin has also the function to deliver electrons to different cytochromes P450. A shuttle model is favored, where the oxidized ferredoxin interacts first with the ferredoxin reductase and further undergoes the reduction with the corresponding formation of an Fe^{3+}/Fe^{2+} iron–sulfur cluster. This cluster is then dissociated from the reductase, and interacts with the respective P450, where it delivers this electron before returning to the reductase and the second electron is transferred to the P450. However, the mechanism of electron transfer between the components of different cytochromes P450 is not well understood yet [35]. It has been suggested that posttranslational modifications can regulate these electron transfer reactions [316]. This electron transfer to the P450 seems to be low and rate limiting in P450 catalysis, therefore, engineering of this step leads to significantly improved biocatalysts [317].

Flavin-Dependent Monooxygenases Flavin monooxygenases (FMOs, EC 1.14.13.-) catalyze a wide variety of biochemical processes, including microbial biodegradation of activated aromatic compounds such as phenols, salicylic acid, or *p*-hydroxybenzoic acid, the oxygenation of nitrogen-, sulfur-, phosphorous-, selenium-, and other nucleophilic heteroatoms-containing chemicals, including many drugs, and the Baeyer–Villiger reaction [318]. Depending on the type of enzyme, flavin is either tightly but noncovalently bound and reduced in the monooxygenase moiety itself, or acts as an electron shuttle being reduced by a separate reductase component, and then bound and stabilized by the monooxygenase component. In any case, the electrons are derived from NAD(P)H, thus involving a direct coupling to the redox metabolism of living cells [144]. The overall reaction catalyzed by FMOs involves three general chemical processes: (i) the reduction of the flavin by NAD(P)H; (ii) the reaction of the reduced flavin with O_2 to provide a C-4a-flavin (hydro)peroxide which is the oxygenating agent (the peroxide for electrophilic substrates and the hydroperoxide for nucleophilic substrates); and (iii) binding, orienting, and activating of the substrate for its oxygenation by the C-4a-(hydro)peroxide (Figure 1.10) [145,318,319]. These three processes have unique requirements and usually more than one active site in the enzyme is required.

In single-component FMOs, such as *p*-hydroxybenzoic acid hydroxylase (PHBH, EC 1.14.13.2), there is an open conformation that gives access of both substrates and products to solvent, and a closed conformation for the reaction with oxygen and where the hydroxylation occurs. This closed form prevents from the destabilization of the (hydro)peroxy-flavin intermediate by the solvent. Finally, the so-called "out conformation," is achieved by movement of several angstroms of the isoalloxazine ring of the flavin toward the solvent, which exposes its N-5 for hydride delivery from

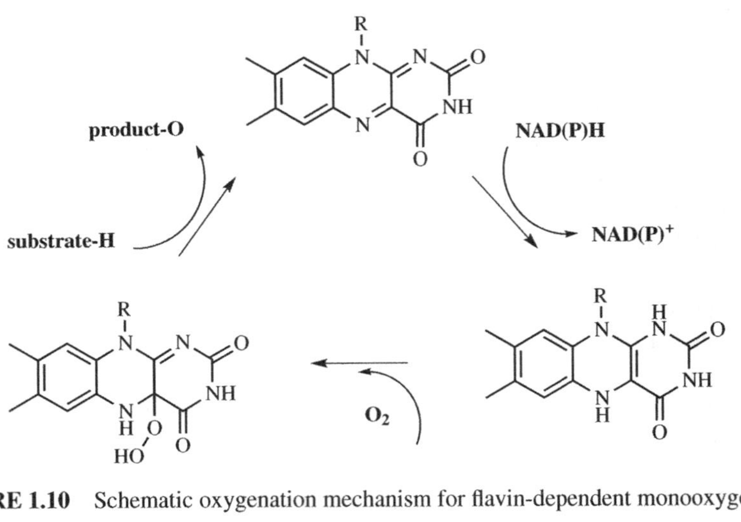

FIGURE 1.10 Schematic oxygenation mechanism for flavin-dependent monooxygenases.

NAD(P)H. The protein undergoes significant rearrangements during the catalysis. This group of enzymes have multiple active sites [318,320–326]. The enzymes PHBH and phenol hydroxylase (EC 1.14.13.7) play important roles in soil detoxification processes, and can be found in prokaryotic microorganisms (*Pseudomonas putida*) as well as in eukaryotes (*Trichosporon cutaneum*, human liver). They are capable to introduce a second hydroxyl group in *ortho-* position with respect to the previously existent one [318,325,327–329].

The group of two-component hydroxylases is characterized by two separated catalytic roles. These enzymes use an oxidoreductase to generate the reduced flavin, and an oxygenase that receives the reduced flavin and reacts with O_2 hydroxylating the substrate. These two-component systems transfer the reduced flavin from the reductase to the oxygenase, stabilizing a C-4a-peroxyflavin until the substrate binds to be hydroxylated, all before flavin oxidation and release of H_2O_2. The two-component systems have no structural or sequence similarities to the one-component enzymes. The mechanisms through which the labile reduced flavin is transferred from the reductase to the oxygenase in the two-component systems is still not well understood, although it is suggested that for most, the reduced flavin diffuses to the oxygenase components before it reacts with oxygen, so that the appropriate hydroxylating species are formed without the production of H_2O_2 [318,330,331].

For all these enzymes, flavin is reduced by NAD(P)H for its reaction with O_2. This reduction occurs stereospecifically by hydride transfer, usually with the *pro-R* hydrogen of the reduced pyridine nucleotide being transferred to the N-5 of the flavin isoalloxazine [332]. The reduction of flavin is often a critical control point in flavoprotein monooxygenase-catalyzed processes [318]. For most of the known single-component flavoprotein aromatic hydroxylases, the reduction is not effective in the absence of substrate, thus preventing the wasteful use of NAD(P)H that would produce reactive oxygen species such as H_2O_2. On the other hand, controlling the rate in the reduction does not regulate the Baeyer–Villiger oxygenases, or mammalian FMOs [333–335]. In these cases, NADPH reduces the enzymes at the same rate in the presence or absence of substrates, and the $NADP^+$ product remains tightly bound. The reduced enzyme-bound flavin then reacts with oxygen to form the C-4a-(hydro)peroxy-flavin, which is stable in the absence of substrates, and the bound $NADP^+$ stabilizes the intermediate. In the presence of substrate, the reaction with the C-4a-(hydro)peroxy-flavin is fast, yielding the oxygenated product [336,337]. Thus, turnover to produce H_2O_2 and oxidized flavin is avoided by a completely different mechanism than in the single-component aromatic hydroxylases. Within this group, Baeyer–Villiger monooxygenases (BVMOs) are NAD(P)H-dependent flavoproteins, which incorporate one atom of molecular oxygen into the substrate, and the other atom is reduced to water. These enzymes can be classified in two groups: type I BVMOs and type II BVMOs. Type I BVMOs are composed of a single chain protein containing FAD as cofactor, and use NADPH as source for electrons. A fingerprint motif was found in characterized type I BVMOs. The most outstanding example of this type of enzymes is cyclohexanone monooxygenase (CHMO, EC 1.14.13.22) from *A. calcoaceticus* NCIMB 9871, which was purified and characterized by Trudgill et al. [338]. The enzyme catalyzes the transformation of cyclohexanone into ε-caprolactone, through the participation of an FAD-4a-hydroperoxyde intermediate, which is the nucleophile that reacts with the carbonyl group [339]. The enzyme has a broad substrate specificity accepting cyclic ketones as well as open chain ketones, aldehydes, and boronic acids (see Chapter 4) [327]. On the other hand, type II BVMOs contain FMN as cofactor, use NADH as electron donor, and are composed of $\alpha_2\beta$ trimers [340]. Examples of this type of enzymes are 2,5-diketocamphane monooxygenase (2,5-DKCMO) and 3,6-DKCMO (3,6-DKCMO) from *P. putida* ATCC 17453 (NCIMB 10007) (see Chapter 4). All BVMOs cloned could be classified as type I enzymes, while no type II BVMO sequence is known. There are no crystal structures available of BVMOs that would disclose the structural features of this class of enzymes [341].

Most of biochemical and mechanistic studies have been performed with type I BVMO [341,342]. The generally accepted mechanism for Baeyer–Villiger oxidation catalyzed by BVMOs was proposed by Walsh and coworkers based on results obtained with CHMO (Figure 1.11) [200,343].

First, the reduced cofactor NADPH binds to the enzyme and the tightly bound FAD prosthetic group (**A**) is reduced to $FADH_2$ (**B**). The cofactor is held in the active site in the oxidized form ($NADP^+$) until the last step of the catalytic cycle [200]. In fact, the release of the oxidized cofactor is the rate limiting step of the catalysis [333]. In the next step, this complex (enzyme-$FADH_2$-$NADP^+$) reacts with molecular oxygen to form the flavin-peroxide species (**C**), equivalent to peracids used in the chemical Baeyer–Villiger oxidation. The C-4a-peroxiflavin intermediate **C** is in equilibrium with the corresponding hydroperoxide **D** as demonstrated by spectroscopic studies [333,344]. It has been proposed that these two forms account for the ambivalent character of the BVMOs in catalyzing both, electrophilic (e.g., sulfoxidation of thioethers) and nucleophilic (e.g., Baeyer–Villiger reaction and epoxidation of alkenes) oxidation processes [344–346]. Attack of C-4a-peroxiflavin intermediate **C** at the carbonyl functional group of the ketone substrate produces a tetrahedral "Criegee-like" intermediate (**E**). This intermediate rearranges to give the C-4a-hydroxyflavin **F** and the corresponding ester or lactone (ε-caprolactone when cyclohexanone is the substrate). Water is spontaneously eliminated from the hydroxy-flavin generating the oxidized prosthetic group FAD (**A**). The catalytic cycle is finally closed by $NADP^+$ release. As in the chemical reaction, the enzyme-catalyzed reaction proceeds with retention of configuration at the migrating carbon atom, as determined by Schwab and coworkers employing isotopically labeled substrates [347,348]. Also, the same stereoelectronic effects (primary and secondary) drive the migratory preference, as long as only one C–C bond is antiperiplanar to the O–O bond in the tetrahedral intermediate **E** [342,349].

Copper-Containing Monooxygenases Copper is a key cofactor in a diverse array of enzymes involved in biological oxidation–reduction reactions [350]. Among monooxygenases, tyrosinase (PPO, EC 1.14.18.1), peptidylglycine α-amidating monooxygenase (PAM), and dopamine β-monooxygenase (DBM, EC 1.14.17.1) are outstanding examples of copper-containing enzymes.

Many bioactive peptides need to be amidated at their carbonyl terminus to exhibit activity, but these amides are not generated through transamidation reactions.

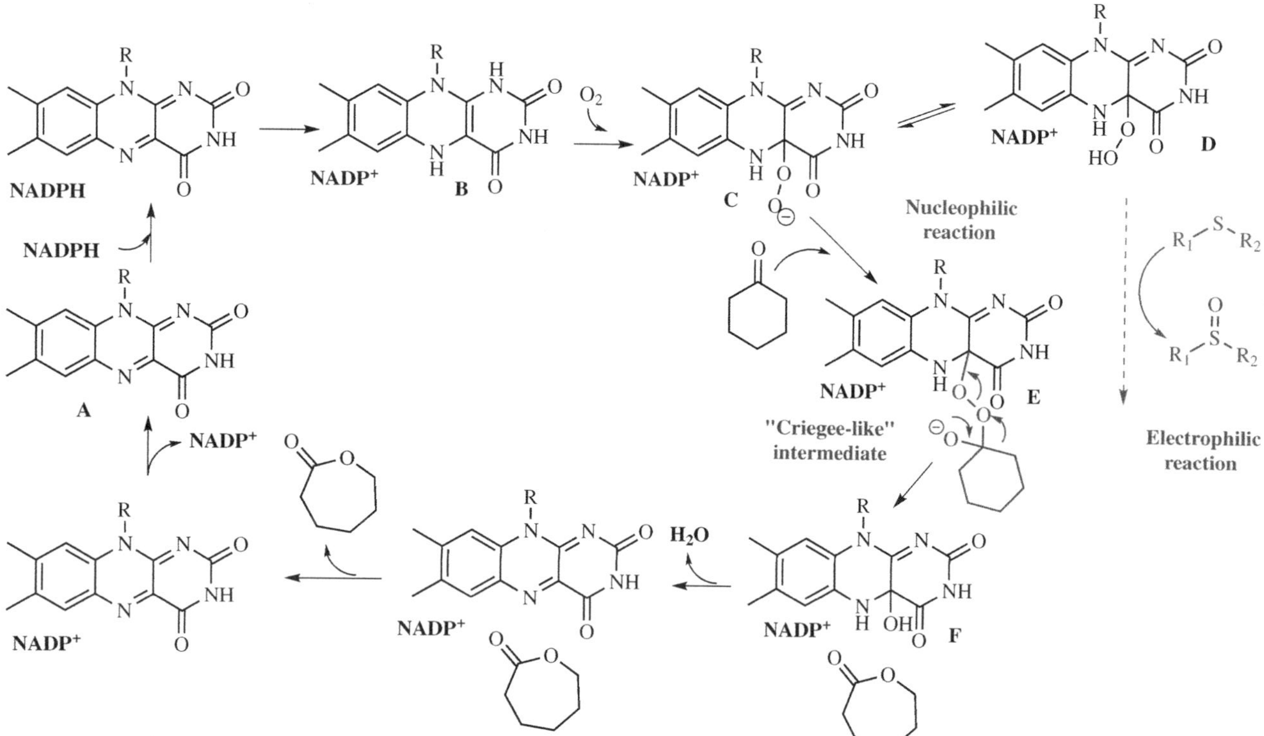

FIGURE 1.11 Proposed mechanism for the Baeyer–Villiger oxidation of cyclohexanone catalyzed with BVMOs of type I.

Instead, these amides are synthesized from glycine-extended intermediates that are converted into the active terminal amides by oxidative cleavage of the glycine N–Cα bond [351]. Although in some organisms this activity is developed by independently expressed enzymes, in higher organisms the process is catalyzed by PAM, a bifunctional copper-containing enzyme [351]. The PAM gene encodes one polypeptide with two enzymes catalyzing the two sequential reactions required for amidation, namely peptidylglycine α-hydroxylating monooxygenase (PHM, EC 1.14.17.3) and peptidylhydroxyglycine α-amidating lyase (PAL, EC 4.3.2.5). The former catalyzes the stereospecific hydroxylation of the glycine α-carbon of all the peptidylglycine substrates and the latter generates the α-amidated peptide product and glyoxylate. PHM is an ascorbate-dependent monooxygenase, containing two redox-active copper atoms that cycle between Cu^{+} and Cu^{2+} during the monooxygenase reaction [352–354]. The catalytic cycle starts with the oxidized resting form of PHM, and the reduction of the two copper atoms from Cu^{2+} to Cu^{+} by two one-electron transfers from two ascorbate molecules, which are oxidized to two semidehydroascorbate species, and then converted into one ascorbate and one dehydroascorbate molecule, producing the catalytically active PHM [354]. Then, the ternary complex of reduced enzyme with peptidylglycine substrate and molecular oxygen is responsible for the reaction progress through an alkyl radical intermediate [355,356], producing the amidated product, water, and the oxidized enzyme [351]. The reaction is stereospecific, involving the abstraction of the *pro-S* hydrogen, thus yielding the (*S*)-configurated product [357,358]. Molecular oxygen is then incorporated into peptidyl-α-hydroxyglycine product [359–361]. PHM is homologous in sequence and mechanism to another copper-containing monooxygenase, dopamine β-monooxygenase (DBM, EC 1.14.17.1), which converts dopamine into norepinephrine during catecholamine biosynthesis, thus the structural and mechanistic insights of PHM can be extended to DBM [351,362].

For these enzymes, two alternative catalytic mechanisms have been proposed differing in the nature of the hydrogen-abstracting species [363,364]. In both schemes, molecular oxygen binds to one of the copper atoms after ascorbate reduces the catalytic coppers. In the presence of substrate, each copper atom donates one electron to molecular oxygen, which undergoes a two-electron reduction to peroxide. This peroxide is protonated, forming a copper-bound hydroperoxide [365]. From this point, the two reaction schemes diverge. In one of the proposed approaches, homolysis of the hydroperoxide is coupled with the abstraction of hydrogen from the substrate [363]. The second approach relies on an active-site residue, which is supposed to be a tyrosine residue, donating a hydrogen atom to form a copper-bound radical [364]. Both mechanisms result in a copper-alkoxide complex, which dissociates to yield the free enzyme and the product.

Certain oxygenases are usually recognized separately from monooxygenases, for example, copper-containing oxidases [149,366]. One of these enzymes is a PPO – also known as tyrosinase or monophenol monooxygenase (EC 1.14.18.1), which is a monooxygenase that uses molecular oxygen as the electron acceptor in the site-specific hydroxylations and oxidations of phenols and catechols. Tyrosinase is a copper monooxygenase that belongs to a larger group or proteins named type 3 copper proteins [367–369]. Together with laccases (EC 1.10.3.2), tyrosinases belong to the group of PPOs, and are able to catalyze the oxidation of aromatic compounds by oxygen. Tyrosinase catalyzes both the *o*-hydroxylation of monophenols (cresolase activity) and the oxidation of *o*-diphenols to the corresponding *o*-quinones (catechol oxidase activity, EC 1.10.3.1) [350,370–372]. The enzyme is widely distributed throughout the phylogenetic scale including bacteria, fungi, plants, and mammals, and usually presents different characteristics in different organs of the same organisms, such as in roots and leaves of higher plants [373]. These enzymes are involved in the melanin pathway and are specifically responsible for the first steps of melanin biosynthesis, which involves the oxidation of L-tyrosine yielding L-3,4-dihydroxyphenylalanine (L-DOPA), and leading to the formation of L-dopaquinone and L-dopachrome [374]. The *o*-quinone products thus produced are spontaneously converted into melanin pigments *via* nonenzymatic reaction pathways [375]. The peculiarity of tyrosinase is that it catalyzes the *o*-hydroxylation of monophenols displaying monooxygenase activity, and also the subsequent oxidation of the resulting *o*-diphenols into reactive *o*-quinones, showing their catalytic oxidase activity [368]. Both reactions proceed using molecular oxygen.

Similar to other PPO, like catechol oxidase (EC 1.10.3.1), which is a ubiquitous plant enzyme catalyzing the oxidation of a broad range of catechols to the corresponding *o*-quinones, and hemocyanin, acting as an oxygen carrier and storage protein in mollusks and arthropods—tyrosinase has a coupled dinuclear active site [367,370]. Despite large differences in primary, tertiary, and quaternary structure, the oxygen binding-site is highly conserved in all these enzymes, suggesting that these proteins have evolved from the same origin. Basically, the catalytic center consists of a hydrophobic pocket inside a helix bundle comprising four densely pocket helices [376]. The structure is maintained by electrostatic and cation-π interactions among the helical segments. Two histidine-rich regions named CuA and CuB are the peptidic segment involved in the two coppers. Both regions contain three perfectly conserved histidine residues [370], and oxygen binds as a side-on peroxide bridge [376]. The catalytic mechanism of tyrosinase has been object of intensive research because of the complexity and peculiarities of the enzyme, such as the existence of two catalytic activities at the same active site and a lag period displayed by the cresolase activity which depends on the presence of catechols—the product of this reaction and the substrate for the oxidase activity [377]. The most accepted proposed reaction mechanism describes a common catalytic site for the two activities with three different forms of the enzyme, called, *met*-, *oxy*-, and *deoxy*-forms, according to the absence or presence of oxygen and the oxidation state of the copper ions [Cu(II)/Cu(I)] [368,372]. The *oxy*-form of tyrosinase and the appropriate substrates initiate either the oxidase- or the hydroxylase-cycles. Native tyrosinase occurs in the inactive *met*-form in which the binuclear copper site is in an inadequate oxidation state [Cu(II)] to bind dioxygen [378]. A two-electron reduction by a catechol converts *met*-tyrosinase into *deoxy*-tyrosinase, which readily binds dioxygen giving *oxy*-tyrosinase (Figure 1.12) [370].

In the *oxy*-form, one dioxygen molecule coordinates between the coppers, each of which is ligated to the protein matrix by three histidine residues provided by an

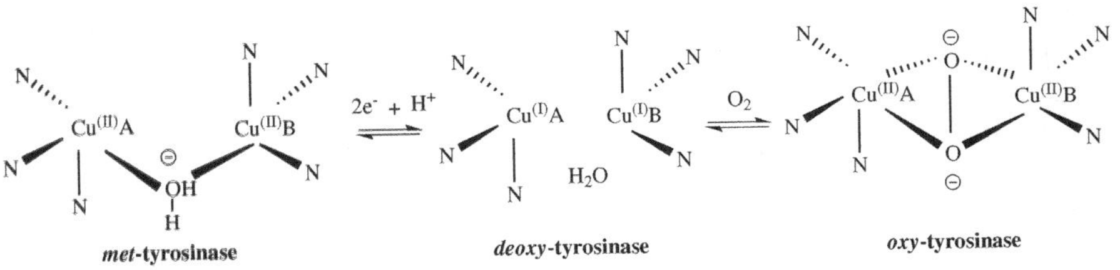

FIGURE 1.12 Binuclear copper-site in the catalytic mechanism of tyrosinase.

antiparallel α-helix pair, except for one His at the CuA motif that comes from a flexible loop. This histidine residue has been proposed as the residue involved in the proton shift from the *ortho-* position of the substrate tyrosine during the catalysis [379]. Oxygen binding induces an oxidative change in the valence of the copper ions that are at the Cu(I) state in the *deoxy*-form and become Cu(II) upon oxygenation. Both phenols and catechols are subsequently oxidized to *o*-quinones by *oxy*-tyrosinase, but the mechanisms of these oxidations are different [370,380]. Oxidation of catechols leads to *met*-tyrosinase, which cannot bind oxygen to regenerate *oxy*-tyrosinase. Only in the presence of a second catechol molecule the *met*-tyrosinase is reduced to *deoxy*-tyrosinase, which is able to regenerate the *oxy*-form (Figure 1.13).

Oxidation of monophenols by *oxy*-tyrosinase gives as final product an *o*-quinone and *deoxy*-tyrosinase, which binds further oxygen, and the oxidation cycle continues to completion (Figure 1.14) [370].

While all enzyme forms oxidize catechols to *o*-quinones, *met*-tyrosinase is inert toward phenols, though they bind reversibly to the active site. Therefore, reaction of phenols with native tyrosinase does not occur unless a catechol (added or formed by an indirect mechanism) leads to *deoxy*-tyrosinase formation. This aspect of phenol oxidation is responsible for the characteristic "lag period" observed in *in vitro* tyrosinase oxidations.

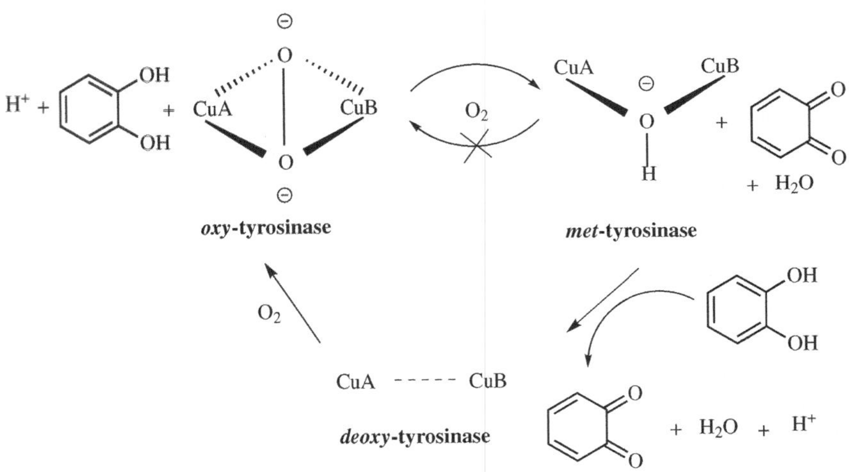

FIGURE 1.13 Catecholase cycle of tyrosinase.

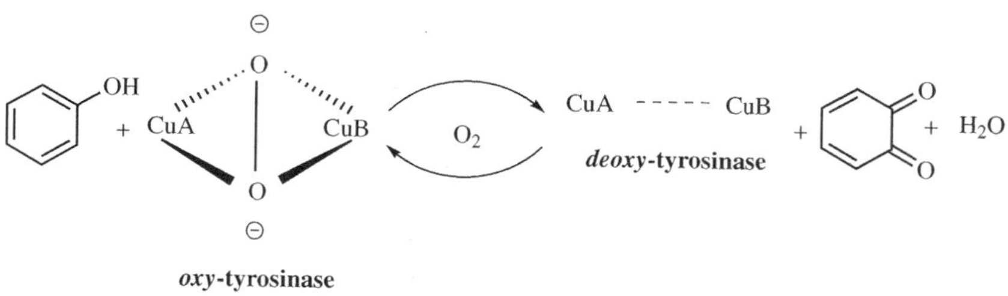

FIGURE 1.14 Cresolase cycle of tyrosinase.

Cofactor-Free Monooxygenases Although the majority of oxygenases require transition metal ions and/or organic cofactors, several enzymes undergo hydroxylations without the apparent need of cofactors [381,382]. Herein, a mechanistically intriguing problem is how triplet dioxygen is activated to react with a singlet organic compound. Among bacterial monooxygenases, quinone-forming monooxygenases involved in the tailoring of polyketides and luciferase monooxygenase from *Renilla* sp. (RLuc) have been identified as cofactor-independent enzymes [382]. Cofactor-independent monooxygenases catalyze very different reactions and belong to several different protein families, reflecting their diverse origin. Nevertheless, they all share the common mechanistic concept of initial base-catalyzed activation of their substrates and "substrate-assisted catalysis." *Renilla* luciferase possesses an α/β-hydrolase architecture for oxygenation reactions. The enzyme catalyzes the monooxygenation of coelenterazine to coelenteramide and CO_2, emitting blue light [382].

The biosynthesis of polyketides involves reactions catalyzed by polyketide synthases, which polymerize simple fatty acids to a polyketide chain, mediate in cyclization steps, and further reactions catalyzed by so-called tailoring enzymes. Type I polyketide synthases are multifunctional proteins that contain multiple active sites and are responsible for one complete cycle of building and modifying of the polyketide backbone. Type II polyketide synthases are constituted of monofunctional proteins that act iteratively [383–386]. Tailoring enzymes comprise reductases, methyltransferases, glycosyltransferases, decarboxylases, and different type of oxygenases. An important tailoring step in the synthesis of a number of aromatic polyketides is the oxidation of naphthacenone- and anthrone-type precursors of aromatic polyketides to the corresponding quinone derivatives. Functional studies on quinone-forming monooxygenases are available since 1990s, when tetracenomycin F1 monooxygenase (TcmH) was isolated from *Streptomyces glaucescens* by Hutchinson and colleagues [387]. The enzyme is a 12.6-kDa protein catalyzing the oxidation of the naphthacenone tetracenomycin F1 at C-5 to 5,12-naphthacenequinone tetracenomycin D3. This quinone-forming monooxygenase requires only dioxygen for activity and uses its substrate as reducing equivalent for the reduction of one oxygen atom of dioxygen to water. It does not possess either prosthetic groups known for monooxygenases, or contain or require metal ions for its catalytic process, albeit cofactors such as flavins or transition metal ions are thought to be required for the "activation" of molecular oxygen or the substrate [381]. It was suggested that sulfhydryl groups and histidine residues could be essential for catalysis. A histidine residue has been proposed to act as a general base, allowing the dehydration of the proposed semiquinone peroxide intermediate [387]. The next quinone-forming monooxygenase to be isolated was ActVA-Orf6 protein, from *Streptomyces coelicolor*, which showed 39% identity to TcmH and catalyzed the oxidation of 6-deoxydihydrokalafungin to dihydrokalafungin [382,388–390]. This enzyme was suggested to have a relaxed substrate specificity, since in addition to its tetracyclic natural substrate, it also oxidizes tricyclic anthrones [389]. The ActVA-Orf6 protein is a homodimer with a ferredoxin-like fold and, contrary to TcmH—which possess catalytically relevant sulfhydryl groups—it does not contain any cysteine residue. Other homologous monooxygenases have been identified in polyketide synthase gene clusters, such as ElmH from *Streptomyces olivaceus* and anthrone monooxygenase AknX from *Streptomyces galilaeus* [391,392]. However, the structure of the quinol oxidizing protein YgiN of *E. coli*, which is similar to that of ActVA-Orf6, suggests that this family of monooxygenases is not restricted to the polyketide biosynthetic pathways of Gram-positive bacteria [393]. This type of monooxygenases may be useful biocatalysts for the modification of polyketide structures [382]. Combination of genetic modules for the synthesis of the "core polyketide" with genes coding for tailoring enzymes that catalyze hydroxylations, methylations, or glycosylations can expand molecular diversity among polyketide antibiotics and other aromatic polyketide-based pharmaceuticals [382,394–396].

1.3.2.2 Dioxygenases Dioxygenases belong to the large family of mononuclear nonheme iron-containing oxygenases, and are a heterogeneous group of enzymes that introduce two oxygen atoms into a C–C double bond. This class of enzymes contains either a high-spin ferrous site involved in O_2 activation, or a high-spin ferric site that activates substrates [144]. The former group includes extradiol dioxygenases and Rieske-type enzymes, whereas the latter comprises LOX and intradiol dioxygenases [208]. The ferric site is coordinated by a variable histidine-rich ligand environment and catalyzes intradiol aromatic ring cleavage and lipoxygenation. In contrast, the ferrous site is invariably coordinated by two histidines and one aspartate or glutamate, which is a recurring motif referred to as the 2-His-1-carboxylate facial triad [144]. The remaining three ligand sites, which in the resting enzyme can be vacant or occupied by water, OH^-, or protein ligands, are available for the binding of substrates, cofactors, and/or O_2 during catalysis.

Iron(III) Dioxygenases: Enzymes that Function by Substrate Activation

LIPOXYGENASE Lipoxygenase (LOX, EC 1.13.11.-) belongs to a group of enzymes found mainly in animals and plants, involved in the regulation of polyunsaturated fatty acid metabolism [397]. These enzymes are nonheme iron dioxygenases that catalyze the regio- and stereospecific hydroperoxidation of polyunsaturated fatty acids by incorporation of molecular oxygen into the 1,4-*cis*,*cis*-diene moiety of fatty acid substrates [398]. Plant LOXs usually act on linoleic acids involved in growth regulation and wound repair, whereas human LOXs catalyze the peroxidation of arachidonic acid in the biosynthesis of leukotrienes and lipoxins [399]. X-ray structures of several soybean LOXs in the iron(II) state reveal a high-spin metal(II) center with distorted octahedral geometry [400–403]. The catalytic domain in LOXs is bigger than the *N*-terminal domain, composed roughly of seven hundred amino acids, organized in twenty-three α-helices and two small antiparallel β-sheets. Their structure reveals a nonheme mononuclear iron site within the helical bundle of the catalytic domain. The iron atom coordinates to five amino acid ligands: the imidazole *N*-atoms of three histidine ligands (His499, His504, and His690 numbering of the soybean LOX), a carboxylate oxygen of the *C*-terminal Ile839 residue, and a carbonyl oxygen of the amide backbone of a weakly bound (3 Å) Asn residue, in addition to a water molecule [397,398]. Rabbit LOX differs from the soybean enzymes in that the asparagine residue is replaced by a fourth, more strongly bound histidine residue, and it is suggested that other mammalian enzymes have similar active sites [398]. The imidazole-rich active sites give rise to a high Fe^{III}/Fe^{II} redox potential, which is proposed to be crucial for the catalytic role of the iron center in the enzyme. The iron(III) form is the catalytically active state [404,405]. The iron(II) form is activated solely by oxidation with the fatty acid hydroperoxide product [406]. The reaction profile of the iron(II)-enzyme and substrate in the presence of air shows an initial lag phase with no activity. This phase is followed by an increase in enzymatic activity as the fatty acid hydroperoxide formed by nonenzymatic auto-oxidation of the substrate, oxidizing the iron(II)-form to active iron(III)-enzyme [398]. Active LOX has a high-spin six-coordinated iron(III) metal center which is axial in the soybean enzyme, but rhombic in the rabbit LOX [406,407]. This difference has been attributed to the substitution of the weak Asn amide ligand in the former with a much stronger His ligand in the latter.

INTRADIOL-CLEAVING CATECHOL DIOXYGENASES Intradiol-cleaving catechol dioxygenases are enzymes found in an ample range of soil bacteria and are responsible for the last step in the biodegradation of aromatic rings, yielding open-chain intermediates [398]. Thus, these enzymes convert dihydroxybenzenes into acyclic compounds, which are then utilized as carbon sources for cell growth [208,398,408]. Catechol dioxygenases catalyze the oxidative cleavage of an aromatic double bond in a catechol molecule, inserting both oxygen atoms from an O_2 molecule into the product. Depending on the position of the cleaved double bond relative to the hydroxyl groups, catechol dioxygenases can be classified in two families: (i) the intradiol-cleaving catechol dioxygenases, which cleave the carbon–carbon bond of the enediol moiety (leading to muconic acids), and (ii) the extradiol-cleaving catechol dioxygenases, which cleave the bond adjacent to the enediol, yielding muconic semialdehyde adducts (Figure 1.15).

These enzymes share similar substrates, but the intradiol- and extradiol-cleaving enzymes exhibit almost complete exclusivity in their oxidative cleavage products, suggesting two different catalytic mechanisms [398]. The intradiol-cleaving catechol dioxygenases use an $[Fe(III)\text{-}(His)_2(Tyr)_2]$ active site, whereas extradiol-cleaving enzymes contain a $[M(II)\text{-}(His)_2(Asp/Glu)]$ motif in the active site, known as 2-His-1-carboxylate facial triad, the metal being typically iron(II), although manganese(II) is found in a few cases [408–414]. The most thoroughly studied enzyme of the class

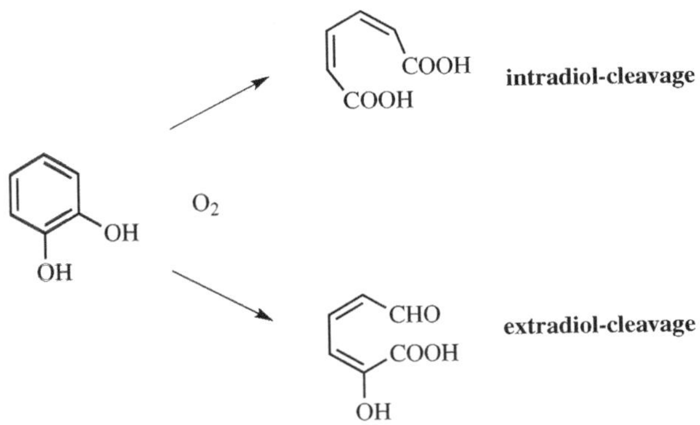

FIGURE 1.15 Intradiol vs. extradiol cleavage reaction mediated by catechol dioxygenase.

of intradiol-cleaving dioxygenases is protocatechuate (3,4-dihydroxybenzoate) 3,4-dioxygenase (3,4-PCD, EC 1.13.11.3) [398]. A number of crystallographic studies of 3,4-PCD from *P. putida* have been carried out, giving structural information of the iron(II)-state, enzyme–substrate complexes and complexes with inhibitors [415–419]. Moreover, crystal structures of the related catechol 1,2-dioxygenase from *Acinetobacter* sp. ADP1 have been solved [420]. Iron(II) 3,4-PCD from *P. putida* shows a trigonal bipyramidal iron center with four endogenous protein ligands (His460, His462, Tyr408, and Tyr447) [415,416]. The fifth coordination position located in the trigonal plane, is occupied by a solvent-derived ligand, identified as an hydroxide, thus giving rise to a charge-neutral iron(III) active site [421]. The structure of the enzyme–substrate complex reveals that the substrate binding to the iron center results in the displacement of the hydroxide and the axial Tyr447 residue [417,418]. The catechol molecule donates its protons to the displaced ligands and chelates to the metal center as a dianionic ligand. The two Fe–O bonds are *trans-* to the His460 and Tyr408 residues, and differ by 0.2 Å in length, presumably due to the distinct effects of the histidine and tyrosinate groups. The proposed catalytic mechanism postulates that the metal center retains its iron(III) character throughout the catalytic cycle (Figure 1.16) [422–424].

The first step in the catalytic mechanism is the binding of the catecholate substrate to the iron(III) center, displacing the hydroxide and the axial Tyr447 residue, and generating a square pyramidal [Fe(His)$_2$(Tyr)$_2$-catecholate] complex (**A**). The covalent character of the interaction iron(III)-catecholate introduces a semiquinonate radical character to the bound substrate and makes it susceptible to O_2 attack, leading to a alkylperoxo-iron(III) intermediate (**C**) *via* the transient species **B**. This intermediate then undergoes a Criegee-type rearrangement (with acyl migration) to form muconic anhydride **D** and an Fe^{III}-OH species which acts as the nucleophile to convert the anhydride into the ring-opened product **E**. Recent calculations suggest that the tridentate alkylperoxo intermediate **C** may evolve to a hydroperoxide bound in a bidentate fashion to the iron(III) center, which undergoes Criegee rearrangement to

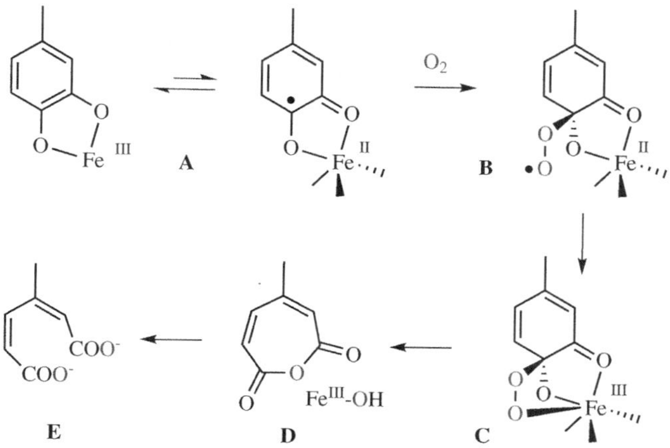

FIGURE 1.16 Proposed catalytic mechanism for intradiol-cleaving catechol dioxygenases.

yield anhydride **D** or, alternatively, the anhydride results from homolytic O–O bond cleavage and further electron transfer [425,426].

Iron(II) Dioxygenases: Enzymes with the 2-His-1-carboxylate Facial Triad Motif. Oxygen Activation The iron(II) center in these enzymes is invariably coordinated by three protein residues—two His and one Asp or Glu—arranged in one face of an octahedron, a recurring motif referred to as the 2-His-1-carboxylate facial triad [233,427]. This type of dioxygenases comprise enzymes for the biodegradation of aromatic molecules catalyzing oxidative ring cleavage, or the *cis*-dihydroxylation of arenes, as well as the superfamily of enzymes that require α-ketoacids as cosubstrates. Sequence comparisons show that the 2-His-1-carboxylate triad is conserved within each group but the sequence motif differs from each other [398]. This facial triad serves as an adequate platform for binding divalent metals. The three remaining sites on the opposite face of the octahedron are available for the binding of exogenous ligands. In the resting enzymes, these sites are usually occupied by solvent molecules but can accommodate both, substrate (or cosubstrate) and O_2 in later steps of the catalytic cycle. A general catalytic mechanism at the iron(II) center in various enzymes belonging to this superfamily has been proposed on the basis of spectroscopic and crystallographic studies [208,427,428]. The iron(II) center is six coordinated at the start of the catalytic cycle, and relatively unreactive toward O_2. Oxygen binding then initiates the oxidative mechanism, which is specific for each subclass. The metal center becomes able to bind O_2 only when substrate and cofactors are present in the active site, promoting strong coupling between the reduction of O_2 and the oxidation of substrate.

EXTRADIOL-CLEAVING CATECHOL DIOXYGENASES A second class of catechol dioxygenases includes the enzymes responsible for the extradiol cleavage of dihydroxyarenes, constituting the more common pathway for the biodegradation of aromatic molecules in soil. While the intradiol-cleaving enzymes utilize an iron(III) active

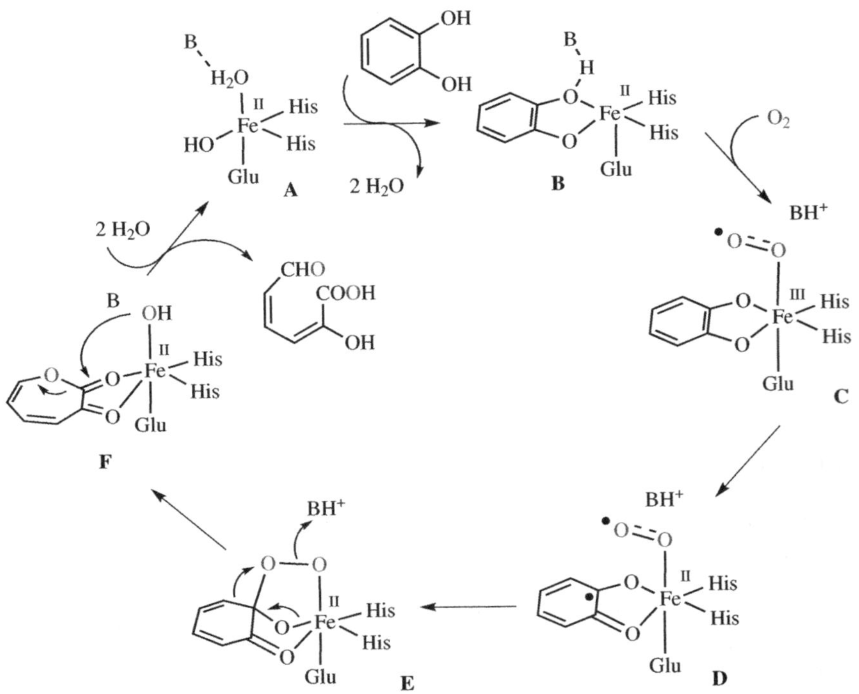

FIGURE 1.17 Proposed mechanism for the extradiol-cleaving catechol dioxygenases.

site, extradiol-cleaving dioxygenases use mainly iron(II) or else Mn(II). Crystallographic data show that the metal center is in a square pyramidal active site with a 2-His-1-carboxylate facial triad [398]. Steady-state kinetic studies revealed that this type of enzymes have an ordered catalytic mechanism with substrate binding prior to O_2 activation [409]. The coordination of the substrate serves as a trigger and significantly increases the affinity of the metal center for oxygen [429]. Crystallographic studies on 2,3-dihydroxybiphenyl 1,2-dioxygenase (BphC, EC 1.13.11.39), 3,4-dihydroxyphenylacetate 2,3-dioxygenase (HPCD, EC 1.13.11.15), and protocatechuate 4,5-dioxygenase (LigAB, EC 1.13.11.8) demonstrate that the first step of the accepted mechanism is catechol binding, resulting in the displacement of the two water ligands of the enzyme in its resting state (**A**) to form a square pyramidal metal center (**B**) (Figure 1.17) [430–434]. In the BphC structure (isolated from *Pseudomonas* sp.), one oxygen atom of the bidentate catecholate occupies the vacant position in the iron(II)-state of the enzyme, while the other occupies the site *trans*- to His210 residue, leaving the site *trans*- to Glu260 vacant [398]. The catechol binds in an asymmetric fashion to the iron(II) center as a monoion, with Fe–O bonds differing by 0.2–0.4 Å [435–438]. This monoanionic nature of the catechol substrate in extradiol dioxygenases contrast with the dianionic catecholate character commonly found in iron(III)complexes [408,409]. Oxygen binding is thus expected to occur at the vacant site *trans*- to Glu260, leading to a six-coordinated species. The change in the covalency of the interaction (substrate-Fe vs. H_2O–Fe) would decrease the redox potential of the metal center, making it more susceptible to react with O_2. The ternary enzyme–substrate–O_2 complex thus formed (**C**) consists of the 2-His-1-carboxylate triad on one face of the metal octahedron and, on the opposite face, a bidentate monoanionic catecholate and O_2. This arrangement juxtaposes the two substrate molecules into the proper orientation for reaction. It remains unclear how the electrons flow from substrate to oxygen [398,426,439].

In one of the proposed mechanisms [398], oxygen binding to the iron(II) center results in one-electron transfer from the metal to O_2, forming the iron(III)-superoxide complex (**C**), and another one-electron transfer from substrate to metal forms a semiquinonate-iron(II)-superoxide species (**D**). The conserved His residue in the second coordination sphere may serve as the base depicted in the proposed mechanism. This residue is supposed to act later as an acid to facilitate O–O bond lysis in the ring cleavage step, and to protonate the oxide. Such oxide becomes the hydroxide ligand needed to hydrolyze the lactone. The next step in the mechanism is the nucleophilic attack of the incipient superoxide on the aromatic ring that generates an alkylperoxo intermediate (**E**), which undergoes a Criegee rearrangement to generate the seven-membered α-oxolactone ring (**F**), which hydrolyzes affording the product [398]. The alkylperoxo intermediate was observed in the crystallization of homoprotocatechuate 2,3-dioxygenase (HPCD, from *Brevibacterium fuscum*) in the presence of the low-rate substrate 4-nitrocatechol [440]. The regiospecificity of the cleavage has been rationalized invoking stereoelectronic factors (the peroxo group is positioned axially with respect to the cyclohexadiene ring providing optimum alignment for the required 1,2-shift), or acid–base catalysis [426]. However, considering the similarities to the intermediates of the intradiol-cleaving enzymes, the found regiospecificity is not fully understood, and other mechanisms have been postulated, including the formation and cleavage of an epoxide before lactone generation [398].

ENZYMES REQUIRING α-KETOACIDS AS COSUBSTRATE Enzymes requiring α-ketoacids as cosubstrates constitute the largest and most diverse family of mononuclear nonheme iron enzymes, catalyzing many pivotal metabolic transformations [441–443]. Substrate oxidation—typically involving (but not limited to) the functionalization of an inactivated C–H bond—occurs with the oxidative decarboxylation of an α-ketocarboxylate, such as 2-oxoglutarate. For the hydroxylation, one atom of dioxygen is incorporated into the product, while the other atom ends up on the carboxylate derived from the ketoacid, constituting an intermolecular variant of the dioxygenase reaction (Figure 1.18) [398].

Moreover, in other oxidative processes catalyzed by these enzymes, only one of the dioxygen atoms is finally incorporated into the secondary substrate (yielding succinate), while the other one is reduced to water. As an example, the primary

$$\mathrm{R{-}H} + \mathrm{R'COCOOH} + O_2 \longrightarrow \mathrm{R{-}OH} + \mathrm{R'COOH} + CO_2$$

FIGURE 1.18 Hydroxylation of unactivated C–H bonds mediated by α-ketoacid-dependent dioxygenases.

substrate undergoes a two-electron oxidation without oxygen incorporation to form a ring (clavaminate synthase, which catalyzes the three steps: (i) hydroxylation, (ii) cyclization, and (iii) desaturation), or a ring expansion during antibiotic biosynthesis (deacetoxycephalosporin C synthase) (Figures 1.19A and B, respectively).

α-Ketoacid-dependent enzymes are found in microorganisms, plant and animals, and play important roles in biochemical pathways that have environmental, pharmaceutical, and biological significance [398]. Enzymes of this family also appear in microbes that use alternative energy sources, such as sulfonates and hypophosphites. For instance, taurine/2-OG dioxygenase (TauD, EC 1.14.11.17) enables *E. coli* to use the aliphatic sulfonate taurine as a sulfur source during periods of sulfate starvation [444]. Despite the vast different transformations these enzymes are able to carry out, the members of this family require one equivalent of iron(II), an α-ketoacid, dioxygen, and ascorbate for full activity. Crystal structures of a number of different 2-oxoglutarate-dependent enzymes show a double-stranded β-helix as common architecture [445–452]. While many of these enzymes show minimal sequence identity, there is a high degree of sequence homology. The conserved residues of the facial triad bind the high-spin iron(II) center, while other conserved arginine, lysine, and serine residues serve to bind the anionic C-5 carboxylate of 2-oxoglutarate [398]. However, several enzymes in this class have substrates with a built-in α-ketoacid function, and thus do not require 2-oxoglutarate as cosubstrate. These enzymes appear to have different topologies from enzymes utilizing 2-oxoglutarate, and

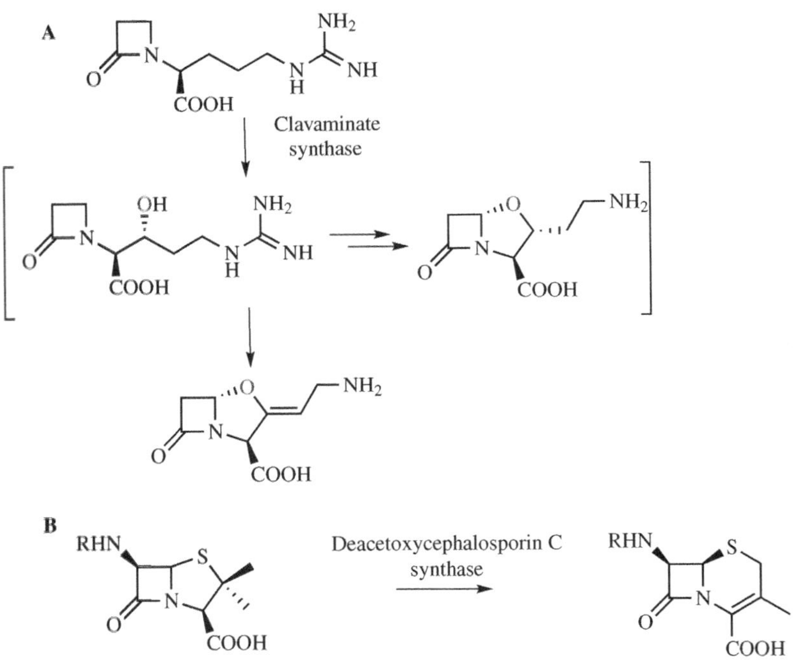

FIGURE 1.19 Two-electron oxidations mediated by α-ketoacid dependent dioxygenases. **A**: Ring formation without oxygen incorporation. **B**: Ring expansion.

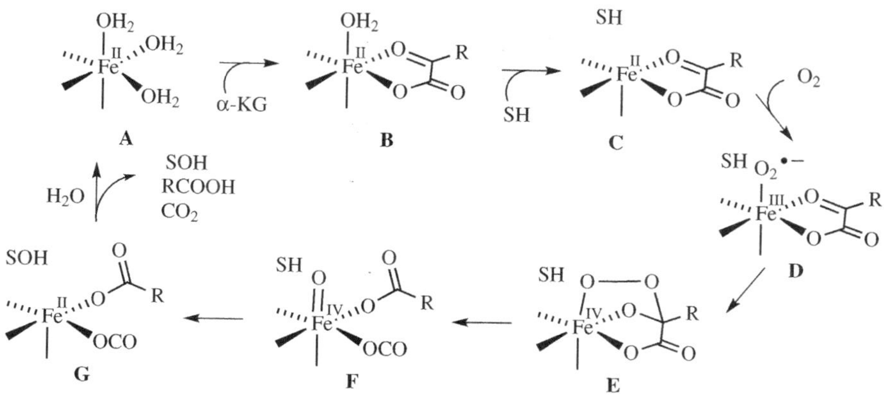

FIGURE 1.20 Proposed reaction mechanism for the α-ketoacid-dependent enzymes.

include 4-hydroxyphenylpyruvate dioxygenase (HPP dioxygenase, EC 1.13.11.27) and 4-hydroxymandelate synthase (EC 1.13.11.46), which utilize the same substrate but catalyze different transformations. A common mechanism was proposed for α-ketoacid-dependent iron oxygenases (Figure 1.20) [208,408,443], in which the iron(II) center in the resting enzyme is ligated to the 2-His-1-carboxylate facial triad and three water molecules complete the coordination sphere (**A**) [453–455].

The α-ketocarboxylate binds orderly, prior to binding of either dioxygen or substrate, by displacement of two water molecules, leading to a six-coordinated iron(II) center with a bidentate α-carboxylate (**B**) [445,446,449,452,456–462]. The equatorial plane of the distorted octahedron is composed of the nearly planar five-membered ring of the chelated α-ketocarboxylate, as well as a histidine and a carboxylate of the facial triad, with the α-keto- group bound in *trans* conformation to the aspartate or glutamate residue. Apical sites are occupied by the remaining histidine residue and one water molecule, which appear to have hydrogen-bonding interaction with the carboxylate residue. The next step in the catalytic cycle is still under discussion. Oxygen binding may occur prior to substrate binding, or the sequence may be the opposite, with substrate binding occurring prior to oxygen [456,457,462–466]. Later studies show that substrate binding results in the formation of a coordinatively unsaturated iron(II) center, presumably by loss of the remaining water ligand, that primes the metal center for oxygen binding and activation [446,448], supporting the proposed substrate-induced conversion of a six-coordinated to a five-coordinated iron center. Substrate binding may induce a shift of the carboxylate residue, removing the hydrogen-bonding interaction with the water ligand in the binary complex, thus facilitating its withdrawal. In this mechanistic proposal, the binding of dioxygen leads to an adduct with significant iron(III) superoxide radical anion character [456,467,468]. The oxidative decarboxylation of the α-ketoacid can be uncoupled from substrate oxidation, particularly when no substrate or a poor substrate analog is employed [456,469–471]. The attack of O_2 on the coordinated α-ketoacid leads to the formation of a high-valent intermediate responsible for the oxidation of the substrate in α-ketocarboxylate-dependent enzymes. Such species has been postulated as an iron(IV)-peroxo species (**E**) that can lose CO_2 to form an iron(II)-peracid adduct and then undergo heterolytic cleavage of the O–O bond to yield an iron(IV)-oxo species (**F**). Substrate hydroxylation is postulated to occur *via* an analogous mechanism to that of cytochrome P450, that is, a two-step process involving hydrogen atom abstraction by the oxo-iron(IV) species followed by oxygen rebound [398]. The difference is that P450 uses a formal Fe(III)/Fe(V)=O couple, while the α-ketoacid-dependent enzymes must use a formally Fe(II)/Fe(IV)=O couple. Carbon dioxide is a product of reactions catalyzed by these classes of enzymes, and it was suggested that CO_2 dissociation follows release of the hydroxylated product [457,459]. For enzymes catalyzing ring cyclization or desaturation instead of hydroxylation, the rebound step is replaced by a step which involves either oxidative ligand transfer to close a ring, or a second hydrogen-atom abstraction to form a double bond [398].

RIESKE-TYPE DIOXYGENASES The hydroxylation of aromatic rings is the most difficult catalytic step in the aerobic degradation of aromatic compounds [472]. Bacterial Rieske-type nonheme dioxygenases catalyze the stereoselective introduction of two hydroxyl groups into the aromatic ring. Rieske-type dioxygenases (NAD(P)H dependent) play important roles in the biosynthesis of secondary metabolites such as flavonoids and alkaloids, and in performing a key step in the natural degradation of aromatic compounds [473–475]. These enzymes display broad substrate specificity catalyzing highly enantiospecific hydroxylations. These features make them attractive tools for the preparation of synthons for the production of industrially useful chiral chemicals [145,476]. Microbial dioxygenases catalyze a key step in the degradation of aromatic compounds pathway, which is the *cis*-dihydroxylation of arenes. Wild-type strains of microorganisms rapidly metabolize *cis*-diols into aromatic compounds, *via* a dihydrodiol dehydrogenase-mediated reaction, with the subsequent loss of chirality. However, in Gibson's blocked mutant *P. putida* 39/D, the expression of dihydrodiol dehydrogenase is knocked out so that the intermediate *cis*-dihydrodiol accumulates in the culture broth [477]. The use of mutant strains with suppressed dehydrogenase activity, allows the high yielding isolation for further use in synthesis of the accumulated *cis*-dihydrodiols [473,478,479]. The most important enzymes in this group are arene dioxygenases (EC 1.14.12.-), which catalyze the conversion of aromatic compounds into vicinal *cis*-diols, and are produced exclusively in bacteria (*Pseudomonas*, *Rhodococcus*, *Sphingomonas*). Their broad substrate specificity enables them to catalyze enantiospecific reactions on a wide variety of aromatic compounds such as benzene, naphthalene, biphenyl derivatives, and phthalic and benzoic acids [478,479]. Over 100 arene dioxygenases have been identified so far [480]. These Rieske nonheme iron oxygenases (ROs) comprise actually two/three components: (i) a reductase domain which obtains electrons from NAD(P)H; (ii) often a Rieske ferredoxin component that shuttles the electrons; (iii) and an oxygenase component that performs catalysis. The oxygenase component structures have all shown to be of the $\alpha3$ or $\alpha3\beta3$ types, containing a mononuclear iron site and a Rieske-type iron–sulfur cluster [2Fe–2S] in each α subunit. The transfer of electrons takes place from the Rieske center to the mononuclear iron of the neighboring subunit *via* a conserved aspartate residue, which is shown to be involved in gating electron transport [208,476,481].

Dioxygenases were initially classified on the basis of the electron transfer components present in the oxygenase system. Two component dioxygenases (reductase and oxygenase) were classified as Class I, and three component enzymes (reductase, ferredoxin and oxygenase) were classified as Class II and III. This classes were further subdivided based on the type of flavin cofactor (FAD or FMN) in the reductase, the presence or absence of an iron–sulfur center in the reductase, the number of proteins in the oxygenase, and if a ferredoxin was involved, the type of iron–sulfur center in the ferredoxin [472]. A second classification system, based on the phylogeny of the oxygenase α subunit was proposed later, and four families were distinguished: Toluene/biphenyl, naphthalene, benzoate, and phthalate dioxygenases [472,476,482]. This classification system was based on the catalytic activity of the enzymes, because α subunit of the oxygenase plays an important role in determining substrate specificity. A new family has emerged recently, consisting of enzymes that catalyze the oxidation at either the 1- or 5-position of salicylate as well as other substrates, and was designated as the salicylate family [483–486]. However, there are still members that do not belong to any of these families, such as enzymes that oxygenate aniline, dibenzodioxins, 3-phenylpropionate, or *o*-benzoate.

Some Gram-positive bacteria such as *Rhodococcus opacus*, are also capable to use polycyclic aromatic hydrocarbons (dibenzofurans, dibenzo-*p*-dioxins) as carbon sources, by producing a unique arene dioxygenase which catalyzes side dioxygenation reactions [487]. Consequently, a new classification was recently put forward [488]. In addition, arene dioxygenases have been classified in many types (namely toluene-, chlorobenzene-, phthalate-dioxygenase, etc.) based on the arene used as a carbon source. The most widely used types in biotransformations include benzene- (BDO), toluene- (TDO), biphenyl- (BPDO), chlorobenzene- (CDO), benzoic acid- (BZDO), nitrobenzene- (NBDO), and naphthalene-dioxygenase (NDO). Because of the multicomponent nature of these ROs and the need of cofactors, the biotransformations are performed using whole-cell biocatalytic systems. The overall reaction stoichiometry for the arene dioxygenation (Figure 1.21) requires molecular dioxygen and two electrons that come from a NAD(P)H reductase and are transferred to the final oxygenase component through a ferredoxin (in the two-component dioxygenase systems, the ferredoxin is absent).

A large amount of structural information is now available in the Protein Data Bank for several members of this family of ROs. The most thoroughly studied dioxygenase is naphthalene 1,2-dioxygenase (NDO), which catalyzes the NAD(P)H and O_2-dependent oxidation of naphthalene to (+)-*cis*-(1*R*,2*S*)-dihydroxy-1,2-dihydronaphthalene and is used as a model and as a basis for comparison to other related enzyme systems [489–496]. NDO consists of three protein components: (i) a flavo-[2Fe–2S] reductase [497], (ii) a Rieske ferredoxin electron transfer protein

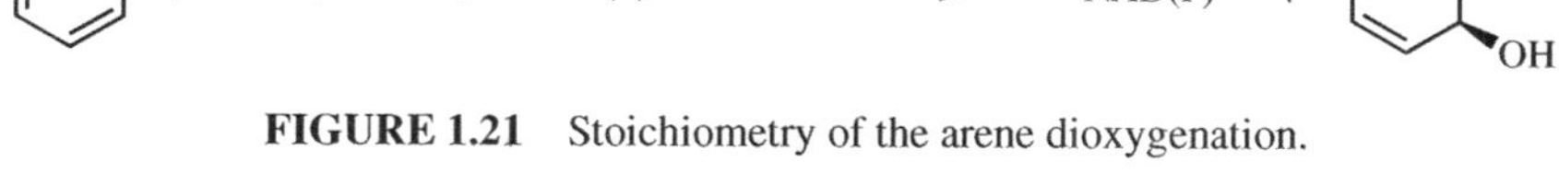

FIGURE 1.21 Stoichiometry of the arene dioxygenation.

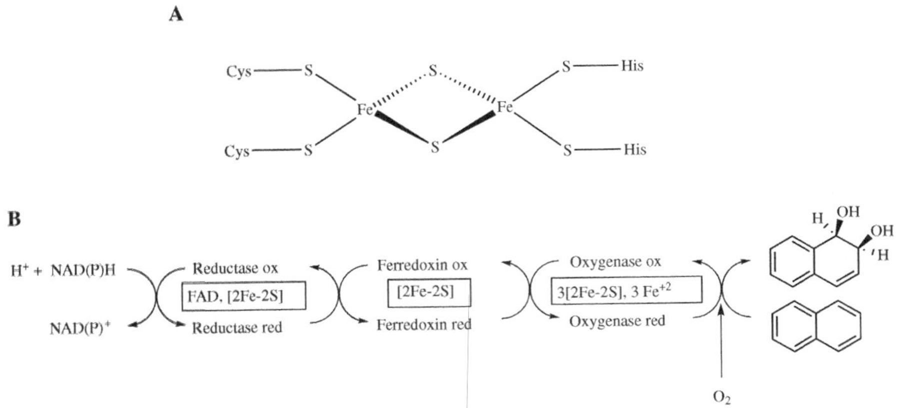

FIGURE 1.22 **A**: Rieske-type [2Fe–2S] cluster in the active site of NDO. **B**: Electron transfer in *P. putida* NDO-mediated dihydroxylation of naphthalene.

[498], and (iii) an $\alpha3\beta3$ oxygenase containing a mononuclear iron(II) center and a Rieske-type [2Fe–2S] cluster in each α-subunit (Figure 1.22A). The active site is built across a subunit–subunit boundary, and each subunit contributes with one type of metal center (Figure 1.22B) [481,495,496,499,500].

The rational basis on how the electrons are transferred from the reductase to the ferredoxin and from this to the oxygenase has advanced substantially with the disclosure of structures of biphenyl- and carbazole-dioxygenases [426,501]. Conformational changes in the flavin ring give rise to a direct electron transfer to the Rieske cluster of the ferredoxin [496]. From studies on carbazole dioxygenase, it is likely that there is a gating mechanism that senses the conformational state of the iron cofactor to regulate the transfer of the electron from the ferredoxin to the Rieske cluster of the neighboring subunit [426]. The active site of NDO from *Pseudomonas* sp. NCIB 9816-4 contains a hydrophobic pocket with a mononuclear iron(II) center, coordinated by His208, His213, and a bidentate Asp362 (the 2-His-1-carboxylate facial triad [233], present in all ROs) [502]. The mononuclear iron(II) center is positioned 12 Å from the [2Fe–2S] cluster of another subunit in the $\alpha3\beta3$ oxygenase domain, which delivers two electrons to the dioxygenase active site during each turnover. Structure determination of the ternary complexes with substrates (naphthalene or indole) and dioxygen have shown that dioxygen is bound side-on to the iron(III) center, an unusual binding mode for dioxygen, and that the aryl substrate is positioned 4.3 Å from the iron center, beyond and parallel to the bound oxygen [496]. An important difference in the geometry of the active sites of cytochrome P450 and NDO is that the active-site iron is highly accessible in NDO, with a large part of its surface available for the binding of both oxygen atoms to iron. Structures of other Rieske dioxygenases have been determined recently, and are similar to that of the NDO enzyme [426,472,481]. Detailed low temperature crystallographic studies of NDO and its complexes with either oxygen or substrates or both, have provided insights into possible catalytic mechanisms [496]. The structure, function, and constitution of the dioxygenase three-component protein, including cofactors and electron-transport chain, are relatively well understood. However, the precise catalytic mechanism for *cis*-dihydroxylation is not clear, and different theories have been developed regarding the oxidation state of iron during catalysis [481,489,490].

The catalytic cycle of arene dioxygenases may involve two steps: the activation of molecular oxygen, and the addition to the substrate. Some studies suggested that the activation of oxygen in the active site of the oxygenase goes through Fe^{4+} and Fe^{5+}. On the other hand, this activation was suggested to take place through an Fe^{3+}–hydroperoxide complex [R–Fe^{3+}–OOH], and currently this is the accepted mechanism [481,503,504]. The involvement of [2Fe–2S] iron–sulfur clusters implies that single electron transfer events take place in the catalytic cycle. Single turnover studies of the NDO reaction have shown that turnover requires reduced enzyme and bound substrate, and that the mononuclear iron(II) center and one Rieske [2Fe–2S] cluster are oxidized during turnover, resulting in iron(III) at the end of the catalytic cycle [489]. The original crystal structure solved for NDO contained an indole hydroperoxide ligand, ligated to the iron center, with the indole ring positioned at about 4 Å from the iron center [504], possibly indicating the existence of a hydroperoxide intermediate in the catalytic mechanism. Hydrogen peroxide is released by NDO if benzene is used as substrate, suggesting that dioxygen is activated *via* reduction to superoxide, which in this case is uncoupled from substrate hydroxylation and is further reduced to peroxide [505]. Recent studies on benzoate 1,2-dioxygenase have shown that enzyme containing a fully oxidized Rieske cluster can use hydrogen peroxide to form *cis*-diol products [506]. In the same way, the observation of monooxygenase activity with certain substrates, suggests that arene *cis*-dihydroxylation is not a concerted process, and that high-valent iron-oxo intermediates may be involved in monohydroxylation [507]. Despite these different mechanistic interpretations [508,509], each of these observations is consistent with an iron(III) hydroperoxy intermediate (Fe^{3+}–OOH), which might either be the active oxidant, reacting directly with the aryl substrate, or undergo O–O bond homolysis to form a reactive O=Fe(V)–OH species [489] (Figure 1.23) [426]. The oxidant could theoretically react with the aryl substrate *via* a cationic or a radical mechanism. Recent radical trap studies showed that the monooxygenation of norcarane using NDO occurs *via* radical intermediates, remaining unclear whether the dioxygenation reaction also proceeds in the same way [510]. It seems that the gathered evidence points to a radical mechanism, but clearly further experimental and theoretical studies are needed to unravel the precise mechanism of the dihydroxylation of arene substrates.

NDO is a relatively specific enzyme, capable to hydroxylated benzene, toluene, and substituted phenols, incorporating the oxygen atoms not only into the *o*-position, but also into the *p*-position [491]. Furthermore, NDO can catalyze monohydroxylations, sulfoxidation, desaturation reactions (formation of C–C double bonds), dehydrogenations, and *O*- and *N*-dealkylations, in a similar way as CYP450 (see Chapter 4) [492,493]. Due to their great versatility, arene dioxygenases are biocatalysts with enormous potential in biotechnological applications, but considering their complexity and cofactor NAD(P)H requirements, their utilization is highly focalized toward the whole-cell biotransformations [37].

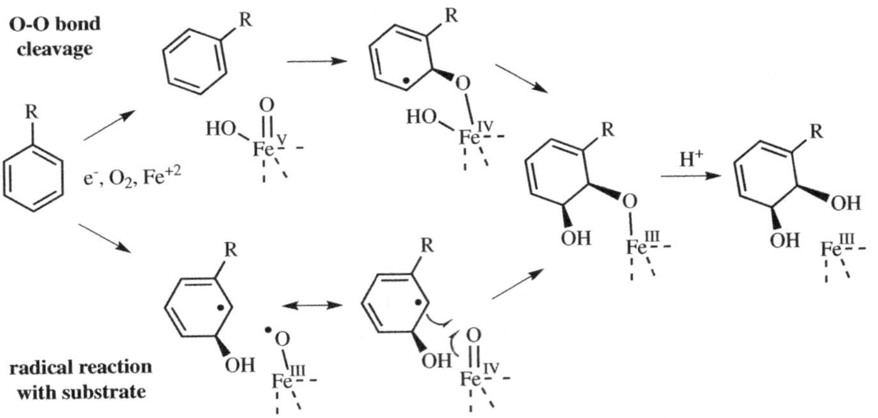

FIGURE 1.23 Proposed radical-based dioxygenase-mediated mechanism for *cis*-hydroxylation of arenes [426].

1.3.3 Oxidases

Oxidases (EC 1.1.3.-, 1.2.3.-, 1.3.3.-, 1.5.3.-, 1.10.-, 1.13.-, and 1.14.-) catalyze oxidation reactions using molecular oxygen as the electron acceptor and reducing it either to hydrogen peroxide or to water, without the incorporation of oxygen atoms into the substrate [149,199]. These enzymes do not require cofactors for developing their catalytic cycle, and use oxygen as electron acceptor in the oxidation of alcohols. This no need of cofactors makes the use of oxidases in biocatalytic processes more straightforward than the use of cofactor dependent enzymes such as dehydrogenases or oxygenases. Most oxidases are not involved in biosynthetic reactions, but play a key role in metabolism serving as terminal electron acceptors in energy storage pathways [144]. However, there are oxidases that make use of the oxidizing potential of O_2 to promote biosynthetic reactions. These enzymes may couple one-, two-, or four-electron oxidation of substrates to the two- or four-electron reduction of O_2 to H_2O_2 or H_2O, not requiring the external addition or regeneration of coenzymes. Oxidases catalyze a wide variety of synthetically interesting oxidations including the oxidation or amines, the oxidation of alcohols including sugars, oxidative ring closures, or oxidative decarboxylations, and can be divided into iron- or copper-containing oxidases, and flavin-dependent enzymes.

1.3.3.1 Iron-Containing Oxidases This type of oxidases belong to the mononuclear nonheme iron enzyme family, featuring the 2-His-1-carboxylate facial triad [439,511]. Mechanisms of oxygen activation proposed for this subgroup of enzymes are diverse, but the general strategy appears to involve the binding of both the substrate and O_2 to the iron, similar to extradiol-cleaving and 2-oxo acid dioxygenases, which belong to the same family of enzymes [144]. This promotes the electron transfer away from the substrate and, in most cases, the iron. The reduced dioxygen species undergoes O–O bond cleavage yielding a water molecule and leaving an Fe(IV)=O species as in 2-oxo acid-dependent oxygenases and pterin-dependent hydroxylases. Finally, the high-valent iron species is used as a reagent completing the second part of the reaction, and accepting two more electrons to form the second molecule of water. Examples of oxidases featuring the 2-His-1-carboxylate facial triad are isopenicillin *N*-synthase (IPNS, EC 1.21.3.1) [512,513], (*S*)-2-hydroxypropylphosphonic acid epoxidase (HppE, EC 1.14.19.7) [514], and ethylene forming enzyme 1-aminocyclopropane-1-carboxylate oxidase (ACCO, EC 1.14.17.4) [515]. These enzymes activate O_2 and ultimately form water using either four electrons from the substrate (IPNS), two from the substrate and two from NADH (HppE), or two from the substrate and two from ascorbate (ACCO).

1.3.3.2 Copper-Containing Oxidases These enzymes catalyze the oxidation of a wide variety of substrates ranging from small molecules (e.g., methane) to large peptides, with concomitant reduction of oxygen to hydrogen peroxide or water. The copper sites of these enzymes have been historically divided into three classes based on their spectroscopic features (reflecting the geometric and electronic structure of the active site): (i) type 1 (T_1) or blue copper; (ii) type 2 (T_2) or normal copper; and (iii) type 3 (T_3) or coupled binuclear copper centers [350]. Along these lines, copper-containing oxidases can be divided in four groups: (i) mononuclear enzymes containing sulfur-ligated copper (type 1); (ii) mononuclear copper oxidases with non-sulfur ligation (type 2); (iii) dinuclear copper oxidases (type 3); and (iv) multicopper oxidases [366].

Type 1 copper oxidases, actually functioning not as oxidases but as electron transfer proteins, are also called "blue-copper proteins" because of their intense color arising from a Cu(II)→S(cysteine) charge transfer transition. Amine oxidases (AOs) can be classified on the basis of their cofactor dependence, as either copper- (formerly EC 1.4.3.6, now EC 1.4.3.21 and EC 1.4.3.22) or flavin-containing (EC 1.4.3.4) enzymes [516]. These enzymes catalyze the oxidation of primary amines to aldehydes, with the subsequent release of ammonia and hydrogen peroxide. In Nature, they catalyze the oxidative deamination of a wide range of biogenic amines including many neurotransmitters, histamine, and xenobiotic amines [366,517,518].

Copper-containing amine oxidases (CuAO) are typical type 2 copper oxidases, occurring in both prokaryotes and eukaryotes. Enzymes in prokaryotes (e.g., *E. coli*) allow the organism to use various amine substrates as source of carbon and nitrogen. In higher eukaryotes their biological functions are less well defined, with suggested roles for detoxification and cell signaling in animals and cell wall cross-linking, lignifications and defense in plants [516]. The CuAO from *E. coli* is a 150–190 kDa homodimer containing two distinct active sites [518–523], each one contains a copper ion and an organic cofactor, 2,4,5-trihydroxyphenylalanine quinone (TPQ), which is derived from the posttranslational modification of an endogenous tyrosine residue [516,524]. The two-electron chemistry is carried out at a redox site consisting of a mononuclear Cu(II) ion and TPQ. The catalytic cycle initiates with the reductive half-reaction in which TPQ in the resting enzyme undergoes nucleophilic attack by the primary amine substrate, thus forming a Schiff-base intermediate (**B**) followed by an aspartate-mediated proton abstraction to generate a carbanionic intermediate (**C**)

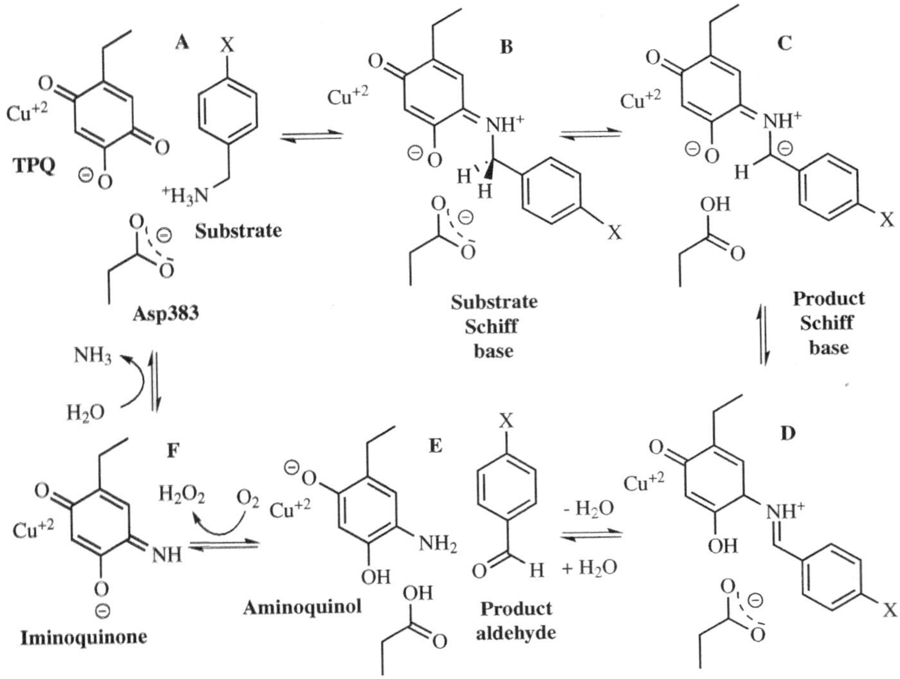

FIGURE 1.24 Catalytic mechanism for *E. coli* CuAO-catalyzed reaction.

(Figure 1.24). Further rearrangement and hydrolysis lead to the release of product aldehyde and the formation of aminoquinol (**E**). The oxidative half-reaction uses O_2 to regenerate the resting enzyme, probably *via* an iminoquinone intermediate (**F**) to release hydrogen peroxide [525]. Further hydrolysis of the mentioned intermediate releases NH_3 and regenerates the cofactor in the resting enzyme (**A**). However, the oxidative half-reaction has remained elusive in terms of oxygen activation and the relative roles of both the copper and TPQ in this process [518]. It seems clear that copper is not required to change its redox state during catalysis, and it has been suggested that its principal role is in electrostatic stabilization [526–529].

Galactose oxidase (GAO, EC 1.1.3.9) is the best characterized enzyme of the family of extracellular mono-copper oxidases of fungal origin. It was first described from *Fusarium* spp. [530–533], and later, its catalytic mechanism [534,535], structure [536,537] and substrate range [538–540], were thoroughly studied. The enzyme has a molecular weight of 68.5 kDa and is composed of 639 amino acidic residues [535]. Its physiological role, still not clear, could be involved in the production of hydrogen peroxide for further use as a co-oxidant for lignin- and cellulose-degrading peroxidases and/or as an antibiotic defense within the rhizosphere [199]. GAO catalyzes the oxidation of D-galactose at the C-6 position in the presence of oxygen to give D-galactohexodialdose and H_2O_2 as by product, which is usually destroyed by catalase to avoid inactivation of the enzyme. These enzymes perform two-electron redox chemistry, coupling the oxidation of the primary alcohol to the corresponding aldehyde with the reduction of O_2 to H_2O_2 [541,542]. The reaction stops at the oxidation stage of the aldehyde, and does not proceed to the corresponding acid. The two-electron oxidation catalyzed by a single copper atom is explained by the presence of a tyrosyl radical unit in the protein active site, acting as a second redox center during the catalytic cycle. The crystal structure of GAO from the fungus *Dactylium dendroides* was published in 1991, and provided new directions for mechanistic studies [543]. A Cu(II) ion is equatorially ligated by His496, His581, and Tyr272 residues and a solvent molecule. An axial Tyr495 completes the distorted square pyramidal geometry. Tyr272 is cross-linked to Cys228 *via* a unique thioether bond to create the radical-forming cofactor in active GAO. In addition, the indole ring of Trp290 is stacked above the thioether bond. As isolated, the enzyme is a mixture of inactive, semireduced Cu(II)-Tyr, and active, oxidized Cu(II)-Tyr• [366]. For oxidized GAO, the visible absorption band at 810 nm arises from a copper-mediated Tyr495 to Tyr272• charge transfer transition. In the semireduced enzyme, the absorption band at 458 nm was attributed to a Tyr272 to Cu(II) ligand-to-metal charge transfer transition. The data also suggested that Trp290 acts to stabilize the radical cofactor [544]. GAO also accepts as substrates carbohydrates containing an accessible D-galactose moiety such as galactosides, melibiose, raffinose, lactose, and lactitol [199]. Furthermore, the stereospecific oxidation of glycerol, 3-halogenopropane-1,2-diols, and polyols to the corresponding aldehydes was described using the commercially available enzyme from *D. dendroides* [545,546]. On the other hand, the enzyme cannot oxidize non-D-galactose-based carbohydrates, whereas it could accept shorter polyol chains, such as glycerol, which could be twisted to form a conformation similar to the D-galactose C4–C6 fragment [547].

Type 3 (dinuclear copper-oxidases) and multicopper oxidases catalyze the four-electron reduction of O_2 to water coupled to the concomitant one-electron oxidation of the substrate [144,149]. Tyrosinases (EC 1.14.18.1) are prominent examples of type 3 copper oxidases (although they can also be classified as monooxygenases), catalyzing the oxidation of both monophenols (known as cresolase or monophenolase activity) and *o*-diphenols (catecholase or diphenolase activity) into *o*-quinones (see Section "Copper-containing monooxygenase") [368]. In the cresolase cycle, the four-electron reduction of one molecule of O_2 is coupled to the four-electron oxidation of one monophenol unit *via* diphenol to an *o*-quinone, whereas in the catecholase cycle, it is coupled to the successive two-electron oxidation of two diphenol molecules [120]. Laccases (EC 1.10.3.2) are conspicuous examples of multicopper oxidases. They act *via* a free-radical mechanism, using molecular oxygen in relatively non-specific reactions [548]. The enzyme couples the monoelectronic oxidation of four equivalents of the substrate, with the transfer of four electrons to molecular oxygen, together with its reduction to water [549]. The enzymes are extracellular glycoproteins secreted by white-rot fungi, usually occurring as isoenzymes that oligomerize to form multimeric complexes, and act on phenolic substrates catalyzing the oxidation of hydroxyl groups to phenoxyl radicals. Further nonenzymatic transformations result in lignin degradation [550]. The molecular weight of the monomer ranges from 50 to 130 kDa. The carbohydrate moiety of laccases, which is believed to contribute to the stability of the enzyme, typically consists of mannose, *N*-acetylglucosamine,

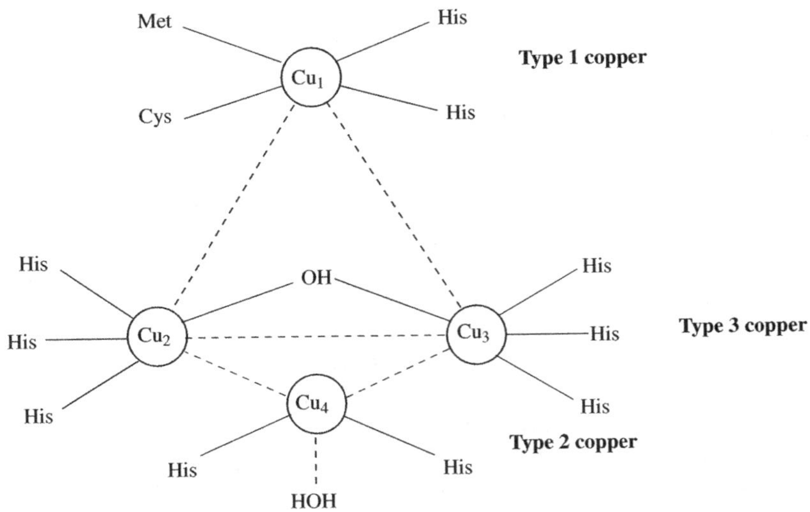

FIGURE 1.25 Schematic representation of the active site of laccase CotA from *Bacillus subtilis*.

and galactose, ranging from 10% to 45% of the protein mass [551]. The active site of laccase contains at least four copper atoms, which are one type 1 (T_1) and a trinuclear cluster (T_2/T_3) as the minimal functional unit, consisting of one type 2 (T_2) and two type 3 (T_3) copper atoms [350,552]. The T_1 copper atom is located at a distance of approximately 12 Å from the T_2/T_3 site, and the T_2 copper atom is located at 4 Å from the T_3 copper atoms (Figure 1.25).

The T_1 copper has a trigonal coordination with two histidines and one cysteine, and the axial ligand is methionine in the bacterial (CotA) enzyme and leucine or phenylalanine in fungal laccases [552]. This copper atom is the one which confers the typical blue color to multicopper proteins (also called "blue copper proteins") due to the strong adsorption around 600 nm, caused by the covalent copper-cysteine bond. Substrate oxidation takes place at this type 1 site due to two contributing factors, namely its high redox potential and the easy accessibility to the active site from the surface (compared to the limited substrate access to the trinuclear cluster). Type 2 copper coordinates with two histidines, while each T_3 atom is coordinated to three histidine ligands. Both take place at the trinuclear cluster (T_2/T_3 site) the reduction of molecular oxygen and release of water. The catalytic properties of laccase have been attributed to three major steps [350,552]: (i) type 1 copper is reduced by accepting electrons from the reducing substrate; (ii) electrons are transferred ~13 Å from type 1 copper to the trinuclear T_2/T_3 cluster; and (iii) molecular oxygen is activated and reduced to water at the T_2/T_3 cluster. During the catalytic cycle of laccase-mediated reactions, the dioxygen molecule interacts with the completely reduced trinuclear cluster T_2/T_3 *via* a two-electron process to produce the peroxide intermediate which contains the dioxygen anion (Figure 1.26) [553]. One oxygen atom of the dioxygen anion is bound to the T_2 and T_3 copper ions and the other oxygen atom is coordinated

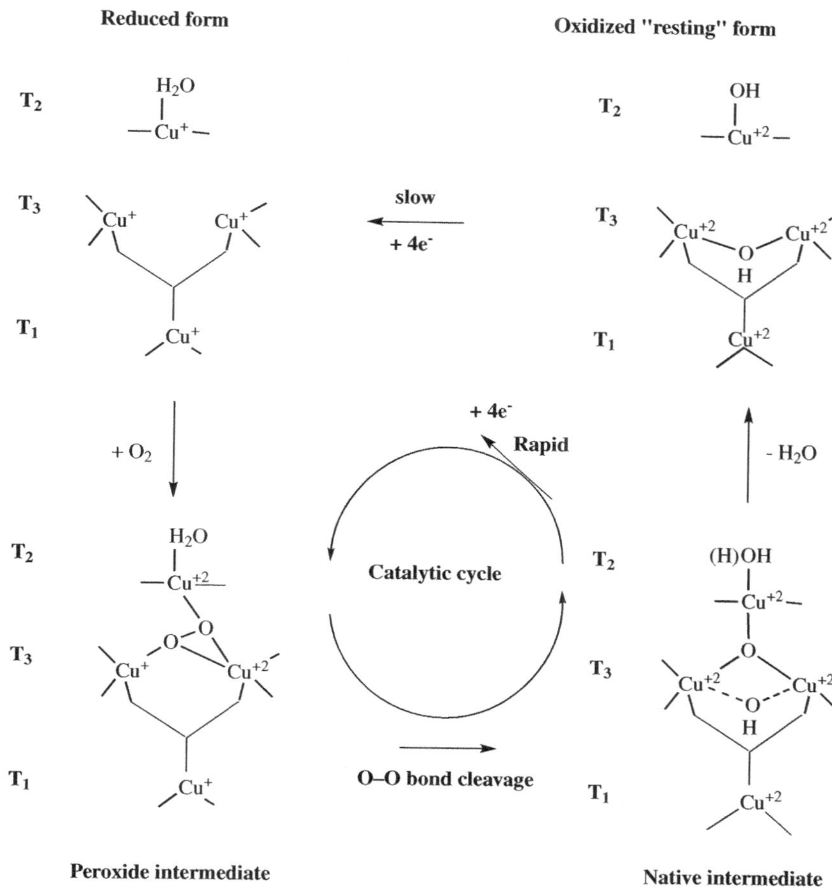

FIGURE 1.26 Catalytic cycle of laccase showing the mechanism of four-electron reduction of a dioxygen molecule to water at the enzyme copper sites.

with the other T_3 copper. Then, the peroxide intermediate undergoes a second two electron process and the peroxide O–O bond breaks to produce an intermediate, termed native intermediate, which is the fully oxidized form with the three copper centers in the trinuclear site bridged by the product of full O_2 reduction with at least one Cu–Cu distance of 3.3 Å. Solomon and coworkers confirmed the structure of this native intermediate by the combination of X-ray spectroscopy and magnetic circular dichroism [554,555].

The three copper centers in the trinuclear cluster are all bridged by a μ_3-oxo ligand at the center of the cluster, with the second oxygen atom from O_2 either remaining bound or dissociated from the trinuclear site. This structure provides relative stability, which is the thermodynamic driving force for the four-electron process for oxygen reduction and provides efficient electron transfer pathways from the T_1 site to the other copper center in the trinuclear cluster [555]. This electron transfer leads to the

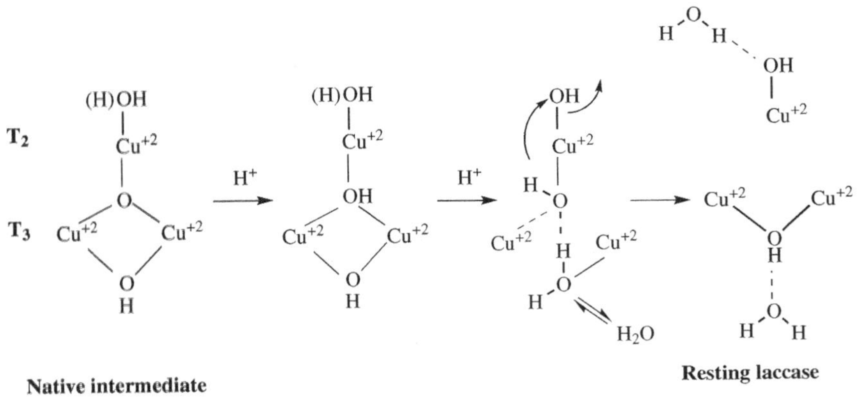

FIGURE 1.27 Proposed decay mechanism of the native intermediate to the resting laccase.

fast reduction of the complete oxidized trinuclear cluster in the native intermediate, thus generating the fully reduced site in the reduced form for further turnover with O_2. The native intermediate slowly converts to a completely oxidized form called "resting" laccase, which has the T_2 copper isolated from the coupled binuclear T_3 centers. This rearrangement only occurs in the absence of reductant and the decay of the native intermediate to the resting enzyme proceeds *via* successive proton-assisted steps (Figure 1.27) [556]. The first proton binds at the μ_3-oxo center and then the second proton binds at the T_3 OH^- bridge. Finally, the three copper centers in the trinuclear cluster are uncoupled to form the resting form of the enzyme.

Since oxidase catalysis is typically not dependent on an additional electron source such as NAD(P)H and thus it does not require the addition of an organic cosubstrate for its regeneration, these enzymes are usually applied *in vitro*, where they are independent from cell metabolism. However, enzyme deactivation by reactive oxygen species and the addition or removal of hydrogen peroxide are most critical issues that have to be considered [144].

1.3.3.3 Flavin-Dependent Oxidases Among flavin-dependent enzymes, the vanillyl alcohol oxidase (VAO, EC 1.1.3.38) family of flavoproteins containing a covalently bound FAD cofactor, constitutes a prominent and synthetically interesting group [557]. As an example, VAO from *Penicillium simplicissimum* catalyzes both the benzylic desaturation and the benzylic hydroxylation of *p*-alkylphenols, depending on the nature of the aliphatic chain [558,559]. The typical VAO-catalyzed reaction is the oxidation of vanillyl alcohol (4-(hydroxymethyl)-2-methoxyphenol) to vainillin. However, the enzyme is also able to catalyze deaminations, hydroxylations, as well as other dehydrogenation reactions, frequently giving flavor compounds. For the benzylic desaturation and hydroxylation of *p*-alkylphenols, the catalytic mechanism involves the substrate oxidation by the flavin to *p*-quinone methides as common intermediates, and differs in whether water attacks the methide. Thus, in contrast with flavin-dependent oxygenases, the flavin is not reduced by NAD(P)H but directly by the substrate and the introduced oxygen atom, in case of benzylic hydroxylation, is derived from water as in the case of hydroxylating dehydrogenases [144]. The reduced flavin transfers the electrons to O_2, and, as another difference to flavin-dependent oxygenases, the resulting hydroperoxyflavin is not stabilized, leading to hydrogen peroxide formation. A remarkable feature of this oxidase family is that it favors the covalent attachment of the flavin cofactor [557].

Alcohol oxidases (AO, EC 1.1.3.13) are flavin-dependent enzymes, where FAD or FMN function as the primary hydride acceptor, and from which the excess reducing equivalents are shuttled to molecular oxygen [145,560,561]. Aliphatic alcohol oxidases from methylotrophic yeasts such as *Candida*, *Hansenula*, or *Pichia* sp. mostly accept primary alcohols as substrates, which are selectively oxidized to the corresponding aldehydes [540,562–565].

Likewise, glucose oxidase (GOX, EC 1.1.3.4) is a prominent flavoenzyme that catalyzes the oxidation of β-D-glucose to D-glucono-δ-lactone, utilizing molecular oxygen as an electron acceptor with simultaneous production of hydrogen peroxide [145,566–568]. In the reductive half reaction, the enzyme catalyzes the oxidation of β-D-glucose to D-glucono-δ-lactone, which is spontaneously and nonenzymatically hydrolyzed to gluconic acid. Subsequently, the FAD ring of GOX is reduced to $FADH_2$ [569]. In the oxidative half-reaction, the reduced enzyme is reoxidized by oxygen to yield H_2O_2 [570]. Usually, the formed hydrogen peroxide is then cleaved by catalase to produce water and oxygen. Its molecular weight ranges from 130 to 175 kDa [571]. This enzyme is receiving much attention due to its wide use in chemical, pharmaceutical, and food and beverage industries, as well as in clinical chemistry and biotechnology, mainly in novel applications for biosensors [572–574]. Microbial wild-type GOX is highly specific for β-D-glucose. GOX from ascomycetes is a dimeric glycoprotein composed of two identical polypeptide chain subunits which are covalently linked together *via* disulfide bonds [571,575]. The structure of *Penicillium amagasakiense* GOX shows each of its subunits to contain one mole of tightly bound, but not covalently linked, FAD moiety as cofactor [569,575]. This enzyme is glycosylated, with a mannose-rich carbohydrate content of nearly 11–13% [566]. The key conserved active-site residues of GOX from *P. amagasakiense* are Tyr73, Phe418, Trp430, Arg516, Asn518, His520, and His563 [569], being Arg516 the most critical amino acid for the efficient binding of β-D-glucose to the enzyme, while Asn518 contributed to a lesser extent. The aromatic residues Tyr73, Phe418, and Trp430 are important for the correct orientation of the substrate as well as for the maximal rate of its oxidation. His520 and His563 establish hydrogen bonds with the 1-OH of glucose during the reaction.

Pyranose oxidases (PyO, EC 1.1.3.10) are flavoproteins expressed in many lignin-degrading white rot fungi [576]. Unlike GOX, which acts on the anomeric carbon, PyO oxidizes in the presence of oxygen, glucose, and a number of common monosaccharides at their hydroxyl group in C-2, giving the corresponding 2-ketosugars and hydrogen peroxide. It is believed that these fungi use this transformation to produce H_2O_2 for their lignin-decomposing peroxidases [199]. The enzyme was first characterized in the 1960s [577], and since then important biotechnological applications have been developed in analytical as well as in synthetic carbohydrate chemistry.

PyO can also use quinones, complexed metal ions, and radicals as electron acceptors [199]. Some of these compounds are even better substrates for the enzyme compared with oxygen, suggesting that PyO can also play a direct role in the reduction of quinones during the process of ligninolysis [578]. Intracellular PyO from *Trametes multicolor* was purified from mycelia extracts and characterized by Haltrich and coworkers [579]. The native enzyme has a molecular mass of 270 kDa and is composed of four identical 68 kDa subunits, each one containing one covalently bound FAD as its prosthetic group.

In addition to the oxidation of the C–O bond to produce carbonyl compounds, oxidases are also well known as catalysts for the oxidation of C–N bonds, including examples in which these reactions occurs enantioselectively [533]. The most relevant oxidases catalyzing the oxidation of primary and secondary amines are flavoproteins, conducting the transfer of a hydride equivalent from the C–N bond to the flavin cofactor. Most of these AOs can be classified into two structural families: (i) the monoamine oxidase family (MAO) and (ii) the amino acid oxidase (AAO) family that comprises L-amino acid oxidases (LAAO, EC 1.4.3.2) and D-amino acid oxidases (DAAO, EC 1.4.3.3) [580]. The mammalian flavin-dependent MAO family includes MAO A and B (EC 1.4.3.4) [581–583] as the most interesting examples regarding synthetic applications. There are also flavin-dependent MAOs of fungal origin with increasing use in synthesis (see Chapter 4). Similar to copper-containing AOs, both flavin-dependent MAO A and B catalyze the oxidation of the primary amino groups of aromatic alkyl amines and are widely distributed in higher eukaryotes. The two isoforms of MAO (MAO A and MAO B) are separate gene products with 70% sequence identity, but exhibit overlapping substrate specificities in catabolism of endogenous amine neurotransmitters such as serotonin and dopamine, as well as a wide range of dietary amines and drugs [581]. Mammalian MAOs are bound to the outer mitochondrial membrane and have a FAD molecule covalently bound to the protein *via* an 8α-thioether linkage to a cysteinyl residue [584]. Although both enzymes have the same overall structure, MAO A has a more open active site, producing differences in substrate specificities for the two forms [581–583,585]. However, their catalytic mechanisms are proposed to be the same, and both have been discussed. Mechanism I involves a radical intermediate, while mechanism II proposes the nucleophilic attack of the amine substrate to the flavin C-4a position (Figure 1.28) [584,585].

Radical-based mechanisms for MAO are supported on studies with inhibitors containing cyclopropyl (or cyclobutyl) rings, since *trans*-2-phenylcyclopropylamine has been reported to be a time-dependent inhibitor for these enzymes [586]. Subsequently, Paech and coworkers showed that this compound is also a substrate, and that the inactivation results in covalent attachment of the inhibitor to the protein rather than to the covalently bound FAD [587]. These authors proposed that inhibition was due to nucleophilic attack of a protein thiol on the resulting cyclopropyl imine or its hydrolysis product. A number of other primary and secondary amines also inactivate the enzymes, through labeling of the protein or the flavin, or both [588]. These findings support a mechanism of inactivation involving one-electron transfers, although it was suggested that a protein rather than a flavin radical reacted with the inhibitor [589].

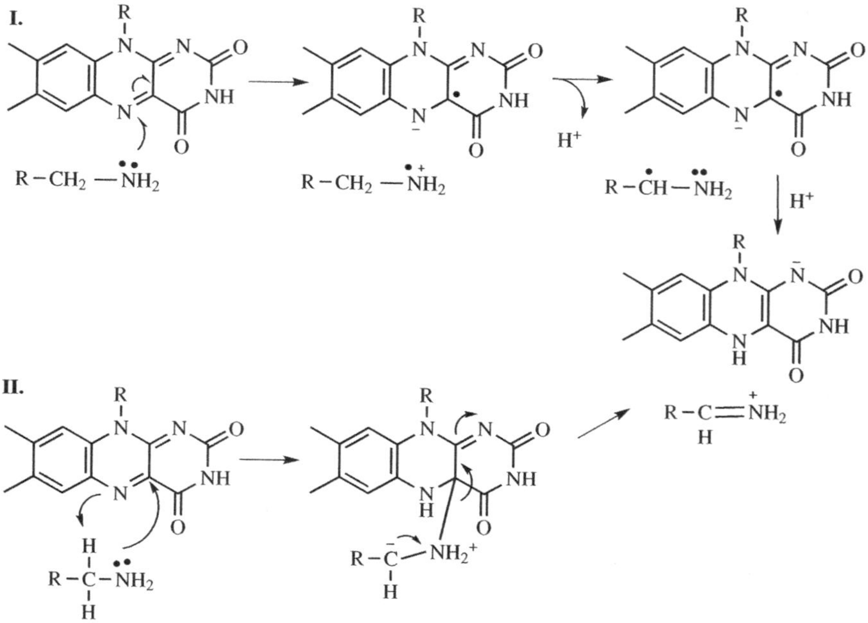

FIGURE 1.28 Proposed catalytic mechanisms for MAO A and B.

Mechanism II for the catalytic mechanism of MAO was developed by Edmondson et al., showing a nucleophilic mechanism for amine oxidation by MAOs and related enzymes [585]. A flavin-amine adduct forms prior to abstraction of the substrate hydrogen as a proton. However, this requires that no irreversible steps precede C–H bond cleavage, so that adduct formation must be reversible. Moreover, there is no obvious base in the active site of the enzyme to abstract a proton from the substrate [581]. This observation led to the mechanism II shown in Figure 1.28, in which the flavin N-5 is the base, with the proton abstraction occurring simultaneously as the substrate nitrogen adds to the flavin rather than in separate steps [585]. The pK_a for the neutral N-5 in reduced flavins has been estimated to be greater than 20 [590].

The other important family of flavin-dependent oxidases with relevant synthetic applications is the AAO family. D-Amino acid oxidases (DAAO) have been more studied and used in preparative applications than the L-selective enzymes [533]. DAAO has been employed for many years as an industrial biocatalyst in the commercial production of 7-aminocephalosporanic acid (7-ACA). The enzyme catalyzes the initial oxidative deamination of the D-amino acid side chain in cephalosporin C [591]. The D-amino acid oxidase family contains DAAO [592] and enzymes that oxidize glycine or *N*-methylated amino acids [593–597]. Thus, both MAO and DAAO families of AOs contain members that oxidize amino acids as well as primary and secondary amines. DAAO was first reported from kidney in 1935, and decades later became a model for the study of FAD-dependent oxidases [591,598]. More recently discovered bacterial glycine oxidases have similar structure to DAAOs from pig kidney and yeasts, oxidizing D-alanine but not L-alanine [593,599], and are proposed to have the same mechanism of catalysis. The catalytic cycle involves a reductive half-reaction in which a hydride equivalent is transferred from the substrate to the flavin yielding the oxidized substrate bound to the reduced enzyme [580]. This step is energetically favorable for most substrates, thus enzyme reduction is effectively irreversible. The product then dissociates from the enzyme, which subsequently reacts with oxygen generating the oxidized enzyme, or else oxygen can react with the reduced enzyme before product dissociates, followed by product dissociation from the oxidized enzyme. The oxidation of an amine group involves the net loss of two electrons, as the C–N single bond is converted to a double bond, and of two protons, one from the carbon and one from the positively charged nitrogen. The ability of the flavin to accept one or two electrons and to form adducts with nucleophiles, in combination with the various possible sequences of loss of protons and electrons from the substrate, allows to propose a variety of catalytic mechanisms [580]. Through kinetic studies of pig kidney DAAO with a series of ring-substituted phenylglycines as substrates, it was concluded that the α-hydrogen of the substrate was removed as a proton [600]. Subsequently, it was proposed that oxidation of an amine or an alcohol could occur by nucleophilic attack of the heteroatom on the flavin C-4a position, followed by loss of the α-hydrogen as a proton (Figure 1.29) [601].

Later, Walsh et al. described the ability of DAAO to catalyze the elimination of HCl from β-Cl-alanine in addition to the normal oxidation of the substrate to β-Cl-pyruvate [602]. This was considered an additional evidence for proton removal to form a substrate carbanion in the initial step of catalysis, consistent with the proposed mechanism (Figure 1.30).

The ratio between oxidation and HCl elimination varied with the oxygen concentration, where higher amounts of O_2 favored the oxidation. To account for the effect of the oxygen concentration on the ratio between oxidation and elimination, it was necessary to propose the reaction of the carbanion intermediate with the flavin as a reversible step. The elimination reaction was subsequently studied with α-amino-β-Cl-butyrate, establishing that the hydrogen originally on the substrate α-carbon was found at the β-carbon in the α-ketoglutarate formed in the elimination reaction [580]. For this suggested mechanism, the tautomerization of the initial enamine to the imino acid and its hydrolysis to the corresponding ketoacid is required to occur in the active

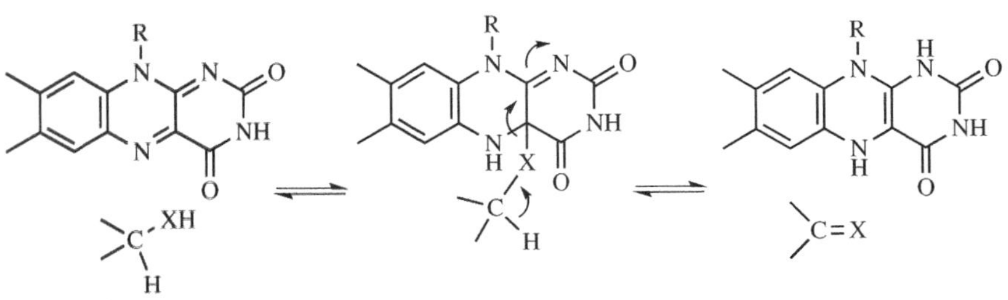

FIGURE 1.29 Proposed mechanism for the oxidation of amines (and alcohols) by flavin-dependent oxidases, on the basis of kinetic studies of pig kidney DAAO [601].

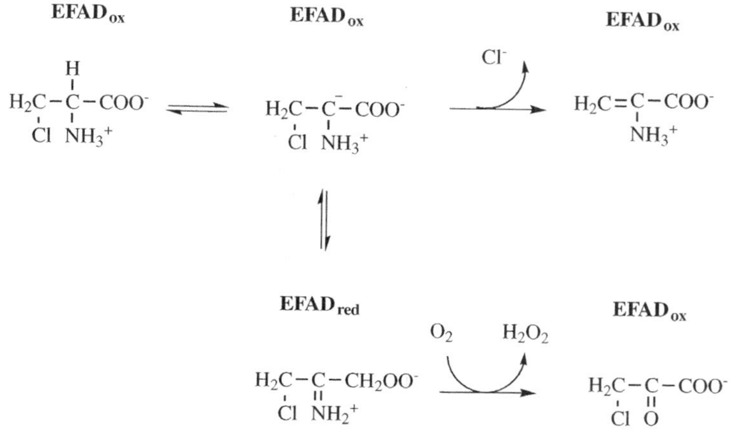

FIGURE 1.30 Proposed mechanism for DAAO-mediated oxidation of amines, consistent with the proton removal to form a substrate carbanion as initial step in the catalytic cycle.

site of the enzyme, in contrast to the behavior of other substrates, where the imino acid product dissociates from the enzyme before the ketoacid is formed [603].

An alternative mechanism was suggested accounting for both the effect of the oxygen concentration on the partitioning between elimination and oxidation, and the retention of the α-hydrogen on the β-carbon (Figure 1.31).

The amino acid substrate is oxidized to an imino acid by the normal catalytic mechanism. The reduced flavin can react with O_2 thus forming the oxidized enzyme, followed by dissociation of the β-Cl-imino acid. Alternatively, the reduced flavin can act as a hydride donor for an S_N2 reaction in which the chloride is the leaving group producing the oxidized enzyme and the β-Cl-imino acid. This mechanism accounts for both the oxygen dependence of the product, and the shift of the α-hydrogen to the β-carbon of the product. If the reduced flavin acts as a donor to the imino acid in the elimination, it is likely to act as a hydride acceptor from the imino acid in the normal reaction.

1.3.4 Peroxidases

Peroxidases are ubiquitous enzymes present in all domains of life, being found in the cytosol, in specific organelles, or are even secreted to the extracellular medium [604]. These enzymes use hydrogen peroxide (H_2O_2) or organic hydroperoxides (ROOH) as electron accepting cosubstrates, oxidizing a variety of compounds and do not require external addition of coenzymes for their regeneration [149,605]. They catalyze a wide variety of reactions under mild and controlled conditions. Among them, olefin epoxidations, sulfoxidation, halogenation, hydroperoxyde reduction, and allylic, propargylic, and benzylic hydroxylations (but no aliphatic hydroxylations) have been reported [37,606–608]. The vast majority of peroxidases are heme-enzymes, containing ferric protoporphyrin IX (protoheme) as prosthetic group in their active sites. In addition, there are several nonheme peroxidases which contain either

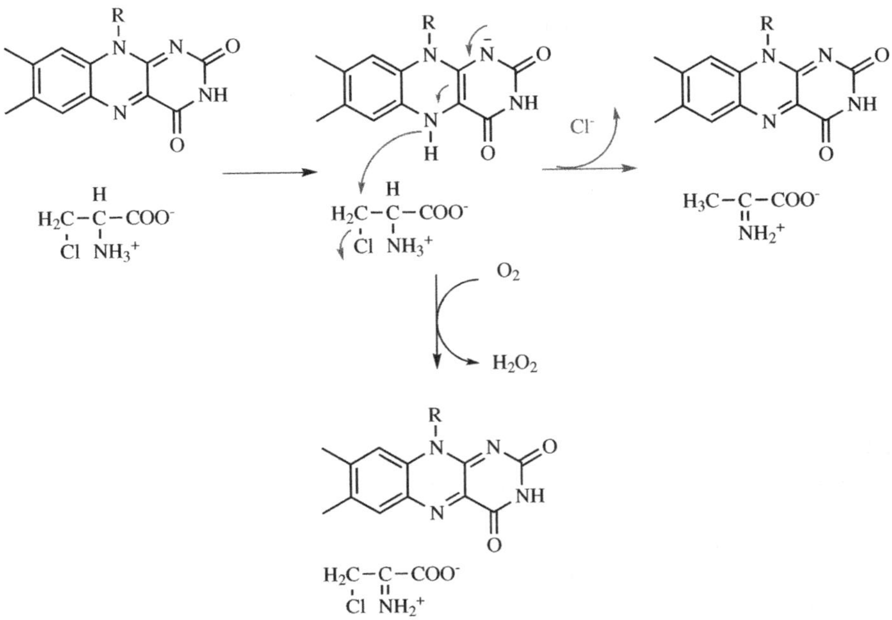

FIGURE 1.31 Proposed mechanism for the oxygen concentration-dependent amine oxidation versus HCl elimination in DAAO-mediated reactions.

other metals instead of iron (Mn, V, Se) such as vanadium haloperoxidases or manganese catalases [117,606], or specific metal-free prosthetic groups like the thiol peroxidases or the alkyl hydroperoxidases [609–611]. Peroxidases are grouped under EC 1.11.- (donor: peroxide oxidoreductases) in the Nomenclature Committee of the International Union of Biochemistry and Molecular Biology (NC-IUBMB) system of enzyme classification [148,612]. Since the classification system is based on the catalyzed reactions, peroxidases with different molecular architecture and/or catalytic mechanism may be classified under the same EC number. For instance, the entry EC 1.11.1.10, corresponding to fungal chloride peroxidases, includes vanadium- (halide peroxidases) and heme-peroxidases (chloroperoxidase), and they are very different regarding their three-dimensional structure. However, peroxidases have also been classified according to other characteristics, such as structural properties and sequence information [611]. Concerning the heme-containing peroxidases, this has led to a classification in two main superfamilies: (i) the peroxidase-cyclooxygenases (animal or mammalian peroxidases) and (ii) the nonanimal peroxidases (plant peroxidases), which include most of the heme-dependent peroxidases. In addition, according to the sequence, host organism, and physiological function, plant peroxidases comprise three main classes, I, II, and III (see Section 1.2.1) [117,609,613–618]. Within the peroxidase-cyclooxygenase superfamily, which includes lactoperoxidase (LPO), thyroid peroxidase, and myeloperoxidase, similar subgrouping according to physiology and enzyme function was established [619]. However, there are already a number of new fungal heme-containing peroxidases which do not fit into the two previous superfamilies: (i) the synthetically relevant heme-thiolate peroxidases, comprising chloroperoxidase from *Caldariomyces fumago* (CPO, EC 1.11.1.10) and aromatic peroxygenases from basidiomycetes; and (ii) the dye-decolorizing peroxidases (DyPs) from fungi and eubacteria [117,611,616,620–623]. Peroxidases are often named according to their natural source, such as horseradish peroxidase (HRP, EC 1.11.1.7) from *Armoracia rusticana*, LPO from bovine milk, or myeloperoxidase (MPO) from human leukocytes. They can also be named according to their natural substrates, namely cytochrome C peroxidase (CcP, EC 1.11.1.5), chloroperoxidase from *C. fumago* (CPO, EC 1.11.1.10), or lignin peroxidase (LiP, EC 1.11.1.14).

The most frequently used peroxidases in organic synthesis are chloroperoxidase (CPO) from *C. fumago* and peroxidase from *A. rusticana* root (HRP). Chloroperoxidase is the most versatile peroxidase, and also behaves as a monooxygenase in sulfoxidations with molecular oxygen and external reducing agent, with substantial increase in enantioselectivity and stability [624]. The classical peroxidase-catalyzed reaction consists of a one-electron oxidation of the substrate with a peroxide as electron acceptor [144]. Heme-dependent peroxidases in turn, also catalyze other oxygen transfer reactions such as olefin epoxidations, sulfoxidations, or hydroxylations [37,606–608]. Such reactions are two-electron oxidations, in which peroxide serves as oxygen donor and electron acceptor, and one molecule of water (or alcohol

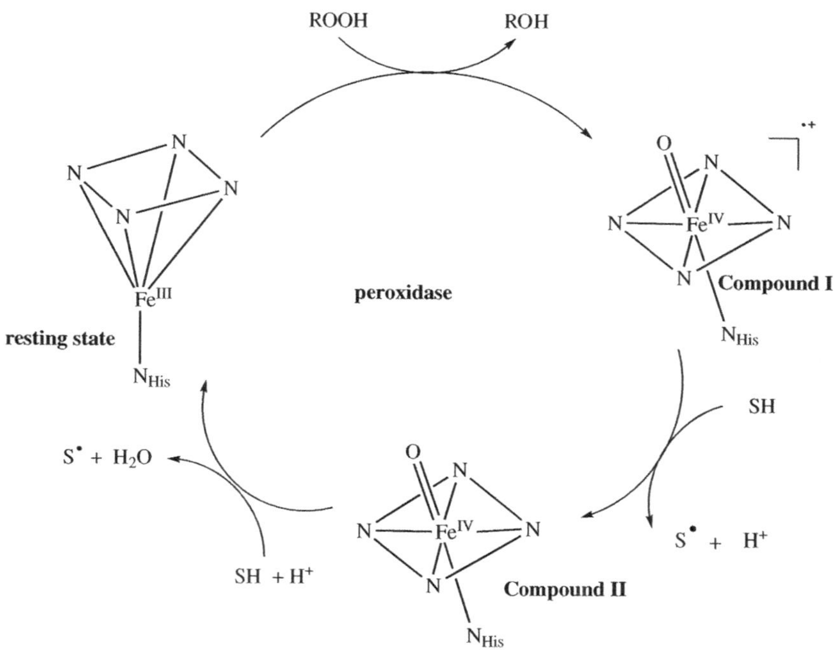

FIGURE 1.32 Catalytic cycle of heme-peroxidases. SH: substrate.

in case of organic peroxide driven reactions) is produced as coproduct. In these oxygen transfer reactions, heme-dependent peroxidases exhibit more typical reactivity for CYP450 than of classical peroxidases, which catalyze oxidative dehydrogenations *via* one-electron processes. This class of peroxidases are not only functionally but also structurally related to P450 monooxygenases, which are also able to catalyze substrate oxyfunctionalizations with hydrogen peroxide as oxygen source in the so-called peroxide shunt pathway. The catalytic mechanism of heme-peroxidases for the classical peroxidase reaction has been studied using HRP as model (Figure 1.32) [606,625].

In heme-dependent peroxidases, hydrogen peroxide enters the catalytic cycle at the hydroperoxo-stage coordinating as a hydroperoxyl anion to Fe(III) and then leading to the pivotal intermediate compound I. In addition to this classical reaction, oxygen transfer from oxidized peroxidase to the substrate can also take place, provided that the heme iron is accessible for the substrates (see Chapter 5, Section 5.2). The fifth ligand of the iron is a histidine, except in CPO, which belongs to the group of heme-thiolate peroxidases [110,626]. With respect to oxygen transfer reactions, this peroxidase with iron ligated to cysteine, similar to P450s, is the most versatile enzyme of the known peroxidases (see Chapter 5, Section 5.2.1.3)

1.4 CONCLUDING REMARKS

As exposed in this chapter, there is a significant diversity of enzymes, reactions, and processes occuring in Nature. Many of these performances may be useful for industrial catalytic processes, once key-aspects of the reactions (e.g., cofactor regeneration) are properly optimized. In a nutshell, enzymatic redox reactions can be gathered in three different groups: (i) dehydrogenase-catalyzed, nicotinamide cofactor-dependent reactions; (ii) oxidations at the expense of molecular oxygen, catalyzed by mono- or dioxygenases, (also nicotinamide cofactor-dependent biocatalysts); (iii) peroxidations, which require hydrogen peroxide and are catalyzed by peroxidases. In Chapters 3–6, different applications of these type of enzymes have been discussed in depth.

REFERENCES

1. Anastas PT, Zimmerman JB. Environ Sci Technol 2003;37:94A–101A.
2. Tang S, Bourne R, Smith R, Poliakoff M. Green Chem 2008;10:268–269.
3. Nestl BM, Nebel BA, Hauer B. Curr Opin Chem Biol 2011;15:187–193.
4. Santacoloma PA, Sin G, Gernaey KV, Woodley JM. Org Process Res Dev 2011;15:203–212.
5. Wohlgemuth R. Curr Opin Biotechnol 2010;21:713–724.
6. Kiełbasiński P, Ostaszewski R, Szymański W. In: Cybulski A, Moulijn JA, Stankiewicz A, editors. Novel Concepts in Catalysis and Chemical Reactors: Improving the Efficiency for the Future. Weinheim: Wiley-VCH; 2010.

7. García Junceda E, García García JF, Bastida A, Fernández Mayoralas A. Bioorg Med Chem 2004;12:1817–1834.
8. Leresche JE, Meyer HP. Org Process Res Dev 2006;10:272–280.
9. Thomas SM, Di Cosimo R, Nagarajan V. Trends Biotechnol 2002;20:238–242.
10. Koeller KM, Wong CH. Nature 2001;409:232–240.
11. Liese A, Villela Filho M. Curr Opin Biotechnol 1999;10:595–603.
12. Loughlin WA. Bioresour Technol 2000;74:49–62.
13. May SW. Curr Opin Biotechnol 1997;8:181–186.
14. Schmid A, Dordick JS, Hauer B, Kiener A, Wubbolts M, Witholt B. Nature 2001;409:258–268.
15. Schoemaker HE, Mink D, Wubbolts MG. Science 2006;299:2912–2926.
16. Hudlicky T, Reed JW. Chem Soc Rev 2009;38:3117–3132.
17. Illanes A. In: Illanes A, editor. Enzyme Biocatalysis. The Netherlands: Springer; 2008, pp. 57–106.
18. Underkofler LA, Barton RR, Rennert SS. Appl Environ Microbiol 1958;6:212–221.
19. Lambert PW, Meers JLB, Best DJ , Phil Trans R Soc Lond B 1983;300:263–282.
20. Yildiz H, Akyilmaz E, Dinçkaya E, Artif Cells, Blood Substitues Immobilization Biotechnol 2004;32:443–452.
21. Oberg CJ, Merrill RK, Brown RJ, Richardson GH. J Dairy Sci 1992;75:669–675.
22. Baladrin MF, Klocke JA, Wurtele ES, Bollinger WH. Science 1985;228:1154–1160.
23. Uhlig H. Industrial Enzymes and Their Applications. New York: Wiley-IEEE; 1998.
24. Pressi G, Toso RD, Monte RD, Carturan G. J Sol-Gel Sci Technol 2003;26:1189–1193.
25. Cimino C, Vairo-Cavalli S, Spina F, Natalucci C, Priolo N. Elect J Biotechnol 2006;9:237–240.
26. Su WW, Arias R. J Biosci Bioeng 2003;95:13–20.
27. Chu L, Robinson DK. Curr Opin Biotechnol 2001;12:180–187.
28. Barnes LM, Bentley CM, Dickson AJ. Cytotechnology 2000;32:109–123.
29. Beloqui A, Domínguez de María P, Golyshin PN, Ferrer M. Curr Opin Microbiol 2008;11:240–248.
30. Buchholz K, Kasche V, Bornscheuer U. Biocatalysis and Enzyme Technology. 1st ed., Weinheim: Wiley-VCH; 2005, p. 448.
31. Leisola M, Jokela J, Pastinen O, Turunen O, Schoemaker H. Industrial Use of Enzymes. Oxford: Eolss Publishers; 2001.
32. Liu Z, Weis R, Gleider A. Food Technol Biotechnol 2004;42:237–249.
33. Liese A, Seebach K, Wandrey C. Industrial Biotransformations. 2nd ed., Weinheim: Wiley-VCH; 2006.
34. Cereghino JL, Cregg JM. FEMS Microbiol Rev 2000;24:45–66.
35. Bernhardt R. J Biotechnol 2006;124:128–145.
36. Sono M, Roach MP, Coulter ED, Dawson JH. Chem Rev 1996;96:2841–2888.
37. Ullrich R, Hofrichter M. Cell Mol Life Sci 2007;64:271–293.
38. Peterson JA, Graham SE. Structure 1998;6:1079–1085.
39. Denisov IG, Makris TM, Sligar SG, Schlichting I. Chem Rev 2005;105:2253–2278.
40. Guengerich FP. Chem Res Toxicol 2001;14:611–650.

41. Guengerich FP, Shimada T. Chem Res Toxicol 1991;4:391–407.
42. González FJ. Trends Pharmacol Sci 1992;13:346–352.
43. Rodrigues AD, Rushmore TH. Curr Drug Metab 2002;3:289–309.
44. Jörnvall H. Eur J Biochem 1970;16:41–49.
45. Jörnvall H. Eur J Biochem 1970;16:25–40.
46. Jörnvall H, Harris JI. Eur J Biochem 1970;13:565–576.
47. Jörnvall H. Eur J Biochem 1970;14:521–534.
48. Pietruszko R, Ringold HJ, Li TK, Valee BL, Akeson A, Theorell H. Nature 1969;221:440–443.
49. Pietruszko R, Theorell H. Arch Biochem Biophys 1969;131:288–299.
50. Lutstorf UM, von Wartburg JP. FEBS Lett 1969;5:202–206.
51. Theorell H, Taniguchi S, Akeson A, Skursky L. Biochem Biophys Res Commun 1966;24:603–610.
52. Bonnichsen RK, Wassen AM. Arch Biochem Biophys 1948;18:361–363.
53. Pietruszko R, Ringold HJ. Biochem Biophys Res Commun 1968;33:497–502.
54. Peterson PE, Smith TJ. Structure 1999;7:769–782.
55. Giri A, Dhingra V, Giri CC, Singh A, Ward OP, Narasu ML. Biotechnol Adv 2001;19:175–199.
56. Orden AA, Magallanes Noguera C, Agostini E, Kurina Sanz M. J Mol Catal B: Enzym 2009;61:216–220.
57. Szejtli J. Carbohydr Polym 1990;12:375–392.
58. Casas DA, Pitta Alvarez SI, Giulietti AM. Appl Biochem Biotechnol 1998;69:127–136.
59. Giri A, Lakshmi Narasu M. Biotechnol Adv 2000;18:1–22.
60. Blanchard N, van de Weghe P. Org Biomol Chem 2006;4:2348–2353.
61. Fonseca AM, Monte FJQ, de Oliveira MCF, de Mattos MC, Cordell GA, Braz Filho R, Lemos TLG. J Mol Catal B: Enzym 2009;57:76–82.
62. Maczka WK, Mironowicz A. Tetrahedron: Asymmetry 2002;13:2299–2302.
63. Maczka WK, Mironowicz A. Tetrahedron: Asymmetry 2004;15:1965–1967.
64. Yadav JS, Nanda S, Thirupathi Reddy P, Bhaskar Rao A. J Org Chem 2002;67:3900–3903.
65. Aldabalde V, Arcia MP, González A, González D. Green Chem Lett Rev 2007;1:25–30.
66. Baldassarre F, Bertoni G, Chiappe C, Marioni F. J Mol Catal B: Enzym 2000;11:55–58.
67. Baskar B, Ganesh S, Lokeswari TS, Chadha A. J Mol Catal B: Enzym 2004;27:13–17.
68. Ravía S, Gamenara D, Schapiro V, Bellomo A, Adum J, Seoane G, González D. J Chem Ed 2006;83:1049–1051.
69. Yadav JS, Reddy GSKK, Sabitha G, Krishna AD, Prasad AR, Rahaman H.-U.-R, Rao KV, Rao AV. Tetrahedron: Asymmetry 2007;18:717–723.
70. Liu X, Pan ZG, Xu JH, Li HX. Chin Chem Lett 2010;21:305–308.
71. Chadha A, Manohar M, Soundararajan T, Lokeswari TS. Tetrahedron: Asymmetry 1996;7:1571–1572.
72. Xie Y, Xu JH, Lu WY, Lin GQ. Bioresour Technol 2009;100:2463–2468.
73. Yadav JS, Reddy BVS, Sreelakshmi C, Rao AB. Synthesis 2009;11:1881–1885.
74. Kumaraswamy G, Ramesh S. Green Chem 2003;5:306–308.

75. Comasseto JV, Omori AT, Porto ALM, Helgueira Andrade L. Tetrahedron Lett 2004;45:473–476.
76. Yadav JS, Reddy PT, Nanda S, Bhaskar Rao A. Tetrahedron: Asymmetry 2001;12:3381–3385.
77. Machado LL, Monte FJQ, de Oliveira MCF, de Mattosa MC, Lemos TLG, Gotor Fernández V, Gotor V. J Mol Catal B: Enzym 2008;54:130–133.
78. Yadav JS, Reddy PT, Nanda S, Rao AB. Tetrahedron: Asymmetry 2002;12:3381–3385.
79. Bruni R, Fantín G, Medici A, Pedrini P, Sacchetti G. Tetrahedron Lett 2002;43:3377–3379.
80. Takemoto M, Achiwa K. Phytochemistry 1998;49:1627–1629.
81. Takemoto M, Achiwa K. Tetrahedron: Asymmetry 1995;6:2925–2928.
82. Andrade LH, Utsunomiya RS, Omori AT, Porto ALM, Comasseto JV. J Mol Catal B: Enzym 2006;38:84–90.
83. Orden AA, Bisogno F, Cifuente DA, Giordano OS, Kurina Sanz M. J Mol Catal B: Enzym 2006;42:71–77.
84. Caron D, Coughlan AP, Simard M, Bernier J, Piché Y, Chevenert R. Biotechnol Lett 2005; 27:713–716.
85. Seip JE, Fager SK, Gavagan JE, Gosser LW, Anton DL, Di Cosimo R. J Org Chem 1993; 58:2253–2259.
86. Adam W, Lazarus M, Boss B, Saha Möller CR, Humpf HU, Schreier P. J Org Chem 1997; 62:7841–7843.
87. Macheroux P, Mulrooney SB, Williams CH, Jr., Massey V. Biochim Biophys Acta 1992; 1132:11–16.
88. Stenberg K, Lindqvist Y. Protein Expr Purif 1996;8:295–298.
89. Macheroux P, Massey V, Thiele DJ, Volokita M. Biochemistry 1991;30:4612–4619.
90. Payne MS, Petrillo KL, Gavagan JE, Wagner LW, Di Cosimo R, Anton DL. Gene 1995;167:215–219.
91. Lindqvist Y. J Mol Biol 1989;209:151–166.
92. Adam W, Boland W, Hartmann Schreier J, Humpf HU, Lazarus M, Saffert A, Saha Möller CR, Schreier P. J Am Chem Soc 1998;120:11044–11048.
93. Azevedo AM, Martins VC, Prazeres DM, Vojinovic V, Cabral JM, Fonseca LP. Biotechnol Annu Rev 2003;9:199–247.
94. Welinder KG. Curr Opin Struct Biol 1992;2:388–393.
95. Schuller DJ, Ban N, van Huystee RB, McPherson A, Poulos TL. Structure 1996;4:311–321.
96. Puppo A, Rigaud J, Job D, Ricard J, Zeba B. Biochim Biophys Acta 1980;614:303–312.
97. McEldoon JP, Pokora AR, Dordick JS. Enzyme Microb Technol 1995;17:359–365.
98. Mäder M, Füssl R. Plant Physiol 1982;70:1132–1134.
99. Evans JJ. Plant Physiol 1968;43:1037–1041.
100. Henriksen A, Welinder KG, Gajhede M. J Biol Chem 1998;273:2241–2248.
101. Tran DN, Balkus KJJ. ACS Catal 2011;1:956–968.
102. Hulst AC, Tramper J. Enzyme Microb Technol 1989;11:546–558.
103. Zhao H, Chen W. Curr Opin Biotechnol 2008;19:541–543.

104. Matsuda T, Yamanaka R, Nakamura K. Tetrahedron: Asymmetry 2009;20:513–557.
105. Prelog V. Pure Appl Chem 1964;9:119–130.
106. Faber K. *Biotransformations in Organic Chemistry*. 5th ed. Berlin: Springer Verlag; 2004.
107. Gamenara D, Domínguez de María P. Biotechnol Adv 2009;27:278–285.
108. Bastos Borges K, de Souza Borges W, Durán Patrón R, Tallarico Pupo M, Sueli Bonato P, González Collado I. Tetrahedron: Asymmetry 2009;20:385–397.
109. Fantin G, Giovannini PP, Guerrini A, Maietti S, Medici A, Pedrini P. Biotechnol Lett 2006;28:805–810.
110. Dembitzky VM. Tetrahedron 2003;9:4701–4720.
111. Fernández I, Khiar N. Chem Rev 2003;103:3651–3705.
112. Kinne M, Poraj Kobielska M, Aranda E, Ullrich R, Hammel KE, Scheibner K, Hofrichter M. Bioorg Med Chem Lett 2009;19:3085–3087.
113. Gochev VK, Krastanov AI. Bulg J of Agric Sci 2007;13:75–83.
114. Baldrian P. FEMS Microbiol Lett 2006;30:215–242.
115. Gianfreda L, Xu F, Bollag JM. Bioremed J 1999;3:1–26.
116. Sakurai T, Kataoka K. Chem Rec 2007;7:220–229.
117. Conesa A, Punt PJ, van den Hondel CAMJJ. J Biotechnol 2002;93:143–158.
118. Xu F. Ind Biotechnol 2005;1:38–50.
119. Martínez AT, Ruiz Dueñas FJ, Martínez MJ, del Río JC, Gutiérrez A. Curr Opin Biotechnol 2009;20:348–357.
120. Carballeira JD, Quezada MA, Hoyos P, Simeó Y, Hernáiz MJ, Alcántara A, Sinisterra JV. Biotechnol Adv 2009;27:686–714.
121. Zeyaullah M, Kamli MR, Islam B, Atif M, Benkhayal FA, Nehal M, Rizvi MA, Ali A. Biotechnol Mol Biol Rev 2009;4:49–54.
122. Cowan D, Meyer Q, Stafford W, Muyanga S, Cameron R, Wittwer P. Trends Biotechnol 2005;23:321–329.
123. Tuffin M, Anderson D, Heath C, Cowan DA. Biotechnol J 2009;4:1671–1683.
124. Singh BK. Trends Biotechnol 2009;28:111–116.
125. Riesenfeld CS, Schloss PD, Handelsman J. Annu Rev Genet 2004;38:525–552.
126. Tringe SG, Rubin EM. Nat Rev Genet 2005;6:805–814.
127. Schloss PD, Handelsman J. Curr Opin Biotechnol 2003;14:303–310.
128. Hugenholtz P, Tyson GW. Nature 2008;455:481–483.
129. De Long EF. Nature 2009;459:200–206.
130. Beja O, Suzuki MT, Heidelberg JF, Nelson WC, Preston CM, Hamada T, Eisen JA, Fraser CM, De Long EF. Nature 2002;415:630–633.
131. Demain AL, Adrio JL. Mol Biotechnol 2008;38:41–53.
132. Lorenz P, Eck J. Nat Rev Microbiol 2005;3:510–516.
133. Streit WR, Daniel R, Jaeger KE. Curr Opin Biotechnol 2004;15:285–290.
134. Steele HL, Jaeger KE, Daniel R, Streit WR. J Mol Microbiol Biotechnol 2009;16:25–37.
135. Henne A, Daniel R, Schmitz RA, Gottschalk G. Appl Environ Microbiol 1999;65:3901–3907.

136. Wang C, Meek DJ, Panchal P, Boruvka N, Archibald FS, Driscoll BT, Charles TC. Appl Environ Microbiol 2006;72:384–391.
137. Knietsch A, Waschkowitz T, Bowien S, Henne A, Daniel R. Appl Environ Microbiol 2003;69:1408–1416.
138. Knietsch A, Waschkowitz T, Bowien SH, Henne A, Daniel R. J Mol Microbiol Biotechnol 2003;5:46–56.
139. Sánchez Amat A, Solano F, Lucas Elío P, Mar Drugs 2010;8:519–541.
140. Xu M, Xiao X, Wang F. Extremophiles 2008;12:255–262.
141. Lim HK, Chung EJ, Kim JC, Choi GJ. Appl Environ Microbiol 2005;71:7768–7777.
142. Wexler M, Bond PL, Richardson DJ, Johnston AWB. Environ Microbiol 2005;7:1917–1926.
143. Wohlgemuth R. In: Fessner WD, Anthonsen T, editors. *Modern Biocatalysis*. Weinheim: Wiley-VCH; 2009. pp. 313–338.
144. Blank LM, Ebert BE, Buehler K, Bühler B. Antioxid Redox Signal 2010;13:349–394.
145. Hollmann F, Arends I, Buehler K, Schallmey A, Bühler B. Green Chem 2011;13:226–265.
146. May SW. Curr Opin Biotechnol 1999;10:370–375.
147. Webb EC. *Enzyme Nomenclature 1992: Recommendations of the Nomenclature Committee of the International Union of Biochemistry and Molecular Biology on the Nomenclature and Classification of Enzymes*. San Diego: Academic Press; 1992.
148. Moss GP. *Recommendations of the Nomenclature Committee of the International Union of Biochemistry and Molecular Biology on the Nomenclature and Classification of Enzymes by the Reactions they Catalyse*. Vol 2011. http://www.chem.qmul.ac.uk/iubmb/enzyme/ ed. (Ed.: N. C. o. t. I. U. o. B. a. M. B. (NC-IUBMB)).
149. Burton SG. Trends Biotechnol 2003;21:543–549.
150. Bühler B, Schmid A. J Biotechnol 2004;113:183–210.
151. Leak DJ, Sheldon RA, Woodley JM, Adlercreutz P. Biocatal Biotransform 2009;27:1–26.
152. Woodley JM. Adv Appl Microbiol 2006;60:1–15.
153. Woodley JM. Biochem Soc Trans 2006;34:301–303.
154. Berenger Murcia A, Fernández Lafuente R. Curr Org Chem 2010;14:1000–1021.
155. Wichmann R, Vasic Racki D. In: Scheper T, editor. Technology Transfer in Biotechnology: From Lab to Industry to Production. Vol. 92. Berlin: Springer-Verlag; 2005.
156. Chenault HK, Whitesides GM. Appl Biochem Biotechnol 1987;14:147–197.
157. De Wildeman SMA, Sonke T, Schoemaker HE, May O. Acc Chem Res 2007;40:1260–1266.
158. Nakamura K, Matsuda T. In: Gotor V, Alfonso I, García-Urdiales E, editors. *Asymmetric Organic Synthesis with Enzymes*. Weinheim: Wiley-VCH; 2008. pp. 193–228.
159. Kula MR, Kragl U. In: Patel RN, editor. *Stereoselective Biocatalysis*. New York: Marcel Dekker, Inc.; 2000. pp. 839–866.
160. Kroutil W, Mang H, Edegger K, Faber K. Curr Opin Chem Biol 2004;8:120–126.
161. Moore JC, Pollard DJ, Kosjek B, Devine PN. Acc Chem Res 2007;40:1412–1419.
162. Jörnvall H, Persson M, Jeffery J. Proc Natl Acad Sci USA 1981;78:4226–4230.

163. Auld DS. Biometals 2001;14:271–313.
164. Eklund H, Plapp BV, Samama JP, Brändén CI. J Biol Chem 1982;257:14349–14358.
165. Baker PJ, Britton KL, Fisher M, Esclapez J, Pire C, Bonete MJ, Ferrer J, Rice DW. Proc Natl Acad Sci USA 2009;106:779–784.
166. Kleifeld O, Frenkel A, Martin JM, Sagi I. Nat Struct Biol 2003;10:98–103.
167. Rao KS, Albro M, Dwyer TM, Frerman FE. Biochemistry 2006;45:15853–16861.
168. Ghisla S, Thorpe C. Eur J Biochem 2004;271:494–508.
169. Lenich A, Goodman S. J Biol Chem 1986;261:4090–4096.
170. Rao KS, Albro M, Zirrolli J, van der Velde D, Jones DN, Frerman F. Biochemistry 2005;44:13932–13940.
171. Cunane LM, Chen ZW, Shamala N, Mathews FS, Cronin CN, McIntire WS. J Mol Biol 2000;295:357–374.
172. Bossert ID, Whited G, Gibson DT, Young LY. J Bacteriol 1989;171:2956–2962.
173. Hopper DJ, Jones MR, Causer MJ. FEBS Lett 1985;182:485–488.
174. Furukawa H, Wieser M, Morita H, Sugio T, Nagasawa T. Arch Microbiol 1998;171:37–43.
175. Okamoto K, Matsumoto K, Hille R, Eger BT, Pai EF, Nishiro T. Proc Natl Acad Sci USA 2004;101:7931–7936.
176. Gladyshev VN, Khangulov SV, Stadtman TC. Biochemistry 1996;35:212–223.
177. Hurh B, Yamane T, Nagasawa T. J Ferment Bioeng 1994;78:19–26.
178. Zhang Y, Gladyshev VN. J Mol Biol 2008;379:881–899.
179. Schwarz G, Mendel RR, Ribbe MW. Nature 2009;460:839–847.
180. Anthony C. Biochem J 1996;320:697–711.
181. Matsushita K, Toyama H, Yamada M, Adachi O. Appl Microbiol Biotechnol 2002;58:13–22.
182. Anthony C. Arch Biochem Biophys 2004;428:2–9.
183. Davidson V. Arch Biochem Biophys 2004;428:32–40.
184. Adachi O, Ano Y, Toyama H, Matsushita K. In: Schmid RD, Urlacher VB, editors. *Modern Biooxidation: Enzymes, Reactions and Applications.* Weinheim: Wiley-VCH; 2007. pp. 1–42.
185. Hauge JG. Biochim Biophys Acta 1960;45:250–262.
186. Hauge JG. Arch Biochem Biophys 1961;94:308–318.
187. Hauge JG. J Biol Chem 1964;239:3630–3639.
188. Duine JA, Frank Jzn J, Van Zeeland JK. FEBS Lett 1979;108:443–446.
189. Afolabi PR, Mohammed F, Amaratunga K, Majekodunmi O, Dales SL, Gill R, Thomposon D, Cooper JB, Wood SP, Goodwin PM, Anthony C. Biochemistry 2001;40:9799–9809.
190. Zheng YJ, Xia ZX, Chen ZW, Matheus F, Bruice T. Proc Natl Acad Sci USA 2001;98:432–434.
191. Cozier GE, Salleh RA, Anthony C. Biochem J 1999;340:639–647.
192. Anthony C, Ghosh A. Prog Biophys Mol Biol 1998;69:1–21.
193. Ghosh A, Anthony C, Harlos K, Goodwin MG, Blake CCF. Structure 1995;3:177–187.

194. Elias MD, Nakamura S, Migita CT, Miyoshi H, Toyama H, Matsushita K, Adachi O, Yamada M. J Biol Chem 2004;279:3078–3083.

195. Tishkov VI, Popov VO. Biochemistry (Moscow) 2004;69:1252–1268.

196. Popov VO, Lamzin VS. Biochem J 1994;301:625–643.

197. Li Z, van Beilen JB, Duetz WA, Schmid A, de Raadt A, Grieng H, Witholt B. Curr Opin Biotechnol 2002;6:136–144.

198. Strohmeier GA, Pichler H, May O, Gruber Khadjawi M. Chem Rev 2011;111:4141–4164.

199. Monti D, Ottolina G, Carrea G, Riva S. Chem Rev 2011;111:4111–4140.

200. Leisch H, Morley K, Lau PCR. Chem Rev 2011;111:4165–4222.

201. Torres Pazmiño DE, Fraaije MW. In: Matsuda T, editor. *Future Directions in Biocatalysis*. The Netherlands: Groningen; 2007. pp. 107–128.

202. Bollinger JM, Broderick JB. Curr Opin Chem Biol 2009;13:51–57.

203. Arisawa A, Agematu H. In: Schmid RD, Urlacher VB, editors. *Modern Biooxidation. Enzymes, Reactions and Applications*. Weinheim: Wiley-VCH; 2007. pp. 177–192.

204. Bureik M, Bernhardt R. In: Schmid RD, Urlacher VB, editors. Modern Biooxidation: *Enzymes, Reactions and Applications*. Weinheim: Wiley-VCH; 2007. pp. 155–176.

205. Fraaije MW, Janssen DB. In: Schmid RD, Urlacher VB, editors. Modern Biooxidation: *Enzymes, Reactions and Applications*. Weinheim: Wiley-VCH; 2007. pp. 77–98.

206. Munro AW, Girvan HM, Mc Vey JP, McLean KJ. In Schmid RD, Urlacher VB, editors. *Modern Biooxidation: Enzymes, Reactions and Applications*. Weinheim: Wiley-VCH; 2007. pp. 123–154.

207. Urlacher VB, Bell SG, Wong LL. In: Schmid RD, Urlacher VB, editors. *Modern Biooxidation: Enzymes, Reactions and Applications*. Weinheim: Wiley-VCH; 2007. pp. 99–122.

208. Solomon EI, Brunold TC, Davis MI, Kemsley JN, Lee SK, Lehnert N, Neese F, Skulan AJ, Yang YS, Zhou J. Chem Rev 2000;100:235–350.

209. Wallar BJ, Lipscomb JD. Chem Rev 1996;96:2625–2658.

210. Leahy JG, Batchelor PJ, Morcomb SM. FEMS Microbiol Rev 2003;27:449–479.

211. Lipscomb JD. Annu Rev Microbiol 1994;48:371–399.

212. Lieberman RL, Rosenzweig AC. Crit Rev Biochem Mol Biol 2004;39:147–164.

213. Blazyk JL, Lippard SJ. J Biol Chem 2004;279:5630–5640.

214. Elango N, Radhakrishnan R, Froland WA, Wallar BJ, Earhart CA, Lipscomb JD, Ohlendorf DH. Protein Sci 1997;6:556–568.

215. Merkx M, Kopp DA, Sazinsky MH, Blazyk JL, Müller J, Lippard SJ. Angew Chem Int Ed 2001;40:2782–2807.

216. Rosenzweig AC, Friederick CA, Lippard SJ, Nordlund P. Nature 1993;366:537–543.

217. Austin RN, Buzzi K, Kim E, Zylstra GJ, Groves JT. J Biol Inorg Chem 2003;8:733–740.

218. Austin RN, Luddy K, Erickson K, Pender Cudlip M, Bertrand E, Deng D, Buzdygon RS, van Beilen JB, Groves JT. Angew Chem Int Ed 2008;47:5232–5234.

219. Hsu HF, Que L, Shanklin J, J Inorg Biochem 1999;74:168.

220. Van Beilen JB, Kingma J, Witholt B. Enzyme Microb Technol 1994;16:904–911.

221. van Beilen JB, Smits THM, Roos FF, Brunner T, Balada SB, Rothlisberger M, Witholt B. J Bacteriol 2005;187:85–91.
222. Fu H, Newcomb M, Wong CH. J Am Chem Soc 1991;113:5878–5880.
223. Katopodis AG, Wimalasena K, Lee J, May SW. J Am Chem Soc 1984;106:7928–7935.
224. Katopodis AG, Smith HA, May SW. J Am Chem Soc 1988;110:897–899.
225. May SW, Katopodis AG. Enzyme Microb Technol 1986;8:17–21.
226. Wubbolts MG, Reuvenkamp P, Witholt B. Enzyme Microb Technol 1994;16:608–615.
227. Bühler B, Schmid A, Hauer B, Witholt B. J Biol Chem 2000;275:10085–10092.
228. Bühler B, Witholt B, Hauer B, Schmid A. Appl Environ Microb 2002;68:560–568.
229. Fitzpatrick PF. Annu Rev Biochem 1999;68:355–381.
230. Kappock TJ, Caradonna JP. Chem Rev 1996;96:2659–2756.
231. Flatmark T, Stevens RC. Chem Rev 1999;99:2137–2160.
232. Erlandsen H, Bjorgo E, Flatmark T, Stevens RC. Biochemistry 2000;39:2208–2217.
233. Hegg EL, Que L. Eur J Biochem 1997;250:625–629.
234. Lange SJ, Que L. Curr Opin Chem Biol 1998;2:159–172.
235. Erlandsen H, Fusetti F, Martínez A, Hough E, Flatmark T, Stevens RC. Nat Struct Biol 1997;4:995–1000.
236. Fusetti F, Erlandsen H, Flatmark T, Stevens RC. J Biol Chem 1998;273:16962–16967.
237. Kobe B, Jennings IG, House CM, Michell BJ, Goodwill KE, Santarsiero BD, Stevens RC, Cotton RGH, Kemp BE. Nat Struct Biol 1999;6:442–448.
238. Goodwill KE, Sabatier C, Marks C, Raag R, Fitzpatrick PF, Stevens RC. Nat Struct Biol 1997;4:578–585.
239. Goodwill KE, Sabatier C, Stevens RC. Biochemistry 1998;37:13437–13445.
240. Ortiz de Montellano PR, Chem Rev 2010;110:932–948.
241. Cytochrome P450 Homepage, http://drnelson.uthsc.edu/CytochromeP450.html.
242. Siglar SG. In: Ballou DP, editor. *Essays in Biochemistry: Metalloproteins*. London: Portland Press; 1999.
243. Hannemann F, Bichet A, Ewen KM, Bernhardt R. Biochim Biophys Acta 2007;1770:330–344.
244. Munro AW, Leys DG, McLean KJ, Marshall KR, Ost TWB, Daff S, Miles CS, Chapman SK, Lysek DA, Moser CC, Page CC, Dutton PL. Trends Biotech Sci 2002;27:250–257.
245. Gotoh O. J Biol Chem 1992;267:83–90.
246. Holland HL. In: Kelly DR, editor. *Biotransformations I*. Vol. 8a. Weinheim: Wiley-VCH; 1998. pp. 475–533.
247. Holland HL, Weber HK. Curr Opin Biotechnol 2000;11:547–553.
248. Lewis DFV. *Cytochrome P450: Structure, Function and Mechanism*. Weinheim: Wiley-VCH; 1996.
249. Lewis DFV. *Guide to Cytochromes P450: Structure and Function*. London: Taylor and Francis; 2001.
250. Degtyarenko KN, Kulikova TA. Biochem Soc Trans 2001;29:139–147.
251. Porter TD, Coon MJ. J Biol Chem 1991;266:13469–13472.

252. Omura T, Sanders E, Estabrook RW, Cooper DY, Rosenthal O. Arch Biochem Biophys 1966;117:660–673.
253. Lu AY, Coon MJ. J Biol Chem 1968;243:1331–1332.
254. Lu AY, Junk KW, Coon MJ. J Biol Chem 1969;244:3714–3721.
255. Meunier B, de Visser SP, Shaik S. Chem Rev 2004;104:3947–3980.
256. McLean KJ, Girvan HM, Munro AW. Expert Opin Drug Metab Toxicol 2007;3:847–863.
257. Bernhardt R, Rev Physiol Biochem Pharmacol 1996;127:137–221.
258. Hasler JA, Estabrook RW, Murray M, Pikuleva I, Waterman MR, Capdevila J, Holla V, Helvig C, Falck J, Farrell G. Mol Aspects Med 1999;20:1–137.
259. Kahn R, Durst F. Recent Adv Phytochem 2000;34:151–189.
260. Gunsalus IC. Hoppe Seylers Z Physiol Chem 1968;349:1610–1613.
261. Katagiri M, Ganguli BN, Gunsalus IC. J Biol Chem 1968;243:3543–3546.
262. Narhi LO, Fulco AJ. J Biol Chem 1986;261:7160–7169.
263. Narhi LO, Fulco AJ. J Biol Chem 1987;262:6683–6690.
264. Hawkes DB, Adam GW, Burlingame AL, Ortiz de Montellano PR, De Voss JJ. J Biol Chem 2002;277:27725–27732.
265. Meharenna YT, Li H, Hawkes DB, Pearson AG, De Voss JJ, Poulos TL. Biochemistry 2004;43:9487–9494.
266. Wright RL, Harris K, Solow B, White RH, Kennelly PJ. FEBS Lett 1996;384:235–239.
267. Nishida CR, Ortiz de Montellano PR. Biochem Biophys Res Commun 2005;338:437–445.
268. Oku Y, Ohtaki A, Kamitori S, Nakamura N, Yohda M, Ohno H, Kawarabayasi Y. J Inorg Biochem 2004;98:1194–1199.
269. Jackson CJ, Lamb DC, Marczylo TH, Warrilow AG, Manning NJ, Lowe DJ, Kelly DE, Kelly SL. J Biol Chem 2002;277:46959–46965.
270. Rylott EL, Jackson RG, Edwards J, Womack GL, Seth Smith HM, Rathbone DA, Strand SE, Bruce NC. Nat Biotechnol 2006;24:216–219.
271. Seth Smith HM, Rosser SJ, Basran A, Travis ER, Dabbs ER, Nicklin S, Bruce NC. Appl Environ Microbiol 2002;68:4764–4771.
272. Roberts GA, Grogan G, Greter A, Flitsch SL, Turner NJ. J Bacteriol 2002;184:3898–3908.
273. Hunter DJ, Roberts GA, Ost TW, White JH, Muller S, Turner NJ, Flitsch SL, Chapman SK. FEBS Lett 2005;579:2215–2220.
274. Warman AJ, Roitel O, Neeli R, Girvan HM, Seward HE, Murray SA, Mc Lean KJ, Joyce MG, Toogood H, Holt RA, Leys D, Scrutton NS, Munro AW. Biochem Soc Trans 2005;33:747–753.
275. Miura Y, Fulco AJ. J Biol Chem 1974;249:1880–1888.
276. Ruettinger RT, Wen LP, Fulco AJ. J Biol Chem 1989;264:10987–10995.
277. Munro AW, Leys DG, Mc Lean KJ, Marshall KR, Ost TW, Daff S, Miles CS, Chapman SK, Lysek DA, Moser CC, Page CC, Dutton PL. Trends Biochem Sci 2002;27:250–257.
278. Fulco AJ. Annu Rev Pharmacol Toxicol 1991;31:177–203.
279. Munro AW, Daff S, Coggins JR, Lindsay JG, Chapman SK. Eur J Biochem 1996;239:403–409.

280. Noble MA, Miles CS, Chapman SK, Lysek DA, MacKay AC, Reid GA, Hanzlik RP, Munro AW. BIochem J 1999;339:371–379.

281. Ost TW, Clark J, Mowat CG, Miles CS, Walkinshaw MD, Reid GA, Chapman SK, Daff S. J Am Chem Soc 2003;125:15010–15020.

282. Kitazume T, Takaya N, Nakayama N, Shoun H. J Biol Chem 2000;275:39734–39740.

283. Nakayama N, Takemae A, Shoun H. J Biochem 1996;119:435–440.

284. Kizawa H, Tomura D, Oda M, Fukamizu A, Hoshino T, Gotoh O, Yasui T, Shoun H. J Biol Chem 1991;266:10632–10637.

285. Takaya N, Suzuki S, Kuwazaki S, Shoun H, Maruo F, Yamaguchi M, Takeo K. Arch Biochem Biophys 1999;372:340–346.

286. Daiber A, Shoun H, Ullrich V. J Inorg Biochem 2005;99:185–193.

287. Nakahara K, Tanimoto T, Hatano K, Usuda K, Shoun H. J Biol Chem 1993;268:8350–8355.

288. Froehlich JE, Itoh A, Howe GA. Plant Physiol 2001;125:306–317.

289. Itoh A, Howe GA. J Biol Chem 2001;276:3620–3627.

290. Lau SM, Harder PA, O'Keefe DP. Biochemistry 1993;32:1945–1950.

291. Shibata Y, Matsui K, Kajiwara T, Hatanaka A. Biochem Biophys Res Commun 1995; 207:438–443.

292. Schoch GA, Yano JK, Wester MR, Griffin KJ, Stout CD, Johnson EF. J Biol Chem 2004; 279:9497–9503.

293. Scott EE, He YA, Wester MR, White MA, Chin CC, Halpert JR, Johnson EF, Stout CD. Proc Natl Acad Sci USA 2003;100:13196–13201.

294. Wester MR, Johnson EF, Marques Soares C, Dijols S, Dansette PM, Mansuy D, Stout CD. Biochemistry 2003;42:9335–9345.

295. Yano JK, Wester MR, Schoch GA, Griffin KJ, Stout CD, Johnson EF. J Biol Chem 2004; 279:38091–38094.

296. Williams PA, Cosme J, Matak Vinkovic D, Ward A, Angove HC, Day PJ, Vohrhein C, Tickle IJ, Jhoti H. Science 2004;305:683–686.

297. Williams PA, Cosme J, Ward A, Angove HC, Matak Vinkovic D, Jhoti H. Nature 2003; 424:464.

298. Wester MR, Yano JK, Schoch GA, Yang C, Griffin KJ, Stout CD, Johnson EF. J Biol Chem 2004;279:35630–35637.

299. Presnell SR, Cohen FE. Proc Natl Acad Sci USA 1989;86:6592–6596.

300. Imai M, Shimada H, Watanabe Y, Matsushima Hibiya Y, Makino R, Koga H, Horiuchi T, Ishimura Y. Proc Natl Acad Sci USA 1989;86:7823–7827.

301. Martinis SA, Atkins WM, Stayton PS, Sligar SG. J Am Chem Soc 1989;11:9252–9253.

302. Groves JT. In: Ortiz de Montellano PR, editor. Cytochrome P450. Structure, Mechanism, and Biochemistry. 3rd ed.New York: Kluwer Academic-Plenum Publishers; 2005. pp. 1–44.

303. Poulos TL, Johnson EF. In: Ortiz de Montellano PR, editor. *Cytochrome P450: Structure, Mechanism, and Biochemistry*. New York: Kluwer Elsevier; 2005. pp. 87–114.

304. Rowland P, Blaney FE, Smyth MG, Jones JJ, Leydon VR, Oxbrow AK, Lewis CJ, Tennant MG, Modi S, Eggleston FS, Chenery RJ, Bridges AM. J Biol Chem 2006;281:7614–7622.

305. Guengerich FP, Miller GP, Hanna IH, Saato H, Martin MV. J Biol Chem 2002;277:33711–33719.

306. Makris TM, Denisov I, Schilichting I, Sligar SG. In: Ortiz de Montellano PR, editor. *Cytochrome P450: Structure, Function, and Mechanism*. New York: Kluwer Elsevier; 2005. pp. 149–182.

307. Davydov R, Kappl R, Hütterman J, Peterson JA. FEBS Lett 1991;295:113–115.

308. Davydov R, Macdonald IDG, Makris TM, Sligar SG, Hoffman BM. J Am Chem Soc 1999;121:10654–10655.

309. Davydov R, Chemerisov W, Werst DE, Rajh T, Matsui T, Ikeda Saito M, Hoffman BM. J Am Chem Soc 2004;126:15960–15961.

310. Makris TM, von Koenig K, Schlichting I, Sligar SG. Biochemistry 2007;46:14129–14140.

311. Davydov R, Makris TM, Kofman V, Werst DE, Sligar SG, Hoffman BM. J Am Chem Soc 2001;123:1403–1415.

312. Paine MJI, Scrutton NS, Munro AW, Gutiérrez A, Roberts GCK, Wolf CR. In: Ortiz de Montellano PR, editors. *Cytochrome P450: Structure, Function, and Mechanism*. New York: Kluwer Elsevier; 2005. pp. 115–148.

313. Puchkaev AV, Ortiz de Montellano PR. Arch Biochem Biophys 2005;434:169–177.

314. Roberts GA, Celik A, Hunter DJB, Ost TWB, White JH, Chapman SK, Turner NJ, Flitsch SL. J Biol Chem 2003;278:48914–48920.

315. Grinberg AV, Hannemann F, Schiffler B, Muller J, Heinemann U, Bernhardt R. Proteins 2000;40:590–612.

316. Bureik M, Zollner A, Schuster N, Montenarh M, Bernhardt R. Biochemistry 2005;44:3821–3830.

317. Guengerich FP. Biol Chem 2002;383:1553–1564.

318. Ballou DP, Entsch B, Cole LJ. Biochem Biophys Res Commun 2005;338:590–598.

319. Massey V. J Biol Chem 1994;269:22459–22462.

320. Entsch B, Palfey BA, Ballou DP, Massey V. J Biol Chem 1991;266:17341–17349.

321. Gatti DL, Palfey BA, Lah MS, Entsch B, Massey V, Ballou DP, Ludwig ML. Science 1994;266:110–114.

322. Moran GR, Entsch B, Palfey BA, Ballou DP. Biochemistry 1997;36:7548–7556.

323. Palfey BA, Entsch B, Ballou DP, Massey V. Biochemistry 1994;33:1545–1554.

324. Palfey BA, Basu R, Frederick KK, Entsch B, Ballou DP. Biochemistry 2002;41:8438–8446.

325. Entsch B, Cole LJ, Ballou DP. Arch Biochem Biophys 2005;433:297–311.

326. Entsch B, van Berkel WJH. FASEB J 1995;9:476–483.

327. Cole LJ, Entsch B, Ortiz-Maldonado M, Ballou DP. Biochemistry 2005;44:14807–14817.

328. Hosokawa K, Stanier RY. J Biol Chem 1966;241:2453–2460.

329. Jadan AP, van Berkel WJH, Golovleva LA, Golovlev EL. Biochemistry (Moscow) 2001;66:898–903.

330. Galán B, Díaz E, Prieto MA, García JL. J Bacteriol 2000;182:627–636.

331. Otto K, Hofstetter K, Rothlisberger M, Witholt B, Schmid A. J Bacteriol 2004;186:5292–5302.

332. Manstein DJ, Pai EF, Schopfer LM, Massey V. Biochemistry 1986;25:6807–6816.
333. Sheng D, Ballou DP, Massey V. Biochemistry 2001;40:11156–11167.
334. Jones KC, Ballou DP. J Biol Chem 1986;261:2553–2559.
335. Beaty NB, Ballou DP. J Biol Chem 1981;256:4619–4625.
336. Palfey BA, Ballou DP, Massey V. In: Valentine JS, Foote CS, Greenberg JF, Liebman JF, editors. *Active Oxygen in Biochemistry*. New York: Blackie Academic and Professional; 1995. pp. 37–83.
337. Palfey BA, Massey V. In: Sinnott M, editor.*Comprehensive Biological Catalysis*. New York: Academic Press; 1998. pp. 83–154.
338. Donoghue NA, Norris DB, Trudgill PW. Eur J Biochem 1976;63:175–192.
339. Wong CH, Whitesides G. In: *Enzymes in Synthetic Organic Chemistry*. Oxford: Pergamon; 1994. pp. 131–194.
340. Willets A. Trends Biotechnol 1997;15:55–62.
341. Kamerbeek NM, Jansen DB, Berkel WJH, Fraaije MW. Adv Synth Catal 2003;345:667–678.
342. Mihovilovic MD, Müller B, Stanetty P. Eur J Org Chem 2002;22:3711–3730.
343. Ryerson CC, Ballou DP, Walsh C. Biochemistry 1982;21:2644–2655.
344. Walsh CT, Jack Chen YC. Angew Chem Int Ed 1988;27:333–343.
345. Chen G, Kayser MM, Mihovilovic MD, Mrstik ME, Martinez CA, Stewart JD. New J Chem 1999;23:827–832.
346. Colonna S, Pironti V, Pasta P, Zambianchi F. Tetrahedron Lett 2003;44:869–871.
347. Schwab JM, J Am Chem Soc 1981;103:1876–1878.
348. Schwab JM, Li W, Thomas LP, J Am Chem Soc 1983;105:4800–4808.
349. Mihovilovic MD. Curr Org Chem 2006;10:1265–1287.
350. Solomon EI, Sundaram UM, Machonkin TE, Chem Rev 1996;96:2563–2606.
351. Prigge ST, Mains RE, Eipper BA, Amzel LM. Cell Mol Life Sci 2000;57:1236–1259.
352. Eipper BA, Quon AS, Mains RE, Boswell JS, Blackburn NJ. Biochemistry 1995;34:2857–2865.
353. Kulathila R, Merkler KA, Merkler DJ. Nat Prod Rep 1999;16:145–154.
354. Freeman JC, Villafranca JJ, Merkler DJ. J Am Chem Soc 1993;115:4923–4924.
355. Katopodis AG, May SW. Biochemistry 1990;29:4541–4548.
356. Bradbury AF, Mistry J, Roos BA, Smyth DG. Eur J Biochem 1990;189:363–368.
357. Ramer SE, Cheng H, Palcic MM, Vederas JC. J Am Chem Soc 1988;110:8526–8532.
358. Ping D, Katopodis AG, May SW. J Am Chem Soc 1992;114:3998–4000.
359. Merkler DJ, Kulathila R, Consalvo AP, Young SD, Ash DE. Biochemistry 1992;31:7282–7288.
360. Zabriskie TM, Cheng H, Vederas JC. J Chem Soc Chem Commun 1991;8:571–572.
361. Noguchi M, Seino H, Kochi H, Okamoto H, Tanaka T, Hirama M. Biochem J 1992;283:883–888.
362. Que L, Watanabe Y. Science 2001;292:651–653.
363. Brenner MC, Klinman JP. Biochemistry 1989;28:4664–4670.
364. Tian G, Berry JA, Klinman JP. Biochemistry 1994;33:226–234.
365. Ahn N, Klinman JP. Biochemistry 1983;22:3096–3106.

366. McGuirl MA, Dooley DM. Curr Opin Chem Biol 1999;3:138–144.
367. Yamazaki S, Morioka C, Itoh S. Biochemistry 2004;43:11546–11553.
368. Halaouli S, Asther M, Sigoillot JC, Hamdi M, Lomascolo A. J Appl Microbiol 2006;100:219–232.
369. Sendovski M, Kanteev M, Ben Yosef VS, Adir N, Fishman A. J Mol Biol 2011;405:227–237.
370. Land EJ, Ramsden CA, Riley PA. Acc Chem Res 2003;36:300–308.
371. Decker H, Schweikardt T, Tuczek F. Angew Chem Int Ed 2006;45:4546–4550.
372. Itoh S, Fukuzumi S. Acc Chem Res 2007;40:592–600.
373. Durán N, Rosa MA, D'Annibale A, Gianfreda L. Enzyme Microb Technol 2002;31:907–931.
374. Sánchez Ferrer A, Rodríguez López JN, García Cánovas F, García Carmona F. Biochim Biophys Acta 1995;1247:1–11.
375. Fenoll LG, Rodríguez López JN, García Sevilla F, García Ruiz PA, Varón R, García Cánovas F, Tudela J. Biochim Biophys Acta 2001;1548:1–22.
376. Olivares C, Solano F. Pigment Cell Melanoma Res 2009;22:750–760.
377. Riley PA. J Theor Biol 2000;203:1–12.
378. Solomon EI, Lowery MD. Science 1993;259:1575–1581.
379. Inoue T, Shiota Y, Yoshizawa K. J Am Chem Soc 2008;130:16890–16897.
380. Fenoll LG, Peñalver MJ, Rodríguez López JN, Varón R, García Cánovas F, Tudela J. Int J Biochem Cell Biol 2004;36:235–246.
381. Fetzner S. Appl Microbiol Biotechnol 2002;60:243–257.
382. Fetzner S, Steiner RA. Appl Microbiol Biotechnol 2010;86:791–804.
383. McDaniel R, Ebert Khosla S, Hopwood DA, Khosla C. Nature 1995;375:549–554.
384. Hopwood DA. Chem Rev 1997;97:2465–2497.
385. Cane DE, Walsh CT, Khosla C. Science 1998;282:63–68.
386. Khosla C, Gokhale RS, Jacobsen JR, Cane DE. Annu Rev Biochem 1999;68:219–253.
387. Shen B, Hutchinson CR. Biochemistry 1993;32:6656–6663.
388. Caballero JL, Martínez E, Malpartida F, Hopwood DA. Mol Gen Genet 1991;230:401–412.
389. Kendrew SG, Hopwood DA, Marsh ENG. J Bacteriol 1997;179:4305–4310.
390. Sciara G, Kendrew SG, Miele AE, Marsh NG, Federici LM, Malatesta F, Schimperna G, Savino C, Vallone B. EMBO J 2003;22:201–215.
391. Chung JY, Fujii I, Harada S, Sankawa U, Ebizuka Y. J Bacteriol 2002;184:6115–6122.
392. Rafanan ER, Le L, Zhao L, Decker H, Shen B. J Nat Prod 2001;64:444–449.
393. Adams MA, Jia J. J Biol Chem 2005;280:8358–8363.
394. Das A, Khosla C. Acc Chem Res 2009;42:631–639.
395. Lombó F, Abdelfattah MS, Braña AF, Salas JA, Rohr H, Méndez C. ChemBioChem 2009;10:296–303.
396. Walsh CT. Acc Chem Res 2008;41:4–10.
397. Abu-Omar MM, Loaiza A, Hontzeas N. Chem Rev 2005;105:2227–2252.
398. Costas M, Mehn MP, Jensen MP, Que L. Chem Rev 2004;104:939–986.

399. Samuelsson B, Dahlen SE, Lindgren JA, Rouzer CA, Serhan CN. Science 1987;237:1171–1176.
400. Boyington JC, Gaffney BJ, Amzel LM. Science 1993;260:1482–1486.
401. Minor W, Steczko J, Bolin JT, Otwinowski Z, Axelrod B. Biochemistry 1993;32:6320–6323.
402. Minor W, Steczko J, Stec R, Otwinowski Z, Bolin JT, Walter R, Axelrod B. Biochemistry 1996;35:10687–10701.
403. Tomchick DR, Phan P, Cymborowski M, Minor W, Holman TR. Biochemistry 2001;40:7509–7517.
404. Nelson MJ, Seitz SP, Cowling RA. Biochemistry 1990;29:6897–6903.
405. Jonsson T, Glickman MH, Sun S, Klinman JP. J Am Chem Soc 1996;118:10319–10320.
406. Zhang Y, Gebhard MS, Solomon EI. J Am Chem Soc 1991;113:5162–5175.
407. Holman TR, Zhou J, Solomon EI. J Am Chem Soc 1998;120:12564–12572.
408. Que L, Ho RYN. Chem Rev 1996;96:2607–2624.
409. Lipscomb JD, Orville AM. In: Sigel H, Sigel A, editors. *Metal Ions in Biological Systems*. Vol. 28. New York: Marcel Dekker; 1992. pp. 243–298.
410. Bugg TDH, Sanvoisin J, Spence EL. Biochem Soc Trans 1997;25:81–85.
411. Que L, Widom J, Crawford RL. J Biol Chem 1981;256:10941–10944.
412. Boldt YR, Sadowsky MJ, Ellis LBM, Que L, Wackett LP. J Bacteriol 1995;177:1225–1232.
413. Whiting AK, Boldt YR, Hendrich MP, Wackett LP, Que L. Biochemistry 1996;35:160–170.
414. Hatta T, Mukherjee Dhar G, Damborsky J, Kiyohara H, Kimbara K. J Biol Chem 2003;278:21483–21492.
415. Ohlendorf DH, Lipscomb JD, Weber PC. Nature 1988;336:403–405.
416. Ohlendorf DH, Orville AM, Lipscomb JD. J Mol Biol 1994;244:586–608.
417. Orville AM, Lipscomb JD, Ohlendorf DH. Biochemistry 1997;36:10052–10066.
418. Elgren TE, Orville AM, Kelly KA, Lipscomb JD, Ohlendorf DH, Que L. Biochemistry 1997;36:11504–11513.
419. Orville AM, Elango N, Lipscomb JD, Ohlendorf DH. Biochemistry 1997;36:10039–10051.
420. Vetting MW, Ohlendorf DH. Structure 2000;8:429–440.
421. True AE, Orville AM, Pearce LL, Lipscomb JD, Que L. Biochemistry 1990;29:10847–10854.
422. Que L, Lipscomb JD, Zimmermann R, Münck E, Orme Johnson NR, Orme Johnson WH. Biochim Biophys Acta 1976;452:320–334.
423. Bull C, Ballou DP, Otsuka S. J Biol Chem 1981;256:12681–12686.
424. Walsh TA, Ballou DP, Mayer R, Que L. J Biol Chem 1983;258:14422–14427.
425. Borowski T, Siegbahn PE. J Am Chem Soc 2006;128:12941–12953.
426. Bugg TD, Ramaswamy S. Curr Opin Chem Biol 2008;12:134–140.
427. Que L. Nat Struct Biol 2000;7:182–184.
428. Solomon EI, Decker A, Lehnert N. Proc Natl Acad Sci USA 2003;100:3589–3594.
429. Arciero DM, Orville AM, Lipscomb JD. J Biol Chem 1985;60:14035–14044.

430. Vetting MW, Wackett LP, Que L, Lipscomb JD, Ohlendorf DH. J Bacteriol 2004;186:1945–1958.
431. Sugimoto K, Senda T, Aoshima H, Masai E, Fukuda M, Mitsui Y. Structure 1999;7:953–965.
432. Vaillancourt FH, Barbosa CJ, Spiro TG, Bolin JT, Blades MW, Turner RFB, Eltis LD. J Am Chem Soc 2002;124:2485–2496.
433. Uragami Y, Senda T, Sugimoto K, Sato N, Nagarajan V, Masai E, Fukuda M, Mitsui Y. J Inorg Biochem 2001;83:269–279.
434. Sato N, Uragami Y, Nishizaki T, Takahashi Y, Sazaki G, Sugimoto K, Nonaka T, Masai E, Fukuda M, Senda T. J Mol Biol 2002;321:621–636.
435. Jo DHC, Chiou YM, Que L. Inorg Chem 2001;40:3181–3190.
436. Chiou YM, Que L. Inorg Chem 1995;34:3577–3578.
437. Reynolds MF, Costas MI, Ito M, Jo DH, Tipton AA, Whiting AK, Que L. J Biol Inorg Chem 2003;8:263-.
438. Shu L, Chiou YM, Orville AM, Miller MA, Lipscomb JD, Que L. Biochemistry 1995; 34:6649–6659.
439. Kovaleva EG, Lipscomb JD. Nature Chem Biol 2008;4:186–193.
440. Kovaleva EG, Lipscomb JD. Science 2007;316:453–457.
441. Schofield CJ, Zhang Z. Curr Opin Struct Biol 1999;9:722–731.
442. Prescott AG, Lloyd MD. Nat Prod Rep 2000;17:367–383.
443. Hanauske Abel HM, Popowicz AM. Curr Med Chem 2003;10:1005–1020.
444. van der Ploeg JR, Weiss MA, Saller E, Nashimoto H, Saito N, Kertesz M, Leisinger T. J Bacteriol 1996;178:5438–5446.
445. Valegård K, van Scheltinga ACTL, Lloyd MD , Hara T, Ramaswamy S, Perrakis A, Thompson A, Lee HJ, Baldwin JE, Schofield CJ, Hajdu J, Andersson S. Nature 1998;394:805–809.
446. Zhang Z, Ren J, Stammers DK, Baldwin JE, Harlos K, Schofield CJ. Nat Struct Biol 2000;7:127–133.
447. Clifton IJ, Hsueh LC, Baldwin JE, Harlos K, Schofield CJ. Eur J Biochem 2001;268:6625–6636.
448. Elkins JM, Ryle MJ, Clifton IJ, Hotopp JCD, Lloyd JS, Burzlaff NI, Baldwin JE, Hausinger RP, Roach PL. Biochemistry 2002;41:5185–5192.
449. Wilmouth RC, Turnbull JJ, Welford RWD, Clifton IJ, Prescott AG, Schofield CJ. Structure 2002;10:93–103.
450. Dann CJ. III, Bruick RK, Deisenhofer J. Proc Natl Acad Sci USA 2002;99:15351–15356.
451. Elkins JM, Hewitson KS, McNeill LA, Seibel JF, Schlemminger I, Pugh CW, Ratcliffe PJ, Schofield CJ. J Biol Chem 2003;278:1802–1806.
452. Clifton IJ, Doan LX, Sleeman MC, Topf M, Suzuki H, Wilmouth RC, Schofield CJ. J Biol Chem 2003;278:20843–20850.
453. Cosper NJ, Stalhandske CMV, Saari RE, Hausinger RP, Scott RA. J Biol Inorg Chem 1999;4:122–129.
454. Lloyd MD, Lee HJ, Harlos K, Zhang ZH, Baldwin JE, Schofield CJ, Charnock JM, Garner CD, Hara T, van Scheltinga ACT, Valegård K, Viklund JAC, Hajdu J, Andersson I, Danielsson A, Bhikhabhai R. J Mol Biol 1999;287:943–960.

455. Pavel EG, Kitajima N, Solomon EI. J Am Chem Soc 1998;120:3949–3962.
456. Tuderman L, Myllylä R, Kivirikko KI. Eur J Biochem 1977;80:341–348.
457. Myllylä R, Tuderman L, Kivirikko KI. Eur J Biochem 1977;80:349–357.
458. Holme E. Biochemistry 1975;14:4999–5003.
459. De Carolis E, De Luca V. J Biol Chem 1993;268:5504–5511.
460. Rundgren M. J Biol Chem 1977;252:5094–5099.
461. Johnson Winters K, Purpero VM, Kavana M, Nelson T, Moran GR. Biochemistry 2003;42:2072–2080.
462. Ho RYN, Mehn MP, Hegg EL, Liu A, Ryle MJ, Hausinger RP, Que L. J Am Chem Soc 2001;123:5022–5029.
463. Ryle MJ, Padmakumar R, Hausinger RP. Biochemistry 1999;38:15278–15286.
464. Pavel EG, Zhou J, Busby RW, Gunsior M, Townsend CA, Solomon EI. Biochemistry 1998;120:743–753.
465. Zhou J, Gunsior M, Bachmann BO, Townsend CA, Solomon EI. J Am Chem Soc 1998; 120:13539–13540.
466. Zhou J, Kelly WL, Backmann BO, Gunsior M, Townsend CA, Solomon EI. J Am Chem Soc 2001;123:7388–7398.
467. San Filippo JC, Chern C, Valentine JS. J Org Chem 1976;41:1077–1078.
468. Myllylä R, Schubotz LM, Wesser U, Kivirikko KI. Biochem Biophys Res Commun 1979;89:98–102.
469. Holme E, Lindstedt S, Nordin O. Biochem Biophys Res Commun 1982;107:518–534.
470. Wehbie RS, Punekar NS, Lardy HA. Biochemistry 1988;27:2222–2228.
471. Myllylä R, Majamaa KG, Günzler V, Hanauske-Abel HM, Kivirikko KI. J Biol Chem 1984;259:5403–5405.
472. Parales RE, Resnick SM. In: Ramos J.-L, Levesque RC, editors.Pseudomonas. Vol. 4, Netherlands: Springer; 2006.
473. Boyd D, Sharma N, Allen C. Curr Opin Biotechnol 2001;12:564–573.
474. van der Geize R, Dijkhuizen L. Curr Opin Microbiol 2004;7:255–261.
475. Pieper DH. Appl Microbiol Biotechnol 2005;67:170–191.
476. Gibson DT, Parales RE. Curr Opin Biotechnol 2000;11:236–243.
477. Gibson DT, Hensley M, Yoshioka H, Mabry TJ. Biochemistry 1970;9:1626–1630.
478. Boyd DR, Sheldrake GN. Nat Prod Rep 1998;15:309–324.
479. Hudlicky T, González D, Gibson DT. Aldrichimica Acta 1999;32:35–62.
480. Parales RE, Resnick SM. In: Ramos JL, Levesque RC, editors.Pseudomonas. Vol. 4, The Netherlands: Springer; 2006.
481. Ferraro DJ, Gakhar L, Ramaswamy S. Biochem Biophys Res Commun 2005;338:175–190.
482. Werlen C, Kohler HP, van der Meer JR. J Biol Chem 1996;271:4009–4016.
483. Cho O, Choi KY, Zylstra GJ, Kim YS, Kim SK, Lee JH, Sohn HY, Kwon GS, Kim YM, Kim E. Biochem Biophys Res Commun 2005;327:656–662.
484. Demaneche S, Meyer C, Micoud J, Louwagie M, Willison JC, Jouanneau Y. Appl Environ Microbiol 2004;70:6714–6725.

485. Pinyakong O, Habe H, Yoshida T, Nojiri H, Omori T. Biochem Biophys Res Commun 2003;301:350–357.
486. Zhou NY, Al Dulayymi J, Baird MS, Williams PA. J Bacteriol 2002;184:1547–1555.
487. Kimura N, Kitagawa W, Mori T, Nakashima N, Tamura T, Kamagata Y. Appl Microbiol Biotechnol 2006;73:474–484.
488. Kweon O, Kim SJ, Baek S, Chae JC, Adjei MD, Baek DH, Kim YC, Cerniglia CE. BMC Biochemistry 2008;9:1–20.
489. Wolfe MD, Paralesi JV, Gibson DT, Lipscomb JD. J Biol Chem 2001;276:1945–1953.
490. Bassan A, Blomberg MR, Siegbahn PE. J Biol Inorg Chem 2004;9:439–452.
491. Lee K. FEMS Microbiol Lett 2006;255:316–320.
492. Gibson DT, Resnick SM, Lee K, Brand JM, Torok DS, Wackett LP, Schocken MJ, Haigler BE, J Bacteriol 1995;177:2615–2621.
493. Resnick SM, Lee K, Gibson DT. J Ind Microbiol Biotechnol 1996;17:438–457.
494. Hage A, Schoemaker HE, Field JA. Appl Microbiol Biotechnol 2001;57:79–84.
495. Ensley BD, Gibson DT. J Bacteriol 1983;155:505–511.
496. Karlsson A, Parales JV, Parales RE, Gibson DT, Eklund H, Ramaswamy S. Science 2003;299:1039–1042.
497. Haigler BE, Gibson DT. J Bacteriol 1990;172:457–464.
498. Haigler BE, Gibson DT. J Bacteriol 1990;172:465–468.
499. Wolfe MD, Lipscomb JD. J Biol Chem 2003;278:829–835.
500. Suen WC, Gibson DT. J Bacteriol 1993;175:5877–5881.
501. Senda M, Kishigami S, Kimura S, Fukuda M, Ishida T, Senda T. J Mol Biol 2007;373:382–400.
502. Kauppi B, Lee K, Carredano E, Parales RE, Gibson DT, Eklund H, Ramaswamy S. Structure 1998;6:571–586.
503. Que LJ. J Biol Inorg Chem 2004;9:684–690.
504. Carredano E, Karlsson A, Kauppi B, Choudhury D, Parales RE, Parales JV, Lee K, Gibson DT, Eklund H, Ramaswamy S. J Mol Biol 2000;296:701–712.
505. Lee K. J Bacteriol 1999;181:2719–2725.
506. Neibergall MB, Stubna A, Mekmouche Y, Munck E, Lipscomb JD. Biochemistry 2007;46:8004–8016.
507. Wackett LP, Kwart LD, Gibson DT. Biochemistry 1988;27:1360–1367.
508. Hudlicky T, Reed JW. Synlett 2009;2009:685–703.
509. Boyd DR, Bugg TD. Org Biomol Chem 2006;4:181–192.
510. Chakrabarty S, Austin R, Deng D, Groves JT, Lipscomb JD. J Am Chem Soc 2007;129:3514–3515.
511. Koehntop KD, Emerson JP, Que L. J Biol Inorg Chem 2005;10:87–93.
512. Brown CD, Neidig ML, Neibergall MB, Lipscomb JD, Solomon EI. J Am Chem Soc 2007;129:7427–7438.
513. Roach MP, Clifton IJ, Hensgens CMH, Shibata N, Schofield CJ, Hajdu J, Baldwin JE. Nature 1997;387:827–830.
514. Yan F, Moon SJ, Liu P, Zhao Z, Lipscomb JD, Liu A, Liu HW. Biochemistry 2007;46:12628–12638.

515. Mirica LM, Klinman JP. Proc Natl Acad Sci USA 2008;105:1814–1819.
516. Wilmot CM, Murray JM, Alton G, Parsons MR, Convery MA, Blakeley V, Corner AS, Palcic MM, Knowles PF, McPherson MJ, Philllips SEV. Biochemistry 1997;36, 1608–1620.
517. Brazeau BJ, Johnson BJ, Wilmot CM. Arch Biochem Biophys 2004;428:22–31.
518. Wilmot CM. Biochem Soc Trans 2003;31:493–496.
519. Chang CM, Klema VJ, Johnson BJ, Mure M, Klinman JP, Wilmot CM. Biochemistry 2010;49:2540–2550.
520. Kumar V, Dooley DM, Freeman HC, Guss JM, Harvey I, McGuirl MA, Wilce MC, Zubak VM. Structure 1996;4:943–955.
521. Li R, Klinman JP, Matheus FS. Structure 1998;6:293–307.
522. Parsons MR, Convery MA, Wilmot CM, Yadav KDS, Blakeley V, Corner AS, Phillips SEV, McPherson MJ, Knowles PF. Structure 1995;3:1171–1184.
523. Wilce MC, Dooley DM, Freeman HC, Guss JM, Matsunami J, McIntire WS, Ruggiero CE, Tanizawa K, Yamaguchi H. Biochemistry 1997;36:16116–16133.
524. Janes SM, Mu D, Wemmer D, Smith AJ, Kaur S, Maltby D, Burlingame AL, Klinman JP. Science 1990;248:981–987.
525. Klinman JP. Chem Rev 1996;96:2541–2562.
526. Mills SAG, Goto Y, Su Q, Plastino J, Klinman JP. Biochemistry 2002;41:10577–10584.
527. Su Q, Klinman JP. Biochemistry 1998;37:12513–12525.
528. Wilmot CM, Hajdu J, McPherson MJ, Knowles PF, Phillips SEV. Science 1999;286:1724–1728.
529. Hajdu J, Neutze R, Sjogren T, Edman K, Szoke A, Wilmouth RC, Wilmot C. Nat Struct Biol 2000;7:1006–1012.
530. Cooper AD, Smith W, Bacila M, Medina H. J Biol Chem 1959;234:445–448.
531. Ogel ZB, Brayford D, McPherson MJ. Mycol Res 1994;98:474–480.
532. Pereira AM, Kemmelmeier C. Braz J Microbiol 2000;31:129–134.
533. Turner NJ. Chem Rev 2011;111:4073–4087.
534. Whittaker MM, Whittaker JW. Biochemistry 2001;40:7140–7148.
535. Borman CD, Saysell CG, Sokolowski A, Twitchett MB, Wright C, Sykes AG, Coord Chem Rev 1999;190–192:771–779.
536. Ito N, Phillips SEV, Stevens C, Ogel ZB, McPherson MJ, Keen JN, Yadav KDS, Knowles PF. Nature 1991;350:87–89.
537. Ito N, Phillips SEV, Yadav KDS, Knowles PF. J Mol Biol 1994;238:794–814.
538. Asafu M, Perlin AS. Carbohydrate Res 1974;32:93–99.
539. Avigad G, Amaral D, Asencio C, Horecker BL. J Biol Chem 1962;237:2736–2743.
540. Siebum A, van Wijk A, Schoevaart R, Kieboom T. J Mol Catal B: Enzym 2006;41:141–145.
541. Whittaker JW. Chem Rev 2003;103:2347–2363.
542. Whittaker JW. Arch Biochem Biophys 2005;433:227–239.
543. Ito N, Philllips SEV, Stevens C, Ogel ZB, McPherson MJ, Keen JN, Yadav KDS, Knowles PF. Nature 1991;350:87–90.

544. Baron AJ, Stevens C, Wilmot C, Seneviratne KD, Blakeley V, Dooley DM, Phillips SE, Knowles PF, McPherson MJ. J Biol Chem 1994;269:25095–25105.
545. Klibanov AM, Alberti BN, Marletta MA. Biochem Biophys Res Commun 1982;108:804–808.
546. Root RL, Durrwachter JR, Wong CH. J Am Chem Soc 1985;107:2997–2999.
547. Franke D, Machajewski T, Hsu CC, Wong CH. J Org Chem 2003;68:6828–6831.
548. Burton SG. Curr Org Chem 2003;7:1317–1331.
549. Quintanar L, Stoj C, Taylor AB, Hart PJ, Kosman DJ, Solomon EI. Acc Chem Res 2007;40:445–452.
550. Widsten P, Kandelbauer A. Enzyme Microb Technol 2008;42:293–307.
551. Morozova OV, Shumakovich GP, Gorbacheva MA, Shleev SV, Yaropolov AI. Biochemistry (Moscow) 2007;72:1136–1150.
552. Witayakran S, Ragauskas AJ. Adv Synth Catal 2009;351:1187–1209.
553. Cole JL, Ballou DP, Solomon EI. J Am Chem Soc 1991;113:8544–8546.
554. Lee S-K, DeBeer George S, Antholine WE, Hedman B, Hodgson KO, Solomon EI. J Am Chem Soc 2002;124:6180–6193.
555. Yoon J, Mirica LM, Stack TDP, Solomon EI. J Am Chem Soc 2005;127:13680–13693.
556. Yoon J, Liboiron BD, Sarangi R, Hodgson KO, Hedman B, Solomon EI. Proc Natl Acad Sci USA 2007;104:13609–13614.
557. Leferink NGH, Heuts DPHM, Fraaije MW, van Berkel WJH. Arch Biochem Biophys 2008;474:292–301.
558. van der Heuvel RHH, Laane C, van Berkel WJH. Adv Synth Catal 2001;343:515–520.
559. van den Heuvel RHH, Fraaije MW, Mattevi A, Laane C, Van Berkel WJH. J Mol Catal B: Enzym 2001;11:185–188.
560. Menon V, Hsieh CT, Fitzpatrick PF. Bioorg Chem 1995;23:42–53.
561. Ozimek P, Veenhuis M, van der Klei IJ. Yeast research 2005;5:975–983.
562. Duff SJB, Murray WD. Biotechnol Bioeng 1989;34:153–159.
563. Karra Chaabouni M, Pulvin S, Meziani A, Thomas D, Touraud D, Kunz W. Biotechnol Bioeng 2003;81:27–32.
564. Borzeix F, Monot F, Vandecasteele JP. Enzyme Microb Technol 1995;17:615–622.
565. Dienys G, Jarmalavicius S, Budriene S, Citavicius D, Sereikaite J. J Mol Catal B: Enzym 2003;21:47–49.
566. Bankar SB, Bule MV, Singhal RS, Ananthanarayan L. Biotechnol Adv 2009;27:489–501.
567. Pluschkell S, Hellmuth K, Rinas U. Biotechnol Bioeng 1996;51:215–220.
568. Hatzinikolaou DG, Hansen OC, Macris BJ, Tingey A, Kekos D, Goodenough P, Stougaard P. Appl Microbiol Biotechnol 1996;46:371–381.
569. Witt S, Wohlfahrt G, Schomburg D, Hecht H, Kalisz H. J Biochem 2000;347:553–559.
570. Witteveen C, Veenhuis M, Visser J. Appl Environ Microbiol 1992;58:1190–1194.
571. Kalisz H, Hendle J, Schmid R. Appl Microbiol Biotechnol 1997;47:502–507.
572. Santoni T, Santianni D, Manzoni A, Zanardi S, Mascini M. Talanta 1997;44:1573–1580.

573. Vodopivec M, Berovic M, Jancar J, Podgornik A, Strancar A. Anal Chim Acta 2000;407:105–110.
574. Chudobová I, Vrbová E, Kodícek M, Janovcová J, Kás J, Chudobov'a E. V. a. I, Kod'ıcek M, Janovcov'a J, K'as J. Anal Chim Acta 1996;319:103–110.
575. Rando D, Kohring G, Giffhorn F. Appl Microbiol Biotechnol 1997;48:34–40.
576. Giffhorn F. Appl Microbiol Biotechnol 2000;54:727–740.
577. Ruelius HW, Kerwin RM, Janssen FW. Biochim Biophys Acta 1968;167:493–500.
578. Ander P, Marzullo LJ. J Biotechnol 1997;53:115–131.
579. Leitner C, Volc J, Haltrich D. Appl Environ Microb 2001;67:3636–3644.
580. Fitzpatrick PF. Arch Biochem Biophys 2010;493:13–25.
581. Binda C, Mattevi A, Edmondson DE. J Biol Chem 2002;277:23973–23976.
582. Binda C, Newton Vinson P, Hubalek F, Edmondson DE, Mattevi A. Nat Struct Biol 2002;9:22–26.
583. De Colibus L, Li M, BInda C, Lustig A, Edmondson DE, Mattevi A. Proc Natl Acad Sci USA 2005;102:12684–12689.
584. Kearney EB, Salach JI, Walker WH, Seng RL, Kenney W, Zeszotek E, Singer TP. Eur J Biochem 1971;24:321–327.
585. Edmondson DE, Binda C, Mattevi A. Arch Biochem Biophys 2007;464:269–276.
586. McEwen CM, Sasaki G, Jones DC. Biochemistry 1969;8:3952–3962.
587. Paech C, Salach JI, Singer TP. J Biol Chem 1980;255:2700–2704.
588. Silverman RB. Acc Chem Res 1995;28:335–342.
589. Silverman RB. J Biol Chem 1983;258:14766–14769.
590. Macheroux P, Ghisla S, Sanner C, Rüterjans H, Müller F, BMC Biochem 2005;6:26.
591. Tishkov V, Khoronenkova SV. Biochemistry (Moscow) 2005;70:40–54.
592. Mattevi A, Vanoni MA, Todone F, Rizzi M, Teplyakov A, Coda M, Bolognesi B, Curti B. Proc Natl Acad Sci USA 1996;93:7496–7501.
593. Settembre EC, Dorrestein PC, Park J, Agustine AM, Begley TP. Biochemistry 2003;42: 2971–2981.
594. Khanna P, Schuman Jorns M. Biochemistry 2001;40:1441–1450.
595. Trickey P, Wagner MA, Schuman Jorns M, Mathews FS. Structure 1999;7:331–345.
596. Leys D. EMBO J 2003;22:4038–4048.
597. Ilari A, Bonamore A, Franceschini S, Fiorillo A, Boffi A, Colotti G. Proteins Struct Funct Bioinform 2008;71:2065–2075.
598. Krebs HA. Biochem J 1935;29:1620–1644.
599. Job V, Marcone GL, Pilone MS, Pollegioni L. J Biol Chem 2002;277:6985–6993.
600. Neims AH, De Luca DC, Hellerman L. Biochemistry 1966;5:203–213.
601. Brown LE, Hamilton GA. J Am Chem Soc 1970;92:7225–7227.
602. Walsh CT, Schonbrunn A, Abeles RH. J Biol Chem 1971;246:6855–6866.
603. Porter DJT, Bright HJ. Biochem Biophys Res Commun 1972;46:571–577.
604. Hofrichter M, Ullrich R, Pecyna MJ, Liers C, Lundell T. Appl Microbiol Biotechnol 2010;87:871–897.
605. Adam W, Heckel F, Saha-Möller CR, Schreier P. J Organomet Chem 2002;661: 17–19.

606. Adam W, Lazarus M, Saha-Möller CR, Weichold O, Hoch U, Häring D, Schreier P. Adv Biochem/Eng Biotech 1999;63:73–108.
607. van Deurzen MPJ, van Rantwijk F, Sheldon RA. Tetrahedron 1997;53:13183–13220.
608. Van Rantwijk F, Sheldon RA. Curr Opin Biotechnol 2000;11:554–564.
609. Welinder KG. In: Penel C, Gaspar T, Greppin H, editors. *Plant Peroxidases, Topics and Detailed Literature on Molecular, Biochemical and Physiological Aspects*. Geneve: Université de Geneve; 1992. pp. 1–24.
610. Poulos TL, Edwards SL, Wariishi H, Gold MH. J Biol Chem 1993;268:4429–4440.
611. Passardi F, Theiler G, Zamocky M, Cosio C, Rouhier N, Teixera F, Margis Pinheiro M, Ioannidis V, Penel C, Falquet L, Dunand C. Phytochemistry 2007;68:1605–1611.
612. NC-IUBMB, http://www.chem.qmul.ac.uk/iubmb/, International Union of Biochemistry and Molecular Biology.
613. Everse J, Everse KE, Grisham MB. *Peroxidases in Chemistry and Biology*. Vol. I, 1st ed. Boca Raton, Ann Arbor, and Boston: CRC Press; 1991.
614. Everse J, Everse KE, Grisham MB. Peroxidases in Chemistry and Biology. Vol. II, 1st ed. Boston: CRC Press; 1991.
615. Petrides PE, Nauseef WM. *The Peroxidase Multigene Family of Enzymes: Biochemical Basis and Clinical Applications*. 1st ed. Berlin: Springer-Verlag; 2000.
616. Passardi F, Bakalovic N, Teixeira FK, Margis Pinheiro M, Penel CD, Dunand C, Genomics 2007;89:567–579.
617. Zámocký M, Furtmüller PG, Obinger C. Biochem Soc Trans 2009;37:772–777.
618. Lundell TK, Mäkelä MR, Hildén K. J Basic Microbiol 2010;50:5–20.
619. Zámocký M, Jakopitsch C, Furtmüller PG, Dunand C, Obinger C. Proteins 2008;72:589–605.
620. Faraco V, Piscitelli A, Sannia G, Giardina P. World J Microbiol Biotechnol 2007;23: 889–893.
621. Sugano Y, Muramatsu R, Ichiyanagi A, Sato T, Shoda M, J Biol Chem 2007;282:36652–36658.
622. Sugano Y. Cell Mol Life Sci 2009;66:1387–1403.
623. Pecyna MJ, Ullrich R, Bittner B, Clemens A, Scheibner K, Schubert R, Hofrichter M. Appl Microbiol Biotechnol 2009;84:885–897.
624. Colonna S, Gaggero N, Richelmi C, Pasta P. Trends Biotechnol 1999;17:163–168.
625. Dawson JH, Sono M. Chem Rev 1987;87:1255–1276.
626. Hofrichter M, Ullrich R. Appl Environ Microbiol 2006;71:276–288.

CHAPTER 2

Natural Cofactors and Their Regeneration Strategies

2.1 TYPES OF NATURAL COFACTORS—MECHANISMS

Chapter 1 has shown the enormous diversity of redox enzymes that can be found in Nature, which actually enables an ample number of important natural transformations. In addition, many of those enzymes can be used out-of-the-box for practical purposes (see following chapters), thus making them promising biocatalysts for environment friendly processes.

For the electron transfer of these redox reactions, often the aid of several chemical compounds—the so-called cofactors—is needed. Generally speaking, cofactors are small organic molecules (e.g., coenzymes) or metal ions that are required by an enzyme for its activity. They may be either movable or tightly attached (e.g., a prosthetic group, such as flavins or cytochromes) to the enzyme [1]. In particular, the great majority of redox enzymes require cofactors for their catalytic activity. These cofactors transfer the needed chemical equivalents (such as hydride or oxygen) [2]. Most of these enzymes use nicotinamide adenine dinucleotide (NAD(H)), its 2′-phosphate form (NADP(H)), pyrroloquinoline quinone (PQQ), heme groups, or flavin mononucleotide and flavin adenine dinucleotide (FMN and FAD, respectively) (Figure 2.1) [3].

As stated above, the biological role of these cofactors is essentially the transfer of hydrogen species, as well as electron transfer, including formation of radical species. The nicotinamide derivatives, NAD(H) or NADP(H), are loosely bound to the enzyme, existing as a soluble component of the enzymatic system. Catalytic activity in the majority of oxidoreductases (oxygenases and dehydrogenases) is directly dependent on these nicotinamide cofactors [4]. Flavins (such as FMN and FAD, mainly present in flavin-dependent monooxygenases) and heme groups (e.g., in cytochrome P450 monooxygenases) are tightly bound to the enzymes and are self-regenerating in most cases (see Chapter 1, Section 1.3) [5]. Even though the transfer mechanism of the electrons to those prosthetic groups is not unique, these redox equivalents are typically obtained from the coenzymes NAD(P)H [6]. PQQ was discovered by Hauge

Redox Biocatalysis: Fundamentals and Applications, First Edition.
Daniela Gamenara, Gustavo A. Seoane, Patricia Saenz-Méndez, and Pablo Domínguez de María.

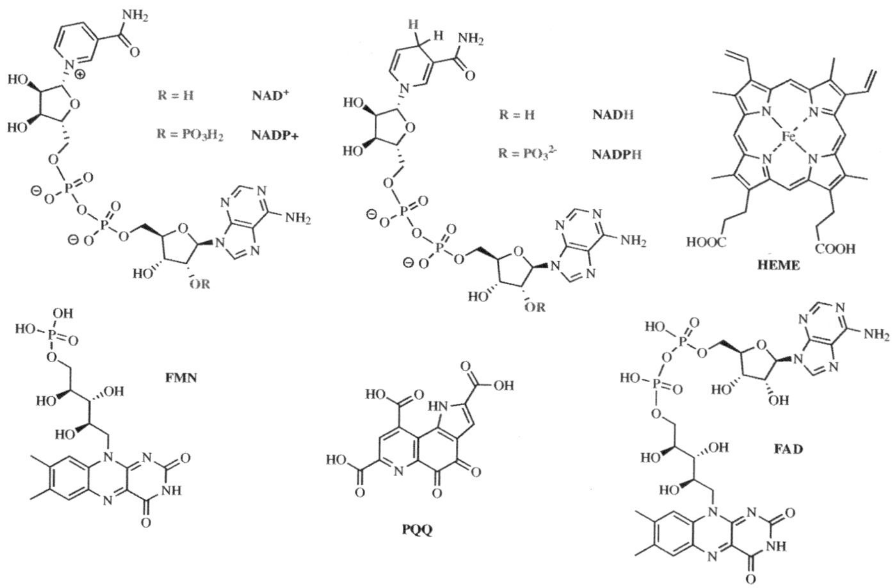

FIGURE 2.1 Structure of cofactors utilized by the majority of redox enzymes.

as the third nonheme redox cofactor after nicotinamide and flavin in bacteria [7], and is also likely to be important in mammals [8]. PQQ is strongly bound to the enzyme, being then a prosthetic group [9,10]. Regardless the particular cofactor, the enzymes require the natural molecules for activity, meaning that they cannot be replaced for synthetic analogs [11,12].

As discussed in Chapter 1, when whole-cell biocatalysts are applied, the system provides the needed enzymes and cofactors. Conversely, when isolated enzymes are employed to catalyze redox reactions, cofactors must be externally provided, either in stoichiometric proportion or in catalytic amounts. From a practical viewpoint cofactors must be regenerated *in situ* by means of a second system, thus allowing them to reenter again in the reaction cycle [13]. In most biocatalytic redox processes, one equivalent of nicotinamide cofactor is required for the reaction to take place. Recycling of NAD(H) or NADP(H) is possible due to the fact that only their oxidation state is modified during the reaction (see Figure 2.1), and therefore, no significant structural changes are involved. Nicotinamide cofactors are highly expensive chemicals to be used in a stoichiometric ratio with the substrate. Thus, it becomes evident that an efficient cofactor recycling system must be employed. Cofactor regeneration can also simplify product purification, since it prevents the accumulation of inhibitory cofactor by-products, and drive the reaction thermodynamically to completion [14]. From an economical perspective, the availability and stability of the regenerative system (enzyme and reagents) under process conditions is an obvious prerequisite. Moreover, the regeneration step should be favorable both kinetically and thermodynamically, and the involved reagents and by-products must be compatible with the rest of the reaction system (see also Chapters 7 and 8) [15,16]. With regard to efficiency, the

TABLE 2.1 Relationship Between TTN and Scope of the Redox Reaction [17]

$TTN = \frac{n_{product}}{n_{cofactor}}$	
TTN < 100	Insufficient
1.000 < TTN < 100.000	Sufficient for reaction on a laboratory scale
100.000 < TTN	Required for production at high scale

total turnover number (TTN) must be high enough (TTN, the total number of moles of product formed per mole of cofactor during the course of a complete reaction) [17]. In this respect, the scale within the redox reaction is practicable is directly related to the TTN (Table 2.1). The TTN value to make a process economically viable actually depends on the cost relationship between the cofactor and the reaction product. If the value of the product is high, the TTN may be lower [18]. Usually, decisions must be made on a case-by-case scenario.

2.2 COFACTOR REGENERATION

For cofactor regeneration issues, a variety of approaches are available, including enzymatic, chemical, electrochemical and photochemical regeneration of reduced or oxidized nicotinamide cofactors [19–21]. However, compared with enzymatic approaches, the last three methods are not competitive for commercial applications, due to the lower TTN obtained, side reactions, or the requirement of expensive mediators [12].

2.2.1 Enzymatic Regeneration of Reduced Cofactors

An important approach in this area is the use of enzymes for the *in situ* cofactor regeneration. Herein, there are two different options for cofactor recycling, namely the substrate-assisted and the enzyme-assisted methods.

2.2.1.1 Substrate-Assisted Method This approach involves the regeneration of reduced cofactors employing the same enzyme responsible of the substrate conversion (Enzyme, Figure 2.2), together with a sacrificial auxiliary substrate. This method

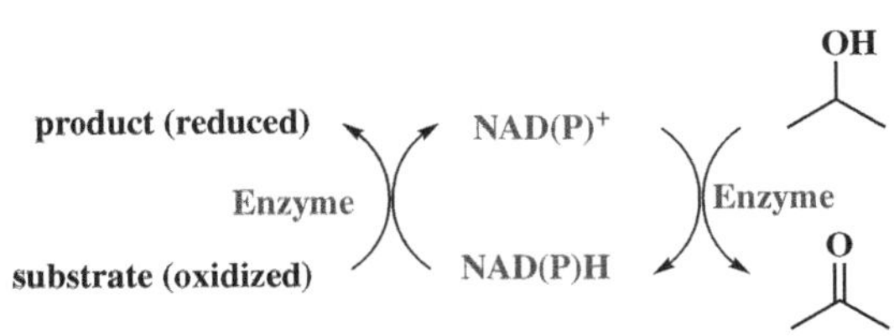

FIGURE 2.2 Cofactor recycling by the substrate-assisted process.

is usually restricted to an alcohol dehydrogenase, which accepts small aliphatic alcohols, and therefore, the cosubstrate is typically 2-propanol. The oxidation of the cosubstrate regenerates the required cofactor [22].

In order to shift the equilibrium of the reaction toward the desired product, high concentrations of the auxiliary substrate are normally required. As a consequence, in such systems the enzyme frequently displays inactivation by high concentrations of cosubstrate or coproduct. Moreover, the isolation of the desired product can be difficult or, in some cases, virtually impossible. Finally, the activity of the enzyme(s) is distributed between both substrates, compromising in most cases the efficiency of the process [12]. In fact, the medium engineering of oxidoreductases is a critical and crucial aspect to be tackled if a practical biotransformation involving these enzymes is envisaged (see Chapters 7 and 8). In recent years, some dehydrogenases that tolerate high concentrations of cosubstrate have been described. For example, alcohol dehydrogenase from *Thermoanaerobium brockii* tolerates concentrations of 2-propanol of up to 20% (v/v) [23,24], and the alcohol dehydrogenase from *Rhodococcus ruber* DSM 44541 (termed ADH-'A') accepts the cosubstrate in amounts up to 50–80% (v/v) [25–29]. Likewise, use of recombinant *Escherichia coli* whole-cells with overexpressed *Candida parapsilopsis* carbonyl reductase allows the enantioselective reduction of ketones in neat substrates at a controlled water activity [30].

2.2.1.2 *Enzyme-Assisted Method* Another approach involves two different enzymes to catalyze two parallel redox reactions. The reaction enzyme ($Enzyme_1$) catalyzes the conversion of substrate into the desired product. Subsequently, a regenerating enzyme ($Enzyme_2$) is responsible for the recycling of the cofactor by oxidizing the ancillary substrate (Figure 2.3).

If the specificity for the substrate is sufficiently different, then the principal substrate and the auxiliary one are not competing for the active site of a single enzyme. As a consequence, there is no inhibition of the production enzymatic system. This also avoids the thermodynamic restrictions mentioned in the previous method. By far, this is the most common approach employed to regenerate needed cofactors. In this area several oxidoreductases have been used as second enzyme to regenerate the cofactor. All of these available enzymatic systems for cofactor recycling present advantages and disadvantages, as shown in Table 2.2 [12].

As observed, systems with lower enzyme cost are related to higher cosubstrate cost and viceversa, meaning that there is no ideal system to achieve desired regeneration. The most widely used method for NADH regeneration uses formate dehydrogenase (FDH) isolated from *Candida boidinii* to convert formate into carbon dioxide [31]. Advantages of this method include that formate is inexpensive, stable and innocuous

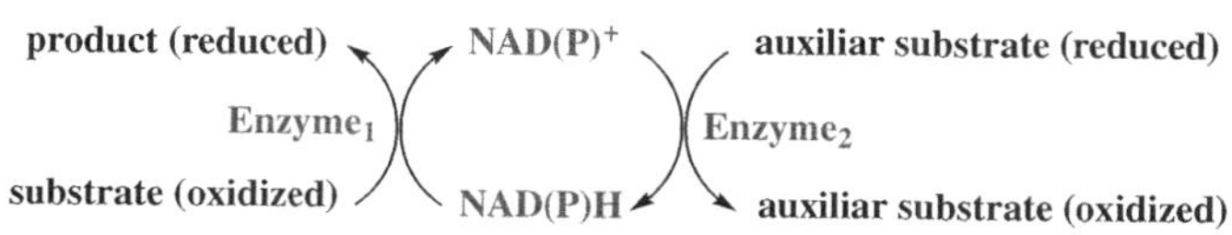

FIGURE 2.3 Cofactor recycling by the enzyme-assisted process.

TABLE 2.2 Comparison of Enzyme Systems for Cofactor Recycling

		Cost	
Enzyme	Activity	Cosubstrate	Enzyme
FDH	−	−	+ +
G6PDH	+ +	+ +	−
GDH	+ +	−	+ +
Aldehyde DH	+	+	+

FDH, formate dehydrogenase; G6PDH, glucose-6-phosphate dehydrogenase; GDH, glucose dehydrogenase; aldehyde DH, aldehyde dehydrogenase.

for the enzyme. The product carbon dioxide is also innocuous and can be easily stripped from the reaction. The main disadvantage is the high cost of the enzyme production and its low specific activity (approximately 6 U/mg) [18]. Likewise, the enzymatic systems for cofactor regeneration based on FDH are somehow limited because these enzymes are specific for NADH. This limitation has been overcome by using a genetically modified FDH from *Pseudomonas* sp. 101, which has high affinity toward $NADP^+$ [32,33]. The mutant FDH is actually one of the best catalysts for NADPH regeneration, and this represents a major achievement considering the fact that the NADPH is more expensive than NADH [34].

All the mentioned factors made that the first enzyme-assisted cofactor generation process implemented at multiton scale was based on the use of yeast FDH from *C. boidinii*. The commercial process was the production of L-*tert*-leucine with an amino acid dehydrogenase by Degussa (Figure 2.4) (see Chapter 8) [35–38].

FDH is efficiently produced by the yeast *C. boidinii*, and the corresponding gene has been successfully cloned and overexpressed in *E. coli* in order to perform enzymatic reductions on a large scale. The recombinant enzyme has been stabilized by site-directed mutagenesis as well, resulting in a very valuable regenerating system for reduced nicotinamide cofactors [39]. Thus, above-reported drawbacks with regard to FDH from *C. boidinii* can now be overcome. Due to this, the number of published works dealing with FDH applications for cofactor regeneration in organic synthesis employing NAD(P)H-dependent oxidoreductases has been increasing in recent years [40–45].

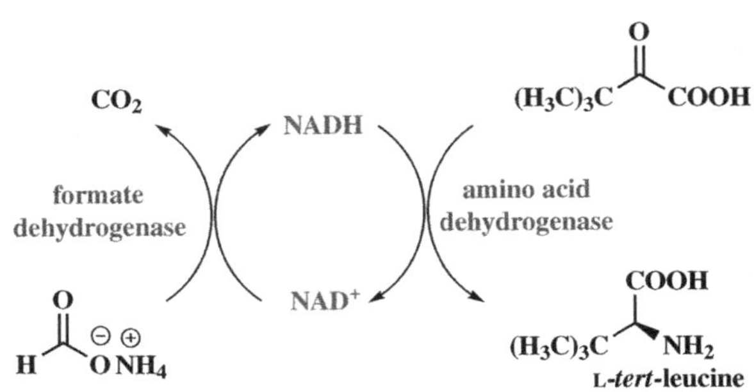

FIGURE 2.4 Production of L-*tert*-leucine at industrial scale.

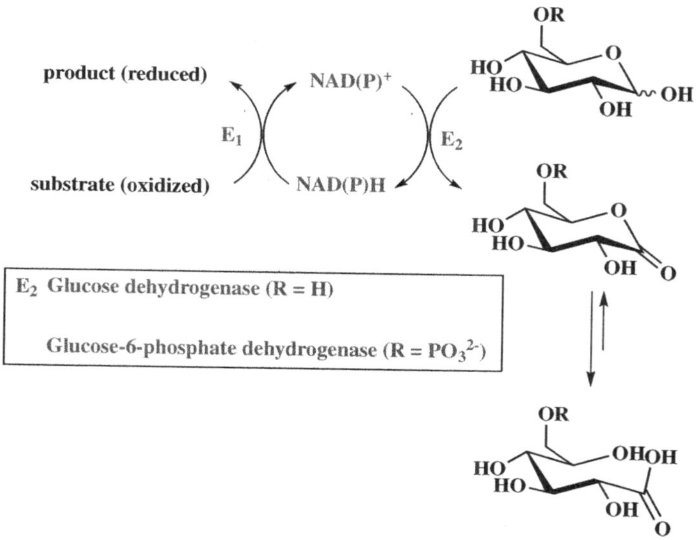

FIGURE 2.5 Regeneration of NAD(P)H using GDH and G6PDH.

Another enzymatic system for recycling both NADH and NADPH involves the oxidation of glucose catalyzed by glucose dehydrogenase (GDH) or by glucose-6-phosphate dehydrogenase (Figure 2.5) [46,47].

One advantage of this approach is that the equilibrium is shifted toward products because the gluconolactone is spontaneously hydrolyzed, yielding gluconic acid (when glucose is the substrate) or phosphogluconic acid (for glucose-6-phosphate). The GDH from *Bacillus cereus* is very stable, accepts both NAD^+ and $NADP^+$, and displays high specific activity. G6PDH from *Leuconostoc mesenteroides* is also stable and works well for both nicotinamide cofactors. The main drawback for these systems is the high cost of GDH in one case, and glucose-6-phosphate in the other one [22,48–50].

A different enzymatic approach for NAD(P)H regeneration is using ethanol and alcohol dehydrogenase. This is a very simple system for lab-scale purposes. However, due to the low redox potential, only activated carbonyl groups are reduced in good yields. With less reactive substrates, the equilibrium can be shifted to the products by removing obtained acetaldehyde by further oxidation to acetate using aldehyde dehydrogenase (Figure 2.6).

Yet, this strategy provides low TTN and therefore its practical use has been rather testimonial as regenerating system for nicotinamide cofactors [51]. To improve the strategy, the successful enantioselective asymmetric reduction of prochiral ketones yielding the corresponding optically active alcohols was described with an alcohol dehydrogenase from *Thermoanaerobacter ethanolicus* using high concentrations of 2-propanol (30% v/v) as cosolvent and cosubstrate. This concentration of 2-propanol was key both to enhance the solubility of substrates and to shift the equilibrium toward products [52].

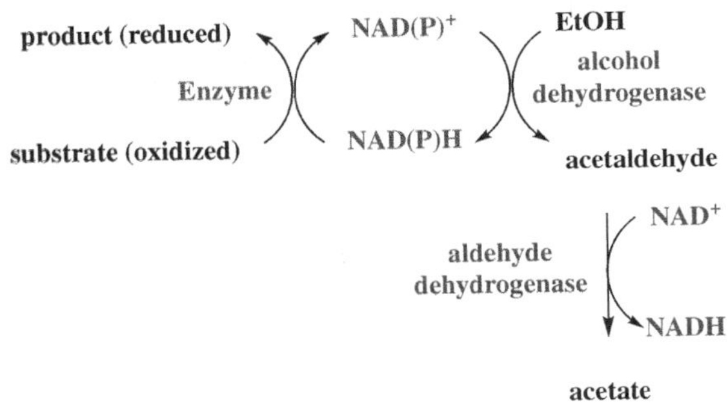

FIGURE 2.6 Regeneration of NAD(P)H using alcohol and aldehyde dehydrogenases.

A new and promising method for NAD(P)H regeneration uses phosphite dehydrogenase (PTDH) (Figure 2.7) [53–55]. This enzyme catalyzed the oxidation of phosphite to phosphate. Even though the wild-type PTDH has a 100-fold preference for NAD^+ over $NADP^+$, a mutant enzyme accepting both cofactors was developed by rational design [16,56]. This system involves an inexpensive substrate (phosphite), an easily removable by-product, and both substrate and by-products are innocuous to the enzyme [57,58].

A totally different alternative for regeneration of NADPH is the use of a hydrogenase from the thermophilic organism *Pyrococcus furiosus*. The enzyme is not only reasonably stable but it also needs high temperatures to achieve an acceptable activity, which in turns leads to a reduction in cofactor stability [6,22]. The main advantage of using this biocatalyst is the fact that the enzyme uses molecular hydrogen as substrate, which is completely converted for cofactor recycling, thus no by-products are generated. Yet, the enzyme is sensitive to molecular oxygen and therefore it is not applicable to monooxygenases, being this an important drawback for further practical applications of this biocatalyst [59] (Figure 2.8).

2.2.2 Enzymatic Regeneration of Oxidized Cofactors

While the regeneration of reduced nicotinamide cofactors has been intensively studied, recycling of $NAD(P)^+$ is comparatively underdeveloped, mainly due to the

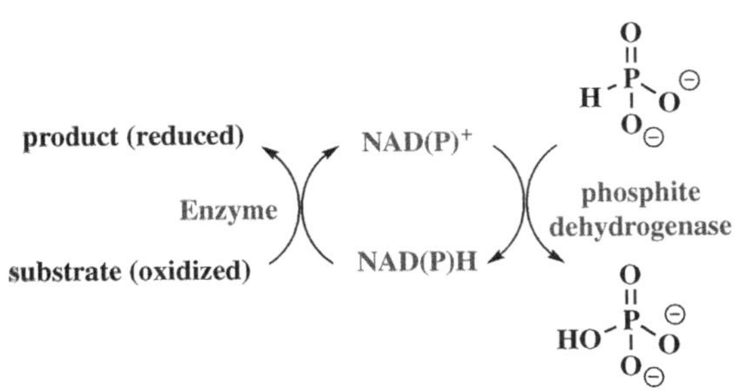

FIGURE 2.7 Regeneration of NAD(P)H with phosphite dehydrogenase.

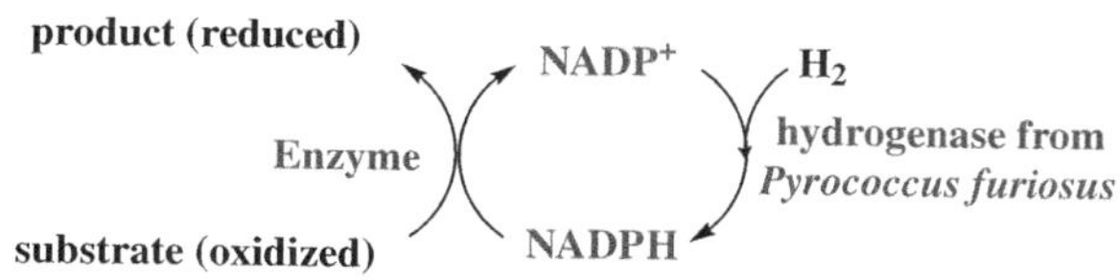

FIGURE 2.8 Regeneration of NADPH with hydrogenase from *Pyrococcus furiosus*.

narrower scope of the involved reactions. Normally enzymatic alcohol oxidations are restricted to kinetic resolution of racemic mixtures of alcohols, as well as the synthesis of ketones (particularly those being difficult to prepare chemically). The most widely used method for recycling of $NAD(P)^+$ involves the use of glutamate dehydrogenase (GluDH) to catalyze the reductive amination of α-ketoglutarate to give L-glutamate (Figure 2.9) [16,18].

The main disadvantage of GluDH system is its moderate specific activity. This may be one of the reasons why the enantioselective enzymatic oxidation of alcohols using ADHs attracted little attention. In the past few years, the regeneration of oxidized cofactors has been significantly improved [59]. NADH or NADPH are oxidized at the expense of the cheapest and harmless oxidant, molecular oxygen, to produce hydrogen peroxide or water as by-product (Figure 2.10).

The NADH oxidase from *Lactobacillus brevis* catalyzes the oxidation of NADH to NAD^+ with molecular oxygen, leading to water formation. This enzymatic system has been recently employed for the oxidative kinetic resolution of racemic 1-phenylethanol [59,60]. Also, the regeneration of NAD^+ by NADH oxidase with molecular oxygen enabled to oxidize alcohols to carboxylic acids under very mild conditions and good yields [61]. Another example comprises the water-forming oxidase from *Lactobacillus sanfranciscensis*, which is able to utilize not only NADH but also NADPH as a cofactor. The ability of this NAD(P)H oxidase to accept both cofactors turns the enzyme into a highly useful catalyst for coupled enzymatically catalyzed oxidation reactions [62].

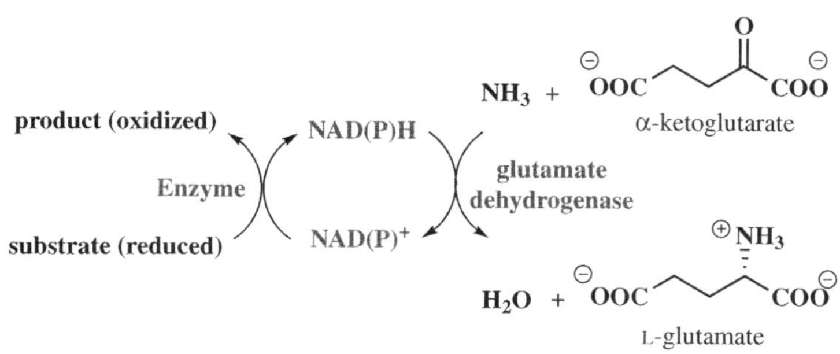

FIGURE 2.9 Regeneration of oxidized nicotinamide cofactors with glutamate dehydrogenase.

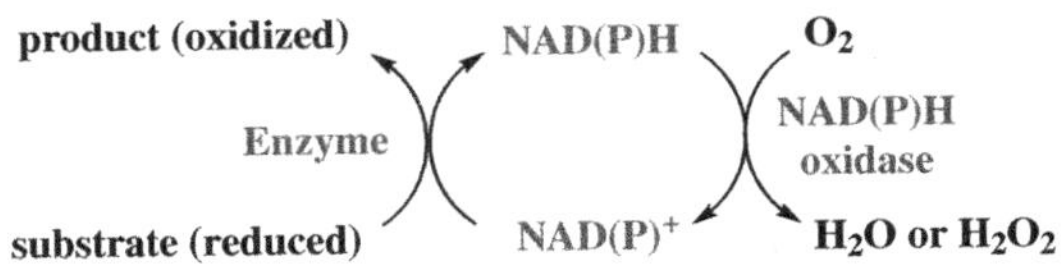

FIGURE 2.10 Regeneration of oxidized nicotinamide cofactors using NAD(P)H oxidase.

2.2.3 Chemical Regeneration of Cofactors

The first developed method for the regeneration of NAD^+ was the chemical reduction of NADH using sodium dithionite ($Na_2S_2O_4$) [63]. Corresponding TTN were lower than 100 and also deactivation of the enzyme occurred due to modification of thiol groups. Therefore, this method is not currently used. Chemical regeneration employing molecular hydrogen is preferred, because hydrogen is the most economic reducing agent and as both atoms of hydrogen are transferred to the products, by-products are not generated. However, this method usually involves complex systems where molecular hydrogen reduces an intermediate hydride carrier, which subsequently reduces the cofactor [13].

More recently, transition metal catalysts for the reaction between H_2 and $NAD(P)^+$, capable to be coupled to the main enzymatic reaction, have been developed. As an example, the reduction of 2-heptanone to (*S*)-2-heptanol using ADH from *T. brockii* was performed with catalytic amounts of $NADP^+$ and $[RuCl_2(TPPTS)_2]_2$ [TPPTS = tris(*m*-sulfonatophenyl)-phosphine] [64]. Although the TTN was low (approximately 10), this work demonstrates that it is possible the chemical regeneration of nicotinamide cofactors, thus opening a new research avenue for the future (Figure 2.11).

The organometallic complex $[Cp^*Rh(bpy)(H_2O)]^{2+}$ (Cp^*, pentamethylcyclopentadienyl; bpy, bipyridine) is the only nonenzymatic catalyst for NADH regeneration that has been successfully coupled to a monooxygenase reaction. However, on a preparative scale, this complex showed a relatively low specific activity (approximately 2 U/mg) [21,65–67].

A novel water-soluble 1,4-dihydropyridine could function in biological processes through NADH regeneration, driving efficiently the horse liver alcohol dehydrogenase reduction of cyclohexanone [68]. Likewise, very recently some iron(III) porphyrins were successfully used as biomimetic NAD(P)H oxidases, using molecular oxygen and forming water to regenerate the cofactor within enzymatic cycles [69]. However, chemical regeneration of cofactors is still not effective enough to substitute the enzymatic approaches as the first choice methods when practical biotransformations are envisaged.

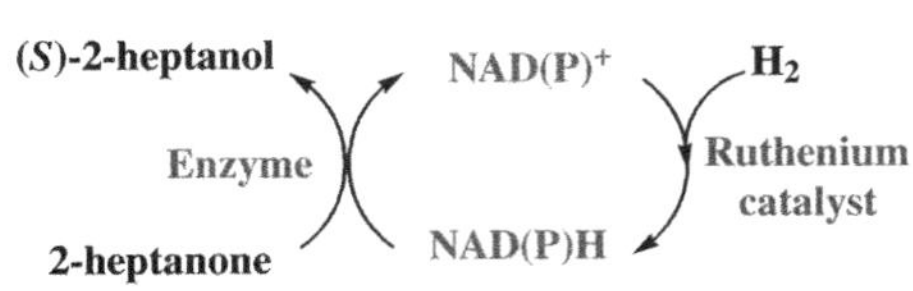

FIGURE 2.11 Chemical regeneration of nicotinamide cofactors.

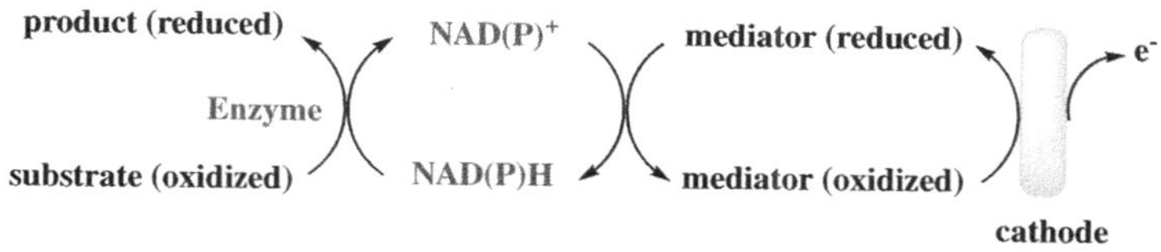

FIGURE 2.12 Electrochemical regeneration of nicotinamide cofactors.

2.2.4 Electrochemical Regeneration of Cofactors

Electrochemical regeneration of cofactors has been described as a promising alternative to the above-described strategies. However, only a few examples combining electrochemical reactions and enzymatic catalysis with synthetic purposes are available so far (Figure 2.12) [70].

Main advantages are that the supply of redox equivalents is essentially mass free, since only electron transfer reactions are involved, no auxiliary substrate is required and thus, no byproducts are obtained. Finally, electrons are one of the most economical redox equivalents available. However, for this electrochemical set-up special bioreactors are required—and normally high potentials are needed—often leading to cofactor dimerization. Moreover, only enzymes located in the immediately vicinity of the electrode are productive, and electrode deposition is frequently observed [18,70]. These drawbacks can be partially overcome by the addition of mediators. In this case, cofactor regeneration can take place at lower potential, thus minimizing the mentioned side effects (Figure 2.13) [70,71].

In this latter approach, electrons generated at the cathode are transferred through the mediator to the cofactor. The rhodium complex showed in Figure 2.13 is employed for regeneration of the reduced cofactor, and $[Cp^{*}Rh(bpy)(H_2O)]^{2+}$ for regeneration of the oxidized one (Figure 2.14). The mediator resulted inert to molecular oxygen, and could be applied for cofactor regeneration in monooxygenase catalyzed reactions [66,72].

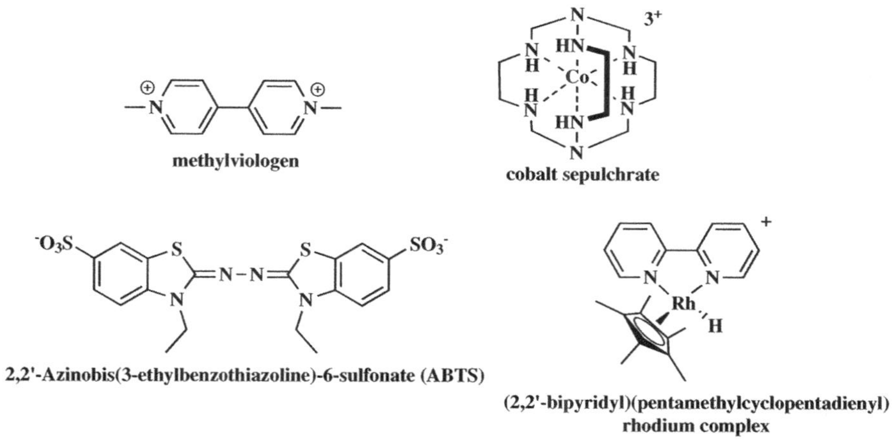

FIGURE 2.13 Examples of mediators used in electrochemical regeneration of cofactors.

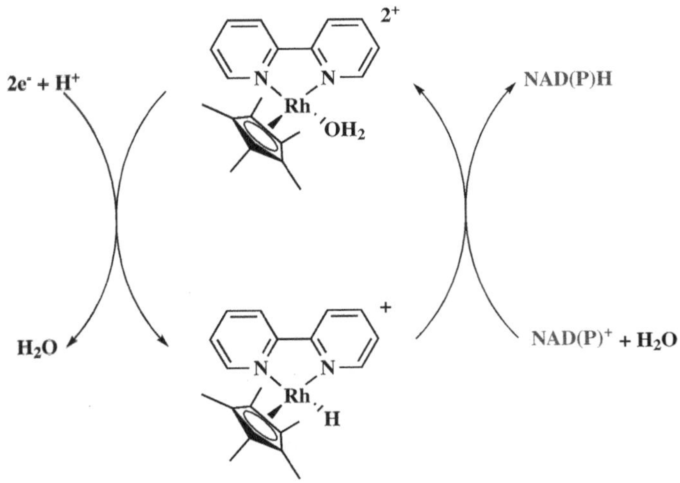

FIGURE 2.14 Electrochemical regeneration of NAD(P)H using $[Cp^{*}Rh(bpy)(H_2O)]^{2+}$ as mediator.

Another recent application of electroenzymatic synthesis is the reduction of acetophenone to (*R*)-phenylethanol catalyzed by the alcohol dehydrogenase from *L. brevis* in combination with electrochemical regeneration of NADPH mediated by a rhodium complex [73]. Even though the indirect electrochemical regeneration of cofactors using mediators has been a more successful strategy, recently the use of special electrodes allowed the efficient mediator-free electrochemical regeneration of nicotinamide cofactors. The electroenzymatic oxidation of 2-propanol with a tin oxide electrode showed efficient regeneration of $NADP^+$ [74]. Finally, an interesting approach involves the immobilization of both the mediator and the cofactor on the electrode surface by using a layer-by-layer adsorption technique [75]. The production of L-malic acid with fixation of HCO_3^- by malic enzyme catalyzed reaction was successfully combined with regeneration of NADH on a layer-by-layer modified electrode [76].

2.2.5 Photochemical Regeneration of Cofactors

Visible light (sun light) is not only cheap, but also environment friendly, and can be used to trigger electron transfer reactions. Therefore, the irradiation of light-active substances (photosensitizers) with visible light causes the regeneration of reduced nicotinamide cofactors as well (Figure 2.15) [12].

Initially, photosensitizer dyes such as methylene blue were used for this purpose [77]. Also, organometallic compounds such as $Ru(bpy)_3^{2+}$ or metalloporphyrins such as Zn-$TMPyP^{4+}$ (Zn-*meso*-tetramethylpyridinium porphyrin), flavin dyes, and semiconductors such as TiO_2 or Fe_2O_3 have been applied as photosensitizers that induce electron transfer reactions [78]. Light-mediated cofactor recycling using living cells of a cyanobacteria (*Synechococcus elongatus* PCC 7942) was successfully

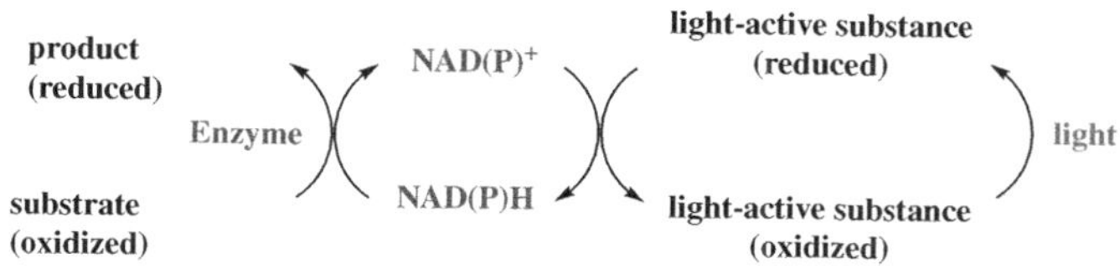

FIGURE 2.15 Photochemical regeneration of nicotinamide cofactors.

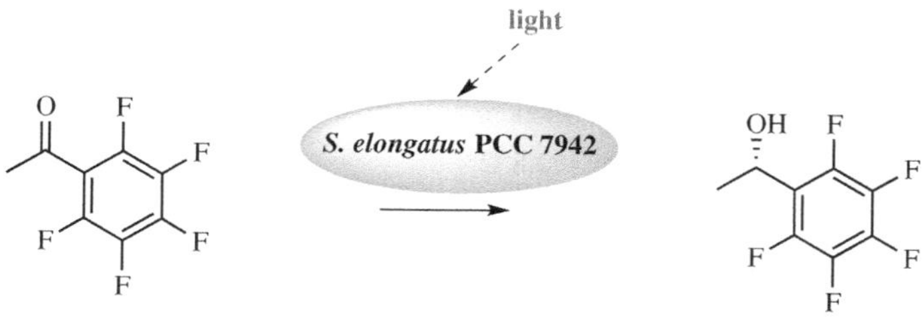

FIGURE 2.16 Enantioselective reduction of pentafluoroacetophenone with *Synechococcus elongatus* and light.

employed as well (Figure 2.16) [79]. In that work, the NADPH generated through photosynthesis was used for reducing an artificial ketone. Cofactor recycling is performed by the cyanobacteria, which captures light in order to generate NADPH through photosynthetic electron transfer reactions.

More recently, a photoregeneration of NADH using carbon-containing TiO_2 as photosensitizer was described [80]. In this report, $[Cp^*Rh(bpy)(H_2O)]^{2+}$ was employed as catalyst. The carbon-containing TiO_2 has a strong absorption toward visible light, and in the presence of the rhodium complex and electron donors, such as ethylendiaminetetraacetic acid (EDTA), ascorbic acid, or mercaptoethanol, it is an appropriate system for NADH regeneration [80].

2.3 CONCLUDING REMARKS

Most redox enzymes require expensive cofactors, which are too expensive to be added in stoichiometric amounts. Therefore, an efficient cofactor regeneration system has to be provided for practical biocatalytic approaches. Over the years, progress made in several coenzyme regeneration approaches allowed to use oxidoreductases in an efficient and inexpensive manner. In this chapter regeneration systems for nicotinamide cofactors, such as enzymatic, chemical, photochemical and electrochemical, have been described. Even though substantial progress has been accomplished in nonenzymatic approaches, still best methods available for cofactor regeneration are the enzymatic ones, either involving an ancillary substrate, or a second enzyme. Chemical, photochemical and electrochemical regeneration systems are still in their infancy, and further research is needed to convert them in more generalized applicable alternatives for cofactor regeneration cycles.

REFERENCES

1. IUPAC. Pure Appl Chem 1992;64:143–168.
2. NC-IUBMB, http://www.chem.qmul.ac.uk/iubmb/, International Union of Biochemistry and Molecular Biology.
3. Fischer JD, Holliday GL, Rahman SA, Thornton JM. J Mol Biol 2010;403:803–824.
4. Molinari F. Curr Org Chem 2006;10:1247–1263.
5. Legrand Y-M, Gray M, Cooke G, Rotello VM. J Am Chem Soc 2003;125:15789–15795.
6. Torres Pazmiño DE, Winkler M, Glieder A, Fraaije MW. J Biotechnol 2010;146:9–24.
7. Hauge JG. J Biol Chem 1964;239:3630–3639.
8. Kasahara T, Kato T. Nature 2003;422:832.
9. Kay CWM, Mennenga B, Görisch H, Bittl R. FEBS Lett 2004;564:69–72.
10. Kay CWM, Mennenga B, Görisch H, Bittl R. J Biol Chem 2006;281:1470–1476.
11. Liu W, Wang P. Biotechnol Adv 2007;25:369–384.
12. De Wildeman SMA, Sonke T, Schoemaker HE, May O. Acc Chem Res 2007;40:1260–1266.
13. Wichmann R, Vasic-Racki D. Cofactor regeneration at lab scale. In: Scheper T, editor. *Technology Transfer in Biotechnology: From Lab to Industry to Production*. Volume 92, Berlin: Springer-Verlag; 2005.
14. Koeller KM, Wong C-H. Nature 2001;409:232–240.
15. Zhao H, van der Donk WA. Curr Opin Biotechnol 2003;14:583–589.
16. Woodyer RD, Johannes TW, Zhao H. In: Pandey A, Webb C, Soccol CS, Larroche C, editors. *Enzyme Technology*. New Delhi: Asiatech Publishers Inc.; 2004. pp. 83–101.
17. Chenault HK, Whitesides GM. Appl Biochem Biotechnol 1987;14:147–197.
18. van der Donk WA, Zao H. Curr Opin Biotechnol 2003;14:421–426.
19. Huisman GW, Gray D. Curr Opin Biotechnol 2002;13:352–358.
20. Burton SG. Trends Biotechnol 2003;21:543–549.
21. Hollman F, Hofstetter K, Schmid A. Trends Biotechnol 2006;24:163–171.
22. Eckstein M, Daubmann T, Kragl U. Biocatal Biotransform 2004;22:89–96.
23. Schubert T, Hummel W, Muller M. Angew Chem Int Ed 2002;41:634–637.
24. Villela FM, Stillger T, Mueller M, Liese A, Wandrey C. Angew Chem Int Ed 2003;42:2993–2996.
25. Stampfer W, Kosjek B, Faber K, Kroutil W. J Org Chem 2003;68:402–406.
26. Edegger K, Mang H, Faber K, Gross J, Kroutil W. J Mol Catal A: Chem 2006;251:66–70.
27. de Gonzalo G, Lavandera I, Faber K, Kroutil W. Org Lett 2007;9:2163–2166.
28. Edegger K, Gruber CC, Poesl TM, Wallner SR, Lavandera I, Faber K, Niehaus F, Eck J, Oehrlein R, Hafnerd A, Kroutil W. Chem Commun 2006;22:2402–2404.
29. Stampfer W, Kosjek B, Kroutil W, Faber K. Biotechnol Bioeng 2003;81:865–869.
30. Jakoblinnert A, Mladenov R, Paul A, Sibilla F, Schwaneberg U, Ansorge-Schumacher MB, Domínguez de María P. Chem Commun 2011;47:12230–12232.
31. Gröger H, Hummel W, Rollman C, Chamouleau F, Husken H, Werner H, Wunderlich C, Abokitse K, Drauz K, Buchholz S. Tetrahedron 2004;60:633–640.

32. Tishkov VI, Galkin AG, Fedorchuk W, Savitsky PA, Rojkova AM, Gieren H, Kula MR. Biotechnol Bioeng 1999;64:187–193.
33. Tishkov VI, Popov VO. Biomol Eng 2006;23:89–110.
34. Hummel W. Trends Biotechnol 1999;17:487–492.
35. Bommarius AS, Drauz K, Hummel W, Kula MR, Wandrey C. Biocatalysis 1994;10:37–47.
36. Bommarius AS, Schwarm M, Stingl K, Kottenhahn M, Huthmacher K, Drauz K. Tetrahedron: Asymmetry 1995;6:2851–2888.
37. Bommarius AS, Schwarm M, Drauz K. J Mol Catal B: Enzym 1998;5:1–11.
38. Evonik, http://corporate.evonik.com/en/Pages/default.aspx, Evonik Industries.
39. Slusarczyk H, Felber S, Kula M-R, Pohl M. Eur J Biochem 2000;267:1280–1289.
40. Tao J, McGee K. Org Proc Res Dev 2002;6:520–524.
41. Maurer SC, Schulze H, Schmid RD, Urlacher V. Adv Synth Catal 2003;345:802–810.
42. Gröger H, May O, Werner H, Menzel A, Altenbuchner J. Org Proc Res Dev 2006;10:666–669.
43. Bolivar JM, Wilson L, Ferrarotti SA, Fernandez-Lafuente R, Guisan JM, Mateo C. Enzyme Microb Technol 2007;40:540–546.
44. Wu X, Wang Y, Ju J, Chen C, Liu N, Chen Y. Tetrahedron: Asymmetry 2009;20:2504–2509.
45. Hölsch K, Weuster-Botz D. Biotechnol Appl Biochem 2010;56:131–140.
46. Weckbecker A, Hummel W. Methods Biotechnol 2005;17:225–237.
47. Zhang J, Witholt B, Li Z. Adv Synth Catal 2006;348:429–433.
48. Betancor L, Berne C, Luckarift HR, Spain JC. Chem Commun 2006;34:3640–3642.
49. Kosjek B, Nti-Gyabaah J, Telari K, Dunne L, Moore JC. Org Proc Res Dev 2008;12:584–588.
50. Truppo MD, Pollard D, Devine P. Org Lett 2007;9:335–338.
51. Lutz J, Mozhaev VV, Khmelnitsky YL, Witholt B, Schmid A. J Mol Catal B: Enzym 2002;19-20:177–187.
52. Musa MM, Ziegelmann-Fjeld KI, Vieille C, Zeikus JG, Phillips RS. J Org Chem 2007;72:30–34.
53. Costas AMG, White AK. J Biol Chem 2001;276:17429–17436.
54. Vrtis JM, White AK. J Am Chem Soc 2001;123:2672–2673.
55. Vrtis JM, White AK, Metcalf WW, van der Donk WA. Angew Chem Int Ed 2002;41:3257–3259.
56. Woodyer R, van der Donk WA. Biochemistry 2003;42:11604–11614.
57. Johannes TW, Woodyer RD, Zhao H. Biotechnol Bioeng 2007:96:18–26.
58. McLachlan MJ, Johannes TW, Zhao H. Biotechnol Bioeng 2008;99:268–274.
59. Kroutil W, Mang H, Edegger K, Faber K. Curr Opin Biotechnol 2004;8:120–126.
60. Geueke B, Riebel B, Hummel W. Enzyme Microb Technol 2003;32:205–211.
61. Hirano J-i, Miyamoto K, Ohta H. Tetrahedron Lett 2008;49:1217–1219.
62. Riebel B, Gibbs PR, Wellborn WB, Bommarius AS. Adv Synth Catal 2003;345:707–712.
63. Jones JB, Sneddon DW, Higgins W, Lewis AJ. J Chem Soc, Chem Commun 1972, 856–857.
64. Wagenknecht PS, Penney JM, Hembre RT. Organometallics 2003;22:1180–1182.

65. Hollmann F, Lin P-C, Witholt B, Schmid A. J Am Chem Soc 2003;125:8209–8217.
66. Hollmann F, Witholt B, Schmid A. J Mol Catal B: Enzym 2003;19-20:167–176.
67. Lutz J, Hollmann F, Vinh Ho T, Schnyder A, Fish RH, Schmid A. J Organomet Chem 2004;689:4783–4790.
68. Sambongi Y, Nitta H, Ichihashi K, Futai M, Ueda I. J Org Chem 2002;67:3499–3501.
69. Maid H, Böhm P, Huber SM, Bauer W, Hummel W, Jux N, Gröger H. Angew Chem Int Ed 2011;50:2397–2400.
70. Kohlmann C, Markle W, Lutz S. J Mol Catal B: Enzym 2008;51:57–72.
71. Udit AK, Gray HB. Biochem Biophys Res Commun 2005;338:470–476.
72. Hollmann F, Schmid A, Steckhan E. Angew Chem Int Ed 2001;40:169–171.
73. Hildebrand F, Lütz S. Tetrahedron: Asymmetry 2007;18:1187–1193.
74. Kim YH, Yoo Yj. Enzyme Microb Technol 2009;44:129–134.
75. Suye S-i, Zheng H, Okada H, Hori T. Sens Actuators B 2005;108:671–675.
76. Zheng H, Ohno Y, Nakamori T, Suye S-i. J Biosci Bioeng 2009;107:16–20.
77. Julliard M, Le Petit J. Photochem Photobiol 1982;36:283.
78. Willner I, Mandler D. Enzyme Microb Technol 1989;11:467–483.
79. Nakamura K, Yamanaka R. Chem Commun 2002;21:1782–1783.
80. Jiang Z, Lü C, Wu H. Ind Eng Chem Res 2005;44:4165–4170.

CHAPTER 3

Reactions Involving Dehydrogenases

Dehydrogenases comprise a large group of enzymes that catalyze reversible redox reactions, thus being used as oxidative or reductive biocatalysts. They are the most widely used oxidoreductases for synthetic purposes, finding application in carbonyl reductions, C–C and C–N double-bond reductions, reductive aminations, and the oxidation of alcohols, aldehydes, and amines. All these reactions are performed through a two-electron transfer mechanism using nicotinamide cofactors. Historically, these enzymes were employed mainly for reduction processes, which can be performed with high stereoselectivity, but their use in oxidation reactions has increased recently, since they provide an environmentally benign alternative to the conventional oxidation reagents. In addition, dehydrogenases are used in the industrial production of enantioenriched compounds, through the selective reduction of prochiral substrates or *via* several deracemization procedures. In this chapter, dehydrogenase-catalyzed reactions will be discussed in depth describing the type of reactions, substrates involved, and products' distribution for individual enzymes.

3.1 GENERAL CONSIDERATIONS

When practical applications are envisaged, enzymes in general, and dehydrogenases in particular, are classified according to the general reaction involved, and not with regard to the biological source. Thus, enzymes under the same classification type may derive from different organisms, displaying different profiles in terms of substrate specificity, stereochemistry, kinetic constants, or stability. This often provides several enzymes for a given reaction, and the choice for practical purposes is made based on the particular substrate profile of each biocatalyst [1]. For instance, horse liver alcohol dehydrogenase (HLADH) can be used for the reduction of a broad range of cyclic ketones and α- or β-keto esters [2], while open-chain methyl and ethyl ketones are preferred substrates for *Thermoanaerobium brockii* alcohol dehydrogenase (TBADH) [3]. For synthetic purposes, there are currently an ample number of commercially available dehydrogenases and ketoreductases (KREDs) [4].

Redox Biocatalysis: Fundamentals and Applications, First Edition.
Daniela Gamenara, Gustavo A. Seoane, Patricia Saenz-Méndez, and Pablo Domínguez de María.

Furthermore, the application of new screening methods (metagenomic) [5–9], enzyme directed evolution, and rational design [10,11], is favoring a rapid incorporation of enzymatic redox methodologies into industrial processes [12] (see Chapters 1 and 7).

In a broad sense, dehydrogenases catalyze the stereoselective transfer of a hydride from the reduced cofactor NAD(P)H to the *Si*- or *Re*- face of a prochiral carbonyl group, resulting in the formation of the corresponding (*R*)- or (*S*)-alcohol. Most alcohol dehydrogenases (ADHs) catalyze ketone reduction reactions according to the Prelog's rule (see Chapter 1, Section 1.3.1) [13], furnishing predominantly (*S*)-secondary alcohols, in the assumption that the smaller substituent in the ketone substrate has also lower priority according to the Cahn, Ingold and Prelog's rules [14,15]. Remarkably wild-type and recombinant [16–30] microbial dehydrogenases with exclusive *anti*-Prelog selectivity are known as well, for example, *Lactobacillus kefir* alcohol dehydrogenase (LKADH) [31–34], which displays a broad substrate spectrum and has been extensively used for synthetic purposes either as isolated enzymes or as (immobilized) whole-cells (see Chapter 8) [35–38]. Besides its interesting stereoselectivity, LKADH also displays a broad substrate specificity, encompassing acetophenone and halogenated derivatives, aliphatic open-chain and cyclic ketones and α- and β-keto esters. Likewise, alcohol dehydrogenases from *Leifsonia* sp. (LSADH) [39], *Pseudomonas fluorescens* (PFADH) [40] or *Pseudomonas* sp. [31] display *anti*-Prelog selectivity for the reduction of structurally different ketones. As an outstanding example of this, Kroutil and coworkers described the asymmetric reduction of different azido-, bromo-, and nitroketones (up to substrate concentrations of 16 g L^{-1}) in a "substrate coupled" approach using 20% (v/v) 2-propanol as hydrogen donor, with lyophilized cells of *Comamonas testosteroni* DSM 1455, obtaining excellent *anti*-Prelog stereoselectivities (Figure 3.1) [17].

Moreover, the opposite enantiopreference of different ADHs, namely ADH-'A' from *Rhodococcus ruber* (a Prelog enzyme) and *Lactobacillus brevis* alcohol dehydrogenase (LBADH) (an *anti*-Prelog enzyme) can be capitalized by combining activated ketones with racemic secondary alcohols. Through this strategy, two optically enriched alcohols—one derived from the prochiral ketone and the other from the racemic sacrificial cosubstrate—can concurrently be obtained, thus turning the sacrificial reaction into a highly valuable transformation [41]. In this way, several α-chloro, α-azido, and α-cyano ketones combined with racemic 2-octanol were reduced in a one-pot process in which enantiocomplementary products could be obtained by simply changing the enzyme (Figure 3.2). In this reaction system, the selective reduction of the prochiral ketone plus the kinetic resolution of the racemate could be achieved *via* hydrogen transfer by a single enzyme, and a catalytic amount of the pyrimidinic cofactor, *in situ* recycled. Due to the high selectivity of each enzyme, the hydride was abstracted from a single enantiomer of the racemic alcohol and transferred to one stereoface of the prochiral ketone resulting in high enantiomeric excesses (>99% *ee*) of each alcohol.

Although Nature's toolbox for asymmetric reductions is vast, enzyme-catalyzed reductions for practical purposes were not considered as a real competitive alternative to chemical methods until recent years [4]. Despite the mild conditions required for

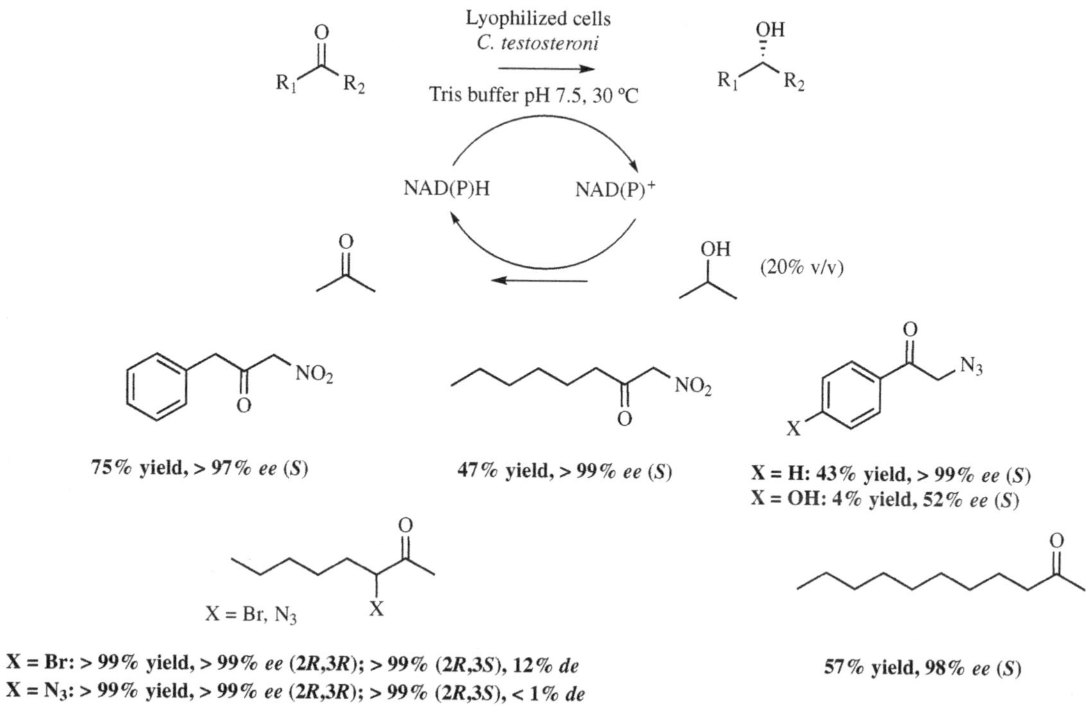

FIGURE 3.1 Substrate-coupled approach for the stereoselective *anti*-Prelog reduction of ketones using *C. testosteroni* DSM 1455 as catalyst.

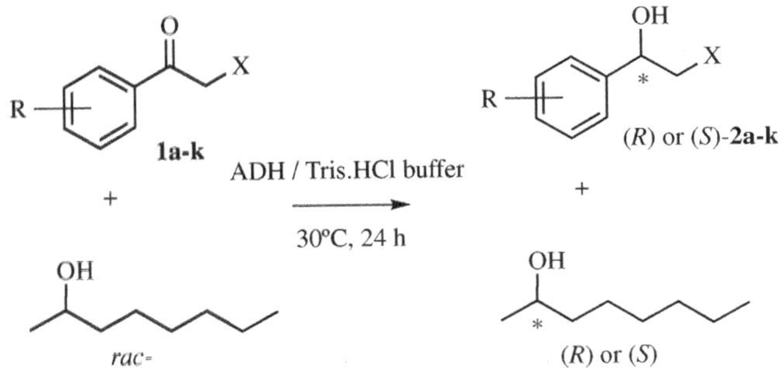

1a: R = H, X = Cl
1b: R = *p*-OH, X = Cl
1c: R = *p*-Me, X = Cl
1d: R = *p*-Cl, X = Cl
1e: R = *m,p*-Cl, X = Cl
1f: R = *m*-NO_2, X = Cl
1g: R = *p*-NO_2, X = Cl
1h: R = H, X = N_3
1i: R = H, X = CN

FIGURE 3.2 ADH-'A' (Prelog selectivity) and LBADH (*anti*-Prelog selectivity)-catalyzed tandem concurrent hydrogen transfer by using activated aromatic ketones and 2-octanol. ADH: either ADH-'A' from *Rhodococcus ruber* or *Lactobacillus brevis* alcohol dehydrogenase (LBADH).

enzymatic reductions, other industrial relevant parameters such as catalyst costs, productivities, final product concentrations, and/or complex downstream processing were not attractive for industrial applications. The main cause of these limitations was the use of growing or resting wild-type cells in earlier applications, which typically grew slowly, were affected by high substrate concentrations, and contained other enzymes that competitively catalyzed side reactions. Additionally, the specific equipment and know-how for microbiological processes were not standard in chemical industries, limiting their use in bioreductive processes. With the use of isolated enzymes [42] and the optimization of cofactor regeneration systems in the 1980s, many of these drawbacks could be avoided (see Chapter 8).

Up to this point, the potential of dehydrogenase-catalyzed processes as a tool for synthetic organic chemistry has not been yet fully described. The modification of the reaction conditions, like the choice of the enzyme and the cofactor recycling strategy (in case of isolated enzymes), the culture medium in microorganisms—sources of C, use of inhibitors—the use of mono- or biphasic systems, or biocatalyst immobilization, is often not enough to improve the process selectivity. Moreover, all these methodologies are experience based, and therefore the prediction of the experiment's behavior often becomes difficult. Thus, the need for "rational design" and biocatalyst improvement for a practical application emerges as an outstanding goal (see Chapter 7). This often involves the use of genetic engineering strategies, such as random or site directed mutagenesis to improve the overall efficiency of the process, which leads to the concept of "designer cells"—for example, overexpression of several enzymes in the same host—offering a high synthetic potential, unexplored until recent years [43]. During the last decades, a wide variety of microbial keto- and aldoreductases have been isolated, purified, characterized, and cloned [18,27,44–56], while many mutants have been designed through genetic engineering procedures. Remarkably, genetic engineering has driven enzymatic reductions either with recombinant enzymes [57,58] or microorganisms [59–61] to an exponential development. Excellent yields and selectivities were achieved with a wide range of substrates such as alkyl and aryl ketones [57,59,61–68], α- and β-keto esters and derivatives [53,57,59,60,62,63,66,68–72], α-haloketones [57,62,63,73], diketones [61–64,74,75], heterocyclic- [76], or α,β-unsaturated ketones [63,69]. Consequently, the first industrial enzymatic reductive amination process with *in situ* cofactor regeneration, mediated by formate dehydrogenase (FDH) from *Candida boidinii*, was developed for the production of L-*tert*-leucine, starting from the corresponding keto acid [77]. Since then, the number of industrial bioreduction processes for the preparation of secondary alcohols, (*R*)-amino- as well as (*R*)-hydroxy-acids has increased substantially [78,79]. Dehydrogenase-catalyzed reductions, either using native or mutant whole-cells or isolated enzymes, have been used for the preparation of a broad range of bioactive compounds, such as synthetic intermediates of steroids, terpenes and related compounds [80], carbohydrates [42,81–88], alkaloids [42,89,90], and pharmaceuticals, namely 3-hydroxy-β-lactams and derivatives as intermediates in the preparation of phenylisoserine [91], cyclic hydroxyketones as building blocks for Zoapatanol [92], or (*R*)-allylic alcohols as precursors of nonanoic acid substituted derivatives [93]. Currently, the synthetic applications of dehydrogenase-catalyzed redox reactions are rapidly increasing, encompassing not only the enantioselective reduction of carbonyl compounds, but also new enzymatic redox processes like racemizations, deracemizations, and oxidations. Meanwhile, the biocatalytic characterization of novel enzymes provides innovative entries for efficient new processes. For instance, the use of enone- and enoate reductases (ERs) for the enantioselective reduction of C–C double bonds, or the use of amine dehydrogenases and D-selective amino acid dehydrogenases (AADHs), are outstanding applications for the production of chiral amines and other high added value products [4].

3.2 REDUCTION OF CARBONYL GROUPS

The biocatalytic reduction of prochiral carbonyl groups is an increasingly important approach for the preparation of chiral pharmaceuticals, synthetic intermediates, and for the synthesis of natural products, motivated by the use of mild reaction conditions as well as by the chemo-, regio-, and stereoselectivity of these biocatalysts [1,94–97]. This high chemo- and regioselectivity of dehydrogenase-catalyzed reductions leads to the reduction of simple or multifunctionalized substrates, such as aliphatic ketones [95,96,98–103], aryl [28,39,102,104,105], diaryl [22], halo-aryl [106], hetero-arylketones [107], α- or β-keto esters and derivatives [39,108–120], diketo esters [121–126], α,β-unsaturated ketones [100,127,128], haloketones [129–133], β-ketonitriles [134], or diketones [135,136]. The biocatalytic procedures can be carried out using a variety of biocatalyst's sources including wild type yeast, fungal or bacterial strains, crude plants fragments, and isolated KREDs with the corresponding cofactor recycling system, as well as recombinant microorganisms (see Chapters 1 and 2) [4,137–140].

Usually high enantioselectivities of the ADH-catalyzed carbonyl reductions are governed by several factors affecting the energetics of substrate binding, such as steric requirements, temperature, medium polarity, and so on. For instance, the enantioselectivity of TBADH-mediated reductions depends on the steric bulk of the substrate, and thus (*S*)-secondary alcohols can be obtained when using bulky ketones as substrates, while (*R*)-alcohols result from the reduction of small ketones [3]. Keinan et al. proposed the existence of two hydrophobic sites of different size within the active site, the smaller one having greater affinity toward alkyl groups. As an example, for 2-butanone the ethyl group could fit properly in the smaller site, affording (*R*)-2-butanol as the major product. In the case of larger ketones, the larger alkyl group might be forced to enter the larger site with the smaller alkyl group being accommodated into the smaller pocket, to afford the corresponding (*S*)-configured alcohols. For some carbonyl reductases the enantioselectivity was temperature dependent [141]. For instance, a secondary ADH (SADH) from *Thermoanaerobacter ethanolicus* and its mutant Ser39Thr, showed an increase in (*R*)-specificity when rising the temperature [142,143]. In addition, water activity also affected the stereoselectivity of ketone reductions catalyzed by TBADH [3,144]. For the reduction of 2-pentanone, the (*S*)/(*R*)-enantiomeric ratio increases from 2.6 to 4.6 when the water activity in hexane rises from 0.53 to 0.97. In these conditions, the reaction medium becomes biphasic, as the water amount increases so that the solubility limit is exceeded. The effect on the enantioselectivity of changing the reaction medium from hexane containing 2.5 % (v/v) of water to pure water is negligible, because of the minimal, related change of water content in the active site [144]. The enantioselectivity of TBADH also remains unchanged in water-miscible organic solvents [145], but it can be modified by the incorporation of cyclodextrins as additives in the reaction medium [146]. Other examples of tuning enantioselectivity of ketone reduction by changing the reaction medium have been reported for ADHs from *Thermoanaerobacterium* sp. KET4B1 [141] *L. brevis* [147], and *T. ethanolicus* mutant [148,149].

3.2.1 Reduction of Aliphatic and Aromatic Ketones

A wide variety of aliphatic ketones can be reduced to the corresponding alcohols, often with excellent enantioselectivities using dried whole-cells of *Geotrichum candidum* (APG4), NADH, and 2-propanol for cofactor regeneration (Table 3.1, entries 1–6) [96,98,99]. The microorganism's ability to distinguish two alkyl groups with a difference of only one methylene unit is noteworthy (Table 3.1, entry 1). Likewise, TBADH is also highly suitable for reducing aliphatic ketones [96] as well as α-haloketones [150,151]. The enzyme displays an interesting substrate size-induced reversal of enantioselectivity, the smaller substrates are reduced to the corresponding (*R*)-enantiomers, while bulky ketones furnish (*S*)-alcohols (Table 3.1, entries 7–12) [96]. Other aliphatic ketones, such as α-chloro- and α-bromopropargylic ones were efficiently reduced by commercially available HLADH, TBADH, and LBADH, yielding the corresponding alcohols with excellent enantiomeric excesses (up to >99%) (Table 3.1, entry 13) [100].

The reduction of structurally different cyclic aliphatic ketones was successfully achieved by Moore and coworkers using commercially available KREDs [95]. Five- to seven-membered rings with an alkyl group in α or β position to the carbonyl group, often with a heteroatom in the ring, and occasionally with a C–C double bond in conjugation with the ketone, were used as substrates (Table 3.2). In this

TABLE 3.1 Enantioselective Reduction of Open-Chain Aliphatic Ketones

Entry	Product	Absolute Configuration	Biocatalyst	*ee* (%)	Reference
1	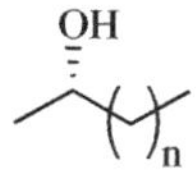 n = 1 - 8	*S*	*G. candidum* (dried whole-cells)	94– >99	[98]
2	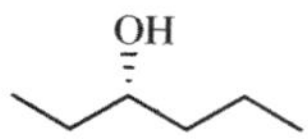	*S*	*G. candidum* (dried whole-cells)	98	[98]
3	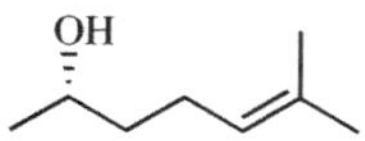	*S*	*G. candidum* (dried whole-cells)	>99	[99]
4	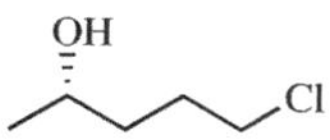	*S*	*G. candidum* (dried whole-cells)	>99	[99]
5	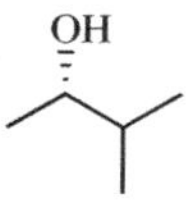	*S*	*G. candidum* (dried whole-cells)	98	[98]
6	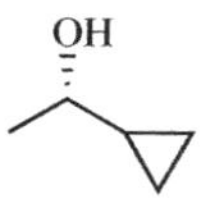	*S*	*G. candidum* (dried whole-cells)	>99	[98]
7	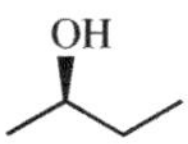	*R*	TBADH	48	[96]
8	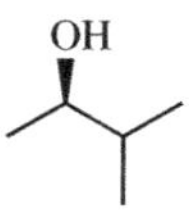	*R*	TBADH	86	[96]
9	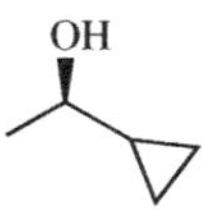	*R*	TBADH	44	[96]
10	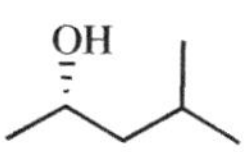	*S*	TBADH	95	[96]
11	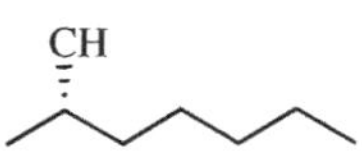	*S*	TBADH	98	[96]

(Continued)

TABLE 3.1 (*Continued*)

Entry	Product	Absolute Configuration	Biocatalyst	*ee* (%)	Reference
12	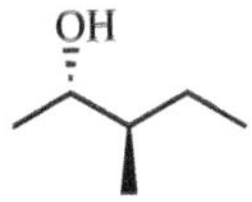	*S*	TBADH	81	[96]
13	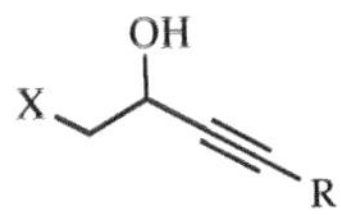 X = Cl, Br R = Ph, TBS, TMS	HLADH: *R* TBADH: *R* rLBADH: *S*	rLBADH	98.5– >99	[100]

rLBADH, *L. brevis* ADH overexpressed in recombinant *E. coli* strains; TBADH, *Thermoanaerobium brockii* alcohol dehydrogenase; HLADH, horse liver alcohol dehydrogenase.

TABLE 3.2 Enantioselective Reduction of Cyclic Aliphatic Ketones [95]

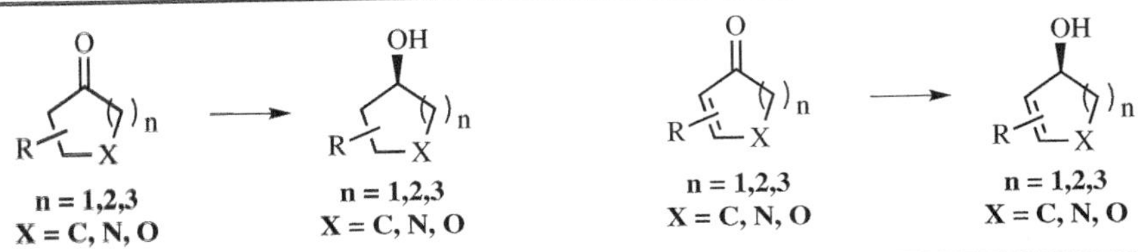

Entry	n	Substrate	*ee* (%)	Enzyme
1	1	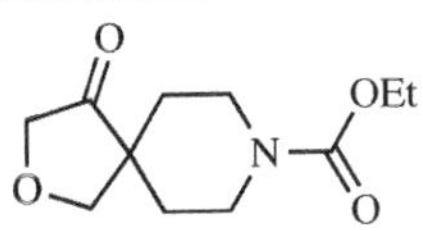	98 (*R*) >99 (*S*)	KRED 119 ADH RE
2	1	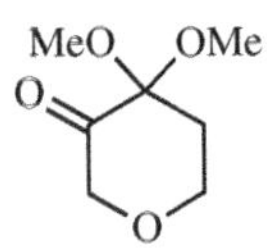	87 (+) 89 (−)	KRED 101 KRED 108
3	2	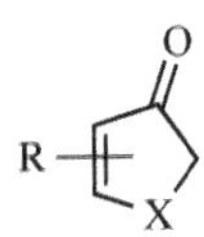	98 (*R*) >99 (*S*)	KRED 119 ADH RE
4	3	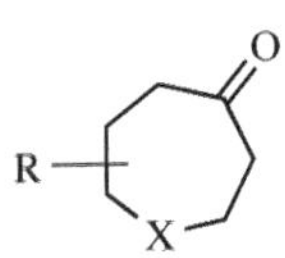	94 (*R*) 92 (*S*)	KRED 112 KRED 124

X, Cl, N or O; R, alkyl groups in α or β position to the carbonyl group; KRED, ketoreductase; ADH RE, alcohol dehydrogenase from *Rhodococcus erythropolis*.

case, glucose dehydrogenase (GDH) was not suitable for cofactor recycling, since the substrates were also accepted by this enzyme, and thus the FDH system was successfully implemented.

Similarly, aromatic ketones can be enantioselectively reduced using isolated enzymes or whole-cells. Using dried whole-cells or immobilized resting cells of *G. candidum*, moderate-to-high yields and excellent enantiomeric excesses could be obtained (Table 3.3, entries 1–4) [99,152,153]. Likewise, 1,2-diols were prepared from aromatic α-hydroxyketones in high yields and enantioselectivities (Table 3.3, entries 5 and 6) [154]. However, microbial reductions usually do not afford the corresponding alcohols with satisfactory optical purity, because of the presence of other enzymes displaying different (enantio)selectivities within the cell. Therefore, several strategies such as inhibition of undesired enzymes [60,155,156], as well as the addition of additives [157–161], polymers [160,161], or organic solvents [159,162,163], have been usually used to improve the enantioselectivity of the biocatalytic processes. In this respect, Nakamura and coworkers reported that the stereoselectivity of the *G. candidum*-catalyzed reduction of aromatic ketones or keto esters could be significantly improved (up to >99%) by using an acetone powder of the cells [164–166]. Apart from these strategies, the above-mentioned cloning and overexpression of a desired enzyme in a heterologous host must also be considered as a useful tool to avoid other enzymatic background reactions.

Studies on the influence of halogens in *ortho*-, *meta*-, or *para*-positions of prochiral aromatic ketones in *Rhizopus arrhizus*-catalyzed reductions allowed to establish that the reaction course was driven by the relative position of the substituent with respect to the carbonyl group, as well as by its size (Table 3.3, entry 7) [167]. The presence of the electron-withdrawing halogens increased the reactivity of the acetophenone derivatives compared to 1-phenylethanol. However, the presence of halogen substituents in *m*-position dramatically reduced the enantioselectivity of the reaction. The *o*-substituted substrates afforded the best yields and enantiomeric excesses, regardless the nature of the halogen. Concerning the effect of different halogens with the same substitution pattern, the *ee* increased with the halogen size (F < Cl < Br). Nevertheless, the effect of the *o*-substitution was so predominant that the nature of the halogen atom did not have significant influence on the course of the reaction. Thus, the effect of the nature of the halogen was more evident in the *m*- and *p*-substituted substrates. In the same area, the enantioselective reduction of acetophenone derivatives using growing cells under fluorescent light of the phototrophic cyanobacteria *Synechococcus* sp. was also reported (Table 3.3, entry 8) [168]. The process was highly stereoselective, though with dissimilar results in terms of conversion. The use of plant whole-cells as ketone reducing agents has also experienced a continuous growth during the past decade [116,169–180]. Acetophenone derivatives were reduced to the corresponding chiral alcohols in moderated-to-high yields and high-to-excellent enantiomeric excesses, according to the Prelog's rule, using plant whole-cells as catalysts (Table 3.3, entries 9–12) [181–184], whereas aromatic and heteroaromatic ketones were reduced with *anti*-Prelog bias using *Raphanus sativus* hairy roots (Table 3.3, entries 13–14) [185]. Besides baker's yeast *Saccharomyces cerevisiae* [186], other yeasts have proven to be useful biocatalysts for the reduction

TABLE 3.3 Selected Examples of Enantioselective Reduction of Aromatic Ketones using Microorganisms or Plant Whole-cells as Biocatalysts

Entry	Product	Substituents	Biocatalyst/Conditions	Yield (%)	*ee* (%)	Reference
1	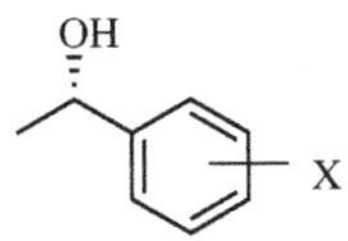	X (*o*, *m*, *p*): F, Cl, Br, Me, OMe, CF_3	*G. candidum* (dried cells)	73–99	>99 (*S*)	[99]
2	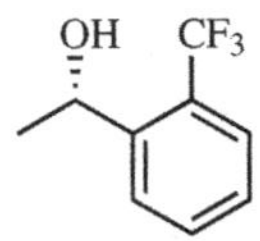	—	*G. candidum* (dried cells)	6	97 (*S*)	[99]
3	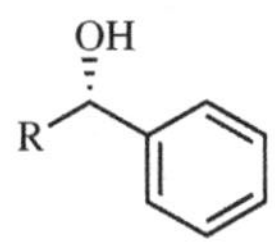	R = Et R = Pr R = *i*-Pr R = *t*-Bu R = $MeOCH_2$- R = $ClCH_2$-	*G. candidum* (dried cells)	41 0 12 1 8 80	>99 (*S*) – >99 (*S*) – >99 (*R*) 98 (*R*)	[153] [153] [153] [153] [152] [152]
4	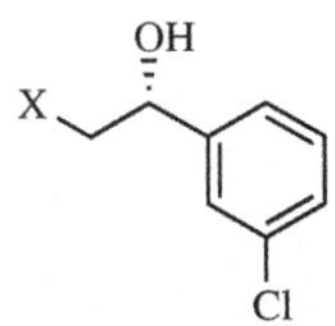	X = Cl, X = Br	*G. candidum* (dried cells)	94 95	98 (*R*) 93 (*S*)	[152]

5	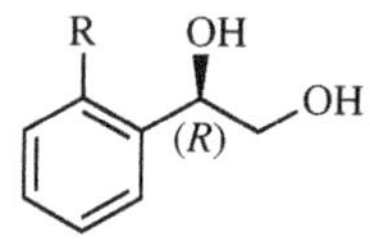	R: Cl, NO_2	*G. candidum* (resting cells)	37–77	47–60	[154]
6	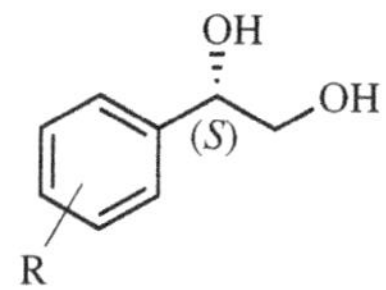	R (*m*, *p*): H, Cl, NO_2, $-NHSO_2CH_3$	*G. candidum* (resting cells)	62–95	47–100	[154]
7	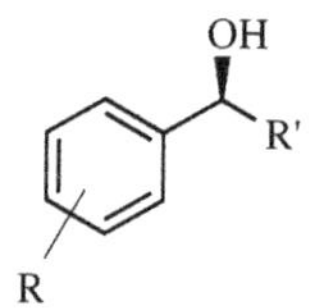	R (*o*, *m*, *p*): F, Cl, Br, Me, NO_2 R': Me, Et.	*R. arrhizus* (fermentation)	48–98	10–99	[167]
8	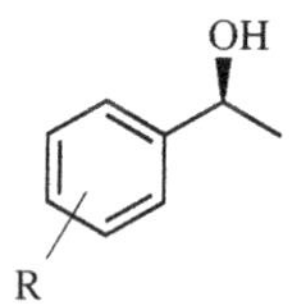	R (*o*, *m*, *p*): H, Cl, F, Me, –OMe	*Synechococcus* sp. (fermentation, fluorescent light)	3–84	96–100	[168]
9	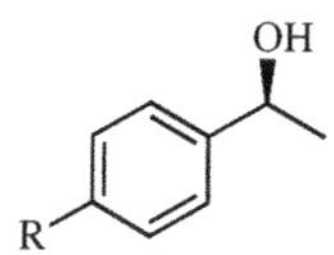	R: H, F, Cl, Me, OH, –OMe	*Phaseolus aureus* L. (whole-cells)		72–95	[181]

(*Continued*)

TABLE 3.3 (*Continued*)

Entry	Product	Substituents	Biocatalyst/Conditions	Yield (%)	*ee* (%)	Reference
10	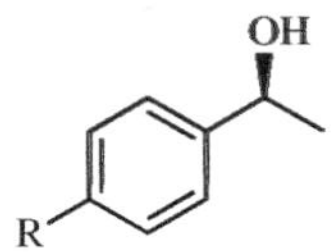	R: H, F, Cl, Br, Me, OH, –OMe, NO_2	*Daucus carota* (whole-cells)	61–82	90–96	[182]
11	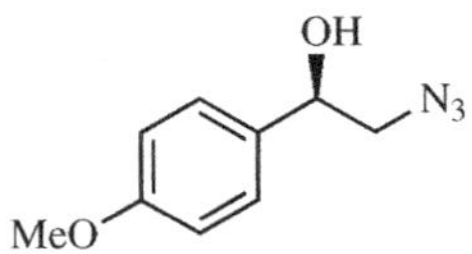	—	*D. carota* (whole-cells)	92	99	[184]
12	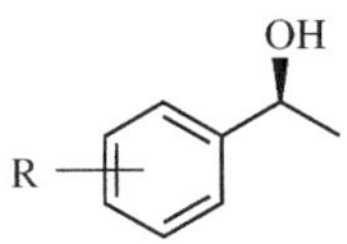	R (*o*, *m*, *p*): -SeMe, -SePh, –SMe	*D. carota* (whole-cells)	<97	>99	[183]
13	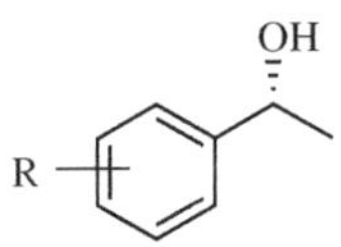	R: H, *p*-OCH_3, *p*-CH_3, *p*-*i*Pr, *p*-NO_2, *p*-Cl, *p*-Br, *p*-F, *m*-F, *o*-F	*R. sativus* hairy roots	52–100	Up to 99	[185]
14	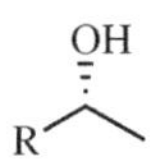	R: 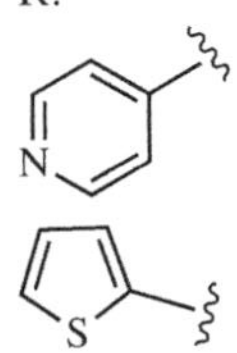	*R. sativus* hairy roots	100 14	47 96	[185]

15	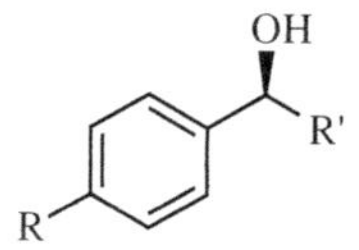	R: H, Cl, Br, NH_2, NO_2, –OMe R’: Me, Et, $–CH_2Br$, $–CH_2Cl$, $–C_2H_4Cl$	*Rhodotorula* sp. (resting cells)	35–84	97–99	[112]
16	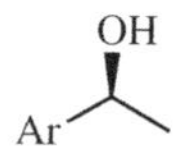	Ar: 2-piridil-, 3-piridil-, 4-piridil	*Rhodotorula* sp. (resting cells)	70–81	89–99	[112]
17	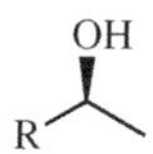	R: 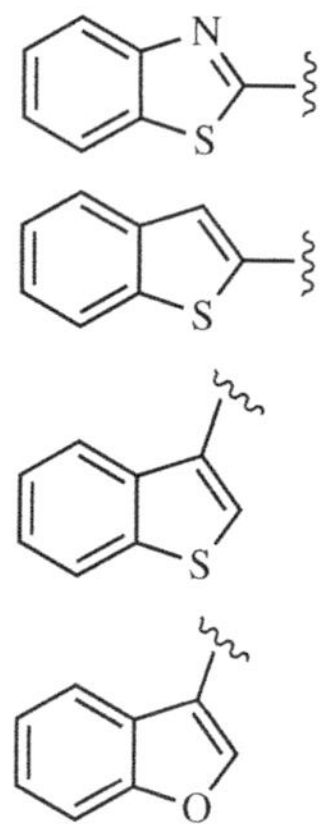	*S. cerevisiae* (resting cells and fermentation)	92–94	98–99	[186]
18	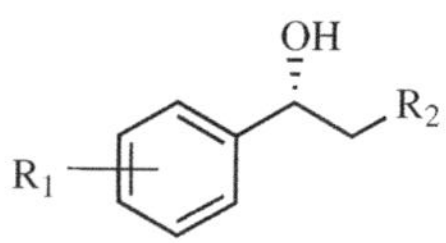	R_1: H, *m*-Cl, *p*-Cl, *p*-CH_3, *p*-OCH_3 R_2: OH, Br, H	*E. coli* BL21(DE3)(pETSCR1)	Up to 90	>99	[65]

of aromatic and heteroaromatic prochiral ketones with moderate-to-high yields and excellent enantiomeric excesses (Table 3.3, entries 15–17) [44,112,186–190]. Yan et al. constructed the recombinant *Escherichia coli* BL21(DE3)(pETSCR1), overexpressing the NADPH-dependent carbonyl reductase SCR1 from *Candida parapsilosis* CCTCCM203011, and the recombinant cell-free extract (total soluble protein 5 gL^{-1}) was successfully applied for asymmetric reduction of prochiral aromatic ketones, obtaining the corresponding alcohols with >99% *ee*, and up to 90% yield (Table 3.3, entry 18) [65].

In this area, enantiodivergent results can be obtained by just modifying the reaction conditions. Nakamura and coworkers described the *G. candidum* IFO 5767-mediated reduction of aromatic ketones yielding the corresponding (*S*)-alcohols with excellent enantioselectivities when adding the hydrophobic polymer amberlite XAD-7, and producing the (*R*)-isomer also in excellent enantiomeric excess, when the reaction was carried out in anaerobic conditions (Figure 3.3) [161]. The influence of the hydrophobic polymer amberlite XAD-7 is noteworthy, since the stereochemical course of the reduction was changed into up to 99% *ee* of the (*S*)-alcohols starting from 85 to >99% *ee* (*R*) in the absence of polymer.

Likewise, all stereoisomers of 2-allyl-3-hydroxy-2-methylcyclopentanone were prepared by Fujii et al. in high conversions and optically pure forms by manipulation of the biotransformation conditions of the microbial reductions and oxidations [191]. The reduction of symmetric diketone **8** by *G. candidum* NBRC 4597 under anaerobic conditions gave hydroxyketone **9** in 98% conversion and >99% *ee* and diastereomeric excess (*de*), whereas its reduction by *G. candidum* NBRC 5767 under aerobic conditions gave **10** with a conversion of 99% and >99% *ee*. In addition, oxidation of *meso*-diol **11** by the same strain of *G. candidum* afforded hydroxyketone **12** with 99% conversion and >99% *ee*, and oxidation of *meso*-diol **13** by *Mucor hiemalis* IAM 6095 in the presence of cyclohexanone afforded **14** with 75% conversion and >99% *ee* (Figure 3.4).

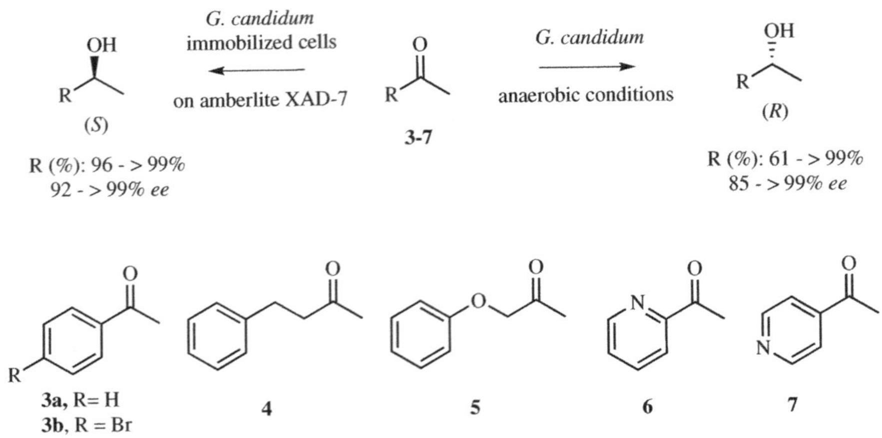

FIGURE 3.3 Enantiodivergent ketone reduction using *Geotrichum candidum* IFO 5767.

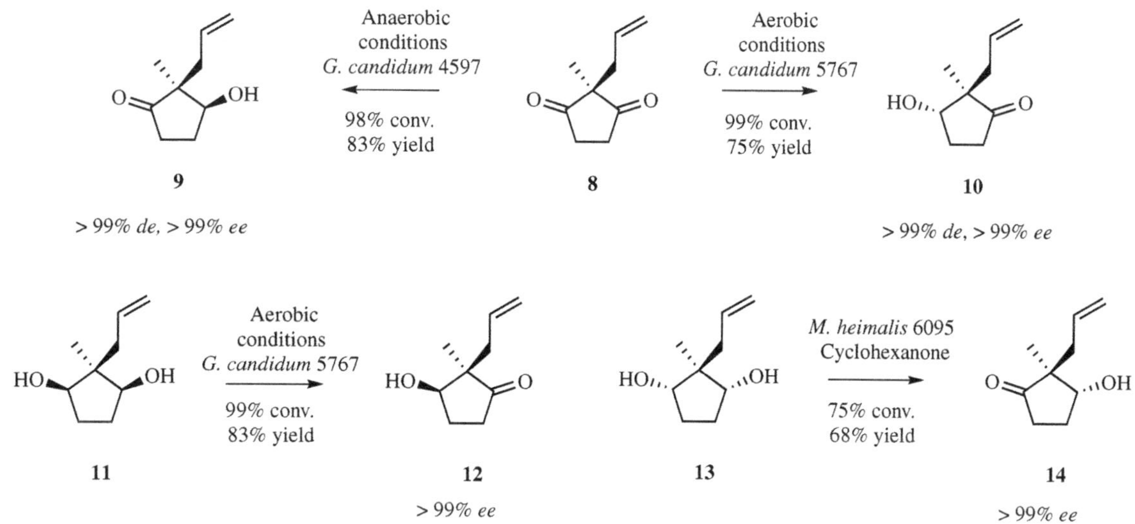

FIGURE 3.4 All stereoisomers of 2-allyl-3-hydroxy-2-methylcyclopentanone prepared by microbial reduction and oxidation from the corresponding *meso*-compounds.

Aromatic and aliphatic ketones are often poorly soluble in aqueous media, making it necessary to work with substrate concentrations <5–10 mM. This fact is an important drawback when scale-up processes at industrial level are envisaged. To overcome this issue, a well-known strategy employed in the field of hydrolases [35,192], is the use of organic solvents [163]. However, the use of organic solvents in redox-biocatalytic processes has been scarce so far. As examples, HLADH [193,194] (Table 3.4, entry 1), as well as whole-cells of *S. cerevisiae* (Table 3.4, entry 2)[195–197] for the reduction of cyclic aliphatic or aromatic ketones, or *G. candidum* (Table 3.4, entries 3–4) [198] in the stereoselective oxidation of racemic 1-phenylethanol, showed only moderate activity in biphasic systems. By using purified alcohol dehydrogenases in biphasic systems, usually the amount of the organic layer must be low, due to the instability of the ADHs in that medium [23,199,200]. Nevertheless, while reducing aromatic or aliphatic ketones with recombinant ADH-'A' from *Rhodococcus rubber* and lyophilized whole-cells of *E. coli* harboring the overexpressed enzyme (*E. coli*/ADH-'A') in microaqueous systems (99% v/v) of hydrophobic solvents, the corresponding alcohols were obtained with excellent enantiomeric excesses (>99%) and moderated to good yields (Table 3.4, entries 5–7) [62]. Water miscible organic solvents (dimethyl sulfoxide (DMSO), acetonitrile, 1,4-dioxane or ethanol) led to complete deactivation of both biocatalysts. Low activity was retained in 2-propanol, which also acted as cosubstrate. In contrast, using water-immiscible organic solvents, full activity was retained with *E. coli*/ADH-'A,' and only modest deactivation was observed with the recombinant enzyme when using less than 0.5% v/v of water. Likewise, high biocompatibility was found for systems involving hydrophobic solvents (logP >2) such as toluene, cyclohexane, and hexane. Moreover, substrate concentrations close to 2.0 M were tolerated by recombinant ADH-'A,' which is an important starting point for scaling-up processes. Stereoselectivity remained completely unchanged even at these elevated substrate concentrations, obtaining the corresponding (*S*)-alcohols in enantiopure form. In addition, the enzyme was evaluated in the two-phase system Tris buffer-ionic liquid (IL), finding [BMIM][NTf_2] (1-butyl-3-methylimidazolium bis(trifluoromethylsulfonyl)imide) as the optimal cosolvent (up to 30% v/v) [201]. Initially, *E. coli* cells containing overexpressed ADH-'A' were employed for the reduction of 4-phenylacetophenone in the presence of 2-propanol for the recycling of the nicotinamide cofactor (NADH). The enzyme catalyzed the reduction in the biphasic system Tris buffer-[BMIM][NTf_2] even at high concentrations of the IL (70% v/v), but the reaction slowed down with increasing amounts of IL, and the enantioselectivity dropped significantly above 30% v/v of the IL. Then, different diaryl acetophenone derivatives were reduced employing the optimized reaction system, thus furnishing the corresponding (*S*)-configured diaryl alcohols in good to excellent yields (55–94%) and >99% *ee*. Recently, recombinant *E. coli* cells overexpressing *C. parapsilosis* carbonyl reductases proved to be active in neat "water-free" substrates, thus providing a promising entry to combine biotransformations and productivity, while avoiding wastewater formation at the same time [202].

Apart from the above-mentioned *classic* NAD(P)H-dependent alcohol dehydrogenases, displaying a broad substrate range, other dehydrogenase types deserve

TABLE 3.4 Enantioselective Ketone Reduction in Systems Involving Organic Media

Entry	Substrate	Product	Biocatalyst/Conditions	Yield (%)	*ee* (%)	Reference
1	(+/-) 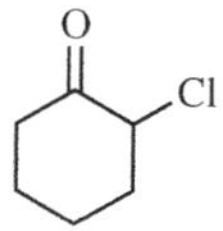	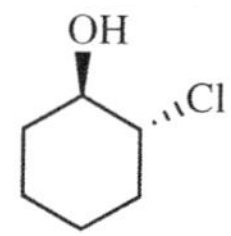	HLADH isopropyl ether	8.3	98	[193]
2	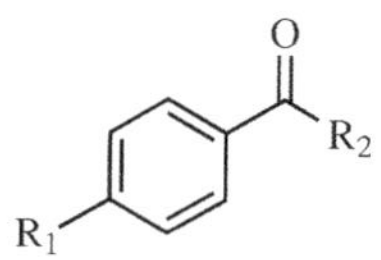 R_1: H, CH_3, -OCH_3, Br R_2: CH_3, -C_2H_5, -CH_2Br	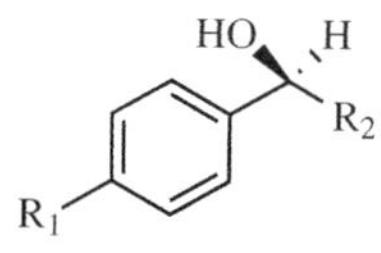	*S. cerevisiae* (dried cells) hexane	51–63	82–90	[195]
3	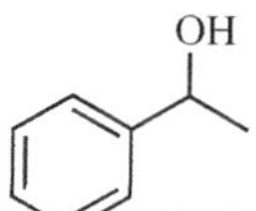 (*rac*)	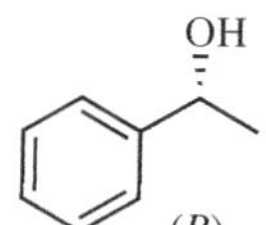 (*R*)	*G. candidum* (immobilized cells) hexane	44	86	[198]
4	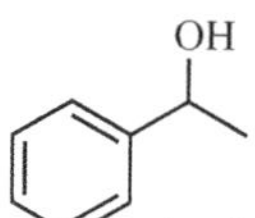 (*rac*)	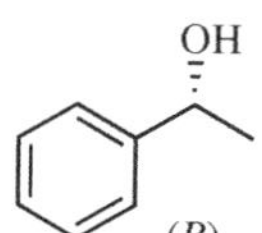 (*R*)	*G. candidum* (immobilized cells) hexane-cyclohexanone	51	99	[198]

(*Continued*)

TABLE 3.4 (*Continued*)

Entry	Substrate	Product	Biocatalyst/Conditions	Yield (%)	*ee* (%)	Reference
5	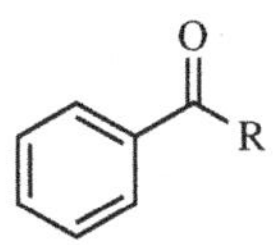 15: R = CH_3, 16: R = $-CH_2Cl$	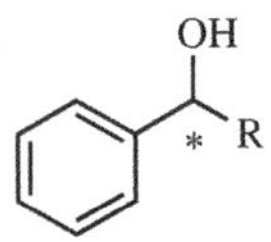	*E. coli*/ADH-'A' Microaqueous system-hexane (99% v/v)	15: 67 16: 96	15: >99 (*S*) 16: >99 (*R*)	[62]
6	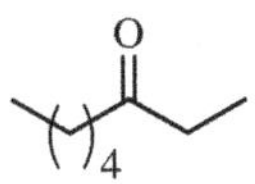	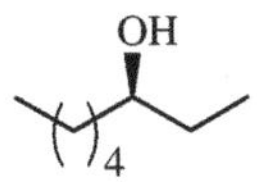	*E. coli*/ADH-'A' Microaqueous system-hexane (99% v/v)	14	>99 (*S*)	[62]
7	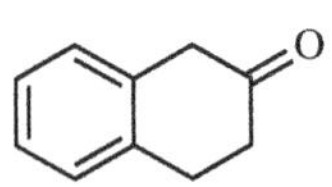	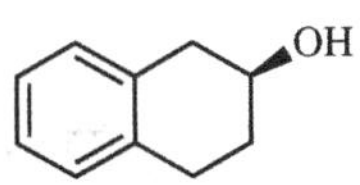	*E. coli*/ADH-'A' Microaqueous system-hexane (99% v/v)	35	>99 (*S*)	[62]

consideration as well. For instance, the family of nicotinoprotein alcohol dehydrogenases (EC 1.1.99.-) differs from classic alcohol dehydrogenases in having a nondissociable NAD(P)H in their active site, representing an alternative cofactor-independent system [203–207]. In this case, the hydride acceptor is a nitrosoaniline, phenolindophenol, or carbonyl compound. In the catalytic mechanism of nicotinoprotein-ADHs, the electron donor is converted into the oxidized product by reducing the enzyme-bound NAD(P)$^+$ cofactor *via* hydride transfer. The reduced cofactor is then reoxidized by the cosubstrate—the electron acceptor that is converted into the reduced coproduct. The nicotinamide cofactor acts as a temporary internal storage of reducing equivalents without leaving the active site [208]. When using these enzymes, difficulties associated to cofactor regeneration may be circumvented, and it can be considered as a coupled-substrate system, lacking the cofactor dissociation step. The optimization of biocatalytic processes catalyzed by this class of enzymes would constitute a major advance in industrial enzymatic reduction processes. As an example, the nicotinoprotein alcohol dehydrogenase from *Rhodococcus erythropolis* DSM 1069 was assessed for asymmetric reductions, kinetic resolutions, and racemizations. Yet, the enzyme catalyzes the asymmetric reduction of ketones with high stereoselectivity, albeit at low reaction rates for a high-performance process. In addition, the enzyme is not efficient in catalyzing kinetic resolution of alcohols, due to dismutation of the aldehyde used as cosubstrate, and also the racemization of secondary alcohols *via* the corresponding ketone is obstructed by side reactions. Thus, these processes are still not suitable for performing efficient biotransformations, unless these drawbacks could be overcome.

Phenylacetaldehyde reductase (PAR), produced by styrene assimilating *Corynebacterium* strain ST-10, has a broad substrate range including various prochiral aromatic ketones and β-keto esters [59,209–211]. Recombinant cells of *E. coli* expressing the *par* gene catalyze the high-yield production of pharmaceutical intermediates such as (*R*)-2-chloro-1-(3-chlorophenyl)ethanol, ethyl (*R*)-4-chloro-3-hydroxybutanoate, and (*S*)-*N*-*tert*-butoxycarbonyl-3-pyrrolidinol from the corresponding carbonyl compounds. The high performance of this enzyme is explained by its capacity to concomitantly regenerate NADH in the presence of 2-propanol (up to 7% v/v). The enzyme showed strong activity toward substitute acetophenones (**17–19**) (Figure 3.5), as well as toward 1-(3,4-dimethoxyphenyl)-propan-2-one (**20**), yielding the corresponding alcohols according to the Prelog's rule with >99% *ee*.

3.2.2 Reduction of α- and β-keto Esters and Derivatives

As described in the previous section, the ample structural diversity of compounds accepted as substrates by dehydrogenases turns these processes into powerful tools for synthetic purposes. Particularly, enzymatic reductions of α- or β-keto esters have been widely explored involving different enzymatic systems, among which isolated enzymes [110,135,136,212,213], wild-type strains of yeast and fungi [111,117,214–219], crude plant cells [116,179], and recombinant microorganisms are worth mentioning [59,60,68,70,71,115,117,220–225]. A multilevel engineering approach is usually needed to significantly improve dehydrogenase-catalyzed

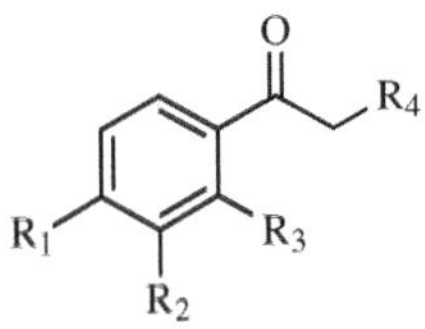

17: $R_1 = R_3 = H$; $R_2 = R_4 = Cl$: > 99% *ee* (*R*); A = 258 %
18: $R_1 = R_3 = R_4 = H$; $R_2 = OCH_3$: > 99% *ee* (*S*); A = 744 %
19: $R_2 = R_4 = H$; $R_1 = R_3 = Cl$: > 99% *ee* (*S*); A = 546 %

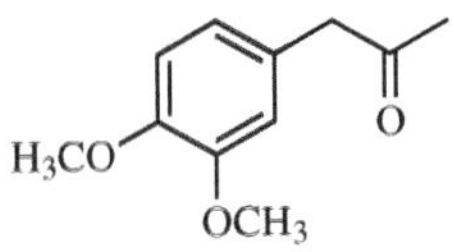

> 99% *ee* (*S*); A = 70 %

20

FIGURE 3.5 Selectivity and activity of PAR toward aromatic ketones and chloroketones. A, relative activity toward acetophenone (100%).

processes at *on-spec* conversions and optical purities. Enzyme immobilization [56,149,226–230]; variations in substrate, cosubstrate, and enzyme loadings [225]; use of organic solvents, biphasic systems [62,149,157,196,199,230–233]; supercritical fluids or ILs [234–238]; and design of membrane reactors [74,239,240], are often suitable strategies (see Chapter 7). As an example, the use of perfluorinated solvents in the reduction of β-keto esters with immobilized whole-cells of *S. cerevisiae* allowed for simplest work-up processes, thus recovering and reusing the solvent (Table 3.5, entries 1-3). Nevertheless, although almost 100% of conversion could be observed in most cases, the recovery was not successfully achieved when volatile products were formed, due to their evaporation during the downstream processing [230]. The use of immobilized cells largely contributed to the improvement of biocatalytic activity, compared to the use of free cells of *S. cerevisiae* under the same conditions.

Kroutil and coworkers used bi- and monophasic IL/buffer systems for ketone and β-keto ester reductions using the partially purified alcohol dehydrogenase ADH-'A' from *R. rubber* (Table 3.5, entry 4) [235]. The enzyme was a suitable catalyst for the reduction of ketones, diketones, and keto esters, as well as for cofactor recycling according to a coupled-substrate approach [241,242]. The use of water-miscible second-generation hydroxyfunctionalyzed "Tris-like" ILs resulted (in some cases) in successful enzymatic outcomes (excellent yields and enantioselectivities) in monophasic systems, even with up to 90% v/v of IL.

NADH-dependent xilose reductase from *Candida tenuis* (CTXR) [47,243], belonging to the aldo-ketoreductases (AKRs) superfamily, was overexpressed and cloned in *S. cerevisiae* by Nidetzky and coworkers [244,245] and used for the reduction of a variety of α-keto esters [225]. Ethanol was the cosubstrate of choice, and anaerobic conditions were employed aiming at suppressing the NADPH-dependent enantiodivergent reductase background displayed by the host genome [156,246–248]. Under these conditions, reducing equivalents could be regenerated mainly in the form of NADH. Several α-keto esters with aromatic and aliphatic side chains were assessed as substrates of the recombinant strain *S. cerevisiae* WF2μ (r*S. cerevisiae*) (Table 3.5, entries 5–7) [225]. This reaction system led to significant improvements in optical

TABLE 3.5 Medium Engineering for the Enantioselective Reduction of α- or β-Keto Esters

Entry	Substrate	Product	Biocatalyst/Conditions	Yield (%)	*ee* (%)	Reference
1	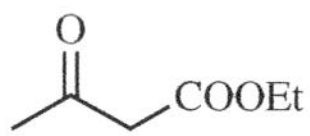	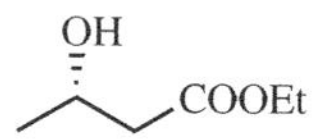	*S. cerevisiae* (immobilized cells) perfluorooctane	25	95	[230]
2	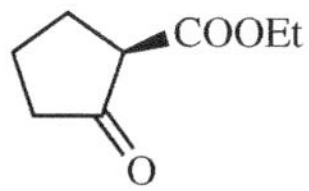	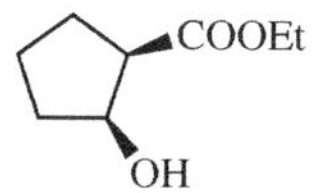	*S. cerevisiae* (immobilized cells) perfluorooctane	28	99	[230]
3	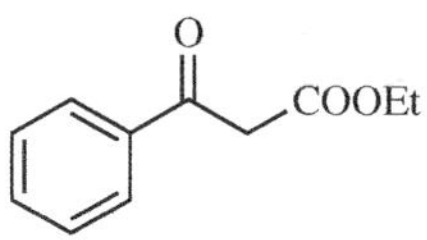	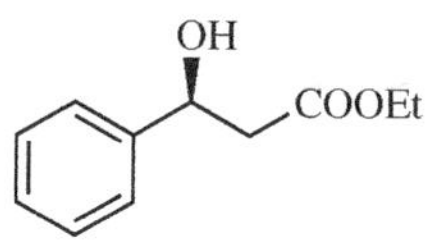	*S. cerevisiae* (immobilized cells) perfluorooctane	66	94	[230]
4	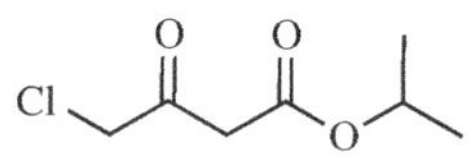	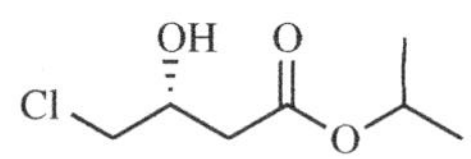	ADH-'A' 90% v/v IL[a] 10% v/v aqueous buffer	19	>99%	[235]

(Continued)

TABLE 3.5 (*Continued*)

Entry	Substrate	Product	Biocatalyst/Conditions	Yield (%)	*ee* (%)	Reference
5	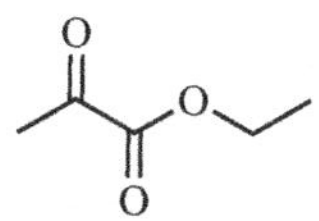	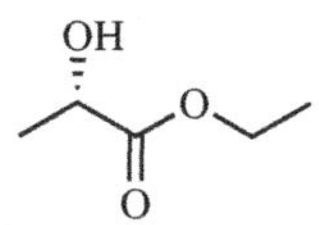	r*S. cerevisiae* WF2μ EtOH 1M anaerobic conditions	35	99.9	[225]
6	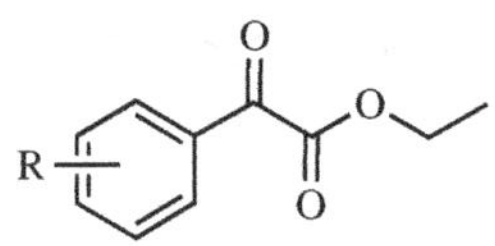	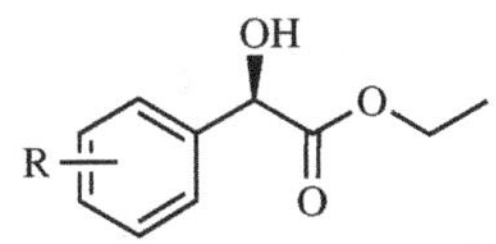 R: H; *o*-Cl; *m*-Cl; *p*-Cl	r*S. cerevisiae* WF2μ EtOH 1M anaerobic conditions	14–80	>99	[225]
7	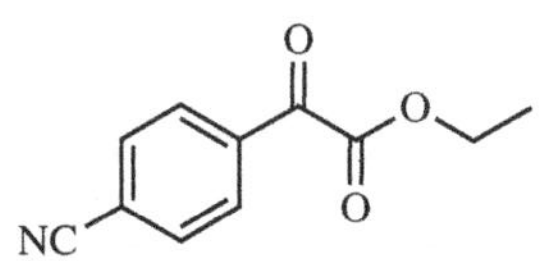	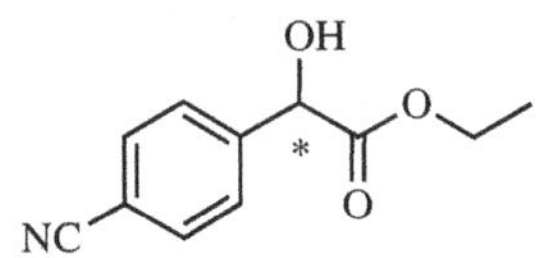	r*S. cerevisiae* WF2μ EtOH 1M anaerobic conditions	100	99.4 (n.d)[b]	[225]

[a] [MTEOA][$MeSO_4$], tris-(2-hydroxyethyl)-methylammonium methylsulfate; r*S. cerevisiae*, recombinant strain of *S. cerevisiae* WF2μ.
[b] Absolute configuration of the alcohol was not determined.

purity of the products as well as increased efficiency of α-keto ester reduction and co-substrate yield (expressed as the molar ratio of formed α-hydroxy ester to consumed ethanol). The corresponding reduction products were obtained with >50% yield for most aromatic substrates and $\geq$ 99% *ee*.

As stated before, current genetic engineering and molecular biology techniques allow the cloning of genes involved in enzyme production in microbial hosts (bacteria or yeasts), as well as their overexpression (see Chapter 7) [43]. The seminal work of Stewart about the overexpression of *S. cerevisiae* KREDs in *E. coli* or *S. cerevisiae* has boosted the expansion of the use of these enzymes and their applications in synthetic organic processes [247]. In the 1990s, Stewart and coworkers developed recombinant strains of *S. cerevisiae* and *E. coli* [60,156], by the isolation and overexpression of yeast ketoreductases and dehydrogenases [220,221,223]. Overexpression and gene knockout are molecular biology tools available to modify the expression level of the enzyme in a certain microorganism. The design of mutant strains lacking competitive enzymes to those involved in a certain process often results in a significant improvement in the biotransformation selectivity. An example of genetic manipulation of *S. cerevisiae* to improve its selectivity in β-keto ester reductions was the construction of recombinant strains in which two of the most important KREDs were alternatively overexpressed and blocked, leading to outstanding improvements in process selectivity (Figure 3.6) [156,223].

Recombinant strains were constructed to improve the poor selectivity achieved for certain substrates by unmodified bakers' yeast. Model β-keto esters **21a–c** were used to assess the improvements of the engineered strains (Figure 3.7).

The alteration of the fatty acid synthetase concentration within the cell resulted in a dramatic effect on the enantioselectivity of the reduction of **21a** (Table 3.6). The strain overexpressing this (R)-selective enzyme afforded the corresponding alcohol in 88% *ee*, showing a significant increase compared to the unmodified strain (70% *ee*). By knocking out fatty acid synthetase, the enantioselectivity was totally switched, leading to the (S)-hydroxyester in high optical purity (88% *ee*).

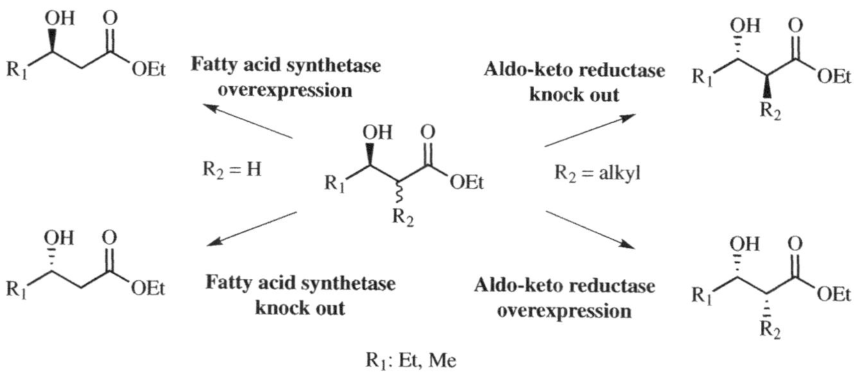

FIGURE 3.6 Enantio- and diastereoselectivity of the engineered *S. cerevisiae* strains, overexpressing or blocking fatty acid synthetase and aldo-ketoreductase, the major β-keto ester reductases in *S. cerevisiae*.

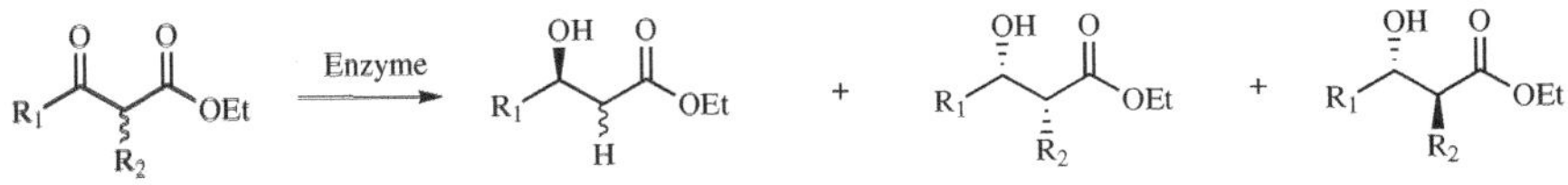

21a: R_1 = Et, R_2 = H
21b: R_1 = Me, R_2 = alkyl
21c: R_1 = Me, R_2 = propargyl

22a (**22b** and **22c** not observed)
Fatty acid synthetase product

23a, 23b, 23c (2*R*,3*S*-*syn*)
Aldo-keto reductase product

24b, 24c (2*S*,3*S*-*anti*)
α-acetoxy keto reductase product

FIGURE 3.7 Enantio- and diastereoselectivity of engineered yeast strains [156].

When compound **21a** was reduced by the engineered strain lacking AKR, only a small decrease in the amount of (*S*)-hydroxy ester was observed, thus demonstrating that this enzyme has a minor contribution to the reduction of this substrate in wild-type cells. On the other hand, the strain overexpressing AKR produced significantly more (*S*)-hydroxyester than the unmodified strain. These effects were demonstrated clearly when α-substituted-β-keto esters **21b** and **21c** were used as substrates, since the enzyme produced the *syn-* diastereomers, whereas the α-acetoxy KRED afforded the corresponding *anti-* isomers (Figure 3.7). Expectedly, the (2*S*,3*S*)-*anti* diastereomers **24b** and **24c** were the major products obtained with the strain lacking the AKR. However, (2*R*,3*S*)-*syn*-diastereomers **23b** and **23c** were still produced, indicating the presence of at least another enzyme with the same stereospecificity. Conversely, the AKR overexpressing strain resulted in a high production of the *syn*-diastereomer, and reduction of **21c** with this strain yielded only one of the four possible diastereomers. These results indicate the potential of "designed cells" approaches to tailor the stereospecificity of *S. cerevisiae*, providing useful biocatalysts for highly stereoselective transformations, by combining overexpression and knock out strategies through the manipulation of single genes (see Chapter 7 for a detailed discussion on the recombinant oxidoreductases present in baker's yeast).

TABLE 3.6 Enantio- and Diastereoselectivity of Engineered Yeast Strains Expressing or Blocking Fatty Acid Synthetase and Aldo-ketoreductase [156]

		Products		
Substrate	Enzyme Modified	Unmodified Strain	Overexpressing Strain	Knocked Out Strain
21a	Fatty acid synthetase	**22a** 70% *ee* (*R*)	**22a** 88% *ee* (*R*)	**23a** 88% *ee* (*S*)[a]
21a	Aldo-keto reductase	**22a** 70% *ee* (*R*)	**22a** 36% *ee* (*R*)	**22a** 78% *ee* (*R*)
21b	Aldo-keto reductase	**24b** 60% *de* (*anti*)[b]	**23b** 65% *de* (*syn*)[b]	**24b** 83% *de* (*anti*)[b]
21c	Aldo-keto reductase	**23c** 28% *de* (*syn*)[b]	**23c** >99% *de* (*syn*)[c]	**24c** 42% *de* (*anti*)[b]

[a]Only the 2*R*,3*S*-enantiomer was observed.
[b]Knock out based on *S. cerevisiae* X 2180.
[c]*ee* >99% at C3.

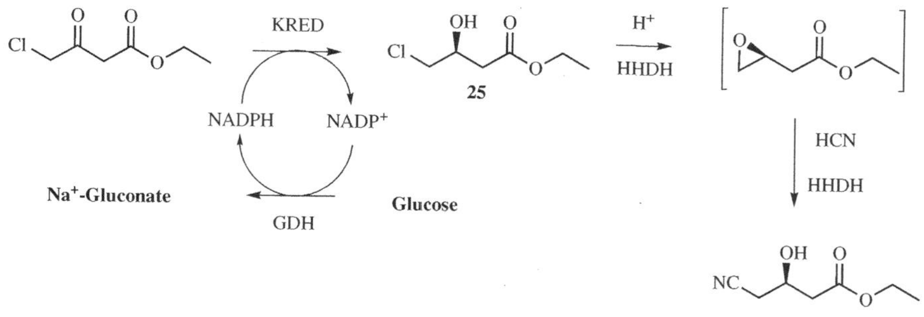

FIGURE 3.8 Two-step-three-enzyme process for the synthesis of a key intermediate in the production of atorvastatin.

Recently, a green approach involving a two-step-three-enzyme process was developed for the synthesis of a key intermediate in the production of atorvastatin, the active component of the cholesterol lowering drug Lipitor® [113]. The first step involved the biocatalytic reduction of ethyl 4-chloroacetoacetate using a KRED in combination with glucose and NADP-dependent GDH for cofactor regeneration. The ethyl (*S*)-4-chloro-3-hydroxybutyrate (**25**) was obtained in 96% isolated yield and >99.5% *ee*. In the second step, the replacement of the chloro substituent with a cyano group with HCN was catalyzed by a halohydrin dehalogenase (HHDH) (Figure 3.8). Both natural enzymes were highly selective but exhibited productivities that were not acceptable for large-scale use. Consequently, *in vitro* enzyme evolution using gene-shuffling technologies on the enzymes expressed in *E. coli* was employed to optimize the process performance (see Chapters 7 and 8 for further details of this process).

The enantioselective reduction of 2-chlorophenylbutyrate to ethyl (*R*)-2-hydroxy-4-phenylbutyrate **26**, a key intermediate in the synthesis of angiotensin-converting enzyme (ACE) inhibitors was reported by Chen et al. The reaction was performed using *C. boidinii* cells, in plain aqueous medium with high yield (92%) and >99% *ee*. (Figure 3.9).

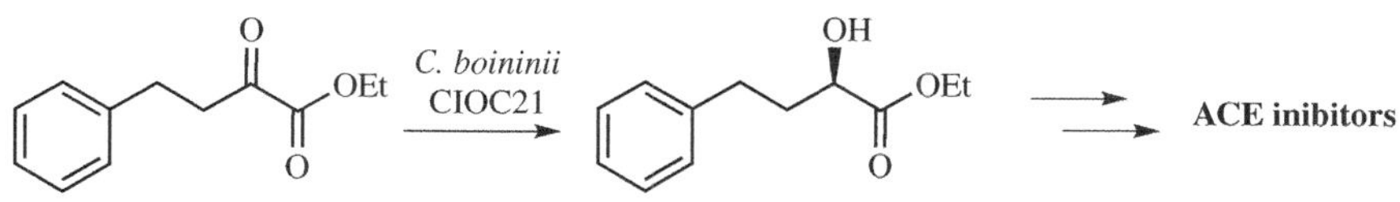

26

Yield 92%
> 99% *ee*

FIGURE 3.9 Redox process catalyzed by *C. boidinii* cells to afford a precursor for angiotensin-converting enzyme (ACE) inhibitors.

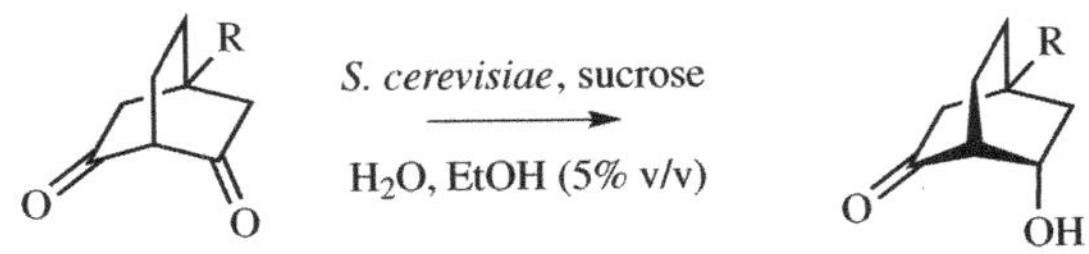

27: R = OTBS: (87%, 46% *ee*)
28: R = OAc: (73%, 10% *ee*)
29: R = OBn: (86%, 69% *ee*)
30: R = O-*p*-BrBn: (36%, 47% *ee*)
31: R = OSEM (59%, 68% *ee*)
32: R = OCH_2CHCH_2 (80%, 82% *ee*)

FIGURE 3.10 Asymmetric reduction with baker's yeast of bridgehead-substituted bicyclo [2.2.2]octanes.

3.2.3 Reduction of Diketones

S. cerevisiae has also been largely used for the reduction of diketones. As an example, baker's yeast-catalyzed enantioselective reductions were used by Frejd and coworkers for the desymmetrization of 4-substituted bicyclo[2.2.2]octane-2,6-diones, achieving the corresponding hydroxyketones with moderate-to-high yields (up to 87%) and enantiomeric excesses (up to 82%) in a reaction medium containing sucrose and 5% v/v ethanol (EtOH) (Figure 3.10) [249]. The enantioselectivity was strongly influenced by the size and shape of the substituent in C4.

Smonou and coworkers designed a biocatalytic cascade reaction for the synthesis of optically pure 2-alkyl-1,3-diols employing two KREDs and glucose/GDH for NADPH recycling [250]. The process consisted of two consecutive steps: (i) a stereoselective diketone reduction and (ii) a hydroxy ketone reduction. In all cases, the first enzymatic reduction led to the β-hydroxyketone, and the corresponding diol was formed by the addition of a second KRED without the isolation of the intermediate product (Figure 3.11A). Chiral diols were obtained by the addition of both KREDs in the same vessel, affording the final products with high stereoselectivity and

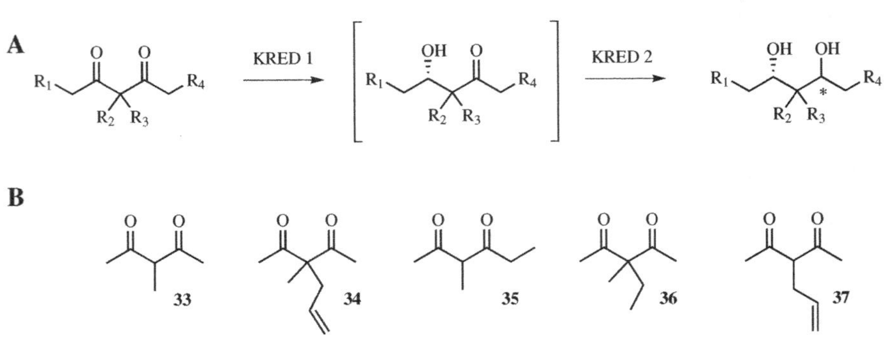

FIGURE 3.11 **A**, one pot stereoselective reduction of 1,3-diketones by using isolated ketoreductases; **B**, 1,3-diketones used as substrates in the ketoreductase-catalyzed cascade reductions to obtain 1,3-diols.

chemical yield. The authors synthesized single diastereomers of the product diols by a judicious choice of commercially available enzymes. The substrates for the biotransformation were three monoalkyl- and two dialkyl-substituted 1,3-diketones (Figure 3.11B). The reductions were carried out in a phosphate buffer (200 mM, pH 6.5-6.9) in the presence of GDH, glucose and NADPH.

In all cases 2-substituted-1,3-diols were obtained in high chemical and stereochemical purity, and isolated without chromatographic purification due to the absence of any detectable by-product (Table 3.7). The first reductions were always quantitative and highly stereoselective, with no further reduction to the corresponding 1,3-diol, thus producing only one of the four possible stereoisomers. The second reaction was also highly stereoselective yielding only one stereoisomer in 24 h.

TABLE 3.7 Enzymatic Reduction of Diketones 33–37

Substrate	KRED 1	KRED 2	Yield (%)	Time (h)	*ee* (%)[a] *de* (%)[b]	Product
33	KRED102	KRED101	88	48	>99 >99	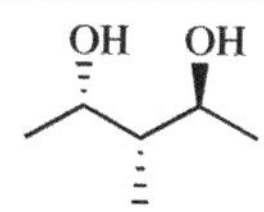
34	KRED112	KRED114	91	36	>99 98	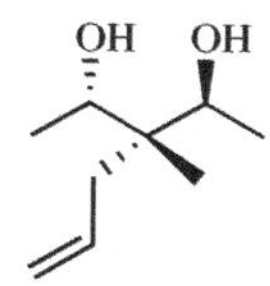
35	KRED102	KRED101	90	48	>99 >99	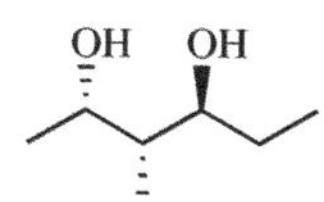
35	KRED102	KREDA1B	85	48	>99 96	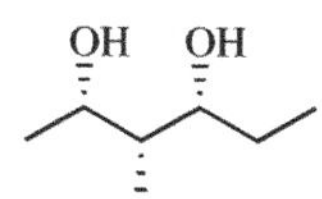
36	KRED18	KRED101	90	36	>99 >99	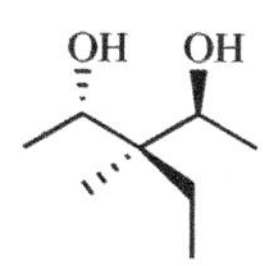
37	KRED108	KRED111	92	48	>99 98	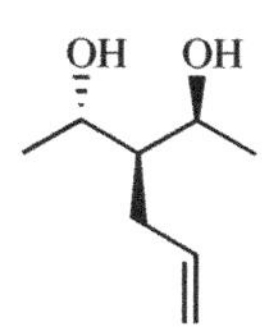

[a]*ee* (%) measured by chiral GC analysis.
[b]*de* (%) determined by chiral GC analysis and ^{1}H NMR spectra.
KRED, Ketoreductase.

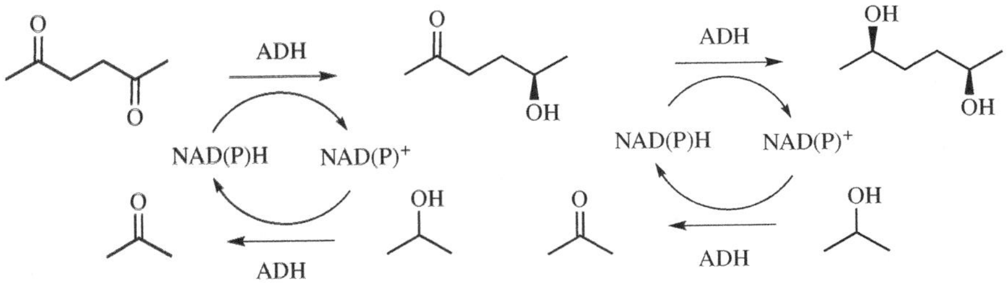

FIGURE 3.12 Biocatalytic reduction of 2,5-hexanedione with substrate-coupled cofactor regeneration. ADH: LBADH, *Lactobacillus brevis* alcohol dehydrogenase.

Lütz and coworkers designed a biocatalytic approach for the reduction of γ-diketones, using a new reactor concept that enables a continuous enzymatic process with concurrent *in situ* pervaporation of the generated (cofactor regeneration). Thus, the thermodynamic inhibition of complete conversion caused by the equilibrium between all involved reactants and products can be overcome (Figure 3.12) (see Chapter 7) [74]. In such bimembrane reactor systems, recombinant *E. coli* cells overexpressing LBADH were applied for the reduction of 2,5-hexanedione. This enzyme is highly effective and robust, especially at elevated (co)substrate concentrations. The biotransformation was carried out in 0.05 $molL^{-1}$ potassium phosphate buffer (pH 6.0) with varying concentrations of 2,5-hexanedione, 2-propanol and biomass, and a $NADP^+$ concentration of 0.01 $mmolL^{-1}$ during the whole process, under a constant vacuum of 100 mbar and a temperature of 40°C in order to enable a high acetone removal rate. Due to this *in situ* acetone removal technique, the acetone concentration did not exceed 50 $mmolL^{-1}$, which corresponds to 90% of acetone removal. This value was calculated for the time period of 10–13 days, where the highest yield of the corresponding hexanediol was achieved. In the experimental conditions reported, high yields (77%) and excellent enantiomeric and diastereomeric excesses (>99%) of the product (2*R*,5*R*)-hexanediol were achieved. This represents an exceedingly high space–time yield of 179 $gL^{-1}d^{-1}$ and a catalyst usage of 17.9 $g_{product}/g_{wet\ cell\ weight}$.

Furthermore, dehydrogenases are important in the production of acyloins (α-hydroxyketones), valuable building blocks for the manufacture of fine chemicals, as well as pharmaceuticals [251]. These versatile compounds can be obtained by dehydrogenase-catalyzed reduction of the corresponding α-diketones (Figure 3.13), as well as by oxidation of 1,2-diols, or deracemization of α-hydroxyketones. In particular, selective bioreduction of α-diketones has been performed by means of different isolated enzymes and also microorganisms as whole-cell biocatalysts [252–256].

3.2.4 Reduction of Aldehydes

Chemical reduction of aldehydes to the corresponding alcohols can be easily carried out in different conditions, but some other functional groups such as ketones, imines, or epoxides may interfere, making the process scarcely selective. Selective

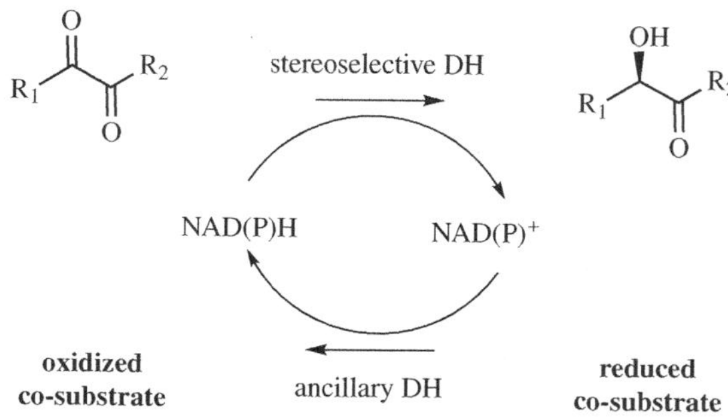

FIGURE 3.13 Bioreductive approach for the synthesis of enantio-enriched α-hydroxy ketones. DH, dehydrogenase.

dehydrogenase-catalyzed aldehyde reductions are less studied than ketone reductions, presumably because aldehyde reductions cannot generate asymmetric centers. Not withstanding this consideration, this strategy can be used for kinetic resolutions in remote centers. For instance, enantiopure glyceraldehyde has been successfully used for the preparation of antibiotics [257], anticancer drugs [258], carbohydrates, and natural products [259,260]. Hummel et al. reported the cloning, heterologous expression and biochemical characterization of the glycerol dehydrogenase (GlyDH) from *Gluconobacter oxydans*, and its use in the kinetic resolution of racemic glyceraldehyde for the production of enantiopure L-glyceraldehyde in preparative scale [261,262]. The reaction was performed with simultaneous regeneration of NADPH by using glucose and GDH to obtain a conversion of 50% after 26 h to give L-glyceraldehyde with an enantiomeric excess greater than 99%. Subsequently, a whole-cell catalyst coexpressing GlyDH from *G. oxydans* and GDH from *Bacillus subtilis* was constructed. Compared to the cell-free system, the recombinant *E. coli* showed higher thermal and similar operational stability, together with the catalyst recycling without any loss of activity, leading to a more efficient production of L-glyceraldehyde with 98% *ee* and 54% conversion on a 4 g scale (Figure 3.14).

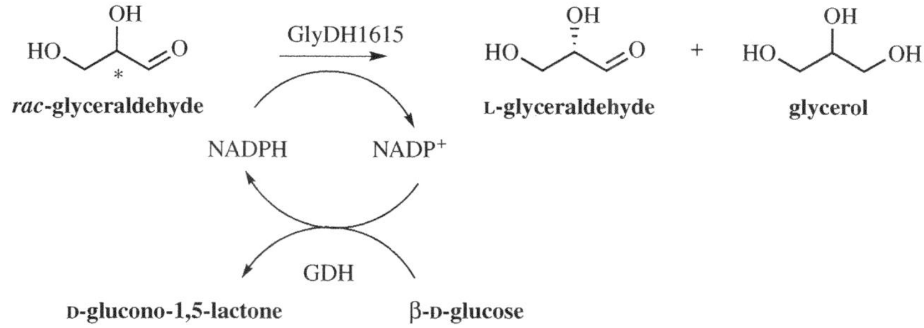

FIGURE 3.14 Kinetic resolution of glyceraldehyde by using a recombinant *E. coli* coexpressing GlyDH and GDH. GlyDH, glycerol dehydrogenase; GDH, glucose dehydrogenase.

Furthermore, Gröger and coworkers reported the synthesis of the aroma chemical cinnamyl alcohol by means of a whole-cell-catalyzed reduction of cinnamyl aldehyde (see Chapter 8) [263]. Recombinant whole-cells overexpressing LKADH and GDH from *Thermoplasma acidophilum* (TA-GDH) were used with D-glucose as co-substrate. Conversions of 98% (yields of 77%) of cinnamyl alcohol, even at a high substrate loadings (up to 166 gL^{-1}) were achieved. The developed approach offered important advantages such as the use of a low-cost whole-cell biocatalyst and the high purity of the product, as well as the no need for addition of external cofactor in the biotransformation step (Figure 3.15).

The yeast strain *Rhodotorula* sp. AS2.2241, isolated from soil samples by supplying acetophenone in vapor state, was found to be an efficient reducing catalyst for ketones and aldehydes [264]. The enzyme showed high activity toward aldehydes including aliphatic and aromatic ones. The highest activity was observed with acetaldehyde, followed in the sequence by propionaldehyde and *n*-butyraldehyde. Regarding aromatic aldehydes, benzaldehyde was a suitable substrate for the enzyme. Based on the exposed results, short-chain alkyl aldehydes and benzaldehyde were suitable substrates (Table 3.8).

Likewise, several aromatic aldehydes were reduced to alcohols with moderate-to-excellent yields using plant cell preparations from *Passiflora edulis*, using freshly cut slices of fruit barks in aqueous solutions (Figure 3.16) [265]. Except for aldehyde **41** all compounds were reduced with moderate-to-high conversions. The reducing process for **42** and **43** was chemoselective, with the transformation occurring only in the carbonyl group, keeping unaltered side-chain double bond (Figure 3.16).

Highly chemoselective enzymatic reductions of aldehydes to the corresponding alcohols were also successfully performed using *E. coli* JM109 whole-cells [266]. Biotransformations were carried out in a two-phase system consisting of an aqueous buffer and an organic phase (9:1 isooctane-isopropyl ether) in a 7:3 ratio, with structurally different aldehydes. Reactions were highly chemoselective since none of the potential interfering groups were affected. Methyl ketones and aldoximes were not converted (Figure 3.17). On the contrary, the biocatalyst accepted a wide range of aldehydic substrates, including aromatic and aliphatic aldehydes, unsaturated aldehydes as well as dialdehydes.

3.3 RACEMIZATION AND DERACEMIZATION REACTIONS

In contrast with enzymatic (or chemical) reductions of prochiral ketones, able to furnish the corresponding alcohols in quantitative yields, resolution of racemic alcohols *via* selective oxidation or esterification cannot obviously exceed conversions of 50%. To this end lipase-catalyzed (trans)esterifications of alcohols is a widely used methodology, successfully implemented for many substrates. As an alternative to hydrolases, oxidoreductases have also been assessed as biocatalysts for the synthesis of enantiomerically pure alcohols in high yields from the corresponding racemic mixtures. In this regard, deracemization of racemic mixtures leads to the corresponding enantiopure alcohols with conversions up to 100% [267–271].

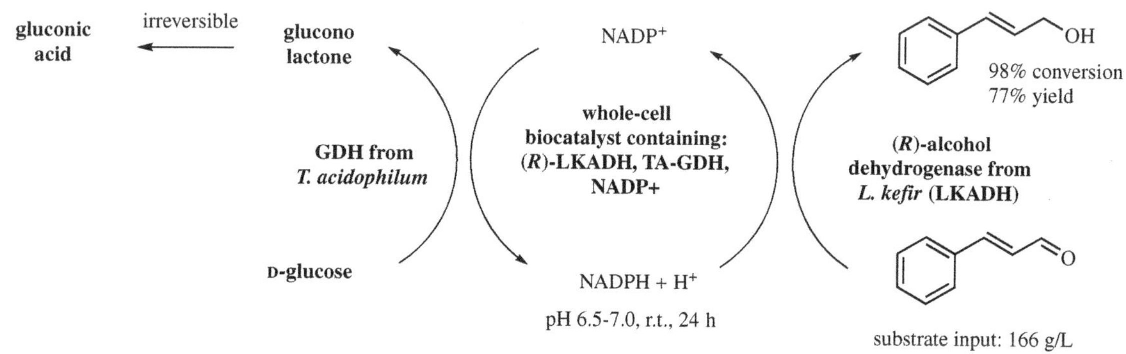

FIGURE 3.15 Biocatalytic approach for the synthesis of cinnamyl alcohol from cinnamyl aldehyde. GDH, glucose dehydrogenase; TA-GDH, *T. acidophilum* glucose dehydrogenase.

TABLE 3.8 Relative Activities of the Partially Purified Enzyme from *Rhodotorula* sp. AS2.2241 in the Reduction of Aldehydes

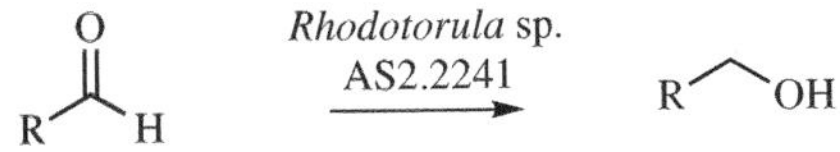

R	Relative Activity (%)[a]
H	367
CH_3-	1667
CH_3CH_2-	483
$CH_3(CH_2)_2$-	400
$CH_3(CH_2)_5$-	167
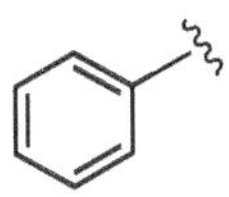	383
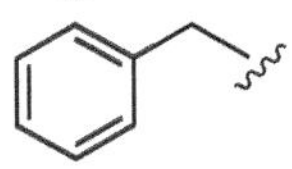	0

[a]Relative activity toward α-bromoacetophenone (100%).

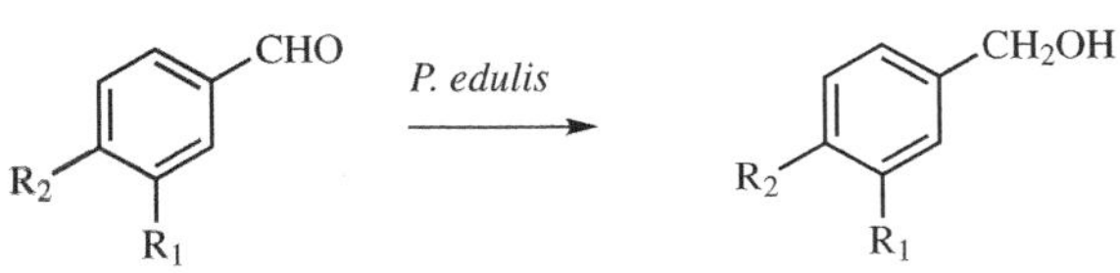

38: $R_1 = R_2 = H$
39: $R_1 = H$; $R_2 = OCH_3$
40: $R_1 = OCH_3$; $R_2 = H$
41: $R_1 = OH$; $R_2 = OCH_3$

38a: $R_1 = R_2 = H$; 96%
39a: $R_1 = H$; $R_2 = OCH_3$; 84%
40a: $R_1 = OCH_3$; $R_2 = H$; 67%
41a: $R_1 = OH$; $R_2 = OCH_3$; 0%

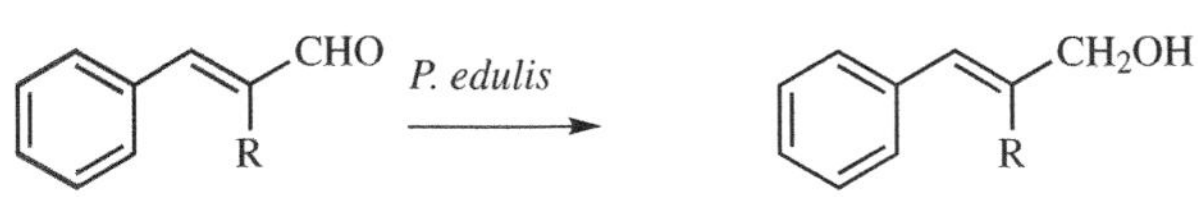

42: R = H;
43: $R = CH_3$

42a: R = H; 41%
43a: $R = CH_3$; 35%

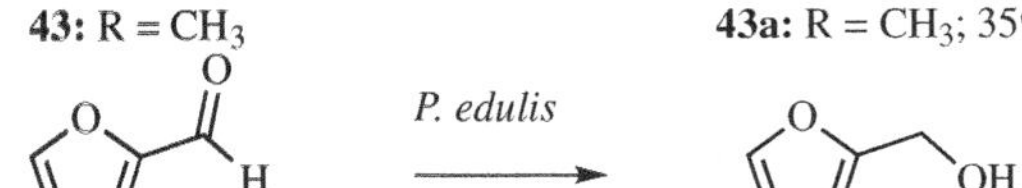

44

44a: 95%

FIGURE 3.16 Conversion of aldehydes **38–44** to the corresponding alcohols **38a–44a** using fruit barks of *P. edulis*.

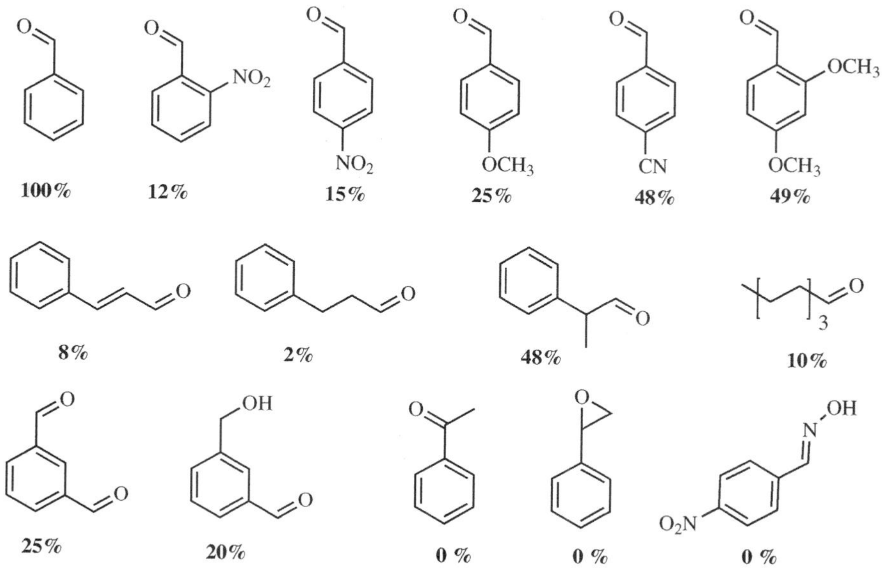

FIGURE 3.17 Relative conversions compared to benzaldehyde (100%) for the reduction of aldehydes to the corresponding alcohols, using *E. coli* JM109 as biocatalyst.

Deracemization of secondary alcohols can be performed either by stereoinversion (Figure 3.18A) [268] or dynamic kinetic resolution (DKR) (Figure 3.18B) [269,272]. Enzymatic stereoinversion of secondary alcohols consists in the selective oxidation of one of the enantiomers to the corresponding ketone, and its further selective reduction to the other enantiomer (Figure 3.18A). Thus, chiral diols [273–278], hydroxyketones [279,280], hydroxyacid derivatives [281–287], or aliphatic [288–292], and aromatic alcohols [291,293–303], can be obtained with high enantiomeric purity.

Stereoinversion reactions can be grouped in two classes, depending on if one or two different biocatalysts are used [268]. Examples of both strategies, the use of two different biocatalysts [281,282,288,291,304,305], or only one [273–275,278,289,290, 292,294,296,300,306–310], have been widely documented. An example of stereoinversion involving two microorganisms is the deracemization of (±)-mandelic acid [305], by using *Alcaligenes bronchisepticus* and *Streptococcus faecalis* as biocatalysts. *A. bronchisepticus* KU 1201 selectively oxidized (*S*)-mandelic acid to benzoylformate which, in turn, was selectively reduced by cell free extracts of *S. faecalis* IFO 12964 containing a NADH-dependent benzoylformate reductase, yielding 80% overall of (*R*)-mandelic acid with >99% *ee* (Figure 3.19A). Conversely, strategies comprising only one biocatalyst were used in the enantioselective preparation of (2*R*)-3-pentyn-2-ol from the corresponding racemic alcohol, in a *Nocardia fusca* AKU 2123-catalyzed reaction (Figure 3.19B) [289,290]. Further examples include arylethanol deracemizations, which have been described using *G. candidum* [198,294,295], *Sphingomonas paucimobilis* [296], and whole plant cells [302,303] as biocatalysts.

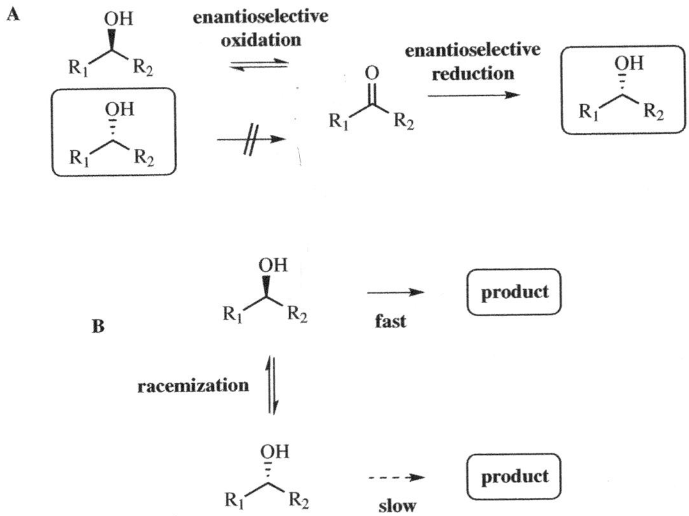

FIGURE 3.18 **A**, deracemization of secondary alcohols *via* stereoinversion; **B**, deracemization of secondary alcohols *via* dynamic kinetic resolution (DKR).

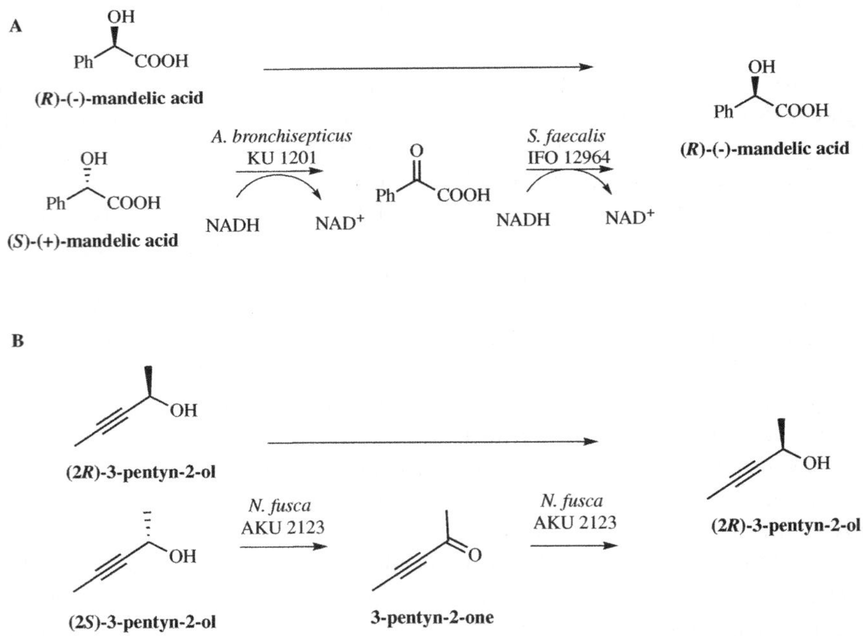

FIGURE 3.19 One- and two-biocatalysts-mediated deracemization of secondary alcohols. **A**, *A. bronchisepticus* KU 1201 selective oxidation of (*S*)-mandelic acid to benzoylformate and subsequent *S. faecalis* IFO 12964-mediated selective reduction to (*R*)-mandelic acid; **B**, (2*R*)-3-pentyn-2-ol preparation from the corresponding racemic mixture in a *Nepenthes fusca*-AKU 2123 mediated deracemization *via* stereoinversion.

The first report on deracemization by stereoinversion was described by Shimizu and coworkers [311], using whole-cells of *R. erythropolis* IFO 12540 for the conversion of L-(-)-pantoyl lactone into its corresponding D-isomer, starting from a racemic mixture. The desired isomer was obtained with an enantiomeric excess of 94.4% and molar yield of 90.5%. From this seminal work, microbial stereoinversions have been widely investigated, being currently a general method for the preparation of single enantiomers from racemates. Many examples can be found employing whole-cell biocatalysts such as bacteria, fungi, yeasts, and plants, as well as isolated enzymes (Table 3.9). As can be observed, the applicability of stereoinversion protocols is remarkably broad and includes α-and β-hydroxyesters, aliphatic and aromatic alcohols, as well as *vic*-diols.

On the other hand, racemizations, meaning the equilibrated interconversion of enantiomers, are usually not considered a synthetic tool, but a nondesired secondary reaction. As a consequence, either chemical or enzymatic controlled racemization of organic compounds has been scarcely explored [312–318]. The impact of racemization processes on synthetic organic chemistry lies in its key role in DKRs [319,320], allowing to convert racemates into one stereoisomer with a theoretical yield of 100% [321]. In this context, the DKR of secondary alcohols (Figure 3.18B) has attracted ample attention, given the possibility of combination of enantioselective lipase or protease-catalyzed acylations or deacylations with racemization protocols based on transition metals [313,314,322,323], aluminum complexes [324] or acid resins [325]. However, most of chemical racemizations require severe reaction conditions, not always compatible with *in situ* enzymatic transformations. Instead, the coupling of two enzymatic steps catalyzed by different enzymes is feasible. Thus, enzymatic racemizations are presented as a valuable promising and underdeveloped tool for mild synthetic procedures [326].

3.4 PREPARATION OF AMINES

The reductive amination of carbonylic compounds is an efficient method for the preparation of amines. A useful synthetic application of this reaction is the reductive amination of α-keto acids, to yield the corresponding (*S*)-amino acids [327]. Amino acid dehydrogenases (EC 1.4.1.-) catalyze the reaction with the concomitant oxidation of NAD(P)H (Figure 3.20) [4]. The reaction requires ammonia and NADH. The oxidized cofactor NAD^+ produced during the reaction is recycled to NADH by oxidation of formate to CO_2 using FDH. The corresponding amino acids can be obtained almost quantitatively, since for thermodynamic reasons in AADH-catalyzed reactions the equilibrium is often shifted to the formation or the amine.

Among the large number of AADHs identified in nature, only a few are relevant in industrial processes for the preparation of enantiopure amino acids [328,329]. The most used for synthetic applications are alanine dehydrogenase (AlaDH, EC 1.4.1.1), glutamate dehydrogenase (GluDH, EC 1.4.1.2-4), and especially phenylalanine dehydrogenase (PheDH, EC 1.4.1.20), with high stability and activity, and leucine dehydrogenase (LeuDH, EC 1.4.1.9). All of them catalyze the conversion of

TABLE 3.9 Selected Examples of Chiral Alcohols Obtained through Stereoinversion from the Corresponding Racemic Mixtures

Product	Biocatalyst	*ee* (%)	Yield (%)	Reference
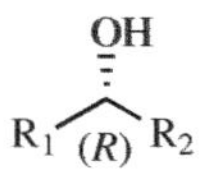 R_1: Ar, Het-Ar, Ar-CH_2- R_2: Me, Et, *n*-Bu, hexyl, vinyl	*S. paucimobilis*	99	90	[296]
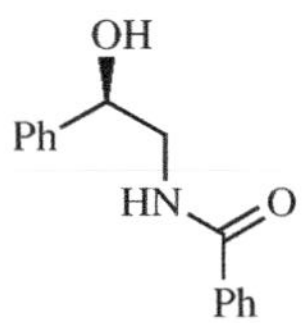	*C. echinulata*	92	57	[297]
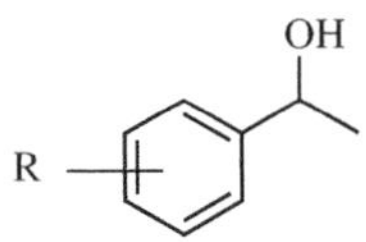 (*R*) for 3320 (*S*) for 4038 R: *m*-F, *p*-NO_2, Br, Cl	*A. terreus* CCT 3320 and CCT 4083	>99	Up to 86	[298,300]
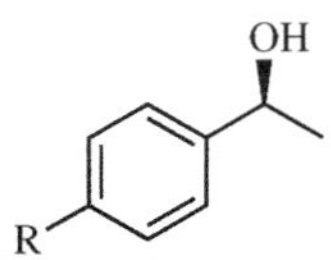 R: H, Cl, Br, CH_3, NO_2	*M. oxydans* ECU2010 (*R* selective oxidation) *Rhodotorula* sp. AS 2.2241 (*S* selective reduction)	Up to 99%	Up to 99	[293]
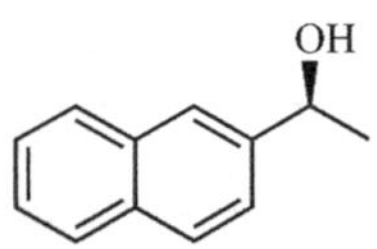	*M. oxydans* ECU2010 (*R* selective oxidation) *Rhodotorula* sp. AS 2.2241 (*S* selective reduction)	>99	86	[293]

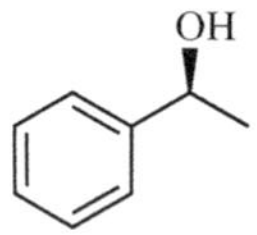	*M. oxydans* ECU2010 (*R* selective oxidation) *Rhodotorula* sp. AS 2.2241 (*S* selective reduction)	>99	>99	[293]
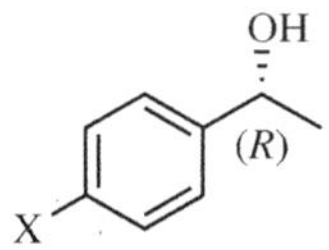 X: H, Cl, OMe	*G. candidum*	Up to 99	55–99	[294,295]
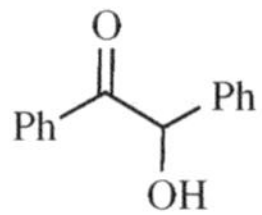 (*S*) for pH 4-5 (*R*) for pH 7.5-8	*R. oryzae*	97 (*R*) 85 (*S*)	73–76 71	[279]
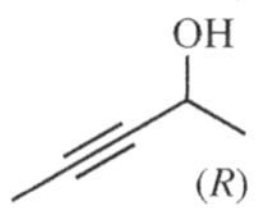 (*R*) or (*S*)	(*R*): *N. fusca*, *Nocardia globerula*, *N. erythropolis* (*S*): *N. pseudosporangifera*	100 (*R*) 98 (*S*)	83 70	[289,290]
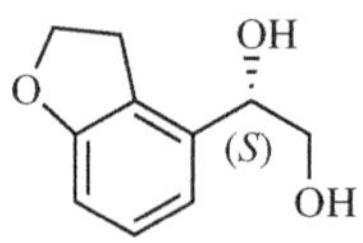	*C. boidinii* *Pichia methanolica* *Hansenula polymorpha*	Up to 100 Up to 100 Up to 60	72 88 99	[275]

(*Continued*)

TABLE 3.9 (*Continued*)

Product	Biocatalyst	*ee* (%)	Yield (%)	Reference
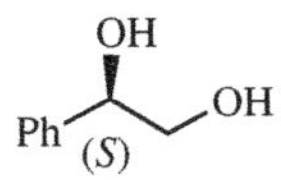	*C. parapsilosis*	98	92	[277,278]
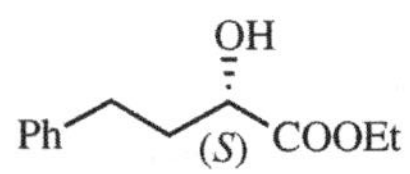	*C. parapsilosis*	>99	85–90	[283]
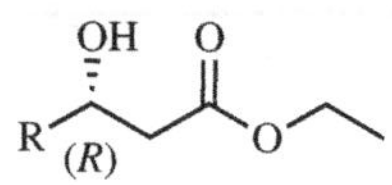 R: Ph, *o*-MeC_6H_4, *p*-MeC_6H_4, *p*-ClC_6H_4, *p*-$NO_2C_6H_4$, *o*-ClC_6H_4 PhCH=CH	*C. parapsilosis* (immobilized)	>99	28–68	[284]
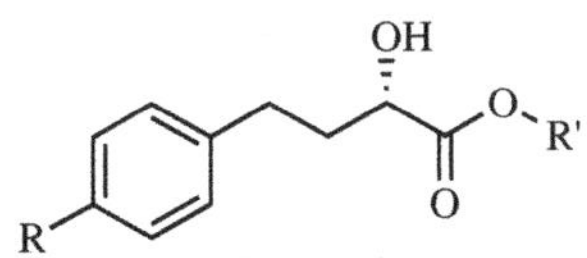 R: Me, Cl R': Me, Et	*C. parapsilosis*	95	69–70	[285]

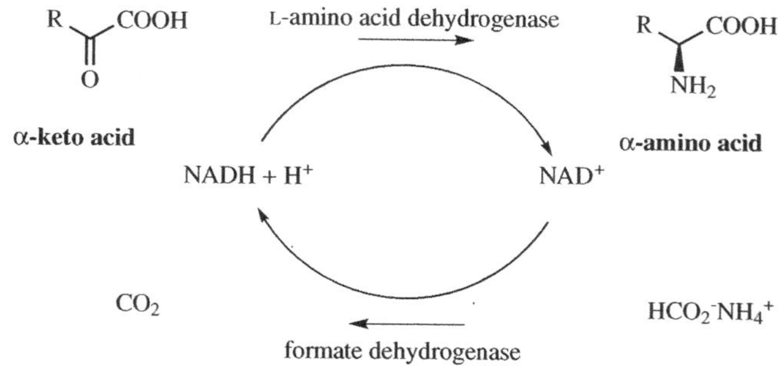

FIGURE 3.20 Reductive amination process to enantiomerically pure L-amino acids with FDH-based cofactor regeneration.

α-keto acids to the corresponding L-amino acids, exclusively with (*S*)-configuration (see Chapter 8 for more details on their industrial applications). Most of the synthetic work developed with NADH-dependent LeuDH has been carried out using enzymes belonging to *Thermoactinomyces intermedius* and different species of *Bacillus* [330]. LeuDH from *B. stearothermophilus*, *B. cereus,* and *Bacillus sphaericus*, display similar substrate specificity, accepting α-keto acid with hydrophobic, aliphatic, hindered, or not hindered side chains with up to six carbon atoms. Furthermore, they also accept cyclic α-keto acids, but not aromatic substrates [331]. Moreover, PheDH (from *T. intermedius*, *Rhodococcus* sp. and *Brevibacterium* spp.) displays a wider substrate specificity, not only comparable to that of LeuDH, but also accepting substrates with aromatic side chains [4]. Actually, the substrate specificity of LeuDH and PheDH are complementary. Different methodologies have been designed at industrial level to efficiently perform reductive amination procedures, including the use of isolated enzymes in membrane reactors or whole-cell processes [78,332]. As an example, recombinant *Pichia pastoris* coexpressing a *T. intermedius* PheDH (EC 1.4.1.20) and also FDH for cofactor regeneration was used for the production of (*S*)-2-amino-5-(1,3-dioxolan-2-yl)pentanoic acid (**45**) (allysine ethylene acetal, building block in the synthesis of the vasopeptidase inhibitor Omapatrilat, BMS 186716) at 15 kg scale, with 97% yield and >98% *ee* (Figure 3.21) [333].

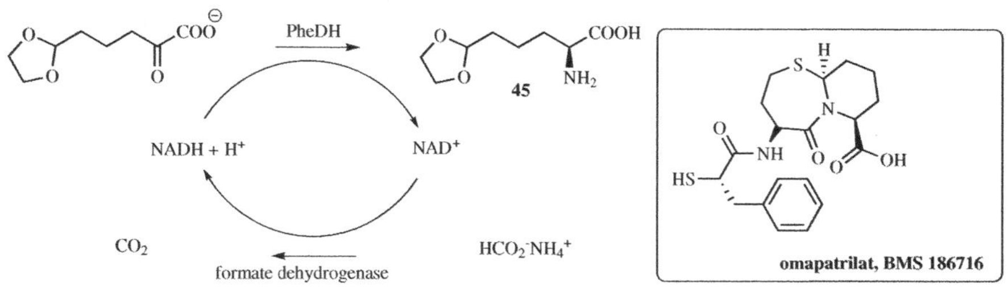

FIGURE 3.21 Production of allysine ethylene acetal (**45**). PheDH catalyzes the reductive amination of the corresponding keto acid and formate dehydrogenase is used for cofactor regeneration.

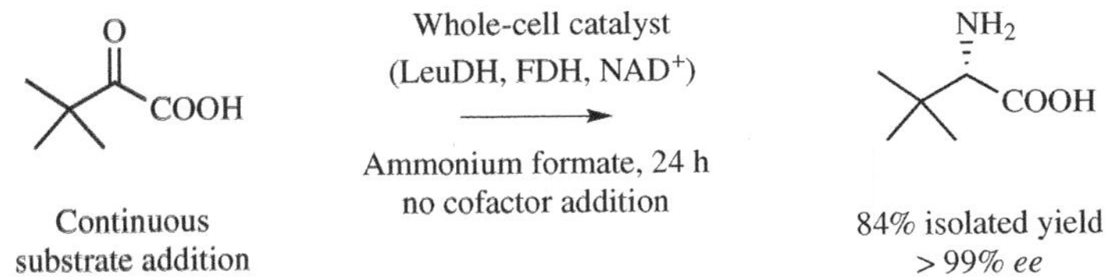

FIGURE 3.22 Whole-cell-catalyzed synthesis of L-*tert*-leucine on preparative scale.

Moreover, two different recombinant *E. coli* strains were constructed, both expressing FDH from *Mycobacterium vaccae* [334] in combination with either LeuDH [335] or PheDH [336] from *T. intermedius*. With these whole-cell-biocatalysts, L-leucine, L-valine, L-norvaline, and L-methionine were obtained in high yields (>88%) and excellent enantioselectivities (>99.9%) by combining LeuDH-FDH, L-phenylalanine, and L-tyrosine with PheDH-FDH. Amino acids were produced in concentrations up to 0.3 M, yet higher concentrations could not be obtained, presumably due to intracellular degradation of the cofactor [327].

In this area, enzymatic reductive amination of α-keto acids through the construction of a whole-cell biocatalyst was used by Degussa, as a cost-attractive alternative to their classic route to L-*tert*-Leucine, based on the application of expensive isolated and purified enzymes [337]. The reaction proceeded even in the absence of externally added NAD^+ cofactor, but substrate concentrations higher than 500 mM led to significant inhibitory effects with incomplete conversions. This was overcome by the addition of a low amount of cofactor (1–10 μM), or by fed-bath operation. The addition of 1.25 M solution of trimethylpyruvate resulted in an overall conversion of 95% after 24 h without the need of external cofactor addition. L-*tert*-Leucine was obtained in 84% yield and >99% *ee* (Figure 3.22).

The same whole-cell biocatalyst was used for the synthesis of L-neopentylglycine, with >95% conversion and >99% *ee* at substrate concentrations up to 88 gL^{-1}, also without the addition of external cofactor (Figure 3.23) [337].

Likewise, a biocatalytic cascade reaction was designed by Kroutil et al. for deracemization of mandelic acid to yield optically pure L-phenylglycine employing three enzymes [338]. The first step in the process was the oxidation of mandelic acid to the corresponding α-keto acid, which was transformed to the corresponding α-amino acid *via* asymmetric reductive amination. The enantiomers of the starting material were interconverted by a racemase (mandelate racemase) enabling that (at least theoretically) 100% of the substrate could be transformed into D-mandelic acid.

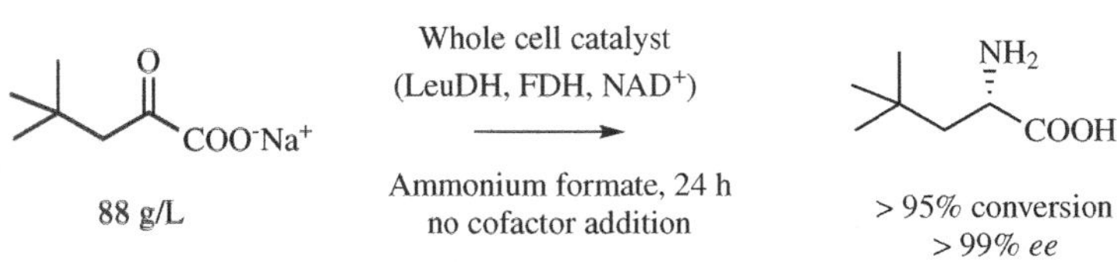

FIGURE 3.23 Whole-cell-based synthesis of L-neopentylglycine.

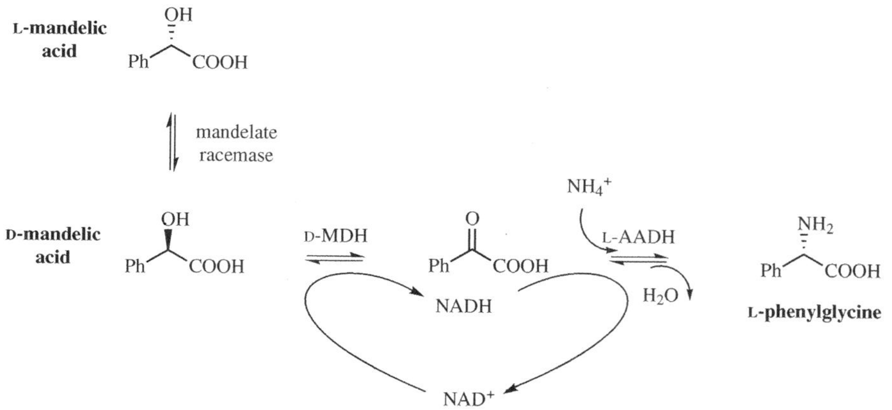

FIGURE 3.24 Deracemization of mandelic acid *via* a concurrent redox-cascade process to yield L-phenylglycine.

Thus, D-mandelic acid was transformed in the presence of D-mandelate dehydrogenase (D-MDH), L-AADH, adequate cofactor, and ammonium to yield the desired L-phenylglycine. Various L-AADHs were screened at substrate concentrations of 5 gL^{-1} on analytical scale. At substrate concentrations of 10 gL^{-1}, a conversion up to 76% was reached, whereby the amount of the intermediate keto acid was below 0.3% (Figure 3.24).

The first synthesis of D-amino acids *via* enzymatic reductive amination was reported in 2006. A highly stereoselective D-AADH was constructed by engineering of the enzyme *meso*-2,6-diaminopimelic acid D-hydrogenase from *Corynebacterium glutamicum*, using both, rational and random mutagenesis [339]. After three rounds of mutagenesis, mutants had a wide substrate range including linear-, branched-aliphatic, and aromatic amino acids. The very high selectivity of these enzymes toward the D-enantiomer (from 95 to >99% *ee*) resulted in a suitable alternative to classic methods for D-amino acid synthesis. Seven mutants displayed high activity, and particularly BC621—which showed an additional mutation, Asp155Gly—was the most active and displayed the broadest substrate range. BC621 was successfully assessed using cyclohexylpyruvate and GDH for cofactor regeneration (Figure 3.25).

A different strategy to obtain enantiopure amines through enzymatic methodologies is the reduction of imines [340,341]. New imine reductase activity was discovered in the anaerobic bacterium *Acetobacterium woodii* when screening a dynamic combinatorial library of virtual imine substrates, using a biphasic water-tetradecane solvent system [340]. Strains were grown in the presence and absence of caffeate, to produce cell preparations containing and lacking the enzyme needed for C–C bond reduction, respectively. This strategy allowed to assess whether or not the C–C reductase could be responsible for any C–N double bond reductions detected. The work targeted the reduction of aldimines as a model for establishing the principle of the screened approach and the identification of new biocatalytic C–N double bond reductions. Benzylidine aniline and butylidine aniline were reduced to the corresponding

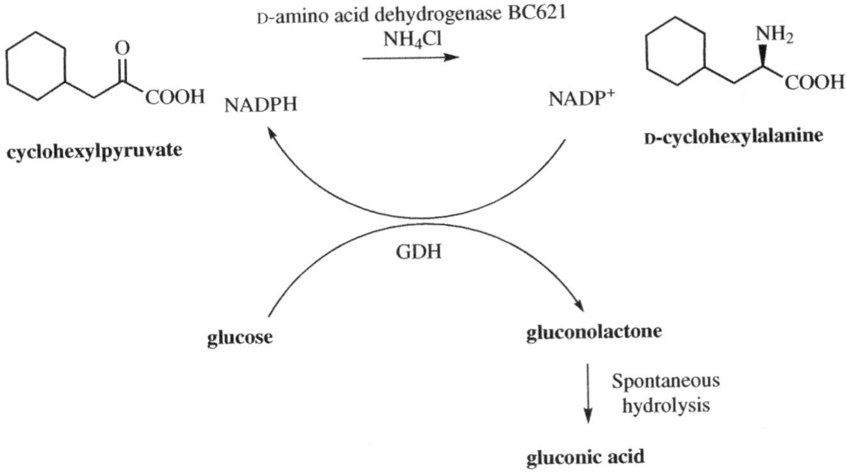

FIGURE 3.25 Reaction scheme for the gram-scale synthesis of D-cyclohexylalanine. GDH, glucose dehydrogenase.

amines by caffeate-induced cells, whereas noninduced cells only reduced butylidine aniline. In the same area, the reduction of arylimines using whole-cells of *C. parapsilosis* ATCC 7330 was described by Chadha and coworkers for the production of secondary (*R*)-arylamines from the corresponding arylimines in aqueous medium, with moderate-to-good yields (up to 80%) and excellent enantiomeric excesses (up to >99% *ee*) (Figure 3.26) [342]. This biocatalyst is well known for the asymmetric reduction of α- and β-keto esters and deracemization of α- and β-hydroxyesters [187,214,283,285,287,307,309,343]. The chemical analogy between the reduction of a carbonyl group and an imine encouraged the group to extend the scope of the biocatalyst, providing evidence of the catalytic promiscuity of *C. parapsilosis* carbonyl reductase.

3.5 REDUCTION OF C–C DOUBLE BONDS

An emergent area in the field of enzymatic redox-reactions is the reduction of C–C double bonds. The enzymes involved in these processes are NAD(P)H-dependent

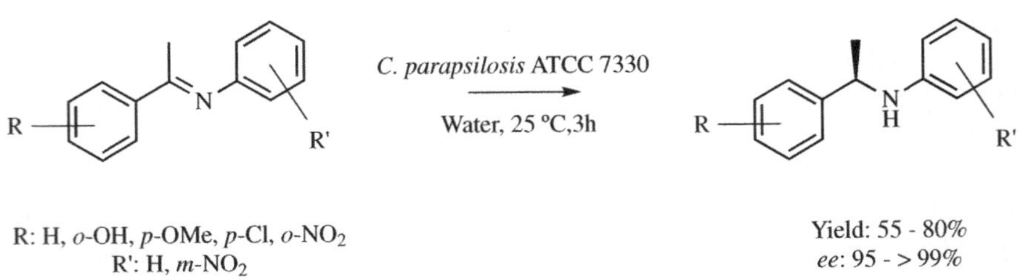

FIGURE 3.26 Asymmetric reduction of aromatic imines using *C. parapsilosis* whole-cells.

ERs, which selectively reduce C–C double bonds in α,β-unsaturated carbonyl compounds or nitriles, and thus can be employed to prepare enantiomerically pure aldehydes, ketones, esters, and nitriles [344]. Although ER from *Saccharomyces carlsbergensis*, the so-called old yellow enzyme, was isolated in 1933 [345], the use of ERs as biocatalysts in stereoselective C–C double bond reduction is still incipient [346–355]. The vast majority of biotransformations using ERs has been performed using growing cells, the most prominent being baker's yeast [356–360] and bacteria [355,361,362], to solve cofactor recycling issues. Albeit the achieved stereoselectivities were often good, the chemoselectivity of whole-cell bioreductions regarding C–C versus C–O double bonds reduction were usually poor, due to the presence of other competitive ADHs within the cell [363]. Since ERs and ADHs depend on the same nicotinamide cofactors, redox decoupling of both enzymatic activities is hardly feasible [363]. To overcome this, Bommarius and coworkers cloned and expressed three ERs in *E. coli*: XenA from *Pseudomonas putida*, KYE1 from *Kluyveromyces lactis* and Yers-ER from *Yersinia bercovieri* [346]. All three enzymes showed broad ER specificity, as well as temperature range and pH optima, but different specificity patterns. The enzymes preferred NADPH as cofactor over NADH, and were stable up to 40°C. Glucose and GDH was the system of choice for cofactor recycling during the cell-free biotransformations. The coupled system revealed that full conversions can be achieved when using high amounts of enzyme and short incubation times. For Yers-ER, conversions of >99% were achieved in 1 h for the reduction of 2-cyclohexenone. However, low enzyme loadings led to biocatalyst deactivation in the presence of the enoate substrate, being a drawback to be overcome when large-scale processes are envisaged. A wide range of α,β-unsaturated carbonyl compounds were tested as substrates for the three enzymes, including aliphatic and aromatic aldehydes and ketones (Figure 3.27).

Mechanistically, ER-catalyzed reactions proceed by a ping–pong mechanism consisting of an oxidative and a reductive half reaction [348,364–366], in which a hydride (derived from the flavin cofactor) is transferred stereoselectively onto C-β, while a Tyr residue adds a proton, ultimately derived from the solvent, onto C-α from the opposite side [363], resulting in the *anti*-addition of hydrogen atoms to the double bond (*trans*-hydrogenation) (Figure 3.28) [367–369]. The catalytic cycle of ERs is completed by the reduction of the oxidized flavin at the expense of NAD(P)H, which is derived from an external H-donor *via* another redox-reaction.

Herein, a rare case of reduction occurring in a *syn*-mode was described with a reductase isolated from cultured cells of *Nicotiana tabacum* (tobacco), which was used in the reduction of the C–C bond adjacent to the carbonyl group of the monoterpenic enones verberone and carvone [370]. The reduction proceeded stereospecifically by the *syn*-addition of hydrogen from the *Re*-face at the α- and β-positions to the carbonyl group. The hydrogen atoms participating in the enzymatic reduction at the α- and β-positions originated from the aqueous medium and the *pro*-4*S*-hydrogen of NADPH, respectively.

Three factors play important roles to determine the enantiospecificity of ERs-catalyzed reactions: (i) the location of substituents in the C–C double bond, for example, α-methyl-2-cyclohexenone yields the (*R*)-enantiomer, while the reduction

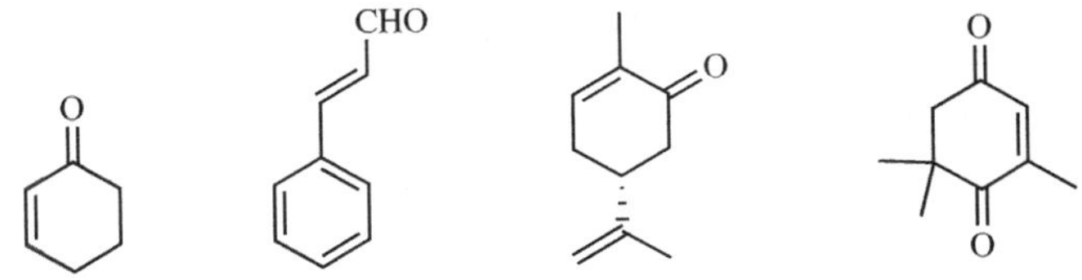

KYE1	**1.54**	**2.30**	**0.73**	**1.33**
XenA	**2.74**	**1.56**	**n.d.**	**1.84**
Yers-ER	**4.22**	**2.91**	**2.54**	**10.4**

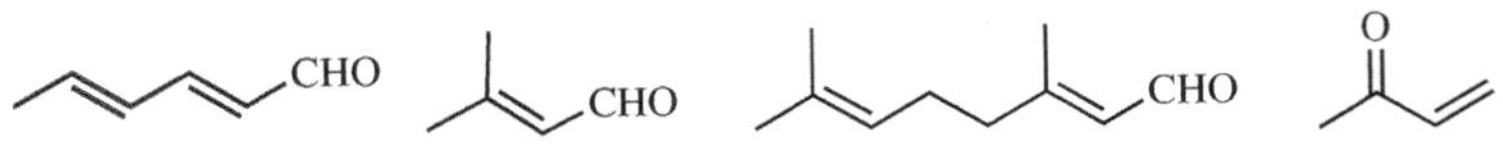

KYE1	**1.51**	**0.45**	**0.67**	**2.14**
XenA	**3.18**	**-0.43**	**1.75**	**22.69**
Yers-ER	**2.19**	**0.39**	**1.97**	**14.85**

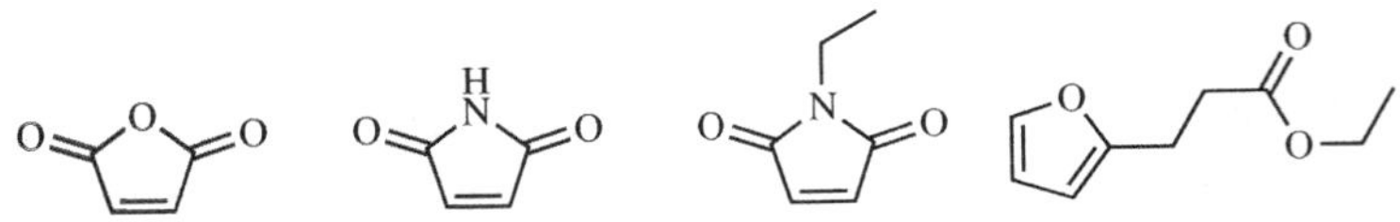

KYE1	**0.86**	**2.04**	**2.46**	**2.36**
XenA	**1.91**	**25.82**	**14.87**	**9.96**
Yers-ER	**5.20**	**18.88**	**15.81**	**12.87**

FIGURE 3.27 Specific activity (U/mg) of enoate reductases XenA from *P. putida*, KYE1 from *K. lactis* and Yers-ER from *Y. bercovieri* expressed in *E. coli* toward different α,β-unsaturated carbonyl compounds. n.d., not detected.

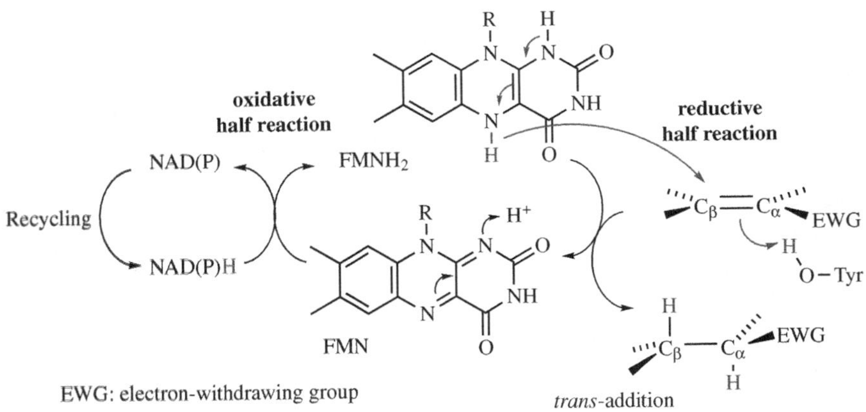

FIGURE 3.28 Catalytic mechanism of enoate reductases.

of β-methyl-2-cyclohexenone affords the (*S*)-enantiomer (Table 3.10, entries 1 and 2) [348,352]; (ii) the *E*- or *Z*- configuration of the double bond, which can be illustrated by different enantio-specificities observed for the reduction of (*E*)- or (*Z*)-citral (Table 3.10, entry 3) [351,355]; and (iii) the substitution on the double bond, because opposite enantiomers could be obtained from (*E*)-2-methyl-2-pentenal and (*E*)-2-methyl-3-phenyl-2-propenal (Table 3.10, entries 4 and 5) [351].

Three cloned ERs from the OYE family were investigated by Faber et al. in the asymmetric bioreduction of activated alkenes [347,349,371]. 12-Oxophytodienoate reductase isoenzymes OPR1 and OPR3 from *Lycopersicon esculentum* (tomato) and YqjM from *B. subtilis* displayed a remarkably broad substrate tolerance, reducing α,β-unsaturated aldehydes, ketones, maleimides, and nitroalkenes with absolute chemoselectivity, being only reduced the conjugated C–C double bond with excellent stereoselectivities (*ee* up to >99%), whereas isolated olefins and carbonyl groups remained intact (Table 3.11, entry 1). Both enantiomers of nitroalkenes were afforded with excellent stereochemical purity by choosing the appropriate enzyme: OPR1 and OPR3 or YqjM (Table 3.11, entry 2). Molecular modeling studies suggested that this enzyme-based stereocontrol is caused by subtle differences within the active site. The same stereoselectivity was shown by ER from *Zymomonas mobilis* and OYE1-3 from yeasts, achieving stereocontrol by variation of the substrate structure, by switching the (*E*/*Z*) geometry of the alkene or by choosing the adequate enzyme, thus allowing the access to the opposite enantiomeric products [348].

O-Protected cyclic acyloins represent a novel type of compounds obtainable by reduction of α-alkoxy enones using flavin-dependent ERs from the OYE family [371]. A series of α-alkoxy cyclopentenones and cyclohexenones were enzymatically reduced to afford the corresponding acyloins with enantiomeric excesses up to >97% (Table 3.11, entries 4 and 5), whereas β-analogs were not converted. The stereochemical outcome of the biotransformation was switched by variation of the *O*-protecting group (methyl, allyl, propyl, benzyl) or of the ring size of the substrate. Furthermore, cloned and overexpressed OYE1 from *Saccharomyces pastorianus* (formerly *carlsbergensis*) and OYE2-3 from *S. cerevisiae* were recently evaluated in the reduction of unsaturated α-chloroesters, yielding the corresponding saturated products in excellent conversions and high enantiomeric excesses (Table 3.11, entries 6 and 7) [372].

A practical enzymatic synthesis of a double chiral compound, (4*R*,6*R*)-4-hydroxy-2,2,6-trimethylcyclohexanone (**46**), starting from 2,6,6-trimethyl-2-cyclohexen-1,4-dione (**47**) was described by Wada et al. [353]. Chirality was induced at the former C-2 position (C-6 of the product) through a stereoselective reduction of the double bond by OYE2 from *S. cerevisiae* expressed in *E. coli*. Thereafter, the carbonyl group at the C-4 position was selective- and stereospecifically reduced by levodione reductase (LVR) of *Corynebacterium aquaticum* M-13 expressed in *E. coli*, to the corresponding alcohol (Figure 3.29). The designed two-step enzymatic asymmetric reduction system produced almost stoichiometric amounts of the desired final product with 94% *ee*.

Two applications for the asymmetric reduction of activated alkenes employing a library of fourteen commercially available ERs were reported by Kosjek and

TABLE 3.10 Enantiospecificity of ERs-Catalized Reductions of α,β-Unsaturated Carbonyl Compounds Regarding Substitution Patterns and Stereochemistry in the Double Bond

Entry	Substrate	Product	Conversion (%)	*ee* (%)	Enzyme	Reference
1	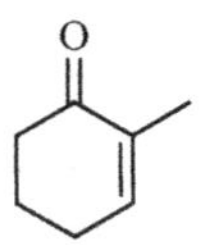	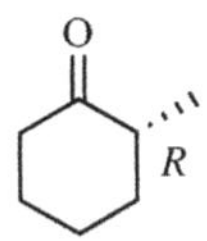	89	88	NCR	[348]
			78	89	OYE1	
			81	91	OYE2	
			78	98	OYE3	
			100	96	*E. coli* BL21(DE3) (pOYE-pET3b)	[352]
2	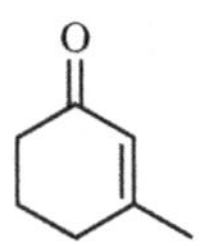	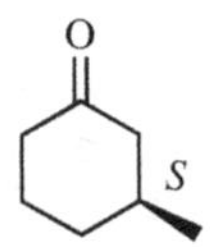	58	>99	NCR	[348]
			74	>99	OYE1	
			60	>99	OYE2	
			37	>99	OYE3	
			100	94	*E. coli* BL21(DE3) (pOYE-pET3b)	[352]
3	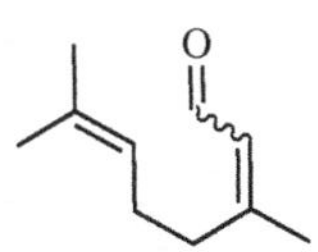 99.4% (*Z*) + 0.6% (*E*)	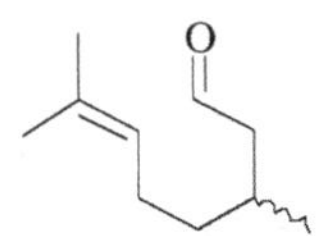	n.d.	>99 (*S*)	NCR	[351]
				10 (*S*)	OYE1	
				15 (*S*)	OYE2	
				75 (*S*)	OYE3	

	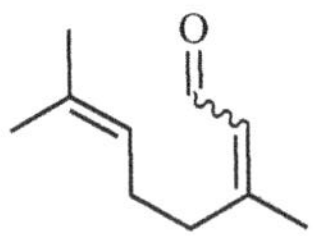 40.5% (Z) + 59.5% (*E*)		n.d.	>99 (*S*)	NCR	
				85 (*R*)	OYE1	
				85 (*R*)	OYE2	
				45 (*R*)	OYE3	
	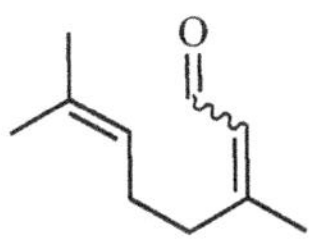 10.5% (Z) + 89.5% (*E*)		n.d.	>99 (*S*)	NCR	
				95 (*R*)	OYE1	
				95 (*R*)	OYE2	
				70 (*R*)	OYE3	
4	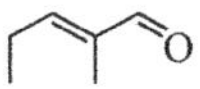	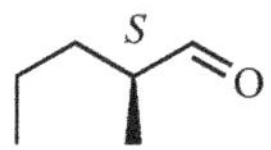	n.d.	75	NCR	[351]
				>99	OYE1	
				>99	OYE2	
				>99	OYE3	
5	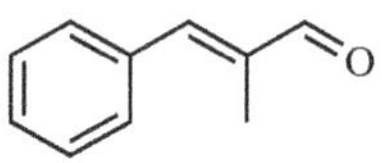	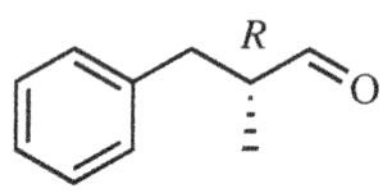	n.d.	50	NCR	[351]
				75	OYE1	
				70	OYE2	
				65	OYE3	

n.d., not determined; n.c., no conversion; NCR, NAD(P)H-dependent 2-cyclohexen-1-one reductase from *Zymomonas mobilis* expressed in *E. coli*; OYE1-3, old yellow enzymes from yeasts; BL21(DE3) (pOYE-pET3b), old yellow enzyme from *S. carlsbergensis* expressed in *E. coli*.

TABLE 3.11 Selected Examples of Substrate Specificity and Stereoselectivity in ERs-Catalyzed Reductions

Entry	Substrate	Product	Conversion (%)	*ee* (%)	Enzyme	Reference
1	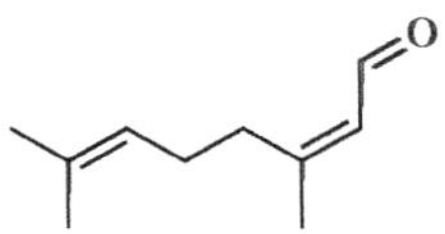	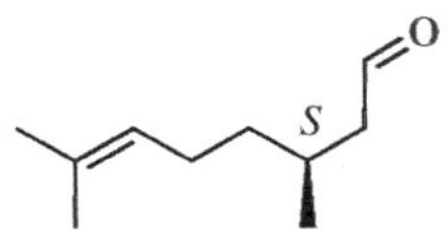	>99 90 70	>95 >95 >95	OPR1 OPR3 YqjM	[347]
2	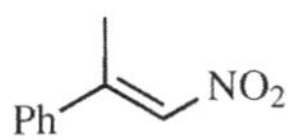	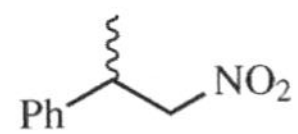	>99 69 94	97 (*R*) 82 (*S*) 92 (*S*)	OPR1 OPR3 YqjM	[347]
3	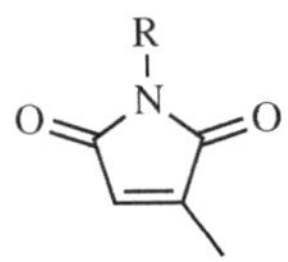	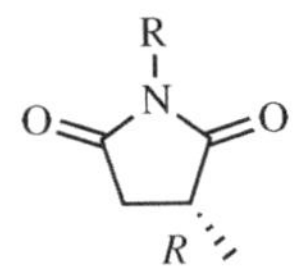	R=H: >99 R=Ph: 99	R=H: 99 R=Ph: >99	OPR1	[347]
			R=H: >99 R=Ph: 99	R=H: 99 R=Ph: >99	OPR3	
			R=H: >99 R=Ph: 99	R=H: 99 R=Ph: >99	YqjM	
4	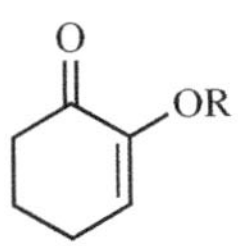	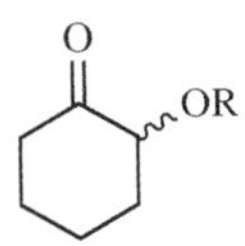	R=OMe: 22 R=O-allyl: 91 R=O-*n*Pr: 99 R=OBn: 98	R=OMe: <5 R=O-allyl: 77 (*S*) R=O-*n*Pr: 83 (*S*) R=OBn: 70 (*S*)	OPR1	[371]
			R=OMe: 9 R=O-allyl: 11 R=O-*n*Pr: 15 R=OBn: 56	R=OMe: 59 (*R*) R=O-allyl: 64 (*S*) R=O-*n*Pr: 38 (*S*) R=OBn: 64 (*S*)	OPR3	

			R=OMe: 31 R=O-allyl: 26 R=O-*n*Pr: 31 R=OBn: 91	R=OMe: 24 (*R*) R=O-allyl: 64 (*S*) R=O-*n*Pr: 62 (*S*) R=OBn: 71 (*S*)	YqjM	
			R=OMe: 37 R=O-allyl: <3 R=O-*n*Pr: <3 R=OBn: 13	R=OMe: 92 (*R*) R=O-allyl: n.d. R=O-*n*Pr: n.d. R=OBn: 87 (*S*)	OYE1	
			R=OMe: 26 R=O-allyl: <3 R=O-*n*Pr: <3 R=OBn: 12	R=OMe: 90 (*R*) R=O-allyl: n.d. R=O-*n*Pr: n.d. R=OBn: 87 (*S*)	OYE2	
			R=OMe: 25 R=O-allyl: <3 R=O-*n*Pr: <3 R=OBn: 6	R=OMe: 84 (*R*) R=O-allyl: n.d. R=O-*n*Pr: n.d. R=OBn: 67 (*S*)	OYE3	
5	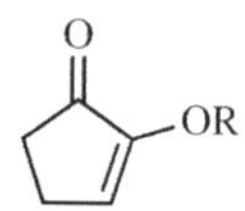	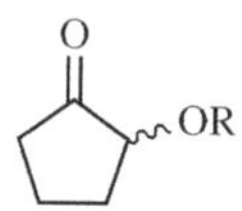	R=OMe: 25 R=OBn: 98	R=OMe: 66 (*S*) R=OBn: 95 (*S*)	OPR1	[371]
			R=OMe: n.c. R=OBn: 5	R=OMe: n.d. R=OBn: 99 (*S*)	OPR3	
			R=OMe: <3 R=OBn: 3	R=OMe: 58 (*S*) R=OBn: 92 (*S*)	YqjM	

(*Continued*)

TABLE 3.11 (*Continued*)

Entry	Substrate	Product	Conversion (%)	*ee* (%)	Enzyme	Reference
			R=OMe: 10 R=OBn: <3	R=OMe: 58 (*S*) R=OBn: n.d.	OYE1	
			R=OMe: <3 R=OBn: <3	R=OMe: n.d. R=OBn: n.d.	OYE2	
			R=OMe: <3 R=OBn: <3	R=OMe: n.d. R=OBn: n.d.	OYE3	
6	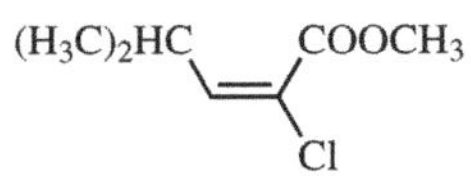	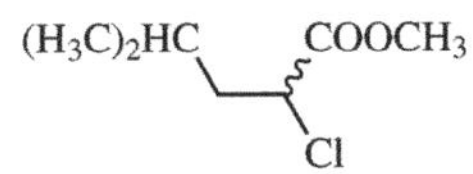	13 6 92	n.d. n.d. 79 (*R*)	OYE1 OYE2 OYE3	[372]
7	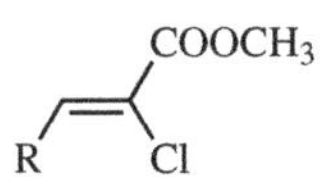	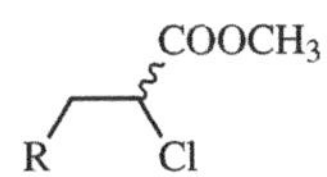	R=$CHCl_2$: 100 R=$CH(CH_3)_2$: 90 R=CCl_3: 78	R=$CHCl_2$: n.d. R=$CH(CH_3)_2$: 91 (*S*) R=CCl_3: 95 (*S*)	OYE1	[372]
			R=$CHCl_2$: 100 R=$CH(CH_3)_2$: 91 R=CCl_3: 95	R=$CHCl_2$: n.d. R=$CH(CH_3)_2$: 82 (*S*) R=CCl_3: 94 (*S*)	OYE2	
			R=$CHCl_2$: 100 R=$CH(CH_3)_2$: 89 R=CCl_3: 95	R=$CHCl_2$: n.d. R=$CH(CH_3)_2$: 83 (*S*) R=CCl_3: 94 (*S*)	OYE3	

n.d., not determined; n.c., no conversion.
OPR1 and OPR3 from *L. esculentum* (tomato).
YqjM from *B. subtilis*.
OYE1 from *S. carlsberensis*.
OYE2 and OYE3 from *S. cerevisiae*.

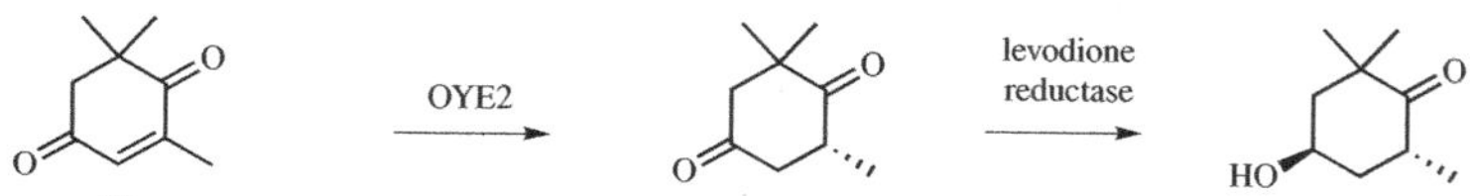

47 **46**

2,6,6-trimethyl-2-cyclohexen-1,4-dione (ketoisophorone) **(6*R*)-2,2,6-trimethylcyclohexane-1,4-dione ((6*R*)-levodione)** **(4*R*,6*S*)-4-hydroxy-2,2,6-trimethyl-cyclohexanone (actinol)**

FIGURE 3.29 Two-step conversion of 2,6,6-trimethyl-2-cyclohexen-1,4-dione (**47**) to (4*R*,6*R*)-4-hydroxy-2,2,6-trimethylcyclohexanone (**46**) using OYE and LVR expressed in *E. coli* as biocatalysts.

coworkers [350]. A series of α,β-unsaturated nitriles were converted to the optically active nitrile products in high yields and excellent enantioselectivities (up to 99% *ee*). Likewise, the reduction of 2,3-disubstituted cyclopentenones provided almost exclusively *trans*-2,3-disubstituted cyclopentanones in high yields and with enantiomeric excesses (Figure 3.30). All enzymes consistently displayed the same facial selectivity giving the corresponding (*R*)-enantiomer. These results pointed out the applicability of ERs for the reduction of structurally complex α,β-unsaturated nitriles and ketones. The process was performed using phosphite/phosphite dehydrogenase for cofactor recycling, providing scope for efficient and scalable processes for the asymmetric reduction of conjugated olefins.

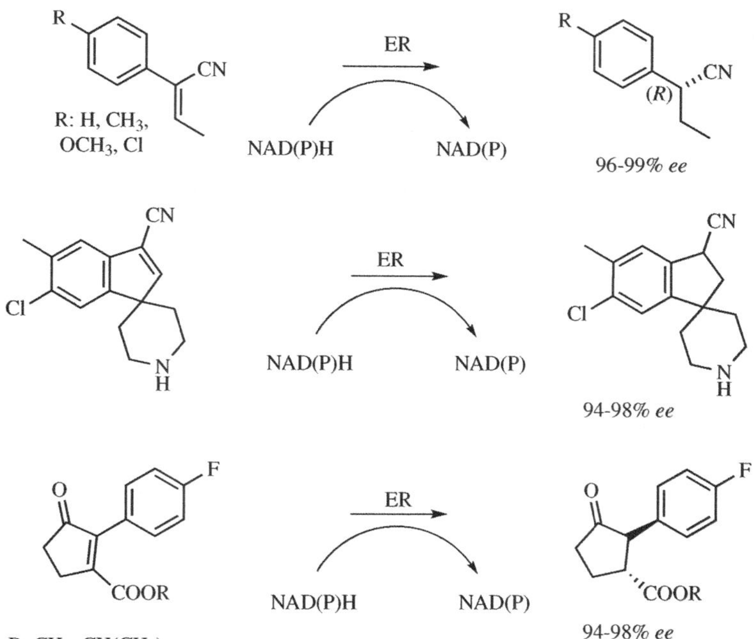

FIGURE 3.30 Asymmetric reduction of α,β-unsaturated nitriles and 2,3-disubstituted ketones using commercially available ERs [350].

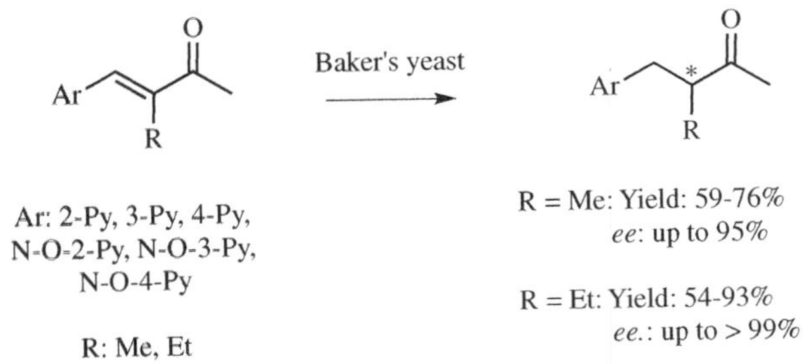

FIGURE 3.31 Stereoselective BY-mediated reduction of α-alkyl-β-heteroarylenones.

The use of whole-cells of yeast and fungal strains for enantioselective reductions of C–C double bonds of enones is well known [355,360,373,374]. The reduction of α-alkyl-β-heteroarylenones was reported with baker's yeast [360]. Although enones have two functional groups susceptible to be reduced—a C–C double bond and a carbonyl group—baker's yeast chemoselectively catalyzed the reduction of the C–C double bond of enones having a pyridyl ring or their *N*-oxide-derivatives to afford the corresponding chiral ketones with excellent stereoselectivity (Figure 3.31). The major enantiomers of yeast reduction products had (*S*)-configuration, except for β-(2-pyridyl-*N*-oxide)-α-methyl enone, which gave the (*R*) ketone in 71% *ee*. The reduction of a series of pyridyl-enones revealed the crucial factors on the stereoselectivity. With regard to β-substituents, the selectivity was induced by a specific interaction, such as an electrostatic or hydrogen bonding interactions, as well as steric factors, such as the length of the α-substituent or the position of the substituent in the β-phenyl ring. For α-substituents, steric factors were overridingly important.

Finally, fungal strains were also used by Zacchino et al. for the chemoselective reduction of nonsubstituted and substituted maleimides to the corresponding succinimides [374]. For instance, 2- and 2,3-methylated *N*-phenylmaleimides were enantioselectively reduced to (*R*)-*N*-phenyl-3-methylsuccinimide and to *trans*-(*R*,*R*)-*N*-phenyl-2,3-dimethylsuccinimide, respectively, with 99% *ee* by *Aspergillus niger*, *A. flavus* and *A. fumigatus*, with conversions from 96% to 99%, and by *Fusarium graminearum* and *Penicillium* sp. with conversions of 37–39%.

3.6 OXIDATION REACTIONS

Dehydrogenases catalyze the reversible oxidation of alcohols and amines, being the primary oxidant the oxidized nicotinamide cofactor $NAD(P)^+$ [375]. Historically, their use in oxidations was not especially relevant, though in the last decade, in the quest for greener and more selective alternatives to traditional oxidation methods, the use of dehydrogenases in the oxidation of alcohols has markedly increased [94,376–380]. The enzymatic alcohol oxidation proceeds under mild conditions that may avoid secondary reactions such as aldol condensations, and without using toxic reagents such as metal salts and oxides.

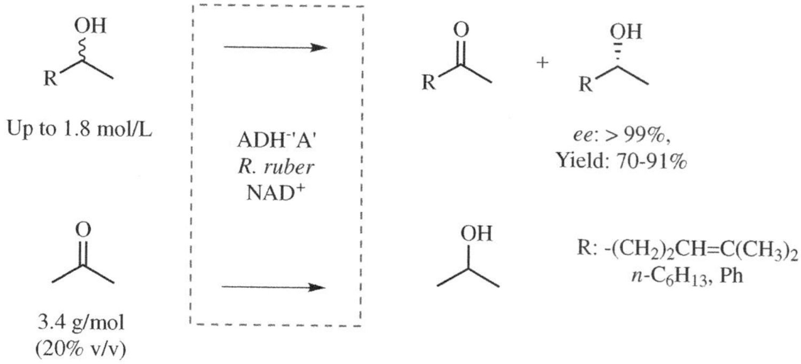

FIGURE 3.32 Preparative-scale alcohol oxidation and simultaneous NAD^+ regeneration by ADH"A" from *R. ruber*.

As stated before, HLADH is one of the most widely used oxidoreductases. This NADH-dependent enzyme exhibits a broad tolerance for primary or secondary alcohols as substrates, and an almost invariable stereospecificity, which turns the enzymatic process highly predictable [193,376,381,382]. The enzyme was used for desymmetrization of prochiral *meso*-diols due to its ability to discriminate between enantiotopic hydroxyl groups [383]. In combination with HLADH, GluDH [384,385] and monooxygenases [386] are commonly used for enzymatic NAD^+ regeneration. Yeast alcohol dehydrogenase (YADH) is structurally related to HLADH, although it is less stable and shows a low resistance toward organic solvents. Its substrate spectrum is limited to primary alcohols and 2-alkanols [376]. Conversely, TBADH is a thermostable NADP-dependent dehydrogenase, showing high activity toward open-chain secondary alcohols, but low activity toward primary alcohols [381]. Alcohol dehydrogenases from *R. erythropolis*, *Pseudomonas* sp., and *L. kefir* accept a broad range of open-chain substrates, including a variety of useful compounds for synthetic purposes [376].

Enzyme-catalyzed alcohol oxidations are mostly used in kinetic resolution of racemic mixtures. The use of single ADHs together with an ancillary ketone as the hydrogen acceptor for cofactor recycling [94] was not feasible on a preparative scale until recently, because of the instability of ADHs in the presence of acetone. However, the $NADP^+$-dependent ADH from *R. ruber* DSM 44541 converts a number of *sec*-alcohols into the corresponding ketones, and also accepts acetone as cosubstrate for $NADP^+$ regeneration, by producing 2-propanol as coproduct (Figure 3.32) [242,387]. The enzyme is highly stable toward elevated cosubstrate concentrations, allowing the use of acetone 20% v/v with whole-cells, and up to 50% v/v with the partially purified enzyme [388,389]. Moreover, high acetone concentrations not only shift the equilibrium toward oxidation, but also increase the solubility of lipophilic substrates, thus enhancing reaction rates.

Excellent productivities were achieved with ADH-'A' in the oxidation mode. Substrate loadings of *rac*-2-octanol of up to 111 gL^{-1} provided a space–time yield of 51 $gL^{-1}d^{-1}$ in batch conditions, with exclusive regioselectivity for secondary over primary alcohols, and preference for the (*S*)-isomer. Preferred substrates for the enzyme include secondary alcohols containing a small substituent (methyl, ethyl) and a large group in the stereocenter (up to C-10 saturated or unsaturated, cyclohexyl, heteroaryl, alkylaryl, or *p*- or *m*-substituted phenyl) [390,391].

Enantioselective oxidation of phenylethanol derivatives were performed by Nakamura and coworkers with immobilized resting cells of *G. candidum* on a water-absorbing polymer, and in a hexane–water biphasic system [99,198], being (*S*)-alcohols selectively oxidized to the corresponding ketones. The biotransformations were carried out immobilizing the cells on a water-absorbing polymer, and in a hexane–water biphasic system; cyclohexanone was used as additive to improve the stereoselectivity. Furthermore, Molinari et al. described the use of acetic acid bacteria (strains of *Acetobacter* and *Gluconobacter* sp.) for the biotransformation of different primary alcohols (2-chloropropanol and 2-phenylpropanol) and diols (1,3- and 2,3-butanediol and 1,4-nonanediol) into the corresponding oxidized products (Table 3.12) [392]. 2-Chloropropanol and 1,3-butanediol were oxidized with good rates but low enantioselectivities (*ee* 18–46% of the (*R*)-acid from 2-chloropropanol, entry 1, and up to 22% *ee* from 1,3-butanediol of the (*S*)-acid, entry 2), while microbial oxidation of 2-phenylpropanol furnished (*S*)-2-phenylpropionic acid with >90% *ee* (Table 3.12, entry 3). The oxidation of 2,3-butanediol strongly depended on the stereochemistry of the substrate. The *meso*- form gave (*S*)-acetoin with all tested strains (Table 3.12, entry 4), with the only exception being a *Gluconobacter cerinus* strain, which provided the (*R*)-isomer (Table 3.12, entry 5). The (*S*)-acetoin was also obtained through the oxidation of (2*S*,3*S*)-butanediol with up to 84% conversion (Table 3.12, entry 6). The exhaustive oxidation to 2,3-butanedione was only obtained from the (2*R*,3*R*)-isomer with *A. hansenii* and *Acetobacter xylinus* with conversions of 30% and 5%, respectively (Table 3.12, entry 7).

A series of microorganisms that assimilate 2-phenylethanol were screened from soil, selecting strains that enantioselectively oxidized 2-phenyl-1-propanol to 2-phenylpropionic acid [393]. By optimizing the culture conditions as well as the reaction system, various enantiomerically enriched 2-aryl-1-propanols and the corresponding propanoic acids were obtained in moderate-to-good yields and good-to-excellent enantiomeric excesses using *Brevibacterium* sp. KU1320 (Table 3.12, entry 8). The (*R*)-specific alcohol dehydrogenase from *L. brevis* was employed to perform the enantioselective oxidation of racemic phenylethanol to acetophenone and (*S*)-phenylethanol with NAD^+ or $NADP^+$ regeneration-mediated by NADH oxidase from *Lactobacillus sanfranciscensis*, which accepts both nicotinamide cofactors as substrates [394,395]. Wild-type (*R*)-ADH only accepts $NADP^+$ whereas its Gly37Asp mutant strongly prefers NAD^+. Purified NADH oxidase was coupled with wild-type ADH from *L. brevis* on NADP(H) and mutant ADH on NAD(H) to achieve 50% conversion, leaving the (*S*)-alcohol unreacted (Table 3.12, entry 9). Depending on the relative concentration of alcohol and cofactor, more than 100 turnovers were achieved. Isolated ADHs as well as whole microbial cells were tested in a hydrogen transfer reaction employing acetaldehyde, chloroacetone, and acetone as hydrogen acceptors, for the oxidation of benzyl alcohol to benzaldehyde. Excellent conversions (95%) were obtained employing lyophilized cells of *Janibacter terrae*

TABLE 3.12 Dehydrogenase-Mediated Oxidation of Primary and Secondary Alcohols and Diols

Entry	Substrate	Product	Conversion (%)	*ee* (%)	Reference
1	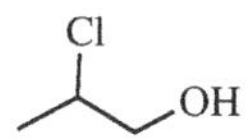	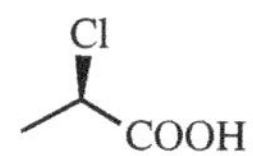	20–50	18–46 (*R*)	[392]
2	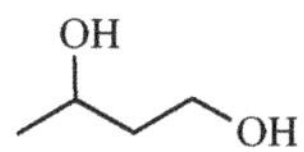	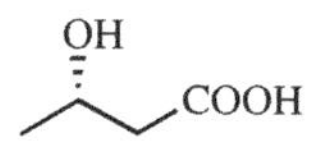	Up to 55	Up to 22 (*S*)	[392]
3	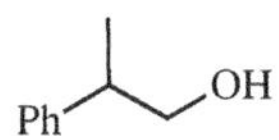	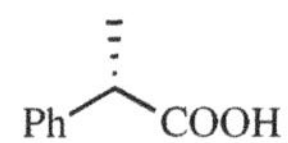	Up to 45	90–97 (*S*)	[392]
4	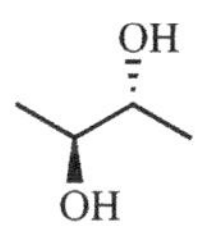	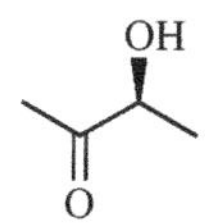	Up to 100	40–97 (*S*)	[392]
5	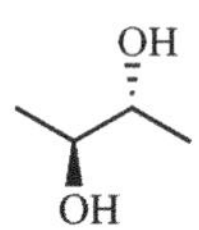	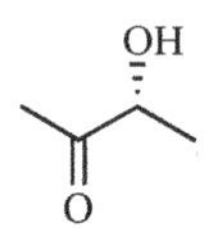	15	72 (*R*)	[392]

(Continued)

TABLE 3.12 (*Continued*)

Entry	Substrate	Product	Conversion (%)	*ee* (%)	Reference
6	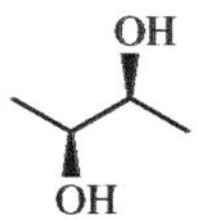	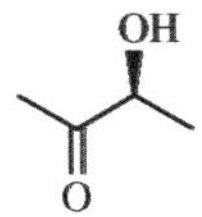	Up to 84		[392]
7	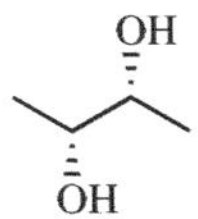	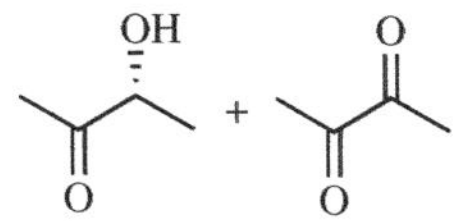	Acetoin: up to 97 2,3-butanedione: 30 (*A. hansenii*) 5 (*A. xylinus*)		[392]
8	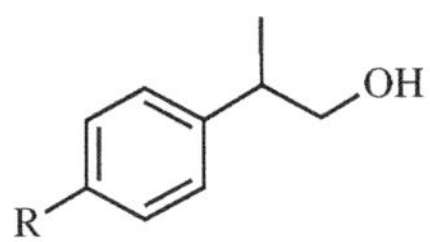 R: H, CH_3, OCH_3, Cl R: H, CH_3, OCH_3, Cl	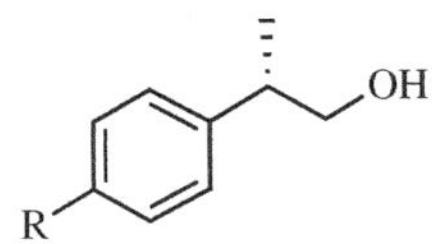	Up to 40	Up to 95 (*S*)	[393]
9	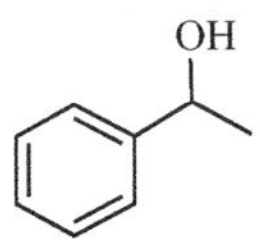	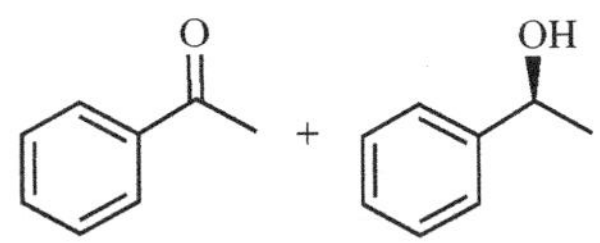	50	>99	[394]
10	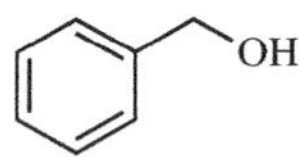	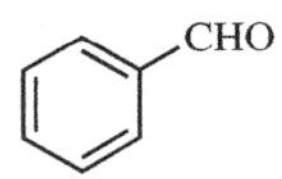	95		[395]

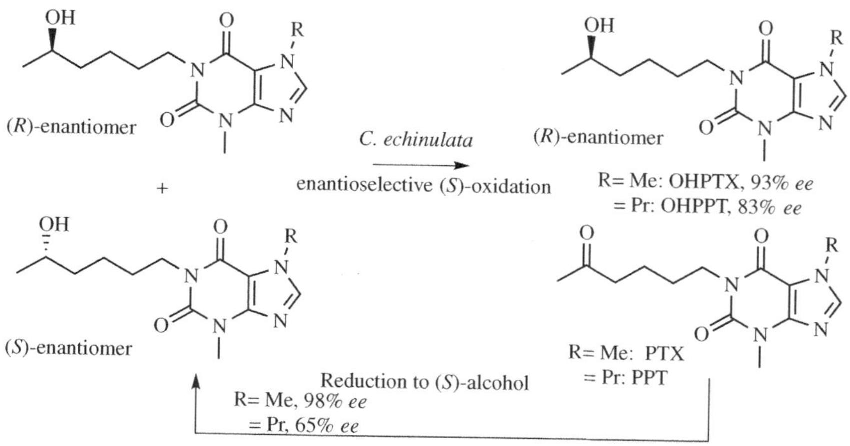

FIGURE 3.33 Biotransformation of PTX, PPT, (±)-OHPTX, and (±)-OHPPT. PTX, 1-(5-oxohexyl)-3,7-dimethylxanthine; PPT, 1-(5-oxohexyl)-3-methyl-7-propylxanthine.

DSM 13953 at a substrate concentration of 97 mM (Table 3.12, entry 10) [396]. The filamentous fungus *Cunninghamella echinulata* NRRL 1384 was used as catalyst for the (*S*)-reduction of pentoxifylline (PTX: 1-(5-oxohexyl)-3,7-dimethylxanthine) and propentofylline (PPT: 1-(5-oxohexyl)-3-methyl-7-propylxanthine), phosphodiesterase inhibiting derivatives of theobromine, and for the (*S*)-oxidation of their racemic hydroxy metabolites (±)-OHPTX and (±)-OHPPT (Figure 3.33) [397]. The oxidation step was selective and relatively fast, while the reduction step was slower and more selective for PTX than for PPT. The bioconversion of (±)-OHPTX gave the (*R*)-enantiomer (lisofylline) with higher enantiopurity (up to 93% *ee*) compared to the bioconversion of (±)-OHPPT, with maximum *ee* values recorded at 83%, and similar yields.

Usually, due to the intrinsic chirality of enzymes (high enantioselectivity), a kinetic resolution occurs, delaying or even disabling the oxidation of the second isomer. Performing the oxidation with two enzymes showing opposite stereo-preference could be a suitable alternative, yet being enzymes with opposite enantio-preferences not always accessible. Herein, for the oxidation of both isomers of alcohols with at least one small alkyl group, a single nonstereoselective short-chain dehydrogenase/reductase from *Sphingobium yanoikuyae* was identified and overexpressed in *E. coli*. The enzyme oxidizes the substrate and also recycles the cofactor, thus working in a coupled substrate approach. It showed high versatility accepting different types of cosubstrates in the active site, such as acetone, methylisobutyl ketone (MIBK), acetaldehyde, and pyruvate. Employing acetone as hydrogen acceptor in large scale (10 eq.), *rac*-1-octanol was efficiently oxidized (99% conversion). Substrate concentrations up to 400 gL^{-1} were tolerated, without significant deviations of the space–time yield with acetone. Moreover, the use of selected ketones as hydrogen acceptors—which can only be reduced but not oxidized, such as α-haloketones, since the resulting halohydrins are not substrates for ADHs—turns the process irreversible (Table 3.13) [398]. Thus, only one equivalent of oxidant (or 1.5) was required instead of a large excess.

TABLE 3.13 Biocatalytic Hydrogen Transfer for Oxidation of Secondary Alcohols to the Corresponding Ketones

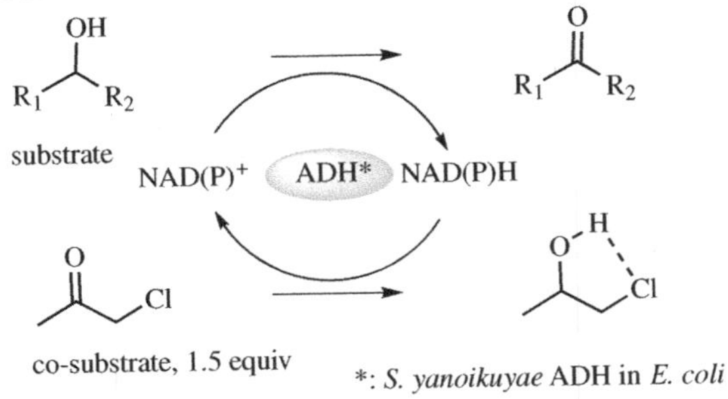

R_1	R_2	Conversion (%)
$(CH_2)_3CH_3$	CH_3	>99
$(CH_2)_4CH_3$	CH_3	95
$(CH_2)_5CH_3$	CH_3	>99
$(CH_2)_6CH_3$	CH_3	97
$(CH_2)_7CH_3$	CH_3	70
$(CH_2)_2CH{=}C(CH_3)_2$	CH_3	90
Phenyl	CH_2CH_3	55

The additional stabilization gained by the formation of an intramolecular hydrogen bond between the newly formed alcohol and the electronegative group (3–4 kcal mol^{-1}) was invoked to account for the quasi-irreversible oxidation process. With this strategy, the use of a nonstereoselective short chain ADH (in lyophilized *E. coli* cells) and chloroacetone as hydrogen acceptor (1.5 eq.) allows high yields in the oxidation of racemic secondary alcohols.

Regiospecific oxidation of polyols can be accomplished enzymatically using ADHs [377,399–404]. As examples, regioselective oxidation at C-5 of L-glucitol to give the rare sugar D-sorbose was catalyzed by L-glucitol dehydrogenase harbored by *Pseudomonas* sp. and the oxidation product was obtained in 35 h with 97% yield [402]. Similarly, L-tagatose was prepared with an overall yield of 78% by enzymatic C-5 oxidation of galactitol with L-lactate dehydrogenase-catalyzed cosubstrate regeneration, using partially purified galactitol dehydrogenase from the phototrophic bacterium *Rhodobacter sphaeroides* D, which is a mutant of the wild type *R. sphaeroides* Si4 (DSM 8371) [403]. The optimization of L-sorbose synthesis by regiospecific C-5 dehydrogenation of D-sorbitol using *G. oxidans* was also reported [404]. The process led to a theoretically maximal L-sorbose productivity of 200 gL^{-1}.

The microorganisms *P. pastoris*, *Pichia membranaefaciens*, *C. boidinii* and *Aureobasidium* sp. were screened for their ability to use amines as a nitrogen source, with different amine acceptance depending on the organism [405]. While *P. pastoris* used α-unsubstituted amines, the other three strains used α-substituted amines like cyclohexyl, 2-butyl and 1-phenylethyl amines. Biotransformations occurred *via* an oxidative pathway leading to the production of the corresponding ketones. Kinetic resolution of racemic 1-phenylethylamine showed that the optically pure (*S*)-enantiomer was obtained with *Aureobasidium* sp., and *P. membranaefaciens* or *C. boidinii* produced the (*R*)-enantiomer, but with lower stereoselectivity (52% < *ee* < 70%).

3.7 DEHYDROGENASE-CATALYZED REDOX REACTIONS IN NATURAL PRODUCTS

Among the complex chemical structures that can be found in Nature, steroids are probably the most widely investigated family of bioactive compounds. Their rigid tetracyclic structure as well as their sensitive functional groups has lured synthetic organic chemists as attractive targets to develop new synthetic reactions [42]. In Nature, the regio- and stereoselective interconversion of secondary alcohols to ketones on the steroid nucleus, as well as on the side chains, is catalyzed by hydroxysteroid dehydrogenases (HSDHs), a group of NAD(P)H-dependent redox enzymes [406]. The most extended synthetic applications of these enzymes have been related to the selective transformations of bile acids, such as cholic acid (**48**), chenodeoxycholic acid (**49**), and ursodeoxycholic acid (**50**) (Figure 3.34) [407].

Due to their carboxylic group, bile acids are water soluble at mild alkaline pH, and therefore enzymatic redox reactions on these substrates have been mainly performed in aqueous solutions [408]. A number of regio- and stereospecific reductions of steroids have been performed with selected HSDHs [409–421]. Dehydrocholic acid (**51**) was selectively reduced by different HSDH with *in situ* cofactor regeneration by either the GDH or the FDH system (Figure 3.35) [409,410]. The reduced products were obtained quantitatively and, by using deuterated substrate donors for cofactor regeneration such as a system involving glucose-1-D and GDH, deuterated steroids could be selectively synthesized [422]. Similarly, cholic acid was regioselective and quantitatively oxidized at each of the three possible hydroxyl groups, using GluDH for cofactor recycling [409]. The same regio- and stereospecific oxidations and reductions were more recently carried out with HSDHs isolated from *Pseudomonas paucimobilis* and *Xanthomonas maltophilia* [411,412].

The enzymatic oxidation of cholic acid to 12-ketochenodeoxycholic acid, a key intermediate in the industrial chemical route to bioactive ursodeoxycholic acid, was performed in one step using 12α-HSDH as biocatalyst, on a multi-kilogram scale in a membrane reactor containing 4% w/v solution of cholic acid [409,423,424]. Enzymes

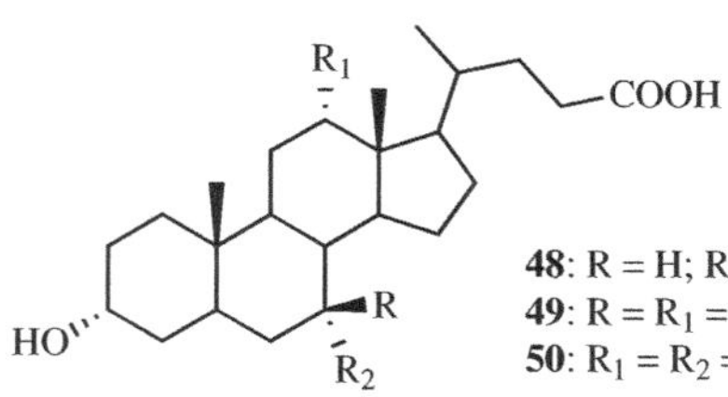

48: R = H; $R_1 = R_2$ = OH
49: R = R_1 = H; R_2 = OH
50: $R_1 = R_2$ = H; R = OH

FIGURE 3.34 Structure of bile acids.

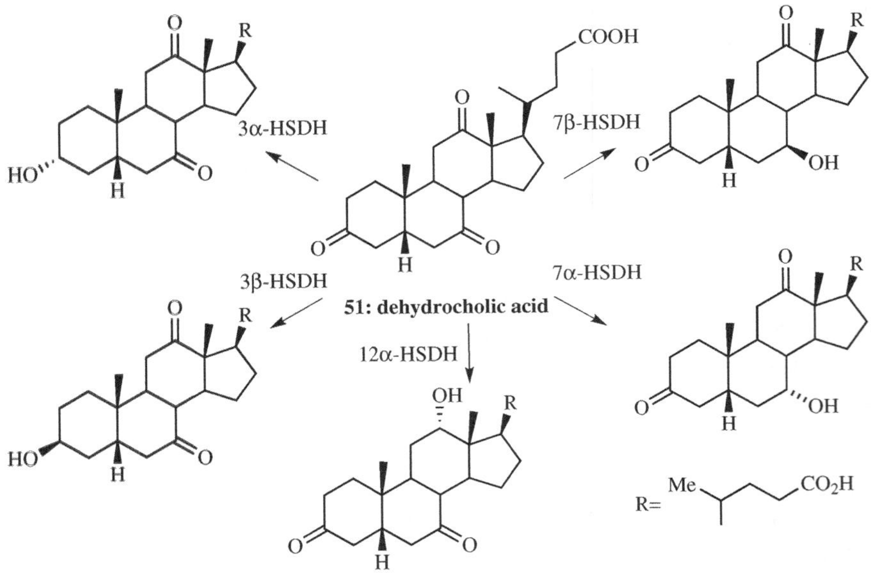

FIGURE 3.35 Regio- and stereospecific reduction of dehydrocholic acid **51** catalyzed by different HSDHs.

were retained inside the reactor and reused up to 50 cycles. Despite the moderate stability of HSDHs in two-phase systems, even at low concentrations of the organic solvent in the aqueous phase, biphasic systems have been used to overcome the very low water solubility of neutral steroids [425], allowing the specific and complete oxidation of hydroxyl groups of steroids or the efficient reduction of keto-steroids [426,427]. Besides their natural steroidal substrates, HSDHs also accept a broad variety of structurally different compounds [42]. Racemic muscarine precursors **52** and **53** were stereoselectively reduced by 20β-HSDH and β-HSDH with high-to-excellent enantiomeric excesses (Figure 3.36) [428].

Other examples are the $NADP^+$-dependent morphine-6-dehydrogenase (MDH, EC 1.1.1.218) and the NADH-dependent morphinone reductase (MR), which have been used in the preparation of semisynthetic analgesic opiates from the naturally occurring alkaloids morphine (**54**) and codeine (**55**) [89]. MDH and MR from *P. putida* M10 were cloned and coexpressed in *E. coli* to transform morphine and codeine into hydromorphone (**56**) and hydrocodone (**57**). However, further reduction of the keto group in **56** and **57** gave significant amounts of the corresponding dihydro derivatives. The use of a third enzyme to improve cofactor recycling, such as GluDH, which accepts NADH and NADPH for the regeneration of both cofactors, was proposed [89]. Finally, the efficient regeneration of the cofactors was accomplished by using a soluble pyridine nucleotide transhydrogenase (STH), which can transfer reducing equivalents between NAD^+ and $NADP^+$. One of the recombinant *E. coli* mutants expressing the three enzymes at different levels converted morphine into hydromorphone (**56**) in 67% yield over several cycles of reuse (Figure 3.37) [90].

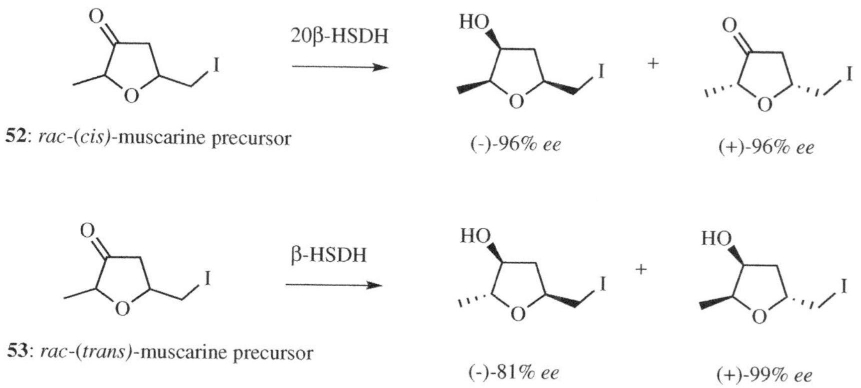

FIGURE 3.36 Stereoselective reduction of racemic muscarine precursors **52** and **53** by HSDHs.

Several examples related to the dehydrogenase-catalyzed selective oxidoreduction of sugars can be found in the literature [42]. UDP-glucuronic acid (**58**) (UDP-GlcA) is the activated sugar donor of glucuronyltransferases, which in Nature catalyzes the glucuronidation of xenobiotics, and of hyaluronic synthase, the enzymes responsible for the synthesis of the polysaccharide hyaluronic acids. Nevertheless, UDP-GlcA is quite expensive, and has strong substrate inhibitory effects on the enzyme, in the

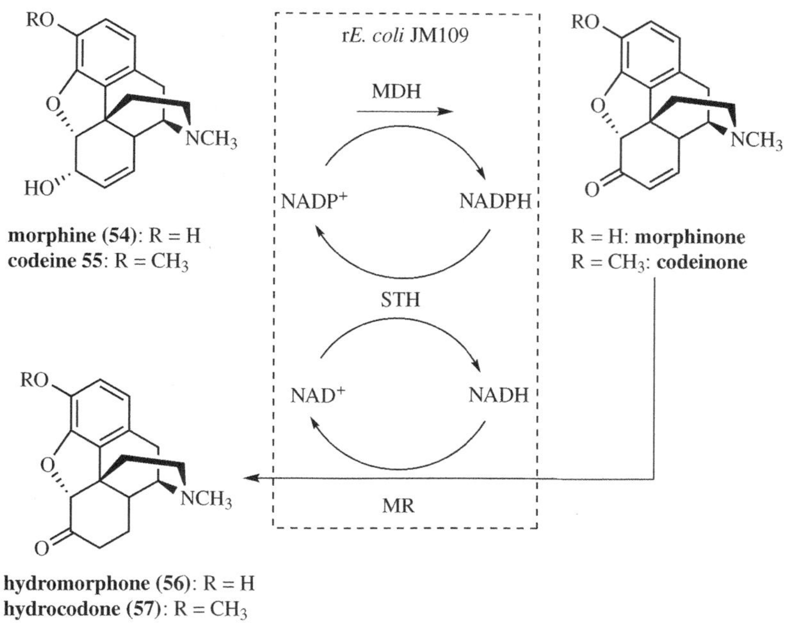

FIGURE 3.37 Whole-cell synthesis of hydromorphone (**56**) and hydrocodone (**57**). MDH: morphine-6-dehydrogenase; MR: morphinone reductase; STH: soluble pyridine nucleotide transhydrogenase.

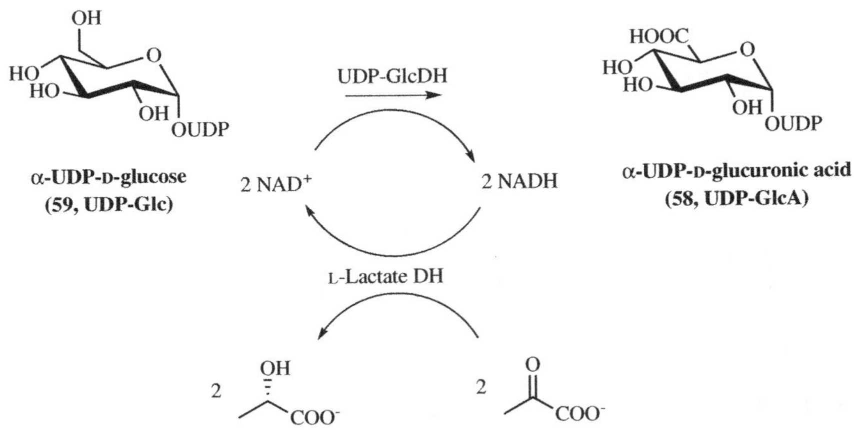

FIGURE 3.38 Preparative-scale synthesis of UDP-glucuronic acid. UDP-GlcDH: Uridine-diphosphate glucose dehydrogenase.

modulation of the *in vivo* synthesis of glucuronates. Uridine diphosphate GDH (UDP-GlcDH, EC 1.1.1.22) provides an interesting tool for the oxidation UDP-glucose (**59**) (UDP-Glc) to UDP-GlcA, and it was used *in vitro* to catalyze the reaction in multigram scale, coupled with lactate dehydrogenase for the *in situ* regeneration of NAD^+ (Figure 3.38) [81,82].

The *in situ* regeneration of UDP-GlcA in combination with glucuronyltransferases has also been reported [83]. β-D-glucuronides **60** and **61** (Figure 3.39) were stereoselectively synthesized by a multienzyme system with *in situ* regeneration of uridine-5′-diphosphoglucuronic acid in a one-pot reaction, using crude liver homogenates containing all enzymes involved. The glucuronidation started with catalytic amounts of UDP-glucose, which was oxidized to UDP-glucuronic acid by the NAD-dependent UDP-GDH. Two products were generated during UDP-glucuronyltransferase-catalyzed transfer of glucuronic acid to the corresponding aglycon, namely the glucuronide and UDP.

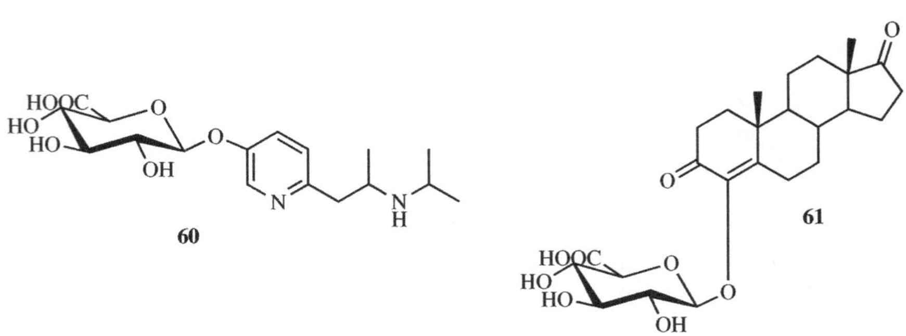

FIGURE 3.39 Glucuronides preparation using multienzyme-mediated *in situ* regeneration of UDP-GlcA.

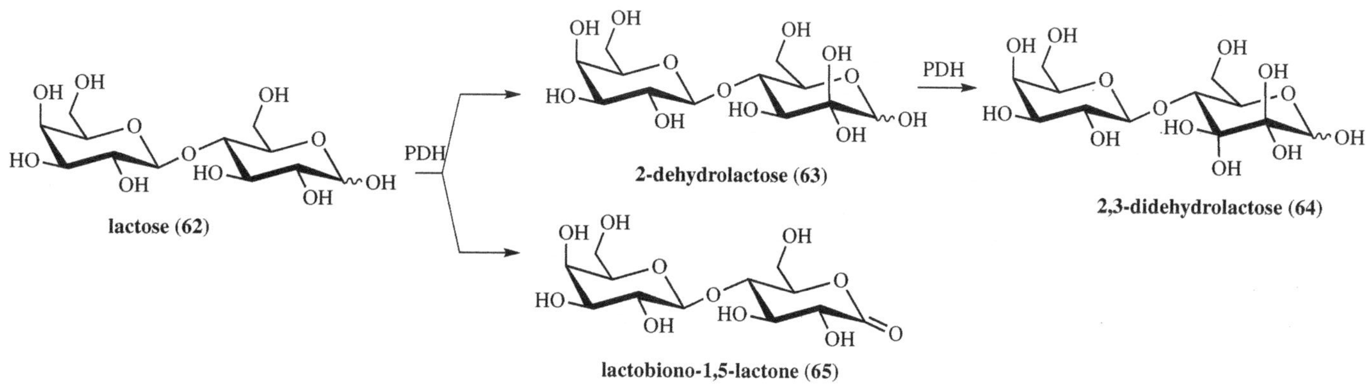

FIGURE 3.40 Reaction scheme for oxidation of lactose (**62**) by PDH from *A. xanthoderma*.

Pyranose dehydrogenase (PDH, EC 1.1.99.29) is a fungal flavin-dependent sugar oxidoreductase, structurally and catalytically related to pyranose oxidase (EC 1.1.3.10) and cellobiose dehydrogenase (EC 1.1.99.18), and probably playing similar biological roles in lignocellulose breakdown [84]. It displays a broad substrate specificity and a variable regioselectivity, being unable to utilize oxygen as electron acceptor, using instead substituted benzoquinones and (organo)metallic ions. Depending on the structure of the sugar in the pyranose form (mono/di/oligosaccharide or glycoside) and the enzyme source, selective monooxidations at C-1, C-2, C-3, or dioxidations at C-2,3 or C-3,4 of the molecule to the corresponding aldonolactones (C-1) or (di)dehydrosugars (aldos(di)uloses) can be performed. These features convert PDH into a promising and versatile biocatalyst for production of highly reactive di- and tri-carbonyl sugar derivatives which are outstanding chiral intermediates for the synthesis of rare sugars, novel drugs, and fine chemicals. The purified fungal PDH catalyzed the substrate-dependent C-1, C2, C-3, C1,3′, or C-2,3(′) (di)oxidation of a number of mono- and disaccharides with 1,4-benzoquinone as an electron acceptor [85]. D-Ribose, D-allose, D-gulose, and D-talose were oxidized to the corresponding aldonic acids. L-Arabinose was exclusively converted to 2-dehydro-L-arabinose (L-*erythro*-pentos-2-ulose) whereas D-xylose underwent competing C-2 and C-3 oxidation followed by di-oxidation to 2,3-didehydro-L-arabinose (D-*glycero*-pentos-2,3-diulose). The major final oxidation products of maltose and cellobiose were 2,3′-didehydromaltose and 2,3′-didehydrocellobiose (α- and β-D-*ribo*-hexos-3-ulopyranosyl-(1→4)-D-arabino-hexos-2-ulose), formed *via* 2- and 3′-monooxidation intermediates. Minor concomitant (di)-oxidation at C-1,(3′) to the corresponding acids also took place. Maltotriose was preferentially oxidized at C-3″ of the terminal glycopyranosyl unit and at C-1 of the reducing moiety. Conversions of (glycosyl)aldoses in nonbuffered systems were nearly quantitative, and the enzyme were allowed to an easy access to highly reactive di- or tricarbonyl sugars. PDH purified from *Agaricus* sp. catalyzed the simultaneous oxidation of lactose (**62**) to β-D-galactopyranosyl-(1→4)-D-*arabino*-hexos-2-ulose (**63,** 2-dehydrolactose) and lactobiono-1,5-lactone (**65**), with the latter undergoing spontaneous hydrolysis to lactobionic acid (Figure 3.40) [88]. The C-1 versus C-2 oxidation ratio varied from almost exclusive formation of lactobionic acid (C-1 oxidation) to 2-dehydrolactose as a major product (C-2 oxidation), according to the source of PDH. The highest selectivity for C-2 oxidation was exhibited by PDH from *Agaricus xanthoderma* and *Arothron meleagris* [429], with yields of approximately 75% of **63**. Upon prolonged incubation with excess of oxidant, minor amounts of 2,3-didehydrolactose (**64**) (β-D-galactopyranosyl-(1→4)-D-*threo*-arabino-hexos-2,3-diulose) were identified.

3.8 CONCLUDING REMARKS

From the identification of different types of dehydrogenases in Nature, their use in a broad number of biocatalytic applications has created a useful toolbox of options for chemical industries. This chapter has provided a comprehensive state-of-the-art of substrates, applications, and smart ways to set up dehydrogenase-catalyzed reactions at practical scale. Approaches range from the enantioselective reduction of prochiral substrates to (dynamic) kinetic resolutions, deracemizations, and stereoinversions, covering both aliphatic and aromatic compounds in all cases. Likewise, different strategies have been used to solve the cofactor regeneration problems, regardless whether isolated enzymes or whole-cells are applied. In fact, dehydrogenase-catalyzed asymmetric reactions are currently the first choice in many industrial laboratories, when novel synthetic approaches are assessed [4]. It is therefore expected that more industrial applications of these enzymes will appear in the coming decades.

REFERENCES

1. Kula MR, Kragl U. In: Patel RN, editor. *Stereoselective Biocatalysis*. New York: Marcel Dekker, Inc.; 2000. pp. 839–866.
2. Ryzewski CN, Pietruszko R. Arch Biochem Biophys 1977;183:73–82.
3. Keinan E, Hafeli EK, Seth KK, Lamed R. J Am Chem Soc 1986;108:162–169.
4. De Wildeman SMA, Sonke T, Schoemaker HE, May O. Acc Chem Res 2007;40:1260–1266.
5. Cowan D, Meyer Q, Stafford W, Muyanga S, Cameron R, Wittwer P. Trends Biotechnol 2005;23:321–329.
6. Cowan D, Arslanoglu A, Burton SG, Baker GC, Cameron RA, Smith JJ, Meyer Q. Biochem Soc Trans 2004;32:298–302.
7. Gabor E, Liebeton K, Niehaus F, Eck J, Lorentz P. Biotechnol J 2007;2:201–206.
8. Langer M, Gabor EM, Liebeton K, Meurer G, Niehaus F, Schulze R, Eck J, Lorenz P. Biotechnol J 2006;1:815–821.
9. Short JM. Nat Biotechnol 1997;15:1322–1323.
10. May O, Voigt CA, Arnold FH. In: Drauz K, Waldman W, editors. *Enzyme Catalysis in Organic Synthesis*. Weinheim: Wiley-VCH. pp. 95–138.
11. Reetz MT. Proc Natl Acad Sci USA 2004;101:5716–5722.
12. Broussy S, Cheloha RW, Berkowitz DB. Org Lett 2009;11:305–308.
13. Prelog V. Pure Appl Chem 1964;9:119–130.
14. Cahn RS, Ingold CK. J Chem Soc 1951;1:612–622.
15. Cahn RS, Ingold C, Prelog V. Angew Chem Int Ed 1966;5:385–415.
16. Zhang R, Geng Y, Xu Y, Zhang W, Wang S, Xiao R. Bioresour Technol 2011;102:483–489.
17. Wallner SR, Lavandera I, Mayer SF, Ohrlein R, Hafner A, Edegger K, Faber K, Kroutil W. J Mol Catal B: Enzym 2008;55:126–129.
18. Nie Y, Xu Y, Yang M, Mu XQ. Lett Appl Microbiol 2007;44:555–562.
19. Zhu D, Yang Y, Hua L. J Org Chem 2006;71:4202–4205.
20. Xie Y, Xu JH, Xu Y. Bioresour Technol 2010;101:1054–1059.
21. Gröger H, Rollman C, Chamouleau F, Sebastien I, May O, Wienand W, Drauz K. Adv Synth Catal 2007;349:709–712.
22. Truppo MD, Pollard D, Devine P. Org Lett 2007;9:335–338.

23. Hu J, Xu Y. Biotechnol Lett 2006;28:1115–1119.
24. Zhu D, Hua L. J Org Chem 2006;71:9484–9486.
25. Bergo de Lacerda PS, Ribeiro JB, Leite SGF, Ferrara MA, Coelho RB, Bon EPS, da Silva Lima EL, Antunes OAC. Tetrahedron: Asymmetry 2006;17:1186–1188.
26. Fardelone LC, Rodrigues JAR, Moran PJS. J Mol Catal B: Enzym 2006;39:9–12.
27. Hanson RL, Goldberg S, Goswami A, Tully TP, Patel RN. Adv Synth Catal 2005;347: 1073–1080.
28. Zhu D, Rios BE, Rozzell JD, Hua L. Tetrahedron: Asymmetry 2005;16:1541–1546.
29. Tsujigami T, Sugai T, Ohta H. Tetrahedron: Asymmetry 2001;12:2543–2549.
30. Lou WY, Wang W, Smith TJ, Zong MH. Green Chem 2009;11:1377–1384.
31. Bradshaw CW, Hummel W, Wong CH. J Org Chem 1992;57:1532–1536.
32. Pekala E, Godawska Matysik A, Zelaszczyk D, Biotechnol J 2007;2:492–496.
33. Hummel W. In: New Enzymes for Organic Synthesis. Vol. 58. Berlin: Springer; 1997. pp. 145–184.
34. Weckbecker A, Hummel W. Biocatal Biotransform 2006;24:380–389.
35. Faber K. *Biotransformations in Organic Chemistry*. 5th ed. Berlin: Springer; 2004.
36. Wong CH, Whitesides G. In: *Enzymes in Synthetic Organic Chemistry*. 1st ed. Pergamon: Oxford; 1994. pp. 131–194.
37. Crestia D, Guerard C, Veschambre H, Hecquet L, Demuynck C, Bolte J. Tetrahedron: Asymmetry 2001;12:869–876.
38. D. Metrángolo Ruiz de Temiño, Hartmeier W, Ansorge Schumacher MB. Enzyme Microb Technol 2005;36:3–9.
39. Inoue K, Makino Y, Itoh N. Tetrahedron: Asymmetry 2005;16:2539–2549.
40. Hildebrandt P, Riermeier T, Altenbuchner J, Bornscheuer U, Tetrahedron: Asymmetry 2001;12:1207–1210.
41. Bisogno F, Lavandera I, Kroutil W, Gotor V. J Org Chem 2009;74:1730–1732.
42. Monti D, Ottolina G, Carrea G, Riva S. Chem Rev 2011;111:4111–4140.
43. Strohmeier GA, Pichler H, May O, Khadjawi MG. Chem Rev 2011;111:4141–4164.
44. Soni P, Kansal H, Banerjee UC. Process Biochem. 2007;42:1632–1640.
45. Inoue K, Makino Y, Itoh N. Appl Environ Microbiol 2005;71:3633–3641.
46. Kataoka M, Doi Y, Sim TS, Shimizu S, Yamada H. Arch Biochem Biophys 1992;294:469–474.
47. Kavanagh KL, Klimacek M, Nidetzky B, Wilson DK. Biochemistry 2002;41:8785–8795.
48. Maser E, Möbus E, Xiong G. Biochem. Biophys Res Commun 2000;272:622–628.
49. Muheim A, Waldner R, Sanglard D, Reiser J, Schoemaker HE, Leisola MSA. Eur J Biochem 1991;195:369–375.
50. Patel RN, Hou CT, Laskin AI, Derelanko P, Felix A. Eur J Biochem 1979;101:401–406.
51. Saratani Y, Uheda E, Yamamoto H, Nishimura A, Yoshizako F. Biosci Biotechnol, Biochem 2001;67:1417–1420.
52. Schütte H, Fossdorf J, Sahm H, Kula MR. Eur J Biochem 1976;62:151–160.
53. Wu X, Liu N, He Y, Chen Y. Acta Biochim Biophys Sin 2009;41:163–170.

54. Uzura A, Nomoto F, Sakoda A, Nishimoto Y, Kataoka M, Shimizu S. Appl Microbiol Biotechnol 2009;83:617–626.
55. Simpson PJ, Tantitadapitak C, Reed AM, Mather OC, Bunce CM, White SA, Ride JP. J Mol Biol 2009;392:465–480.
56. Martin JR, Vega DE, Cabrera Z, Bolivar JM, Fernandez Lafuente R., Berenguer J, Guisan JM. Process Biochem 2009;44:1004–1012.
57. Yang Y, Zhu D, Piegat TJ, Hua L. Tetrahedron: Asymmetry 2007;17:1799–1801.
58. Kaluzna IA, Rozzell JD, Kambourakis S. Tetrahedron: Asymmetry 2005;16:3682–3689.
59. Itoh N, Matsuda M, Mabuchi M, Dairi T, Wang J. Eur J Biochem 2002;269:2394–2402.
60. Rodríguez S, Kayser M, Stewart JD. J Am Chem Soc 2001;123:1547–1555.
61. Schroer K, Tacha E, Lütz S. Org Process Res Dev 2007;11:836–841.
62. G. de Gonzalo, Lavandera I, Faber K, Kroutil W. Org Lett 2007;9:2163–2166.
63. Edegger K, Gruber CC, Poesl TM, Wallner SR, Lavandera I, Faber K, Niehaus F, Eck J, Oehrlein R, Hafnerd A, Kroutil W. Chem Commun 2006;2006:2402–2404.
64. Goldberg K, Edegger K, Kroutil W, Liese A. Biotechnol Bioeng 2006;95:192–198.
65. Yan Z, Nie Y, Xu Y, Liu X, Xiao R. Tetrahedron Lett 2011;52:999–1002.
66. Parkot J, Gröger H, Hummel W. Appl Microbiol Biotechnol 2010;86:1813–1820.
67. Choi YH, Choi HJ, Kim D, Uhm KN, Kim HK. Appl Microbiol Biotechnol 2010;87:185–193.
68. Wu X, Wang Y, Ju J, Chen C, Liu N, Chen Y. Tetrahedron: Asymmetry 2009:20:2504–2509.
69. Heiss C, Phillips RS. J Chem Soc, Perkin Trans 1 2000;2000:2821–2825.
70. Ye Q, Cao H, Zang G, Mi L, Yan M, Wang Y, Zhang Y, Li X, Li J, Xu L, Xiong J, Ouyang P, Ying H. Appl Microbiol Biotechnol 2010;88:1277–1285.
71. Ye Q, Cao H, Mi L, Yan M, Wang Y, He Q, Li J, Xu L, Chen Y, Xiong J, Ouyang P, Ying H. Bioresour Technol 2010;101:8911–8914.
72. Asako H, Shimizu M, Makino Y, Itoh N. Tetrahedron Lett 2010;51:2664–2666.
73. Schrittwieser JH, Lavandera I, Seisser B, Mautner B, Kroutil W. Eur J Org Chem 2009;14:2293–2298.
74. Schroer K, Luetz S. Org Process Res Dev 2009;13:1202–1205.
75. Parachin NS, Carlquist M, Grauslund MFG. Appl Microbiol Biotechnol 2009;84:487–497.
76. Liang J, Mundorff E, Voladri R, Jenne S, Gilson L, Conway A, Krebber A, Wong J, Huisman G, Truesdell S, Lalonde J. Org Process Res Dev 2010;14:188–192.
77. Bommarius AS, Drauz K, Hummel W, Kula MR, Wandrey C. Biocatalysis 1994;10:37–47.
78. Liese A, Seebach K, Wandrey C. *Industrial Biotransformations*. 2nd ed. Weinheim: Wiley-VCH; 2006.
79. Liu Z, Weis R, Gleider A. Food Technol Biotechnol 2004;42:237–249.
80. Zhi LW, Li ZY, Lin GQ. Synthesis 2000;12:1673–1676.
81. Bülter T, Elling L. Glycoconjugate J 1999;16:147–159.
82. Toone EJ, Simon ES, Whitesides GM. J Org Chem 1991;56:5603–5606.
83. Gygax D, Spies P, Winkler T, Pfaar U. Tetrahedron 1991;47:5119–5122.

84. Peterbauer CK, Volc J. Appl Microbiol Biotechnol 2010;85:837–848.
85. Sedmera P, Halada P, Kubátová E, Haltrich D, Prkrylová V, Volc J. J Mol Catal B: Enzym 2006;41:32–42.
86. Volc J, Sedmera P, Halada P, Daniel G, Prikrylova V, Haltrich D. J Mol Catal B: Enzym 2002;17:91–100.
87. Volc J, Sedmera P, Halada P, Dwivedi P, Ferreira MC. J Carbohydrate Chem 2003;22:217–216.
88. Volc J, Sedmera P, Kujawa M, Halada P, Kubátová E, Haltrich D. J Mol Catal B: Enzym 2004;30:177–184.
89. Bruce NC, French CE, Hailes AM, Long MT, Rathbone DA. Trends Biotechnol 1995;13:200–205.
90. Boonstra B, Rathbone DA, French CE, Walker EH, Bruce NC. Appl Environ Microb 2000;66:5161–5166.
91. Yang Y, Wang J, Kayser M. Tetrahedron: Asymmetry 2007;18:2021–2029.
92. Renouf P, Poirier JM, Duhamel P. J Org Chem 1999;64:2513–2515.
93. Pollard DJ, Telari K, Lane J, Humphrey G, McWilliams C, Nidositko S, Salmon P, Moore J. Biotechnol Bioeng 2006;93:674–686.
94. Kroutil W, Mang H, Edegger K, Faber K. Curr Opin Chem Biol 2004;8:120–126.
95. Moore JC, Pollard DJ, Kosjek B, Devine PN. Acc Chem Res 2007;40:1412–1419.
96. Nakamura K, Yamanaka R, Matsuda T, Harada T. Tetrahedron: Asymmetry 2003;14:2659–2681.
97. Patel RN. Curr Opin Biotechnol 2001;12:587–604.
98. Matsuda T, Nakajima Y, Harada T, Nakamura K. Tetrahedron: Asymmetry 2002;13:971–974.
99. Nakamura K, Matsuda T, Harada T. Chirality 2002;14:703–708.
100. Schubert T, Hummel W, Muller M. Angew Chem Int Ed 2002;41:634–637.
101. Carballeira JD, Valmaseda M, Alvarez A, Sinisterra JV. Enzyme Microb Technol 2004;34:611–623.
102. Pedrini P, Giovannini PP, Mantovani EA, Colalongo C. J Mol Catal B: Enzym 2009;60:145–150.
103. Rivilla CP, Cafeu MC, Casatejada JA, Araujo AR, Collado IG. Tetrahedron: Asymmetry 2009;20:2666–2672.
104. Musa MM, Fjeld KIZ, Vieille C, Zeikus JG, Phillips RS. J Org Chem 2007;72:30–34.
105. Rocha LC, Ferreira HV, Pimenta EF, Berlinck RGS, Seleghim MHR, Javaroti DCD, Sette LD, Bonugli RC, Porto ALM. Biotechnol Lett 2009;31:1559–1563.
106. Pollard DJ, Truppo M, Pollard J, Chen CW, Moore J. Tetrahedron: Asymmetry 2006;17:554–559.
107. Yang W, Xu JH, Pan J, Wang ZL. Biochem Eng J 2008;42:1–5.
108. Zhang J, Witholt B, Li Z. Adv Synth Catal 2006;348:429–433.
109. Zhang J, Witholt B, Li Z. Chem Commun 2006;4:398–400.
110. Kalaitzakis D, Rozzell JD, Kambourakis S, Smonou I. Org Lett 2005;7:4799–4801.
111. Ravía S, Carrera I, Seoane G, Vero S, Gamenara D. Tetrahedron: Asymmetry 2009;20:1393–1397.

112. Yang W, Xu JH, Xie Y, Zhao G, Lin GQ. Tetrahedron: Asymmetry 2006;17:1769–1774.
113. Ma SK, Gruber J, Davis C, Newman L, Gray D, Wang A, Grate J, Huisman GH, Sheldon RA. Green Chem 2010;12:81–86.
114. Zhu D, Stearns JE, Ramírez M, Hua L. Tetrahedron 2006;62:4535–4539.
115. Zhu D, Mukherjee C, Rozzell JD, Kambourakis S, Hua L. Tetrahedron 2006;62:901–905.
116. Baskar B, Ganesh S, Lokeswari TS, Chadha A. J Mol Catal B: Enzym 2004;27:13–17.
117. Padhi SK, Kaluzna IA, Buisson D, Azerad R, Stewart JD. Tetrahedron: Asymmetry 2007;18:2133–2138.
118. Danchet S, Buisson D, Azerad R. J Mol Catal B: Enzym 1998;5:255–259.
119. Danchet S, Bigot C, Buisson D, Azerad R. Tetrahedron: Asymmetry 1997;8:1735–1739.
120. Cabon O, Buisson D, Larcheveque M, Azerad R. Tetrahedron: Asymmetry 1995;6:2199–2210.
121. Wolberg M, Kaluzna IA, Müller M, Stewart JD. Tetrahedron: Asymmetry 2004;15:2825–2828.
122. Wolberg M, Ji A, Hummel W, Müller M. Synthesis 2001;6:937–942.
123. Wolberg M, Hummel W, Müller M. Chem Eur J 2001;7:4562–4571.
124. Wolberg M, Filho MV, Bode S, Geilenkirchen P, Feldmann R, Liese A, Hummel W, Müller M. Bioprocess Biosyst Eng 2008;31:183–191.
125. Patel JM. J Mol Catal B: Enzym 2009;61:123–128.
126. Lüdeke S, Richter M, Mueller M. Adv Synth Catal 2009;351:253–259.
127. Schubert T, Hummel W, Kula MR, Müller M. Eur J Org Chem 2001;22:4181–4187.
128. Soni P, Kaur G, Chakraborti AK, Banerjee UC. Tetrahedron: Asymmetry 2005;16:2425–2428.
129. Goswami A, Mirfakhrae KD, Totleben MJ, Swaminathan S, Patel RN. J Ind Microbiol Biotechnol 2001;26:259–262.
130. Poessl TM, Kosjek B, Ellmer U, Gruber CC, Edegger K, Faber K, Hildebrandt P, Bornscheuer U, Kroutil W. Adv Synth Catal 2005;347:1827–1834.
131. Rosen TC, Feldmann R, Dünkelmann P, Daußmann T. Tetrahedron Lett 2006;47:4803–4806.
132. Zhu D, Hyatt BA, Hua L. J Mol Catal B: Enzym 2009;56:272–276.
133. Zhu D, Mukherjee C, Hua L. Tetrahedron: Asymmetry 2005;16:3275–3278.
134. Zhu D, Ankati H, Mukherjee C, Yan Y, Biehl ER, Hua L. Org Lett 2007;9:2561–2563.
135. Kalaitzakis D, Rozzell JD, Kambourakis S, Smonou I. Eur J Org Chem 2006;10:2309–2313.
136. Kalaitzakis D, Rozzell JD, Smonou I, Kambourakis S. Adv Synth Catal 2006;348:1958–1969.
137. Goldberg K, Schroer K, Lütz S, Liese A. Appl Microbiol Biotechnol 2007;76:249–255.
138. Goldberg K, Schroer K, Lütz S, Liese A. Appl Microbiol Biotechnol 2007;76:237–248.
139. Matsuda T, Yamanaka R, Nakamura K. Tetrahedron: Asymmetry 2009;20:513–557.
140. Nakamura K, Matsuda T. In: Gotor V, Alfonso I, García-Urdiales E, editors. *Asymmetric Organic Synthesis with Enzymes*. Weinheim: Wiley-VCH; 2008. pp. 193–228.
141. Zhu D, Hua L. Pure Appl Chem 2010;82:117–128.

142. Pham VT, Phillips RS. J Am Chem Soc 1990;112:3629–3632.
143. Tripp AE, Burdette DS, Zeikus JG, Phillips RS. J Am Chem Soc 1998;120:5137–5141.
144. Jönsson A, Wehtje E, Adlercreutz P, Mattiasson B. Biochim Biophys Acta 1999;1430:313–322.
145. Olofsson L, Nicholls IA, Wikman S. Org Biomol Chem 2005;3:750–755.
146. Petkova GA, Král V. Bioorg Med Chem 2010;18:6651–6656.
147. Schumacher J, Eckstein M, Kragl U. Biotechnol J 2006;1:574–581.
148. Musa MM, Fjeld KIZ, Vieille C, Phillips RS. Org Biomol Chem 2008;6:887–892.
149. Musa MM, Fjeld KIZ, Vieille C, Zeikus JG, Phillips RS. Angew Chem Int Ed 2007;46:3091–3094.
150. Bisogno F, Cuetos A, Orden AA, Sanz MK, Lavandera I, Gotor V. Adv Synth Catal 2010;352:1657–1661.
151. Bisogno F, Urdiales EG, Valdés H, Lavandera I, Kroutil W, Suárez D, Gotor V. Chem Eur J 2010;16:11012–11019.
152. Nakamura K, Matsuda T. Curr Org Chem 2006;10:1217–1246.
153. Nakamura K, Matsuda T, Ohno A. Tetrahedron: Asymmetry 1996;7:3021–3024.
154. Wei ZL, Lin GQ, Li ZY. Bioorg Med Chem 2000;8:1129–1137.
155. Nakamura K, Kawai Y, Nakajima N, Ohno A. J Org Chem 1991;56:4778–4783.
156. Rodríguez S, Kayser M, Stewart JD. Org Lett 1999;1:1153–1155.
157. Nakamura K, Takano S, Ohno A. Tetrahedron Lett 1993;34:6087–6090.
158. Nakamura K, Kawai Y, Oka S, Ohno A. Bull Chem Soc Jpn 1989;62:875–879.
159. M. Brzezinska-Rodak, Zymanczyk-Duda E, Kafarski P, Lejczak B. Biotechnol Progr 2002;18:1287–1291.
160. Nakamura K, Fujii M, Ida Y. J Chem Soc, Perkin Trans 1 2000;19:3205–3211.
161. Nakamura K, Takenaka K, Fujii M, Ida Y. Tetrahedron: Asymmetry 2002;43:3629–3631.
162. Klibanov AM. Nature 2001;409:241–246.
163. Klibanov AM. Curr Opin Biotechnol 2003;14:427–431.
164. Matsuda T, Harada T, Nakajima N, Nakamura K. Tetrahedron Lett 2000;41:4135–4138.
165. Nakamura K, Kitano K, Matsuda T, Ohno A. Tetrahedron Lett 1996;37:1629–1632.
166. Nakamura K, Matsuda T. J Org Chem 1998;63:8957–8964.
167. Salvi NA, Chattopadhyay I. Tetrahedron: Asymmetry 2008;19:1992–1997.
168. Nakamura K, Yamanaka R, Tohi K, Hamada H. Tetrahedron Lett 2000;41:6799–6802.
169. Maczka WK, Mironowicz A. Tetrahedron: Asymmetry 2002;13:2299–2302.
170. Maczka WK, Mironowicz A. Tetrahedron: Asymmetry 2004;15:1965–1967.
171. Aldabalde V, Arcia MP, González A, González D. Green Chem Lett Rev 2007;1:25–30.
172. Baldassarre F, Bertoni G, Chiappe C, Marioni F. J Mol Catal B: Enzym 2000;11:55–58.
173. Blanchard N, van de Weghe P. Org Biomol Chem 2006;4:2348–2353.
174. Ravía S, Gamenara D, Schapiro V, Bellomo A, Adum J, Seoane G, González D. J Chem Ed 2006;83:1049–1051.
175. Yadav JS, Reddy GSKK, Sabitha G, Krishna AD, Prasad AR, Rahaman H-U-R, Rao KV, Rao AV. Tetrahedron: Asymmetry 2007;18:717–723.
176. Liu X, Pan ZG, Xu JH, Li HX. Chin Chem Lett 2010;21:305–308.

177. Chadha A, Manohar M, Soundararajan T, Lokeswari TS. Tetrahedron: Asymmetry 1996;7:1571–1572.
178. Xie Y, Xu JH, Lu WY, Lin GQ. Bioresour Technol 2009;100:2463–2468.
179. Fonseca AM, Monte FJQ, de Oliveira MCF, de Mattos MC, Cordell GA, Filho RB, Lemos TLG. J Mol Catal B: Enzym 2009;57:76–82.
180. Yadav JS, Reddy BVS, Sreelakshmi C, Rao AB. Synthesis 2009;11:1881–1885.
181. Kumaraswamy G, Ramesh S. Green Chem. 2003;5:306–308.
182. Yadav JS, Nanda S, Reddy PT, Rao AB. JOrg Chem 2002;67:3900–3903.
183. Comasseto JV, Omori AT, Porto ALM, Andrade LH. Tetrahedron Lett. 2004;45:473–476.
184. Yadav JS, Reddy PT, Nanda S, Rao AB. Tetrahedron: Asymmetry 2001;12:3381–3385.
185. Orden AA, Noguera CM, Agostini E, Sanz MK. J Mol Catal B: Enzym 2009;61:216–220.
186. Tosa MI, Podea PV, Paizs C, Irimie FD. Tetrahedron: Asymmetry 2008;17:2068–2071.
187. Gamenara D, de María PD. Biotechnol Adv 2009;27:278–285.
188. Fatima Y, Kansal H, Soni P, Banerjee UC. Process Biochem 2007;42:1412–1418.
189. Xie Q, Wu J, Xu G, Yang L. Biotechnol Progr 2006;22:1301–1304.
190. Zelinski T, Liese A, Wandrey C, Kula MR. Tetrahedron: Asymmetry 1999;10:1681–1687.
191. Fujii M, Takeuchi M, Akita H, Nakamura K. Tetrahedron Lett 2009;50:4941–4944.
192. Carrea G, Riva S. Angew Chem Int Ed 2000;39:2226–2254.
193. Grunwald J, Wirz B, Scollar MP, Klibanov AM. J Am Chem Soc 1986;108:6732–6734.
194. Zaks A, Klibanov AM. J Biol Chem 1988;263:8017–8021.
195. Liu X, Zhu TS, Sun PD, Xu JH. Synth Commun 2001;31:1521–1526.
196. Nakamura K, Inoue K, Ushio K, Oka S, Ohno A. J Org Chem 1988;53:2589–2593.
197. Nakamura K, Kondo S, Nakajima N, Ohno A. Tetrahedron 1995;51:687–694.
198. Nakamura K, Inoue Y, Matsuda T, Misawa I. J Chem Soc, Perkin Trans 1 1999;16:2397–2402.
199. Gröger H, Hummel W, Buchholz S, Drauz K, Van Nguyen T, Rollmann C, Hüsken H, Abokitse K. Org Lett 2003;5:173–176.
200. Lu J, Lazzaroni MJ, Hallet JP, Bommarius AS, Liotta CL, Eckert CA. Ind Eng Chem Res 2004;43:1586–1590.
201. Gauchot V, Kroutil W, Schmitzer AR. Chem Eur J 2010;2010:6748–6751.
202. Jakoblinnert A, Mladenov R, Paul A, Sibilla F, Schwaneberg U, Ansorge-Schumacher MB, de María PD. Chem Commun 2011;47:12230–12232.
203. Bystrykh LV, Vonck J, van Bruggen EF, van Beeumen J, Samyn B, Govorukhina NI, Arfman N, Duine JA, Dijkhuizen L. J Bacteriol 1993;175:1814–1822.
204. Daussmann T, Aivasidis A, Wandrey C. Eur J Biochem 1997;248:889–896.
205. Kato N, Yamagami T, Shimao M, Sakazawa C. Eur J Biochem 1986;156:59–64.
206. van der Werf MJ, van der Ven C, Barbirato F, Eppink MHM, de Bont JAM, van Berkel WJH. J Biol Chem 1999;274:26296–26304.
207. van Ophem PW, van Beeumen J, Duine JA. Eur J Biochem 1993;212:819–826.
208. Schenkels P, De Vries S, Straathof AJJ. Biocatal Biotransform 2001;19:191–212.
209. Makino Y, Inoue K, Dairi T, Itoh N. Appl Environ Microbiol 2005;71:4713–4720.

210. Itoh N, Mizuguchi N, Mabuchi M. J Mol Catal B: Enzym 1999;6:41–50.
211. Itoh N, Moriharma R, Wang J, Okada K, Mizuguchi N. Appl Environ Microbiol 1997;63:3783–3788.
212. Kalaitzakis D, Kambourakis S, Rozzell JD, Smonou I. Tetrahedron: Asymmetry 2007;18:2418–2426.
213. Peters J, Minuth T, Kula MR. Enzyme Microb Technol 1993;15:950–958.
214. Baskar B, Pandian NG, Priya K, Chadha A. Tetrahedron: Asymmetry 2004;15:3961–3966.
215. Mangone CP, Pereyra EN, Argimón S, Moreno S, Baldessari A. Enzyme Microb Technol 2002;30:596–601.
216. He JY, Zhou LM, Wang P, Zu L. Process Biochem 2009;44:316–321.
217. Cui J, Zhang MJ, Wang XQ, Cui NJ. Chin Chem Lett 2008;19:311–313.
218. Ishihara K, Yamaguchi H, Hamada H, Nakamura K, Nakajima N. J Mol Catal B: Enzym 2000;10:419–428.
219. de Souza Ramos A, Ribeiro JB, Vazquez L, Fiaux SB, Leite SGF, de Andrade Cruz R, Ramos MCK, de Aquino Neto FR, Antunes OAC. Tetrahedron: Asymmetry 2009;20: 559–561.
220. Kaluzna IA, Andrew AA, Bonilla M, Martzen MR, Stewart JD. J Mol Catal B: Enzym 2002;17:101–105.
221. Kaluzna IA, Feske BD, Wittayanan W, Ghiviriga I, Stewart JD. J Org Chem 2005;70:342–345.
222. Panizza P, Onetto S, Rodríguez S. Biocatal Biotransform 2007;25:414–417.
223. Rodríguez S, Schroeder KT, Kayser M, Stewart JD. J Org Chem 2000;65:2586–2587.
224. Kratzer R, Pukl M, Egger S, Nidetzky B. Microbial Cell Factories 2008;7. DOI: 10.1186/1475-2859-7-37.
225. Kratzer R, Egger S, Nidetzky B. Biotechnol Bioeng 2008;101:1094–1101.
226. Carballeira JD, Quezada MA, Hoyos P, Simeó Y, Hernáiz MJ, Alcántara A, Sinisterra JV. Biotechnol Adv 2009;27:686–714.
227. Goldberg K, Krueger A, Meinhard T, Kroutil W, Mautner B, Liese A. Tetrahedron: Asymmetry 2008;19:1171–1173.
228. Cuetos A, Valenzuela ML, Lavandera I, Gotor V, Carriedo GA, Biomacromolecules 2010;11:1291–1297.
229. Bhattacharyya MS, Singh A, Banerjee UC. Bioresour Technol 2010;101:1581–1586.
230. Yajima A, Naka K, Yabuta G. Tetrahedron Lett 2004;45:4577–4579.
231. Gröger H, Hummel W, Rollman C, Chamouleau F, Husken H, Werner H, Wunderlich C, Abokitse K, Drauz K, Buchholz S. Tetrahedron 2004;60:633–640.
232. Villela FM, Stillger T, Mueller M, Liese A, Wandrey C. Angew Chem Int Ed 2003;42:2993–2996.
233. He JY, Sun ZH, Ruan WQ, Xu Y. Process Biochem 2006;41:244–249.
234. Harada T, Kubota Y, Kamitanaka T, Nakamura K, Matsuda T. Tetrahedron Lett 2009;50:4934–4936.
235. G. de Gonzalo, Lavandera I, Durchschein K, Wurm D, Faber K, Kroutil W. Tetrahedron: Asymmetry 2007;18:2541–2546.
236. Braütigam S, Meyer SB, Weuster-Botz D. Tetrahedron: Asymmetry 2007;18:1883–1887.

237. Bräutigam S, Dennewald D, Schürmann M, Lutje-Spelberg J, Pitner WR, Weuster-Botz D. Enzyme Microb Technol 2009;45:310–316.
238. Musa MM, Fjeld KIZ, Vieille C, Phillips RS. Org Biomol Chem 2008;6:887–892.
239. Liese A, Zelinski T, Kula MR, Kierkels H, Karutz M, Kragl U, Wandrey C. J Mol Catal B: Enzym 1998;4:91–99.
240. Müller D, Liauw MA, Greiner L. Chem Eng Technol 2005;28:1569–1571.
241. Kosjek B, Stampfer W, Pogorevc M, Goessler C, Faber K, Kroutil W. Biotechnol Bioeng 2004;86:55–62.
242. Stampfer W, Kosjek B, Moitzi C, Kroutil W, Faber K. Angew Chem Int Ed 2002;41:1014–1017.
243. Kavanagh KL, Klimacek M, Nidetzky B, Wilson DK. Biochem J 2003;373:319–326.
244. Kratzer R, Nidetzky B. Chem Commun 2007;10:1047–1049.
245. Kratzer R, Leitgeb S, Wilson DK, Nidetzky B. Biochem J 2006;393:51–58.
246. Johanson T, Katz M, Grauslund MFG. FEMS Yeast Res 2005;5:513–525.
247. Kaluzna IA, Matsuda T, Sewell AK, Stewart JD. J Am Chem Soc 2004;126:12827–12832.
248. Sybesma WFH, Straathof AJJ, Jongejan JA, Pronk JT, Heijnen JJ. Biocatal Biotransform 1998;16:95–134.
249. Manner S, Hansson C, Larsson JM, Oltner VT, Frejd T. Tetrahedron: Asymmetry 2010;21:1374–1381.
250. Kalaitzakis D, Smonou I. J Org Chem 2010;75:8658–8661.
251. Hoyos P, Sinisterra JV, Molinari F, Alcántara A, de María PD. Acc Chem Res 2010;43:288–299.
252. R. Bel-Rhlid, Fauve A, Veschambre H. J Org Chem 1989;54:3221–3223.
253. Hoyos P, Sansottera G, Fernández M, Molinari F, Sinisterra JV, Alcántara A. Tetrahedron 2008;64:7929–7936.
254. Demir AS, Ayhan P, Demirtas U, Erkilic U. J Mol Catal B: Enzym 2008;55:164–168.
255. Saito T, Maruyama R, Ono S, Yasukawa N, Kodaira K, Nishizawa M, Ito S, Inoue M. Appl Biochem Biotechnol 2003;111:185–190.
256. Bortolini O, Fantín G, Fogagnolo M, Giovannini PP, Guerrini A, Medici A. J Org Chem 1997;62:1854–1856.
257. Areces P, Carrasco E, Light ME, Santos M, Plumet J. Synlett 2007;20:3180–3182.
258. Bi J, Aggarwal VK. Chem Commun 2008;1:120–122.
259. Jurczak J, Pikul S, Bauer T. Tetrahedron 1986;42:447–488.
260. Baer E, Fishcher HOL. J Biol Chem 1939;128:463–473.
261. Richter N, Neumann M, Liese A, Wohlgemuth R, Eggert T, Hummel W. ChemBioChem 2009;10:1888–1896.
262. Richter N, Neumann M, Liese A, Wohlgemuth R, Weckbecker A, Eggert T, Hummel W. Biotechnol Bioeng 2010;106:541–552.
263. Chamouleau F, Hagedorn C, May O, Groeger H. Flavour Fragance J 2007;22:169–172.
264. Ni Y, Xu JH. J Mol Catal B: Enzym 2002;18:233–241.
265. Machado LL, Monte FJQ, de Oliveira MCF, de Mattosa MC, Lemos TLG, Fernández VG, Gotor V. J Mol Catal B: Enzym 2008;54:130–133.

266. Sello G, Orsini F, Bernasconi S, Di Gennaro P. Molecules 2006;11:365–369.
267. Azerad R, Buisson D. Curr Opin Biotechnol 2000;11:565–571.
268. Gruber CC, Lavandera I, Faber K, Kroutil W. Adv Synth Catal 2006;348:1789–1805.
269. Matute BM, Backvall JE. In: Carrea G, Riva S, editors. *Organic Synthesis with Enzymes in Non-Aqueous Media*. Weinheim: Wiley-VCH; 2008. pp. 113–144.
270. Tessaro D, Molla G, Pollegioni L, Servi S. In: Fessner WD, Anthonsen T, editors. *Modern Biocatalysis*. Weinheim: Wiley-VCH; 2009. pp. 195–228.
271. Turner NJ. In: Gotor V, Alfonso I, Urdiales EG, editors. *Asymmetric Organic Synthesis with Enzymes*. Weinheim: Wiley-VCH; 2008. pp. 115–132.
272. Matute BM, Backvall JE. Curr Opin Chem Biol 2007;11:226–232.
273. Hasegawa J, Ogura M, Tsuda S, Maemoto S, Kutsuki H, Ohashi T. Agric Biol Chem 1990;54:1819–1827.
274. Carnell AJ, Iacazio G, Roberts SM, Willets A. Tetrahedron Lett 1994;35:331–334.
275. Goswami A, Mirfakhrae KD, Patel RN. Tetrahedron: Asymmetry 1999;10:4239–4244.
276. Matsumura S, Kawai Y, Takahashi Y, Toshima K. Biotechnol Lett 1994;16:485–490.
277. Nie Y, Xu Y, Lv TF, Xiao R. J Chem Technol Biotechnol 2009;84:468–472.
278. Nie Y, Xu Y, Mu XQ. Org Process Res Dev 2004;8:246–251.
279. Demir AS, Hamamci H, Sesenoglu O, Neslihanoglu R, Asikoglu B, Capanoglu D. Tetrahedron Lett 2002;43:6447–6449.
280. Demir AS, Hamamci H, Ayhan P, Duygu N, Igdir AC, Capanoglu D. Tetrahedron: Asymmetry 2004;15:2579–2582.
281. Shimizu S, Hattori S, Hata H, Yamada H. Enzyme Microb Technol 1987;9:411–416.
282. Takahashi E, Nakamichi K, Furui M. J Ferment Bioeng 1995;80:247–250.
283. Chadha A, Baskar B. Tetrahedron: Asymmetry 2002;13:1461–1464.
284. Padhi SK, Chadha A. Tetrahedron: Asymmetry 2005;16:2790–2798.
285. Baskar B, Pandian NG, Priya K, Chadha A. Tetrahedron 2005;61:12296–12306.
286. Buisson D, Azerad R, Sanner C, Larcheveque M. Biocatalysis 1992;5:249–265.
287. Padhi SK, Pandian NG, Chadha A. J Mol Catal B: Enzym 2004;29:25–29.
288. Fantín G, Fogagnolo M, Giovannini PP, Medici A, Pedrini P. Tetrahedron: Asymmetry 1995;6:3047–3053.
289. Xie SX, Ogawa J, Shimizu S. Appl Microbiol Biotechnol 1999;52:327–331.
290. Ogawa J, Xie SX, Shimizu S, Biotechnol Lett 1999;21:331–335.
291. Voss CV, Gruber CC, Faber K, Knaus T, Macheroux P, Kroutil W. J Am Chem Soc 2008;130:13969–13972.
292. Voss CV, Gruber CC, Kroutil W. Tetrahedron: Asymmetry 2007;18:276–281.
293. Li YL, Xu JH, Xu Y. J Mol Catal B: Enzym 2010;64:48–52.
294. Nakamura K, Fujii M, Ida Y. Tetrahedron: Asymmetry 2001;12:3147–3153.
295. Nakamura K, Inoue Y, Matsuda T, Ohno A. Tetrahedron Lett 1995;36:6263–6266.
296. Allan GR, Carnell AJ. J Org Chem 2001;66:6495–6497.
297. Cardus GJ, Carnell AJ, Trauthwein H, Riermeir T. Tetrahedron: Asymmetry 2004;15:239–243.
298. Comasseto JV, Omori ÁT, Andrade LH, Porto ALM. Tetrahedron: Asymmetry 2003;14:711–715.

299. Comasseto JV, Assis LF, Andrade LH, Schoenlein-Crusius IH, Porto ALM. J Mol Catal B: Enzym. 2006;39:24–30.

300. Comasseto JV, Andrade LH, Omori AT, Assis LF, Porto ALM. J Mol Catal B: Enzym 2004;29:55–61.

301. Andrade LH, Utsunomiya RS, Omori AT, Porto ALM, Comasseto JV. J Mol Catal B: Enzym 2006;38:84–90.

302. Takemoto M, Achiwa K. Tetrahedron: Asymmetry 1995;6:2925–2928.

303. Takemoto M, Achiwa K. Phytochemistry 1998;49:1627–1629.

304. Voss CV, Gruber CC, Kroutil W. Angew Chem Int Ed 2008;47:741–745.

305. Tsuchiya S, Miyamoto K, Ohta H. Biotechnol Lett 1992;14:1137–1142.

306. Takemoto M, Matsuoka Y, Achiwa K, Kutney JP. Tetrahedron Lett 2000;41:499–502.

307. Vaijayanthi T, Chadha A. Tetrahedron: Asymmetry 2007;18:1077–1084.

308. Titu D, Chadha A. J Mol Catal B: Enzym 2008;52-53:168–172.

309. Padhi SK, Titu D, Pandian NG, Chadha A. Tetrahedron 2006;62:5133–5140.

310. Titu D, Chadha A. Tetrahedron: Asymmetry 2008;19:1698–1701.

311. Shimizu S, Hattori S, Hata H, Yamada H. Appl Environ Microbiol 1987;53:519–522.

312. Ebbers EJ, Ariaans GJA, Houbiers JPM, Bruggink A, Zwanenburg B. Tetrahedron 1997;53:9417–9476.

313. van Nispen SFGM, van Buijtenen J, Vekemans JAJM, Meuldijk J, Hulshof LA. Tetrahedron: Asymmetry 2006;17:2299–2305.

314. Kim N, Ko SB, Kwon MS, Kim MJ, Park J. Org Lett 2005;7:4523–4526.

315. Nestl BM, Voss CV, Bodlenner A, Schaumberger UE, Kroutil W, Faber K. Appl Microbiol Biotechnol 2007;76:1001–1008.

316. Nestl BM, Bodlenner A, Stuermer R, Hauer B, Kroutil W, Faber K. Tetrahedron: Asymmetry 2007;18:1465–1474.

317. Nestl BM, Glueck SM, Hall M, Kroutil W, Stuermer R, Hauer B, Faber K. Eur J Org Chem 2006;20:4573–4577.

318. Gruber CC, Nestl BM, Gross J, Hildebrandt P, Bornscheuer U, Faber K, Kroutil W. Chem Eur J 2007;13:8271–8276.

319. Pamies O, Bäckvall JE. Chem Rev 2003;103:3247–3261.

320. Pàmies O, Bäckvall JE. Curr Opin Biotechnol 2003;14:407–413.

321. Faber K. Chem Eur J 2001;7:5004–5010.

322. Hoyos P, Fernández M, Sinisterra JV, Alcántara AR. J Org Chem 2006;71:7632–7637.

323. Borén L, Martín-Matute B, Xu Y, Córdova A, Bäckvall JE. Chem Eur J 2006;12:225–232.

324. Berkessel A, Sebastian-Ibarz ML, Mueller T. Angew Chem Int Ed 2006;45:6567–6570.

325. Ödman P, Wessjohann LA, Bornscheuer UT. J Org Chem 2005;70:9551–9555.

326. Pellissier H. Tetrahedron 2011;67:3769–3802.

327. Galkin A, Kulakova L, Yoshimura T, Soda K, Esaki N. Appl Environ Microbiol 1997;63:4651–4656.

328. Bommarius A. In: Drauz K, Waldman W, editors. *Enzyme Catalysis in Organic Synthesis*. 2nd ed. Weinheim: Wiley-VCH; 2002. pp. 1047–1063.

329. Oshima T, Soda K. In: Patel RN, editor. *Stereoselective Biocatalysis*. New York: Marcel Dekker, Inc.; 2000. pp. 877–902.
330. Oshima T, Nishida N, Bakthavatsalam S, Kataoka K, Takada H, Yoshimura T, Esaki N, Soda K. Eur J Biochem 1994;222:305–312.
331. Brunhuber NMW, Thoden JB, Blanchard JS, Vanhooke JL. Biochemistry 2000;39:9174–9187.
332. Lütz S, Rao N, Wandrey C. Chem Eng Technol 2006;29:1404–1415.
333. Hanson RL, Howell JM, La Porte TL, Donovan MJ, Cazzulino DL, Zannella V, Montana MA, Nanduri VB, Schwarz SR, Eiring RF, Durand SC, Wasylyk JM, Parker WL, Lui MS, Okuniewicz FJ, Chen BC, Harris JC, Natalie KJ, Ramig K, Swaminathan S, Rosso VW, Pack SK, Lotz BT, Bernot PJ, Rusowicz A, Lust DA, Tse KS, Venit JJ, Szarka LJ, Patel RN. Enzyme Microb Technol 2000;26:348–358.
334. Galkin A, Kulakova L, Tishkov V, Esaki N, Soda K. Appl Microbiol Biotechnol 1995;44:479–483.
335. Ohshima T, Nishida N, Bakthavatsalam S, Kataoka K, Takada H, Yoshimura T, Esaki N, Soda K. Eur J Biochem 1994;222:305–312.
336. Takada H, Yoshimura T, Ohshima T, Esaki N, Soda K. J Biochem 1991;109:371–376.
337. Gröger H, May O, Werner H, Menzel A, Altenbuchner J. Org Process Res Dev 2006;10:666–669.
338. Resch V, Fabian WMF, Kroutil W. Adv Synth Catal 2010;352:993–997.
339. Peters KV, Gunawardana M, Rozzell JD, Novick SJ. J Am Chem Soc 2006;128:10923–10929.
340. Li H, Williams P, Micklefield J, Gardiner JM, Stephens G. Tetrahedron 2004;60:753–758.
341. Muramatsu H, Mihara H, Kakutani R, Yasuda M, Ueda M, Kurihara T, Esaki N. J Biol Chem 2005;280:5329–5335.
342. Vaijayanthi T, Chadha A. Tetrahedron: Asymmetry 2008;19:93–96.
343. Thangavel V, Chadha A. Tetrahedron 2007;63:4126–4133.
344. Hall M, Bommarius A. Chem Rev 2011;111:4088–4110.
345. Warburg O, Christian W. Biochem Zeit 1933;266:377–411.
346. Riggers JFC, Roggers TA, Figueroa EV, Polizzi KM, Bommarius A. Adv Synth Catal 2007;349:1521–1531.
347. Hall M, Stuecker C, Ehammer H, Pointner E, Oberdorfer G, Gruber K, Hauer B, Stuermer R, Kroutil W, Macheroux P, Faber K. Adv Synth Catal 2008;350:411–418.
348. Hall M, Stuecker C, Hauer B, Stuermer R, Friedrich T, Breuer M, Kroutil W, Faber K. Eur J Org Chem 2008;9:1511–1516.
349. Hall M, Stuecker C, Kroutil W, Macheroux P, Faber K. Angew Chem Int Ed 2007;46:3934–3937.
350. Kosjek B, Fleitz FJ, Dormer PG, Kuethe JT, Devine PN. Tetrahedron: Asymmetry 2008;19:1403–1406.
351. Müller A, Hauer B, Rosche B. Biotechnol Bioeng 2007;98:22–29.
352. Swiderska MA, Stewart JD. J Mol Catal B: Enzym. 2006;42:52–54.
353. Wada M, Yoshizumi A, Noda Y, Kataoka M, Shimizu S, Takagi H, Nakamori S. Appl Environ Microbiol 2003;69:933–937.

354. Richter N, Gröger H, Hummel W. Appl Microbiol Biotechnol 2011;89:79–89.
355. Muller A, Hauer B, Rosche B. J Mol Catal B: Enzym 2006;38:126–130.
356. Gatfield IL. Adv Biochem Eng/Biotechnol 1997;55:221–238.
357. Högberg HE, Berglund P, Edlund H, Fägerhag J, Hedenström E, Lundh M, Nordin O, Servi S, Vörde C. Catal Today 1994;22:591–606.
358. Servi S. Synthesis 1990;1:1–25.
359. Fuganti C. Pure Appl Chem 1990;62:1449–1452.
360. Kawai Y, Hayashi M, Tokitoh N. Tetrahedron: Asymmetry 2001;12:3007–3013.
361. Rambeck B, Simon H. Angew Chem Int Ed 1974;13:609.
362. Günther H, Simon H. Biocatal Biotransform 1995;12:1–26.
363. Hall M, Hauer B, Stuermer R, Kroutil W, Faber K. Tetrahedron: Asymmetry 2006;17:3058–3062.
364. French CE, Nicklin S, Bruce NC. J Bacteriol 1996;178:6623–6627.
365. Niino YS, Chakraborty S, Brown BJ, Massey V. J Biol Chem 1995;270:1983–1991.
366. Stott K, Saito K, Thiele DJ, Massey V. J Biol Chem 1993;268:6097–6106.
367. Brown BJ, Hyun JW, Duwuri S, Karplus PA, Massey V. J Biol Chem 2002;277:2138–2145.
368. Kohli RM, Massey V. J Biol Chem 1998;273:32763–32770.
369. Xu D, Kohli RM, Massey V. Proc Natl Acad Sci USA 1999;96:3556–3561.
370. Shimoda K, Ito DI, Izumi S, Hirata T. J Chem Soc, Perkin Trans 1 1996;4:355–358.
371. Winkler CK, Stuecker C, Mueller NJ, Pressnitz D, Faber K, Eur. J. Org. Chem. 2010;33:6354–6358.
372. Brenna E, Fronza G, Fuganti C, Monti D, Parmeggiani F. J Mol Catal B: Enzym 2011;73:17–21.
373. Koul S, Crout DHG, Errington W, Tax J. J Chem Soc, Perkin Trans 1 1995;23:2969–2988.
374. Sortino MA, Filho VC, Zacchino SA. Tetrahedron: Asymmetry 2009;20:1106–1108.
375. Hollmann F, Arends I, Buehler K, Schallmey A, Bühler B. Green Chem 2011;13:226–265.
376. Kroutil W, Mang H, Edegger K, Faber K. Adv Synth Catal 2004;346:125–142.
377. Arterburn JB. Tetrahedron 2001;57:9765–9788.
378. Thomas SM, Di Cosimo R, Nagarajan V. Trends Biotechnol 2002;20:238–242.
379. Mihovilovic MD, Bianchi DA. In: Gotor V, Alfonso I, Urdiales EG, editor. *Asymmetric Organic Synthesis with Enzymes*. Weinheim: Wiley-VCH; 2008. pp. 229–274.
380. Sello G, Orsini F. Mini-Rev Org Chem 2006;3:11–21.
381. Hummel W. Adv Biochem Eng Biotechnol, 1997;58:145–184.
382. Boratynski F, Kiełbowicz G, Wawrzenczyk C. J Mol Catal B: Enzym 2010;65:30–36.
383. Lok KP, Jakovac IJ, Jones JB. J Am Chem Soc 1985;107:2521–2526.
384. Wong CH, Matos JR. J Org Chem 1985;50:1992–1994.
385. Matos JR, Wong CH. J Org Chem 1986;51:2388–2389.
386. Willets A, Knowles CJ, Levitt MS, Roberts SM, Sandey H, Shipston NF. J Chem Soc, Perkin Trans 1 1991;6:1608–1610.
387. Edegger K, Mang H, Faber K, Gross J, Kroutil W. J Mol Catal A: Chem 2006;251:66–70.

388. Stampfer W, Kosjek B, Kroutil W, Faber K. Biotechnol Bioeng 2003;81:865–869.
389. Kosjek B, Stampfer W, Pogorevc M, Goessler W, Faber K, Kroutil W. Biotechnol Bioeng 2004;86:55–62.
390. Stampfer W, Kosjek B, Faber K, Kroutil W. J Org Chem 2003;68:402–406.
391. Stampfer W, Kosjek B, Faber K, Kroutil W. Tetrahedron: Asymmetry 2003;14:275–280.
392. Romano A, Gandolfi R, Nitti P, Rollini M, Molinari F. J Mol Catal B: Enzym 2002;17:235–240.
393. Miyamoto K, Fujimori K, Hirano JI, Ohta H. Biocatal Biotransform. 2009;27:66–70.
394. Geueke B, Riebel B, Hummel W, Enzyme Microb Technol 2003;32:205–211.
395. Riebel BR, Gibbs PR, Wellborn WB, Bommarius A. Adv Synth Catal 2003;345:707–712.
396. Orbegozo T, Lavandera I, Fabian WMF, Mautner B, de Vries JG, Kroutil W. Tetrahedron 2009;65:6805–6809.
397. Pekala E, Kochan M, Carnell AJ. Lett Appl Microbiol 2009;48:19–24.
398. Lavandera I, Kern A, Resch V, Silva BF, Glieder A, Fabian WMF, De Wildeman S, Kroutil W, Org Lett 2008;10:2155–2158.
399. Woodyer RD, Wymer NJ, Racine M, Khan SN, Saha BC. Appl Environ Microbiol 2008;74:2967–2975.
400. Molinari F. Curr Org Chem 2006;10:1247–1263.
401. Landis BH, McLaughlin JK, Heeren R, Grabner RW, Wang PT. Org Process Res Dev 2002;6:547–552.
402. Huwig A, Emmel S, Giffhorn F. Carbohydrate Res 1996;281:183–186.
403. Huwig A, Emmel S, Jägel G, Giffhorn F. Carbohydrate Res 1998;305:337–339.
404. De Wulf P, Soetaert W, Vandamme EJ. Biotechnol Bioeng 2000;69:339–343.
405. Ferré E, N'guyen M, Le Petti J, Gil G. Biocatal Biotransform 2002;20:311–317.
406. Maser E. Biochem Pharmacol 1995;49:421–440.
407. Bortolini O, Medici A, Poli S. Steroids 1997;62:564–577.
408. Chenault HK, Whitesides GM. Appl Biochem Biotechnol 1987;14:147–197.
409. Riva S, Bovara R, Pasta P, Carrea G. J Org Chem 1986;51:2902–2906.
410. Carrea G, Bovara R, Longhi R, Barani R. Enzyme Microb Technol 1984;6:307–311.
411. Bianchini E, Chinaglia N, Dean M, Giovannini PP, Medici A, Pedrini P, Poli S. Tetrahedron 1999;55:1391–1398.
412. Pedrini P, Andreotti E, Guerrini A, Dean M, Fantin G, Giovannini PP. Steroids 2006;71:189–198.
413. Anelli PL, Brocchetta M, Morosini P, Palano D, Carrea G, Falcone L, Pasta P, Sartore D. Biocatal Biotransform 2002;20:29–34.
414. Riva S, Bovara R, Zetta L, Pasta P, Ottolina G, Carrea G. J Org Chem 1988;53:88–92.
415. Bovara R, Canzi E, Carrea G, Pilotti A, Riva S. J Org Chem 1993;58:499–501.
416. Medici A, Pedrini P, Bianchini E, Fantin G, Guerrini A, Natalini B, Pellicciari R. Steroids 2002;67:51–56.
417. Carrea G, Pilotti A, Riva S, Canzi E, Ferrari A. Biotechnol Lett 1992;14:1131–1134.
418. Bovara R, Carrea G, Riva S, Secundo F. Biotechnol Lett 1996;18:305–308.
419. Giovannini PP, Grandini A, Perrone D, Pedrini P, Fantin G, Fogagnolo M. Steroids 2008;73:1385–1390.

420. Monti D, Ferrandi EE, Zanellato I, Hua L, Polentini F, Carrea G, Riva S. Adv Synth Catal 2009;351:1303–1311.
421. Cho YS, Kim CH, Cheon HG. Anal Biochem 2009;392:110–116.
422. Riva S, Ottolina G, Carrea G, Danieli B. J Chem Soc, Perkin Trans 1 1989;11:2073–2074.
423. Carrea G, Bovara R, Cremonesi P, Lodi R. Biotechnol Bioeng 1984;26:560–563.
424. Carrea G, Bovara R, Longhi R, Riva S. Enzyme Microb Technol 1985;7:597–600.
425. Carrea G. Trends Biotechnol 1984;2:102–106.
426. Cremonesi P, Carrea G, Sportoletti G, Antonini E. Arch Biochem Biophys 1973;159: 7–10.
427. Carrea G, Riva S, Bovara R, Pasta P. Enzyme Microb Technol 1988;10:333–340.
428. De Amici M, De Micheli C, Molteni G, Pitre D, Carrea G, Riva S, Spezia S, Zetta L. J Org Chem 1991;56:67–72.
429. Pisanelli I, Kujawa M, Gschnitzer D, Spadiut O, Seigoth B, Peterbauer CK. Appl Microbiol Biotechnol 2010;86:599–606.

CHAPTER 4

Reactions Involving Oxygenases

Oxidation is a central process in organic chemistry, and oxidative biocatalysts are very attractive for synthetic purposes due to unsolved limitations of chemical oxidative performances. These problems include the lack of control (e.g., loss of chemo-, regio-, and stereoselectivity), requiring complicated protection-deprotection sequences. Also, the use of molecular oxygen (the cheapest and the most environmentally friendly oxidant) is difficult to implement due to the difficulties associated with the activation of dioxygen by conventional means. Chemical oxidizing agents based on transition metal ions or peroxo-compounds are used. Such oxidizing agents are often costly and toxic, constituting an important environmental burden [1,2]. However, the high specificity of enzymes performing oxidation reactions, and the fact that they use molecular oxygen as oxidizing agent, make this class of enzymes a great alternative as catalysts in organic synthesis [3,4]. Oxygenase-catalyzed biotransformations include a variety of useful reactions, such as sulfoxidation reactions (together with selenide and amine oxidations), Baeyer–Villiger oxidations, epoxidations, hydroxylations, and aromatic dihydroxylations (Figure 4.1) [5–7].

4.1 MONOOXYGENASE-CATALYZED REACTIONS

Monooxygenases catalyze the reductive activation of molecular oxygen (see Section 1.3.2), requiring reducing equivalents (directly or indirectly) from reduced nicotinamide cofactors. As stated in Chapter 2, this requirement of expensive cofactors represents an important constraint for the development of biocatalytic applications involving monooxygenases. Therefore, a regeneration system is an essential aspect that has to be addressed, most of the monooxygenase-catalyzed reactions being performed by using whole-cell systems [1,7].

One of the best known monooxygenases is the family of cytochrome P450 monooxygenases. These enzymes contain a heme group as cofactor, and are ubiquitously distributed enzymes protecting organisms from xenobiotics, toxins, and performing the biosynthesis of important compounds such as steroids, fatty acids, and

Redox Biocatalysis: Fundamentals and Applications, First Edition.
Daniela Gamenara, Gustavo A. Seoane, Patricia Saenz-Méndez, and Pablo Domínguez de María.

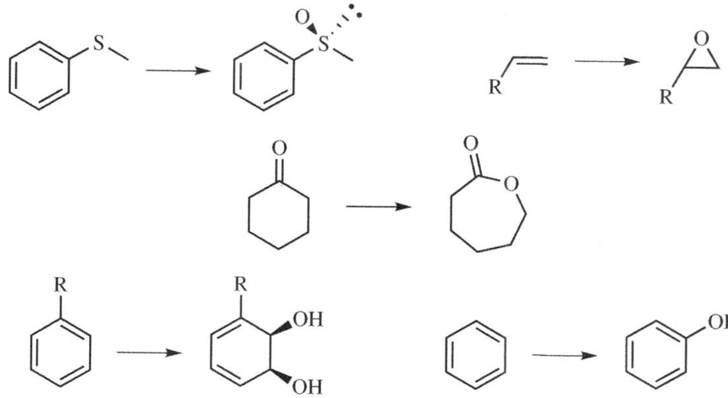

FIGURE 4.1 Oxidative reactions catalyzed by mono- and dioxygenases.

eicosanoids. Monooxygenase-catalyzed reactions include hydroxylation of alkanes, aromatic hydroxylations, epoxidation reactions, heteroatom oxygenations, and oxidative dehalogenation reactions [8–10]. Besides cytochrome P450 enzymes, flavin-dependent monooxygenases are of increasing interest in organic synthesis, covering a broad range of oxidation reactions, including epoxidation reactions, halogenations of phenols, and Baeyer–Villiger oxidations [7,11].

4.1.1 Hydroxylation of Aliphatic Compounds

The specific hydroxylation of nonactivated substrates is particularly important because their chemical alternatives are rather scarce or simply not available. Considering that oxyfunctionalized compounds are valuable intermediates or final products in pharmaceutical industry, biocatalytic oxygen transfer has gained attention as an efficient way to achieve selective hydroxylation reactions [12]. In this area, site specific hydroxylation of terpenoids catalyzed by bacterial and fungal enzymes has been extensively investigated. As a result, regioselective hydroxylation of steroids can now be achieved at industrial scale, for example, the hydroxylation of progesterone in the 11α-position (nonactivated methylene) using *Rhizopus arrhizus* or *Aspergillus niger*, leading to 11α-hydroxyprogesterone (Figure 4.2) [1,13].

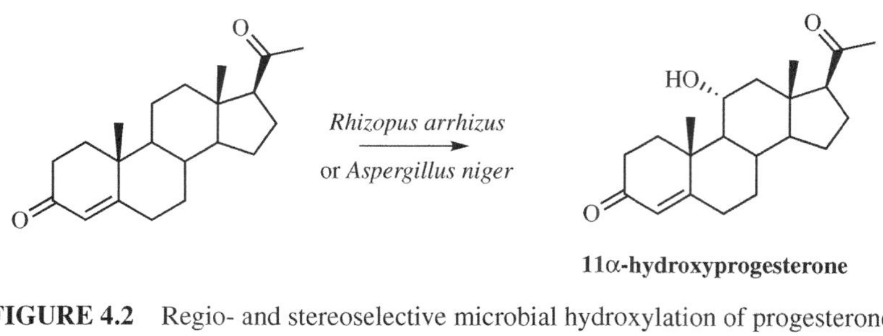

FIGURE 4.2 Regio- and stereoselective microbial hydroxylation of progesterone.

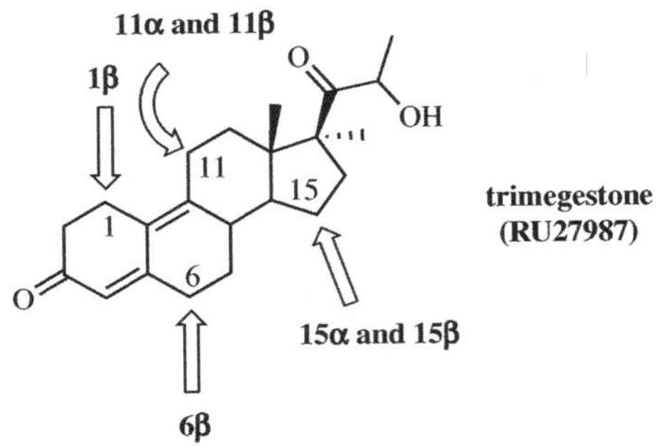

FIGURE 4.3 Microbial hydroxylation of trimegestone.

Likewise, several microorganisms were found to convert Trimegestone® (RU27987) to hydroxyderivatives in several positions (Figure 4.3) [14]. Trimegestone is used in osteoporosis therapy as a regulator of calcium assimilation. In that work, 41 fungal strains and 10 *Streptomyces* strains were tested, being most of the strains able to metabolize Trimegestone, producing six major hydroxyl derivatives, namely C-1β, C-6β, C-11α, C-11β, C-15α, and C-15β (Figure 4.3).

Bacillus megaterium P450 (BM3) (CYP106A2) was also identified as a steroid hydroxylase [15]. An *E. coli* system composed of a mitochondrial electron transfer system and the bacterial enzyme CYP102A2, was designed an employed in the regio- and stereoselective 15β-hydroxylation of 11-deoxycorticosterone (Figure 4.4) [16].

Besides hydroxylation of steroids, bioconversion of terpenoids is of importance for flavor and fragrance compounds, as well as for their potential antibacterial, antiviral, or cytotoxic activity. Several systematic studies aiming at discovering new oxyfunctionalized derivatives have been carried out. Agrawal and coworkers reported the induced mutations of various *Aspergillus* sp. and *Penicillium* sp. for the improved conversion of α-pinene to verbenol, the mint flavor ingredient (Figure 4.5) [17]. Even though some fungal species are active, the conversion is economically unattractive.

Verbenol and verbenone were the major products of the biotransformation (Figure 4.5), due to hydroxylase and dehydrogenase activities, respectively. A mutant strain of *Aspergillus* sp. showed a 15-fold improved activity compared to the wild-type strain, whereas the verbenone formation was substantially reduced. In the case of *Penicillium*, an 8-fold increase was found. More recently, the design of active-site

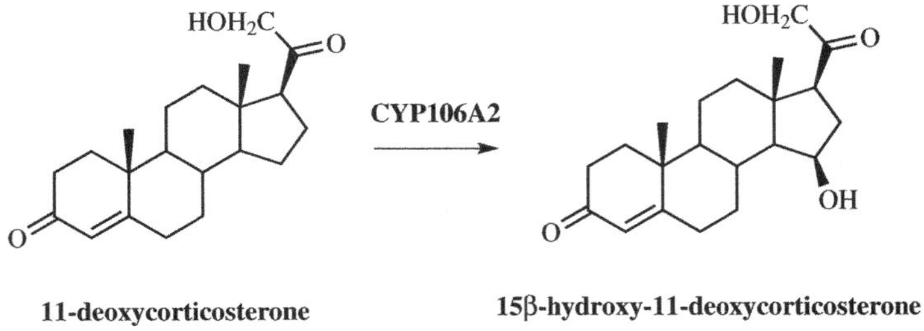

FIGURE 4.4 11-Deoxycorticosterone hydroxylation catalyzed by CYP106A2.

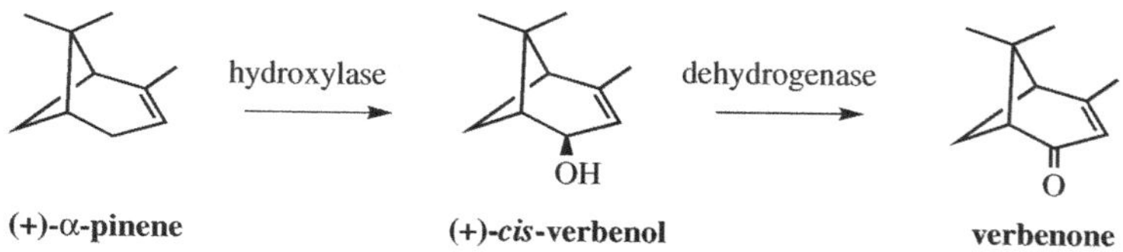

FIGURE 4.5 α-Pinene hydroxylation.

mutations on cytochrome $P450_{cam}$ (CYP101) based on a comparison of the structures of camphor and (+)-α-pinene was described. The engineered $P450_{cam}$ mutants exhibited enhanced activity and product selectivity for (+)-α-pinene oxidation to verbenol [18]. In the same field of terpenoids, (-)-perillyl alcohol is an interesting target for microbial biotransformations, due to its anticarcinogenic properties and the limited availability of the alcohol in nature. The oxidation of L-limonene might be a synthetic route for that, yet until 2005 the only microbial enzyme enabling such biotransformation was found in *Bacillus stearothermophilus* BR388. However, the need of expensive purification methods due to the low regioselectivity observed precluded the industrial production of perillyl alcohol employing this microorganism. In 2005, the biocatalytic production of (-)-perillyl alcohol from limonene was reported using a *Mycobacterium* sp. strain HXN-1500 cytochrome P450 alkane hydroxylase expressed in *Pseudomonas putida* (Figure 4.6) [19]. So far, several cytochrome P450 alkane hydroxylases of the CYP153 family were identified and expressed in *P. putida* GPo12 [20].

Nonoxygenated terpenes are frequently responsible for the unpleasant scent of essential oils, while oxygenated derivatives impart floral notes. Taking this into account, Pinheiro and Marsaioli reported the transformation of fragrant compounds by *Trichosporum cutaneum* CCT 1903 whole-cells [21]. The biotransformation of *cis*-jasmone using *T. cutaneum* CCT 1906 gave 4-hydroxyjasmone, 7,8-epoxyjasmone, and 7,8-dihydroxyjasmone (Figure 4.7).

Ionones are aroma compounds found in a variety of essential oils. In particular, β-ionone is a sesquiterpene that contributes to the aroma of roses. Also, 4-hydroxy-β-ionone is a flavor component. Wild-type P450(BM3) can hydroxylate β-ionone, producing a mixture of different hydroxylated products. Therefore, the selective biooxidation of β-ionone was conducted with a P450(BM3) triple mutant (Ala74Glu/Phe87Val/Pro386Ser), producing 4-hydroxy-β-ionone with modest stereoselectivity (39% *ee*) but completely regioselective (Figure 4.8) [22]. The same triple mutant was

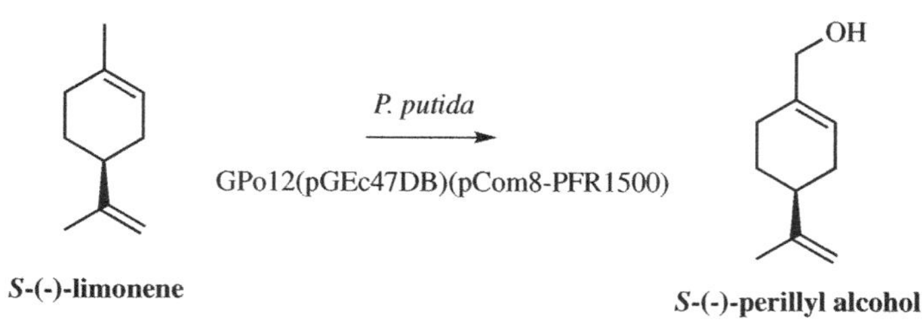

FIGURE 4.6 Production of perillyl alcohol from limonene.

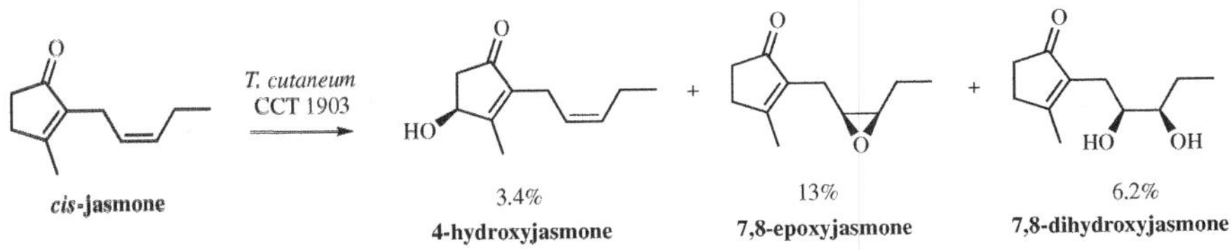

FIGURE 4.7 Biooxidation of *cis*-jasmone using *T. cutaneum* CCT 1906 whole-cells.

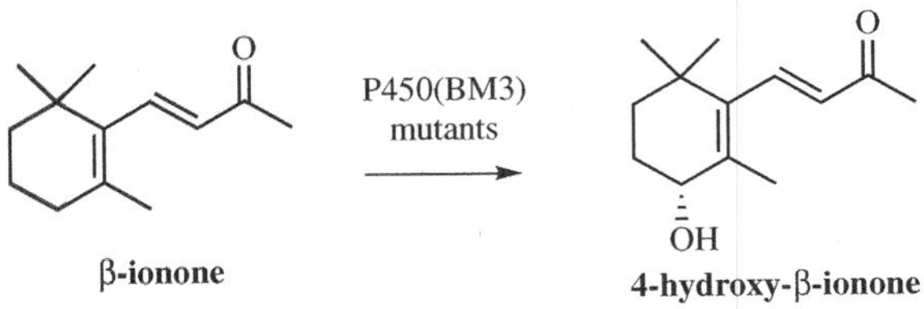

FIGURE 4.8 Bioconversion of *β*-ionone to 4-hydroxy-*β*-ionone catalyzed by P450(BM3) mutants.

immobilized jointly with a cofactor recycling system based on formate dehydrogenase (FDH) from *Pseudomonas* sp. 101, envisaging a technical application of P450 monooxygenases in fine chemical synthesis. The system was successfully tested in the oxidation of *β*-ionone to 4-hydroxy-*β*-ionone [23].

In a different area, Sun and coworkers found that a taxadiene derivative was converted to hydroxylated derivatives by the fungus *Cunninghamella elegans* (Figure 4.9) [24]. The results showed the potential use of filamentous fungi in the production of new taxoids with potential biological activity.

Pravastatin, a current therapeutic agent for hypercholesterolemia, was originally isolated from the urine of a canine fed with compactin. Because of the high cost and

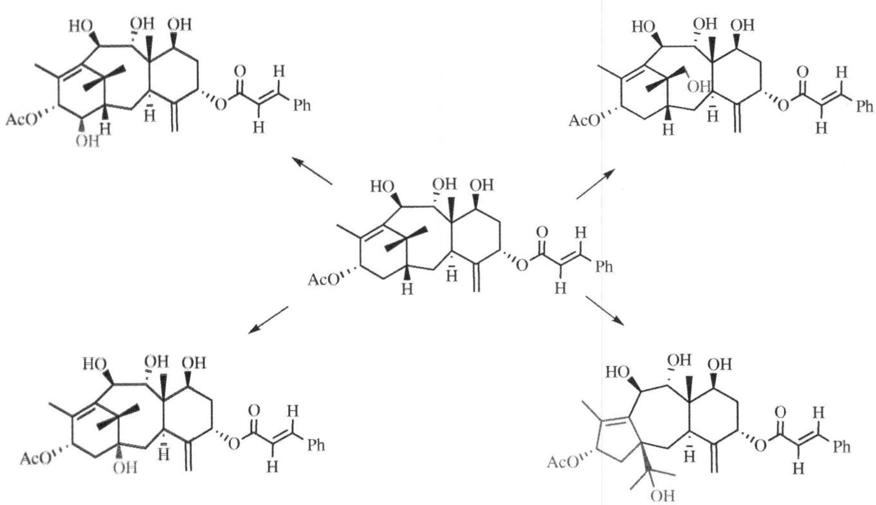

FIGURE 4.9 Microbial transformation of a taxadiene derivative.

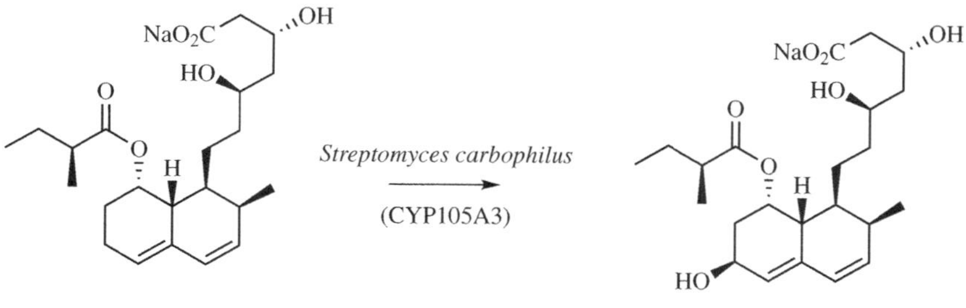

FIGURE 4.10 Stereoselective hydroxylation of compactin to Pravastatin.

nonselective chemical synthesis, the enzymatic hydroxylation of compactin is the preferred methodology for the preparation of pravastatin (Figure 4.10) [25,26].

Epothilones isolated from *Sorangium cellulosum* possess high cytotoxic activity by promoting tubulin polymerization [27]. Epothilone F is the hydroxylation product of epothilone B at the 21 position (Figure 4.11), and is employed as an intermediate in the synthesis of more bioavailable derivatives displaying antitumor activity. Therefore, its efficient synthesis is of great importance and to this end, several mutations were introduced in the gene of epothilone B hydroxylase based on previous knowledge on mutagenesis of P450 enzymes [28].

The hydroxylation of the great majority of organic compounds is not easily predicted or not even described. Therefore, it is often necessary to perform an extensive screening to find the appropriate biocatalytic system for preparing desired derivatives. Moreover, as mentioned before, sometimes strains displaying the particular hydroxylation activity are not suitable for a large-scale (or even a laboratory-scale) process, and therefore protein engineering approaches are needed (see Chapters 7–8). One different and more general way that allows the transformation of different substrate classes by a limited number of microorganisms is the concept of docking/protecting (d/p) group in biohydroxylation [29,30]. Some representative examples are shown in Figure 4.12.

The term protecting, as usual, means that the group has the ability to prevent undesired side reactions. The term docking group suggests that the d/p group promotes the interaction of the substrate with the active site of the enzyme, leading to a very selective hydroxylation pattern. Without the d/p group the reaction is not observed or proceeds without selectivity. Finally, after the biotransformation, the d/p group is removed (Figure 4.13).

An interesting case of enzyme engineering is illustrated by the use of bulky amino acids substitutions in the $P450_{cam}$ active site to promote the binding and oxidation of gaseous alkanes [31]. So far, the most active $P450_{cam}$ mutant for propane oxidation is the EB mutant (Phe87Trp/Tyr96Phe/Thr101Leu/Leu1244Met/Val247Leu). Xu et al. introduced additional mutations to EB to decrease the active site volume with bulky substitutions. Thus, the EB/Leu294M/Tyr185Met/Leu1358Pro/Gly248Ala mutant was successfully employed for ethane oxidation to ethanol.

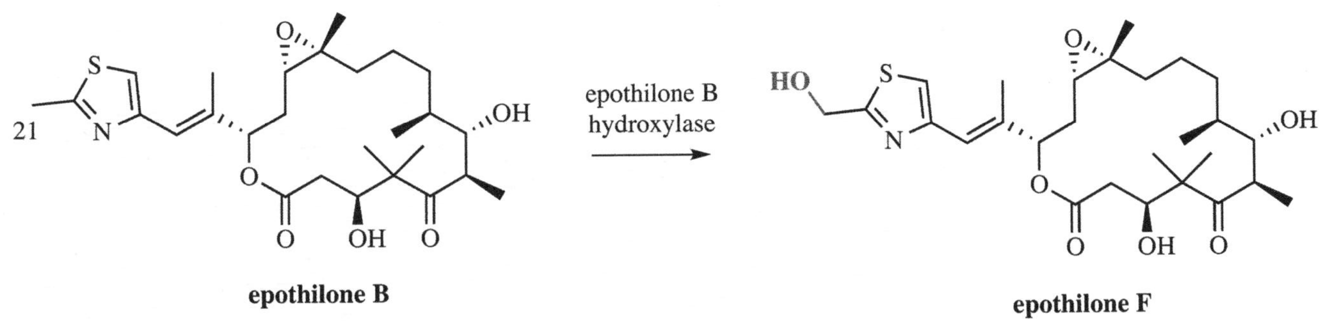

FIGURE 4.11 Regioselective enzymatic hydroxylation of epothilone B to afford epothilone F.

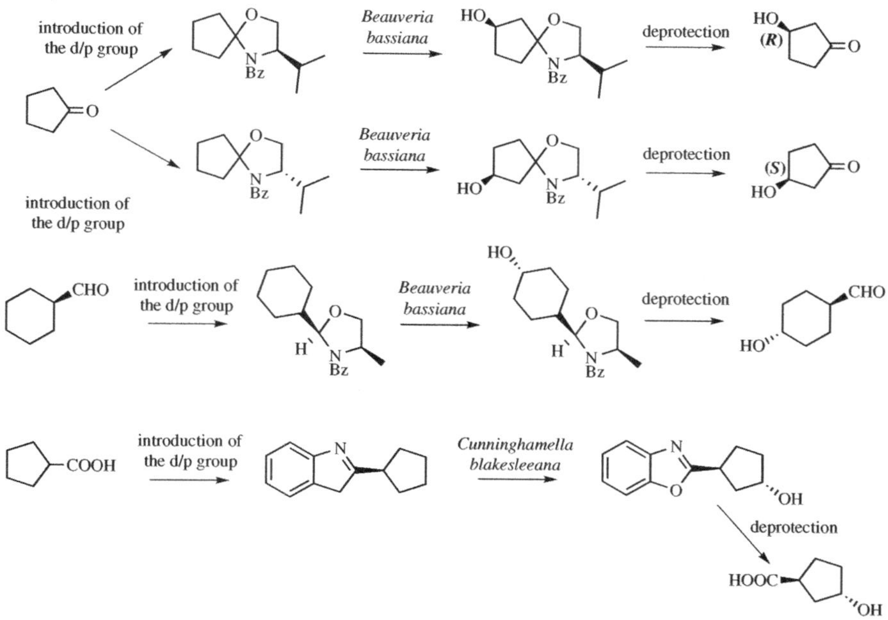

FIGURE 4.12 Examples of the docking/protecting group approach.

4.1.2 Hydroxylation of Aromatic Compounds

Selective enzymatic hydroxylation of aromatic compounds is an interesting alternative to produce important building blocks, especially considering the intrinsic difficulty of performing the reaction by chemical methods [32]. Monooxygenases catalyze aromatic hydroxylations, in particular the cytochrome P450 family introduce an oxygen atom into monoaromatic, polyaromatic, and heterocyclic compounds (Figure 4.14) [8,32].

The oxidation of aromatic compounds involves the epoxidation of the ring, affording an arene oxide. This unstable intermediate rapidly undergoes the epoxide ring opening associated to migration of a hydride anion, yielding a ketone intermediate. Final tautomerization of the ketone leads to the final phenolic product. This hydride

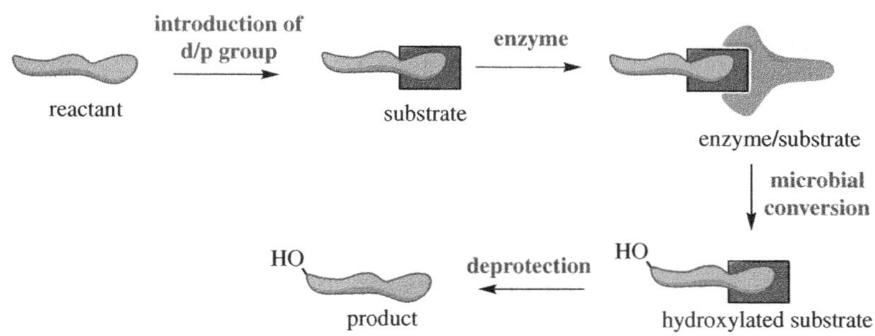

FIGURE 4.13 General procedure for the docking/protecting groups approach [30].

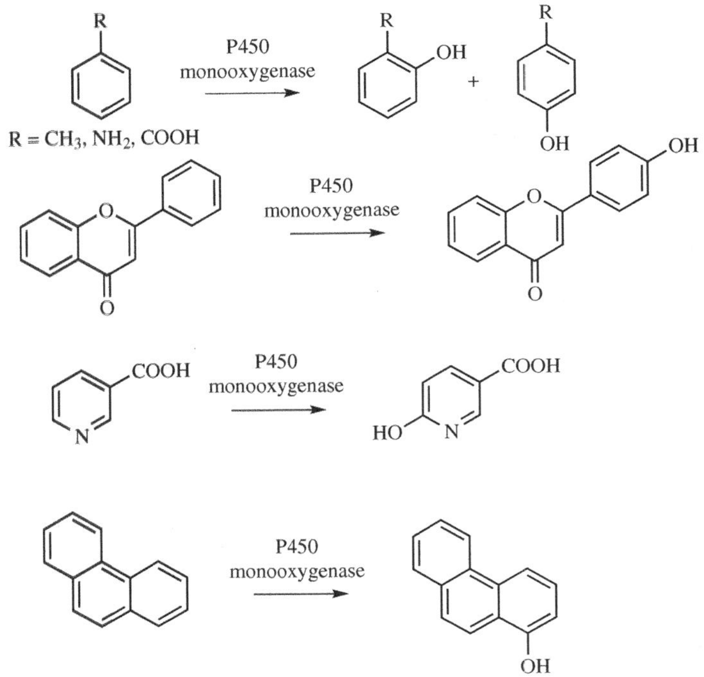

FIGURE 4.14 Examples of hydroxylation of aromatic compounds catalyzed by monooxygenases [32,33].

shift was observed for the first time in the National Institute of Health, Bethesda, the United States of America and for that is known as "NIH shift" (Figure 4.15) [32,34].

The oxidation of polycyclic aromatic hydrocarbons (PAHs) is of utmost importance from an environmental viewpoint. PAHs are persistent environmental contaminants considered as procarcinogens. The rate limiting step for biodegradation by microorganisms is the initial oxidation by P450 monooxygenases [35]. $P450_{cam}$ has been engineered for the oxidation of PAHs. The double mutants Phe87Ala/Tyr96Phe and Phe87Leu/Tyr96Phe were employed in the oxidation of phenanthrene, fluoranthene, pyrene and benzo[*a*]pyrene (Figure 4.16) [36].

Wild-type P450(BM3) was also tested for the biooxidation of phenanthrene, fluoranthene and benzo[*a*]pyrene, showing a low activity. Mutation of amino acids in the

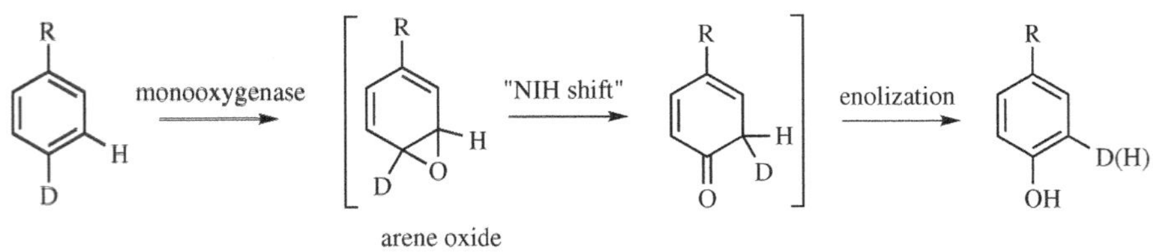

FIGURE 4.15 Accepted mechanism for the monooxygenase catalyzed hydroxylation of aromatic compounds.

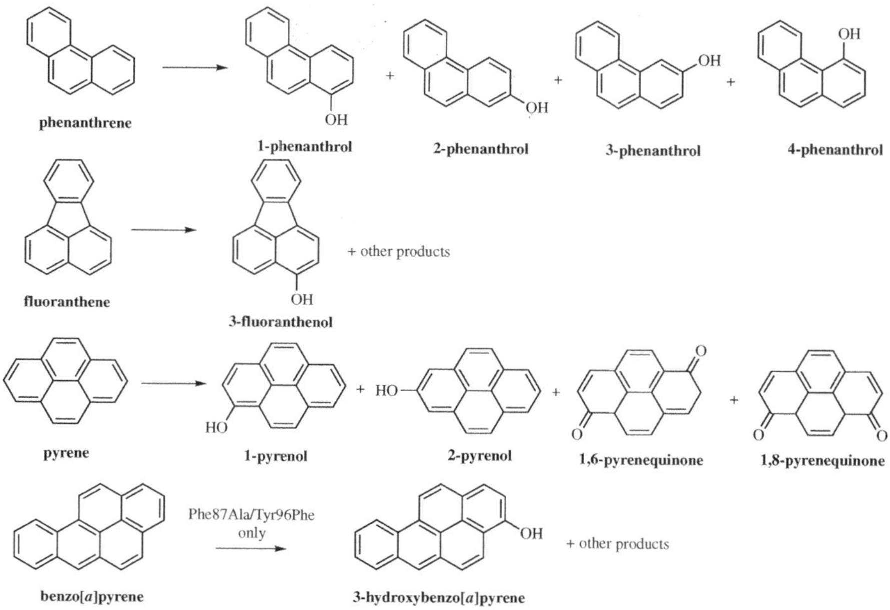

FIGURE 4.16 Oxidation of PAHs catalyzed by $P450_{cam}$ mutants.

active site was performed in order to improve the PAHs biodegradation by CYP102. The Arg47Leu/Tyr51Phe mutation greatly promoted the oxidation of the hydrophobic PAHs substrates, while the Ala264Gly mutation had the greatest effect in increasing the PAH oxidation rate (Figure 4.17) [37].

Among recalcitrant pollutants, polychlorinated benzenes are important targets that had triggered enzyme engineering for their efficient biodegradation. Jones and coworkers developed several genetically engineered $P450_{cam}$ enzymes capable of performing the oxidation of polychlorinated benzenes into their phenol derivatives (Figure 4.18) [38]. The product distribution with different enzymes was evaluated during the oxidation of 1,3-dichlorobenzene. 2,5-dichlorophenol and 2,4-dichlorophenol can be derived from rearrangement.

4.1.3 Baeyer–Villiger Reactions

The Baeyer–Villiger reaction of ketones affording esters or lactones is a powerful synthetic methodology possessing a wide range of applications, for example, synthesis of pheromones, antibiotics, steroids, and high-value building blocks. The reaction was described for the first time by Adolf von Baeyer (Nobel laureate in chemistry in 1905) [39] and Victor Villiger in 1899 for the oxidation of menthone to the corresponding lactone, employing a mixture of sodium persulfate and concentrate sulfuric acid (Figure 4.19) [40]. The persulfuric acid (also known as Caro's acid) was subsequently replaced by an organic peracid, such as *m*-chloroperbenzoic acid (*m*CPBA) or trifluoroperacetic acid.

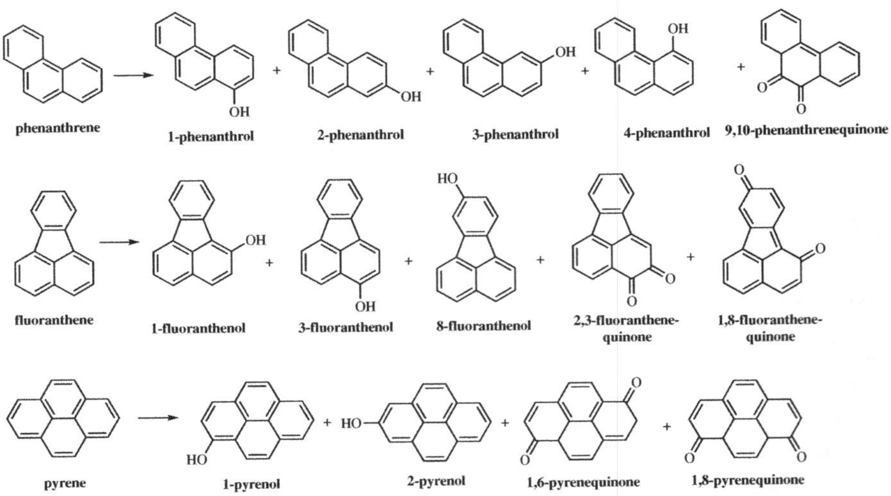

FIGURE 4.17 Oxidation of PAHs catalyzed by P450(BM3) mutants.

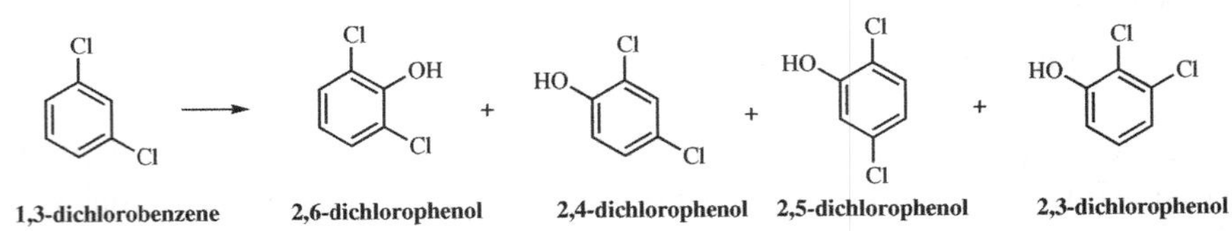

	2,6-dichlorophenol	2,4-dichlorophenol	2,5-dichlorophenol	2,3-dichlorophenol
wild-type	4	55	18	23
Tyr96Phe	4	63	8	25
Tyr96Phe/Phe87Trp	3	47	40	10
Tyr96Phe/Phe87Trp/Phe98Trp	5	54	31	10
Tyr96Phe/Phe87Trp/Val247Leu	2	60	30	8

FIGURE 4.18 Enzymatic oxidation of 1,3-dichlorobenzene with $P450_{cam}$ wild-type and mutants, including the percentage distribution for each product.

$$K_2S_2O_8 + H_2SO_4 + H_2O \longrightarrow 2\ KHSO_4 + \boxed{H_2SO_5}$$

Caro's acid

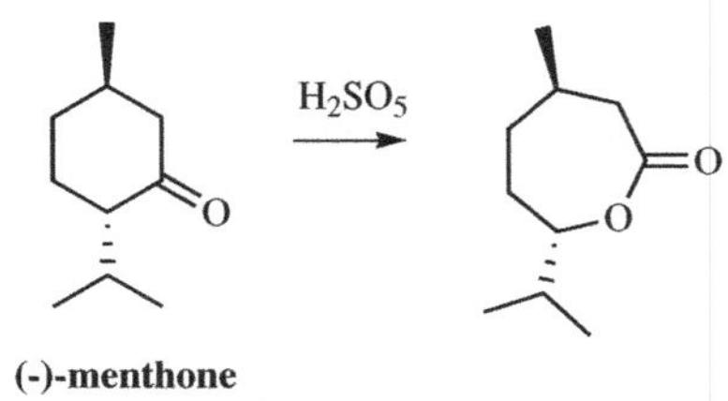

FIGURE 4.19 Baeyer–Villiger oxidation of menthone with Caro's acid.

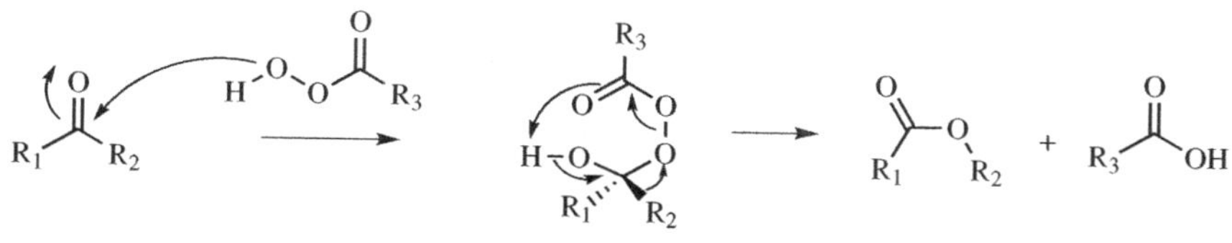

FIGURE 4.20 Accepted mechanism for the Baeyer–Villiger oxidation by peracids.

More than a century later, the reaction became a widely applied synthetic methodology in organic synthesis; firstly because it represents a versatile method for breaking carbon–carbon bonds under relatively mild conditions. Secondly, it tolerates a large number of functional groups in the substrate (i.e., chemoselective). The regiochemistry of the reaction is highly controlled by stereoelectronic aspects and therefore it is highly predictable. Finally, the Baeyer–Villiger reaction is generally stereoselective, meaning that the migrating group retains its configuration [41]. The currently accepted mechanism of the Baeyer–Villiger oxidation was proposed by Criegee in 1948 and involves a two-step reaction (Figure 4.20) [42].

In the first step, the nucleophilic attack of a peracid to the carbonyl group leads to the formation of a tetrahedral intermediate, known as "Criegee intermediate." In the second step, the Criegee intermediate undergoes a concerted rearrangement by migration of one of the groups (either R_1 or R_2) onto the peroxygen, affording the corresponding ester and carboxylic acid. It is normally accepted that the last step is rate limiting and the most important in determining the properties of the reaction [41]. As mentioned before, the regiochemistry of the reaction is highly controlled by stereoelectronic aspects. When different substituents are attached to the carbonyl group (R_1 and R_2), the one that can stabilize better a partial positive charge (i.e., more electron rich group) migrates to the oxygen atom (R_2 in Figure 4.20). Consequently, the usual migratory preference in the Baeyer–Villiger oxidation follows the trend *t*-alkyl > cyclohexyl > *sec*-alkyl > benzyl > phenyl > *n*-alkyl > methyl > hydride. However, it should be stated that the migratory aptitude can be overridden by the so-called primary and secondary stereoelectronic effects. In the Criegee intermediate, a particular alignment is required for the rearrangement achievement—the migrating group bond (R_M) has to be in an antiperiplanar position with respect to the O–O bond of the leaving group (primary stereoelectronic effect) and antiperiplanar to a lone pair of the hydroxyl group (secondary stereoelectronic effect) (Figure 4.21) [43–47].

Even though the Baeyer–Villiger reaction holds properties such as high chemo-, regio- and stereoselectivity, it still bears several disadvantages. Organic peracids are shock sensitive, being hazardous chemicals with strongly restricted transportation and storage. Furthermore, the Baeyer–Villiger reaction results in the formation of one equivalent of the corresponding carboxylic acid, which has to be properly disposed. As a consequence, the development of alternative approaches, such as the use of hydrogen peroxide in the presence of a catalyst or the *in situ* peracid generation has received increasing attention (see Chapter 6) [41,45]. Also, in order to match with the principles of green chemistry [48], metal-free organocatalytic Baeyer–Villiger

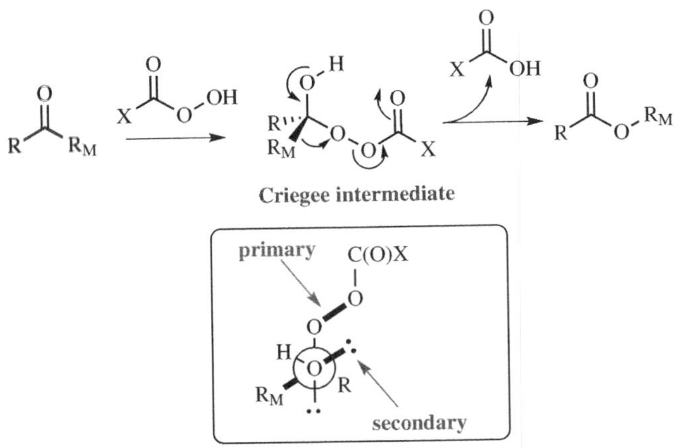

FIGURE 4.21 Primary and secondary stereoelectronic requirements for migration.

reactions employing hydrogen peroxide as primary oxidant have been developed [49,50]. An interesting approach exploiting the benefits of compressed (liquefied or supercritical) CO_2 as a green reaction media was developed by Bolm et al. In this work, the authors performed the Baeyer–Villiger oxidation of several ketones using oxygen as primary oxidant and benzaldehyde or pivalaldehyde as coreductant [51]. However, such approaches not always allow the access to asymmetric esters and lactones with high enantiomeric purity. Considering that the stereoselective outcome is one of the main strengths of biocatalytic methods, the development of an enzymatic Baeyer–Villiger oxidation is therefore highly attractive. In this area, Turfitt described the first biocatalytic Baeyer–Villiger reaction in 1948, studying the degradation of the steroid A-ring by fungi [52]. Since then, Baeyer–Villiger reactions have been found in biosynthetic pathways in several organisms [53–55]. The enzymatic Baeyer–Villiger oxidation is catalyzed by flavin-dependent monooxygenases, referred to as Baeyer–Villiger monooxygenases (BVMOs) [56].

4.1.3.1 Classification and Metabolic Role of BVMOs All BVMOs characterized so far depend on a flavin cofactor for catalysis. These enzymes incorporate one atom of molecular oxygen into the substrate and the other atom is reduced to water at the expense of NAD(P)H. BVMOs are essential enzymes in degradative metabolic pathways, allowing the cells to use alcohols as sources of carbon and energy [57]. Based on biochemical data facilitated by the increased number of BVMOs characterized, Willets has classified BVMOs in two types: Type I and Type II, [58] belonging to the class B and C flavoprotein monooxygenases, respectively [7].

Type I BVMOs are composed of a single chain protein, contain flavin adenine dinucleotide (FAD) as cofactor, and use NADPH as electron donor. Also, sequence similarity has shown that a fingerprint motif enables the screening and identification of Type I BVMOs (see Section 4.1.3.2). The prototype of Type I BVMOs is cyclohexanone monooxygenase (CHMO) from *Acinetobacter calcoaceticus* NCIMB 9871 (EC 1.14.13.22), purified and characterized by Trudgill and coworkers in 1976

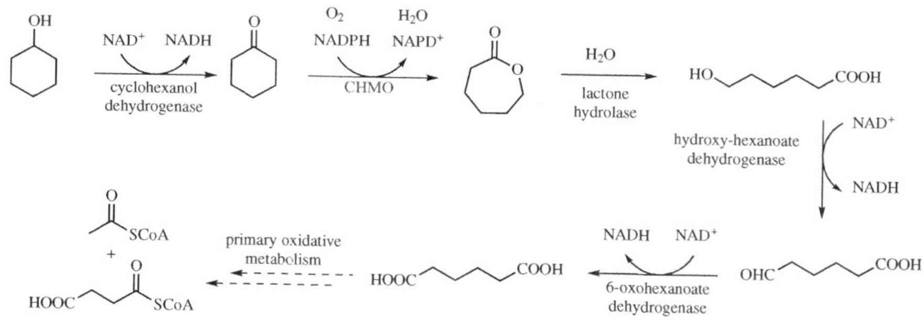

FIGURE 4.22 Metabolic role of CHMO in the degradation of cyclohexanol in *Acinetobacter*.

[59]. So far, most of the known BVMOs belong to this type. CHMO is the enzyme responsible for the oxygen insertion in cyclohexanone, which allows bacteria to degrade cyclohexanol to adipic acid (Figure 4.22). Further oxidation of adipic acid to acetyl-coenzyme A takes place, thus entering in the primary metabolism [60–63].

Type II BVMOs are composed of $\alpha_2\beta$ trimers, contain flavin mononucleotide (FMN) as cofactor and use NADH as electron donor [64]. The best known enzymes of this type are 2,5-diketocamphane monooxygenase (2,5-DKCMO) and 3,6-diketocamphane monooxygenase (3,6-DKCMO) from *P. putida* ATCC 17453 (NCIMB 10007) [65,66]. Degradation of camphor by *P. putida* involves several BVMOs [67–74]. The first step is the oxidation of each stereoisomer of camphor catalyzed by a cytochrome P450 enzyme (Figure 4.23). A subsequent oxidation by a dehydrogenase produces 2,5-diketocamphane **1** (from (+)-camphor) and 3,6-diketocamphane **2** (from (−)-camphor). These bicyclic ketones are the substrates for 2,5-DKCMO and 3,6-DKCMO (a mixture originally named as MO1), respectively [65,66]. The resulting lactones (**3** and **4**, respectively) undergo elimination to give

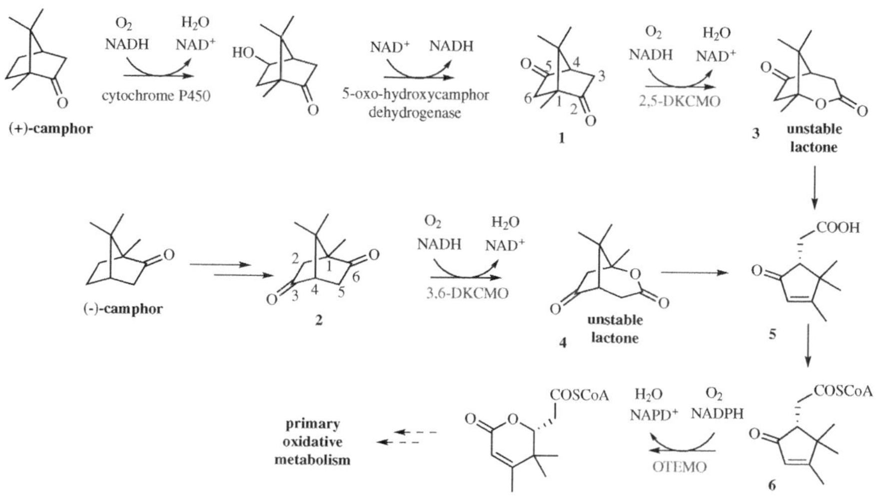

FIGURE 4.23 Camphor degradation involving BVMOs of Type II.

the cyclopentenylacetic acid derivative **5**, which in turn is activated to the coenzyme A derivative **6**. This activated thioester is oxidized by a third BVMO, namely 2-oxo-Δ3-4,5,5-trymethylcylcopentenylacetic acid monooxygenase (OTEMO, previously called MO2) [75].

4.1.3.2 Isolated Enzymes versus Whole-Cell Systems (Wild-Type and Recombinant Microorganisms)

Monooxygenases are NAD(P)H-dependent, and therefore cofactor regeneration issues need to be tackled. Whole-cell biocatalysis provide the needed enzymes and cofactors. However, if isolated enzymes are employed to catalyze redox reactions, nicotinamide cofactors must be externally provided. Nicotinamide cofactors are highly expensive chemicals to be used in equimolar amounts, and therefore an efficient cofactor recycling system must be designed (see Chapter 2). Both approaches have advantages and drawbacks (see Chapters 7–8).

Isolated BVMOs In BVMOs, the flavin reduction (FAD or FMN) proceeds at the expense of NADPH (BVMOs of Type I) or NADH (BVMOs of Type II). For enzymatic regeneration of oxidized $NAD(P)^+$, a second enzyme and an auxiliary substrate are often added to the biocatalytic system. There is no ideal approach to achieve the cofactor regeneration, as systems with low enzyme costs proceed with high auxiliary substrate costs and vice versa (see Chapter 2) [76,77]. As mentioned before, the most extensively studied and used BVMOs are those of Type I, particularly CHMO. Therefore, NADPH regeneration systems are of especial interest within the biocatalyzed Baeyer–Villiger reaction. As examples, FDH from *Pseudomonas* sp. 101 was engineered by multiple site-directed mutagenesis to accept $NADP^+$ [78–80], allowing the efficient regeneration of NADPH in biotransformations with CHMO [81]. In addition, the glucose 6-phosphate/glucose 6-phosphate dehydrogenase system is commonly used in BVMO catalyzed reactions with isolated enzymes [82–86].

An alternative approach developed by Willets and coworkers employs an alcohol dehydrogenase from *Thermoanaerobium brockii* and several secondary alcohols (**7–9**) as cosubstrates. In this closed-loop recycling system, the alcohol is transformed into the corresponding ketone producing NADPH. In a subsequent step, the ketone becomes the substrate for the BVMO (Figure 4.24) [87,88].

Finally, taking advantage of the favorable thermodynamics for the conversion of phosphite to phosphate catalyzed by phosphite dehydrogenase (PTDH), this system

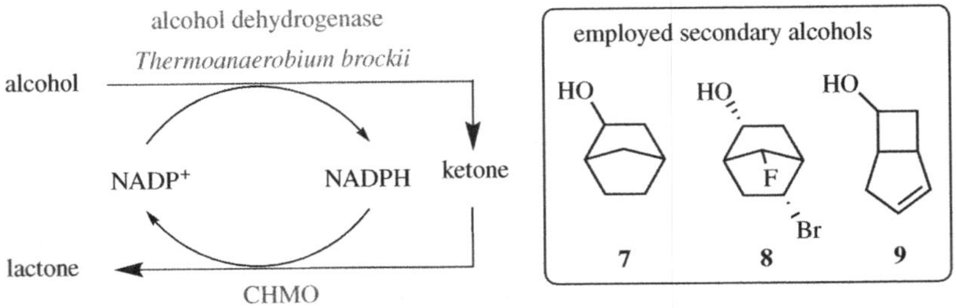

FIGURE 4.24 Closed-loop cofactor recycling.

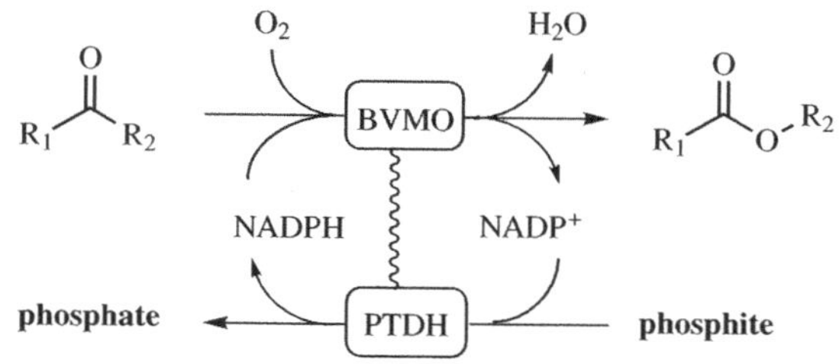

FIGURE 4.25 Cofactor regeneration by fusion enzymes.

was also employed for cofactor regeneration [89,90]. This approach was further studied by developing a fusion protein, being the BVMO covalently connected to PTDH to produce a self-sufficient fusion protein. This protein is able to perform both a Baeyer–Villiger oxidation and cofactor regeneration (Figure 4.25) [91]. PTDH converts phosphite into phosphate while $NADP^+$ is reduced to NADPH. By this approach, CHMO from *Acinetobacter* sp., cyclopentanone monooxygenase (CPMO) from *Comamonas* sp., and phenylacetone monooxygenase (PAMO) from *Thermobifida fusca* were successfully linked to PTDH.

Whole-Cell Systems Whole-cell conversions are attractive mainly because of the efficient intracellular coenzyme regeneration, yet with the existence of some unwanted reactions in some cases [92]. Particularly, in whole-cell biotransformations with native *Acinetobacter* strains, the Baeyer–Villiger lactone product can be further metabolized by a lactone hydrolase [57]. An inhibition strategy of the lactone hydrolase, employing tetraethyl pyrophosphate in the reaction media, was successfully developed [93]. Also, hydrolase-deficient mutants can be employed for Baeyer–Villiger oxidations [94]. Besides the potential side reactions, *Acinetobacter* strain is a class II pathogen, which requires microbiological expertise in cell handling, hampering its application in organic synthesis. Finally, the enzymes required to metabolize cyclohexanol, which includes CHMO, are produced only when this substrate is present. As a consequence, the presence of cyclohexanol and its metabolic products makes target lactone product isolation and purification difficult [57]. To overcome this, molecular biology techniques enabled the overexpression of the gene for BVMOs in both eukaryotic (*Saccharomyces cerevisiae*) [57,95] and prokaryotic (*Escherichia coli*) [96] host organisms. The engineered microorganisms are nonpathogenic and can be easily used as a biocatalytic tool for Baeyer–Villiger oxidations by organic chemists. Also, the enzyme can be produced in high quantities by using strong promoter systems. Finally, control of protein production is achieved by addition of selective inducers such as galactose or IPTG (isopropyl-β-D-thiogalactopyranoside) (see Chapters 7–8 for further details on designer cells for biocatalysis).

4.1.3.3 Substrate Profile of Available Baeyer–Villiger Monooxygenases

Table 4.1 provides a selection of the most relevant BVMOs together with information on the original strain, the natural substrate and the year of cloning when applicable.

TABLE 4.1 BVMOs Available with Known Potential for Synthetic Applications

BVMO	Original Strain	Natural Substrate	Year of Cloning	Reference
Type I BVMOs				
Cyclododecanone monooxygenase (CDMO$_{Rhodo}$)	*Rhodococcus ruber* SC1	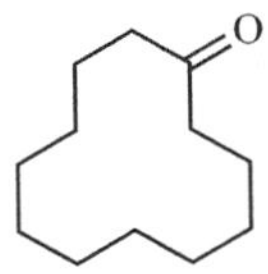	2001	[97]
Cyclohexanone monooxygenase (CHMO$_{Acineto}$)	*Acinetobacter* sp. NCIMB 9781	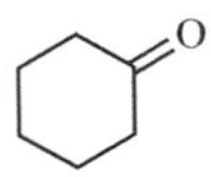	1988	[59,98]
Cyclohexanone monooxygenase (CHMO$_{Arthro}$)	*Arthrobacter* sp. BP2	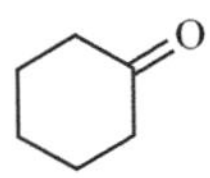	2003	[62,99]
Cyclohexanone monooxygenase (CHMO$_{Brevi1\&2}$)	*Brevibacterium* HCU	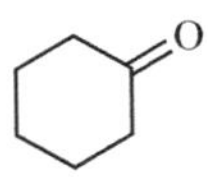	2000	[100]
Cyclohexanone monooxygenase (CHMO$_{Rhodo1\&2}$)	*Rhodococcus* Phi1 and Phi2	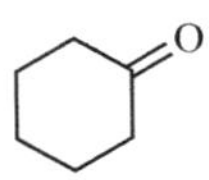	2003	[99]
Cyclopentadecanone monooxygenase (CPDMO)	*Pseudomonas* sp. HI-70	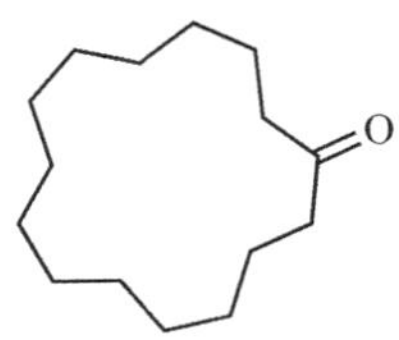	2006	[101]
Cyclopentanone monooxygenase (CPMO$_{Coma}$)	*Comamonas* NCIMB 9872	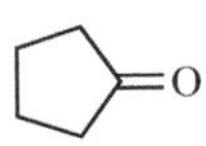	2002	[63,102]
4-Hydroxyacetophenone monooxygenase (HAPMO$_{Pseudo}$)	*Pseudomonas fluorescences* ACB	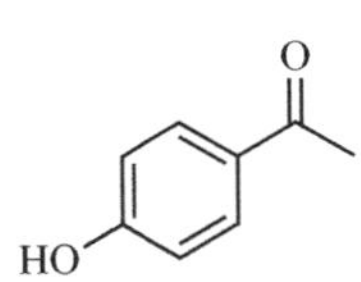	2001	[103]

TABLE 4.1 ***(Continued)***

BVMO	Original Strain	Natural Substrate	Year of Cloning	Reference
Phenylacetone monooxygenase (PAMO)	*T. fusca*	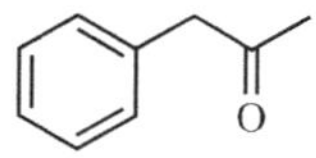	2005	[104,105]
Steroid monooxygenase (SMO_{Rhodo})	*Rhodococcus rhodochrous* IFO 3338	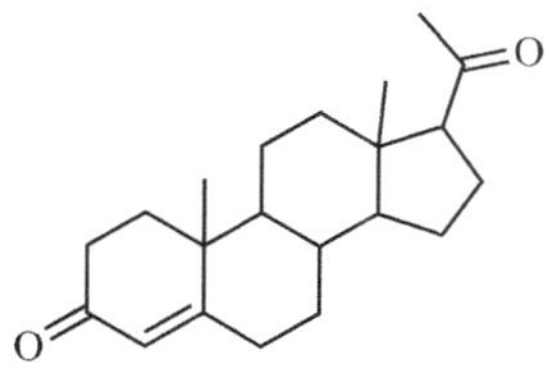	1999	[106]
Type II BVMOs				
2,5-Diketocamphane monooxygenase (2,5-DKCMO)	*P. putida* ATCC 17453 (NCIMB 10007)		n.a.[a]	[65]
3,6-Diketocamphane monooxygenase (3,6-DKCMO)	*P. putida* ATCC 17453 (NCIMB 10007)	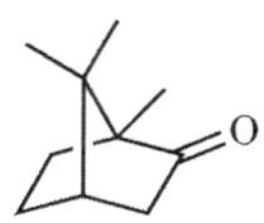	n.a.[a]	[66]

[a]n.a., not applicable.

Cyclohexanone Monooxygenase CHMO from *Acinetobacter* sp. NCIMB 9871 is so far the most studied Type I BVMO regarding its mechanism, biocatalytic properties and substrate profile. CHMO was the first BVMO which the gene had been cloned [98]. However, the current 543 amino acid sequence of CHMO was deposited later by Iwaki and coworkers [61]. CHMO displays an impressive broad substrate specificity allowing the oxidation of ketones, including cyclic, bycyclic, heterocyclic with different ring sizes and substitution patterns [57,107–109]. Some representative substrates for CHMO are included in Figure 4.26.

Besides CHMO from *Acinetobacter* sp., more recently some other CHMO were studied (Table 4.1). Systematic studies employing 2-, 3- and 4-substituted cyclohexanones showed that substrate profile and enantioselectivities for the new BVMO are similar to that of CHMO from *Acinetobacter* sp. [110].

Cyclopentanone Monooxygenase CPMO was originally purified by Trudgill in 1976 from *Pseudomonas* sp. NCIMB 9872 [102]. More recently, this microorganism was reclassified as belonging to genus *Comamonas* and in 2002 the gene encoding

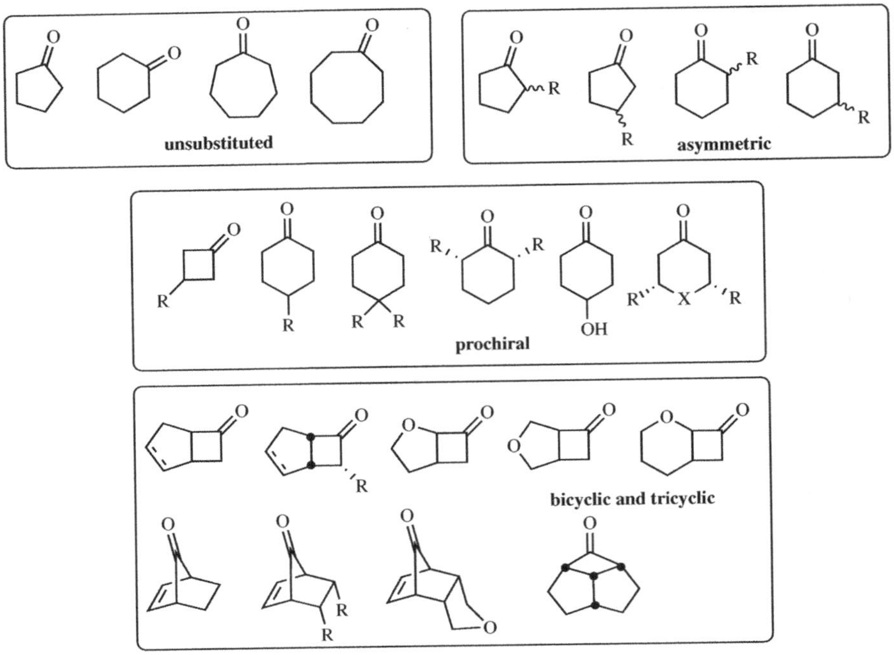

FIGURE 4.26 Representative substrates accepted by CHMO.

CPMO was cloned and expressed in *E. coli* [63]. Most of the studies with CPMO were conducted employing whole-cells, enabling transformation of C_4–C_8 ketones which are also oxidized by CHMO. Moreover, CPMO can oxidize aromatic and conjugated ketones that are not accepted by CHMO (Figure 4.27) [63,64,111].

Even though several substrates are converted by both CHMO and CPMO, the stereoselectivity differs. The oxidation of 2-substituted cyclopentanones and cyclohexanones with both CHMO and CPMO produces the (*S*)-lactone, however with lower optical purities with the latter. Prochiral 4-methylcyclohexanone is oxidized by CHMO to the (*S*)-lactone whereas CPMO affords the (*R*)-lactone [63]. CPMO is able to transform prochiral fused systems such as bicycle[4.3.0]ketones forming the opposite enantiomer to that obtained with CHMO [112]. Finally, CPMO accepts substrates that are not accepted by CHMO, such as indanone derivatives, broadening the scope of enzymatic Baeyer–Villiger reactions [63,113].

Cyclododecanone Monooxygenase Alicyclic hydrocarbons (e.g., cyclopentane, cyclohexane and cyclodecane) are important components of petroleum, and BVMOs of several microorganisms are able to metabolize them through oxidation reactions [114]. In particular, cyclododecanone monooxygenase (CDMO), the enzyme that oxidizes cyclododecanone to lauryl ketone, was first isolated and purified in 1999 from *Rhodococcus ruber* CD4 [115]. Subsequently the gene from another strain (*R. ruber* SC1) was cloned and expressed in *E. coli*. This recombinant microorganism is active toward cyclic ketones (C_{11}–C_{15}) producing the corresponding lactones, but is

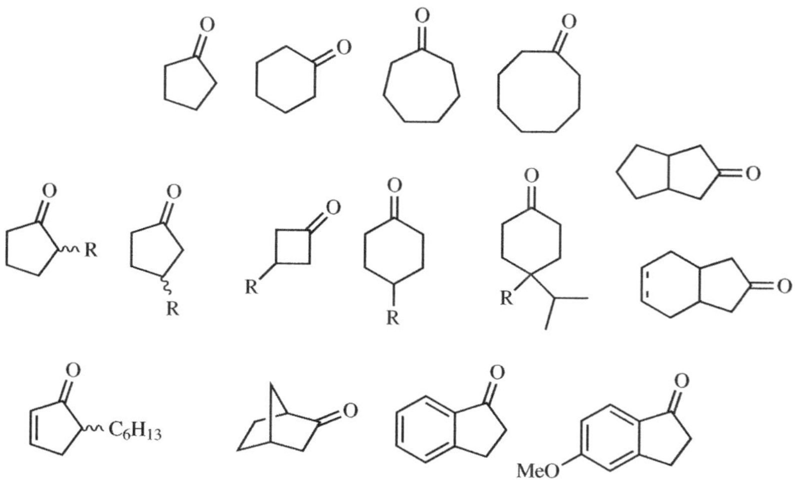

FIGURE 4.27 Representative substrates accepted by CPMO.

not able to oxidize smaller ring systems (C_6–C_{10}) [97]. The ability to convert large ring systems that are not accepted by CHMO makes CDMO a valuable biocatalyst.

Cyclopentadecanone Monooxygenase Cyclopentadecanone monooxygenase (CPDMO), originally derived from *Pseudomonas* sp. HI-70 was cloned in 2006 [101]. This enzyme displays an unusual substrate profile, showing that CPDMO is able to accommodate and oxidize both large and small cyclic (C_5, C_6, C_{11}–C_{15}) and bicyclic ketones (such as decalone and β-tetralone), often with outstanding enantioselectivities.

Steroid Monooxygenase The first reports on the biocatalyzed Baeyer–Villiger oxidation involved studies of steroid metabolism [52], and steroid monooxygenases (SMOs) from *Rhodococcus rhodochrous* have been isolated and characterized [116]. Moreover, steroid metabolism in different fungi was investigated [117]. The overexpression of SMO in *E. coli* turned it a valuable tool for the bioconversion of steroids [106]. However, the substrate specificity has not been thoroughly established and further studies are needed to assess the ability of SMO to convert compounds somehow unrelated to its natural substrate.

Phenylacetone Monooxygenase PAMO, originally discovered by searching for a more stable BVMO, was obtained from the thermophilic actinomycete *T. fusca* employing genome mining [105]. Biochemical characterization of PAMO shows a NADPH-dependent and FAD-containing BVMO, active at higher temperatures than these typical for BVMO (approximately 50°C). PAMO mostly accepts aromatic ketones (phenylacetone, benzylacetone, 4-hydroxyacetophenone), but it is also able to convert 2-dodecanone and bicyclohept-2-en-6-one into their corresponding esters (or lactones). However, PAMO is not able to oxidize cyclohexanone [105,118].

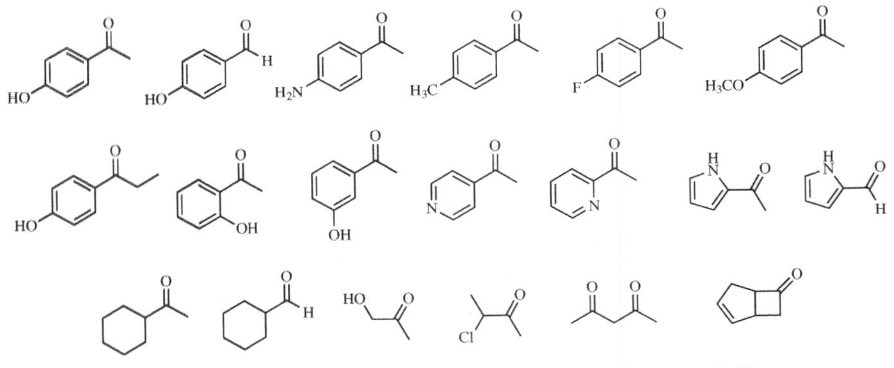

FIGURE 4.28 Representative substrates accepted by HAPMO.

4-Hydroxyacetophenone Monooxygenase Aryl ketones can be enzymatically degraded through a metabolic pathway involving a Baeyer–Villiger oxidation, first reported in 1970 [119], and further studied for several substrates such as acetophenones [120,121], 4-ethylphenol [122], and fluorene [123]. The first BVMO converting such substrate profile was described in 1999, with the isolation of 4-hydroxyacetophenone monooxygenase (HAPMO) from *Pseudomonas fluorescens* ACB [124]. In 2001, the encoding gene was cloned and overexpressed in *E. coli* [103]. HAPMO displays a broad substrate specificity, accepting a wide range of aryl, heteroaromatic, and aliphatic ketones (Figure 4.28) besides its natural substrate 4-hydroxyacetophenone [103,125,126]. Although the best results are observed with acetophenones and propiophenones, with heteroaromatic ketones, the enzymatic activity is in general high, except for aliphatic ketones, which display moderate conversion rates.

2,5- and 3,6-Diketocamphane Monooxygenase 2,5-DKCMO and 3,6-DKCMO are BVMO of Type II, identified as involved in the microbial degradation of camphor by *P. putida* [65,66]. As mentioned before, most of the systematic studies involving BVMOs are with Type I enzymes. However, conversion of racemic fused bicyclic ketones with BVMOs of Type II was performed. The treatment of the ketone substrate with 2,5-DKCMO led to a regiodivergent oxidation to the frequently called "normal" and "abnormal" ketones (Figure 4.29) with high enantiomeric excess. 3,5-DKCMO showed similar enantiopreference but lower selectivity [127–129].

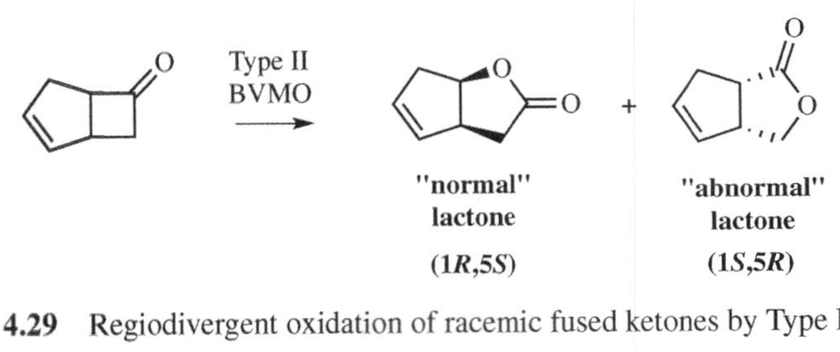

FIGURE 4.29 Regiodivergent oxidation of racemic fused ketones by Type II BVMOs.

4.1.3.4 Synthetic Applications of BVMOs Besides the efforts devoted to elucidation of the enzymatic mechanism and the evaluation of substrate specificities of BVMOs, the impressive potential of these enzymes has been fully demonstrated in the production of chiral compounds, valuable intermediates, building blocks, natural products, and bioactive compounds. In the following sections most relevant biotransformations in organic synthesis will be outlined.

Classical and Dynamic Kinetic Resolutions BVMO-catalyzed kinetic racemic resolutions have been thoroughly studied [130]. Representative examples for the biooxidation of racemic 2-cyclopentanones are included in Table 4.2.

An early and important example [93] is the oxidation of 2-undecylcyclopentanone (**10h**) with *Acinetobacter* sp. NCIMB 9781 in the presence of the lactone hydrolase inhibitor tetraethyl pyrophosphate, affording the (*S*)-lactone **11h**, with good optical purity. The recovered (*R*)-ketone could be chemically oxidized through a Baeyer–Villiger reaction, affording the (*R*)-lactone 5-hexadecanolide, a pheromone isolated from the oriental hornet *Vespa orientalis* (Figure 4.30, Table 4.2) [93].

The regioselectivity for the enzymatic Baeyer–Villiger oxidation follows the expected migration of the more electron rich (higher substituted) α-carbon, affording the frequently called "normal" lactone (being the "abnormal" lactone the product obtained when the less electron rich group migrates during the rearrangement). Moreover, a preference for the generation of the (*S*)-lactone is observed, meaning that the enzyme recognizes the chiral center at the substrate, converting only the (*S*)-ketones [107,135]. The enzymatic Baeyer–Villiger oxidation is also chemoselective, meaning that it tolerates a large number of functional groups in the substrate. An example that fully illustrates this observation involves the bioconversion of 5-hexylcyclopent-2-enone (**12**) by CPMO from *Comamonas* sp. NCIMB 9872 (Figure 4.31) [111].

BVMOs have been successfully employed (mainly for CHMO) to catalyze the oxidation of a large number of 2-substituted cyclohexanones (Table 4.3) [57,59,83,95,110,130,136–140].

Optically active 6-substituted ε-caprolactones are valuable building blocks in asymmetric synthesis, being the kinetic resolution of 2-substituted cyclohexanones employing a Baeyer–Villiger oxidation an attractive strategy to afford those products [137]. As shown in Table 4.3, enantioselectivities are high, being the lactone and the residual ketone recovered in high optical purities through a single reaction. Again, the same preference for generation of (*S*)-lactone is observed, leaving the (*R*)-ketone unchanged. Racemic 2-methylcyclohexanone (**14a**) is oxidized by CHMO to the corresponding lactone with only modest selectivity for the (*S*)-ketone [130]. A beneficial effect on optical purities is observed when the chain size is increased [136,137,139]. The enzymatic preparation of lactone **15l**, which was converted into (*R*)-(+)-lipoic acid (see Section "Synthesis of natural product and biologically active compounds"), produced antipode isomers when using different biocatalysts [111,140,141]. The MO2 from *Pseudomonas* sp. NCIMB 10007 catalyzed biooxidation of **15l**, yielded the (*R*)-lactone, whereas the Baeyer–Villiger reaction performed by CPMO from *Comamonas* sp. NCIMB 9872 afforded the opposite enantiomer in moderate

TABLE 4.2 Kinetic Resolution of Racemic 2-Substituted Cyclopentanones

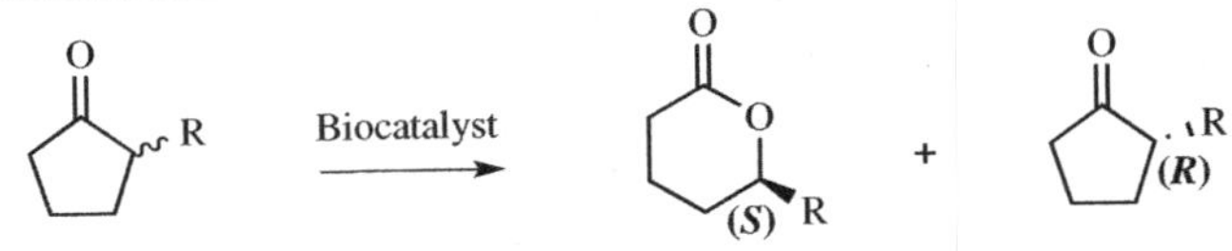

Substrate	R	Biocatalyst	*ee* (%)[a,b]	Yield (%)	Reference
10a	Me	$CHMO_{Acineto}$ (recombinant cells)	**11a**: 32 (−) **10a**: 44 (+)	**11a**: 36 **10a**: 34	[95,131]
		CPDMO (recombinant cells)	**11a**: 87 (−) **10a**: n.r.	**11a**: 10 **10a**: n.r.	[101]
10b	Et	$CHMO_{Acineto}$ (recombinant cells)	**11b**: 39 (n.r.) **10b**: 46 (n.r.)	**11b**: 44 **10b**: 37	[95,131]
10c	*n*-Pr	$CHMO_{Acineto}$ (recombinant cells)	**11c**: 67 (−) **10c**: 72 (−)	**11c**: 51 **10c**: 21	[95,131]
10d	Allyl	$CHMO_{Acineto}$ (recombinant cells)	**11d**: 32 (−) **10d**: 51 (−)	**11d**: 76 **10d**: 10	[95,131]
10e	*n*-Bu	$CHMO_{Acineto}$ (recombinant cells)	**11e**: >98 (−) **10e**: >98 (−)	**11e**: 18 **10e**: 32	[95,131]
10f	*n*-Hex	$CHMO_{Acineto}$ (recombinant cells)	**11f**: >98 (−) **10f**: >98 (−)	**11f**: 32 **10f**: 42	[95,131]
10g	*n*-Oct	$CHMO_{Acineto}$ (recombinant cells)	**11g**: >98 (−) **10g**: >98 (−)	**11g**: 25 **10g**: 14	[95,131]
10h	*n*-Undec	$CHMO_{Acineto}$ (recombinant cells)	**11h**: >98 (−) **10h**: >98 (−)	**11h**: 39 **10h**: 37	[95,131]
		$CHMO_{Acineto}$ (*Acinetobacter* NCIMB 9871, wild-type cells)	**11h**: 74 (−) **10h**: 95 (−)	**11h**: 25 **10h**: 30	[93]
10i	$(CH_2)_3Br$	$CHMO_{Acineto}$ (recombinant cells)	**11i**: 95 (n.r.) **10i**: 92 (n.r.)	**11i**: 47 **10i**: 47	[132]
10j	CH_2Oallyl	$CHMO_{Acineto}$ (recombinant cells)	**11j**: 93 (n.r.) **10j**: n.r.	**11j**: 33 **10j**: 19	[132]
10k	CH_2COOEt	$MO2_{Pseudo}$ (*P. putida* NCIMB 10007, isolated enzyme)	**11k**: 98 (n.r.) **10k**: 75 (n.r.)	**11k**: 27 **10k**: 37	[133]
10l	CH_2OBn	$CHMO_{Acineto}$ (recombinant cells)	**11l**: 97 (−) **10l**: 43 (n.r.)	**11l**: 43 **10l**: 48	[134]

[a] Sign of specific rotation or absolute configuration is given in parenthesis.

[b] Some substituents require change in the Cahn–Ingold–Prelog priority numbering, leading to a reversal of the (*R*)/(*S*) assignment; however, the sense of chirality is the same.

n.r., not reported.

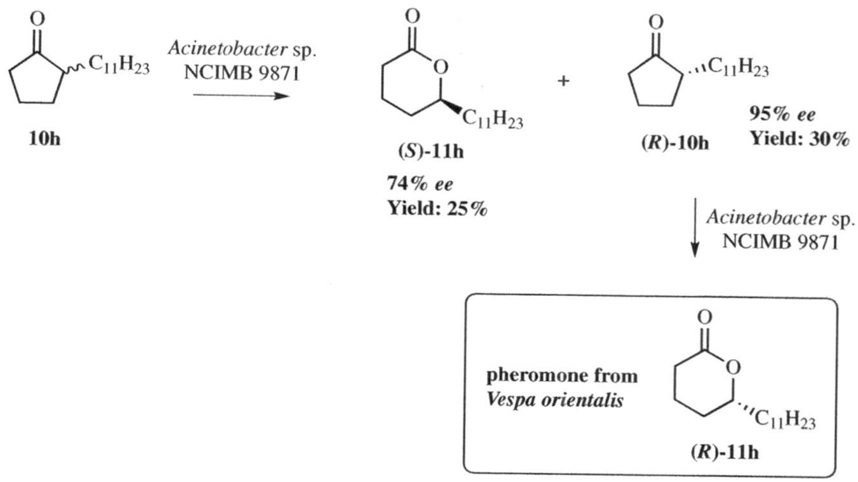

FIGURE 4.30 Kinetic resolution of racemic 2-undecylcyclopentanone.

enantiomeric excess. Likewise, the BVMO-catalyzed kinetic resolution of 4-hydroxy-2-aliphatic ketones was described by Kirschner and Bornscheuer in 2006 [140,142]. Recombinant BVMO from *P. fluorescens* DSM 50106 expressed in *E. coli* converted the aliphatic ketones **16a**, **16c**, and **16e** into the corresponding (*S*)-hydroxyl acetates **17** with good enantioselectivities (observed enantiomeric excesses higher than 90%) (Table 4.4). Also, the other possible product ("abnormal" ester, **18**) was detected in small proportions (less than 5%). This biotransformation is particularly useful because an optically active acetate of a 1,2-diol is afforded, broadening the scope of BVMOs. Directed evolution of the enzyme led to an improvement of the enantioselectivity [143]. More recently, several additional enzymes of various bacterial origins were tested as biocatalysts for the kinetic resolution of β-hydroxyketones. A BVMO from *Pseudomonas veronii* MEK700 was used with the same three substrates, yielding the (*S*)-hydroxyl acetates, yet with low enantioselectivities [144]. Furthermore, CHMOs from different bacteria (*A. calcoaceticus* NCIMB 9871, *Brevibacterium* sp., and *Rhodococcus* sp.) as well as CPMO from *Comamonas* sp. 9875 were tested with several substrates. Such substrates included 4-hydroxy-2-ketones (**16a–g**) and 5-hydroxy-3-ketones (**16h–i**) [145]. In the case of a protection of the hydroxyl group

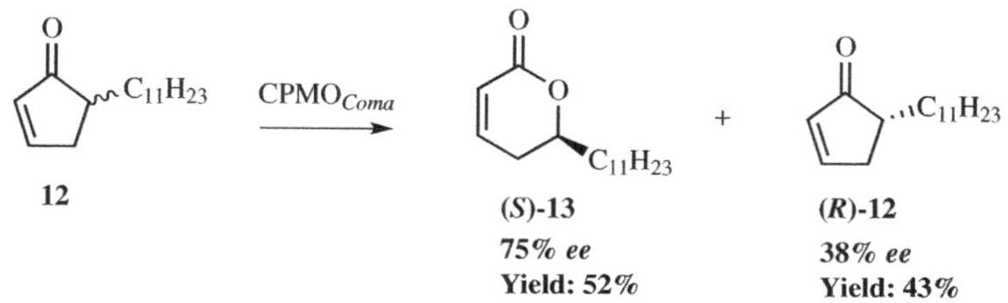

FIGURE 4.31 Kinetic resolution of racemic 5-hexylcyclopent-2-enone by cyclopentanone monooxygenase (isolated enzyme).

TABLE 4.3 Kinetic Resolution of Racemic 2-Substituted Cyclohexanones

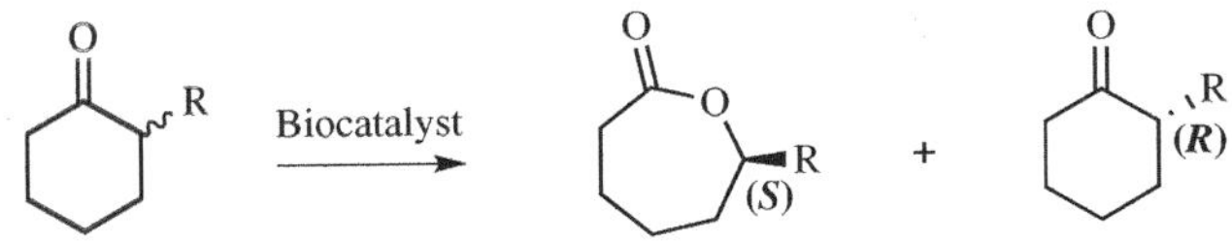

14a-l 15a-l 14a-l

Substrate	R	Biocatalyst	*ee* (%)[a,b]	Yield (%)	Reference
14a	Me	CHMO$_{Acineto}$ (*Acinetobacter* NCIMB 9871, wild-type cells)	**15a**: 61 (−) **14a**: 35 (+)	**15a**: 35 **14a**: 52	[136]
		CHMO$_{Acineto}$ (recombinant cells)	**15a**: 49 (−) **14a**: n.r.	**15a**: 50 **14a**: n.r.	[137,139]
		CHMO$_{Xantho}$ (*Xanthobacter* sp. ZL5, recombinant cells)	**15a**: 62 (−) **14a**: 29 (+)	**15a**: 24 **14a**: n.r	[138]
14b	Et	CHMO$_{Acineto}$ (*Acinetobacter* NCIMB 9871, wild-type cells)	**15b**: 38 (−) **14b**: >98 (−)	**15b**: 10 **14b**: 60	[136]
		CHMO$_{Acineto}$ (recombinant cells)	**15b**: 95 (−) **14b**: >98 (−)	**15b**: 79 **14b**: 69	[137,139]
14c	*n*-Pr	CHMO$_{Acineto}$ (recombinant cells)	**15c**: 97 (−) **14c**: 92 (−)	**15c**: 54 **14c**: 66	[137,139]
14d	*i*-Pr	CHMO$_{Acineto}$ (recombinant cells)	**15d**: >98 (−) **14d**: 96 (−)	**15d**: 41 **14d**: 46	[137,139]
14e	Allyl	CHMO$_{Acineto}$ (recombinant cells)	**15e**: >98 (−) **14e**: >98 (−)	**15e**: 59 **14e**: 58	[137,139]
		CHMO$_{Xantho}$ (*Xanthobacter* sp. ZL5, recombinant cells)	**15e**: 90 (−) **14e**: >99 (−)	**15e**: 59 **14e**: n.r.	[138]
14f	*n*-Bu	CHMO$_{Acineto}$ (*Acinetobacter* NCIMB 9871, wild-type cells)	**15f**: >98 (−) **14f**: 10	**15f**: 6 **14f**: 37	[136]
		CHMO$_{Acineto}$ (recombinant cells)	**15f**: >98 (−) **14f**: 98 (−)	**15f**: 59 **14f**: 64	[137,139]
14g	*n*-Hex	CHMO$_{Acineto}$ (*Acinetobacter* NCIMB 9871, wild-type cells)	**15g**: 10 (−) **14g**: 98 (−)	**15g**: 23 **14g**: 25	[136]

TABLE 4.3 (*Continued*)

Substrate	R	Biocatalyst	*ee* (%)[a,b]	Yield (%)	Reference
14h	*n*-Non	$CHMO_{Acineto}$ (*Acinetobacter* NCIMB 9871, wild-type cells)	**15h**: 85 (−) **14h**: 42 (−)	**15h**: 26 **14h**: 32	[136]
14i	Ph	$CHMO_{Acineto}$ (*Acinetobacter* NCIMB 9871, wild-type cells)	**15i**: >98 (+) **14i**: 86 (−)	**15i**: 40 **14i**: 48	[136]
		$CHMO_{Xantho}$ (*Xanthobacter* sp. ZL5, recombinant cells)	**15i**: >99 (+) **14i**: 76 (−)	**15i**: 48 **14i**: n.r.	[138]
14j	Bn	$CHMO_{Acineto}$ (*Acinetobacter* NCIMB 9871, wild-type cells)	**15j**: >96 (−) **14j**: 78 (−)	**15j**: 28 **14j**: 22	[136]
		$CHMO_{Xantho}$ (*Xanthobacter* sp. ZL5, recombinant cells)	**15j**: >99 (−) **14j**: 71 (−)	**15j**: 45 **14j**: n.r.	[138]
14k	CH_2COOEt	$CHMO_{Acineto}$ (*Acinetobacter* NCIMB 9871, isolated enzyme)	**15k**: >99 (−) **14k**: 64 (−)	**15k**: 39 **14k**: 60	[81]
14l	CH_2CH_2OAc	$CHMO_{Acineto}$ (*Acinetobacter* NCIMB 9871, isolated enzyme)	**15l**: >99 (n.r.) **14l**: 76 (n.r.)	**15l**: 34 **14l**: 66	[81]
		$MO2_{Pseudo}$ (*P. putida* NCIMB 10007, isolated enzyme)	**15l**: 83 (−) **14l**: 75 (+)	**15l**: 36 **14l**: 13	[141]
		$CPMO_{Coma}$ (*Comamonas* sp. NCIMB 9872 isolated enzyme)	**15l**: 42 (+) **14l**: 68 (−)	**15l**: 59 **14l**: 37	[141]

[a]Sign of specific rotation or absolute configuration is given in parenthesis.

[b]Some substituents require change in the Cahn–Ingold–Prelog priority numbering, leading to a reversal of the (*R*)/(*S*) assignment; however, the sense of chirality is the same.

n.r., not reported.

TABLE 4.4 Kinetic Resolution of Aliphatic Ketones

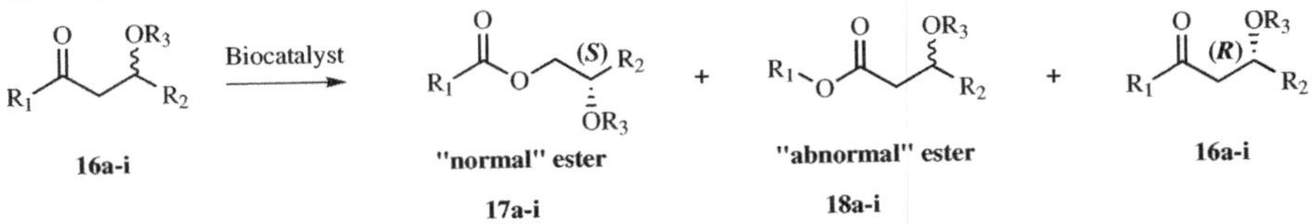

Substrate	R_1	R_2	R_3	Biocatalyst	*ee* (%)[a]	Reference
16a	Me	Bu	H	BVMOF1 (*P. fluorescens* DSM 50106, recombinant cells)	**17a**: 93 (*S*) **16a**: 61 (*R*)	[142]
				CHMO$_{Acineto}$ (recombinant cells)	**17a**: 93 (*S*) **16a**: 91 (*R*)	[145]
				CPMO$_{Coma}$ (*Comamonas* sp. NCIMB 9872, recombinant cells)	**17a**: 56 (*R*) **16a**: 64 (n.r.)	[145]
				MeKMO (*P. veronii* MEK700, isolated enzyme)	**17a**: 82 (*S*) **16a**: 72 (*R*)	[144]
16b	Me	Pent	H	BVMOF1 (*P. fluorescens* DSM 50106, recombinant cells)	**17b**: 98 (*S*) **16b**: 89 (*R*)	[145]
				CHMO$_{Acineto}$ (recombinant cells)	**17b**: >99 (*S*) **16b**: 96 (*R*)	[145]
16c	Me	Hex	H	BVMOF1 (*P. fluorescens* DSM 50106, recombinant cells)	**17c**: 91 (*S*) **16c**: 84 (*R*)	[142]
				CHMO$_{Acineto}$ (recombinant cells)	**17c**: 93 (*S*) **16c**: 92 (*R*)	[145]
				MeKMO (*P. veronii* MEK700, isolated enzyme)	**17c**: 15 (*R*) **16c**: 95 (n.r.)	[144]
16d	Me	Hept	H	CHMO$_{Acineto}$ (recombinant cells)	**17d**: 52 (*S*) **16d**: 13 (*R*)	[145]
				CHMO$_{Rhodo2}$ (*Rhodococcus* sp., recombinant cells)	**17d**: 80 (*S*) **16d**: 17 (*R*)	[145]
16e	Me	Oct	H	BVMOF1 (*P. fluorescens* DSM 50106, recombinant cells)	**17e**: 90 (*S*) **16e**: 74 (*R*)	[142]
				MeKMO (*P. veronii* MEK700, isolated enzyme)	**17e**: 53 (*R*) **16e**: 27 (n.r.)	[144]
16f	Me	Hex	Ac	CPMO$_{Coma}$ (*Comamonas* sp. NCIMB 987,2 recombinant cells)	**17f**: >99 (*R*) **16f**: 45 (n.r.)	[145]
				CHMO$_{Brevi2}$ (*Brevibacterium* HCU, recombinant cells)	**17f**: 91 (*R*) **16f**: 41 (n.r.)	[145]

with an acetyl (**16f**) and a formyl (**16g**) group, enantiocomplementary esters of those obtained in the former biotransformations were afforded. Finally, the location of the keto group modifies the regioselectivity of the enzymes. For 5-hydroxy-3-ketones, the "abnormal" ester was generated in 64% (**16h**) or 40% (**16i**) in favor over the "normal" one.

TABLE 4.4 (*Continued*)

Substrate	R_1	R_2	R_3	Biocatalyst	*ee* (%)[a]	Reference
16g	Me	Bu	CHO	$CPMO_{Coma}$ (*Comamonas* sp. NCIMB 9872, recombinant cells)	**17g**: 99 (*R*) **16g**: 99 (n.r.)	[145]
				$CHMO_{Brevi2}$ (*Brevibacterium* HCU, recombinant cells)	**17g**: >99 (*R*) **16g**: 27 (n.r.)	[145]
16h	Et	Bu	H	$CPMO_{Coma}$ (*Comamonas* sp. NCIMB 9872, recombinant cells)	**17h**: 96 (n.r.) **16h**: >99 (n.r.)	[145]
				BVMOF1 (*P. fluorescens* DSM 50106, recombinant cells)	**17h**: 53 (n.r.) **16h**: 25 (n.r.) **re**: 64	[145]
16i	Et	Pent	H	$CPMO_{Coma}$ (*Comamonas* sp. NCIMB 9872, recombinant cells)	**17i**: 87 (n.r.) **16i**: 81 (n.r.)	[145]
				$CHMO_{Brevi2}$ (*Brevibacterium* HCU, recombinant cells)	**17i**: 82 (n.r.) **16i**: >99 (n.r.) **re**: 40	[145]

[a]absolute configuration is given in parenthesis.
n.r., not reported; re, regioisomeric excess in favor of "abnormal" ester determined by chiral phase GC.

The kinetic resolution of 3-aryl-2-ketones was first reported in 2007, employing four different BVMOs (CHMO from *A. calcoaceticus* NCIMB 9871, CPMO from *Comamonas* sp. NCIMB 9872, BVMOF1 from *P. fluorescens* DSM 50106, and a BVMO from *P. putida* KT2440) expressed in *E. coli* [146]. Even though arylaliphatic ketones were accepted as substrates by different BVMOs, only BVMOF1 gave satisfactory enantioselectivity in the kinetic resolution of 3-phenyl-2-butanone affording the corresponding (*S*)-acetate with 90% *ee*. BVMO from *P. putida* KT2440 showed opposite enantiopreference toward **1a**, giving the (*R*)-acetate (Table 4.5). Also, purified PAMO and HAPMO were used for the Baeyer–Villiger oxidation of racemic ketones **19a–d** and 2-phenylpropionaldehyde **19e** [147]. In further systematic studies, purified PAMO, HAPMO, and a PAMO mutant M446G were employed in the kinetic resolution of 3-aryl-2-ketones substituted in the aromatic ring **19f–n** and racemic benzylketones with different chain types R_1 and R_2 (**19o–s**) with good enantioselectivities [148]. These kinetic resolutions enable the preparation of different (*R*)-α-alkyl-benzylketones and (*S*)-α-alkyl-benzylesters.

As a well-known fact, the major limitation of classis kinetic resolutions is the maximal theoretical yield of 50%. The development of dynamic processes combining *in situ* substrate racemization and kinetic resolution to increase the yield are of utmost importance [5,135,149]. α-Substituted ketones are substrates prone to undergo racemization through a keto-enol tautomerization. In 2002, Furtoss and coworkers reported the first example of a dynamic kinetic resolution applied to a Baeyer–Villiger oxidation (Figure 4.32) [134]. Classical kinetic resolution using recombinant

TABLE 4.5 Kinetic Resolution of Alkyl Benzyl Ketones

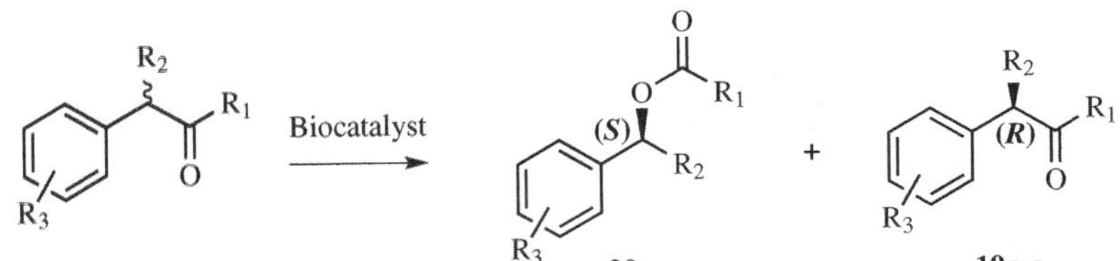

Substrate	R_1	R_2	R_3	Biocatalyst	*ee* (%)[a]	Reference
19a	Me	Me	H	BVMOF1 (*P. fluorescens* DSM 50106, recombinant cells)	**20a**: 90 (*S*) **19a**: 75 (n.r.)	[146]
				BVMO (*P. putida* KT2440, recombinant cells)	**20a**: 71 (*R*) **19a**: 72 (n.r.)	[146]
				PAMO (*T. fusca*, isolated enzyme)	**20a**: 98 (*S*) **19a**: 36 (*R*)	[147]
				HAPMO (*P. fluorescens* ACB, isolated enzyme)	**20a**: 97 (*S*) **19a**: 43 (*R*)	[147]
19b	Me	Et	H	$CHMO_{Acineto}$ (recombinant cells)	**20b**: 73 (*S*) **19b**: n.r.	[146]
				$CPMO_{Coma}$ (*Comamonas* sp. NCIMB 9872, recombinant cells)	**20b**: 48 (*R*) **19b**: n.r.	[146]
				PAMO (*T. fusca*, isolated enzyme)	**20b**: 99 (*S*) **19b**: 45 (*R*)	[147]
				HAPMO (*P. fluorescens* ACB, isolated enzyme)	**20b**: 96 (*S*) **19b**: 78 (*R*)	[147]
19c	Et	Me	H	PAMO (*T. fusca*, isolated enzyme)	**20c**: 99 (*S*) **19c**: 26 (*R*)	[147]
				HAPMO (*P. fluorescens* ACB, isolated enzyme)	**20c**: 98 (*S*) **19c**: 93 (*R*)	[147]
19d	Et	Et	H	PAMO (*T. fusca*, isolated enzyme)	**20d**: 95 (*S*) **19d**: 98 (*R*)	[147]
				HAPMO (*P. fluorescens* ACB, isolated enzyme)	**20d**: 90 (*S*) **19d**: 98 (*R*)	[147]
19e	H	Me	H	PAMO (*T. fusca*, isolated enzyme)	**20e**: 92 (*S*) **19e**: 9 (*R*)	[147]
				HAPMO (*P. fluorescens* ACB, isolated enzyme)	**20e**: 88 (*S*) **19e**: 9 (*R*)	[147]
19f	Me	Me	4-OMe	PAMO (*T. fusca*, isolated enzyme)	**20f**: 91 (*S*) **19f**: 44 (*R*)	[148]
19f	Me	Me	4-OMe	HAPMO (*P. fluorescens* ACB, isolated enzyme)	**20f**: 98 (*S*) **19f**: 33 (*R*)	[148]

TABLE 4.5 ***(Continued)***

Substrate	R_1	R_2	R_3	Biocatalyst	*ee* (%)[a]	Reference
19g	Me	Me	4-Et	PAMO (*T. fusca*, isolated enzyme)	**20g**: 86 (*S*) **19g**: 36 (*R*)	[148]
19h	Me	Me	3-Me	PAMO M446G (mutant isolated enzyme)	**20h**: 98 (*S*) **19h**: 10 (*R*)	[148]
19h	Me	Me	3-Me	HAPMO (*P. fluorescens* ACB, isolated enzyme)	**20h**: 98 (*S*) **19h**: 25 (*R*)	[148]
19i	Me	Me	3-OMe	PAMO M446G (mutant isolated enzyme)	**20i**: 96 (*S*) **19i**: 11 (*R*)	[148]
19j	Me	Me	4-Br	PAMO M446G (mutant isolated enzyme)	**20j**: 95 (*S*) **19j**: 11 (*R*)	[148]
19k	Me	Me	4-Cl	PAMO M446G (mutant isolated enzyme)	**20k**: 96 (*S*) **19k**: 11 (*R*)	[148]
19l	Me	Me	3-Cl	PAMO M446G (mutant isolated enzyme)	**20l**: 96 (*S*) **19l**: 13 (*R*)	[148]
19l	Me	Me	3-Cl	PAMO (*T. fusca*, isolated enzyme)	**20l**: 92 (*S*) **19l**: 27 (*R*)	[148]
19m	Me	Me	3-CF_3	HAPMO (*P. fluorescens* ACB, isolated enzyme)	**20m**: 97 (*S*) **19m**: 62 (*R*)	[148]
19m	Me	Me	3-CF_3	PAMO M446G (mutant isolated enzyme)	**20m**: 92 (*S*) **19m**: 39 (*R*)	[148]
19n	Me	Me	4-NO_2	PAMO M446G (mutant isolated enzyme)	**20n**: 93 (*S*) **19n**: 98 (*R*)	[148]
19o	Me	Bu	H	PAMO M446G (mutant isolated enzyme)	**20o**: 86 (*S*) **19o**: 9 (*R*)	[148]
19p	Me	Allyl	H	PAMO (*T. fusca*, isolated enzyme)	**20p**: 64 (*S*) **19p**: 26 (*R*)	[148]
19q	Et	Prop	H	PAMO M446G (mutant isolated enzyme)	**20q**: 98 (*S*) **19q**: 5 (*R*)	[148]
19r	Et	Bu	H	PAMO M446G (mutant isolated enzyme)	**20r**: 95 (*S*) **19r**: 10 (*R*)	[148]
19s	Et	Allyl	H	HAPMO (*P. fluorescens* ACB, isolated enzyme)	**20s**: 98 (*S*) **19s**: 5 (*R*)	[148]

[a] Absolute configuration is given in parenthesis.
n.r., not reported; re, regioisomeric excess in favor of "abnormal" ester determined by chiral phase GC.

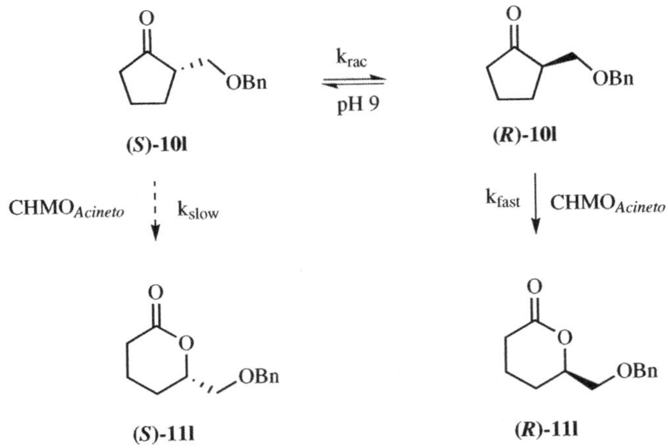

FIGURE 4.32 Dynamic kinetic resolution of racemic 2-benzyloxymethylcyclopentanone by recombinant cells expressing CHMO from *Acinetobacter calcoaceticus*.

whole-cells gave lactone (*R*)-**111** in 97% enantiomeric excess (43% yield), while the remaining ketone (*S*)-**101** was recovered with low stereoselectivity (43% *ee*). It was observed that racemization was faster at pH 9, and thus complete conversion to (*R*)-**111** was achieved in 85% overall yield and in excellent enantiomeric purity (96% *ee*). Efficient dynamic kinetic resolutions were carried out employing wild-type and mutant enzymes, and several substrates such as racemic 2-alkyl-1-indanones and alkyl benzyl ketones (Figures 4.33 and 4.34; Table 4.6).

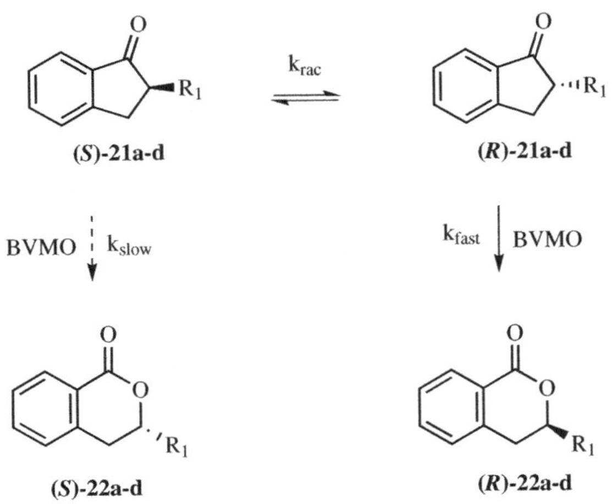

FIGURE 4.33 Dynamic kinetic resolution of racemic 2-alkyl-1-indanones by BVMOs.

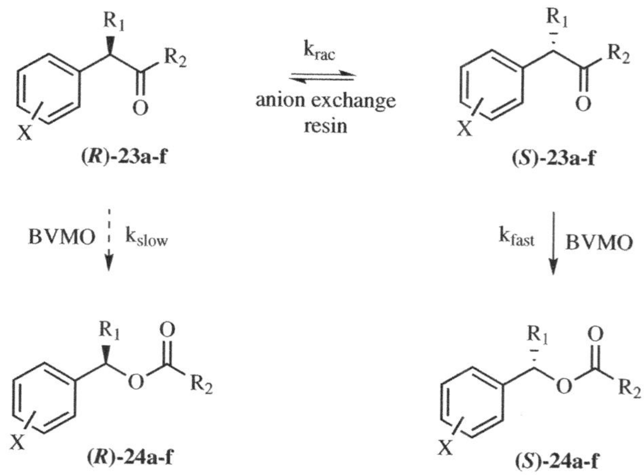

FIGURE 4.34 Dynamic kinetic resolution of racemic alkyl benzyl ketones by BVMOs in the presence of anion exchange resins.

When employing a mutant PAMO (M446G) under basic conditions it was possible to achieve 3-substituted 3,4-dihydroisocoumarins with high yields and optical purities through a dynamic kinetic resolution process. Oxidation of racemic 2-methyl-1-indanone (**21a**) was performed in a biphasic system, Tris-HCl buffer pH 10 containing 5% hexane, leading to (*R*)-**22** with 84% conversion and 82% *ee*. Likewise, this dynamic kinetic resolution was extended to other racemic 1-indanones obtaining desired products with high enantiomeric excess [150]. Recently, the preparation of optically active benzylesters through dynamic kinetic resolution of racemic benzylketones employing isolated HAPMO was reported. When racemic 3-phenylbutan-2-one (**23a**) was oxidized in the presence of HAPMO at pH 10, the racemization of the substrate was inefficient, producing (*S*)-1-phenylethyl acetate (**24a**) with low enantiomeric excess (21%). Therefore, the enzymatic resolution of **23a** was tested in the presence of anion-exchange resins. Strong anion resins led to a significant deactivation of the enzyme. The best result was obtained when using Dowex MWA-1, demonstrating that a dynamic kinetic resolution can be performed to obtain the chiral ester. The same biocatalytic system was extended to other benzyl ketones (**23b–f**), affording the corresponding (*S*)-benzylesters with good enantiomeric excesses [151].

Desymmetrization of Prochiral Substrates Enantioselective enzymatic desymmetrizations of prochiral substrates represent a powerful tool in synthetic organic chemistry, with theoretical yields of 100% [152]. In this context, the potential of BVMOs in desymmetrization reactions has been tested, for example, the oxidation of mesomeric 4-cyclohexanones first reported in 1988 [84]. 4-Methyl- (**25a**) and 4-methoxycyclohexanone (**25b**) (as well as polysubstituted prochiral cyclohexanones) were oxidized with purified CHMO from *Acinetobacter* sp. NCIMB 9871, providing the (*S*)-lactones with excellent enantiomeric excesses (Table 4.7). Carbon chains up to C_3 were oxidized by the enzyme, but larger chains than C_4 results in diminished

TABLE 4.6 BVMOs-Catalyzed Dynamic Kinetic Resolution of 2-Alkyl-1-Indanones and Alkyl Benzyl Ketones

Substrate	R_1	R_2	X	Biocatalyst	Racemization Conditions	Yield (%)[a]	*ee* (%)[b]	Reference
21a	Me	–	–	PAMO M446G (mutant isolated enzyme)	Tris-HCl, pH 10, 5% hexane	84	82 (*R*)	[150]
21b	Et	–	–	PAMO M446G (mutant isolated enzyme)	Tris-HCl, pH 10	72	91 (*R*)	[150]
21c	*i*-Pr	–	–	PAMO M446G (mutant isolated enzyme)	Tris-HCl, pH 10, 5% hexane	53	>97 (*R*)	[150]
21d	*n*-Bu	–	–	PAMO M446G (mutant isolated enzyme)	Tris-HCl, pH 10	80	92 (*R*)	[150]
23a	Me	Me	H	HAPMO (*P. fluorescens* ACB, isolated enzyme)	Dowex MWA-1, pH 10	84	86 (*S*)	[151]
23b	Me	Me	*m*-Me	HAPMO (*P. fluorescens* ACB, isolated enzyme)	Dowex MWA-1, pH 10	69	81 (*S*)	[151]
23c	Me	Me	*m*-CF_3	HAPMO (*P. fluorescens* ACB, isolated enzyme)	Dowex MWA-1, pH 10	88	79 (*S*)	[151]
23d	Me	Me	*p*-Cl	HAPMO (*P. fluorescens* ACB, isolated enzyme)	Dowex MWA-1, pH 10	69	58 (*S*)	[151]
23e	Me	Et	H	HAPMO (*P. fluorescens* ACB, isolated enzyme)	Dowex MWA-1, pH 10	66	83 (*S*)	[151]
23f	Et	Me	H	HAPMO (*P. fluorescens* ACB, isolated enzyme)	Dowex MWA-1, pH 10	72	80 (*S*)	[151]

[a]Conversion determined by GC.

[b]Enantiomeric excess of lactone or ester product, absolute configuration is given in parenthesis.

TABLE 4.7 Desymmetrization of 4-Substituted Cyclohexanones

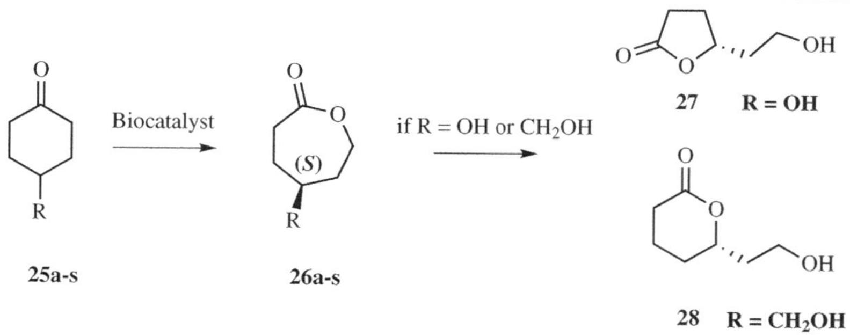

Substrate	R	Biocatalyst	Yield (%)	*ee* (%)[a]	Reference
25a	Me	$CHMO_{Acineto}$ (*Acinetobacter* NCIMB 9871, isolated enzyme)	80	>98 (−)	[84]
		$CHMO_{Acineto}$ (recombinant cells)	83	>98 (−)	[95,139,153]
		$CHMO_{Acineto}$ (recombinant cells)	61	>98 (−)	[154]
		$CHMO_{Xantho}$ (*Xanthobacter* sp. ZL5, recombinant cells)	>90 conv.	>98 (−)	[155]
		$CPMO_{Coma}$ (*Comamonas* sp. NCIMB 9872, recombinant cells)	68	46 (+) (*R*)	[63]
25b	OMe	$CHMO_{Acineto}$ (*Acinetobacter* NCIMB 9871, isolated enzyme)	76	75 (−)	[84]
		$CHMO_{Acineto}$ (recombinant cells)	84	78 (+)	[154]
		$CPMO_{Coma}$ (*Comamonas* sp. NCIMB 9872, recombinant cells)	71	28 (+)	[156]
25c	Et	$CHMO_{Acineto}$ (*Acinetobacter* NCIMB 9871, isolated enzyme)	84	>98 (−)	[85]
		$CHMO_{Acineto}$ (recombinant cells)	74	>98 (−)	[95,139,153]
		$CHMO_{Acineto}$ (recombinant cells)	91	97 (−)	[154]
25d	*i*-Pr	$CHMO_{Acineto}$ (*Acinetobacter* NCIMB 9871, isolated enzyme)	60	>98 (−)	[85]
		$CHMO_{Acineto}$ (recombinant cells)	60	>98 (−)	[95,139,153]
		$CPMO_{Coma}$ (*Comamonas* sp. NCIMB 9872, recombinant cells)	79	33 (−) (*S*)	[63]
25e	*n*-Pr	$CHMO_{Acineto}$ (*Acinetobacter* NCIMB 9871, isolated enzyme)	80	>98 (−)	[85]
		$CHMO_{Acineto}$ (recombinant cells)	63	92 (−)	[95,139,153]
		$CPMO_{Coma}$ (*Comamonas* sp. NCIMB 9872, recombinant cells)	68	36 (*S*)	[63]
25f	Allyl	$CHMO_{Acineto}$ (recombinant cells)	62	95 (−)	[95,139]
25g	*t*-Bu	$CHMO_{Acineto}$ (*Acinetobacter* NCIMB 9871, isolated enzyme)	17	>98 (−)	[85]

(Continued)

TABLE 4.7 ***(Continued)***

Substrate	R	Biocatalyst	Yield (%)	*ee* (%)[a]	Reference
		$CHMO_{Xantho}$ (*Xanthobacter* sp. ZL5, recombinant cells)	>90 conv.	>99 (−)	[155]
25h	OH	$CHMO_{Acineto}$ (*Acinetobacter* NCIMB 9871, isolated enzyme)	73	9.6 (−)	[85]
		$CHMO_{Acineto}$ (recombinant cells)	61	9.1 (−)	[154]
		$CPMO_{Coma}$ (*Comamonas* sp. NCIMB 9872, recombinant cells)	73	85 (+)	[156]
25i	CH_2OH	$CHMO_{Acineto}$ (*Acinetobacter* NCIMB 9871, isolated enzyme)	80	98 (−)	[85]
25j	*n*-Bu	$CHMO_{Acineto}$ (*Acinetobacter* NCIMB 9871, isolated enzyme)	70	52 (+) (*R*)	[85]
25k	I	$CHMO_{Acineto}$ (recombinant cells)	60	97 (−)	[154]
		$CPMO_{Coma}$ (*Comamonas* sp. NCIMB 9872, recombinant cells)	65	82 (+)	[156]
		$CHMO_{Xantho}$ (*Xanthobacter* sp. ZL5, recombinant cells)	>90 conv.	>99 (−)	[155]
25l	Br	$CHMO_{Acineto}$ (recombinant cells)	63	97 (−)	[154]
		$CPMO_{Coma}$ (*Comamonas* sp. NCIMB 9872, recombinant cells)	70	64 (+)	[156]
		$CHMO_{Xantho}$ (*Xanthobacter* sp. ZL5, recombinant cells)	>90 conv.	99 (−)	[155]
25m	Cl	$CHMO_{Acineto}$ (recombinant cells)	56	95 (−)	[156]
		$CPMO_{Coma}$ (*Comamonas* sp. NCIMB 9872, recombinant cells)	64	34 (+)	[156]
		$CHMO_{Xantho}$ (*Xanthobacter* sp. ZL5, recombinant cells)	>90 conv.	>97 (−)	[155]
25n	COOEt	$CPMO_{Coma}$ (*Comamonas* sp. NCIMB 9872, recombinant cells)	83	64 (+)	[156]
		$CHMO_{Xantho}$ (*Xanthobacter* sp. ZL5, recombinant cells)	>90 conv.	98 (−)	[155]
25o	OEt	$CPMO_{Coma}$ (*Comamonas* sp. NCIMB 9872, recombinant cells)	90	37 (−)	[156]
25p	OAllyl	$CPMO_{Coma}$ (*Comamonas* sp. NCIMB 9872, recombinant cells)	80	45 (+)	[156]
25q	OBn	$CPMO_{Coma}$ (*Comamonas* sp. NCIMB 9872, recombinant cells)	95	75 (−)	[156]
25r	OAc	$CPMO_{Coma}$ (*Comamonas* sp. NCIMB 9872, recombinant cells)	81	5 (n.r.)	[156]
25s	Ph	$CHMO_{Xantho}$ (*Xanthobacter* sp. ZL5, recombinant cells)	>90 conv.	98 (−)	[155]

[a]Sign of specific rotation is given in parenthesis.
n.r., not reported.

stereoselectivity [85]. Comparable results were obtained employing recombinant whole-cells [153,154]. The imposed steric hindrance by a branched substituent, such as 4-*t*-butyl- (**25g**), leads to the corresponding lactone with significant decreased yields [85]. Interestingly, when the Baeyer–Villiger oxidation was catalyzed by a BVMO from *Xanthobacter* sp. ZL5, 4-*t*-butylcyclohexanone was an excellent substrate and a preparative-scale biotransformation was performed. The same enzyme was employed to catalyze the Baeyer–Villiger oxidation of several monosubstituted cyclohexanones (**25a**, **25k–n**, **25s**) [155]. Other functional groups are tolerated by the enzyme, such as olefinic substituents (**25f**) which are not converted to the corresponding epoxide, yielding only the Baeyer–Villiger oxidation lactone [95,139]. The oxidation of prochiral ketones with CHMO from *Acinetobacter* sp. NCIMB 9871 and CPMO from *Comamonas* sp. NCIMB 9872, displayed an enantiodivergent trend [63,156]. The comparative use of the two monooxygenases in oxidation of several mesomeric 4-substituted cyclohexanones showed that CPMO readily accepts cyclohexanones with larger substituents in 4-position than CHMO, also improving the enantioselectivity with an increasing size of the group (the series OMe, OEt, OAllyl, OBn; **25b**, **25o–q**, respectively). Moreover, the 4-halocyclohexanones (**25k–m**) were converted by the two enzymes (CHMO and CPMO) to the antipode lactones [156]. Interestingly, for the bioconversion of prochiral 4-substituted cyclohexanones bearing hydroxyl-functionalities OH (**25h**) and CH_2OH (**25i**), the initially formed ε-caprolactone rearranges to give the thermodynamically more stable five- or six-membered ring lactones (**27–28**) [85,154,156].

In the same area, desymmetrization of 4,4-disubstituted cyclohexanones was extensively explored for several substrates with different biocatalytic systems (Table 4.8). Disubstitution with methyl and ethyl groups (**29a–c**) does not affect the catalytic activity of the BVMOs [154,155,157]. Also, 4-methyl-4-phenylcyclohexanone (**29g**) was completely oxidized with excellent enantiomeric excess with CHMO from *Xanthobacter* sp. [155]. However, this substrate is not tolerated by other enzymes such as CHMO from *Acinetobacter* sp. and CPMO from *Comamonas* sp. [157]. As observed previously for 4-hydroxycyclohexanone, the oxidation of **29d–e**, **29h–i** spontaneously rearranges to give the more stable five-membered ring system [84,155,157–159]. Position 4 is highly tolerant toward different substituents and even spiro-disubstitution (**29f**, **29n**) at this position does not affect the efficiency or stereoselectivity of the Baeyer–Villiger oxidation [154,159]. The active site also accepts a sp^2 methylene group (**29l**), as well as a carbonyl group [84,154,159].

Heteroatom-containing ketones represent a challenge for the chemoselectivity, since the heteroatom can be oxidized to a *N*-oxide or a sulfoxide. Interestingly, during systematic studies employing such substrates and BVMOs, high chemoselectivity was observed in the biooxidation (Table 4.9) [155,161]. Thus, the ability of CHMO from *Acinetobacter* sp. to convert heterocyclic ketones to lactones was investigated for substrates **32a–f**. Ketones substituted with oxygen and sulfur, as well as several *N*-protected 4-piperidones, were accepted by the enzyme [161]. In addition, the ability of the recombinant CHMO from *Xanthobacter* sp. to catalyze the Bayer–Villiger oxidation of several heterocyclic ketones was evaluated, and all tested ketones resulted good substrates for this enzyme [155].

TABLE 4.8 Desymmetrization of Prochiral 4,4-Disubstituted Cyclohexanones

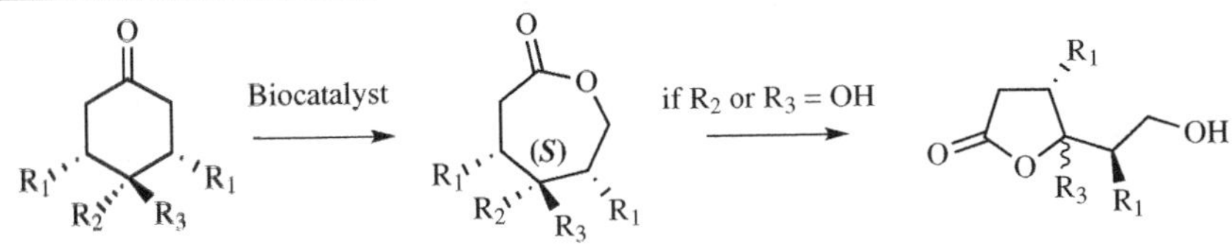

29a-s **30a-s** **31d, 31e, 31h, 31i**

Substrate	R_1	R_2	R_3	Biocatalyst	Yield (%)	*ee* (%)[a]	Reference
29a	H	Me	Me	$CHMO_{Acineto}$ (recombinant cells)	61	n.a.	[154]
				$CHMO_{Xantho}$ (*Xanthobacter* sp. ZL5, recombinant cells)	>90 conv.	n.a.	[155]
29b	H	Et	Me	$CHMO_{Acineto}$ (recombinant cells)	91	75 (−)	[154]
				$CHMO_{Xantho}$ (*Xanthobacter* sp. ZL5, recombinant cells)	>90 conv.	66 (−)	[155]
				$CPMO_{Coma}$ (*Comamonas* sp. NCIMB 9872, recombinant cells)	56	21 (−)	[157]
				$CHMO_{Artrho}$ (*Arthrobacter* sp., recombinant cells)	62	88 (−)	[157]
29c	H	Et	Et	$CHMO_{Acineto}$ (recombinant cells)	60	n.a.	[154]
29d	H	OH	Me	$CHMO_{Acineto}$ (recombinant cells)	59	86 (−)	[158]
				$CHMO_{Brachy}$ (*Brachymonas* sp., recombinant cells)	48	97 (−)	[158]
				$CPMO_{Coma}$ (*Comamonas* sp. NCIMB 9872, recombinant cells)	54	76 (+)	[158]
				$CHMO_{Xantho}$ (*Xanthobacter* sp. ZL5, recombinant cells)	>90 conv.	>99 (−)	[155]
29e	H	OH	Et	$CHMO_{Acineto}$ (recombinant cells)	54	94(−)	[154]
				$CPMO_{Coma}$ (*Comamonas* sp. NCIMB 9872, recombinant cells)	52	42 (+)	[63]

TABLE 4.8 (*Continued*)

Substrate	R_1	R_2	R_3	Biocatalyst	Yield (%)	*ee* (%)[a]	Reference
29f	H	*cyclo*-CH_2CH_2		$CHMO_{Acineto}$ (recombinant cells)	74	n.a.	[154]
29g	H	Me	Ph	$CHMO_{Xantho}$ (*Xanthobacter* sp. ZL5, recombinant cells)	>90 conv.	95 (−)	[155]
29h	Me	OH	H	$CHMO_{Acineto}$ (*Acinetobacter* NCIMB 9871, isolated enzyme)	88	>98 (−)	[84]
				$CHMO_{Acineto}$ (recombinant cells)	77	99 (−)	[159]
				$CHMO_{Xantho}$ (*Xanthobacter* sp. ZL5, recombinant cells)	>90 conv.	>95% (−)	[155]
29i	Me	H	OH	$CHMO_{Acineto}$ (recombinant cells)	80	96 (+)	[159]
				$CHMO_{Xantho}$ (*Xanthobacter* sp. ZL5, recombinant cells)	>90 conv.	>95% (+)	[155]
29j	Me	Cl	H	$CHMO_{Acineto}$ (recombinant cells)	53	99 (−)	[159]
29k	Me	H	Cl	$CHMO_{Acineto}$ (recombinant cells)	40	>99 (−)	[159]
29l	Me	=CH_2		$CHMO_{Acineto}$ (recombinant cells)	54	92 (+)	[159]
				$CPMO_{Coma}$ (*Comamonas* sp. NCIMB 9872, recombinant cells)	63	99 (−)	[159]
				$CHMO_{Xantho}$ (*Xanthobacter* sp. ZL5, recombinant cells)	>90 conv.	>99 (+)	[155]
				$CHMO_{Brevi1}$ (*Brevibacterium* sp., recombinant cells)	70	>99 (+)	[160]
				$CHMO_{Brevi2}$ (*Brevibacterium* sp., recombinant cells)	44	>99 (−)	[160]

(*Continued*)

TABLE 4.8 (*Continued*)

Substrate	R_1	R_2	R_3	Biocatalyst	Yield (%)	*ee* (%)[a]	Reference
29m	Me	=O		$CHMO_{Acineto}$ (*Acinetobacter* NCIMB 9871, isolated enzyme)	25	>98 (−)	[84]
29n	Me	*cyclo*-CH_2CH_2		$CHMO_{Acineto}$ (recombinant cells)	57	>99 (+)	[159]
				$CHMO_{Xantho}$ (*Xanthobacter* sp. ZL5, recombinant cells)	>90 conv.	>99 (+)	[155]

[a]Sign of specific rotation is given in parenthesis.
n.a., not applicable.

Chiral β-substituted γ-butyrolactones are versatile building blocks for the synthesis of bioactive compounds and natural products [135,140]. Prochiral 3-substituted cyclobutanones can be substrates for a biocatalyzed Bayer–Villiger oxidation (Table 4.10). Based on the original work of Taschner on the Baeyer–Villiger oxidation of cyclic ketones [84], several 3-substituted cyclobutanones (**34d**, **34f**, **34h–i**) were oxidized using two different types of biocatalysts, namely whole-cells of *A. calcoaceticus* NCIMB 9871 and a partially purified BVMO from *P. putida* NCIMB 10007 [162]. The oxidation provided the corresponding lactones with moderate enantioselectivity, and in some cases both enzymes showed complementary stereoselectivities. Several systematic studies (**34a–g**) were carried out using CHMO from different sources, such as *Brevibacterium* sp. [158] and *Xanthobacter* sp.[155]. Interestingly, enantiocomplementary reactions were observed when some substrates were oxidized with a fungal strain of *Cunninghamella echinulata* [163,164]. Likewise, biotransformations of **34e–f** with HAPMO from *P. fluorescens* produced the corresponding lactones carrying the aromatic ring in antipodal form to previous oxidations with CHMO from *Acinetobacter* sp. [165].

CHMO from *Acinetobacter* sp. NCIMB 9871 was employed in the desymmetrization of a series of substituted bicyclo[2.2.1]hept-2-en-7-ones (**36a–h**) (Table 4.11). These bridged cycloketones were accepted by the enzyme, resulting in the corresponding lactones in good-to-excellent yields and with high optical purity [166]. Interestingly, the peracid-catalyzed chemical Baeyer–Villiger oxidation of **36a** leads exclusively to the epoxide. However, the enzymatic Baeyer–Villiger oxidation produced the corresponding lactone, thus being fully chemoselective. Mihovilovic and coworkers evaluated different BVMOs, namely CHMO from *Acinetobacter* NCIMB 9871, two CHMOs from *Brevibacterium* sp., a CHMO from *Rhodococcus* sp., and CPMO from *Comamonas* sp. [167,168]. The Baeyer–Villiger oxidation of bridged cycloketone **36i** leads to the formation of enantiocomplementary lactone products depending on the enzyme used, in accordance with the identification of two family

TABLE 4.9 Desymmetrization of Prochiral Heterocyclic Cyclohexanones

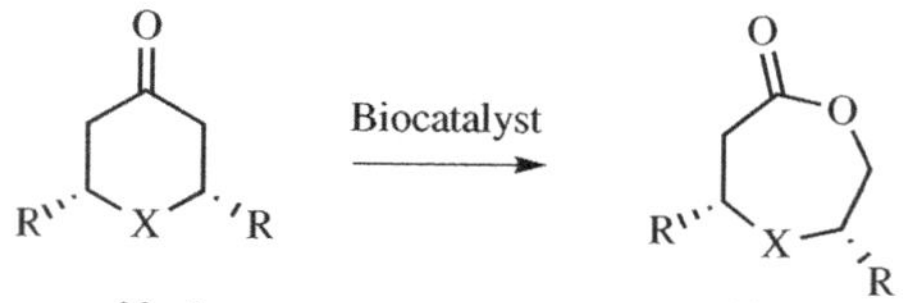

Substrate	R	X	Biocatalyst	Yield (%)	*ee* (%)[a]	Reference
32a	H	S	$CHMO_{Acineto}$ (recombinant cells)	48	n.a.	[161]
			$CHMO_{Xantho}$ (*Xanthobacter* sp. ZL5, recombinant cells)	>90 conv.	n.a.	[155]
32b	H	O	$CHMO_{Acineto}$ (recombinant cells)	79	n.a.	[161]
			$CHMO_{Xantho}$ (*Xanthobacter* sp. ZL5, recombinant cells)	>90 conv.	n.a.	[155]
32c	H	NMe	$CHMO_{Acineto}$ (recombinant cells)	50	n.a.	[161]
			$CHMO_{Xantho}$ (*Xanthobacter* sp. ZL5, recombinant cells)	>90 conv.	n.a.	[155]
32d	H	NAllyl	$CHMO_{Acineto}$ (recombinant cells)	10[b]	n.a.	[161]
32e	H	NAc	$CHMO_{Acineto}$ (recombinant cells)	39	n.a.	[161]
32f	H	NCOOMe	$CHMO_{Acineto}$ (recombinant cells)	40	n.a.	[161]
			$CHMO_{Xantho}$ (*Xanthobacter* sp. ZL5, recombinant cells)	>90 conv.	n.a.	[155]
32g	Me	O	$CHMO_{Xantho}$ (*Xanthobacter* sp. ZL5, recombinant cells)	>90 conv.	>99 (−)	[155]
32h	Me	$CH{=}CH_2$	$CHMO_{Xantho}$ (*Xanthobacter* sp. ZL5, recombinant cells)	>90 conv.	>99 (−)	[155]

[a] Sign of specific rotation is given in parenthesis.
[b] Determined by GC, mainly starting material recovered.
n.a., not applicable.

TABLE 4.10 Desymmetrization of Prochiral 3-Substituted Cyclobutanones

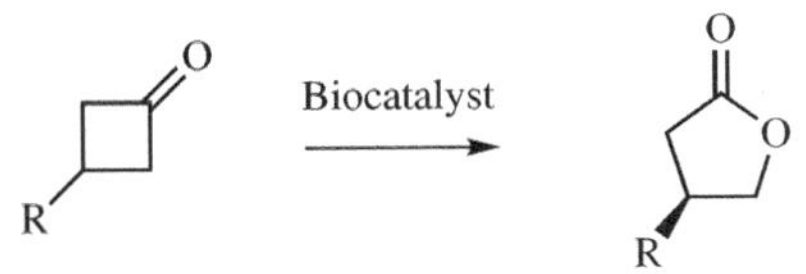

34a-i 35a-i

Substrate	R	Biocatalyst	Yield (%)	*ee* (%)[a]	Reference
34a	Ph	$CHMO_{Brevi1}$ (*Brevibacterium* sp., recombinant cells)	73	98 (−)	[158]
		$CHMO_{Brevi2}$ (*Brevibacterium* sp., recombinant cells)	50	39 (−)	[158]
		Cunninghamella echinulata NRRl 3655 (whole cell)	71	>98 (−)	[163]
		$CHMO_{Xantho}$ (*Xanthobacter* sp. ZL5, recombinant cells)	>90 conv.	83 (−)	[155]
		Acinetobacter NCIMB 9871 (whole cell)	70	43 (−)	[164]
34b	4-Cl-Ph	$CHMO_{Xantho}$ (*Xanthobacter* sp. ZL5, recombinant cells)	>90 conv.	81 (+)	[155]
		Acinetobacter NCIMB 9871 (whole cell)	88	85 (+)	[164]
		Cunninghamella echinulata NRRl 3655 (whole cell)	30	98 (−)	[164]
34c	4-F-Ph	*Acinetobacter* NCIMB 9871 (whole cell)	89	19 (−)	[164]
		Cunninghamella echinulata NRRl 3655 (whole cell)	80	>98 (−)	[164]
34d	Bn	$CHMO_{Xantho}$ (*Xanthobacter* sp. ZL5, recombinant cells)	>90 conv.	88 (−)	[155]
		Acinetobacter NCIMB 9871 (whole cell)	57	82 (*S*)	[162]
		HAPMO (*Pseudomonas fluorescens* ACB, recombinant cells)	26	44 (+)	[165]
34e	3-MeO-Bn	$CHMO_{Xantho}$ (*Xanthobacter* sp. ZL5, recombinant cells)	>90 conv.	97 (−)	[155]
		Cunninghamella echinulata NRRl 3655 (whole cell)	68	77 (−)	[164]
		Acinetobacter NCIMB 9871 (whole cell)	83	96 (−)	[164]
		HAPMO (*Pseudomonas fluorescens* ACB, recombinant cells)	42	56 (+)	[165]

TABLE 4.10 (*Continued*)

Substrate	R	Biocatalyst	Yield (%)	*ee* (%)[a]	Reference
34f	3,4-(OCH_2O)-Bn (piperonyl)	CHMO$_{Xantho}$ (*Xanthobacter* sp. ZL5, recombinant cells)	>90 conv.	99 (−)	[155]
		Cunninghamella echinulata NRRl 3655 (whole cell)	68	91 (−)	[164]
		Acinetobacter NCIMB 9871 (whole cell)	83	95 (−)	[162]
		HAPMO (*Pseudomonas fluorescens* ACB, recombinant cells)	63	66 (+)	[165]
34g	3,4,5-$(MeO)_3$-Bn	CHMO$_{Xantho}$ (*Xanthobacter* sp. ZL5, recombinant cells)	>90 conv.	95 (−)	[155]
34h	Bu	*Acinetobacter* NCIMB 9871 (whole cell)	68	17 (*S*)	[162]
		MO1 (*Pseudomonas putida* NCIMB 10007, crude enzyme)	n.r.	69 (*R*)	[162]
34i	*i*-Bu	CHMO$_{Xantho}$ (*Xanthobacter* sp. ZL5, recombinant cells)	>90 conv.	76 (−)	[155]
		Acinetobacter NCIMB 9871 (whole cell)	56	84 (*S*)	[162]
		MO1 (*Pseudomonas putida* NCIMB 10007, crude enzyme)	n.r.	91 (*R*)	[162]

[a] Sign of specific rotation or absolute configuration is given in parenthesis.
n.a., not applicable; n.r., not reported.

clusters of BVMOs [158]. The enantiomer of **37i** (*ent*-**37i**) was obtained when using CHMO$_{Brevi1}$ *Brevibacterium* sp., whereas lactone **37i** was the main product when the biotransformation was performed with CPMO$_{Coma}$ *Comamonas* sp. [167]. Other bridged cycloketones (**36j–l**) served as substrates for BVMO-mediated oxidations, and again antipodal lactones could be isolated with excellent enantiomeric excesses [168].

The Baeyer–Villiger oxidation of fused bicyclic ketones catalyzed by different bacterial CHMOs and CPMO from *Comamonas* sp., showed an interesting trend regarding the influence of ring size, rigidity of the system, and polarity of the functional groups attached to the bicyclic core (Table 4.12) [112,154,158,160,169]. The enantiodivergent trend in the Baeyer–villiger oxidation was again observed for fused ketones **38a**, **38e** and **38f** when employing CHMO from *Acinetobacter* NCIMB 9871 or CPMO from *Comamonas* sp. [158]. The rigidity in the fused system is essential for the high enantioselectivity; bicyclo[4.3.0]ketones (**38f–g**) were oxidized by CHMO$_{Acineto}$ with very low enantiomeric excesses values, while the corresponding bicylo[3.3.0] system **38a** displayed high optical selectivity. The opposite

TABLE 4.11 Desymmetrization of Bicyclo[2.2.1]heptenones and Bicyclo[2.2.1]heptanones

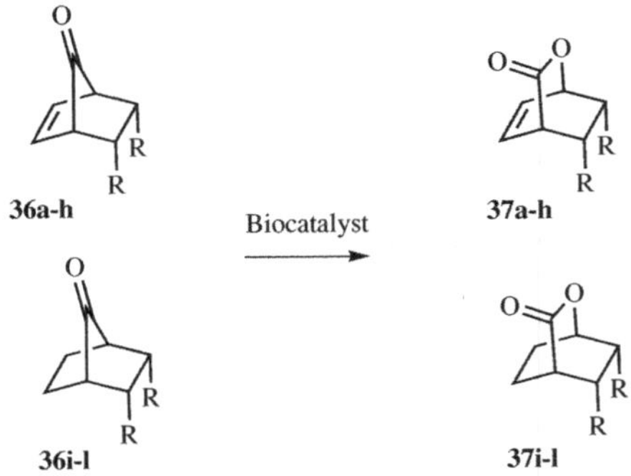

Substrate	R	Biocatalyst	Yield (%)	*ee* (%)[a]	Reference
36a	H	$CHMO_{Acineto}$ (*Acinetobacter* NCIMB 9871, isolated enzyme)	62	80 (+)	[166]
36b	Me	$CHMO_{Acineto}$ (*Acinetobacter* NCIMB 9871, isolated enzyme)	70	>98 (+)	[166]
36c	Et	$CHMO_{Acineto}$ (*Acinetobacter* NCIMB 9871, isolated enzyme)	83	93 (+)	[166]
36d	$-CH_2OCH_2-$	$CHMO_{Acineto}$ (*Acinetobacter* NCIMB 9871, isolated enzyme)	74	>98 (+)	[166]
36e	$-(CH_2)_3-$	$CHMO_{Acineto}$ (*Acinetobacter* NCIMB 9871, isolated enzyme)	80	97 (+)	[166]
36f	$-(CH_2)_4-$	$CHMO_{Acineto}$ (*Acinetobacter* NCIMB 9871, isolated enzyme)	78	>98 (−)	[166]
36g	$-(CH_2)_5-$	$CHMO_{Acineto}$ (*Acinetobacter* NCIMB 9871, isolated enzyme)	57	>98 (+)	[166]
36h	$-(CH_2)_6-$	$CHMO_{Acineto}$ (*Acinetobacter* NCIMB 9871, isolated enzyme)	55	87 (+)	[166]
36i	$-(CH_2)_2-$	$CHMO_{Brevi1}$ (*Brevibacterium* sp., recombinant cells)	100 conv.	96 (−) *er*: 2:98	[167]
		$CPMO_{Coma}$ (*Comamonas* sp. NCIMB 9872, recombinant cells)	56 conv.	86 (+) *er*: 93:7	[167]
36j	$-(CH_2)_3-$	$CHMO_{Rhodo1}$ (*Rhodococcus* sp., recombinant cells)	63	99 (−)	[168]
		$CHMO_{Brevi2}$ (*Brevibacterium* sp., recombinant cells)	67	92 (+)	[168]
36k	$-(CH_2)_4-$	$CHMO_{Rhodo1}$ (*Rhodococcus* sp., recombinant cells)	58	99 (−)	[168]
		$CHMO_{Brevi2}$ (*Brevibacterium* sp., recombinant cells)	78	94 (+)	[168]
36l	$-CH_2OCH_2-$	$CHMO_{Acineto}$ (recombinant cells)	53	92 (−)	[168]
		$CPMO_{Coma}$ (*Comamonas* sp. NCIMB 9872, recombinant cells)	49	71 (+)	[168]

[a]Sign of specific rotation is given in parenthesis.
er, enantiomeric ratio **37i**/*ent*-**37i**.

TABLE 4.12 Desymmetrization of Fused Bicyclic Ketones

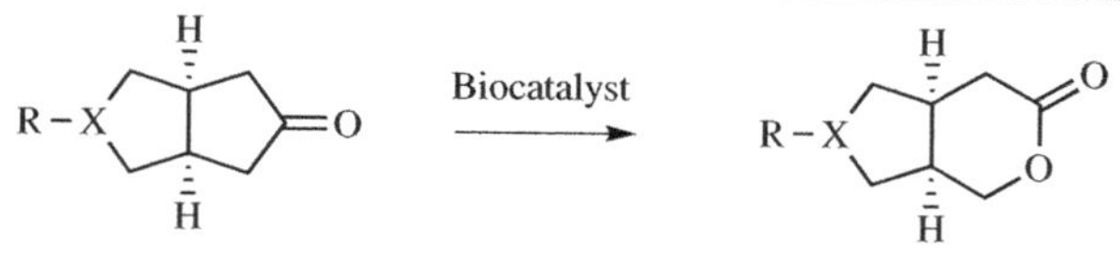

38a-g **39a-g**

Substrate	X	R	Biocatalyst	Yield (%)	*ee* (%)[a]	Reference
38a	–CH_2–	–	CHMO$_{Acineto}$ (recombinant cells)	50	89 (−)	[158]
			CHMO$_{Brachy}$ (*Brachymonas* sp., recombinant cells)	71	91 (−)	[158]
			CPMO$_{Coma}$ (*Comamonas* sp. NCIMB 9872, recombinant cells)	89	9 (+)	[158]
38b	–CH–	*endo*-OMe	CHMO$_{Acineto}$ (recombinant cells)	24	9 (−)	[169]
			CPMO$_{Coma}$ (*Comamonas* sp. NCIMB 9872, recombinant cells)	81	34 (+)	[169]
38c	–CH–	*exo*-OMe	CHMO$_{Acineto}$ (recombinant cells)	40	96 (−)	[169]
			CPMO$_{Coma}$ (*Comamonas* sp. NCIMB 9872, recombinant cells)	75	11 (−)	[169]
38d	–CH–	*endo*-Cl	CHMO$_{Acineto}$ (recombinant cells)	75	80 (−)	[169]
			CPMO$_{Coma}$ (*Comamonas* sp. NCIMB 9872, recombinant cells)	79	60 (−)	[169]
38e	–CH–	*exo*-Cl	CHMO$_{Acineto}$ (recombinant cells)	78	>99 (−)	[158]
			CHMO$_{Brachy}$ (*Brachymonas* sp., recombinant cells)	45	>99 (−)	[158]
			CPMO$_{Coma}$ (*Comamonas* sp. NCIMB 9872, recombinant cells)	92	48 (+)	[158]

(Continued)

TABLE 4.12 (*Continued*)

Substrate	X	R	Biocatalyst	Yield (%)	*ee* (%)[a]	Reference
			CHMO$_{Brevi1}$ (*Brevibacterium* sp., recombinant cells)	55	>99 (−)	[160]
			CHMO$_{Brevi2}$ (*Brevibacterium* sp., recombinant cells)	59	>99 (+)	[160]
			CHMO$_{Xantho}$ (*Xanthobacter* sp. ZL5, recombinant cells)	>90[b] conv.	>99 (−)	[155]
38f	–CH=CH–	–	CHMO$_{Acineto}$ (recombinant cells)	33	5 (−)	[158]
			CHMO$_{Brachy}$ (*Brachymonas* sp., recombinant cells)	56	85 (−)	[158]
			CPMO$_{Coma}$ (*Comamonas* sp. NCIMB 9872, recombinant cells)	76	>99 (+)	[158]
			CHMO$_{Brevi1}$ (*Brevibacterium* sp., recombinant cells)	10	71 (n.r.)	[160]
			CHMO$_{Brevi1}$ (*Brevibacterium* sp., recombinant cells)	92	94 (+)	[160]
			CHMO$_{Xantho}$ (*Xanthobacter* sp. ZL5, recombinant cells)	>90[b] conv.	88 (−)	[155]
38g	–CH_2CH_2–	–	CHMO$_{Acineto}$ (recombinant cells)	21	3 (−)	[112]
			CPMO$_{Coma}$ (*Comamonas* sp. NCIMB 9872, recombinant cells)	83	99 (+)	[112]

[a]Sign of specific rotation is given in parenthesis.
[b]Conversion was reported.
n.r., not reported.

behavior is observed when employing CPMO from *Comamonas* sp. [107,112,158]. Thus, CHMO$_{Acineto}$ is an excellent biocatalyst for prochiral bicylo[3.3.0] systems, while CHMO$_{Coma}$ leads to higher enantiomeric excesses for bicyclo[4.3.0]ketones [135]. When comparing the polarity of functional groups attached to the ring system, CHMO$_{Acineto}$ gave better yields with less polar chlorinated ketones (**38d–e**) than with methoxylated ones (**38b–c**). However, $CHMO_{Coma}$ displayed good conversions for these substrates. Furthermore, a significant influence of the conformation of the ring system was observed. The stereochemistry of the 5-position in the substrate had a pronounced influence in the selectivity of the Baeyer–Villiger oxidation. Generally, *endo-* compounds (sterically more demanding conformation) were converted into the corresponding lactones with low enantiomeric excesses (**38b**, **38d**). On the other hand, the enzyme displays an increased enantioselectivity for the oxidation of *exo*-isomers (**38c**, **38e**) [160,169].

Regioselective and Regiodivergent Transformations The regioselectivity of the chemical Baeyer–Villiger oxidation is governed by stereoelectronic effects. The rearrangement of the Criegee intermediate takes place through the migration of the group antiperiplanar to the O–O bond of the leaving group (primary stereoelectronic effect), and antiperiplanar to a lone pair of the hydroxyl group (secondary stereoelectronic effect) [43–47]. Thus, when the conformation of the Criegee intermediate is not fixed, the more nucleophilic (more substituted) carbon atom migrates preferentially. Yet, when the conformation of the intermediate places the less substituted carbon atom antiperiplanar to the O–O bond, this group migrates preferentially. Migration of the more substituted carbon atom generates the expected "normal" ester or lactone, while the "abnormal" one is formed by the migration of the less substituted carbon atom. The same preferences are observed in the enzymatic Baeyer–Villiger oxidation. However, in contrast to chemical oxidations, BVMO-catalyzed oxidations afford the two regioisomeric products in high enantioselectivity. This means that one enantiomer is positioned in the active site of the enzyme holding the configuration where the more substituted group is antiperiplanar and migrates, while the antipodal substrate is assuming a configuration that leads to the migration of the less substituted carbon atom. This regiodivergent oxidation of ketones constitutes a very useful property of the enzymatic Baeyer–Villiger reaction.

In 1965, the first regiodivergent BVMO-catalyzed oxidation was reported [170], the microbial oxidation of racemic fenchone affording a mixture of 1,2- and 2,3-fencholide using *Corynebacterium* sp. cells (Figure 4.35).

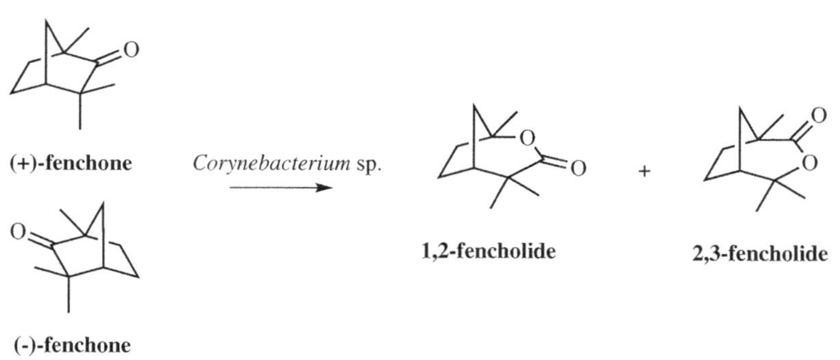

FIGURE 4.35 Regiodivergent biooxidation of racemic fenchone.

Several bicyclic ketones were subjected to Baeyer–Villiger oxidation employing different biocatalysts (Table 4.13). The regiodivergent oxidation of (*S*,*S*)-bicyclo[3.2.0]hept-2-en-6-one (**40a**) with wild-type cells of *Acinetobacter* NCIMB 9871 generates the "normal" lactone **41a**, that is, migration of the more substituted carbon atom. The antipode (*R*,*R*) enantiomer produces the "abnormal" lactone. Both products are produced in high enantiomeric purity [171]. The same regiodivergent and enantioselective Baeyer–Villiger oxidation was observed when the mutant strain *Acinetobacter* TD 63 was employed [172]. Both products are valuable building blocks for the preparation of natural products. The "normal" lactone is a synthon used in prostaglandin synthesis [172], while the "abnormal" lactone can be employed in the preparation of brown algae pheromones [173]. Interestingly, the biotransformation of **40a** with *C. echinulata* only produces the "abnormal" lactone with excellent enantioselectivity [173]. However, the "normal" lactone is afforded when the reaction is performed with CPMO from *Comamonas* sp. NCIMB 9872 [158]. Likewise enzymatic oxidation of mono- and dimethyl ketones **40b–c** [174], and of the saturated analog **40d** [171,175], were explored. These five-membered fused cyclic ketones were converted by $CHMO_{Brevi1}$ to the corresponding regioisomeric lactones (50:50) with excellent enantioselectivities. However, the six-membered ring ketone **40g** was converted with higher enantioselectivity by $CHMO_{Brevi1}$ [175]. Also, several heterocyclic fused ketones substrates were investigated (**40h–m**) [175–177]. These studies showed that the presence of an oxygen atom (for bicyclic furofuran and pyropyran ketones, **40h–l**) does not modify the outcome of the reaction. The ketone **40m**, containing a nitrogen atom was also accepted as substrate by CHMO (both wild-type and mutant strain lacking of lactone hydrolase), affording the *N*-protected Geisman–Waiss lactone **41m** ("normal" lactone) and its regioisomer **42m** ("abnormal" lactone) [176]. The Geisman–Waiss lactone is a versatile intermediate for the preparation of pyrrolizidine alkaloids.

The fact that "normal" and "abnormal" lactones are derived from one enantiomeric ketone means that those antiopodal substrates are fitted into the active site of the enzyme in a different position. As a consequence, different bonds satisfy the stereoelectronic prerequisites for migration. In this respect, Taschner developed a model to explain the regio- and stereochemical outcome of the CHMO-catalyzed oxidation of **40a**. This model shows that only one bond in each enantiomeric substrate is able to occupy the antiperiplanar position in the Criegee intermediate, leading to different regioisomeric lactones (Figure 4.36) [178].

In 1995, Kelly and coworkers proposed an alternative model that explained the observed regiodivergence, and confirmed the validity of the proposal by oxidizing the tricyclic ketone **43** (which reproduces the superposition of the enantiomeric ketones **40a**) to the corresponding lactone (Figure 4.37) [179–181].

The biooxidation of **43** gave lactone (−)-**44** when using CHMO from *Acinetobacter* NCIMB 9871 (Type I BVMO) and (+)-**44** when using 2,5-DKCMO from *Pseudomonas* NCIMB 10007 (Type II BVMO). The model classifies the reactions according to the arrangement of groups in the catalytic site, applying the Cahn–Ingold–Prelog rules by increasing the priority of the migrating group over the nonmigrating one. With the priority peroxide > hydroxyl group > migrating group > nonmigrating

TABLE 4.13 Regiodivergent Baeyer–Villiger Oxidation of Fused Bicyclobutanones

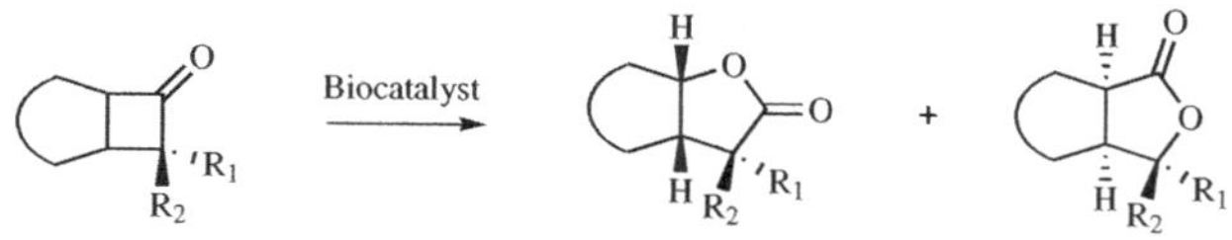

40a-m "normal" lactone **41a-m** "abnormal" lactone **42a-m**

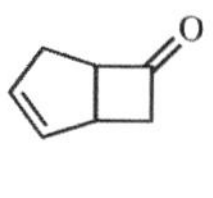
40a

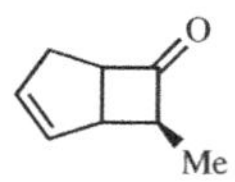
40b

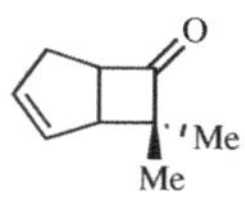
40c

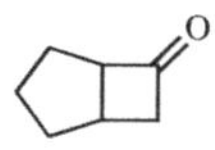
40d

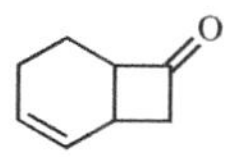
40e

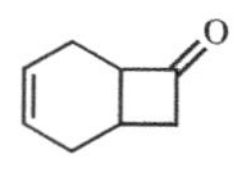
40f

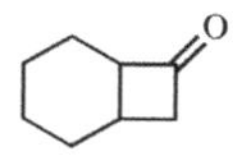
40g

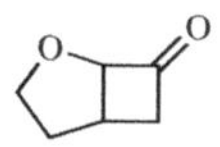
40h

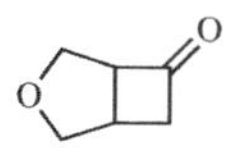
40i

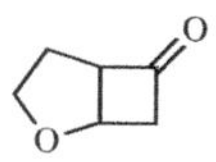
40j

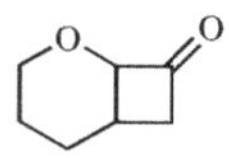
40k

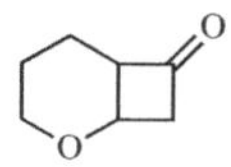
40l

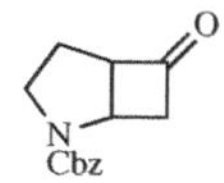
40m

Substrate	Biocatalyst	Yield (%)	Ratio 41:42	*ee* (%)[a] 41	*ee* (%)[a] 42	Reference
40a	CHMO$_{Acineto}$ (*Acinetobacter* NCIMB 9871, wild-type cells)	86	51:49	>95 (1*S*,5*R*)	>95 (1*R*,5*S*)	[171]
	CHMO$_{Acineto}$ (*Acinetobacter* TD 63, whole-cells)	86	52:48	>97 (1*S*,5*R*)	>97 (1*R*,5*S*)	[172]
	C. echinulata NRRl 3655 (whole cell)	30	0:100	n.a.	>98 (1*R*,5*S*)	[173]
	CPMO$_{Coma}$ (*Comamonas* sp. NCIMB 9872, recombinant cells)	61	97:3	0	>99 (1*R*,5*S*)	[158]
	CHMO$_{Brevi1}$ (*Brevibacterium* sp., recombinant cells)	85	51:49	96 (1*S*,5*R*)	>99 (1*R*,5*S*)	[175]
	CHMO$_{Brevi2}$ (*Brevibacterium* sp., recombinant cells)	61	98:2	0	>99 (1*R*,5*S*)	[175]
40b	CHMO$_{Acineto}$ (*Acinetobacter* NCIMB 9871, wild-type cells)	55	50:50	>98 (1*R*,4*R*,5*S*)	>95 (1*R*,4*S*,5*S*)	[174]
40c	CHMO$_{Acineto}$ (*Acinetobacter* NCIMB 9871, wild-type cells)	63	0:100	n.a.	29 (1*R*,5*S*)	[174]

(Continued)

TABLE 4.13 (*Continued*)

Substrate	Biocatalyst	Yield (%)	Ratio 41:42	*ee* (%)[a] 41	*ee* (%)[a] 42	Reference
40d	$CHMO_{Acineto}$ (*Acinetobacter* NCIMB 9871, wild-type cells)	67	54:46	>95 (1*S*,5*S*)	>95 (1*R*,5*S*)	[171]
	$CHMO_{Acineto}$ (*Acinetobacter* TD 63, whole-cells)	68	49:51	>95 (1*S*,5*S*)	>95 (1*R*,5*S*)	[171]
	$CHMO_{Brevi1}$ (*Brevibacterium* sp., recombinant cells)	78	50:50	>99 (1*S*,5*S*)	>99 (1*R*,5*S*)	[175]
	$CHMO_{Brevi2}$ (*Brevibacterium* sp., recombinant cells)	78	55:45	76 (1*S*,5*S*)	91 (1*R*,5*S*)	[175]
40e	$CHMO_{Acineto}$ (*Acinetobacter* NCIMB 9871, wild-type cells)	80	54:46	>95 (1*R*,6*S*)	>95 (1*S*,6*R*)	[171]
	$CHMO_{Acineto}$ (*Acinetobacter* TD 63, whole-cells)	79	51:49	>95 (1*R*,6*S*)	>95 (1*S*,6*R*)	[171]
40f	$CHMO_{Acineto}$ (*Acinetobacter* NCIMB 9871, wild-type cells)	77	53:47	90 (1*S*,6*S*)	>95 (1*S*,6*R*)	[171]
	$CHMO_{Acineto}$ (*Acinetobacter* TD 63, whole-cells)	84	51:49	92 (1*S*,6*S*)	>95 (1*S*,6*R*)	[171]
40g	$CHMO_{Acineto}$ (*Acinetobacter* NCIMB 9871, wild-type cells)	80	65:35	61 (1*S*,6*S*)	>95 (1*S*,6*R*)	[171]
	$CHMO_{Acineto}$ (*Acinetobacter* TD 63, whole-cells)	77	61:39	53 (1*S*,6*S*)	>95 (1*S*,6*R*)	[171]
	$CHMO_{Brevi1}$ (*Brevibacterium* sp., recombinant cells)	83	60:40	78 (1*S*,6*S*)	>99 (1*S*,6*R*)	[175]
	$CHMO_{Brevi2}$ (*Brevibacterium* sp., recombinant cells)	79	96:4	2 (1*S*,6*S*)	>99 (1*S*,6*R*)	[175]
40h	$CHMO_{Acineto}$ (*Acinetobacter* NCIMB 9871, wild-type cells)	67	52:48	90 (1*R*,5*S*)	>98 (1*R*,5*S*)	[177]
40i	$CHMO_{Acineto}$ (*Acinetobacter* NCIMB 9871, wild-type cells)	70	50:50	97 (1*R*,5*R*)	>98 (1*S*,5*R*)	[177]

TABLE 4.13 (*Continued*)

Substrate	Biocatalyst	Yield (%)	Ratio 41:42	*ee* (%)[a] 41	*ee* (%)[a] 42	Reference
	CHMO$_{Brevi1}$ (*Brevibacterium* sp., recombinant cells)	74	65:35	99 (1*R*,5*R*)	>99 (1*S*,5*R*)	[175]
	CHMO$_{Brevi2}$ (*Brevibacterium* sp., recombinant cells)	79	98:2	6 (1*R*,5*R*)	84 (1*S*,5*R*)	[175]
40j	CHMO$_{Acineto}$ (*Acinetobacter* NCIMB 9871, wild-type cells)	74	45:55	>98 (1*S*,5*S*)	>98 (1*R*,5*S*)	[177]
	CHMO$_{Brevi1}$ (*Brevibacterium* sp., recombinant cells)	72	46:54	97 (1*S*,5*S*)	>99 (1*R*,5*S*)	[175]
	CHMO$_{Brevi2}$ (*Brevibacterium* sp., recombinant cells)	83	62:38	57 (1*S*,5*S*)	97 (1*R*,5*S*)	[175]
40k	CHMO$_{Acineto}$ (*Acinetobacter* NCIMB 9871, wild-type cells)	66	50:50	70 (1*R*,6*S*)	>98 (1*R*,6*S*)	[177]
40l	CHMO$_{Acineto}$ (*Acinetobacter* NCIMB 9871, wild-type cells)	78	77:23	33 (1*S*,6*S*)	>98 (1*S*,6*R*)	[177]
	CHMO$_{Brevi1}$ (*Brevibacterium* sp., recombinant cells)	64	75:25	97 (1*S*,6*S*)	>99 (1*S*,6*R*)	[175]
	CHMO$_{Brevi2}$ (*Brevibacterium* sp., recombinant cells)	71	96:4	0	>99 (1*S*,6*R*)	[175]
40m	CHMO$_{Acineto}$ (recombinant cells)	48	77:23	94 (1*R*,5*R*)	87 (1*R*,5*S*)	[176]
	CHMO$_{Acineto}$ (*Acinetobacter* TD 63, whole-cells)	40	28:72	87 (1*R*,5*R*)	98 (1*R*,5*S*)	[176]

[a] Absolute configuration is given in parenthesis.
n.a., not applicable.

group, the CHMO Criegee intermediate has the (*S*)-configuration. Type II BVMOs (such as 2,5- and 3,6-DKCMO) display antipodal migration and therefore, the Criegee intermediate has the opposite configuration. The biooxidation of 3-substituted cyclic ketones is interesting as well. The chemical Baeyer–Villiger reaction usually leads to mixtures of products, due to the fact that both carbon atoms adjacent to the ketone group have comparable migration capabilities and therefore the reaction can afford four products (a "proximal" enantiomeric pair, and a "distal" one) (Table 4.14).

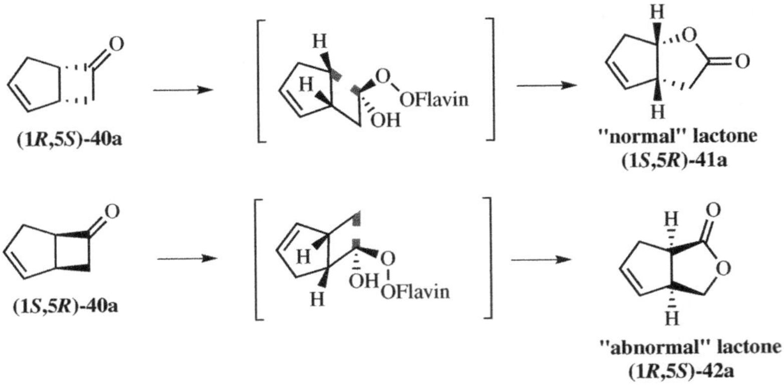

FIGURE 4.36 Model for the oxidation of **40a** by CHMO (migrating bond is shown as dashed lines).

However, the enzymatic reaction displays an interesting regiodivergent outcome. The Baeyer–Villiger oxidation of each enantiomer of 3-methylcyclohexanone (**45a**) afforded different regioisomers [95,110,139,182]. (*R*)-3-Methylcyclohexanone yielded the "proximal" lactone (**46a**), while the (*S*)-enantiomer gave the "distal" isomer, when using CHMO from *Acinetobacter* NCIMB 9871. The oxidation of substrates with increased length of substituents behaved very similar, although the optical purities of lactones were much lower. For the corresponding 3-substituted cyclopentanones, the results were also similar. Again, CHMO and CPMO display regiocomplementary oxidations for tested substrates.

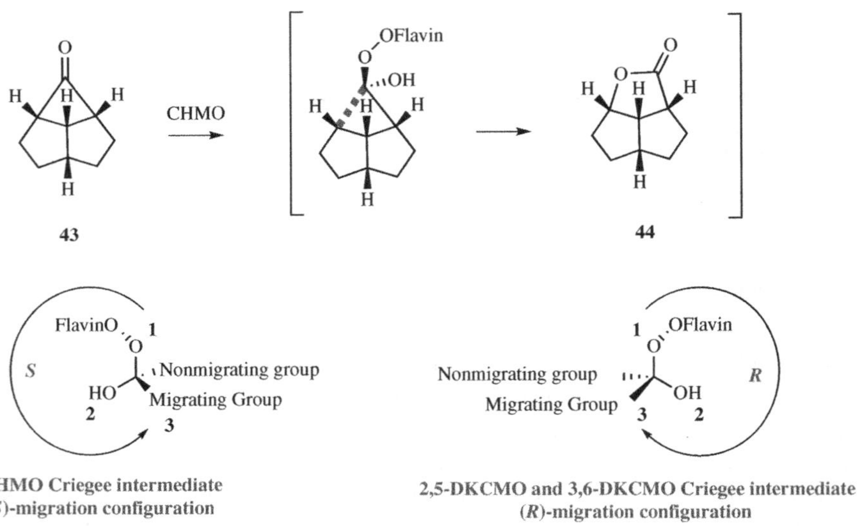

FIGURE 4.37 Absolute configuration of the Criegee intermediates.

TABLE 4.14 Regiodivergent Baeyer–Villiger Oxidation of 3-Substituted Cycloketones

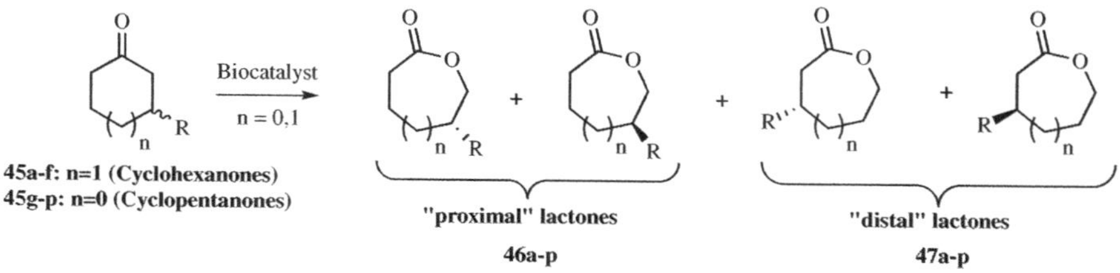

Substrate	R	Biocatalyst	Yield (%)	Ratio 46:47	*ee* (%)[a] 46	*ee* (%)[a] 47	Reference
Cyclohexanones							
45a	*rac*-Me	$CHMO_{Acineto}$ (recombinant cells)	n.r.	n.r.	>98 (−)	>98 (−)	[139]
	(*R*)-Me	$CHMO_{Acineto}$ (recombinant cells)	77	>99:1	>98 (*R*)	n.a.	[182]
		$CPMO_{Coma}$ (*Comamonas* sp. NCIMB 9872, recombinant cells)	75	>99:1	>98 (*R*)	n.a.	[182]
	(*S*)-Me	$CHMO_{Acineto}$ (recombinant cells)	60	1:99	>98 (*S*)	n.a.	[182]
45b	*rac*-Et	$CHMO_{Acineto}$ (recombinant cells)	n.r.	n.r.	70 (−)	70 (−)	[139]
	(*R*)-Et (81% *ee*)	$CHMO_{Acineto}$ (recombinant cells)	89	94:6	94 (*R*)	>99 (*S*)	[182]
45c	*rac*-*n*-Prop	$CHMO_{Acineto}$ (recombinant cells)	80	98:2	>98 (−)	n.r.	[139]
45d	*rac*-*i*-Prop	$CHMO_{Acineto}$ (recombinant cells)	No reaction				[139]
45e	*rac*-Allyl	$CHMO_{Acineto}$ (recombinant cells)	86	93:7	97 (−)	n.r.	[139]
45f	*rac*-*n*-Bu	$CHMO_{Acineto}$ (recombinant cells)	77	98:2	56 (−)	n.r.	[139]

(*Continued*)

TABLE 4.14 ***(Continued)***

Substrate	R	Biocatalyst	Yield (%)	Ratio 46:47	*ee* (%)[a] 46	*ee* (%)[a] 47	Reference
	(*S*)-n-Bu (83% *ee*)	$CHMO_{Acineto}$ (recombinant cells)	n.r.	>99:1	56 (*S*)	n.a.	[182]
		$CPMO_{Coma}$ (*Comamonas* sp. NCIMB 9872, recombinant cells)	85	>99:1	83 (*S*)	n.a.	[182]
Cyclopentanones							
45g	*rac*-Me	$CHMO_{Acineto}$ (recombinant cells)	95	13:87	9 (n.r.)	36 (n.r.)	[131]
		$CHMO_{Acineto}$ (recombinant cells)	95	13:87	9 (*R*)	36 (*S*)	[182]
	(*R*)-Me	$CHMO_{Acineto}$ (recombinant cells)	88	10:90	>99 (*R*)	>99 (*R*)	[182]
		$CPMO_{Coma}$ (*Comamonas* sp. NCIMB 9872, recombinant cells)	62	100:0	>99 (*R*)	n.a.	[182]
45h	*rac*-Et	$CHMO_{Acineto}$ (recombinant cells)	80	80:20	33 (n.r)	19 (n.r)	[131]
		$CPMO_{Coma}$ (*Comamonas* sp. NCIMB 9872, recombinant cells)	78	100:0	*rac.*	n.a.	[182]
	(*R*)-Et	$CPMO_{Coma}$ (*Comamonas* sp. NCIMB 9872, recombinant cells)	69	>99:1	86 (*R*)	n.a.	[182]
45i	*rac*-*n*-Prop	$CHMO_{Acineto}$ (recombinant cells)	44	93:17	33 (n.r)	60 (n.r)	[131]
		$CPMO_{Coma}$ (*Comamonas* sp. NCIMB 9872, recombinant cells)	70	100:0	*rac.*	n.a.	[182]

TABLE 4.14 (*Continued*)

Substrate	R	Biocatalyst	Yield (%)	Ratio 46:47	*ee* (%)[a] 46	*ee* (%)[a] 47	Reference
45j	(*R*)-*i*-Prop (87% *ee*)	$CHMO_{Acineto}$ (recombinant cells)	58	33:67	97 (*R*)	93 (*S*)	[182]
		$CPMO_{Coma}$ (*Comamonas* sp. NCIMB 9872, recombinant cells)	75	>99:1	87 (*R*)	n.a.	[182]
45k	*rac*-Allyl	$CHMO_{Acineto}$ (recombinant cells)	42	44:56	23 (n.r.)	n.r.	[131]
45l	*rac*-*n*-Bu	$CHMO_{Acineto}$ (recombinant cells)	34	>99:1	38 (*S*)	n.a.	[182]
		$CPMO_{Coma}$ (*Comamonas* sp. NCIMB 9872, recombinant cells)	76	96:4	*rac.*	n.a.	[182]
	(*S*)-*n*-Bu (90% *ee*)	$CHMO_{Acineto}$ (recombinant cells)	88	95:5	90 (*S*)	n.a.	[182]
		$CPMO_{Coma}$ (*Comamonas* sp. NCIMB 9872, recombinant cells)	84	>99:1	92 (*S*)	n.a.	[182]
45m	(*S*)-$(CH_2)_2Ph$ (88% *ee*)	$CHMO_{Acineto}$ (recombinant cells)	80	>99:1	99 (*S*)	n.a.	[182]
		$CPMO_{Coma}$ (*Comamonas* sp. NCIMB 9872, recombinant cells)	80	100:0	88 (*S*)	n.a.	[182]
45n	*rac*-*n*-Hex	$CHMO_{Acineto}$ (recombinant cells)	20	99:1	60 (n.r.)	n.a.	[131]
45o	*rac*-*n*-Oct	$CHMO_{Acineto}$ (recombinant cells)	19	99:1	16 (n.r.)	n.a.	[131]
45p	*rac*-*n*-Undec	$CHMO_{Acineto}$ (recombinant cells)	20	99:1	*rac.*	n.a.	[131]

[a] Absolute configuration or sign of specific rotation is given in parenthesis.
n.a., not applicable; n.r., not reported.

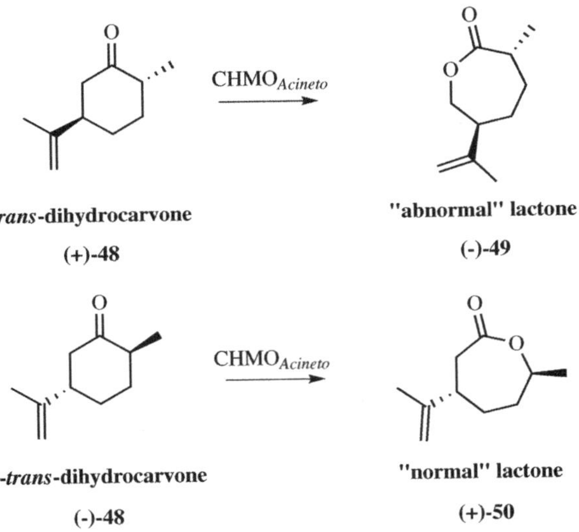

FIGURE 4.38 Regiodivergent enzymatic oxidation of (+) and (−)-dihydrocarvone.

An interesting case is represented by the regiodivergent oxidation of *trans*-dihydrocarvone (**48**). Both enantiomers of a substrate are accepted, and the migratory preference is inverted as well. When employing CHMO from *Acinetobacter* sp., (+)-*trans*-dihydrocarvone gives the "abnormal" lactone, while its enantiomer (-)-*trans*-dihydrocarvone affords the "normal" lactone upon oxidation, both with excellent enantiomeric purities (Figure 4.38) [183].

Several recombinant whole-cell biocatalysts were applied for the oxidation of both enantiomers of *trans*-dihydrocarvone and other optically pure terpenones (Table 4.15) [184]. In most cases, CHMO from *Acinetobacter* NCIMB 9871 gave regiodivergent oxidations to "normal" or "abnormal" lactones depending on the particular enantiomer. When using CPMO from *Comamonas* sp., the "abnormal" lactone was afforded as the major product, regardless of the absolute configuration of the substrate, albeit with poorer conversions. (-)-Menthone was not accepted as substrate by any of the selected BVMOs.

Synthesis of Natural Products and Biologically Active Compounds Chiral seven-membered-lactones are valuable building blocks. In particular, the optically pure lactone **15l** could be converted into (*R*)-(+)-lipoic acid, a growth-promoting enzyme cofactor used for the treatment of hepatitis, pancreatitis, and induced carcinomas [141]. Racemic ketone **14l** was enantioselectively oxidized by CHMO from *Acinetobacter* NCIMB 9871 to the corresponding (*R*)-lactone **15l** (*ee* >99%) [81]. The same lactone was afforded during the enzymatic oxidation with MO2 from *Pseudomonas* NCIMB 10007 [141]. This optically pure lactone could be converted into (*R*)-(+)-lipoic acid through six chemical transformation steps. However, when the racemic ketone **14l** was treated with either whole-cells of *Comamonas* NCIMB 9872 or purified CPMO, the enantiomeric (*S*)-**15l** lactone was afforded. This lactone can be

TABLE 4.15 Regiodivergent Baeyer–Villiger Oxidation of Terpenones [184]

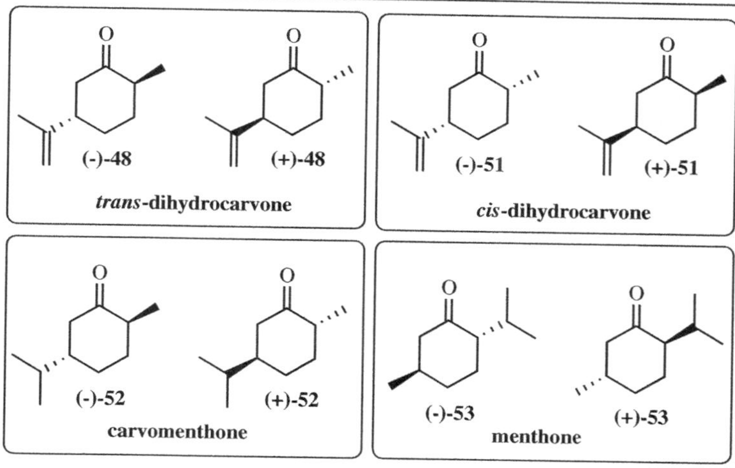

Substrate	Biocatalyst	Conversion[a] (%)	Ratio: normal: abnormal
(−)-48	CHMO$_{Acineto}$ (recombinant cells)	>90	100:0
	CHMO$_{Brevi1}$ (*Brevibacterium* sp., recombinant cells)	>90	100:0
	CPMO$_{Coma}$ (*Comamonas* sp. NCIMB 9872, recombinant cells)	<50	0:100
(+)-48	CHMO$_{Acineto}$ (recombinant cells)	>90	0:100
	CHMO$_{Brevi1}$ (*Brevibacterium* sp., recombinant cells)	50–90	55:45
	CPMO$_{Coma}$ (*Comamonas* sp. NCIMB 9872, recombinant cells)	<50	0:100
(−)-51	CHMO$_{Acineto}$ (recombinant cells)	>90	100:0
	CHMO$_{Brevi1}$ (*Brevibacterium* sp., recombinant cells)	>90	100:0
	CPMO$_{Coma}$ (*Comamonas* sp. NCIMB 9872, recombinant cells)	<50	0:100
(+)-51	CHMO$_{Acineto}$ (recombinant cells)	<50	64:34
	CHMO$_{Brevi1}$ (*Brevibacterium* sp., recombinant cells)	<50	100:0
	CPMO$_{Coma}$ (*Comamonas* sp. NCIMB 9872, recombinant cells)	<50	37:63
(−)-52	CHMO$_{Acineto}$ (recombinant cells)	50–90	100:0
	CHMO$_{Brevi1}$ (*Brevibacterium* sp., recombinant cells)	50–90	100:0
	CPMO$_{Coma}$ (*Comamonas* sp. NCIMB 9872, recombinant cells)	<50	0:100

(*Continued*)

TABLE 4.15 (*Continued*)

Substrate	Biocatalyst	Conversion[a] (%)	Ratio: normal: abnormal
(+)-**52**	$CHMO_{Acineto}$ (recombinant cells)	50–90	0:100
	$CHMO_{Brevi1}$ (*Brevibacterium* sp., recombinant cells)	50–90	41:59
	$CPMO_{Coma}$ (*Comamonas* sp. NCIMB 9872, recombinant cells)	50–90	0:100
(−)-**53**	$CHMO_{Acineto}$ (recombinant cells)	No reaction	
	$CHMO_{Brevi1}$ (*Brevibacterium* sp., recombinant cells)	No reaction	
	$CPMO_{Coma}$ (*Comamonas* sp. NCIMB 9872, recombinant cells)	No reaction	
(+)-**53**	$CHMO_{Acineto}$ (recombinant cells)	>90	100:0
	$CHMO_{Brevi1}$ (*Brevibacterium* sp., recombinant cells)	>90	100:0
	$CPMO_{Coma}$ (*Comamonas* sp. NCIMB 9872, recombinant cells)	No reaction	

[a]Conversion according to GC.

transformed into **54** through a shorter chemical sequence than its enantiomer (Figure 4.39) [111].

The bioxidation of prochiral 4-hydroxy-substituted cyclohexanones, such as **29h** resulted in a rearrangement of the initially formed seven-membered lactone **30h** to furnish the five-membered ring system **31h**. Taschner and coworkers reported this lactone as a potential precursor for the antibiotic tirandamycin (Figure 4.40) [84,185]. In that work, the Baeyer–Villiger oxidation of the prochiral ketone was carried out with isolated CHMO from *Acinetobacter* NCIMB 9871. The same reaction was also reported under different conditions, employing recombinant microorganisms expressing CHMO from *Acinetobacter* NCIMB 9871 [159], as well as CHMO from *Xanthobacter* sp. ZL5 [155]. The same optically pure lactone served as precursor of

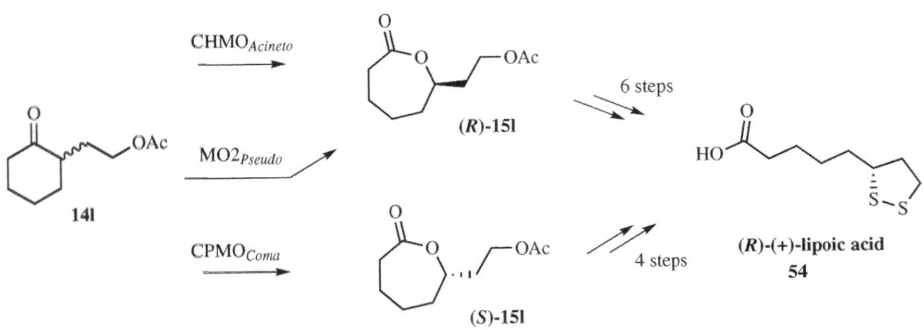

FIGURE 4.39 Application of chiral lactones produced by BVMO-mediated kinetic resolution in (*R*)-(+)-lipoic acid synthesis.

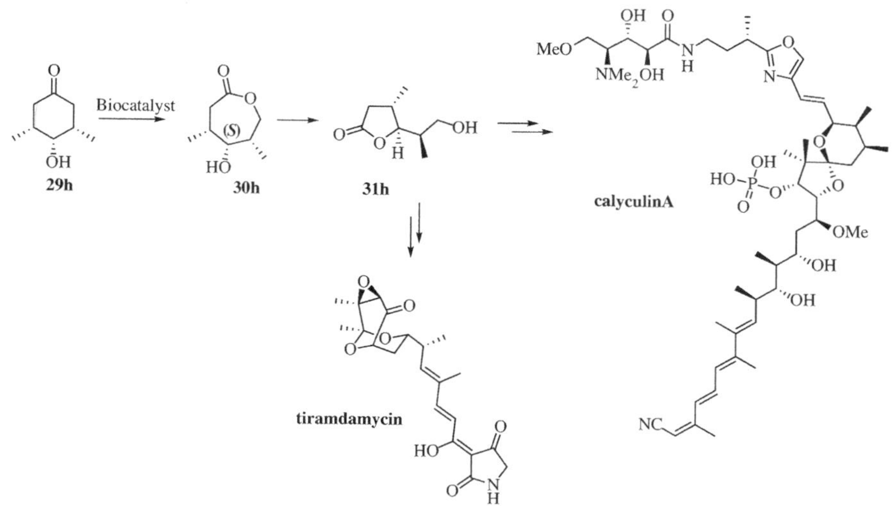

FIGURE 4.40 Lactone **31h** as precursor in natural product synthesis.

Calyculin, an inhibitor of protein phosphatases 1 and 2A, originally isolated from the marine sponge *Discodermia calyx* [186].

Desymmetrization of prochiral ketone **55** through a Baeyer–Villiger oxidation with CHMO isolated from *Acinetobacter* NCIMB 9871 afforded the chiral lactone **56** with excellent optical purity (> 98% *ee*) [84]. This lactone is an important intermediate in the preparation of the C_{11}–C_{16} subunit (**57**) of Ionomycin, a polyether antibiotic (Figure 4.41) [187].

3-Substituted prochiral cyclobutanones can be converted into optically pure 3-substituted γ-butyrolactones by BVMOs. Several cyclobutanones were successfully biotransformed by *C. echinulata* with excellent enantioselectivities [163,164]. A collection of target compounds using γ-butyrolactones as intermediates are included in Figure 4.42.

The oxidation of ketone **58** with whole-cells of *C. echinulata* NRLL 3655 leads to the optically pure lactone (*R*)-**59** with excellent enantiomeric excess (> 97%) [164]. This intermediate can be employed in the synthesis of β-amino acids such as β-proline **60** [188]. Furthermore, GABA (γ-aminobutyric acid) is the major inhibitory neurotransmitter in the central nervous system. Baclofen (**63**), an agonist of GABA,

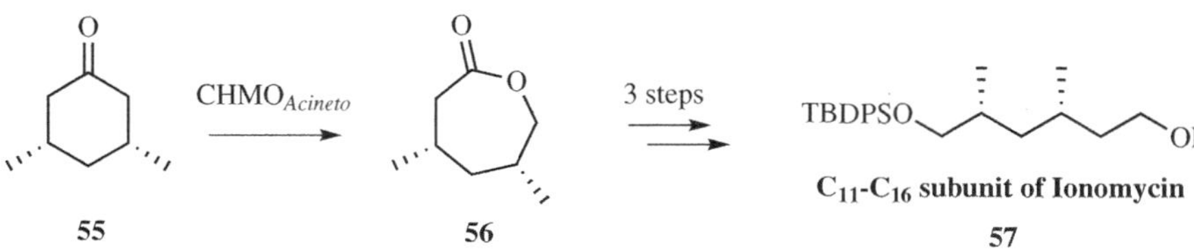

FIGURE 4.41 Prochiral ketone **55** as precursor of C_{11}–C_{16} subunit of Ionomycin.

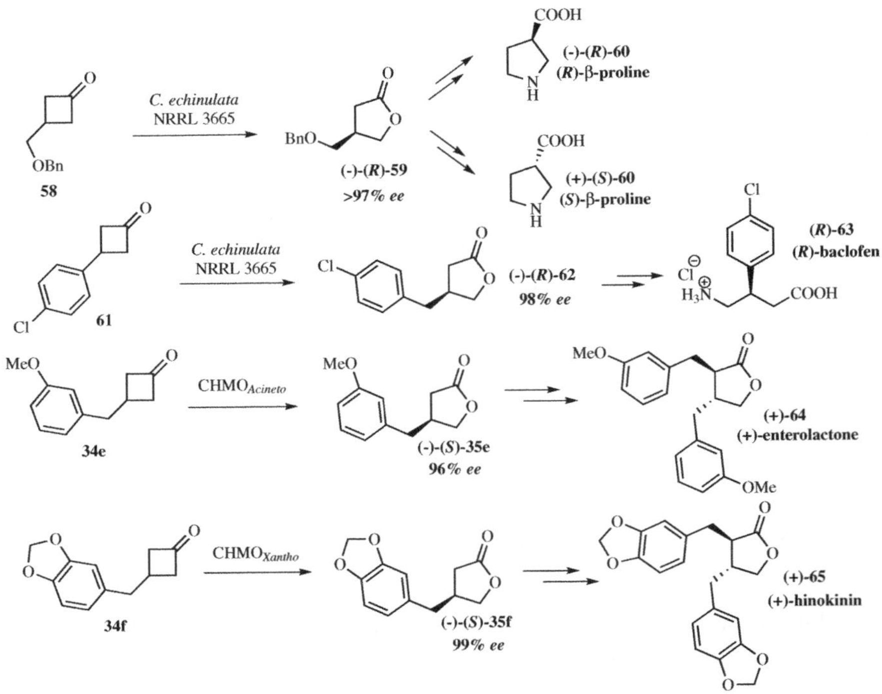

FIGURE 4.42 Desymmetrization of prochiral cyclobutanones to key intermediates for the synthesis of natural products and bioactive compounds.

was prepared from prochiral 3-(4-chlorophenyl)cyclobutanone (**61**) through an enantioselective enzymatic Baeyer–Villiger oxidation [189]. Besides whole-cells of *C. echinulata*, other BVMOs such as CHMO from *Acinetobacter* sp., or *Xanthobacter* sp., and HAPMO were employed in the desymmetrization of prochiral 3-substituted cyclobutanones [164,165]. Chiral butyrolactones (*S*)-**35e–f** are useful intermediates in the synthesis of lignans such as enterolactone (**64**) and hinokinin (**65**) [190,191]. The BVMO-mediated desymmetrization of some fused bicyclic[4.3.0]ketones showed an enantiodivergent trend when employing CHMO from *Brachymonas* sp. or CPMO from *Comamonas* sp. [158]. In particular, the bioconversion of ketone **38f** leads to both antipodes of lactone **39f** when using CHMO or CPMO (Figure 4.43). These lactones are key intermediates for the synthesis of indole alkaloids, such as (−)-alloyohimbane (**66**) and (−)-antirhine (**67**) [112].

Prochiral lactone **68** was selectively oxidized by recombinant cells of *E. coli* expressing CPMO from *Comamonas* sp. leading to the chiral (+)-lactone **69** (Figure 4.44). This chiral intermediate was employed as starting material for the preparation of several tetrahydrofuran-containing natural products, such as the antibiotic (+)-showdomycin (**71**), goniofufurone analogs (**73**), and (+)-*trans*-kumausyne (**75**) [192].

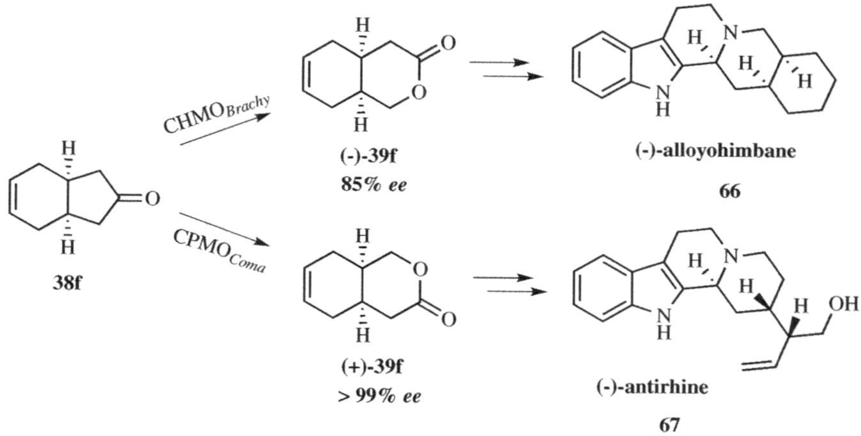

FIGURE 4.43 Desymmetrization of fused bicyclic ketone **38f** to precursors of indole alkaloids.

The kinetic resolution of norcamphor derivatives employing BVMOs represents a key step in the chemoenzymatic synthesis of nucleosides as well as advanced precursors of azadirachtin (Figure 4.45). Kinetic resolution of racemic haloketone 5-bromo-7-fluoro-nornornanone (**76**), employing whole-cells of *Acinetobacter* NCIMB 9871, lead exclusively to the normal lactone (−)-**77** with excellent enantiomeric excess (> 95%). (−)-Ketone **76** remained unchanged and could be recovered with high optical purity (*ee* > 95%). The optically pure ketone seved as key precursor for preparing antiviral fluorocarbocyclic nucleosides (**78**) [87,193,194]. MO1 from

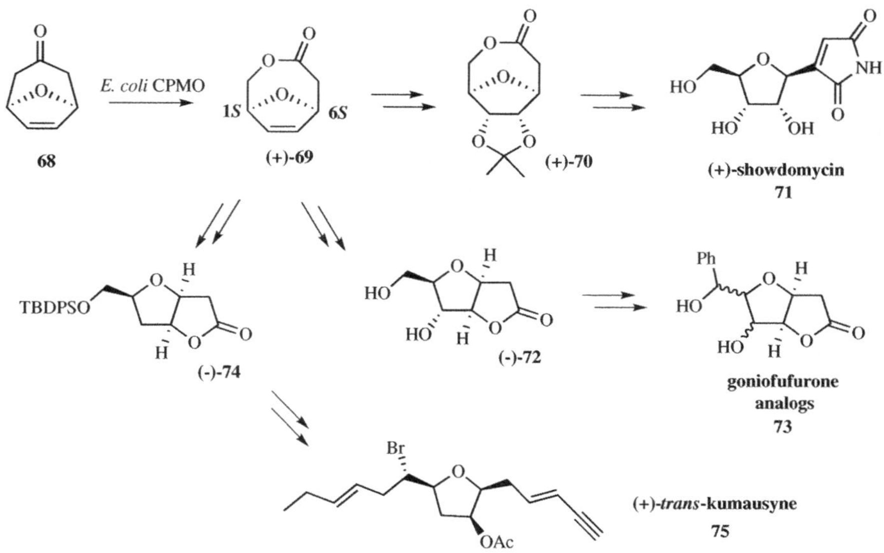

FIGURE 4.44 Desymmetrization of prochiral bicyclic ketone.

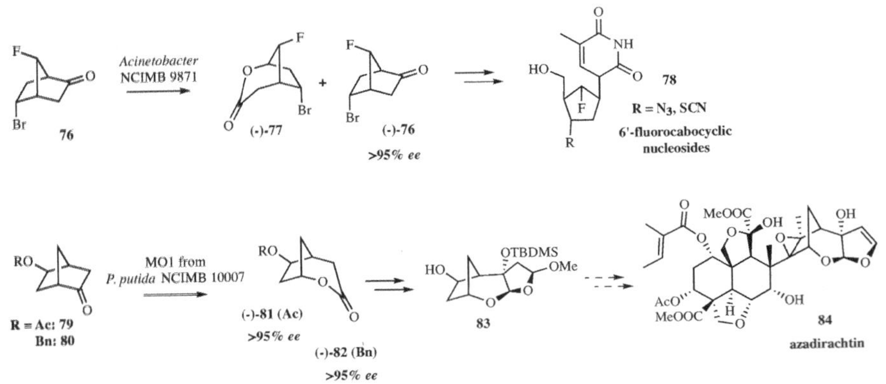

FIGURE 4.45 Kinetic resolution of bicyclic ketones for the production of bioactive compounds.

P. putida NCIMB 10007 was employed in the kinetic resolution of 5-substituted bicycle[2.2.1]heptan-2-ones (**79–80**) to give the normal lactones (−)-**81** and (−)-**82** with excellent enantiomeric excesses (>95%). Afforded lactones were converted into acetal **83**, an intermediate in the preparation of azadirachtin, a potent antifeedant and growth regulator [195].

The regiodivergent BVMO-mediated oxidation of bicyclo[3.2.0]hept-2-en-6-one allows the preparation of antipodal and regioisomeric lactones, depending on the particular selected biocatalyst (Figure 4.46). In particular, the treatment of ketone **40a** with CPMO from *Comamonas* sp. predominantly produced the "normal" lactone **41a** in excellent enantiomeric excess (>99%) [158]. This lactone can be employed in the preparation of various prostaglandins and thromboxanes (**85**) [196–198]. Oxidation of ketone **40a** with *P. putida* NCIMB 10007 produces the enantiomer *ent*-**41a**. This lactone represents a key intermediate in the synthesis of the C_1–C_9 fragment of callipeltoside-A (**86**), a macrocyclic lactone able to protect cells infected with HIV virus [199]. The "abnormal" lactone **42a** was selectively afforded by *C. echinulata*-mediated biooxidation. This lactone proved to be a very versatile building block, serving as starting material for the preparation of brown algae pheromones viridiene (**88**), multifidene (**89**), and caudoxirene (**90**) [173,200]. The same lactone was also employed in the preparation of advanced intermediates in the synthesis of sarkomycin A (**91**), a potent cytostatic [201,202].

Regiodivergent oxidation of prochiral fused ketone **40h** employing whole-cells of *Acinetobacter* sp. NCIMB 9871 led to "normal" furofuran-lactone **41h** (Figure 4.47). This chiral lactone served as precursor for the heterocyclic part of the insect antifeedant clerodin (**92**) [177,203].

4.1.4 Epoxidation of Alkenes

Epoxides are versatile building blocks and useful intermediates in organic synthesis, which can undergo stereospecific ring-opening reactions to afford halohydrins,

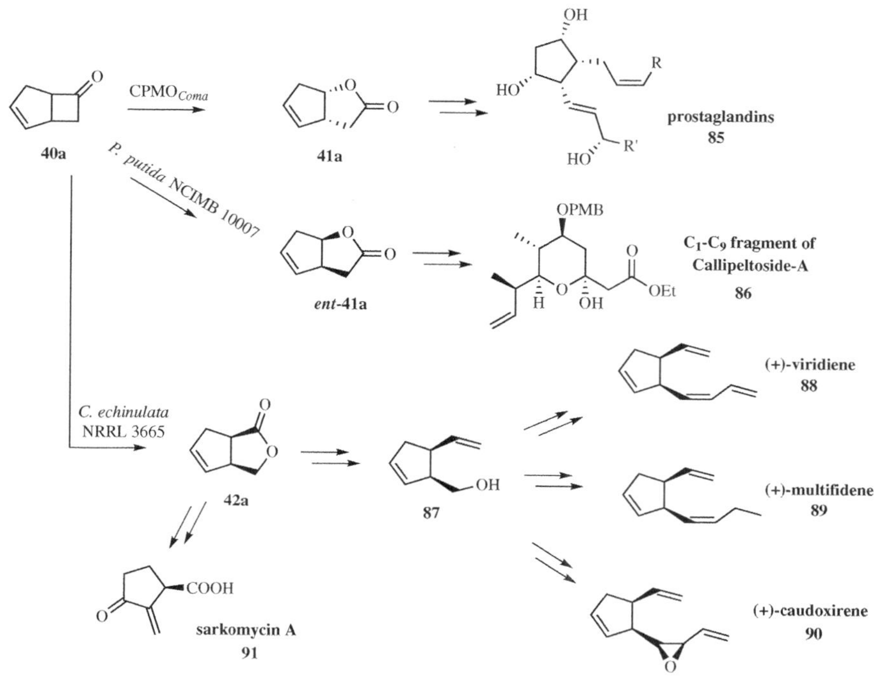

FIGURE 4.46 Regiodivergent biooxidation of fused ketones in the preparation of bioactive compounds and natural products.

aminoalcohols or diols, among others. Therefore, large efforts have been devoted to the development of asymmetric epoxidation reactions [204]. Among them, the Katsuki-Sharpless epoxidation of allylic alcohols constitutes a milestone in enantioselective synthesis [205]. Nonfunctionalized alkenes may be epoxidized employing chiral (salen)Mn complexes developed by Jacobsen [206]. Herein, enzymatic epoxidation catalyzed by monooxygenases constitutes a valuable alternative, since that usually provides good chemo-, regio- and stereoselectivity [207]. Moreover, monooxygenases use oxygen as oxidant and therefore constitute a green alternative

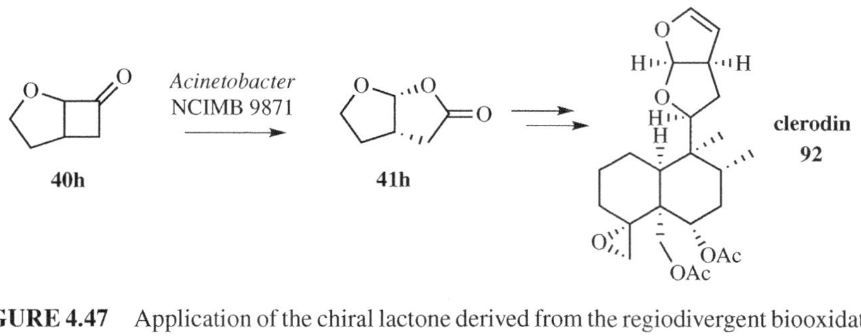

FIGURE 4.47 Application of the chiral lactone derived from the regiodivergent biooxidation of fused bicyclobutanone **40h**.

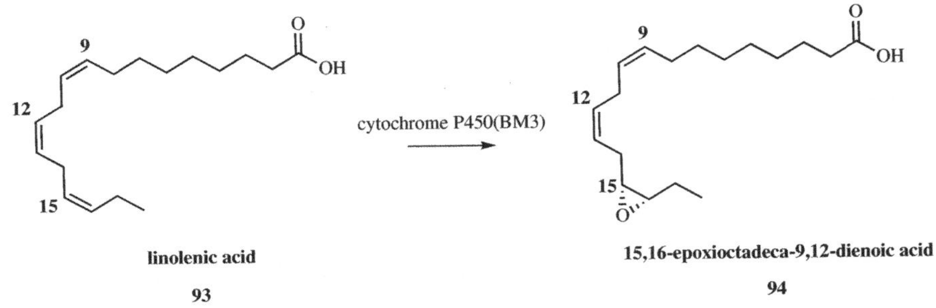

FIGURE 4.48 Cytochrome P450(BM3)-catalyzed epoxidation of linolenic acid.

process for epoxidation of C–C double bonds [3,204,208]. These monooxygenases can be metal-dependent enzymes, such as cytochrome P450 monooxygenases, or flavin-dependent enzymes, such as styrene monooxygenase (StyAB). In most cases, practical applications of P450 monooxygenases as epoxidation catalysts is made difficult by their low enantioselectivity [5,20,209–211]. Among cytochrome P450 monooxygenases, the enzyme from *B. megaterium* (cytochrome P450(BM3), CYP102) was one of the first enzymes reported to catalyze the epoxidation of unsaturated fatty acids, such as palmitoleic acid. In these early examples, in addition to epoxidation, hydroxylation of nonactivated carbon atoms was observed [212,213]. Also, cytochrome P450(BM3) was mutated in order to improve its epoxidizing ability. To this end, the replacement of Phe87 with Val converted cytochrome P450(BM3) into a regio- and stereoselective arachidonic acid epoxygenase [214]. More recently, cytochrome P450(BM3) was employed for the preparative synthesis of enantiomerically enriched epoxy-linolenic acid (**94**), with complete regio- and moderate enantioselectivity (Figure 4.48) [215]. The enzyme responsible for enzymatic epoxidation P450(BM3) was genetically engineered, and variants epoxidizing several substrates are currently available [216–220].

Sesquiterpenoids are naturally occurring compounds of biological importance. Biooxidation of (+)-valencene (**95**) has been investigated as a potential route to (+)-nootkatone (**97**), a fine fragrance. When employing P450(BM3) (wild-type as well as mutants) several products other than nootkatone were obtained (Figure 4.49).

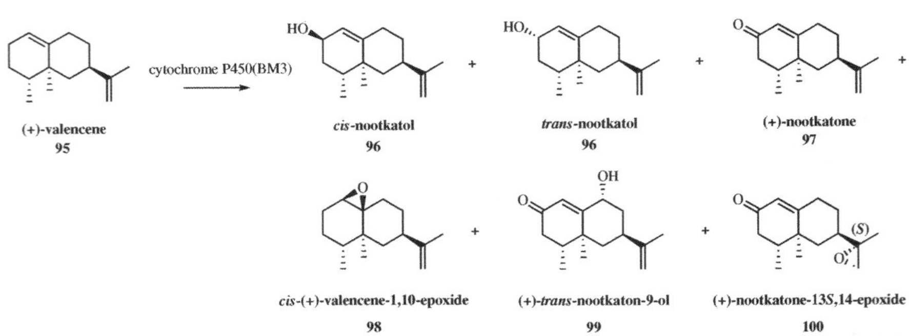

FIGURE 4.49 Cytochrome P450(BM3)-catalyzed epoxidation of (+)-valencene.

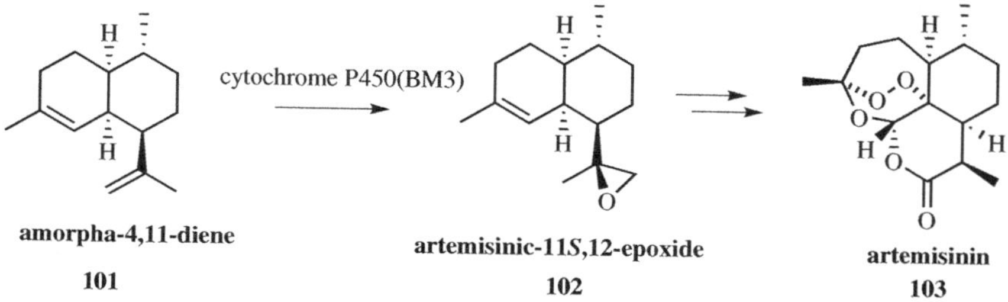

FIGURE 4.50 Cytochrome P450(BM3)-catalyzed epoxide production in the route to artemisinin.

Among those products, two epoxides were isolated and characterized, though in low yields (< 30%) [217].

An engineered substrate-promiscuous P450(BM3) enzyme, accepting amorpha-4,11-diene (**101**) as substrate for the regio- and stereospecific epoxidation, was reported as a key step in the preparation of the antimalarial drug artemisinin (**103**) (Figure 4.50) [219]. The mutant enzyme yielded the epoxide (**102**) at titers higher than 250 mgL^{-1}.

Camphor monooxygenase from *P. putida*, $P450_{cam}$ (CYP101) was also employed as an epoxidation catalyst. Substrates such as styrene derivatives [221], limonene, and α-pinene were tested (Figure 4.51) [18,222]. Conversion of *cis*-β-methylstyrene using $P450_{cam}$ leads to the corresponding epoxide in low enantiomeric excess. For substrates such as α-pinene and limonene, the enzymatic oxidation afforded several products, such as epoxidation, hydroxylation. and overoxidation products. This low

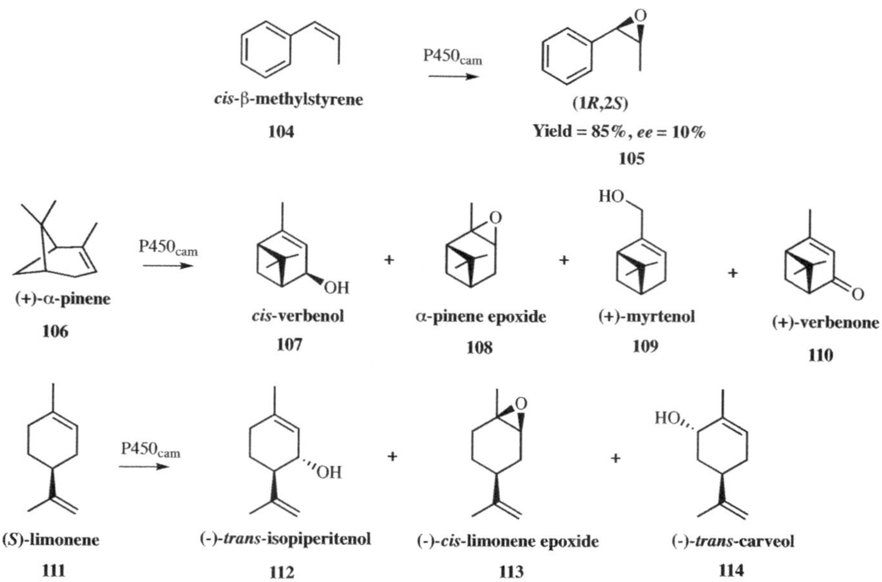

FIGURE 4.51 Cytochrome $P450_{cam}$-catalyzed epoxidaton of different substrates.

TABLE 4.16 Microbial Epoxidation of Alkenes

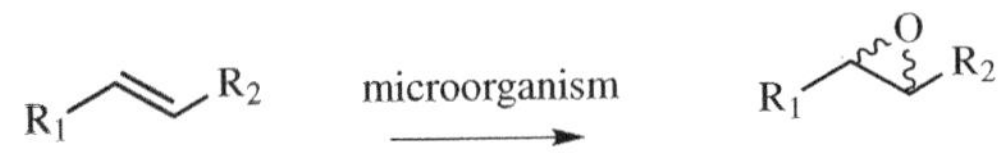

115a-i **116a-i**

Substrate	R_1	R_2	Biocatalyst	*ee* (%)[a]	Reference
115a	CH_3	H	*Mycobacterium* sp.	98 (*R*)	[223]
			Nocardioides sp.	98 (*R*)	[224]
			Xanthobacter sp.	97 (*R*)	[225]
115b	$CH_2\,CH_3$	H	*Nocardia* sp.	98 (*R*)	[223]
			Nocardioides sp.	74 (*R*)	[224]
			Xanthobacter sp.	94 (*R*)	[225]
115c	$CH_2\,Cl$	H	*Mycobacterium* sp.	98 (*S*)	[223]
			Nocardia sp.	98 (*S*)	[223]
115d	CH_3	CH_3	*Mycobacterium* sp.	86 (*R*,*R*)	[223]
			Nocardioides sp.	82 (*R*,*R*)	[224]
115e	$CH_2CH_2\,CH_3$	H	*Burkholderia cepacia*	>99 (*R*)	[226]
115f	n-C_6H_{13}	H	*Pseudomonas oleovorans*	70 (*R*)	[227]
115g	n-C_8H_{17}	H	*Pseudomonas oleovorans*	62 (*R*)	[228]
115h	CH_2O-p-C_6H_4-$CH_2CH_2\,OCH_3$	H	*Pseudomonas oleovorans*	98 (*R*)	[229]
115i	CH_2O-p-C_6H_4-CH_2CONH_2	H	*Pseudomonas oleovorans*	97 (*R*)	[229]

[a] Absolute configuration is given in parenthesis.

chemo-, regio-, and enantioselectivity hampered the use of such enzymatic systems in preparative transformations.

Terminal alkenes are converted into their corresponding epoxides employing *Mycobacterium* sp. and *Nocardia* sp. with good enantiomeric excesses (Table 4.16) [223]. Moreover, *Nocardioides* sp. and *Xanthobacter* sp. catalyze the enantioespecific epoxidation in moderate-to-excellent optical purities [224,225]. Longer-chain alkenes were also metabolized by *Burkholderia cepacia* or *Pseudomonas oleovorans* [226–229]. The epoxide products are generally of (*R*)-configuration.

The alkane monooxygenase (alkB) from *P. oleovorans* has been extensively studied, showing that this ω-hydroxylase is able to epoxidize terminal alkenes [227,230], as well as allylphenyl ethers (**115h–i**) [229]. This epoxidation affords (*R*)-glycidyl ethers in excellent enantiomeric excesses, leading to the industrial synthesis of adrenergic receptor inhibitors Metoprolol and Atenolol (Figure 4.52) [231]. The enzyme has been purified and characterized as a nonheme iron protein [232,233], its genetics has been studied [234], and the *alk* genes has been cloned in *E. coli* [235].

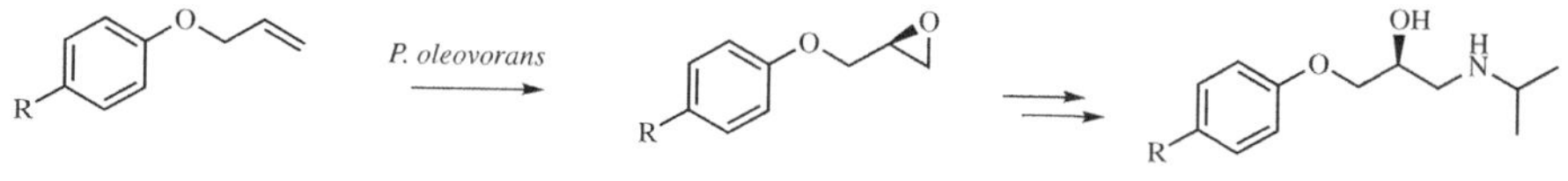

115h $CH_2CH_2OCH_3$
115i CH_2CONH_2

116h 98% *ee*
116i 97% *ee*

117h metoprolol
117i atenolol

FIGURE 4.52 Microbial epoxidation of aryl allyl ethers as a key step in the synthesis of β-blockers.

Other examples and synthetic applications employing whole-cell biocatalysts have been described, such as the enantioselective *trans*-dihydroxylation of C–C double bonds with *Sphingomonas* sp. HXN-200 (Figure 4.53) [236], through a first step epoxidation of the alkene and subsequent hydrolysis, affording the *trans*-dihydroxylation product. Both steps can be performed through enzymatic approaches. In particular, *Sphingomonas* sp. HXN-200 contains an alkane monooxygenase catalyzing the stereoselective epoxidation, and an epoxide hydrolase performing the *trans*-dihydroxylation through a hydrolysis reaction. During the biotransformation of **118a–b** and **121a–b**, the formed epoxides were hydrolyzed to the corresponding *trans*-diols.

Fosfomycin ((-)-(1*R*,2*S*)-1,2-epoxypropylphosphonic acid), a secondary metabolite of many strains of *Streptomyces*, is a broad spectrum antibiotic that has been proven to be effective for the treatment cephalosporin-resistant *Streptococcus pneumoniae* [237]. White and coworkers reported a screening of 20 fungi (18 species of *Penicillium*, 1 of *Paecilomyces* and 1 of *Oidium*) for the enantioselective epoxidation of *cis*-propenylphosphonic acid (**124**) to fosfomycin (**125**) [238]. *Penicillium* sp. produced the desired epoxide in 90% *ee* (Figure 4.54). Recently the strain S101 of

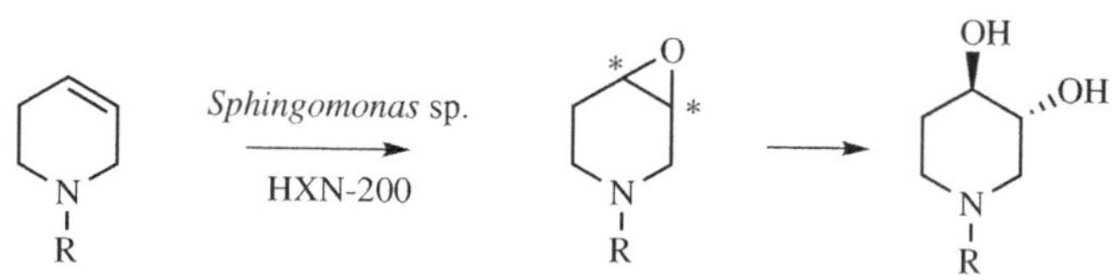

118a **R = CO_2Ph**
118b **R = CO_2CH_2Ph**

119a
116b

120a (*ee* > 90%)
120b (*ee* = 50%)

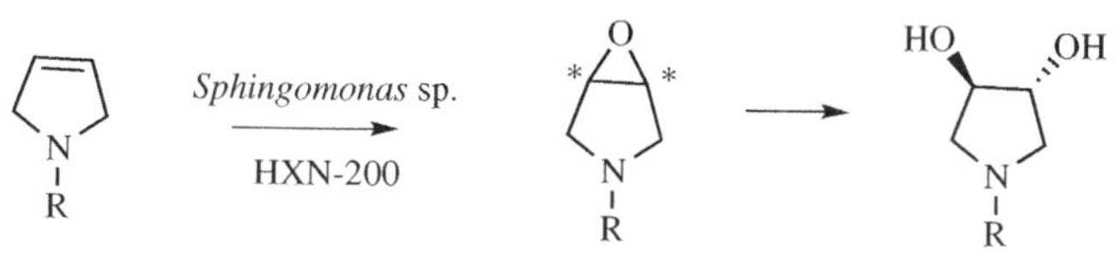

121a **R = CO_2Ph**
121b **R = CO_2CH_2Ph**

122a
122b

123a (*ee* = 80%)
123b (*ee* = 95%)

FIGURE 4.53 *Trans*-dihydroxylation of tetrahydropyridines and pyrrolines with *Sphingomonas* sp. HXN-200.

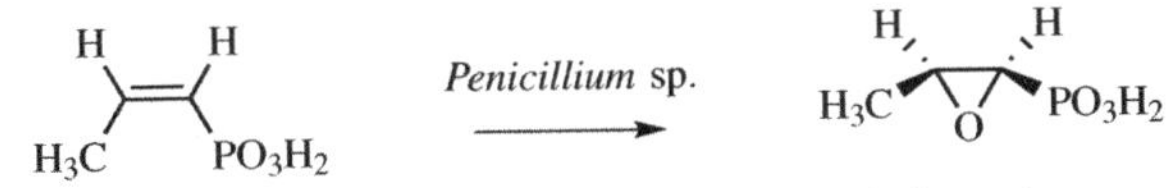

cis-propenylphosphonic acid

124

fosfomycin

125

90% *ee*

FIGURE 4.54 Biotransformation of *cis*-propenylphosphonic acid to fosfomycin.

Bacillus simplex showed the conversion of **124** to fosfomycin without the production of the unwanted (+)-stereoisomer [239].

Another example [240] reports the epoxidation of (*S*)-sulcatol derivatives (**126**) with *A. niger* to afford the corresponding (2*S*,5*S*)-epoxide (**127**), which is a precursor of pityol (**128**), a pheromone of the bark beetle *Pityophthorus pityographus*, with *ee* > 99% (Figure 4.55).

In addition to whole-cell epoxidation reactions, isolated enzymes applications were also evaluated. In particular, StyAB from *Pseudomonas* sp. VLB120 is one of the most studied biocatalysts to epoxidize alkenes. StyAB converts styrene into (*S*)-styrene oxide in > 99% *ee* [241], and is composed of a FAD-specific styrene epoxidase (StyA) and NADH-specific flavin reductase (StyB), encoded by *StyA* and *StyB* genes [242–244]. This enzymatic system was evaluated using catalytic amounts of nicotinamide cofactors and different regeneration systems (Table 4.17). The direct electrochemical regeneration as well as the use of an organometallic complex $[Cp{*}Rh(bpy)(H_2O)]^{2+}$ were applied successfully, allowing the production of enantiopure epoxides *via* enzymatic synthesis [245,246].

To simplify cofactor regeneration, recombinant *E. coli* cells carrying the StyAB genes from *Pseudomonas* sp. VLB120 [247–251], *P. fluorescens* ST [252–254], *P. putida* KT2440 [255], *Pseudomonas* sp. LQ26 [256,257], and *Rhodococcus opacus* 1CP [258,259] have been used in the epoxidation of styrene derivatives. For the natural substrate, those StyAB s afford the optically pure (*S*)-styrene oxide (*ee* >99%), accepting also styrene derivatives with substituents in the aromatic ring (methyl, methoxy, halogen atoms), methyl groups directly attached to the C–C double bond, and also some heteroaromatic systems. By assessing series of ring monosubstituted styrenes, cyclic compounds, and substituted double bonds styrene derivatives, a comprehensive study of the electronic and substitution-position effects was reported (Table 4.18) [253,254].

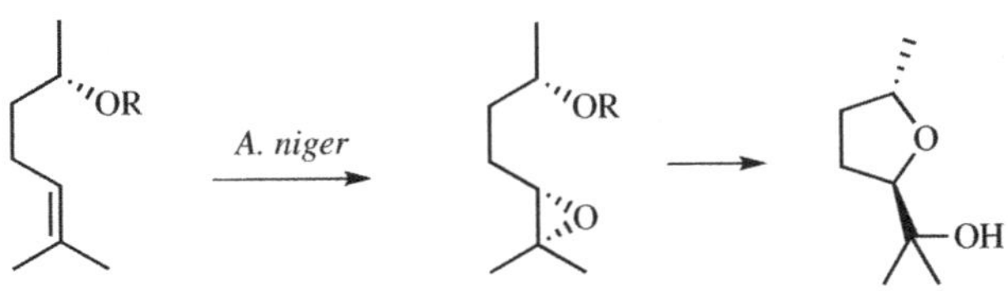

126
R = CONHPh

127
> 99% *ee*

pityol
128

FIGURE 4.55 Asymmetric epoxidation with *Aspergillus niger*.

TABLE 4.17 Epoxidation of Alkenes with Styrene Monooxygenase from *Pseudomonas* sp. VLB120

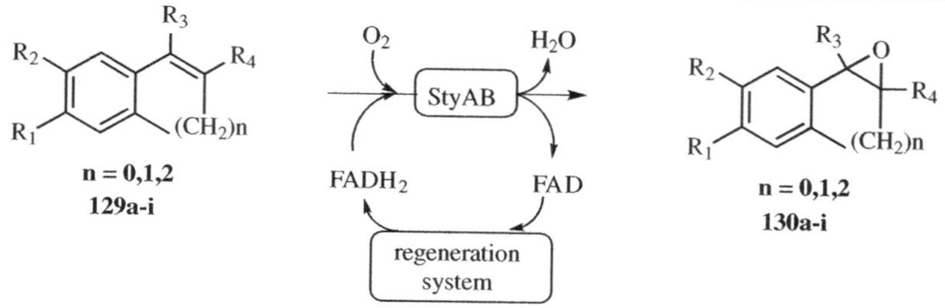

Substrate	n	R_1	R_2	R_3	R_4	*ee* (%)	Reference
129a	0	H	H	H	H	98.5	(245)
129b	0	H	H	CH_3	H	99.5	(245)
129c	0	H	H	H	CH_3	99.9	(245)
129d	0	H	Cl	H	H	98.1	(245)
129e	0	F	H	H	H	99.4	(245)
129f	0	Cl	H	H	H	98.1	(245)
129g	0	Br	H	H	H	95.6	[246]
129h	1	H	H	H	H	97.9	[246]
129i	2	H	H	H	H	99.2	(245)

In this study, an unexpected monotonic increase of the yield along the series *ortho*, *meta*, and *para* was reported. Moreover, the *ortho*-substituted compounds did not show an enhanced influence of the substituent when compared to the *meta*-substituted compounds. These observations evidenced that the main factor in the *ortho*-isomers may be a geometric effect. This became clear when comparing the electronic effect in the *ortho*-substituted compounds ($OCH_3 > CH_3 \approx Cl > CN$), which had indeed almost no influence. For the *para*-isomers (no geometric effect), the expected electronic influence is observed, with a preference for the most electron-donating group (methoxy group) [254].

For substrates containing the C–C double bond substituted by a methyl group, the conversion rates also depended on the substrate, yet a clear correlation could not be established. However, the *Z*-isomers are more reactive than the *E*- ones. All epoxides were obtained in enantiopure form, except the epoxide derived from the only disubstituted alkene, 1-phenyl-2-methyl propene (**135**) [253]. In the same area, a systematic study on the performance of the StyAB from *Pseudomonas* sp. LQ26 was also reported (Table 4.19) [260].

In all cases, (*S*)-epoxides were obtained in enantiomeric excesses over 99%, except for substrates **140** and **151**. On the other hand, the yield of epoxide significantly varied along the different styrene derivatives. When increasing the size of the substituent (from H to propyl) a marked steric effect was observed, lowering the yield of epoxide. Conversely, substrates with aromatic rings larger than styrene (**152–153**) were well accepted, yielding the optically pure (*S*)-epoxides. Substituents in the α-position

TABLE 4.18 Epoxidation of Styrene Derivatives with Recombinant *E. coli* Containing the Styrene Monooxygenase from *P. fluorescens* ST

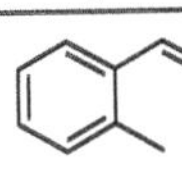
131a

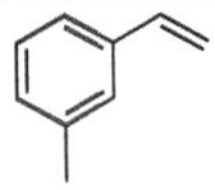
131b

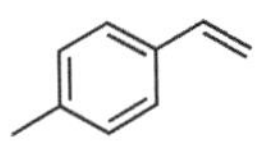
131c

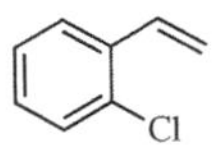

132

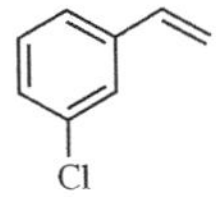

129d

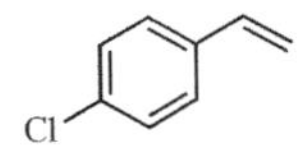

129f

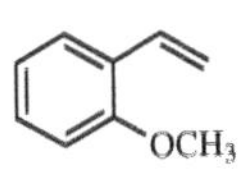

133a

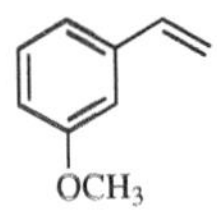

133b

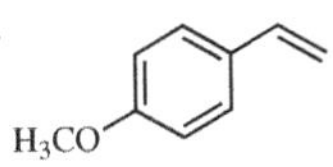

133c

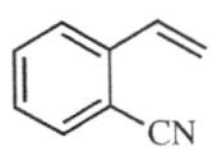

134a

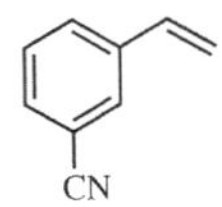

134b

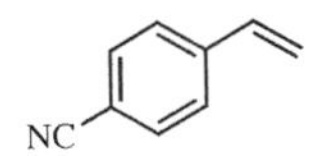

134c

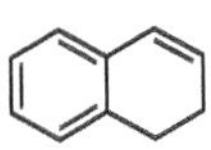
129i

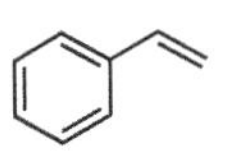
129a

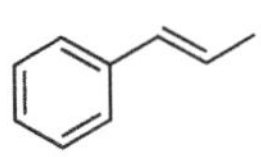
129c

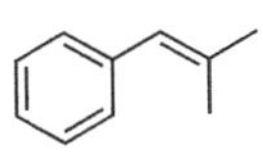
135

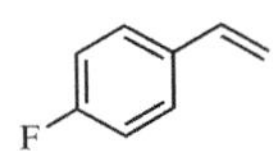

129e

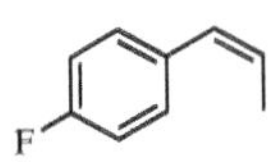

136

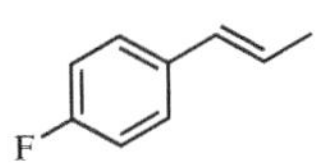

137

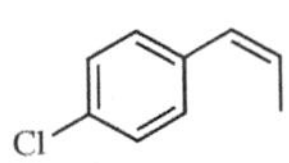

138

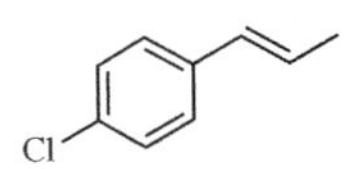

139

Substrate	Relative activity (%)[a]	*ee* (%)	Reference
131a	46	56	[254]
131b	101	>95	[254]
131c	130	>95	[254]
132	60	>95	[254]
129d	107	>95	[254]
129f	145	>95	[254]
133a	55 (racemic diol)	n.d.	[254]
133b	138	>95	[254]
133c	222 (racemic diol)	n.d.	[254]
134a	0	–	[254]
134b	55	>95	[254]
134c	0	–	[254]
129i	43	>95	[254]
129a	100	>95	[253]
129c	166	>95	[253]
135	310	87	[253]
129e	237	>95	[253]
136	38	>95	[253]
137	17	>95	[253]
138	78	>95	[253]
139	16	>95	[253]

[a]Relative activity = [specific activity of substrate (i)]/[specific activity of styrene] × 100; specific activity of (i) = mmoles of product formed in 1h by 1g cells (dry cell weight) in 1 L of broth.
n.d.: not determined.

TABLE 4.19 Epoxidation of α- and β-Substituted Styrene Derivatives with Recombinant *E. coli* Containing the Styrene Monooxygenase from *Pseudomonas* sp. LQ26

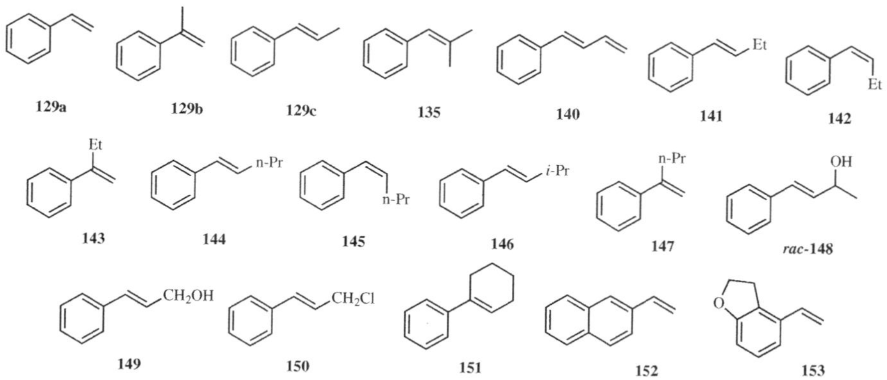

Substrate	Yield (%)[a]	Relative Activity (%)[b]	*ee* (%)
129a	>99%	100	>99%
129b	87	90	>99%
129c	92	97	>99%
135	68	71	>99%
140	58	64	96
141	45	51	>99%
142	6	10	>99%
143	35	39	>99%
144	12	15	>99%
145	1	3	>99%
146	11	15	>99%
147	5	7	>99%
***rac*-148**	>99%	167	>99% (*syn*:*anti* 1:1)
149	>99%	271	>99%
150	73[c]	80	>99%
151	7	9	65
152	76	77	>99%
153	65	70	>99%

[a]HPLC yield after 24 h.

[b]Relative activity = [specific activity of substrate (i)]/[specific activity of styrene] × 100; Specific activity of (i) = mmoles of product formed in 1 h by 1 g of wet cells.

[c]The resulting epoxide is the same obtained for **149** ((2*S*,3*S*)-3-phenyloxiran-2-yl methanol) and *rac*-1-phenylprop-2-en-1-ol was isolated with a yield of 17%.

TABLE 4.20 CHMO$_{Acineto}$-Mediated Epoxidation of Vinyl Phosphonates

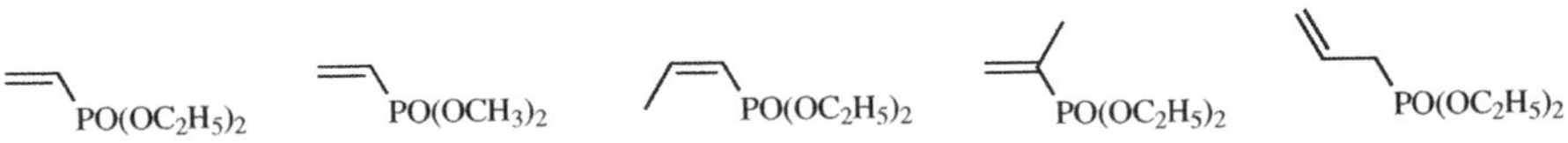

154 **155** **156** **157** **158**

Substrate	Yield (%)	*ee* (%)
154	40	≥98
155	10	≥98
156	No reaction	–
157	No reaction	–
158	No reaction	–

caused greater steric hindrance than those present in the β-position (comparison of **129b**, **143**, and **147** with **129c**, **141**, and **144**). The *E*-isomers were more reactive than the *Z*- ones, contrary to the results observed for StyAB from *P. fluorescens* ST. For styrenes bearing a hydroxyl group at the β-position (**148–149**), StyAB also catalyzed the epoxidation of both substrates. Finally, compound **150** did not afford corresponding epoxide, but yielded the same product than the one obtained for **149**.

Even though BVMOs have been extensively studied, as discussed throughout Section 4.1.3, only two reports on the epoxidation of alkenes with these enzymes are available so far. Colonna and coworkers assessed the olefin epoxidation catalyzed by CHMO from *A. calcoaceticus* NCIMB 9871. Substrates such as styrene, α-methyl styrene, 2-phenylpropene, 2-methyl-2-propen-1-ol, 3-methyl-3-propen-1-ol, 2,3,3-trimethyl-1-butene, among others were tested. Only dimethyl and diethyl vinyl phosphonate were accepted as substrates by CHMO, yielding the optically pure (*R*)-epoxide (Table 4.20) [261]. The selection of such kind of substrates was due to their structural analogy with fosfomycin (**125**).

The second example of a CHMO-mediated epoxidation was found during the substrate profiling of engineered whole-cells overexpressing the BVMO from *Xanthobacter* sp. ZL5. The authors reported that the prochiral bridged ketone **159** was converted into the corresponding epoxide (**160**) in moderate yields (Figure 4.56) [155].

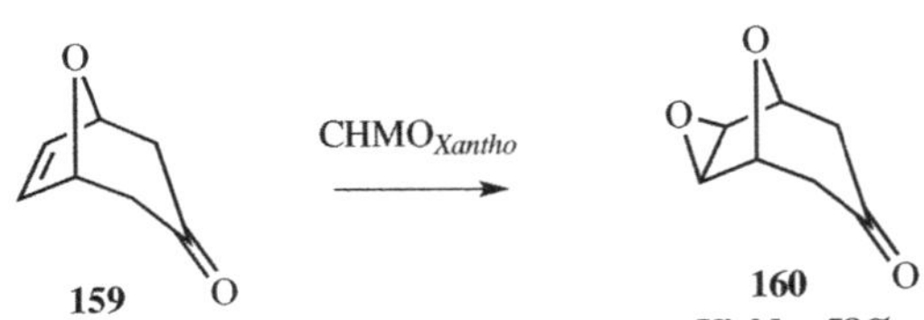

159 **160**
Yield = 58%

FIGURE 4.56 Asymmetric epoxidation of bridged ketones with CHMO from *Xanthobacter* sp. ZL5.

4.2 DIOXYGENASE-CATALYZED REACTIONS

Dioxygenases are a heterogeneous group of enzymes that incorporate both atoms of molecular oxygen into the substrate. These oxidoreductases belong to two subclasses, EC 1.13.-.- and EC 1.14.-.-, depending on the incorporation of molecular oxygen into single or paired donors, respectively. The ability of these enzymes to activate molecular oxygen from the atmosphere and use it in organic reactions has attracted considerable attention, both from a theoretical and a preparative viewpoint, given the little precedent of this type of reactions in organic chemistry. Most dioxygenases require a metal cofactor for oxygen activation, which is usually an iron atom, either as a heme-complex or in a different environment. However, there is a small group of metal-independent dioxygenases [262]. Normally, a double bond of the substrate is transformed into an unstable peroxo-species, such as a hydro- or *endo*-peroxide, which can either be obtained as the final product, or react further to yield the more stable hydroxy derivatives. In this context, alkenes afford allylic hydroperoxides (in a lipoxygenase-mediated dioxygenation reaction) or *endo*-peroxides, whereas arenes are transformed by aromatic dioxygenases into *cis*-dihydrodiols *via* a formal *endo*-peroxide (dioxetane) intermediate. Different groups of dioxygenases exist, catalyzing an ample range of oxidations [262,263]. In this section the most useful enzymes from a preparative point of view will be described. In this regard, aromatic dioxygenases are, arguably, the most used and studied of the group. Also, lipoxygenases, associated to the biocatalytic formation of hydroperoxides, will be discussed.

4.2.1 Aromatic Dioxygenases

These enzymes are crucial for the natural degradation of aromatic compounds by bacteria, performing the first step in the metabolic pathway (namely the oxygenation). The capability of these enzymes to achieve functionalized addition reactions—leading to dearomatization of benzene rings—is fascinating, given the resonance energy deficit involved and the thermodynamic drive to preserve the aromatic sextet. A chemical equivalent of this transformation on a benzene derivative was described in 1995 [264], albeit through a nonselective, uncontrolled, and foul yielding procedure. Moreover, no chemocatalyzed enantioselective modifications have been reported to date. Thus, the synthetic relevance of aromatic dioxygenases has been noticed and widely exploited in synthetic chemistry. The preparative application of aromatic dioxygenases is performed using whole-cell systems. It is worth mentioning that further steps in the degradation of aromatic compounds involve other type of dioxygenases, such as catechol dioxygenases, but the synthetic use of these enzymes is not yet significant from a practical viewpoint.

4.2.1.1 Dihydroxylation of Aromatic Compounds As stated in Chapter 1, the enzymes involved in the dihydroxylation of arenes, the Rieske dioxygenases, belong to the family of Rieske nonheme iron oxygenases. These enzymes are responsible for the generation of *cis*-dihydroxylated metabolites, a common first step

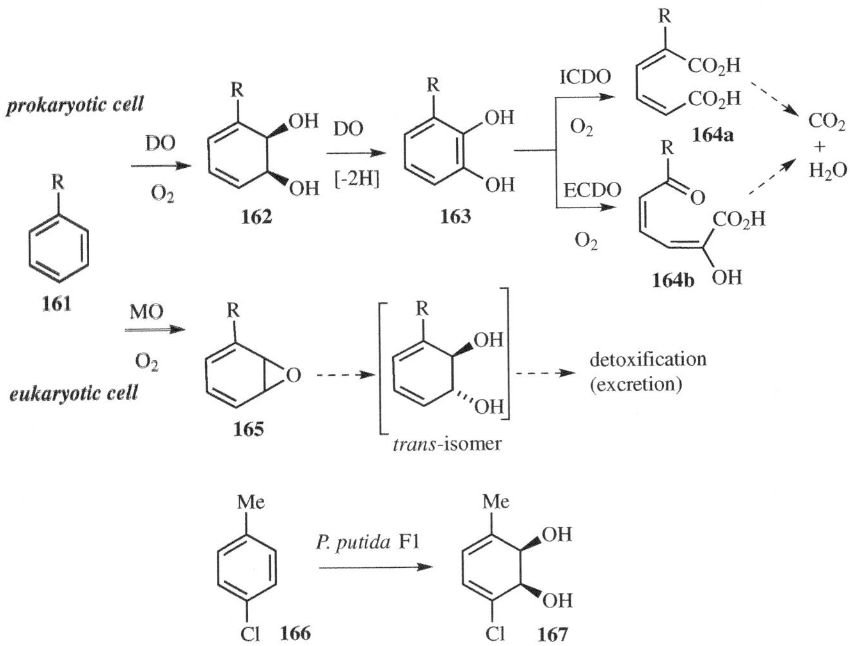

FIGURE 4.57 Degradation pathways for monosubstituted benzene derivatives. MO, monooxygenase; DO, Rieske dioxygenase; ICDO, intradiol-cleaving catechol dioxygenase; ECDO, extradiol-cleaving catechol dioxygenase.

in the bacterial degradation of arenes. The mechanism of *cis*-dihydroxylation in the degradation of arenes was proposed in 1968 by Gibson, who published the first unequivocal identification of an arene *cis*-dihydrodiol as a result of his work with strains of the soil bacterium *P. putida* [265]. He also isolated the first stable arene *cis*-dihydrodiol (**167**), derived from oxidation of *p*-chlorotoluene (**166**) by the action of *P. putida* F1, a mutant strain (Figure 4.57) [266]. This mechanism differed from that of formation of dihydrodiols in eukaryotic cells, usually performed by P450 monooxygenase systems, involving an arene oxide (**165**), which is further cleaved to produce a *trans*-diol.

In the dioxygenase-catalyzed reaction, molecular oxygen was identified as the source of both hydroxy moieties in the products [267,268]. As mentioned, for monosubstituted benzenes, the first step of arene degradation in prokaryotic organisms comprises the dioxygenase-catalyzed dihydroxylation of the arene to give the corresponding *cis*-dihydrodiol **162** as depicted in Figure 4.57 [262,269]. In wild-type bacteria, this diol is rearomatized by a *cis*-diol dehydrogenase to yield a catechol **163** which, in turn, can be degraded by either an intradiol-cleaving catechol dioxygenase (ICDO) to *cis,cis*-muconic acid **164a** or an extradiol-cleaving catechol dioxygenase (ECDO) to muconic semialdehyde **164b** (R = H). Hydration and retro-aldol reaction then provide acetate (and ultimately carbon dioxide) as an energy and carbon

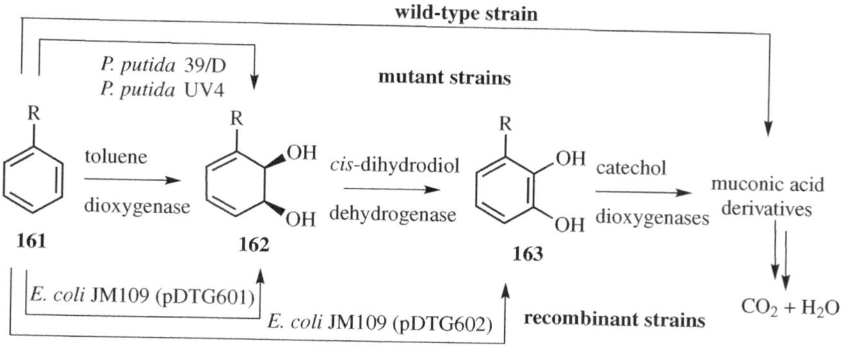

FIGURE 4.58 Examples of recombinant and blocked mutant strains used in the degradation pathway of arenes.

source. The dioxygenation takes place at the 2,3- bond to give monosubstituted *cis*-diol **162**. The only exceptions reported so far are the reaction of nitrobenzene and benzoic acid substrates, where *cis*-dihydroxylation occurs exclusively at the 1,2-bond (*ipso*) to yield stable diols of the parent benzoic acid, and catechols in the case of nitrobenzene, through spontaneous nitrite elimination [270–273].

The *cis*-diols of type **162**, although relatively stable, were difficult to isolate since they are substrates for the following enzyme in the metabolic pathway. However, in Gibson's blocked mutant *P. putida* 39/D the expression of *cis*-diol dehydrogenase is knocked out so that the intermediate *cis*-dihydrodiol **162** accumulates in the fermentation medium [274]. Nowadays, a number of arene *cis*-dihydrodiols are obtained by using whole-cell mutant strains with the *cis*-diol dehydrogenase blocked and by recombinant organisms expressing the dioxygenase genes (see Figure 4.58 and Section "Dioxygenases and bacterial strains for the production of arene *cis*-dihydrodiols"). Moreover, molecular biology techniques such as site-directed mutagenesis and directed-evolution approaches have been used to modify the selectivity and to improve the efficiency of dioxygenases [275,276].

Dioxygenases and Bacterial Strains for the Production of Arene cis-Dihydrodiols The aromatic dioxygenases responsible for the arene *cis*-dihydrodiol formation, Rieske nonheme iron oxygenases, are multicomponent enzyme systems (E.C. 1.14.12.-) that catalyze dihydroxylations, differing from the dioxygenases involved in aromatic ring cleavage, which act on the downstream catechol intermediates in many of the same metabolic pathways (Figure 4.57). Being the first step of a catabolic pathway, these enzymes are typically quite promiscuous, catalyzing the oxidation of a number of compounds in addition to their native substrates. At the same time, however, many of these enzyme systems are highly enantioselective, producing chiral *cis*-dihydrodiols or other chiral compounds in high enantiomeric purities. These properties have made aromatic ring hydroxylating dioxygenases attractive as biocatalysts.

More than one hundred Rieske dioxygenases have been identified so far [277], comprising two or three component systems: (i) a reductase unit that obtains electrons from NAD(P)H; (ii) often a Rieske ferredoxin component that shuttles the electrons; and (iii) an oxygenase component that actually performs catalysis. Currently, they are phylogenetically classified in five families: (i) toluene/biphenyl, (ii) phthalate, (iii) benzoate, (iv) naphthalene, and (v) salicylate, although a new classification was recently put forward (see Chapter 1) [277,278]. Also, they are commonly classified according to the arene used as a carbon source, namely toluene-, chlorobenzene-, phthalate-dioxygenase, and so on. The most widely used types in biotransformations include benzene- (BDO), toluene- (TDO), biphenyl- (BPDO), chlorobenzene- (CDO), benzoic acid- (BZDO), nitrobenzene- (NBDO), and naphthalene-dioxygenase (NDO). Because of the multicomponent nature of these enzymes and the need of cofactors, the biotransformations are performed using whole-cells biocatalytic systems. In this area, the first successful approaches to yield arene *cis*-dihydrodiols used whole-cell mutant strains containing TDO, but blocked in the next enzymatic step (e.g., *P. putida* 39/D or UV4), NDO (e.g., *P. putida* 9816/11), BPDO (e.g., *Sphingomonas yanoikuyae* B8/36), or BZDO (e.g., *P. putida* U103 or *Alcaligenes eutrophus* B9) [277,279]. The development of whole-cell *E. coli* recombinant strains expressing the appropriate encoded dioxygenase genes including TDO (e.g., *E. coli* JM109 [pDTG601] or JM109[pKST11]) [280], NDO (e.g., *E. coli* JM109[DE3][pDTG141]) [279], and BPDO (e.g., *E. coli* pKF2072) [281] and the possibility of using protein engineering techniques, such as random or site-directed mutagenesis and gene shuffling, have become of increasing importance (see Chapter 7). For instance, the substrate range and the activity of a 2,4-dinitrotoluene dioxygenase (DDO) were improved by site-directed mutagenesis [282], and NDO was also subjected to a number of molecular biology procedures to improve its substrate tolerance and activity [275,283,284].

Considering the number of biocatalysts available to perform the aromatic dioxygenation, the size and nature of the substrate are relevant when choosing the catalytic system. As a general rule, BDO and TDO can accommodate suitably substituted monocyclic arene substrates at the active site, yielding the corresponding benzene *cis*-dihydrodiols, whereas they do not accept polycyclic arene substrates larger than naphthalene. Conversely, NDO or BPDO can accommodate larger carbocyclic and heterocyclic arenes (di-, tri-, tetra- and penta-cyclic), but few monocyclic arenes (with the exception of biphenyl analogs) [281] have been dihydroxylated using NDO or BPDO. Indeed, more than 100 substrates are dihydroxylated by TDO (from *P. putida* F1) and more than 60 by NDO [285–288]. Other dioxygenases are much more selective, specifically targeting benzene rings bearing carboxylate (BZDO) [289,290], or nitro groups (NBDO) [272,273]. In addition to the dihydroxylation of arenes, these Rieske dioxygenases catalyze different types of oxidations, showing both dioxygenase- and monooxygenase-behavior (see Section 4.2.1.2) [291].

The most studied Rieske dioxygenase is naphthalene 1,2-dioxygenase (NDO), which catalyzes the NAD(P)H and O_2-dependent oxidation of naphthalene to (+)-*cis*-(1*R*,2*S*)-dihydroxy-1,2-dihydronaphthalene (Figure 4.59). The overall reaction stoichiometry for the arene dioxygenation requires molecular oxygen and two

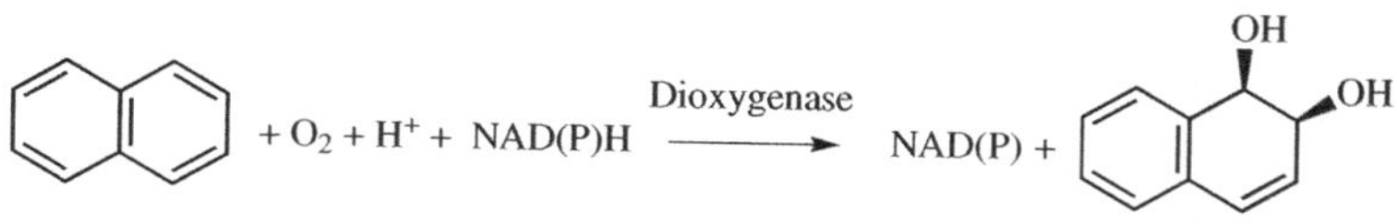

FIGURE 4.59 Stoichiometry of the arene dioxygenation.

electrons coming from a NAD(P)H reductase, and are transferred to the final oxygenase component through a ferredoxin (see Chapter 1, Section "Rieske-type dioxygenases").

Selectivity of the Dioxygenase-Catalyzed Oxidation A wide variety of aromatic compounds are accepted as substrates by aromatic dioxygenases. Actually, more than 300 arene *cis*-dihydrodiol metabolites have been reported, arising from the oxidation of substituted benzenes, naphthalene and higher fused aromatic systems, biphenyl and other linked aromatic systems, heteroaromatics and derivatives. Along the years, several comprehensive compilations of the known metabolites and their use in synthesis have been published [279,285,286,292–297]. The dihydroxylation is remarkably regio-, stereo-, and enantioselective.

REGIOSELECTIVITY OF THE DIHYDROXYLATION OF ARENE *CIS*-DIHYDRODIOLS

Monosubstituted Benzenes The most common *cis*-dihydrodiols, representing more than 50 examples, are of type **162**, resulting from TDO-catalyzed *cis*-dihydroxylation at C2–C3 in monosubstituted benzene substrates **161** (Figures 4.57 and 4.60) (279, 285,286,292–298). When R is a polar group, such as amino, carboxylate, or sulfonate, the corresponding arenes aniline ($R = NH_2$), benzoic acid ($R = CO_2H$), and benzene sulfonic acid ($R = SO_3H$), have not been reported as substrates for TDO. *cis*-Dihydroxylation at other positions have not been found in monosubstituted benzene derivatives using TDO, with the only exception of fluorobenzene, which gives the expected dihydrodiol (**162**, R = F) along with trace amounts (< 3%) of the C3–C4 dihydroxylated regioisomer in 20% *ee* (**169**, R = F) (299).

Dioxygenases different from TDO accomplish *cis*-dihydroxylation at C1–C2 (*ipso*) in monosubstituted substrates. Thus, BZDO-catalyzed reaction of benzoic acid (**161,** $R = CO_2H$) yielded a stable (1*S*,2*R*)-*cis*-dihydrodiol of type **168** ($R = CO_2H$) [270,271,289,290], while the NBDO-catalyzed reaction of nitrobenzene was postulated to give a transient *cis*-dihydrodiol intermediate of the same type before rearomatization (**168**, $R = NO_2$) (Figure 4.60) [272,273].

NDO has not been reported to accept monosubstituted benzenes as substrates, with the exceptions of biphenyl and its aza-analogs [275,299]. In this case, regioisomeric dihydroxylation at C3–C4 yields a mixture of isomers. Thus, NDO-catalyzed *cis*-dihydroxylation of biphenyl affords the expected isomer of type **162** together with the regioisomer of type **169** in a 87:13 ratio (Figure 4.60) [275]. Remarkably, the use of site-directed mutants of NDO caused a major change in the selectivity of the reaction, in particular upon replacement of Phe352 with smaller amino acids

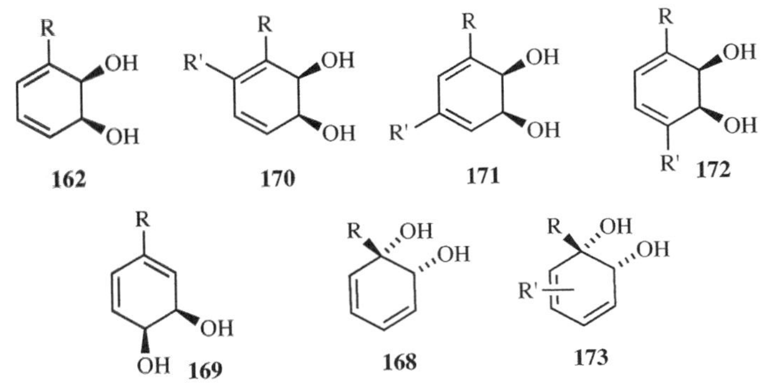

FIGURE 4.60 Dioxygenase-mediated dihydroxylation of simple arenes.

(Table 4.21). For instance, NDO variants Phe352Val and Phe352Thr produced the regioisomer of type **169** in more than 9:1 ratio and at a slightly slower rate than the wild-type enzyme performance. It is interesting to note that also the enantioselectivity could be changed. Thus, whereas the wild-type enzyme produced (+)-biphenyl *cis*-(3*R*,4*S*)-dihydrodiol, the mutants Phe352Val and Phe352Thr yielded (-)-biphenyl *cis*-(3*S*,4*R*)-dihydrodiol in 77% and 60% *ee*, respectively (Table 4.21) [283]. The aza-analogs (benzene rings bearing a pyridine, pyrrole, or pyrazole substituent) afforded mainly diols of type **162**, with trace amounts of regioisomers of type **169**, in high enantiomeric excess [299].

Disubstituted Benzene Derivatives For 1,2- and 1,3-disubstituted benzenes, the regioselectivity observed during TDO-catalyzed *cis*-dihydroxylation was related to the relative size of the substituents (steric effect). Thus, the more sterically demanding substituent R (compared to R') directed the dihydroxylation to the proximate C–C

TABLE 4.21 Product Distribution and Enantioselectivity of Compounds Formed by Wild-Type and Mutant NDO Enzymes

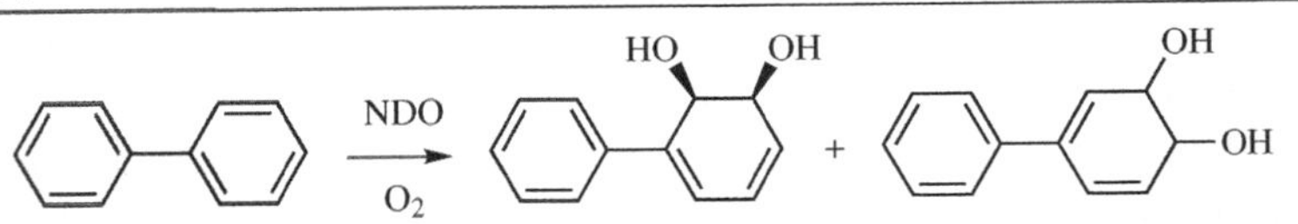

	cis-biphenyl-2,3-dihydrodiol		*cis*-biphenyl-3,4-dihydrodiol		
NDO Enzyme	%	*ee* (%)	%	*ee* (%)	Relative Activity
wild type	87.2	>95	12.8	>98 (3*R*,4*S*)	100
F352G	31.5	>95	68.5	60 (3*R*,4*S*)	<26
F352A	22.6	>95	77.4	65 (3*R*,4*S*)	26
F352T	8.2	>95	91.8	60 (3*S*,4*R*)	77
F352V	4.4	–	95.6	77 (3*S*,4*R*)	99
F352I	14.8	>95	85.2	53 (3*R*,4*S*)	78
F352L	17.0	>95	83.0	70 (3*R*,4*S*)	86

double bond yielding an excess of *cis*-diols of types **170** and **171** [286,300]. The size of the substituents is related to standard steric parameter values (e.g., ν, the Charton steric parameter) and to the preferred conformation adopted within the active site, giving the following sequence: Pr > Et > CF_3 > SMe > I > Br > Cl $\approx$ Me > F > H [301,302]. Alternatively, when using NBDO or BZDO for the *cis*-dihydroxylation of monosubstituted nitrobenzenes or benzoic acids, the R group (R = CO_2H or NO_2) was totally dominant over R', possibly due to H-bonding to the R group at the active site [273], and only *ipso* substituted *cis*-dihydrodiol of type **173** was formed [272,289,290,303]. As TDO did not catalyze *cis*-dihydroxylation at an *ipso*-bond, only *cis*-diols of type **172** could be formed using 1,4-disubstituted benzene substrates [302,304].

Polycyclic Aromatic Hydrocarbons The *cis*-dihydroxylation of PAHs and their heteroarene analogs also shows a strong or exclusive preference for bonds proximate to the sterically hindered "bay" and "fjord" regions. BPDO is often used, given the size of the substrates, and representative examples include the formation of *cis*-dihydrodiols of phenanthrene (**174**) (bay region) [275,283], chrysene (**175**) (two bay regions) [305,306], and benzo[*c*]phenanthrene (**176**) (fjord region) (Figure 4.61) [307]. It is interesting to note that the *cis*-dihydrodiol **175** was also accepted as substrate of BPDO to yield *bis-cis*-dihydrodiol **177**, the first member of a new family of metabolites with four chiral centers resulting from sequential *cis*-dihydroxylation (tetraoxygenation) [305]. Dioxygenases capable to dihydroxylate and ample range of PAHs, made up to five rings have been characterized, for instance the NDO from *Sphingomonas* sp. (strain CHY-1), which yields significant amounts of diols in the bay- and nonbay region, **178** and **179**, respectively. Both compounds are taken as substrates by the enzyme to yield the same tetraoxygenated product [308].

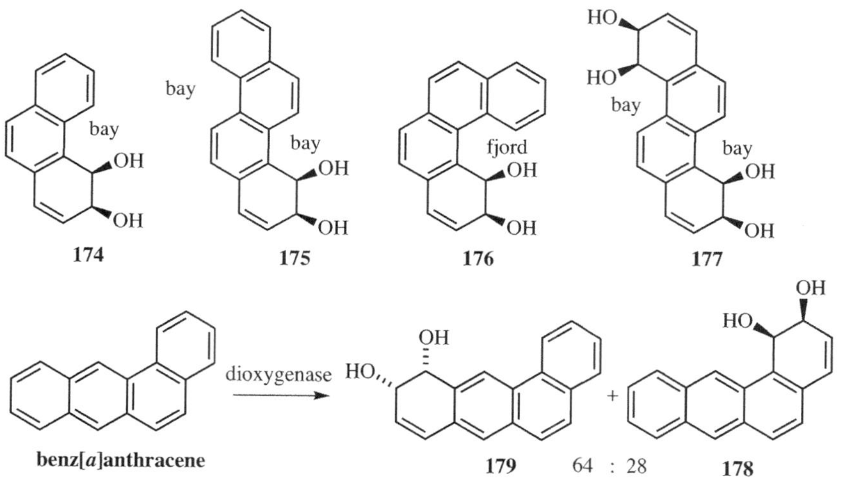

FIGURE 4.61 Dioxygenase-mediated dihydroxylation of polycyclic arenes.

TABLE 4.22 Selectivity in the NDO- and TDO-Mediated Dihydroxylation of 2-Substituted Naphthalenes

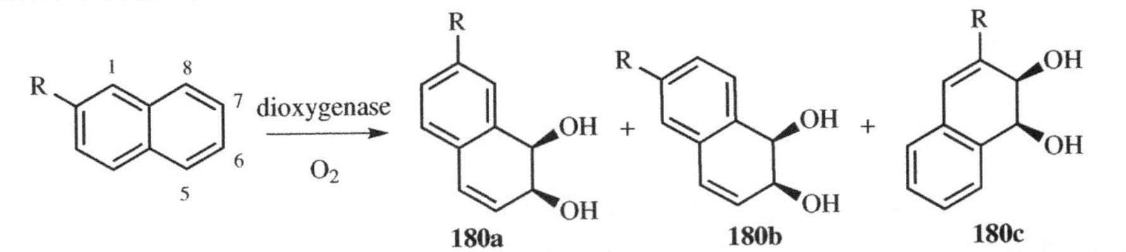

180a 180b 180c

R	Enzyme	Relative Yields (%) 180a	180b	180c	Reference
H	NDO/TDO	100	–	–	[310]
Br	NDO	100	Trace	–	[311]
Br	TDO	33	40	27	[309]
OMe	NDO	93	7	–	[310]
OMe	TDO	17	52	31	[309]
Me	NDO	100	–	–	[310]
Me	TDO	60	10	30	[309]
F	TDO	42	30	38	[309]
CN	TDO	15	57	28	[309]
I	TDO	51	34	15	[309]

The selectivity of the *cis*-dihydroxylation of 2-substituted naphthalenes is very dependent on the enzyme used. As expected, better yields are obtained using NDO, which is also more selective, leading mainly to the 7,8-dihydrodiol (**180a**) and trace amounts of the regioisomer **180b** (Table 4.22). On the contrary, TDO gives ternary mixtures [309].

Heteroarenes TDO catalyzes the *cis*-dihydroxylation of electron-rich aromatic heterocyclic rings (e.g., thiophene, furan, and pyrrole) [312–317], being the thiophene ring the most extensively studied. Similar to the behavior of monosubstituted benzene substrates, where the proximate C2–C3 bond is oxygenated, 3-substituted thiophenes **181** yield the corresponding *cis*-dihydrodiol metabolites of type **182**, whereas the isomeric 2-substituted ring is not accepted by the enzyme. The hemithioacetal **182** isomerizes spontaneously to the corresponding *trans*-dihydrodiol *via* an undetected acyclic aldehyde [316,317]. An alternative TDO-catalyzed *S*-oxidation pathway results in sulfoxide formation and spontaneous dimerization (Figure 4.62) [317]. Biotransformation of 3-phenyl thiophene (**181**, R = Ph) using TDO, indicated a regioselective preference (3:1) for *cis*-dihydroxylation of the carbocyclic over the heterocyclic ring, yielding mainly a type **162** diol [317]. Conversely, no *cis*-dihydrodiol metabolites of monocyclic furan or pyrrole substrates have been reported to date. Probably, the products of the reaction spontaneously open to form the corresponding acyclic hydroxy-aldehydes or dehydrate to yield hydroxyfuran or -pyrrole derivatives. Adding another aromatic ring increases the stability and, in this way, the

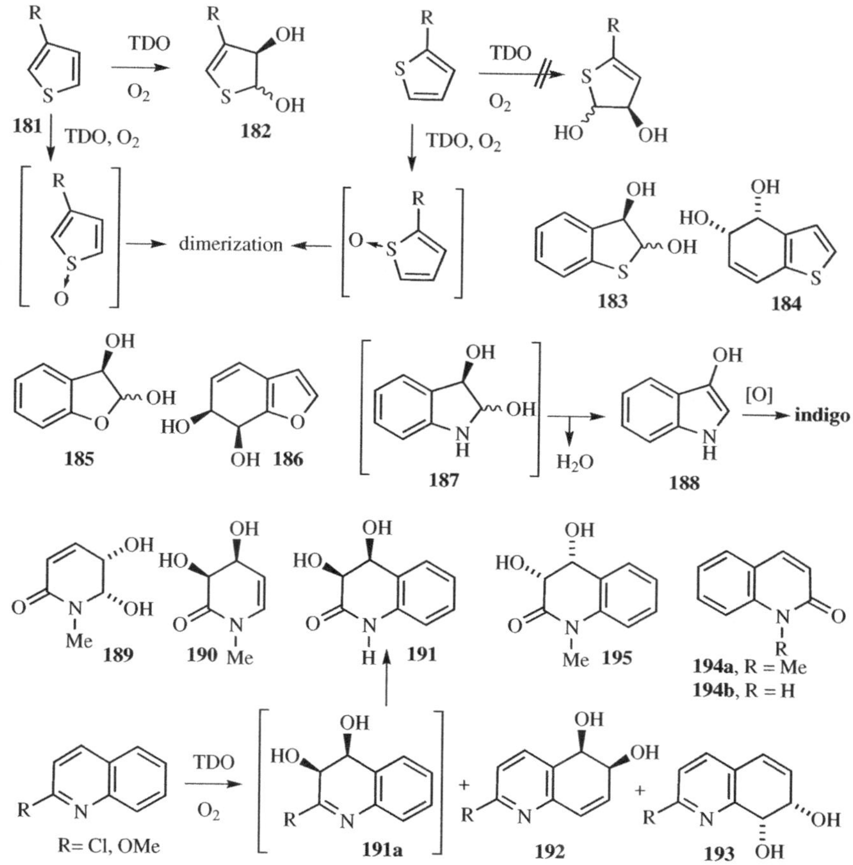

FIGURE 4.62 Dioxygenase-mediated dihydroxylation of heterocyclic compounds.

cis-dihydrodiols resulting from reaction of benzofuran (in both the heterocyclic (**185**) and carbocyclic rings (**186**)) can be isolated (even though they are less stable than the corresponding benzothiophene metabolites **183** and **184**) [312–317]. On the other hand, the indole *cis*-dihydrodiol **187**, being too unstable to detect prior to spontaneous dehydration to indoxyl (**188**) formed indigo dye by autoxidation (Figure 4.62) [318]. *cis*-Diols **183**, **185**, and presumably **187**, are configurationally unstable, because of *cis*/*trans* isomerization. For the benzothiophene ring, the regioselectivity of the *cis*-hydroxylation was controlled by substitution of methyl groups at C-5 (exclusive oxidation in the heterocyclic ring), and C-3 (major oxidation in the carbocyclic ring) [316].

Electron-poor heterobenzene rings, for example, pyridines, are not generally accepted as substrates by dioxygenases and to date no *cis*-dihydrodiol metabolites of monocyclic pyridines have been characterized. Using TDO only alkyl monohydroxylated products were found when 2-, 3-, and 4-alkylpyridines were used as substrates

[319,320]. As pyridine metabolites are likely to be highly unstable, spontaneous dehydration could take place, resulting in the formation of monohydroxylated pyridines. This is observed in biotransformations involving dioxygenase enzymes, for example, 3-hydroxy-4-alkylpyridine from 4-alkylpyridine [320,321], and 3-hydroxyquinoline from quinoline [34]. On the other hand, *N*-methyl-2-pyridone is accepted by TDO, NDO, and BPDO forming the stable monocyclic *cis*-dihydrodiols **189** and **190** (Figure 4.62) [322–324]. Indirect evidence for the intermediacy of an unstable substituted pyridine *cis*-dihydrodiol (**191a**, R = Cl, OMe) was obtained when product **191** was isolated as a minor metabolite from TDO-catalyzed oxidation of 2-chloro- and 2-methoxyquinoline [323,325]. Spontaneous hydrolysis of the postulated pyridine *cis*-dihydrodiol **191a** (R = Cl, OMe) could account for the presence of lactam **191** among the other major metabolites (**192**, **193**, R = Cl, OMe) (Figure 4.62) [323]. Azanaphthalene substrates are in general well accepted for dioxygenases to yield *cis*-dihydrodiols resulting from predominant attack on the carbocyclic ring (with heterocyclic *cis*-diols or phenols also being formed as minor products) [323,326–328]. The NDO-mediated dihydroxylation of quinolones was successful only with *N*-methyl-2-quinolone (**194a**) giving the corresponding diol **195** in 28% yield, along with 34% of the *N*-dealkylated product **194b** [324].

In summary, biotransformations of polycyclic heteroarenes by dioxygenases (e.g., TDO, NDO and BPDO) afford metabolites preferentially *cis*-dihydroxylated at the more sterically demanding bay- or fjord-region, in the carbocyclic ring rather than at a more accessible bond. Both benzene rings and electron-rich five-membered heteroarene rings can be *cis*-dihydroxylated in the presence of electron-deficient azobenzene rings, which remain intact [300].

ENANTIOSELECTIVITY OF THE DIHYDROXYLATION OF ARENE *CIS*-DIHYDRODIOLS As it was mentioned before, the dioxygenase-catalyzed oxidation of arenes to give *cis*-dihydrodiols displays an outstanding enantioselectivity. When using TDO, the configuration of the *cis*-dihydrodiols is always as shown in Figure 4.60. Regarding the enantiomeric purity, the following statements can be made: (i) *cis*-dihydrodiols of all monosubstituted, 1,2-disubstituted, and 1,3-disubstituted benzene substrates formed using TDO biocatalysis are enantiopure (> 98% *ee*) except for those having only the smallest substituents present, that is, F atoms; (ii) *cis*-dihydrodiols of 1,4-disubstituted benzene substrates formed using TDO biocatalysis had *ee* values that varied depending on the relative sizes of substituents (Table 4.23) [329,330]; (iii) *cis*/*trans*-dihydrodiol metabolites isolated from TDO-catalyzed dihydroxylation of thiophene or furan rings were found to have variable *ee* values (10–98%) presumably due to partial racemization *via* the acyclic aldehyde or ketone intermediates.

Dioxygenases may possess different enantiospecificity, providing access to the enantiomeric dihydrodiol. For instance, TDO and NDO are enantiocomplementary in the dihydroxylation of benzocycloalkenes of type **196**. Thus, while TDO oxidizes the alkene to give *cis*-dihydrodiols (and products of benzylic oxidation) with 1*S* configuration [331,332], NDO produces *cis*-diols with opposite configuration in excellent enantiomeric excess (Figure 4.63) [329].

TABLE 4.23 Optical Purity of 1,4-Disubstituted *cis*-Dihydrodiols Obtained from TDO Oxidation

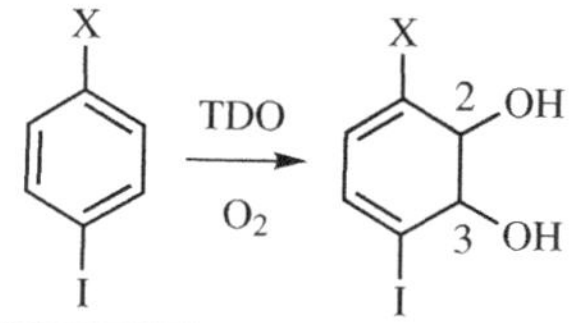

X	*ee* (%)	Absolute Configuration
F	88	2*R*,3*S*
Cl	15	2*R*,3*S*
Br	22	2*R*,3*S*
Me	80	2*S*,3*S*

Other dioxygenases are not so enantioselective, and scalemic mixtures can be obtained. For instance, using recombinant CDO from *Pseudomonas* sp. P51, the biotransformation of monosubstituted benzene derivatives bearing a nitrile group (side chain R = CN, CH_2CN, CH=CHCN) yielded *cis*-dihydrodiols with 66–97% *ee* [333]. To improve the optical purity of those arene *cis*-dihydrodiols that are produced as scalemic mixtures, such as those resulting from the oxidation of fluorobenzene or 1,4-disubstituted benzenes, some methods are available. One approach takes advantage of the high stereoselectivity of the *cis*-dihydrodiol dehydrogenases that constitute the second step in the degradation pathway of arenes. The dehydrogenase only accepts the *cis*-dihydrodiol of "natural" configuration as substrate, leaving the other enantiomer intact. Thus, by feeding the *cis*-diol from oxidation of fluorobenzene (obtained with 60–80% *ee*) to a wild-type *Pseudomonas* containing the (highly enantioselective) dehydrogenase, only one isomer was further metabolized, leaving the other in optical pure form. This procedure has been applied in the preparation of *ent*-diene diols [329], and complex products such as *ent*-7-deoxypancratistatin (see Section "Synthetic applications of arene cis-dihydrodiol metabolites"; Figures 4.65B and 4.75) [334]. Another procedure is based on the lipase-catalyzed kinetic resolution of either the diol or its diacetate to obtain the corresponding isomers after hydrolysis [335].

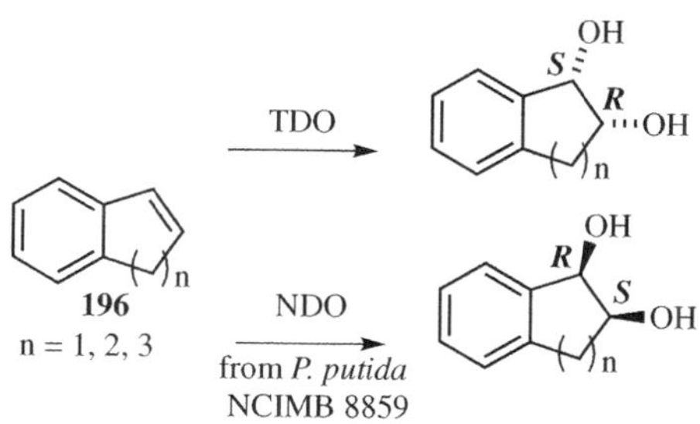

n	*ee* (%)
1,	20-28
2,	>98
3,	>98

n	*ee* (%)
1,	85
2,	>98
3,	>98

FIGURE 4.63 Enantiocomplementary dihydroxylation of benzocycloalkenes.

The enantiomeric purity for *cis*-dihydrodiol derivatives of both mono- and polycyclic arenes have been determined directly by chiral stationary phase HPLC, when both enantiomers are available [302,304]. The enantiomeric excesses (*ee*) of *cis*-dihydrodiols of monocyclic arenes have also been determined indirectly by: (i) GC-MS analysis of the corresponding *n*-butylboronate derivatives on a chiral stationary phase GC column (where both enantiomers are available) [333]; (ii) NMR analysis following formation of cycloadducts using Cookson's dienophile (4-phenyl-1,2,4-triazoline-3,5-dione) and further Mosher's diesterification using (*R*)- and (*S*)-2-methoxy-2-phenyl-2-(trifluoromethyl)acetyl chloride (where one or both enantiomers are present) [336,337]; and (iii) NMR analysis after formation of a boronate derivative by reaction with (*R*)- and (*S*)-2-(1-methoxyethyl)phenyl boronic acids (where one or both enantiomers are present) [302,305,338].

CHEMOSELECTIVITY OF THE DIHYDROXYLATION OF ARENE *CIS*-DIHYDRODIOLS Apart from catalyzing the *cis*-dihydroxylation of arenes, aromatic dioxygenases are known to be involved in other reactions, such as sulfoxidation, benzylic oxidation, desaturation (dehydrogenation), *O*- and *N*-dealkylation (see Section 4.2.1.2) [288]. Therefore, when using susceptible substrates, the possibility of side reactions of this type has to be considered. In alkyl benzenes (e.g., propylbenzene) the sequential TDO-catalyzed benzylic hydroxylation followed by *cis*-dihydroxylation of the ring to yield triols with three new chiral centers (**197**) has been observed (Figure 4.64) [339,340]. In the same way, sequential sulfoxidation and *cis*-dihydroxylation of alkylaryl sulfide substrates

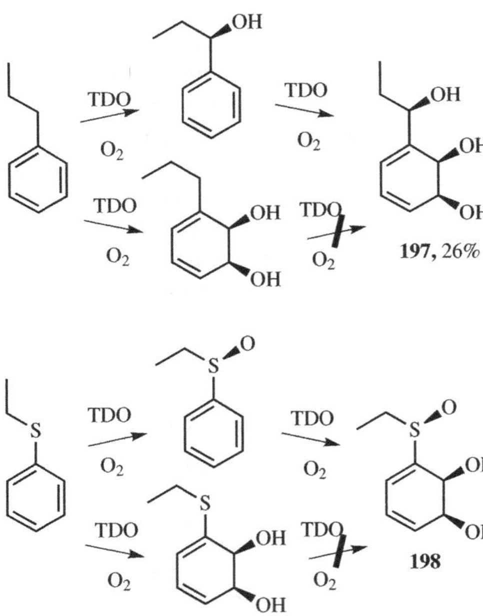

FIGURE 4.64 TDO-catalyzed mono, di-, and trioxygenation pathways for alkyl- and thioalkylbenzenes.

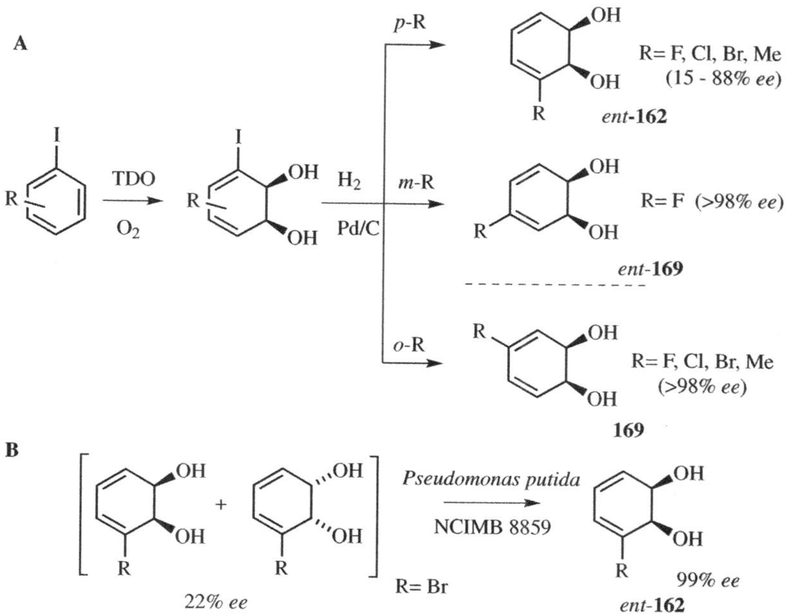

FIGURE 4.65 Chemoenzymatic synthesis of enantiomers and unnatural regioisomers of monosubstituted benzene *cis*-dihydrodiols.

to yield *cis*-diol sulfoxides containing three new chiral centers (**198**) has also been reported (Figure 4.64) [302,340,341]. It is interesting to note that the sulfoxidation and the benzylic oxidation must take place first, because once the more soluble diol is formed; it is not a proper substrate for the enzyme.

The chemoselectivity of the dioxygenases during arene oxidation is dependent on the substitution pattern. In particular, the *meta*-substituted benzenes are less reactive toward *cis*-dihydroxylation than their *ortho*- and *para*-regioisomers (the so-called *meta*-effect) and yield higher proportions of products resulting from alternative dioxygenase-mediated pathways (benzylic hydroxylation, sulfoxidation, and alkene dihydroxylation) [301,302].

In the context of synthetic studies using these *cis*-diols, attempts have been made to rationalize preferences for arene *cis*-dihydroxylation versus benzylic hydroxylation [339,340], arene *cis*-dihydroxylation versus sulfoxidation [302,341,342], and arene *cis*-dihydroxylation versus alkene dihydroxylation [343]. A number of general trends were observed: (i) *cis*-dihydroxylation of monosubstituted monocyclic arenes occurs when using BDO, TDO, or CDO but is less common when using NDO or BPDO; (ii) *cis*-dihydroxylation of disubstituted monocyclic arenes using TDO is generally slower relative to *cis*-dihydroxylation, benzylic hydroxylation, alkylaryl sulfoxidation, or alkene dihydroxylation of monosubstituted arenes; (iii) arene *cis*-dihydroxylation of alkylbenzyl sulfides using TDO is much faster than sulfoxidation, while alkylaryl sulfides prefer the sulfoxidation pathway; (iv) TDO-catalyzed arene *cis*-dihydroxylation of vinylbenzene substrates is faster than alkene dihydroxylation when substituents are present on the alkene group but slower when they are present on the arene ring; (v) benzylic hydroxylation, sulfoxidation, and alkene dihydroxylation using benzocyclo-alkanes, -alkenes, and -hetero analogs as substrates for TDO, are often observed while arene *cis*-dihydroxylation is rarely found; (vi) dialkylsulfide sulfoxidation, allylic hydroxylation and dealkylation are generally observed only when more facile TDO-catalyzed oxidations, for example, arene *cis*-dihydroxylation, are not possible; (vii) desaturation can be detected indirectly when dihydro- or tetrahydroarene substrates gave the corresponding arene *cis*-dihydrodiols, *via* desaturation to the arene and further *cis*-dihydroxylation.

Synthetic Applications of Arene cis*-Dihydrodiol Metabolites* Despite the large number of arene *cis*-dihydrodiols reported (more than three hundred) [5], only about a dozen have been used in synthesis, mostly derived from the oxidation of monosubstituted benzene substrates of type **162** (Figure 4.60). Among them, the most used *cis*-diols are those derived from oxidation of chloro- and bromobenzene, followed by toluene, benzene, *β*-bromoethylbenzene, iodobenzene, styrene, some disubstituted benzenes such as *m*-dibromobenzene, and also some bicyclic arenes, namely naphthalene (Figures 4.70–4.76).

AVAILABILITY OF METABOLITES, REGIOISOMERS AND ENANTIOMERS The earlier difficult access to these metabolites has been a limiting factor for development. However, the wealth of synthetic possibilities offered by these synthons has turned the enzymatic dihydroxylation of arenes into a well-known and respected synthetic methodology. Moreover, this reaction is one of the few natural processes for which an asymmetric chemical equivalent has yet to be designed [344]. In addition to the commercial availability of the most easily prepared *cis*-diols (*cis*-halobenzene dihydrodiols), Sinisterra et al. reported on the preparative-scale transformation of toluene into the corresponding *cis*-diol [345]. One challenge was the high toxicity of the substrate which was overcome by using the two-liquid-phase concept and thereby keeping the aqueous toluene concentration at acceptable values for the microorganism (mutant *P. putida* UV4 immobilized in alginate beads). Overall more than 80 gL^{-1} of enantiopure *cis*-diol could be obtained. After a few years, Hudlicky published an *Organic Syntheses* procedure for the small scale preparation of *cis*-diols using *P. putida* 39/D [346]. This simple protocol produces small amounts (< 3 g per batch) of dioxygenated products, but can be performed in a laboratory not equipped for microbiology and the manipulation is easy. More recently, a number of other *cis*-dihydrodiols have been consistently obtained in yields over 10 gL^{-1} from whole-cell fermentations using recombinant toluene dioxygenase expressed in *E. coli*. Here, substrate concentrations were kept below toxicity levels by continuous feeding. Likewise, removal of the product by an external charcoal column resulted in an additional improvement of the yield [333,347,348].

As for all enzymatic methodologies, the issue of the availability of the unnatural enantiomers has to be considered. In addition, the unavailability of the unnatural *cis*-dihydrodiol regioisomers (e.g., **168** and **169**) is another limitation to be addressed. The earliest chemoenzymatic method to circumvent this was developed by Boyd

[329]. It relies on the directing effect of iodine in the biotransformation of *ortho-*, *meta-*, and *para-*substituted iodobenzene substrates, and then on the greater reactivity of the iodine in the Pd/C hydrogenolysis, to provide either the unnatural regioisomer **169**, its enantiomer (*ent-***169**), or the enantiomer (*ent-***162**) (as an scalemic mixture) of the natural *cis*-dihydrodiol (Figure 4.65A). Unfortunately the biotransformations of substituted *p*-iodobenzenes give *cis*-dihydrodiols of moderate optical purity (Table 4.23), and this is translated to the final product. Boyd overcame this problem by exposing the scalemic mixture of (*ent-***162**) to a second fermentation step using a nonblocked strain of *Pseudomonas* (*P. putida* NCIMB 8859), which is able to completely metabolize the natural enantiomer while accumulating the unnatural one (Figure 4.65B) [329,330]. This strategy has also been applied to obtain the *cis*-dihydrodiol resulting from biotransformation of fluorobenzene in optically pure form.

While this approach produced an unnatural regioisomer (**169**) and two unnatural enantiomers (*ent-***162** and *ent-***169**), in some cases the isolated yields from the substituted iodobenzene substrates were relatively low (e.g., from *meta*-substituted iodobenzenes, where only one isomer gives acceptable yields), the compounds (*ent-***169**) were obtained as scalemic mixtures, and the regioisomers resulting from *ipso* oxygenation (of type **168,** Figure 4.60) were not produced. An alternative approach, based on a monosubstituted benzene *cis*-dihydrodiol (**162**), and involving a four-step synthesis and reduction of benzene dioxide intermediates, has recently been developed [349]. This constitutes an alternative route to *cis*-3,4-dihydrodiol (**169**), and a new route to the previously unavailable 1,2- regioisomer *ent-***168** (Figure 4.66).

The availability of the unnatural isomers can also be addressed by conventional enzymatic techniques. Thus, the recent observation that CDO-catalyzed *cis*-dihydroxylation of several monosubstituted benzene substrates bearing a nitrile group (**162**, side chain R = CN, CH_2CN, CH=CHCN) yield scalemic mixtures (66–95% *ee*) [333] could in principle be used to obtain either enantiomer through an amidase- or nitrilase-mediated kinetic resolution process. The *meso*-diol resulting from biotransformation of benzene has been desymmetrized through kinetic resolution of a conduritol derivative (Figure 4.67) [335].

Regarding the *trans*-analogs **201** and **202**, these isomers were obtained through chemical modification of *cis*-dihydrodiols. Because of the instability of the dienic system toward the conditions for inversion, both *trans*-2,3-dihydrodiol isomers were prepared by inversion in a nondienic system, which then regenerated the desired dienediol (Figure 4.68) [350,351]. Another approach formed the *trans*-3,4-dihydrodiol isomers (**203**) through a six-step sequence involving the reduction of *anti*-dioxide

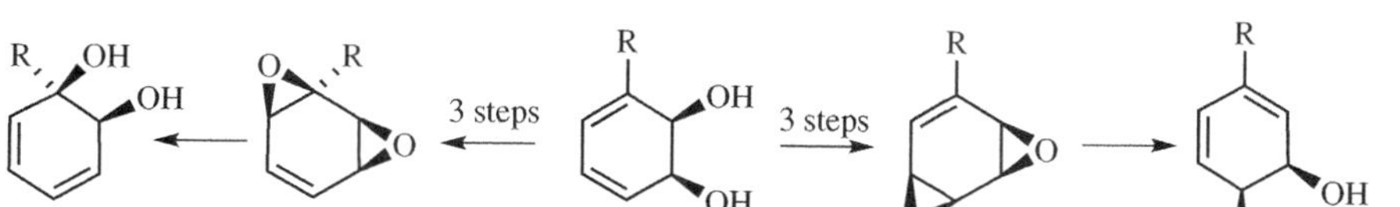

*ent-***168** **162** **169**

FIGURE 4.66 Chemoenzymatic synthesis of unnatural regioisomers of monosubstituted benzene *cis*-dihydrodiols.

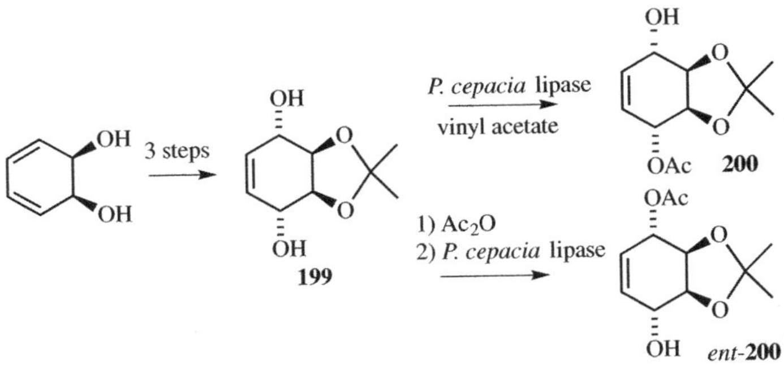

FIGURE 4.67 Desymmetrization of *cis*-1,2-benzenedihydrodiol.

intermediates, similar to that used to prepare the corresponding *cis*-isomers (Figure 4.68) [352]. The *trans*-1,2- dihydrodiol derivatives of toluene were also prepared using this route.

RATIONAL DESIGN AND HISTORICAL DEVELOPMENT OF THE AROMATIC DIHYDROXYLATION METHODOLOGY IN ENANTIOSELECTIVE SYNTHESIS Arene *cis*-dihydrodiol metabolites are densely functionalized compounds providing multiple reactive sites and opportunities for stereocontrol (Figure 4.69). The transfer of chirality from the diol to the periphery of the dienic ring is the most evident transformation, which has been successfully used to prepare a wide range of polyoxygenated compounds, namely cyclitols, conduramines, inositols, and carbohydrates, both natural and unnatural. The diol functionality can be used either as an anchor point for electrophilic tethers (directing effect *syn* to the hydroxy groups) or as a hindering group when

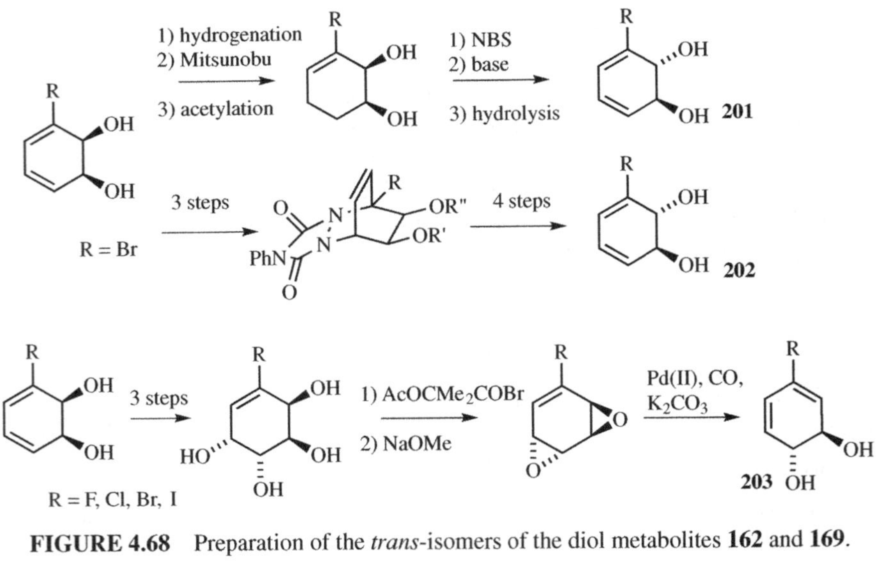

FIGURE 4.68 Preparation of the *trans*-isomers of the diol metabolites **162** and **169**.

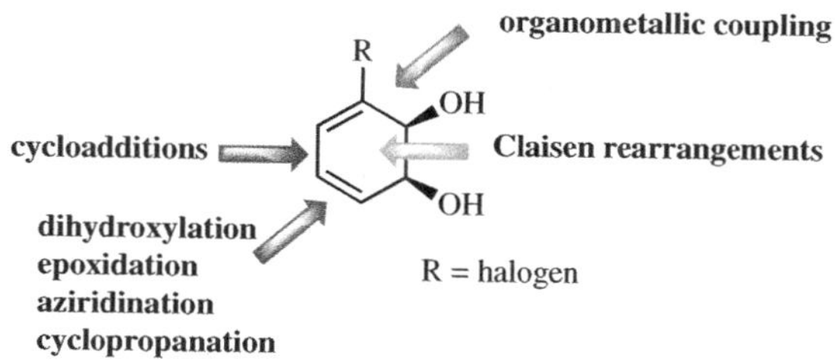

FIGURE 4.69 Reactive options for arene-*cis*-dihydrodiols.

protected, shielding one face of the system. Reaction of the diene in heterocycloadditions leads to bicyclo[2.2.2]octanes, in which the bicyclic system can be cleaved to a monocycle or transformed further into more complex systems through consecutive reaction sequences. The *cis*-diols derived from oxidation of halobenzenes present a polarized dienic system, which allows for complete differentiation of the olefins in dihydroxylations, epoxidations and other electrophilic additions, cycloadditions of the [2 + n] type (cyclopropanations, aziridinations, etc.), and other reactions. Both allylic alcohol residues are differentiated by the proximity of the halogen atom and trigger themselves to Claisen-type rearrangements. The vinyl halide can participate in organometallic coupling reactions, allowing for diversity of substitution on that position of the cyclohexadiene ring or, alternatively, can be easily reduced to an olefin (Figure 4.69). The use of these features in synthesis has been covered in several reviews [279,285,286,292,294–297,344]. It is interesting to note that halobenzene *cis*-dihydrodiols have been used for enantiodivergent synthesis of carbohydrates, inositols, and alkaloids [286]. Reduction of the vinyl halide to an olefin would give a *meso*-compound if not previously functionalized. This feature is involved in the concept of "latent symmetry," introduced to describe the strategy of enantiodivergent design from a single enantiomer of a starting diol. In this strategy, the routes to the enantiomers contain the same number of steps but *either* enantiomer is accessed by merely changing the order of execution of the individual reaction steps [344].

Historically, the first synthetic application of arene *cis*-dihydrodiol metabolites appeared in 1983, when chemists from Imperial Chemical Industries in England reported on the use of the biocatalytically generated *meso-cis*-diol **204**-derived from benzene as a monomer for the synthesis of polyphenylene (**205**) on an industrial scale (Figure 4.70) [353]. The same year Gibson described the synthesis of indigo by dihydroxylation of indole by naphthalene dioxygenase in a process that years later was optimized to an industrial-scale fermentation (see Chapter 8) [318]. In the late 1980s, the number of reports on the use of these metabolites increased sharply. Following Ley's preparation of racemic pinitol in 1987 [354], Hudlicky' synthesis of prostaglandin PGE_{2a} (formal) in 1988 [355], total synthesis of zeylena in 1989 [356], pinitol in 1990 [357], and a fully general method of synthesis for hexoses and other carbohydrates [358,359] were reported. Dihydrodiols have served as starting materials for the synthesis of various cyclitols, inositols [360–365], their oligomers [366–368], and inositol phosphates [369–371]. In the terpene area, specionin [372], hirsutene [373,374], hirsutic acid [375], and approaches to taxanes [376], tricyclic

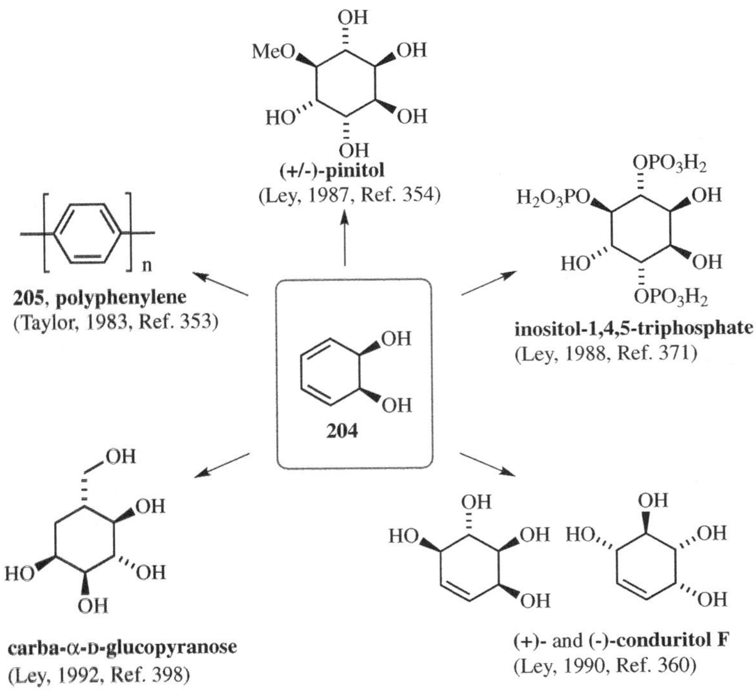

FIGURE 4.70 Representative synthetic targets prepared using the diol derived from benzene.

sesquiterpenes [377], among others, have been performed using *cis*-dihydrodiols. In the alkaloid field, codeine [378], nangustine [379], lycoricidine [380], pancratistatin [381], and narciclasine [382] have been synthesized, along with their enantiomers [383–385]. Many other targets have also been synthesized, among them pyrethroids [386] macrocycles [387] pseudo- [359,388]. and azasugars [389], hydroxylated pyrrolizidine alkaloids [390,391], and sphingosines [392,393]. It is worth mentioning that arene *cis*-dihydrodiols has been recently used as chiral pool molecules. In this way the *cis*-diol derived from bromobenzene has been used, after some manipulation, to prepare chiral inositol polymers [394], and more recently, chiral 2,2′- and 4,4′-bipyridyl derivatives were synthesized from the *cis*-dihydrodiols derived from 2- and 4-chloroquinolines in order to be used as ligands in transition-metal asymmetric catalysis [395–397]. From a historical stand point, Gibson's work opened new ways in different areas, with substantial contributions not only in microbiological and biochemical aspects of the enzymatic dioxygenation of arenes but also reporting many diols and producing recombinant organisms with improved efficiency for synthetic use. Furthermore, the group of T. Hudlicky in USA, the British groups (D. Boyd, H. Dalton, S. Ley, S. Roberts, H. Carless, D. Ribbons, R. Stephenson, D. Crout, among others) and more recently, the group of Banwell (Australia) have made seminal contributions to the development of the field of asymmetric dihydroxylation of arenes.

Alternatively, the impressive array of synthetic applications of the arene *cis*-dihydrodiol metabolites can be viewed from the standpoint of the starting metabolite, instead of focusing in the synthetic targets. In this context, the arene *cis*-dihydrodiol **204**, derived from benzene, was the first to be used in synthesis, both in academic (pinitol) [354] and industrial settings (polyphenylene) [353], but the need for desymmetrization of this *meso*-compound has hampered its use in the last years. It has been used in the preparation of inositol phosphates [369–371], conduritols [360], and sugar derivatives (carbasugars) [398] (Figure 4.70).

Undoubtedly, the most widely used arene *cis*-dihydrodiols for natural product synthesis are those derived from monosubstituted halobenzenes (e.g., chloro- and bromobenzene), which have been continuously used in synthesis since 1990 [357,390,391]. Representative examples of synthetic targets prepared using diols **206** and **207**, derived from chloro- and bromobenzene, respectively, comprise a general methodology for the preparation of hexoses and other carbohydrates (both natural and unnatural, including pseudo- and azasugars) [358,359,389,399], cyclitols and derivatives [357,400], inositols and their oligomers and derivatives [361–368], alkaloids (pyrrolizidine bases) [390], pancratistatin and derivatives [381,401], lycoricidine [380], nangustine [379], macrolides (cladospolide) [387], plant, marine and fungal metabolites (pyrethroids) [386], bengamides [402], epoxyenones [403], and balanol [404], chiral polymers [394], and lipids (sphingosines) [392,393] (Figure 4.71).

The *cis*-dihydrodiol derived from iodobenzene (**208**) has been used more recently, perhaps due to its lower stability and earlier lack of commercial supply. However, since the iodine atom allows for easier halogen substitution reactions, its synthetic applications are rapidly increasing, such as the preparation of carbasugars [388], shikimic acid derivatives [405], and epoxyquinols [406] (Figure 4.72).

The *cis*-dihydrodiol metabolite derived from toluene (**209**) has been used in synthesis since 1988 for the preparation of different types of compounds, mostly prostaglandins [355], sugar derivatives (branched sugars) [407], and terpenes such as specionin [372], *ent*-hirsutene [373,374] hirsutic acid [375] and taxanes [376] (Figure 4.73).

Other *cis*-dihydrodiol metabolites with alkyl side chains have been exploited in synthesis, such as those derived from styrene (**210**) (in the synthesis of the arene oxide derivative zeylena) [356], and from (2-bromoethyl)benzene (**211**) (in the synthesis of codeine and related alkaloids) [378,408]. Recently, the *cis*-dihydrodiol derived from oxidation of ethyl benzoate (**212**) was used in the preparation of the antiviral oseltamivir (Figure 4.74) [409,410].

Once the enzymatic dihydroxylation methodology was established, synthetic applications of *cis*-dihydrodiols derived from disubstituted monocyclic and polycyclic arenes or heteroarenes and of different dioxygenase types appeared. These include the following metabolites derived from oxidation of disubstituted benzenes: achiral diol **213**, derived from *p*-xylene, as precursor of a furanone strawberry flavor compound **214** [276], enantiopure diol **215**, derived from *m*-dibromobenzene, to yield alkaloid narciclasine [411], and a scalemic mixture of diol **216**, derived from

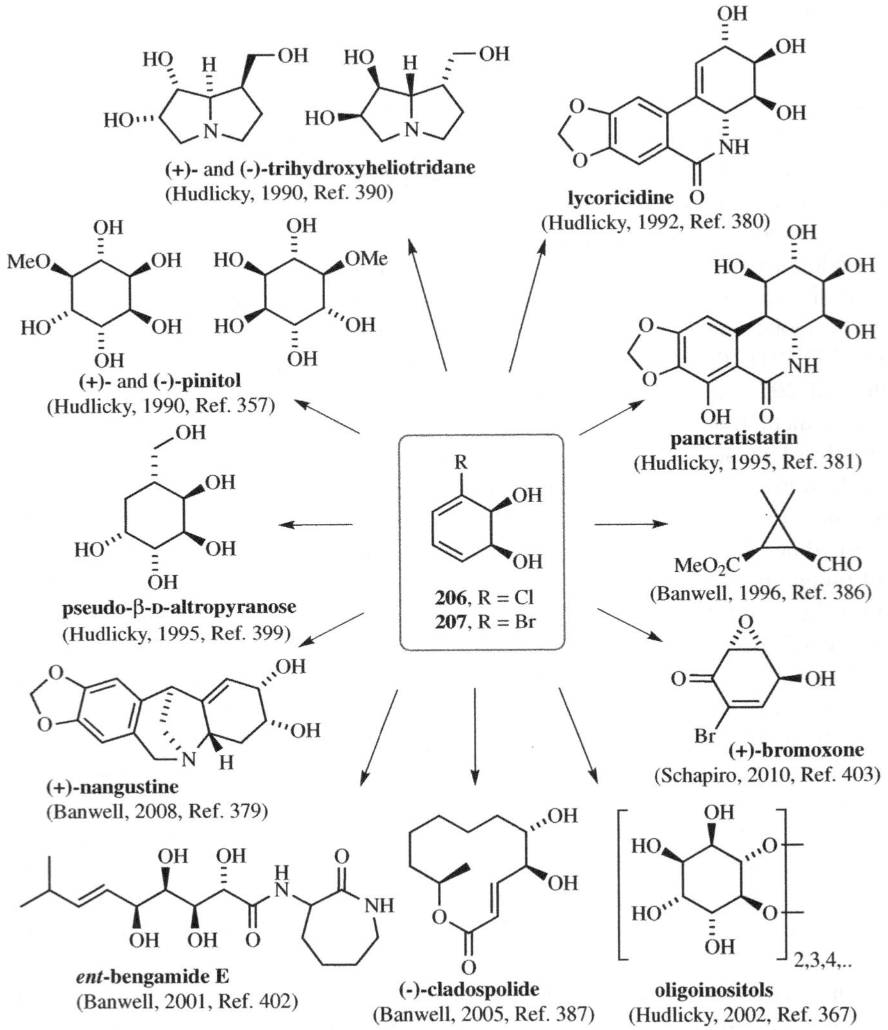

FIGURE 4.71 Representative synthetic targets prepared using the diols derived from chloro- and bromobenzene.

p-bromoiodobenzene, to give 7-deoxypancratistatin and also its enantiomer (Figure 4.75) [411]. It is worth mentioning the high yields of **214** (approximately 20 gL^{-1}) obtained through the application of a directed evolution approach to improve the dioxygenation step [276].

The bicyclic arene diol **217**, derived from naphthalene, has been used in the production of the styryllactone (+)-goniodiol [412]. Similarly, several diols derived from heteroarenes such as quinoline and furoquinoline, have been used in the preparation of furoquinoline alkaloids including 7-isopentenyl-γ-fagarine (**218**) [413] and

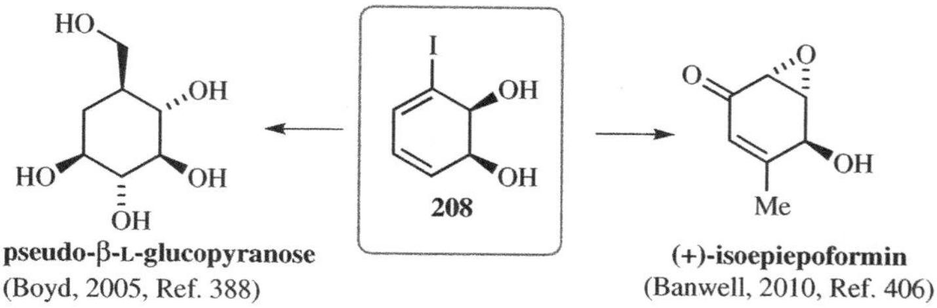

FIGURE 4.72 Representative synthetic targets prepared using the diol derived from iodobenzene.

of a new series of chiral bipyridine ligands, from diols **219** and **220**, giving excellent chiral induction in asymmetric allylic hydroxylation, cyclopropanation, and allylation reactions (Figure 4.76) [395–397]. The 1,2-*cis*-dihydrodiol **221** obtained *via* BZDO-mediated dioxygenation of benzoic acid has been recently used in the preparation of tetracyclines, carba-β-L-fructopyranose, novel polycyclic scaffolds, and the bioactive compound idesolide [414–417].

The range of available dihydrodiols has been increased through the use of efficient coupling techniques, which allowed the preparation of substituted dihydrodiols, whose aromatic precursors were not substrates for the dioxygenases [418,419].

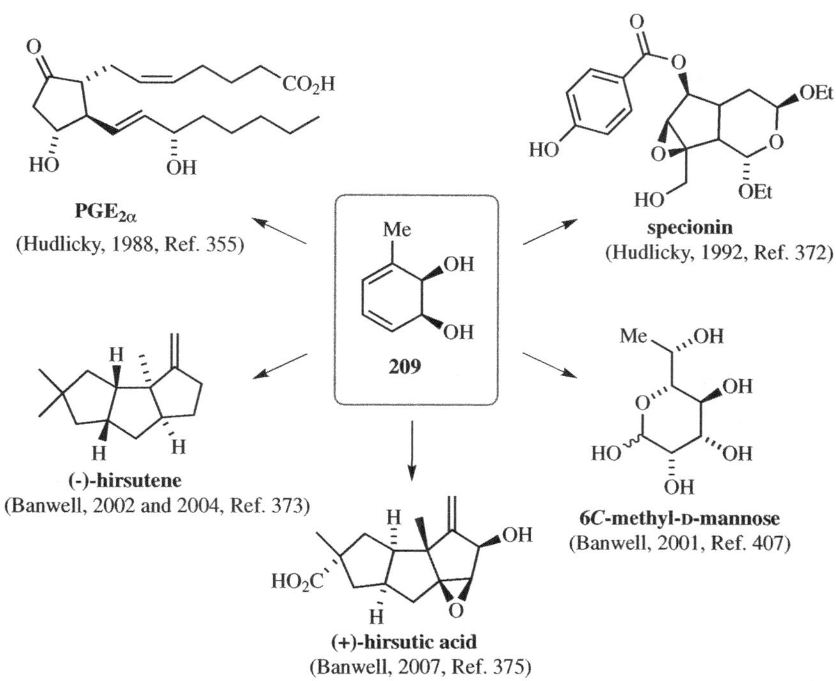

FIGURE 4.73 Representative synthetic targets prepared using the diol derived from toluene.

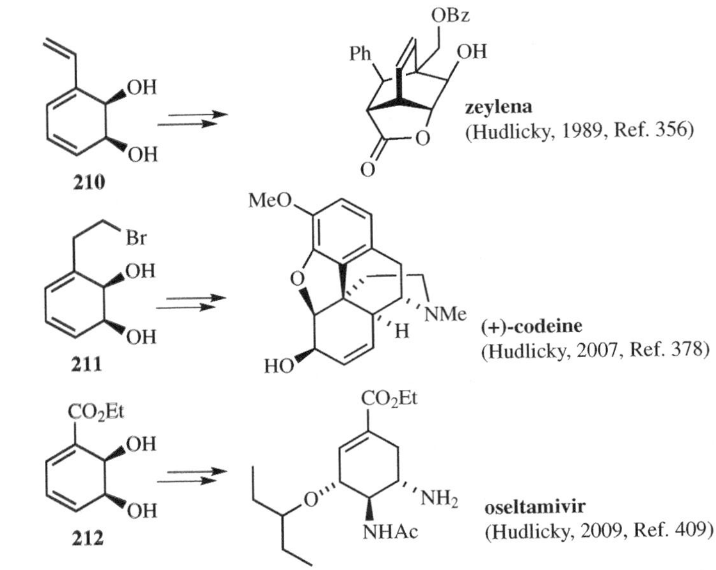

FIGURE 4.74 Representative synthetic targets prepared using the diols derived from styrene, (2-bromoethyl)benzene, and ethyl benzoate.

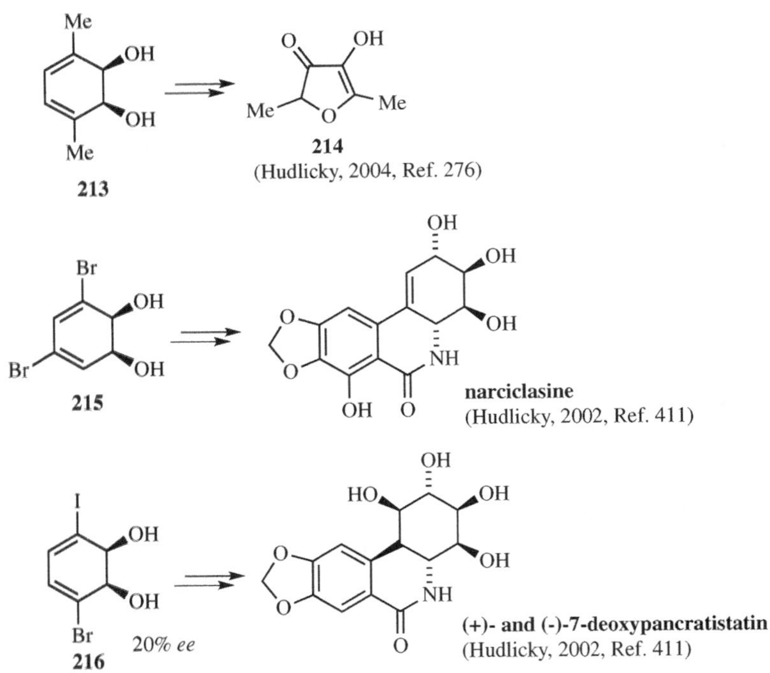

FIGURE 4.75 Miscellaneous diols used in synthesis, derived from disubstituted monocyclic arenes.

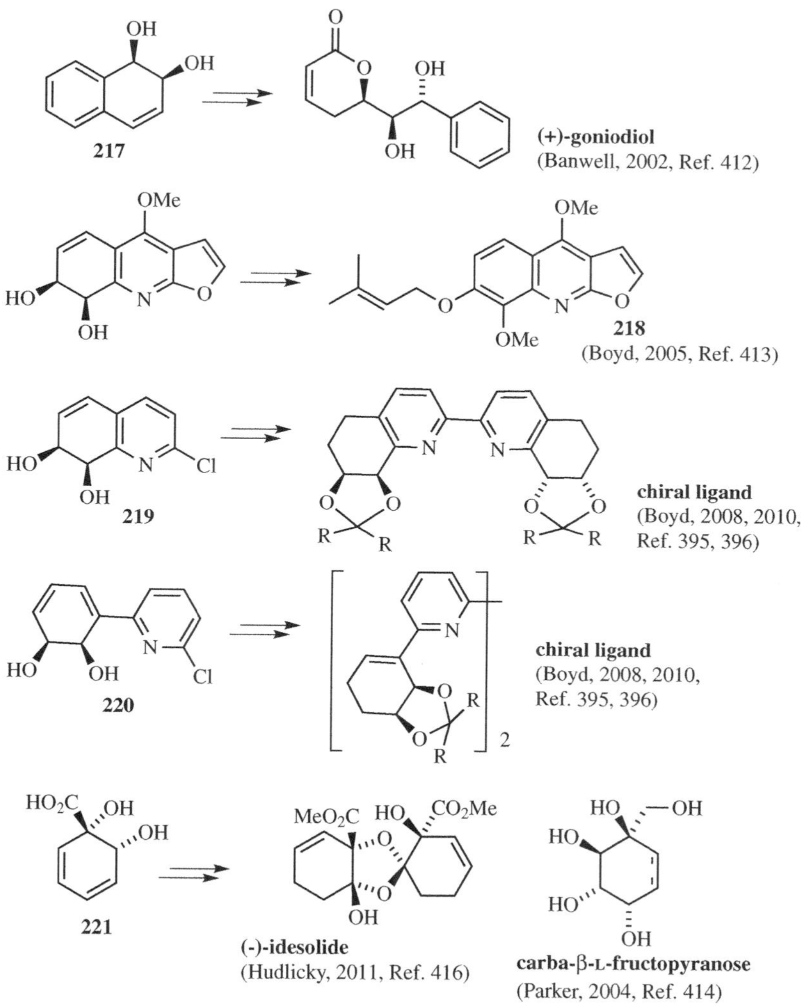

FIGURE 4.76 Miscellaneous diols used in synthesis, derived from polycyclic arenes and heteroarenes.

Additional Biotechnological Applications The broad specificity and high enantioselectivity of aromatic dioxygenases makes them excellent candidates for synthetic applications. Compounds obtained by *cis*-dihydroxylation have been used in the preparation of fine chemicals, natural products, pharmaceutical intermediates, biologically active compounds, and polymers, as mentioned in the previous section. Many of these metabolites are amenable to whole-cell-based large-scale production, not requiring expensive cofactors and relying on the integrity of the multicomponent dioxygenase system, usually overexpressed in recombinant hosts. Not only chiral metabolites, but also final products can be produced using aromatic dioxygenases. A representative example is the production of indigo. A recombinant *E. coli* expressing

NDO was first described as indigo producer, through the oxidation of indole to an unstable *cis*-dihydrodiol, followed by dehydration to indoxyl and spontaneous oxidation to indigo [318]. The brevity of the route led to the use of protein engineering techniques on the recombinant *E. coli* aiming at the commercial production of indigo from glucose [420]. The strategies used include a modification of the tryptophan pathway for high levels of indole production, gene inactivation to eliminate the enzymatic production of byproducts, and gene amplification to achieve a final production of more than 18 gL^{-1} of indigo. However, the process has not been implemented yet at industrial scale.

In addition to their preparative use, dioxygenases find ample application in environmental sciences through their involvement in the degradation of environmental pollutants [421]. Many recalcitrant compounds, mostly halogenated, are oxidized by dioxygenases. For instance, aromatic dioxygenase-containing bacteria have been used in the bioremediation of contamination by trichloroethylene, polychlorobiphenyls (PCBs), nitro-aromatic compounds, and aromatic amines, among others [277,422–424]. PCBs are a major environmental concern, and the study of their degradation pathways aiming at an improved biodegradation is a topic of continuous research. Biphenyl dioxygenase is capable to oxidize a number of PCBs, and genetic engineering has been used to improve its substrate specificity to accept a wider range of PCBs, substituted benzenes, and heterocyclic aromatic compounds [425–429]. Aromatic dioxygenases have also been used to construct new pathways for the degradation of recalcitrant compounds, designing microorganisms capable to degrade polychlorinated compounds [430], and aromatic hydrocarbons in the presence of high levels of radiation [431].

4.2.1.2 Other Oxidation Reactions Performed by Aromatic Dioxygenases

Apart from the above-mentioned arene dihydroxylations, Rieske nonheme iron dioxygenases are involved in different types of oxidations, showing both dioxygenase- and monooxygenase-behavior [291]. Representative examples include (i) monosulfoxidation of alkylaryl or diaryl sulfides [302,432]; (ii) dihydroxylation of cyclic or acyclic conjugated alkenes [298,343]; (iii) monohydroxylation at benzylic or allylic centers [332,343,433]; (iv) desaturation at allylic or benzyl positions [305,434]; and (v) *N*- and *O*-dealkylation [288]. As a general rule, the most common of these transformations are sulfoxidations, allylic/benzylic oxidations, and dihydroxylation of conjugated alkenes, and they only proceed when the arene *cis*-dihydroxylation is not favored. Exceptions are found in the TDO-mediated sulfoxidation of alkylaryl sulfides, which can take place very readily [342,435]. This reaction has been studied in detail using TDO and NDO, for a number of alkyl (C1–C8), aryl (thienyl, pyridyl, *o*- and *p*-substituted phenyl) sulfides, and also cyclic sulfides (Table 4.24) [317,432,435–439]. Similar to the *cis*-dihydroxylation, the yields obtained with TDO were generally higher with sulfides bearing smaller alkyl groups, whereas NDO displayed an opposite behavior, giving higher yields with larger sulfides [302,435]. Moreover, both enzymes are enantiocomplementary in the sulfoxidation, TDO affording mainly the (*R*)-sulfoxide and NDO the optical antipode (Table 4.24). It is

TABLE 4.24 TDO- and NDO-Catalyzed Sulfoxidation

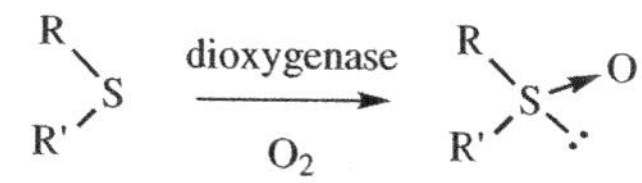

		TDO			NDO			
R	R′	Yield (%)	Absolute Configura-tion	*ee* (%)	Yield (%)	Absolute Configura-tion	*ee* (%)	Reference
Me	Ph	90/2.5 gL^{-1}	*R*	>98/92	33	*S*	91	[435]/[342]
Et	Ph	64	*R*	>98	27	*S*	84	[435]
Butyl	Ph	7	*R*	>98	79	*R*	>98	[435]
Hexyl	Ph	<1	*R*	78	76	*R*	>98	[435]
Me	*p*-FC_6H_4	51	*R*	90	4	*S*	91	[302]
Me	*p*-BrC_6H_4	<1	*S*	73	35	*S*	>98	[302]
Me	*p*-MeC_6H_4	8	*S*	45	17	*S*	>98	[302]
Me	*o*-FC_6H_4	53	*R*	>98	61	*S*	>98	[302]
Me	*o*-ClC_6H_4	20	*R*	41	50	*S*	69	[302]
Me	*o*-BrC_6H_4	17	*R*	18	22	*S*	16	[302]
Me	*m*-FC_6H_4	30	*S*	98	53	*S*	97	[435]
Me	2-thienyl	41	*R*	>98	4	*S*	69	[435]
Me	CH_2Ph	8	Racemic	0				[339]

interesting to note that dialkyl sulfides yielded racemic sulfoxides (only one example is reported) [339].

Another type of dioxygenase-catalyzed transfer of a single oxygen atom is the benzylic/allylic hydroxylation, observed in substituted benzenes, pyridines, and mainly in benzo-fused bicyclic substrates such as indanes [301,332,339,343,433,440]. Much attention was devoted to the hydroxylation of indane and its 2-substituted derivatives, such as 2-indanol, as a means of preparing *cis*-1,2-indanediol in high enantiomeric excess, which is used in the synthesis of the core of indinavir, an HIV-1 protease inhibitor [441–443]. Side-chain oxidation of a number of monosubstituted benzenes (**222**) takes place readily to give transient benzylic alcohols, which are further accepted by the dioxygenase to give the corresponding triol **223**. The triol metabolites were diastereomerically homogeneous (except for **223**, R = $C(CH_3){=}CH_2$) and isolated in variable yields along with the *cis*-dihydrodiols, in some cases being the major products (**223**: R = Et and $C(CH_3){=}CH_2$) (Table 4.25) [339]. It was proposed that the preferential benzylic hydroxylation of propylbenzene may result from a similar conformation to that of indane, which is a good substrate for TDO-mediated benzylic hydroxylation [339,442]. Despite the high enantioselectivity of the reaction to produce the (*R*)-benzylic alcohol, both enantiomers were accepted in the TDO-mediated *cis*-dihydroxylation of racemic benzylic alcohols to yield the diastereomeric triols with little or no kinetic resolution [444,445].

TABLE 4.25 TDO-Mediated Di- and Trihydroxylation of Monosubstituted Benzenes

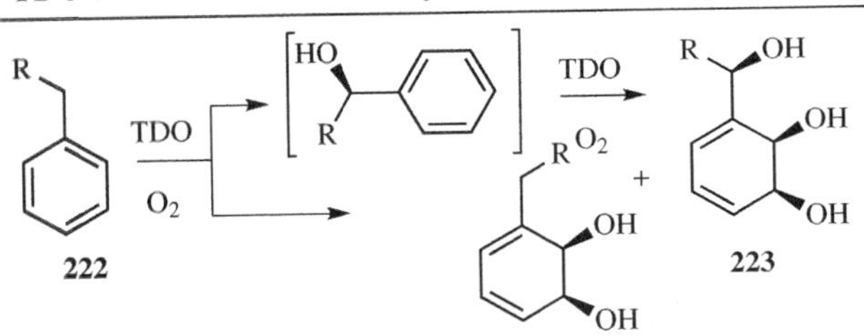

222 223

R	Yield of **223** (%)	diol + **223** (%)	Diastereomeric Purity
H	4	64	Homogeneous
Me	5	65	Homogeneous
Et	26	41[a]	Homogeneous
Pr	9	42	Homogeneous
$CH=CH_2$	15	33	Homogeneous
$CMe=CH_2$	18	27	90%

[a] Also from Ref. [301]

Disubstituted benzenes show a different behavior, with a challenging side chain oxidation. Thus, the *ortho*- and *para*-substituted benzenes yielded the *cis*-dihydrodiols exclusively, and only the *meta*- isomers afforded mixtures of *cis*-dihydrodiols and benzylic alcohols (with (*R*)-configuration), although in very low yields and variable optical purity (Table 4.26) [301]. The oxidation yields of the disubstituted benzenes were low-to-moderate for the *ortho*- and *para*-isomers (2–50%, with no evidence of benzylic hydroxylation or trihydroxylation) and very low for the *meta*-isomers (1–6%), in agreement with the *meta*-effect (*m*-substituent are slower reactant substrates) and with the preference of the *m*-substituted benzenes to be oxidized at the periphery of the aromatic system in the dioxygenase-mediated transformations.

Side chain oxidations also take place upon TDO-mediated oxidation of some monoalkylpyridines [320]. As for the disubstituted benzenes, the enzymatic reaction is highly regioselective: 4-alkylpyridines were hydroxylated exclusively on the ring and 3-alkylpyridines were hydroxylated exclusively on the side-chain α-C, whereas 2-alkylpyridines yielded both types of products (Table 4.27) [320].

In addition to arene *cis*-dihydroxylation, dioxygenases catalyze the dihydroxylation of conjugated alkenes to give the corresponding diols [298,329,331, 332,343,434,446–451]. The dihydroxylation of isolated alkenes is very rare, and only a few cyclic compounds have been reported as substrates [452,453]. However, the dihydroxylation is feasible when the alkene is activated by conjugation with aromatic groups or other alkenes. In this context, the most studied systems as substrates for dioxygenases are benzocycloalkenes, cyclodienes, cyclotrienes, and substituted styrenes (acyclic alkenes). As mentioned previously, the TDO- and NDO-catalyzed dihydroxylation of benzocycloalkenes of type **196** has been studied (Figure 4.63). TDO oxidizes the olefinic moiety to give *cis*-dihydrodiols (and products of benzylic oxidation) with 1*S* configuration [331,332], while NDO produces *cis*-diols with opposite configuration in excellent enantiomeric excess ($\geq$ 85% *ee*) [329,343]. Cyclic

TABLE 4.26 TDO-Mediated Benzylic vs. Aryl Oxidation in Disubstituted Benzenes

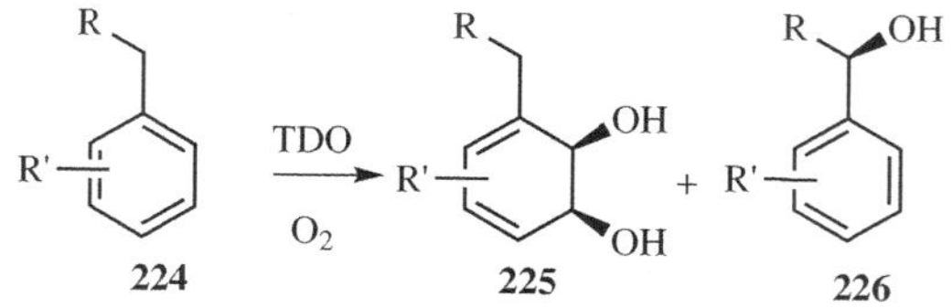

R	R′	225	Yield (%)[a]	ee (%)	226	Yield (%)[a]	ee (%)	Absolute Configuration
Me	2-F	**225a**	100	>98	**226a**	0	–	–
Me	2-Cl	**225b**	100	>98	**226b**	0	–	–
Me	2-Br	**225c**	100	>98	**226c**	0	–	–
Me	2-I	**225d**	100	>98	**226d**	0	–	–
Me	4-F	**225e**	100	>98	**226e**	0	–	–
Me	4-Cl	**225f**	100	>98	**226f**	0	–	–
Me	4-Br	**225g**	100	>98	**226g**	0	–	–
Me	4-I	**225h**	100	>98	**226h**	0	–	–
Me	3-F	**225i**	60,10	>98	**226i**	40, 90	2	*R*
Me	3-Cl	**225j**	14	>98	**226j**	86	6	*R*
Me	3-Br	**225k**	34	>98	**226k**	66	18	*R*
Et	2-F	**225l**	100	>98	**226l**	0	–	–
Et	4-F	**225m**	100	>98	**226m**	0	–	–
Et	4-Cl	**225n**	100	>98	**226n**	0	–	–
Et	4-Br	**225o**	100	>98	**226o**	0	–	–
Et	3-F	**225p**	0	–	**226p**	100	94	*R*
Et	3-Cl	**225q**	68	>98	**226q**	32	46	*R*

[a]Relative yield of **225**:**226**.

C5–C8 dienes (**232**) afforded the corresponding *cis*-diols in low-to-moderate yield but in enantiopure form (except for **233a**) when subjected to either TDO- or NDO-catalyzed dihydroxylation. Substituted fulvenes (**234**) were also accepted by TDO to give enantiopure diols resulting from reaction on the *cis*-double bond (Table 4.28) [343].

The dihydroxylation of acyclic alkenes was assessed in isoprene derivatives [298], and mainly in substituted styrenes [343,446,447]. In the latter case, dioxygenase selection and substitution pattern of the alkene are important factors to determine the preference for arene- versus alkene dihydroxylation, and the occurrence of allylic monohydroxylation. Thus, TDO leads to mixtures of side chain- and arene-dihydroxylation products, while NDO does not catalyze arene-dihydroxylation of substituted benzenes, but some competition from allylic monohydroxylation of methyl groups is observed (**239** and **240**). The enantioselectivity of the arene *cis*-dihydroxylation was complete, while the 1,2-diols from acyclic alkenes displayed

TABLE 4.27 TDO-Mediated Benzylic versus Aryl Oxidation in Monoalkylpyridines

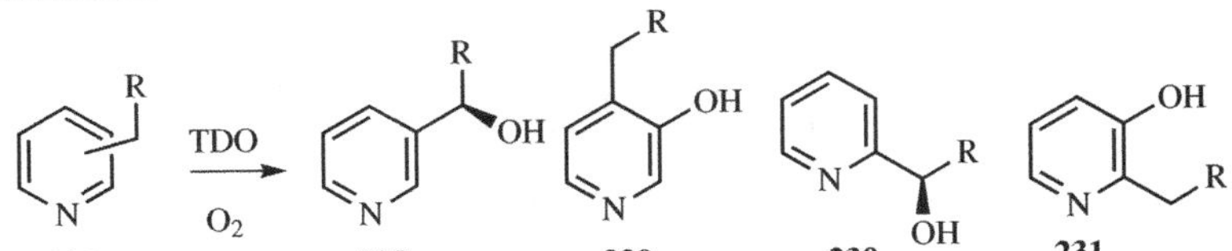

227 228 229 230 231

	Substrate	Hydroxypyridine	Yield (%)	Pyridylalkanol	Yield (%)	*ee* (%)	Absolute Configuration
	227a, R = 2-Me	**231a**	10	**230a**	20	–	–
o-alkyl	**227b**, R = 2-Et	**231b**	11	**230b**	29	*R*	83, >99[a]
	227c, R = 2-Pr	**231c**	10	**230c**	28	*R*	>99
	227d, R = 3-Me	–	–	**228d**	2	–	–
m-alkyl	**227e**, R = 3-Et	–	–	**228e**	30	*R*	90
	227f, R = 3-Pr	–	–	**228f**	70	*R*	>99
	227g, R = 4-Me	**229g**	17	–	–	–	–
p-alkyl	**227h**, R = 4-Et	**229h**	15	–	–	–	–
	227i, R = 4-Pr	**229i**	7	–	–	–	–

[a] Using *E. coli* pKST11.

lower enantiomeric excess values (< 82% *ee*, (1*R*)-diol) (Table 4.29) [343]. It is worth to mention that the TDO-mediated oxidation of 3-chlorostyrene (**236c**), using an enzyme from another strain of *Pseudomonas* (*P. putida* 39D) produced a similar ratio of **237c**:**238c**, whereas the enantiomeric excess values were different, resulting in a much lower enantioselectivity for the ring dihydroxylation (54% *ee* for **237c**) and an increased but reversed selectivity for the side chain-dihydroxylation (95% *ee* for (1*S*)-**238c**) [446].

Dioxygenase-catalyzed desaturations, although less common, have been reported mainly in monocyclic and bicyclic systems. For instance, during the TDO-mediated dihydroxylation of cyclohexadiene (**232b**) the desaturation pathway accounted for almost 50% of the products, yielding benzene, which was further transformed by the enzyme to give *cis*-benzenedihydrodiol (**233b'**) (Table 4.28) [343]. Moreover, TDO-mediated desaturation of 1,2-dihydronaphthalene gave the aromatic hydrocarbon, which yielded *cis*-1,2-dihydroxy-1,2-dihydronaphthalene as the final product *via* arene *cis*-dihydroxylation [434]. NDO has also been shown to catalyze desaturation reactions, for example, the desaturation of indane to indene to give (1*R*, 2*S*)-indanediol after a tandem desaturation/*cis*-dihydroxylation sequence [448]. The less common *N*- and *O*-dealkylation reactions of dioxygenases may turn into the major reaction pathway, for instance, in the biotransformation of *N*-methyl-2-quinolone (**194a**) where the *N*-dealkylated product (**194b**) was formed in 34% and the corresponding *cis*-diol (**195**) in only 28% (Figure 4.62) [324]. In general, these reactions take place when the *cis*-dihydroxylation pathway is not favored [288].

TABLE 4.28 TDO- and NDO-Mediated Dihydroxylation of Cyclic Dienes and Trienes

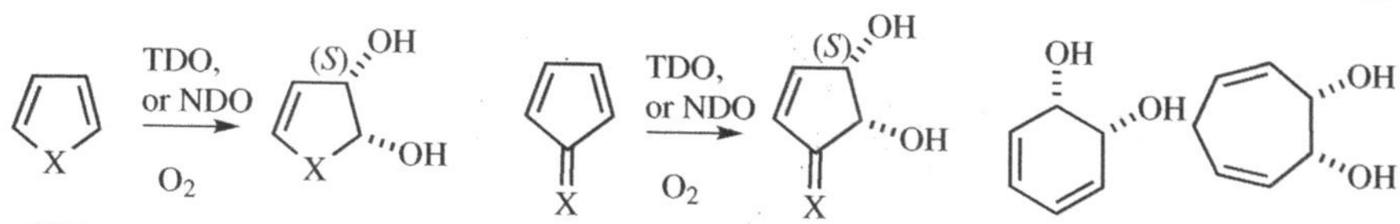

232 233 234 235 233b' 233e'

	TDO (*P. putida* UV4)			NDO (*P. putida* NCIMB 8859)	
Compound	Yield (%)	*ee* (%), Configuration	Compound	Yield (%)	*ee* (%), Configuration
233a, X=CH_2	32	20, (1*R*,2*S*)	**233a**, X=CH_2	14	40, (1*R*,2*S*)
233b, X=$(CH_2)_2$	12 + **233b'**(10)	>98, (1*R*,2*S*)	**233b**, X=$(CH_2)_2$	8	>98, (1*R*,2*S*)
233c, X=$(CH_2)_3$	20	>98, (1*R*,2*S*)	**233c**, X=$(CH_2)_3$	24	>98, (1*R*,2*S*)
233d, X=$(CH_2)_4$	4	>98, (1*R*,2*S*)	**233d**, X=$(CH_2)_4$	4	>98, (1*R*,2*S*)
233e, X=CH=CHCH$_2$	29 + **233e'**(15)	>98, (1*R*,2*S*)	**233e**, X=CH=CHCH$_2$	24	>98, (1*R*,2*S*)
235a, X=CMe_2	19	>98, (1*R*,2*S*)	**235a**, X=CMe_2	7	>98, (1*R*,2*S*)
235b, X=CEt_2	7	>98, (1*R*,2*S*)			
235c, X=C$(CH_2)_4$	5	>98, (1*R*,2*S*)			
235d, X=C$(CH_2)_5$	5	>98, (1*R*,2*S*)			

4.2.2 Miscellaneous Dioxygenases

Of all enzymes capable to activate dioxygen, namely oxidases, mono- and dioxyg nases, the latter enzymes, particularly those which are not aromatic dioxygenas have currently a marginal synthetic use. Yet, the variety of reactions catalyzed by t group deserves further attention from the synthetic community. Within this grou dioxygenases, the potential of lipoxygenases is remarkable and thus will be discu in the following section.

4.2.2.1 Lipoxygenase Lipoxygenases (LOX, EC 1.13.11.-), first descri 1932 [454], are a family of nonheme iron-containing dioxygenases that catal addition of molecular oxygen to polyunsaturated fatty acids with a (*Z,Z*)-1, unit to give an allyl fatty acid hydroperoxide with migrated (*E,Z*) double (Figure 4.77) [455].

LOX are most abundant in plants and mammals, and also occur in fun tebrates and recently have been found in bacteria [456,457]. They are exp some plant and animal tissues in high levels, for example, the soybean lip (SLOX, EC 1.13.11.12), of which at least four isoenzymes are present,

TABLE 4.29 TDO- and NDO-Mediated Dihydroxylation of Acyclic Alkenes

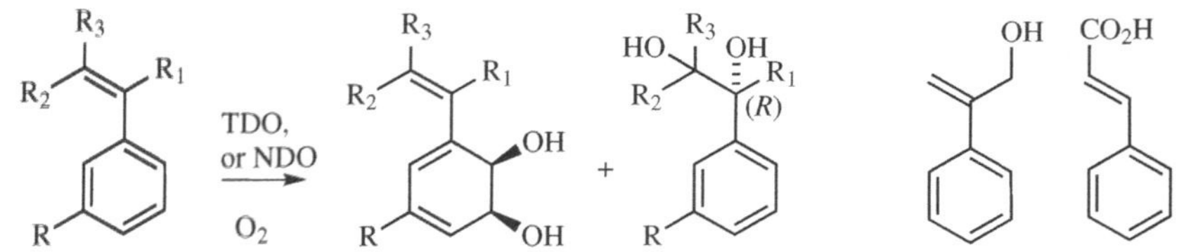

236 237 238 239 240

TDO (*P. putida* UV4)						NDO (*P. putida* NCIMB 8859)		
Arene-*cis*-dihydrodiol			Alkene-1,2-diol			Alkene-1,2-diol		
Compound[a]	Yield (%)	*ee* (%)	Compound[a]	Yield (%)	*ee* (%)	Compound[a]	Yield (%)	*ee* (%)
7a (R = H)	32	>98	**238a** (R = H)	3	88, 1*R*	**238a** (R = H)	60	80, 1*R*
7b (R = F)	27	>98	**238b** (R = F)	22	62, 1*R*	**238b** (R = F)	14	62, 1*R*
(R = Cl)	2	>98	**238c** (R = Cl)	14	42, 1*R*	**238c** (R = Cl)	18	56, 1*R*
= Me)	22	>98	**238d** (R = Me)	8	48, 1*R*	**238d** (R = Me)	12	56, 1*R*
Me)	24	>98				**238e** (R_1 = Me)	20 + **239** (12)	46, 1*R*
'e)	42	>98				**238f** (R_2 = Me)	15	82, 1*R*
	37	>98				**240**	52	–

ss otherwise indicated.

S,
iis
of
sed
ed in
ze the
-diene
bonds
i, inver-
essed in
xygenase
onstitutes

he protein in soybeans [456,458]. Its early discovery, straightfor-
eptable stability, and commercial availability turned SLOX into
used LOX. Historically, these enzymes are designated according
ygenation of their most common substrate (linoleic acid, C-18
ic acid, C-20 in mammals) and, when necessary, the name of
ence is added. In plants, LOXs are involved in plant defense,
tion, plant growth and development, while in mammals the
olved in diverse cell functions related to the maintenance
animal cell, cell maturation, and also in diverse human
on, fever, arthritis, and cancer [459,460]. As shown in
n can take place at either position of the distal vinylic
selectivity depends on the source of the enzyme and on
pH and oxygen concentration) [461]. For instance,
I 9.0 afforded (13*S*)-HPOD (**242**) in a 98:2 ratio,
the (9*S*)-isomer increased to approximately 25%
om corn germ and tomato forms predominantly
n both cases the enantioselectivity is excellent,

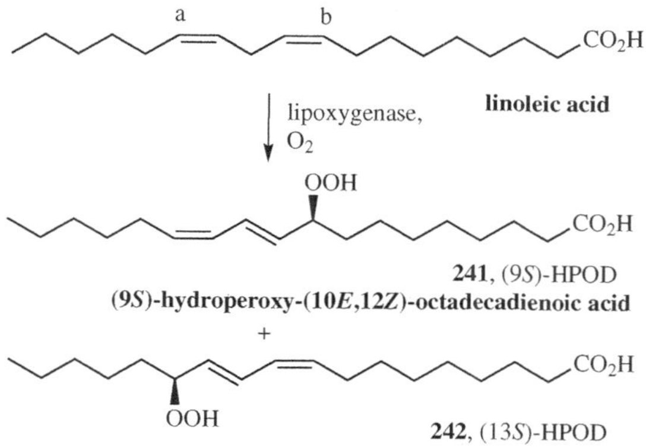

linoleic acid

241, (9*S*)-HPOD
(9*S*)-hydroperoxy-(10*E*,12*Z*)-octadecadienoic acid

+

242, (13*S*)-HPOD
(13*S*)-hydroperoxy-(9*Z*,11*E*)-octadecadienoic acid

FIGURE 4.77 Lipoxygenase-mediated dioxygenation of linoleic acid (the most common substrate for plant LOX).

yielding the *S*-isomers in more than 98% *ee* [463]. When using arachidonic acid, the natural substrate of mammal LOX, the presence of four unsaturations allows for a second LOX-mediated dioxygenation, offering the possibility of producing four dihydroperoxides. For synthetic purposes, the hydroperoxides can be further transformed into alcohols, ketones, or aldehydes (C–C cleavage), representing interesting starting materials for flavors, fragrances, and also for bioactive compounds. In addition to the natural substrates, LOX can accept a number of compounds, provided that they contain the (*Z*,*Z*)-1,4-diene unit and side chains (the proximal and the distal units) of certain length (from C-3 to C-10 for the distal unit) and polarity (a polar group such as a carboxylic acid is needed in the proximal unit), as described by several research groups [464–466]. For instance, Schreier reported the SLOX-catalyzed oxidation of long-chain alkadien(trien)ylphenols, such as cardanol, cardols, and anacardic acids [465]. Monounsaturated fatty acids were substrates of SLOX, yielding enones, *via* the corresponding allylic hydroperoxides (which were detected in a further study). Thus, oleic acid gave mainly 11-oxo-9(*Z*)-octadecenoic acid (**243**) and (12*Z*)-octadecenoic acid was converted to 13-oxo-11(*E*)-octadecenoic acid (**244**), and small amounts of 11-oxo-12(*Z*)-octadecenoic acid (**245**) (Figure 4.78) [466,467]. In both cases the major products resulted from hydrogen abstraction on C-11. The presence of the (*Z*,*Z*)-1,4-diene unit is not an essential requirement for activity and thus, polyenic substrates containing a carbonyl at the position of the central methylene unit were also accepted [468]. Models accounting for the selectivity of the dioxygenation were proposed by Lehman and Hatanaka, where the distal and proximal units of the substrate accommodate into the hydrophobic and the hydrophilic pockets of the enzyme, respectively [469,470].

The enzyme has been used for epoxide formation in polyenes through co-oxidation in the presence of linoleic acid. The co-oxidation of retinol, β-ionone, and 4-hydroxy-β-ionone was regioselective and nonenantioselective, giving the 5,6-epoxides **246** in racemic form (Figure 4.79) [471]. In another interesting application, SLOX was used

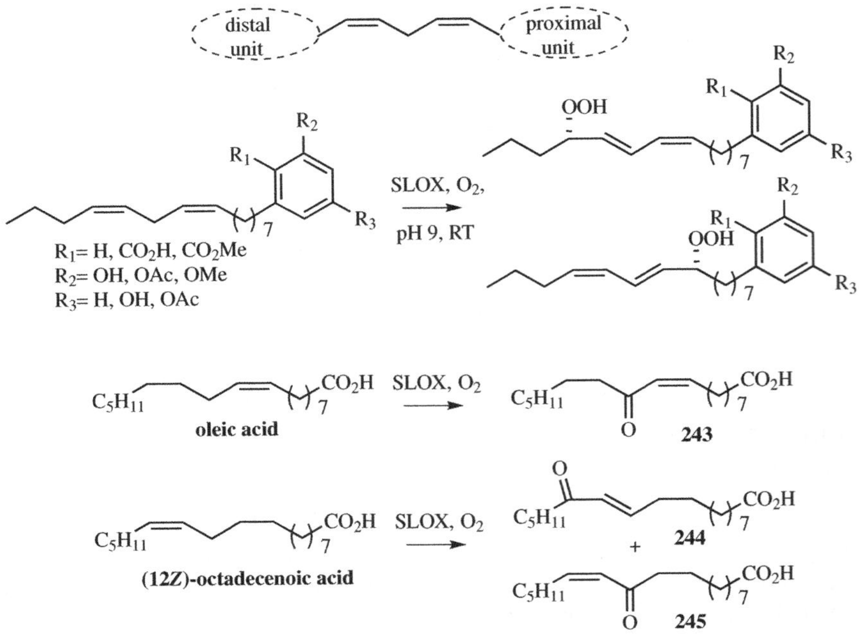

FIGURE 4.78 Miscellaneous substrates for LOX-mediated dioxygenation.

for the enantioselective dehydration of racemic hydroperoxide **247**, yielding 56% of the corresponding ketone **248** and 42% of unreacted chiral hydroperoxide, with 44% *ee* (Figure 4.79) [472].

SLOX has been used in the chemoenzymatic preparation of eicosanoids and related compounds, and also of miscellaneous natural products. A representative example is the seven-step synthesis of (6*R*)-lipoxin A (**249**), from arachidonic acid, using as a key reaction the LOX-catalyzed asymmetric hydroperoxidation of arachidonic acid into (5*S*,15*S*)-dihydroperoxyeicosatetraenoic acid ((5*S*,15*S*)-di-HPETE, **250**)

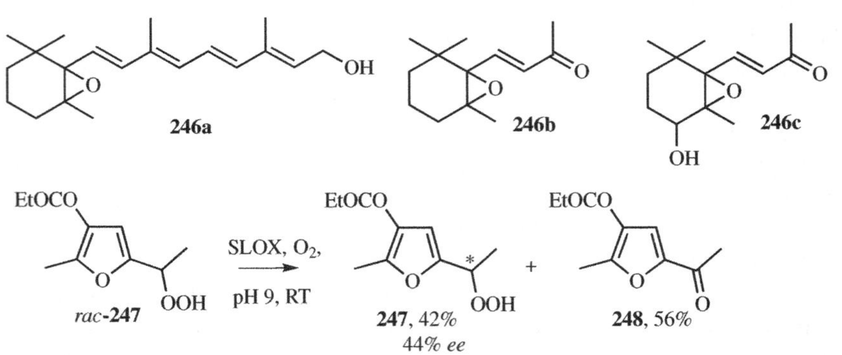

FIGURE 4.79 LOX-mediated co-oxidation and kinetic resolution reactions.

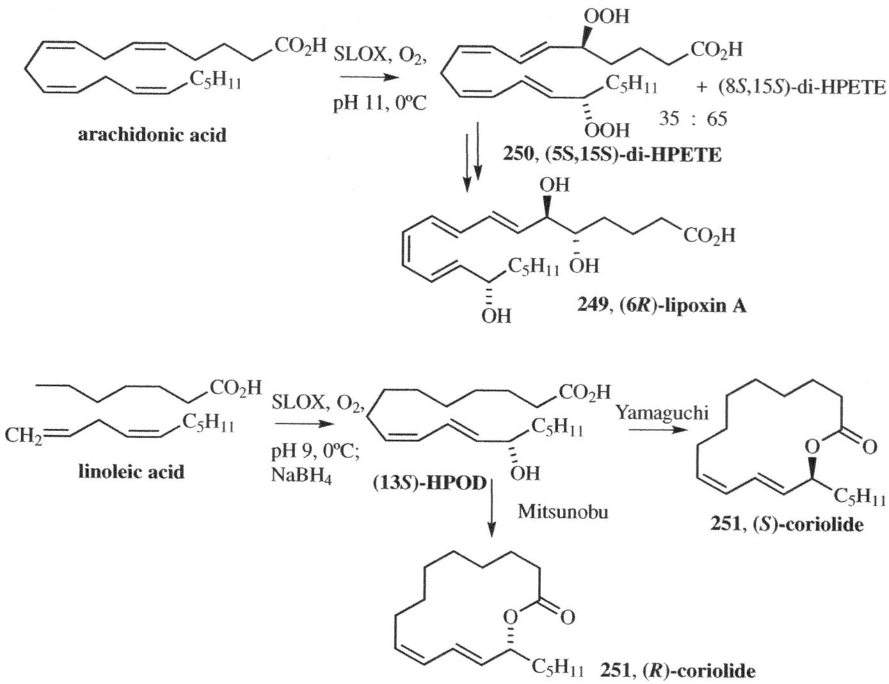

FIGURE 4.80 LOX-catalyzed preparation of natural products.

(Figure 4.80) [473]. Also, the LOX-catalyzed dioxygenation of linoleic acid was used in the preparation of both enantiomers of the macrocycle coriolide (**251**) (Figure 4.80) [474].

A strategy was designed for the enantioselective hydroxylation of short chain nonconjugated dienic systems, through the LOX-catalyzed dioxygenation of pentadienol unit linked to a spacer group that mimics the proximal unit of the natural substrate (linoleic acid). Best results were obtained using adipate as the spacer group, yielding chiral allylic alcohols in good yields with excellent regio- and stereocontrol (Table 4.30) [464,475]. The same strategy was applied for the preparation of methyl (5*S*,15*S*)-di-HETE (**256**) from arachidonic acid. In this strategy, following a regular SLOX-mediated dioxygenation and reduction on C-15, a change in the polarity of the side chains was introduced in order to dioxygenate the proximal double bond of arachidonic acid, which was part of the newly formed hydrophobic unit. This change in polarity was performed by methylation of the acid (becoming the hydrophobic unit) and succinylation of the C-15 alcohol to form the new hydrophilic unit of **257**. The second dioxygenation and chemical reduction produced the desired C-5 hydroxy compound in 78%, which was the desuccinylated to afford methyl (5*S*,15*S*)-di-HETE in 44% from arachidonic acid with a diastereomeric excess greater than 98% (Figure 4.81) [476].

TABLE 4.30 SLOX-Mediated Dioxygenation of Pentadienols Linked to a Spacer Group

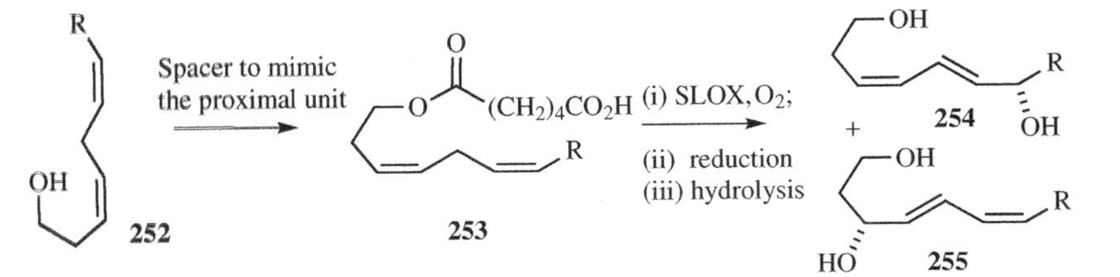

R	Ratio **254**:**255**	Yield of **254** (%)	*ee* (%)
$(CH_2)_4Me$	97:3	92	98
CH_2CHMe_2	82:18	72	96
$CHMe_2$	87:13	53	99
CH_2Ph	89:11	83	98
$(CH_2)_3Ph$	77:23	71	98
CH_2OCH_2Ph	85:15	52	97
$(CH_2)_3COMe$	99:1	83	97
$(CH_2)_3CMe_2OH$	99:1	68	99

From a preparative viewpoint, not only the regio- but also the enantioselectivity of the reaction needs to be considered. In this respect, SLOX and the commonly used LOXs afford hydroperoxides of (*S*)-configuration. Hydroperoxides from natural substrates are obtained with high optical purity, whereas the transformation of nonnatural substrates is achieved with high enantioselectivity only for the products oxygenated at the distal vinylic carbon (position "a" in Figure 4.77), the proximal position showing a very low stereoselectivity. It is interesting to mention that enantiocomplementary LOXs have been found in invertebrates, catalyzing the dioxygenation of polyunsaturated fatty acids with (*R*)-configuration [477,478].

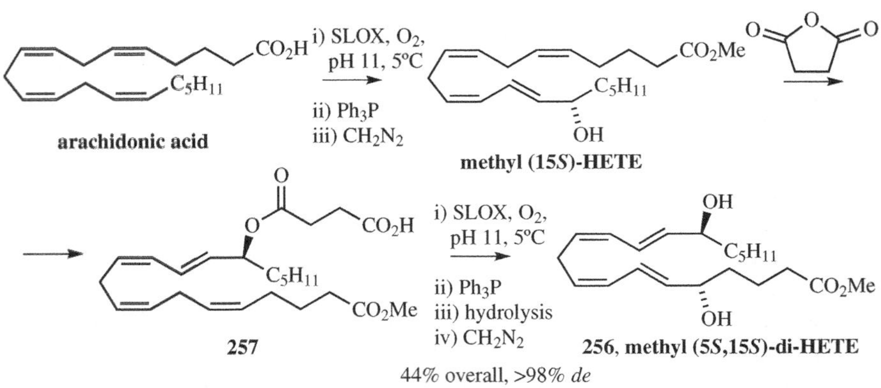

FIGURE 4.81 Use of the spacer strategy to prepare methyl (5*S*, 15*S*)-di-HETE.

Catalyst and Medium Engineering (Immobilization and Molecular Biology Strategies) Since lipids are the natural substrates for LOX, with low solubility in aqueous environments, the development of nonconventional media for LOX catalysis has deserved the attention of several groups. The reaction has been studied in organic solvents, monophasic aqueous-organic systems, biphasic systems, and micellar systems, generally affording better yields and also altered product distribution. For instance, the use of water-soluble cosolvents (sorbitol and other polyols) in the SLOX-catalyzed dioxygenation of linoleic acid produced a significant increase (3.7-fold) in the proportion of 9-HPOD versus 13-HPOD [479]. The same reaction was performed in organic media containing 2% of aqueous buffer, obtaining a 2.6-fold increase in activity compared to that in aqueous buffer when the reaction was done in *iso*-octane. Organic solvents like hexane and octane produced higher activities as well, while halogenated solvents produced an inhibitory effect [480]. In biphasic systems, containing Tris-HCl buffer and organic solvents in a 0–10% concentration range, it was observed again that *iso*-octane (4%) was the best cosolvent, enhancing the enzymatic activity by 4.7-fold [480]. The oxidation of linoleic acid in a ternary micelle system containing *iso*-octane/Tris-HCl buffer/surfactant resulted in enhancement or inhibition of the LOX activity, following an inverse relationship with the concentration of surfactant. In the ternary system, the product distribution was changed to give the 9-HPOD isomer as the major product. Thus the ratio 9-HPOD:13-HPOD shifted from the original 33:67 in aqueous medium to 57.1:42.9 in the ternary system [480]. Usual concerns associated to the work with biocatalysts, such as stability and reusability, can be addressed by enzyme immobilization. Several strategies and supports were used to immobilize SLOX, including adsorption to glutenin [481], glass wool [482], and talc [483], ionic binding to an ion-exchange resin [484] covalent attachment to a polyacrylamide gel [485,486], and gel entrapment in alginate [487]. In addition to improved stability and reusability, the substrate specificity and product distribution of LOX were different depending on the immobilization method used [488,489]. Furthermore, molecular biology approaches have been used to extend the versatility of LOX, particularly related to altering the product distribution [490–493].

4.3 CONCLUDING REMARKS

The ability of oxygenases to use oxygen for their catalytic performances, actually being oxygen the cheapest and more environment friendly oxidating agent, has triggered the development of many biocatalytic applications of these enzymes, usually providing excellent selectivities, and with cases already being performed at industrial scale. Remarkably, in some cases the chemical analogous reactions have not yet been reported, thus reinforcing the interest that these biocatalytic approaches may have. The combination of proper molecular biology techniques (e.g., cloning, directed evolution, generation of knocked-out mutants, etc.) together with the judicious decision on using isolated enzymes or whole-cells (depending on the application), are key aspects for a successful implementation.

REFERENCES

1. Faber K. Pure Appl Chem 1997;69:1613–1632.
2. Burton SG. Trends Biotechnol 2003;21:543–549.
3. Li Z, van Beilen JB, Duetz WA, Schmid A, de Raadt A, Grieng H, Witholt B. Curr Opin Biotechnol 2002;6:136–144.
4. Wohlgemuth R. J Chem Technol Biotechnol 2007;82:1055–1062.
5. Hollman F, Arends IWCE, Buehler K, Schallmey A, Bühler B. Green Chem 2011;13:226–265.
6. Loughlin WA. Bioresour Technol 2000;74:49–62.
7. van Berkel WJH, Kamerbeek NM, Fraaije MW. J Biotechnol 2006;124:670–689.
8. Bernhardt R. J Biotechnol 2006;124:128–145.
9. Torres Pazmiño DE, Winkler M, Glieder A, Fraaije MW. J Biotechnol 2010;146:9–24.
10. Ulracher VB, Lutz-Wahl S, Schmid RD. Appl Microbiol Biotechnol 2004;64:317–325.
11. Joosten V, van Berkel WJH. Curr Opin Chem Biol 2007;11:195–202.
12. Holland HL, Weber HK. Curr Opin Biotechnol 2000;11:547–553.
13. Peterson DH, Murray HC, Eppstein SH, Reineke LM, Weintraub A, Meister PD, Leigh HM. J Am Chem Soc 1952;74:5933–5936.
14. Lacroix I, Biton J, Azerad R. Bioorg Med Chem 1999;7:2329–2341.
15. Ortiz de Montellano PR. Chem Rev 2010;110:932–948.
16. Hannemann F, Virus C, Bernhardt R. J Biotechnol 2006;124:172–181.
17. Agrawal R, Deepika N.-U.-A, Joseph R. Biotechnol Bioeng 1999;63:249–252.
18. Bell SG, Chen X, Sowden RJ, Xu F, Williams JN, Wong L.-L, Rao Z. J Am Chem Soc 2003;125:705–714.
19. van Beilen JB, Holtackers R, Lüscher D, Bauer U, Witholt B, Duetz WA. Appl Environ Microbiol 2005;71:1737–1744.
20. Funhoff EG, Salzmann J, Bauer U, Witholt B, van Beilen JB. Enzyme Microb Technol 2007;40:806–812.
21. Pinheiro L, Marsaioli AJ. J Mol Catal B: Enzym 2007;44:78–86.
22. Urlacher V, Makhsumkhanov A, Schmid RD. Appl Microbiol Biotechnol 2005;70:53–59.
23. Maurer SC, Schulze H, Schmid RD, Urlacher V. Adv Synth Catal 2003;345:802–810.
24. Sun D.-A, Sauriol F, Mamer O, Zamir LO. Bioorg Med Chem 2001;9:793–800.
25. Serizawa N, Serizawa S, Nakagawa K, Kouhei F, Okazaki T, Terahara A. J Antibiotics 1998;36:887–891.
26. Park J.-W, Lee J.-K, Kwon T.-J, Yi D.-H, Kim Y.-J, Moon S.-H, Suh H.-H, Kang S.-M. Park Y.-I. Biotechnol Lett 2003;25:1827–1831.
27. Bollag DM, McQuency PA, Zhu J, Hensens O, Koupal L, Liesch J, Goetz M, Lazarides E, Woods CM. Cancer Res 1995;55:2325–2333.
28. Nayeem A, Chiang S.-J, Liu S.-W, Sun Y, You L, Basch J. Protein Eng Des Sel 2009;22:257–266.
29. Braunegg G, de Raadt A, Feichtenhofer S, Griengl H, Kopper I, Lehmann A, Weber HJ. Angew Chem Int Ed 1999;38:2763–2766.

30. de Raadt A, Griengl H, Weber H. Chem Eur J 2001;7:27–31.
31. Xu F, Bell SG, Lednik J, Insley A, Rao Z, Wong LL. Angew Chem Int Ed 2005;44:4029–4032.
32. Ullrich R, Hofrichter M. Cell Mol Life Sci 2007;64:271–293.
33. Sono M, Roach MP, Coulter ED, Dawson JH. Chem Rev 1996;96:2841–2888.
34. Barr SA, Bowers N, Boyd DR, Sharma ND, Hamilton L, Austin R, McMordie S, Dalton H, J Chem Soc, Perkin Trans 1 1998;20:3443–3451.
35. Cerniglia CE. Biodegradation 1992;3:351–368.
36. Harford-Cross CF, Carmichael AB, Allan FK, England PA, Rouch DA, Wong LL. Protein Eng 2000;13:121–128.
37. Carmichael AB. Wong L.-L. Eur J Biochem 2001;268:3117–3125.
38. Jones JP, O'Hare EJ, Wong LL. Eur J Biochem 2001;268:1460–1467.
39. von Baeyer A, http://nobelprize.org/nobel_prizes/chemistry/laureates/1905/index.html, The Nobel Prize in Chemistry 1905.
40. Renz M, Meunier B, Eur J Org Chem 1999;4:737–750.
41. Strukul G. Angew Chem Int Ed 1998;37:1198–1209.
42. Criegee R. Justus Liebigs Ann Chem 1948;560:127–135.
43. Noyori R, Kobayashi H, Sato T. Tetrahedron Lett 1980;21:2573–2576.
44. Noyori R, Sato T, Kobayashi H. Tetrahedron Lett 1980;21:2569–2572.
45. ten Brink G.-J, Arends IWCE, Sheldon RA. Chem Rev 2004;104:4105–4123.
46. Crudden CM, Chen AC, Calhoun LA. Angew Chem Int Ed 2000;39:2851–2855.
47. Goodman RM, Kishi Y. J Am Chem Soc 1998;120:9392–9393.
48. Anastas PT, Bartlett LB, Kirchhoff MM, Williamson TC. Catal Today 2000;55:11–22.
49. Mazzini C, Lebreton J, Furtoss R. J Org Chem 1996;61:8–9.
50. Murahashi S.-I, Ono S. Imada Y, 2002;41:2366–2368.
51. Bolm C, Palazzi C, Francio G, Leitner W, Chemm Commun 2002;1588–1589.
52. Turfitt GE. Biochem J 1948;42:376–383.
53. Townsend CA, Christensen SB, Davis SG. J Am Chem Soc 1982;104:6152–6153.
54. Damtoft S, Franzyk H, Jensen R. Phytochemistry 1995;40:773–784.
55. Wright JLC, Hu T, McLachlan JL, Needham J, Walter JA. J Am Chem Soc 1996;118:8757–8758.
56. Walsh CT. Jack Chen Y.-C. Angew Chem Int Ed 1988;27:333–343.
57. Stewart JD. Curr Org Chem 1998;2:195–216.
58. Willets A. Trends Biotechnol 1997;15:55–62.
59. Donoghue NA, Norris DB, Trudgill PW. Eur J Biochem 1976;63:175–192.
60. Donoghue NA, Trudgill PW. Eur J Biochem 1975;60:1–7.
61. Iwaki H, Hasegawa Y, Teraoka M, Tokuyama T, Bergeron H, Lau PCK. Appl Environ Microbiol 1999;65:5158–5162.
62. Cheng Q, Thomas SM, Kostichka K, Valentine JR, Nagarajan V. J Bacteriol 2000;182:4744–4751.
63. Iwaki H, Hasegawa Y, Wang S, Kayser MM, Lau PCK. Appl Environ Microbiol 2002;68:5671–5684.

64. Kamerbeek NM, Janssen DB, van Berkel WJH, Fraaije MW. Adv Synth Catal 2003;345:667–678.
65. Taylor DG, Trudgill PW. J Bacteriol 1986;165:489–497.
66. Jones KH, Smith RT, Trudgill PW. J Gen Microbiol 1993;139:797–805.
67. Bradshaw WH, Conrad HE, Corey EJ, Gunsalus IC, Lednicer D. J Am Chem Soc 1959;81:5507–5508.
68. Conrad HE, Hedegaard J, Gunsalus IC. Tetrahedron Lett 1965;6:561–565.
69. Conrad HE, Lieb K, Gunsalus IC. J Biol Chem 1965;240:4029–4037.
70. Hedegaard J, Gunsalus IC. J Biol Chem 1965;240:4038–4043.
71. Trudgill PW, DuBus R, Gunsalus IC. J Biol Chem 1966;241:1194–1205.
72. Trudgill PW, DuBus R, Gunsalus IC. J Biol Chem 1966;241:4288–4290.
73. Yu CA, Gunsalus IC. J Biol Chem 1969;244:6149–6152.
74. Gunsalus IC, Marshall VP, Ribbons DW. Crit Rev Microbiol 1971;1:291–310.
75. Ougham HJ, Taylor DG, Trudgill PW. J Bacteriol 1983;153:140–152.
76. van der Donk WA, Zao H. Curr Opin Biotechnol 2003;14:421–426.
77. Hollman F, Hofstetter K, Schmid A. Trends Biotechnol 2006;24:163–171.
78. Tishkov VI, Galkin AG, Fedorchuk W, Savitsky PA, Rojkova AM, Gieren H, Kula MR. Biotechnol Bioeng 1999;64:187–193.
79. Tishkov VI, Galkin AG, Marchenko GN, Egorova OA, Sheluho DV, Kulakova LB, Dementieva LA, Egorov AM. Biochem Biophys Res Commun 1993;192:976–981.
80. Tishkov VI, Popov VO. Biomol Eng 2006;23:89–110.
81. Schwarz-Linek U, Krödel A, Ludwig F.-A, Schulze A, Rissom S, Kragl U, Tishkov VI, Vogel M. Synthesis 2001;6:947–953.
82. Wong C.-H, Whitesides GM. J Am Chem Soc 1981;103:4890–4899.
83. Abril O, Ryerson CC, Walsh C, Whitesides GM. Bioorg Chem 1989;17:41–52.
84. Taschner MJ, Black DJ. J Am Chem Soc 1988;110:6892–6893.
85. Taschner MJ, Black DJ. Chen Q.-Z. Tetrahedron: Asymmetry 1993;4:1387–1390.
86. Ottolina G, de Gonzalo G, Carrea G, Danieli B. Adv Synth Catal 2006;347:1035–1040.
87. Willets A, Knowles CJ, Levitt MS, Roberts SM, Sandey H, Shipston NF. J Chem Soc Perkin Trans. 1 1991;6:1608–1610.
88. Grogan G, Roberts S, Willets A. Biotechnol. Lett. 1992;14:1125–1130.
89. Vrtis JM, White AK, Metcalf WW, van der Donk WA. Angew Chem Int Ed 2002;41:3257–3259.
90. Woodyer R, van der Donk WA. Biochemistry 2003;42:11604–11614.
91. Torres Pazmiño DE, Snajdrova R, Baas B.-J, Ghobrial M, Mihovilovic MD, Fraaije MW. Angew Chem Int Ed 2008;47:2275–2278.
92. Duetz WA, van Beilen JB, Witholt B. Curr Opin Biotechnol 2001;12:419–425.
93. Alphand V, Archelas A, Furtoss R. J Org Chem 1990;55:347–350.
94. Davey JF, Trudgill PW. Eur J Biochem 1977;74:115–127.
95. Kayser MM, Chen G, Stewart JD. Synlett 1999;1:153–158.
96. Chen G, Kayser MM, Mihovilovic MD, Mrstik ME, Martinez CA, Stewart JD. New J Chem 1999;23:827–832.

97. Kostichka K, Thomas SM, Gibson KJ, Nagarajan V, Cheng Q. J Bacteriol 2001;183:6478–6486.
98. Chen Y.-C. J, Peoples OP, Walsh CT. J Bacteriol 1988;170:781–789.
99. Brzostowicz PC, Walters DM, Thomas SM, Nagarajan V. Rouvière P. Appl Environ Microbiol 2003;69:334–342.
100. Brzostowicz PC, Gibson KL, Thomas SM, Blasko MS. Rouvière P. J Bacteriol 2000;182:4241–4248.
101. Iwaki H, Wang S, Grosse S, Bergeron H, Nagahashi A, Lertvorachon J, Yang J, Konishi Y, Hasegawa Y, Lau PCK. Appl Environ Microbiol 2006;72:2707–2720.
102. Griffin M, Trudgill PW. Eur J Biochem 1976;63:199–209.
103. Kamerbeek NM, Moonen MJH, van der Ven JGM, van Berkel WJH, Fraaije MW, Janssen DB. Eur J Biochem 2001;268:2547–2557.
104. Malito E, Alfieri AF, Marco W, Mattevi A. Proc Natl Acad Sci USA 2004;101:13157–13162.
105. Fraaije MW, Wu J, Heuts DPHM, van Hellemond EW, Lutje Spelberg JH, Janssen DB. Appl Microbiol Biotechnol 2005;66:393–400.
106. Morii S, Sawamoto S, Yamauchi Y, Miyamoto M, Iwami M, Itagaki E. J Biochem 1999;126:624–631.
107. Mihovilovic MD, Müller B, Stanetty P. Eur J Org Chem 2002;3711–3730.
108. Kayser MM. Tetrahedron 2009;65:947–974.
109. Roberts SM, Wan PWH. J Mol Catal B: Enzym 1998;4:111–136.
110. Kyte BG, Rouvière P, Cheng Q, Stewart JD. J Org Chem 2004;69:12–17.
111. Bes MT, Villa R, Roberts SM, Wan PWH, Willets A. J Mol Catal B: Enzym 1996;1:127–134.
112. Mihovilovic MD, Müller B, Kayser MM, Stanetty P Synlett 2002;5:700-702.
113. Gutiérrez MC, Alphand V, Furtoss R. J Mol Catal B: Enzym 2003;21:231–238.
114. Cheng Q, Thomas S. Rouvière P. Appl Microbiol Biotechnol 2002;58:704–711.
115. Schumacher JD, Fakoussa RM. Appl Microbiol Biotechnol 1999;52:85–90.
116. Miyamoto M, Matsumoto J, Iwaya T, Itagaki E. Biochim Biophys Acta 1995;1251:115–124.
117. Liu H.-M, Li H, Shan L, Wu J. Steroids 2006;71:931–934.
118. C. Rodríguez, de Gonzalo G, Torres Pazmiño DE, Fraaije MW, Gotor V. Tetrahedron: Asymmetry 2008;19:197–203.
119. Cripps RE. Biochem J 1975;152:233–241.
120. Higson FK, Focht DD. Appl Environ Microbiol 1990;56:3678–3685.
121. Havel J, Reineke W. Appl Environ Microbiol 1993;59:2706–2712.
122. Jones KH, Trudgill PW, Hopper DJ. Appl Environ Microbiol 1994;60:1978–1983.
123. Casellas M, Grifoll M, Bayona JM, Solanas AM. Appl Environ Microbiol 1997;63:819–826.
124. Moonen MJH, Rietjens IMCM, van Berkel WJH. Flavins and Flavoproteins. Berlin: Scientific; 1999.
125. Tanner A, Hopper DJ. J Bacteriol 2000;182:6565–6569.

126. Kamerbeek NM, Olsthoorn AJJ, Fraaije MW, Janssen DB. Appl Environ Microbiol 2003;69:419–426.
127. Grogan G, Roberts S, Wan P, Willets A. Biotechnol Lett 1993;15:913–918.
128. Grogan G, Roberts S, Willets A J Chem Soc, Chem Commun 1993;8:699–701.
129. Beecher J, Grogan G, Roberts S, Willets A. Biotechnol Lett 1996;18:571–576.
130. Schwab JM, Li W, Thomas LP. J Am Chem Soc 1983;105:4800–4808.
131. Kayser MM, Chen G, Stewart JD. J Org Chem 1998;63:7103–7106.
132. Wang S, Chen G, Kayser MM, Iwaki H, Lau PCK, Hasegawa Y. Can J Chem 2002;80:613–621.
133. Adger B, Bes MT, Grogan G, McCague R, Pedragosa-Moreau S, Roberts S, Villa R, Wan PWH, Willets A. J Chem Soc, Chem Commun 1995;1563–1564.
134. Berezina N, Alphand V, Furtoss R. Tetrahedron: Asymmetry 2002;13:1953–1955.
135. Mihovilovic MD. Curr Org Chem 2006;10:1265–1287.
136. Alphand V, Furtoss R, Pedragosa-Moreau S, Roberts S, Willets A. J Chem Soc, Perkin Trans 1 1996;(15):1867–1872.
137. Stewart JD, Reed KW, Zhu J, Chen G, Kayser MM. J Org Chem 1996;61:7652–7653.
138. Rial DV, Cernuchova P, van Beilen JB, Mihovilovic MD. J Mol Catal B: Enzym 2008;50:61–68.
139. Stewart JD, Reed KW, Martínez CA, Zhu J, Chen G, Kayser MM. J Am Chem Soc 1998;120:3541–3548.
140. Leisch H, Morley K, Lau PCK. Chem Rev 2011;111:4165–4222.
141. Adger B, Bes MT, Grogan G, McCague R, Pedragosa-Moreau S, Roberts S, Villa R, Wan PWH, Willets A. Bioorg Med Chem 1997;5:253–261.
142. Kirschner A, Bornscheuer UT. Angew Chem Int Ed 2006;45:7004–7006.
143. Kirschner A, Bornscheuer UT. Appl Microbiol Biotechnol 2008;81:465–472.
144. A. Völker, Kirschner A, Bornscheuer UT, Altenbuchner J. Appl Microbiol Biotechnol 2008;77:1251–1260.
145. Rehdorf J, Lengar A, Bornscheuer UT, Mihovilovic MD. Bioorg Med Chem Lett 2009;19:3739–3743.
146. Geitner K, Kirschner A, Rehdorf J, Schmidt M, Mihovilovic MD, Bornscheuer UT. Tetrahedron: Asymmetry 2007;18:892–895.
147. C. Rodríguez, de Gonzalo G, Fraaije MW, Gotor V. Tetrahedron: Asymmetry 2007;18:1338–1344.
148. C. Rodríguez, de Gonzalo G, Torres Pazmiño DE, Fraaije MW, Gotor V. Tetrahedron: Asymmetry 2009;20:1168–1173.
149. de Gonzalo G, Mihovilovic MD, Fraaije MW. ChemBioChem 2010;11:2208–2231.
150. A. Rioz-Martínez, de Gonzalo G, Torres Pazmiño DE, Fraaije MW, Gotor V. J Org Chem 2010;75:2073–2076.
151. C. Rodríguez, de Gonzalo G, Rioz-Martínez A, Torres Pazmiño DE, Fraaije MW, Gotor V. Org Biomol Chem 2010;8:1121–1125.
152. García-Urdiales E, Alfonso I, Gotor V. Chem Rev 2005;105:313–354.
153. Stewart JD, Reed KW, Kayser MM J Chem Soc, Perkin Trans 1 1996;8:755–757.

154. Mihovilovic MD, Chen G, Wang S, Kyte B, Rochon F, Kayser MM, Stewart JD. J Org Chem 2001;66:733–738.

155. Rial DV, Bianchi DA, Kapitanova P, Lengar A, van Beilen JB, Mihovilovic MD Eur J Org Chem 2008;7:1203–1213.

156. Wang S, Kayser MM, Iwaki H, Lau PCK. J Mol Catal B: Enzym 2003;22:211–218.

157. Mihovilovic MD, Snajdrova R, Grötzl B. J Mol Catal B: Enzym 2006;39:135–140.

158. Mihovilovic MD, Rudroff F, Grötzl B, Kapitan P, Snajdrova R, Rydz J, Mach R. Angew Chem Int Ed 2005;44:3609–3613.

159. Mihovilovic MD, Rudroff F, Grötzl B, Stanetty P. Eur J Org Chem 2005;5:809–816.

160. Mihovilovic MD, Rudroff F, Müller B, Stanetty P. Bioorg Med Chem Lett 2003;13:1479–1482.

161. Mihovilovic MD, Müller B, Kayser MM, Stewart JD, Fröhlich J, Stanetty P, Spreitzer H. J Mol Catal B: Enzym 2001;11:349–353.

162. Gagnon R, Grogan G, Groussain E, Pedragosa-Moreau S, Richardson PF, Roberts S, Willets A, Alphand V, Lebreton J, Furtoss R. J Chem Soc Perkin Trans 1 1995;20:2527–2528.

163. Alphand V, Furtoss R. J Mol Catal B: Enzym 2000;9:209–217.

164. Alphand V, Mazzini C, Lebreton J, Furtoss R. J Mol Catal B: Enzym 1998;5:219–221.

165. Mihovilovic MD, Kapitan P, Rydz J, Rudroff F, Ogink FH, Fraaije MW. J Mol Catal B: Enzym 2005;32:135–140.

166. Taschner MJ, Peddada L J Chem Soc, Chem Commun 1992;19:1384–1385.

167. Braun I, Rudroff F, Mihovilovic MD, Bach T. Angew Chem Int Ed 2006;45:5541–5543.

168. Snajdrova R, Braun I, Bach T, Mereiter K, Mihovilovic MD. J Org Chem 2007;72:9597–9603.

169. Mihovilovic MD, Müller B, Schulze A, Stanetty P, Kayser MM. Eur J Org Chem 2003;12:2243–2249.

170. Chapman PJ, Meerman G, Gunsalus IC. Biochem Biophys Res Commun 1965;20:104–108.

171. Alphand V, Furtoss R. J Org Chem 1992;57:1306–1309.

172. Alphand V, Archelas A, Furtoss R. Tetrahedron Lett 1989;30:3663–3664.

173. Lebreton J, Alphand V, Furtoss R. Tetrahedron Lett 1996;37:1011–1014.

174. Carnell AJ, Roberts SM, Sik V, Willets AJ J Chem Soc, Perkin Trans 1 1991;10:2385-2389.

175. Mihovilovic MD, Kapitan P. Tetrahedron Lett 2004;45:2751–2754.

176. Luna A, Gutiérrez M.-C, Furtoss R, Alphand V. Tetrahedron: Asymmetry 2005;16:2521–2524.

177. Petit F, Furtoss R. Tetrahedron: Asymmetry 1993;4:1341–1352.

178. Taschner MJ, Peddada L, Cyr P, Chen Q.-Z, Black DJ, The enzymatic Baeyer–Villiger oxidation. In: Serve S, editor. *Microbial Reagents in Organic Synthesis*. London: Kluwer Academic, 1992, pp. 347–360.

179. Kelly DR, Knowles CJ, Mahdi JG, Wright MA, Taylor IN, Hibbs DE, Hursthouse MB, Mish'al AK, Roberts SM, Wan PWH, Grogan G, Willets AJ. J Chem Soc, Perkin Trans 1 1995;(16):2057–2066.

180. Kelly DR, Knowles CJ, Mahdi JG, Wright MA, Taylor IN, Roberts SM, Wan PWH, Grogan G, Pedragosa-Moreau S, Willets AJ Chem Commun 1996;(20):2333–2334.
181. Kelly DR. Tetrahedron: Asymmetry 1996;7:1149–1152.
182. Wang S, Kayser MM. J Org Chem 2003;68:6222–6228.
183. Alphand V, Furtoss R. Tetrahedron: Asymmetry 1992;3:379–382.
184. Cernuchová P, Mihovilovic MD. Org Biomol Chem 2007;5:1715–1719.
185. Taschner MJ, Aminbhavi AS. Tetrahedron Lett 1989;30:1029–1032.
186. Yokokawa F, Hamada Y, Shioiri T, Chem Commun 1996, 871–872.
187. Taschner MJ, Chen Q-Z. Bioorg Med Chem Lett 1991;1:535–538.
188. Mazzini C, Lebreton J, Alphand V, Furtoss R. J Org Chem 1997;62:5215–5218.
189. Mazzini C, Lebreton J, Alphand V, Furtoss R. Tetrahedron Lett 1997;38:1195–1196.
190. Bode JW, Doyle MP, Protopopova MN, Zhou Q.-L. J Org Chem 1996;61:9146–9155.
191. Honda T, Kimura N, Sato S, Kato D, Tominaga H J Chem Soc, Perkin Trans 1 1994;8:1043–1046.
192. Mihovilovic MD, Bianchi DA, Rudroff F. Chem Commun 2006;30:3214–3216.
193. Levitt MS, Newton RF, Roberts SM, Willets AJ. J Chem Soc, Perkin Trans 1 1990;(8):619–620.
194. Gagnon R, Grogan G, Levitt MS, Roberts SM, Wan PWH, Willets AJ. J Chem Soc, Perkin Trans 1 1994;2537–2543.
195. Gagnon R, Grogan G, Roberts SM, Villa R, Willets AJ J Chem Soc, Perkin Trans 1 1995;1:1505–1511.
196. Corey EJ, Noyori R. Tetrahedron Lett 1970;4:311–313.
197. Corey EJ, Su WG. Tetrahedron Lett 1990;31:3833–3836.
198. Newton RF, Roberts SM. Tetrahedron 1980;36:2163–2196.
199. VelázquezF., Olivo HF. Org Lett 2000;2:1931–1933.
200. Lebreton J, Alphand V, Furtoss R. Tetrahedron Lett 1997;53:145–160.
201. Andrau L, Lebreton J, Viazzo P, Alphand V, Furtoss R. Tetrahedron Lett 1997;38:825–826.
202. Königsberger K, Griengl H. Bioorg Med Chem 1994;2:595–604.
203. Petit F, Furtoss R. Synthesis 1995;12:1517–1520.
204. Besse P, Veschambre H. Tetrahedron 1994;50:8885–8927.
205. Katsuki T, Sharpless KB. J Am Chem Soc 1980;102:5974–5976.
206. Zhang W, Loebach JL, Wilson SR, Jacobsen EN. J Am Chem Soc 1990;112:2801–2803.
207. Lin H, Liu J-Y, Wang H-B, Ahmed AAW, Wu Z-L. J Mol Catal B: Enzym 2011;72:77–89.
208. Chang D, Zhang J, Witholt B, Li Z. Biocatal Biotransform 2004;22:113–130.
209. Chefson A, Auclair K. Mol BioSyst 2006;2:462–469.
210. Martinez CA, Stewart JD. Curr Org Chem 2000;4:263–282.
211. Yun C-H, Kim K-H, Kim D-H, Jung H-C Pan J-G. Trends Biotechnol 2007;25:289–298.
212. Ruettinger RT, Fulco AJ. J Biol Chem 1981;256:5728–5734.
213. Ruettinger RT, Wen L-P, Fulco AJ. J Biol Chem 1989;264:10987–10995.
214. Graham-Lorence S, Truan G, Peterson JA, Falck JR, Wei S, Helvig C, Capdevila JH. J Biol Chem 1997;272:1127–1135.

215. Celik A, Sperandio D, Speight RE, Turner NJ. Org Biomol Chem 2005;3:2688–2690.
216. Farinas ET, Alcalde M, Arnold F. Tetrahedron 2004;60:525–528.
217. Sowden RJ, Yasmin S, Rees NH, Bell SG, Wong L.-L. Org Biomol Chem 2005;3:57–64.
218. Whitehouse CJC, Bell SG, Yang W, Yorke JA, Blandford CF, Strong AJF, Morse EJ, Bartlam M, Rao Z, Wong L.-L. ChemBioChem 2009;10:1654–1656.
219. Dietrich JA, Yoshikuni Y, Fisher KJ, Woolard FX, Ockey D, McPhee DJ, Renninger NS, Chang MCY, Baker D, Keasling JD. ACS Chem Biol 2009;4:261–267.
220. Shoji O, Fujishiro T, Nakajima H, Kim M, Nagano S, Shiro Y, Watanabe Y. Angew Chem Int Ed 2007;46:3656–3659.
221. Ortiz de Montellano PR, Fruetel JA, Collins JR, Camper DL, Loew GH. J Am Chem Soc 1991;113:3195–3196.
222. Bell SG, Sowden RJ, Wong L-L. Chem Commun 2001;(7):635–636.
223. Weijers CAGM, van Ginkel CG, de Bont JAM. Enzyme Microb Technol 1988;10:214–218.
224. Owens CR, Karceski JK, Mattes TE. Appl Microbiol Biotechnol 2009;84:685–692.
225. Habets-Crützen AQH, Carlier SJN, de Bont JAM. Enzyme Microb Technol 1985;7:17–21.
226. McClay K, Fox BG, Steffan RJ. Appl Environ Microbiol 2000;66:1877–1882.
227. Abbott BJ, Hou CT. Appl Microbiol 1973;26:86–91.
228. de Smet M-J, Kingma J, Wynberg H, Witholt B. Enzyme Microb Technol 1983;5:352–360.
229. Wubbolts MG, Panke S, Beilen JB, Witholt B. Chimia 1996;50:436–437.
230. May SW, Abbott BJ. J Biol Chem 1973;248:1725–1730.
231. Johnstone SL, Phillips GT, Robertson BW, Watts PD, Bertola MA, Koger HS, Marx AF. In: Laane C, Tramper J, Lilly MD, editors. Biocatalysis in Organic Media. Amsterdam: Elsevier; 1987, pp. 387–392.
232. McKenna EJ, Coon MJ. J Biol Chem 1970;245:3882–3889.
233. Ruettinger RT, Olson ST, Boyer RF, Coon MJ. Biochem Biophys Res Commun 1974;57:1011–1017.
234. van Beilen JB, Wubbolts MG, Witholt B. Biodegradation 1994;5:161–174.
235. Staijen IE, van Beilen JB, Witholt B. Eur J Biochem 2000;267:1957–1965.
236. Chang D, Heringa MF, Witholt B, Li Z. J Org Chem 2003;68:8599–8606.
237. Ribes S, Taberner F, Domenech A, Cabellos C, Tubau F, Liñares J, Fernández Viladrich P, Gudiol F. J Antimicrob Chemother 2006;57:931–936.
238. White RF, Birnbaum J, Meyer RT, Ten Broeke J, Chemerda JM, Demain AL. Appl Microbiol 1971;22:55–60.
239. Xie F, Chao Y, Xue Z, Yang X, Zhang G, Shi J, Qian S. J Ind Microbiol Biotechnol 2009;36:739–746.
240. Archelas A, Furtoss R. Tetrahedron Lett 1992;33:5241–5242.
241. Panke S, Witholt B, Schmid A, Wubbolts MG. Appl Environ Microbiol 1998;64:2032–2043.
242. Otto K, Hofstetter K, Röthlisberger M, Witholt B, Schmid A. J Bacteriol 2004;186:5292–5302.

243. Kantz A, Chin F, Nallamothu N, Nguyen T, Gassner GT. Arch Biochem Biophys 2005;442:102–116.
244. Kantz A, Gassner GT. Biochemistry 2011;50:523–532.
245. Hollmann F, Hofstetter K, Habicher T, Hauer B, Schmid A. J Am Chem Soc 2005;127:6540–6541.
246. Hollmann F, Lin P-C, Witholt B, Schmid A. J Am Chem Soc 2003;125:8209–8217.
247. Panke S, Held M, Wubbolts MG, Witholt B, Schmid A. Biotechnol Bioeng 2002;80:33–41.
248. Panke S, Wubbolts MG, Schmid A, Witholt B. Biotechnol Bioeng 2000;69:91–98.
249. Schmid A, Hofstetter K, Feiten H-J, Hollman F, Witholt B. Adv Synth Catal 2001;343:732–737.
250. Park J-B, Bühler B, Habicher T, Hauer B, Panke S, Witholt B, Schmid A. Biotechnol Bioeng 2006;95:501–512.
251. Kuhn D, Kholiq MA, Heinzle E, Bühler B, Schmid A. Green Chem 2010;12:815–827.
252. Di Gennaro P, Colmegna A, Galli E, Sello G, Pelizzoni F, Bestetti G. Appl Environ Microbiol 1999;65:2794–2797.
253. Sello G, Orsini F, Bernasconi S, Di Gennaro P. Tetrahedron: Asymmetry 2006;17:372–376.
254. Bernasconi S, Orsini F, Sello G, Di Gennaro P. Tetrahedron: Asymmetry 2004;15:1603–1606.
255. Panke S, De Lorenzo V, Kaiser A, Witholt B, Wubbolts MG. Appl Environ Microbiol 1999;65:5319–5623.
256. Lin H, Liu Y, Wu Z-L. Chem Commun 2011;47:2610–2612.
257. Lin H, Qiao J, Liu Y, Wu Z-L. J Mol Catal B: Enzym 2010;67:236–241.
258. Tischler D, Eulberg D, Lakner S, Kaschabek SR, van Berkel WJH, Schlömann M. J Bacteriol 2009;191:4996–5009.
259. Tischler D, Kermer R, Gröning JAD, Kaschabek SR, van Berkel WJH, Schlömann M. J Bacteriol 2010;192:5220–5227.
260. Lin H, Liu Y, Wu Z-L. Tetrahedron: Asymmetry 2011;22:134–137.
261. Colonna S, Gaggero N, Carrea G, Ottolina G, Pasta P, Zambianchi F. Tetrahedron Lett 2002;43:1797–1799.
262. Bugg TDH. Tetrahedron 2003;59:7075–7101.
263. Costas M, Mehn MP, Jensen MP, Que LJ. Chem Rev 2004;104:939–986.
264. Motherwell WB, Williams AS. Angew Chem Int Ed 1995;34:2031–2033.
265. Gibson DT, Koch JR, Kallio RE. Biochemistry 1968;7:2653–2662.
266. Gibson DT, Koch JR, Schuld CL, Kallio RE. Biochemistry 1968;7:3795–3802.
267. Gibson DT, Cardini GE, Maseles FC, Kallio RE. Biochemistry 1970;9:1631–1635.
268. Jeffrey AM, Yeh HJ, Jerina DM, Patel TR, Davey JF, Gibson DT. Biochemistry 1975;14:575–584.
269. Gibson DT, Subramanian V. In: Gibson DT, editor. *Microbial Degradation of Organic Compounds*. New York: Marcel Dekker; 1984. pp. 181–252.
270. Reiner AM. J Bacteriol 1971;108:89–94.
271. Reiner AM, Hegeman GD. Biochemistry 1971;10:2530–2536.

272. Lessner DJ, Johnson GR, Parales RE, Spain JC, Gibson DT. Appl Environ Microbiol 2002;68:634–641.

273. Friemann R, Ivkovic-Jensen MM, Lessner DJ, Yu C.-L, Gibson DT, Parales RE, Eklund H, Ramaswamy S. J Mol Biol 2005;348:1139–1151.

274. Gibson DT, Hensley M, Yoshioka H, Mabry TJ. Biochemistry 1970;9:1626–1630.

275. Parales RE, Lee K, Resnick SM, Jiang H, Lessner DJ, Gibson DT. J Bacteriol 2000;182:1641–1649.

276. Newman LM, Garcia H, Hudlicky T, Selifonov SA. Tetrahedron 2004;60:729–734.

277. Parales RE, Resnick SM. In: Ramos J-L, Levesque RC, editors. *Pseudomonas*. Vol. 4. The Netherlands: Springer; 2006. pp. 287–340.

278. Kweon O, Kim S-J, Baek S, Chae J-C, Adjei MD, Baek D-H, Kim Y-C, Cerniglia CE. In: BMC Biochemistry. Volume 9. BioMed Central Ltd, 2008. DOI:10.1186/1471-2091-9-11.

279. Johnson RA, in Organic Reactions. Vol. 63. New York: John Wiley & Sons; 2004. pp. 117–264.

280. Zylstra GJ, Gibson DT. J Biol Chem 1989;264:14940–14946.

281. Misawa N, Shindo K, Takahashi H, Suenaga H, Iguchi K, Okazaki H, Harayama S, Fukawa K. Tetrahedron 2002;58:9605–9612.

282. Keenan BG, Leungsakul T, Smets BF, Wood TK. Appl Environ Microbiol 2004;70:3222–3231.

283. Parales RE, Resnick SM, Yu CL, Boyd DR, Sharma ND, Gibson DT. J Bacteriol 2000;182:5495–5504.

284. Yu C.-L, Parales RE, Gibson DT. J Ind Microbiol Biotechnol 2001;27:94–103.

285. Boyd DR, Sheldrake GN. Nat Prod Rep 1998;15:309–324.

286. Hudlicky T, Gonzalez D, Gibson DT. Aldrichimica Acta 1999;32:35–62.

287. Lange CC, Wackett LP. J Bacteriol 1997;179:3858–3865.

288. Resnick SM, Lee K, Gibson DT. J Ind Microbiol Biotechnol 1996;17:438–457.

289. Jenkins GN, Ribbons DW, Widdowson DA, Slawin AMZ, Williams DJ. J Chem Soc, Perkin Trans 1 1995;20:2647–2655.

290. Myers AG, Siegel DR, Buzard DJ, Charest MG. Org Lett 2001;3:2923–2926.

291. Gibson DT, Parales RE. Curr Opin Biotechnol 2000;11:236–243.

292. Widdowson DA, Ribbons DW. Janssen Chim Acta 1990;8:3–9.

293. Carless HJ. Tetrahedron: Asymmetry 1992;3:795–826.

294. Sheldrake GN. In: Collins AN, Sheldrake GN, Crosby J, editors. *Chirality in Industry*. Chichester: John Wiley & Sons; 1992. pp. 127–166.

295. Brown SM, Hudlicky T. In: Hudlicky T, editor. *Organic Synthesis: Theory and Applications*. Vol. 2. Greenwich: JAI Press, Inc.; 1993. pp. 113–176.

296. Boyd DR, Sharma ND, Allen CCR. Curr Opin Biotechnol 2001;12:564–573.

297. Hudlicky T, Reed JW. In: Hassner A, editor. *Advances in Asymmetric Synthesis*. Vol. 1. Greenwich: JAI Press; 1995. pp. 271–312.

298. Boyd DR, Clarke D, Cleij MC, Hamilton JTG, Sheldrake GN. Monatshefte für Chemie/Chemical Monthly 2000;131:673–685.

299. Boyd DR, Sharma ND, Coen GP, Hempenstall F, Ljubez V, Malone JF, Allen CCR, Hamilton JTG. Org Biomol Chem 2008;6:3957–3966.

300. Boyd DR, Bugg TDH. Org Biomol Chem 2006;4:181–192.
301. Boyd DR, Sharma ND, Bowers NI, Dalton H, Garrett MD, Harrison JS, Sheldrake GN. Org Biomol Chem 2006;4:3343–3349.
302. Boyd DR, Sharma ND, Byrne BE, Haughey SA, Kennedy MA, Allen CCR. Org Biomol Chem 2004;2:2530–2537.
303. Sun S-Y, Zhang X, Zhou Q, Chen J-C, Chen G-Q. Appl Microbiol Biotechnol 2008;80:977–984.
304. Boyd DR, Sharma ND, Hand MV, Groocock MR, Kerley NA, Dalton H, Chima J, Sheldrake GN. J Chem Soc Chem Commun 1993;11:974–976.
305. Boyd DR, Sharma ND, Hempenstall F, Kennedy MA, Malone JF, Resnick SM, Gibson DT. J Org Chem 1999;64:4005–4011.
306. Boyd DR, Sharma ND, Agarwal R, Resnick SM, Schocken MJ, Gibson DT, Sayer JM, Yagi H, Jerina DM. J Chem Soc Perkin Trans 1 1997;11:1715–1724.
307. Boyd DR, Sharma ND, Harrison JS, Kennedy MA, Allen CCR, Gibson DT. J Chem Soc Perkin Trans 1 2001;11:1264–1269.
308. Jouanneau Y, Meyer C, Jakoncic J, Stojanoff V, Gaillard J. Biochemistry 2006;45:12380–12391.
309. Kwit M, Gawronski J, Boyd DR, Sharma ND, Kaik M, More O'Ferrall RA, Kudavalli JS. Chem Eur J 2008;14:11500–11511.
310. Bestetti G, Bianchi G, Bosetti A, Di Gennaro P, Galli E, Leoni B, Pelizzoni F, Sello G. Appl Microbiol Biotechnol 1995;44:306–313.
311. Hudlicky T, Endoma MA, Butora G. Tetrahedron: Asymmetry 1996;7:61–68.
312. Boyd DR, Sharma ND, Boyle R, Malone JF, Chima J, Dalton H. Tetrahedron: Asymmetry 1993;4:1307–1324.
313. Boyd DR, Sharma ND, Kennedy MA, Shepherd S, Malone JF, Alves-Areias A, Holt R, Allenmark SG, Lemurell MA, Dalton H, Luckarift H. Chem Commun 2002;14:1452–1453.
314. Boyd DR, Sharma ND, Boyle R, McMurray BT, Evans TA, Malone JF, Dalton H, Chima J, Sheldrake GN. J Chem Soc, Chem Commun 1993;(1):49–51.
315. Eaton RW, Nitterauer JD. J Bacteriol 1994;176:3992–4002.
316. Boyd DR, Sharma ND, Brannigan IN, Clarke DA, Dalton H, Haughey SA, Malone JF. Chem Commun 1996;20:2361–2362.
317. Boyd DR, Sharma ND, Gunaratne N, Haughey SA, Kennedy MA, Malone JF, Allen CCR, Dalton H. Org Biomol Chem 2003;1:984–994.
318. Ensley BD, Ratzkin BJ, Osslund TD, Simon MJ, Wackett LP, Gibson DT. Science 1983;222:167–169.
319. Garrett MD, Scott R, Sheldrake GN. Tetrahedron: Asymmetry 2002;13:2201–2204.
320. Garrett MD, Scott R, Sheldrake GN, Dalton H, Goode P. Org Biomol Chem 2006;4:2710–2715.
321. Sakamoto T, Joern JM, Arisawa A, Arnold FH. Appl Environ Microbiol 2001;67:3882–3887.
322. Modyanova L, Azerad R. Tetrahedron Lett 2000;41:3865–3869.
323. Boyd DR, Sharma ND, Modyanova LV, Carroll JG, Malone JF, Allen CCR, Hamilton JTG, Gibson DT, Parales RE, Dalton H. Can J Chem 2002;80:589–600.

324. Chopard C, Azerad R, Prange T. J Mol Catal B: Enzym 2008;50:53–60.

325. Boyd DR, Sharma ND, Carroll JG, Malone JF, Mackerracher DG, Allen CCR. Chem Commun 1998;6:683–684.

326. Boyd DR, McMordie RAS, Porter HP, Dalton H, Jenkins RO, Howarth OW. Chem Commun 1987;(22):1722–1724.

327. Boyd DR, Sharma ND, Dorrity MRJ, Hand MV, McMordie RAS, Malone JF, Porter HP, Chima J, Dalton H, Sheldrake GN. J Chem Soc, Perkin Trans 1 1993;(9):1065–1071.

328. Fetzner S, Vogler B, Lingens F. FEMS Microbiol Lett 1993;112:151–157.

329. Allen CCR, Boyd DR, Dalton H, Sharma ND, Brannigan I, Kerley NA, Sheldrake GN, Taylor SC. J Chem Soc, Chem Commun 1995;(2):117–118.

330. Boyd DR, Sharma ND, Bart SA, Dalton H, Chima J, Whited G, Seemeyer R. J Am Chem Soc 1994;116:1147–1148.

331. Boyd DR, Austin R, McMordie S, Sharma ND, Dalton H, Williams P, Jenkins RO. J Chem Soc, Chem Commun 1989;(6):339–340.

332. Wackett LP, Kwart LD, Gibson DT. Biochemistry 1988;27:1360–1367.

333. Yildirim S, Franko TT, Wohlgemuth R, Kohler HPE, Witholt B, Schmid A. Adv Synth Catal 2005;347:1060–1072.

334. Hudlicky T. Akgün H. Tetrahedron Lett 1999;40:3081–3084.

335. Johnson CR, Adams JP, Collins MA, J Chem Soc Perkin Trans 1 1993;1:1–2.

336. Boyd DR, Sharma ND, Byrne B, Hand MV, Malone JF, Sheldrake GN, Blacker J, Dalton H, J Chem Soc, Perkin Trans 1 1998;1935–1944.

337. Boyd DR, Dorrity MRJ, Hand MV, Malone JF, Sharma ND, Dalton H, Gray DJ, Sheldrake GN. J Am Chem Soc 1991;113:666–667.

338. Resnick SM, Torok DS, Gibson DT. J Org Chem 1995;60:3546–3549.

339. Boyd DR, Sharma ND, Bowers NI, Duffy J, Harrison JS, Dalton H J Chem Soc, Perkin Trans 1 2000;(9):1345–1350.

340. Boyd DR, Sharma ND, Ljubez V, Byrne BE, Shepherd SD, Allen CCR, Kulakov LA, Larkin ML, Dalton H. Chem Commun 2002;17:1914–1915.

341. Boyd DR, Sharma ND, Haughey SA, Malone JF, King A, McMurray BT, Holt R, Dalton H. J Chem Soc Perkin Trans 1 2001;24:3288–3296.

342. Finn KJ, Cankar P, Jones TRB, Hudlicky T. Tetrahedron: Asymmetry 2004;15:2833–2836.

343. Boyd DR, Sharma ND, Brannigan IN, Groocock MR, Malone JF, McConville G, Allen CCR. Adv Synth Catal 2005;347:1081–1089.

344. Hudlicky T, Reed JW. Synlett 2009;5:685–703.

345. Sinisterra J, Dalton H. Influence of the immobilization methodology in the stability and activity of *P. putida* UV4 immobilized whole cells. In: Wijffels R, Buitelaar R, Bucke C, Tramper J, editors. Immobilized Cells: Basics & Applications. Amsterdam: Elsevier Science; 1996. pp. 416–423.

346. Hudlicky T, Stabile MR, Gibson DT, Whited G. Org Synth 1999;76:77–85.

347. Yildirim S, Zezula J, Hudlicky T, Witholt B, Schmid A. Adv Synth Catal 2004;346:933–942.

348. Endoma MA, Bui VP, Hansen J, Hudlicky T. Org Proc Res Dev 2002;6:525–532.

349. Boyd DR, Sharma ND, Llamas NM, O'Dowd CR, Allen CCR. Org Biomol Chem 2007;5:2267–2273.
350. Boyd DR, Sharma ND, Dalton H, Clarke DA. Chem Commun 1996;1:45–46.
351. McKibben BP, Barnosky GS, Hudlicky T. Synlett 1995;8:806–808.
352. Boyd DR, Sharma ND, Llamas NM, O'Dowd CR, Allen CCR. Org Biomol Chem 2006;4:2208–2217.
353. Ballard DGH, Courtis A, Shirley IM, Taylor SC. J Chem Soc Chem Commun 1983;17:954–955.
354. Ley SV, Sternfeld F, Taylor SC. Tetrahedron Lett 1987;28:225–226.
355. Hudlicky T, Luna H, Barbieri G, Kwart LD. J Am Chem Soc 1988;110:4735–4741.
356. Hudlicky T, Seoane G, Pettus T. J Org Chem 1989;54:4239–4243.
357. Hudlicky T, Price JD, Rulin F, Tsunoda T. J Am Chem Soc 1990;112:9439–9440.
358. Hudlicky T. Chem Rev 1996;96:3–30.
359. Hudlicky T, Entwistle DA, Pitzer KK, Thorpe AJ. Chem Rev 1996;96:1195–1220.
360. Ley SV, Redgrave AJ. Synlett 1990;7:393–394.
361. Hudlicky T, Mandel M, Rouden J, Lee RS, Bachmann B, Dudding T, Yost K, Merola JS. J Chem Soc Perkin Trans 1 1994;12:1553–1567.
362. Nguyen BV, York C, Hudlicky T. Tetrahedron 1997;53:8807–8814.
363. Brammer LE, Hudlicky T. Tetrahedron: Asymmetry 1998;9:2011–2014.
364. Vitelio C, Bellomo A, Brovetto M, Seoane G, Gonzalez D. Carbohydr Res 2004;339:1773–1778.
365. Bellomo A, Giacomini C, Brena B, Seoane G, Gonzalez D. Synth Commun 2007;37:3509–3518.
366. Hudlicky T, Abboud KA, Bolonick J, Maurya R, Stanton ML, Thorpe AJ. Chem Commun 1996;15:1717–1718.
367. Paul BJ, Martinot TA, Willis J, Hudlicky T. Synthesis 2001;6:952–956.
368. Paul BJ, Willis J, Martinot TA, Ghiviriga I, Hudlicky T. J Am Chem Soc 2002;124:10416–10426.
369. Ley SV. Pure Appl Chem 1990;62:2031–2034.
370. Ley SV, Parra M, Redgrave AJ, Sternfeld F. Tetrahedron 1990;46:4995–5026.
371. Ley SV, Sternfeld F. Tetrahedron Lett 1988;29:5305–5308.
372. Hudlicky T, Natchus MG. J Org Chem 1992;57:4740–4746.
373. Banwell MG, Edwards AJ, Harfoot GJ, Jolliffe KA, J Chem Soc Perkin Trans 1 2002;22:2439–2441.
374. Banwell MG, Edwards AJ, Harfoot GJ, Jolliffe KA. Tetrahedron 2004;60:535–547.
375. Banwell MG, Austin KAB, Willis AC. Tetrahedron 2007;63:6388–6403.
376. Banwell MG, McLeod MD, Riches AG. Aust J Chem 2004;57:53–66.
377. Dietinger CE, Banwell MG, Garson MJ, Willis AC. Tetrahedron 2010;66:5250–5261.
378. Omori AT, Finn KJ, Leisch H, Carroll RJ, Hudlicky T. Synlett 2007;18:2859–2862.
379. Kokas OJ, Banwell MG, Willis AC. Tetrahedron 2008;64:6444–6451.
380. Hudlicky T, Olivo HF. J Am Chem Soc 1992;114:9694–9696.
381. Tian X, Hudlicky T, Königsberger K. J Am Chem Soc 1995;117:3643–3644.
382. Gonzalez D, Martinot T, Hudlicky T. Tetrahedron Lett 1999;40:3077–3080.

383. Akgün H, Hudlicky T. Tetrahedron Lett 1999;40:3081–3084.
384. Matveenko M, Banwell MG, Willis AC. Tetrahedron 2008;64:4817–4826.
385. Matveenko M, Kokas OJ, Banwell MG, Willis AC. Org Lett 2007;9:3683–3685.
386. Banwell MG, Forman GS. J Chem Soc Perkin Trans 1 1996;21:2565–2566.
387. Banwell MG, Loong DTJ, Willis AC. Aust J Chem 2005;58:511–516.
388. Boyd DR, Sharma ND, Llamas NM, Malone JF, O'Dowd CR, Allen CCR. Org Biomol Chem 2005;3:1953–1963.
389. Hudlicky T, Rouden J, Luna H, Allen S. J Am Chem Soc 1994;116:5099–5107.
390. Hudlicky T, Luna H, Price JD, Rulin F. J Org Chem 1990;55:4683–4687.
391. Hudlicky T, Seoane G, Price JD, Gadamasetti K. Synlett 1990;8:433–440.
392. Hudlicky T, Nugent T, Griffith W. J Org Chem 1994;59:7944–7946.
393. Nugent TC, Hudlicky T. J Org Chem 1998;63:510–520.
394. Bui VP, Hudlicky T. Tetrahedron 2004;60:641–646.
395. Boyd DR, Sharma ND, Sbircea L, Murphy D, Belhocine T, Malone JF, James SL, Allen CCR, Hamilton JTG. Chem Commun 2008;43:5535–5537.
396. Boyd DR, Sharma ND, Sbircea L, Murphy D, Malone JF, James SL, Allen CC, Hamilton JT. Org Biomol Chem 2010;8:1081–1090.
397. Sbircea L, Sharma ND, Clegg W, Harrington RW, Horton PN, Hursthouse MB, Apperley DC, Boyd DR, James SL. Chem Commun 2008;43:5538–5540.
398. Ley SV, Yeung LL. Synlett 1992;(4):291–292.
399. Entwistle DA, Hudlicky T. Tetrahedron Lett 1995;36:2591–2594.
400. Duchek J, Adams D, Hudlicky T. Chem Rev 2011;111:4223–4258.
401. Rinner U, Hudlicky T, Gordon H, Petit GR. Angew Chem Int Ed 2004;43:5342–5346.
402. Banwell MG, McRae KJ. J Org Chem 2001;66:6768–6774.
403. Labora M, Pandolfi EM, Schapiro V. Tetrahedron: Asymmetry 2010;21:153–155.
404. Gilmet J, Sullivan B, Hudlicky T. Tetrahedron 2009;65:212–220.
405. Humphreys JL, Lowes DJ, Wesson KA, Whitehead RC. Tetrahedron Lett 2004;45:3429–3432.
406. White LV, Dietinger CE, Pinkerton DM, Willis AC, Banwell MG. Eur J Org Chem 2010;(23):4365–4367.
407. Banwell MG, Bray AM, Edward AJ, Wong DJ. New J Chem 2001;25:1347–1350.
408. Leisch H, Omori AT, Finn KJ, Gilmet J, Bissett T, Ilceski D, Hudlicky T. Tetrahedron 2009;65:9862–9875.
409. Sullivan B, Carrera I, Drouin M, Hudlicky T. Angew Chem Int Ed 2009;48:4229–4231.
410. Werner L, Machara A, Hudlicky T. Adv Synth Catal 2010;352:195–200.
411. Hudlicky T, Rinner U, Gonzalez D, Akgun H, Schilling S, Siengalewicz P, Martinot TA, Petit GR. J Org Chem 2002;67:8726–8743.
412. Banwell MG, Coster MJ, Karunaratne OP, Smith JA. J Chem Soc Perkin Trans 1 2002;14:1622–1624.
413. Boyd DR, Sharma ND, O'Dowd C, Carroll JG, Loke PL, Allen CCR. J Chem Soc Chem Commun 2005;31:3989–3991.
414. Parker MH, Maryanoff BE, Reitz AB. Synlett 2004;12:2095–2098.
415. Fischer T, Leisch H, Mihovilovic M. Monatsh Chem/Chem Monthly 2010;141:699–707.

416. Adams DR, Aichinger C, Rinner U, Hudlicky T. Synlett 2011;5:725–729.
417. Charest MG, Siegel DR, Myers AG. J Am Chem Soc 2005;127:8292–8293.
418. Bellomo A, Gonzalez D, Stefani HA. J Organomet Chem 2008;693:1136–1142.
419. Heguaburu V, Mandolesi SM, Schapiro V, Pandolfi E. Tetrahedron Lett 2008;49:6787–6790.
420. Berry A, Dodge TC, Pepsin M, Weyler W. J Ind Microbiol Biotechnol 2002;28:127–133.
421. Furukawa K. Curr Opin Biotechnol 2000;11:244–249.
422. Kulkarni M, Chaudhari A. J Environ Manage 2007;85:496–512.
423. Hopkins GD, McCarty PL. Environ Sci Technol 1995;29:1628–1637.
424. Ang EL, Obbard JP, Zhao H. Appl Microbiol Biotechnol 2009;81:1063–1070.
425. Gibson DT, Cruden DL, Haddock JD, Zylstra GJ, Brand JM. J Bacteriol 1993;175:4561–4564.
426. Kimura N, Nishi A, Goto M, Furukawa K. J Bacteriol 1997;179:3936–3943.
427. Kumamaru T, Suenaga H, Mitsuoka M, Watanabe T, Furukawa K. Nat Biotechnol 1998;16:663–666.
428. Suenaga H, Goto M, Furukawa K. J Biol Chem 2001;276:22500–22506.
429. Suenaga H, Mitsuoka M, Ura Y, Watanabe T, Furukawa K. J Bacteriol 2001;183:5441–5444.
430. Wackett LP, Sadowsky MJ, Newman LM, Hur H.-G, Li S. Nature 1994;368:627–629.
431. Lange CC, Wackett LP, Minton KW, Daly MJ. Nat Biotechnol 1998;16:929–933.
432. Boyd DR, Sharma ND, Haughey S, Kennedy MA, Malone JF, Shepherd S. Dalton CCRAH. Tetrahedron 2004;60:549–559.
433. Boyd DR, Sharma ND, Bowers NI, Boyle R, Harrison JS, Lee K, Bugg TDH, Gibson DT. Org Biomol Chem 2003;1:1298–1307.
434. Boyd DR, Sharma ND, Kerley NA, McMordie RAS, Sheldrake GN, Williams P, Dalton H J Chem Soc, Perkin Trans 1 1996;(1):67–74.
435. Boyd DR, Sharma ND, Haughey SA, Kennedy MA, McMurray BT, Sheldrake GN, Allen CCR, Dalton H, Sproule K. J Chem Soc, Perkin Trans 1 1998;(12):1929–1933.
436. Cashman JR, Olsen LD, Boyd DR, McMordie RAS, Dunlop R, Dalton H. J Am Chem Soc 1992;114:8772–8777.
437. Lee K, Brand JM, Gibson DT. Biochem Biophys Res Commun 1995;212:9–15.
438. Boyd DR, Sharma ND, Haughey SA, Malone JF, McMurray BT, Sheldrake GN, Allen CCR, Dalton H. Chem Commun 1996;20:2363–2364.
439. Boyd DR, Sharma ND, King AWT, Shepherd SD, Allen CCR, Holt RA, Luckarift HR, Dalton H. Org Biomol Chem 2004;2:554–561.
440. Brand JM, Cruden DL, Zylstra GJ, Gibson DT. Appl Environ Microbiol 1992;58:3407–3409.
441. Lakshman MK, Chaturvedi S, Zajc B, Gibson DT, Resnick SM, Synthesis 1998;1352–1356.
442. Bowers NI, Boyd DR, Sharma ND, Goodrich PA, Groocock MR, Blacker AJ, Goode P, Dalton H. J Chem Soc, Perkin Trans 1 1999;(11):1453–1461.
443. Boyd DR, Sharma ND, Bowers NI, Goodrich PA, Groocock MR, Blacker AJ, Clarke DA, Howard T, Dalton H. Tetrahedron: Asymmetry 1996;7:1559–1562.

444. Bui V, Hansen TV, Stenstrøm Y, Ribbons DW, Hudlicky T. J Chem Soc, Perkin Trans 1 2000;11:1669-1672.

445. Bui VP, Hansen TV, Stenstrøm Y, Hudlicky T, Ribbons DW. New J Chem 2001;25:116–124.

446. Hudlicky T, Boros EE, Boros CH. Tetrahedron: Asymmetry 1993;4:1365–1386.

447. Konigsberger K, Hudlicky T. Tetrahedron: Asymmetry 1993;4:2469–2474.

448. Gibson DT, Resnick SM, Lee K, Brand JM, Torok DS, Wackett LP, Schocken MJ, Haigler BE. J Bacteriol 1995;177:2615–2621.

449. Torok DS, Resnick SM, Brand JM, Cruden DL, Gibson DT. J Bacteriol 1995;10:5799–5805.

450. Resnick SM, Gibson DT. Appl Environ Microbiol 1996;62:1364–1368.

451. Bowers NI, Boyd DR, Sharma ND, Kennedy MA, Sheldrake GN, Dalton H. Tetrahedron: Asymmetry 1998;9:1831–1834.

452. Geary PJ, Pryce RJ, Roberts SM, Ryback G, Winders JA. J Chem Soc, Chem Commun 1990;204–205.

453. Ziffer H, Kabuto K, Gibson DT, Kobal VM, Jerina DM. Tetrahedron 1977;33:2491–2496.

454. Andre E, Hou K. Comptes rendus de l'academie de sciences 1932;194:645–647.

455. Veldink GA, Vliegenthart JFG. In: Atta-ur-Rahman, editor. Studies in Natural Products Chemistry. Structure and Chemistry (part B). Vol. 9. Amsterdam: Elsevier Science Publishers B.V.; 1991. pp. 559–589.

456. Brash AR. J Biol Chem 1999;274:23679–23682.

457. Porta H, Rocha-Sosa M. Microbiology 2001;147:3199–3200.

458. Axelrod B, Cheesbrough TH, Laakso S. Methods Enzymol 1981;71:441–451.

459. Funk CD. Biochim Biophys Acta 1996;1304:65–84.

460. Grechkin A. Prog Lipid Res 1998;37:317–352.

461. Nanda S, Yadav JS. J Mol Catal B: Enzym 2003;26:3–28.

462. Gardner HW, Biochim Biophys Acta 1989;1001:274.

463. Schneider C, Schreier P. Humpf H.-U. Chirality 1997;9:563–567.

464. Datcheva VK, Kiss K, Solomon L, Kyler KS. J Am Chem Soc 1991;113:270–274.

465. Roth M, Gutsche B, Herderich M, Humpf H.-U, Schreir P. J Agric Food Chem 1998;46:2951–2956.

466. Clapp CH, Senchak SE, Stover TJ, Potter TC, Findeis PM, Novak MJ. J Am Chem Soc 2001;123:747–748.

467. Clapp CH, Strulson M, Rodriguez PC, Lo R, Novak MJ. Biochemistry 2006;45:15884–15892.

468. Wiseman JS, Nichols JS. Biochem Biophys Res Commun 1988;154:544–549.

469. Hatanaka A. Phytochemistry 1993;34:1201–1218.

470. Lehman WD. Free Radical Biol Med 1994;16:241–253.

471. Waldmann D, Schreier P. J Agric Food Chem 1995;43:626–630.

472. Scettri A, Bonadies F, Lattanzi A, Palombi L, Pesci S. Tetrahedron 1997;53:17139–17150.

473. Corey EJ, Su WG. Tetrahedron Lett 1989;30:4181–4184.

474. Matsushita Y, Sugamoto K, Nakama T, Matsui T, Hayashi Y, Uenaki K. Tetrahedron Lett 1997;38:6055–6058.
475. Zhang P, Kyler KS. J Am Chem Soc 1989;111:9241–9242.
476. Martini D, Buono G, Iacazio G. J Org Chem 1996;61:9062–9064.
477. Di Marzo V, Gianfrani C, De Petrocellis L, Milone A, Cimino G. Biochem J 1994;300:501–507.
478. Hawkins DJ, Brash AR. J Biol Chem 1987;262:7629–7634.
479. Pourplanche C, Lambert C, Bejrot M, Marx J, Chopard C, Alix AJP, Larreta-Garde VJ. J Biol Chem 1994;269:31585–31591.
480. Kermasha S, Dioum N, Bisakowski B. J Mol Catal B: Enzym 2001;11:909–919.
481. Allen JC. Eur J Biochem 1968;4:201–208.
482. Graveland A. Biochem Biophys Res Commun 1970;41:427–434.
483. Battu S, H. Rabinovitch-Chable, Beneytout JL. J Agric Food Chem 1994;42:2115–2119.
484. Kermasha S, Dioum N, Bisakowski B, Vega M. J Mol Catal B: Enzym 2002;19–20:305–317.
485. Karadag H, Bilgin R, Tukel S. Biotechnol Biotechnol Equip 2006;20:95–100.
486. Pinto MC, Macias P. Appl Biochem Biotechnol 1996;59:309–318.
487. Hsu AF, Fogila TA, Piazza GJ. Biotechnol Lett 1997;19:71–74.
488. Hsu AF, Shen S, Wu E, Fogila TA. Biotechnol Appl Biochem 1998;28:55–59.
489. Vega M, Karboune S, Kermasha S. Appl Biochem Biotechnol 2005;127:29–42.
490. Boeglin WE, Kim RB, Brash AR. Proc Natl Acad Sci USA 1998;95:6744–6749.
491. Hornung E, Walther M, Kuhn H, Feussner I. Proc Natl Acad Sci USA 1999;96:4192–4197.
492. Hughes RK, Lawson DM, Hornostaj AR, Fairhurst SA, Casey R. Eur J Biochem 2001;268:1030–1040.
493. Hughes RK, West SI, Hornostaj AR, Lawson DM, Fairhurst SA, Sanchez RO, Robinson BH, Casey R. Biochem J 2001;353:345–355.

CHAPTER 5

Reactions Involving Oxidases and Peroxidases

Oxidative biocatalysis is a useful methodology for the oxidation of organic compounds, which complements the transition metal-catalyzed and organocatalyzed oxidations developed in the last decades [1–3]. One of the main advantages of the biocatalytic methods over the chemocatalytic ones is the selectivity and predictability that can be achieved in the oxidation. Another important asset is the ample range of enzymes capable of performing oxidations, ranging from dehydrogenases (see Chapter 3), which catalyze hydrogen removal without the participation of active oxygen intermediates, to those directly reacting with oxygen (either as molecular oxygen or peroxide) to form reactive oxygen intermediates that react further with the substrate [4–7]. Formally, the second group of enzymes is more appealing, since the oxidation can be performed employing the inexpensive and nontoxic molecular oxygen (or hydrogen peroxide) as the oxidant, and sometimes without using expensive cofactors. This group comprises oxygenases (mono- and dioxygenases), oxidases, and peroxidases. Oxygenases, which catalyze the introduction of one or two atoms of molecular oxygen into the substrate, were covered in Chapter 4. Oxidases use oxygen as terminal electron acceptor, without incorporating it into the substrate, and will be described in this chapter, together with peroxidases, which use hydrogen peroxide as oxidant.

Oxidase- and peroxidase-catalyzed reactions include radical formation and coupling, oxidation of alcohols to carbonyl compounds, oxidation of amines to imines, heteroatom oxidation (e.g., sulfoxidation), epoxidation, *N*-dealkylation, active C–H oxidation (benzylic/allylic oxidation), and aromatic hydroxylation, halogenation, and nitration. Such a variety of reactions is accomplished through several mechanisms (including single- and two-electron transfer), employing different metals, such as Fe, Cu, and V; cofactors (e.g., flavin adenine dinucleotide (FAD), flavin mononucleotide (FMN)); and oxidants (oxygen and hydrogen peroxide) (see Chapter 1). As a drawback, the high versatility of oxidase-catalyzed reactions may lead to selectivity issues; for instance, the product of the reaction of the substrate with the reactive oxygen species in the active site may be a radical (without incorporation of oxygen

Redox Biocatalysis: Fundamentals and Applications, First Edition.
Daniela Gamenara, Gustavo A. Seoane, Patricia Saenz-Méndez, and Pablo Domínguez de María.

atoms), which may react further through a nonenzymatic reaction. As a result, this type of reaction can be nonspecific or can produce several different products, being less selective than those mediated by oxygenases, particularly in terms of regioselectivity. On the other hand, the apparent drawback derived from the diminished selectivity is considered an advantage for some biotechnological applications, where an ample, nonselective reactivity is actually sought (e.g., in the environmental field).

5.1 OXIDASE-CATALYZED REACTIONS

As mentioned in Chapter 1, oxidases perform the oxidation of substrates without incorporating atoms from molecular oxygen, which, as the terminal electron acceptor, is reduced to either hydrogen peroxide or water. The aerobic regeneration of the prosthetic group is less costly and simpler than oxidations performed with coenzyme-dependent dehydrogenases. However, this direct electron transfer to oxygen generally results in the formation of hydrogen peroxide, which is highly reactive and rapidly inactivates enzymes unless it is efficiently and rapidly removed from the reaction system. Catalase is generally used to accomplish this task.

Following the four figures code proposed by the Nomenclature Committee of the International Union of Biochemistry and Molecular Biology (NC-IUBMB) [8], the third figure (which usually indicates the type of acceptor involved) corresponds for molecular oxygen to a "3" and thus, oxidases are classified as EC 1.-.3.-. Within this subclass, laccases and alcohol-, sugar-, amine-, amino acid, and phenol oxidases have found significant synthetic or analytical applications. Regarding the wide range of reactions covered and the ample structural diversity of substrates tolerated by these enzymes, the ensuing sections will deal with oxidases individually and not as a group. However, for a more organized discussion they were included in four groups according to their synthetic relevance and type of bond they oxidize.

5.1.1 Oxidases Acting on C–O Bonds

5.1.1.1 Galactose Oxidase Galactose oxidase (GAO, EC 1.1.3.9) belongs to the family of extracellular monocopper oxidases of fungal origin, first described at the end of the 1950s from *Fusarium* spp. [9–11]. To improve the production of enzyme the corresponding gene has been isolated and expressed in *Aspergillus nidulans* [12], *Pichia pastoris* [13], and *Escherichia coli* [14]. The enzyme selectively oxidizes primary alcohols to aldehydes *via* two-electron oxidation, coupled to the reduction of molecular oxygen to hydrogen peroxide. The active site of GAO contains a tyrosine residue and a copper atom, and in its active form the tyrosine is in the radical form and the copper atom as Cu(II) [15,16]. The tyrosyl radical acts as a second redox center during the catalytic cycle, explaining the two-electron redox chemistry with a single cooper atom. In nature, GAO oxidizes the primary alcohol of D-galactose in the presence of molecular oxygen to form D-galactohexodialdose and hydrogen peroxide (Figure 5.1). High concentrations of hydrogen peroxide inactivate GAO, and thus for biocatalytic applications catalase is usually added to decompose it, enhancing the action of GAO [9]. Likewise, the presence of a peroxidase in the reaction medium,

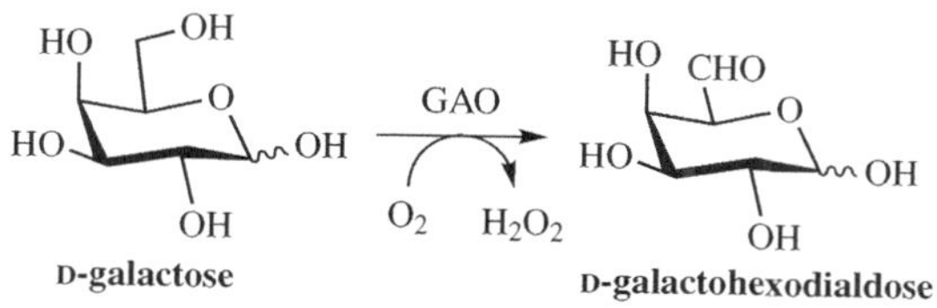

FIGURE 5.1 GAO-catalyzed oxidation of galactose.

such as horseradish peroxidase (HRP), is also known to stimulate the activity of GAO [17].

The selectivity is high for the C-6 hydroxyl of galactose or galactose-containing saccharides, as confirmed by the reported oxidation of galactosides, melibiose, raffinose, lactose, and lactitol and the nonacceptance of glucose [18]. The oxidation can also be conducted on easily accessible galactopyranosyl residues of polysaccharides. In this way, pectic polysaccharides were labeled through the combined use of GAO and tritiated potassium borohydride [19]. Moreover, the use of a multienzyme system with catalase and HRP to enhance the action of GAO, allowed the oxidation of the C-6 hydroxyl groups of terminal galactose units in a series of polysaccharides including spruce galactoglucomannan, guar galactomannan, larch arabinogalactan, corn fiber arabinoxylan, and tamarind seed xyloglucan [20]. Being a dialdehyde, oxidized galactose (galactohexodialdose) is a potential monomer for polymerizations. This has been used in a chemoenzymatic preparation of sugar-based polymers comprising poorly hydrolyzable linkages [21]. The strategy involved the GAO-catalyzed oxidation of galactose, condensation with alkyl diamines, and further reduction of the resulting polyimines with sodium cyanoborohydride to afford open-form AA-BB polymers. Moreover, since β-D-galactosamine is also substrate for GAO, its oxidation and polymerization provides, after chemical reduction, a poly(galactose amine) material that does not contain a hydrocarbon linker, as an example of an A-B polymer (Figure 5.2).

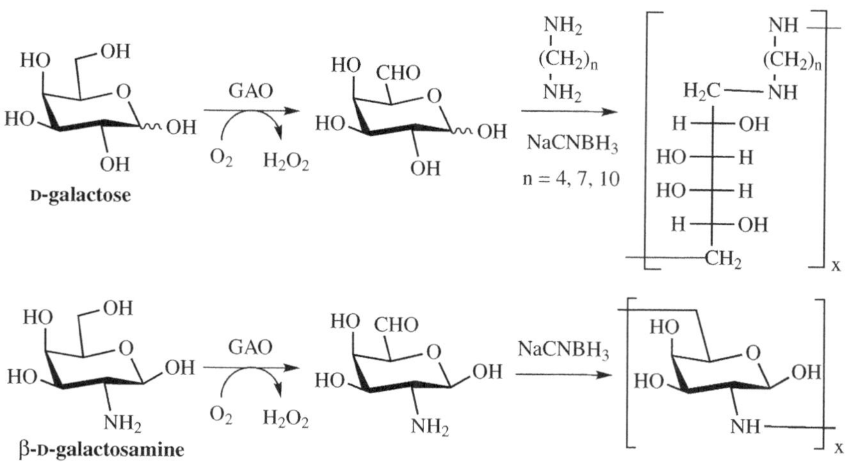

FIGURE 5.2 Sugar-based polymers using GAO catalysis.

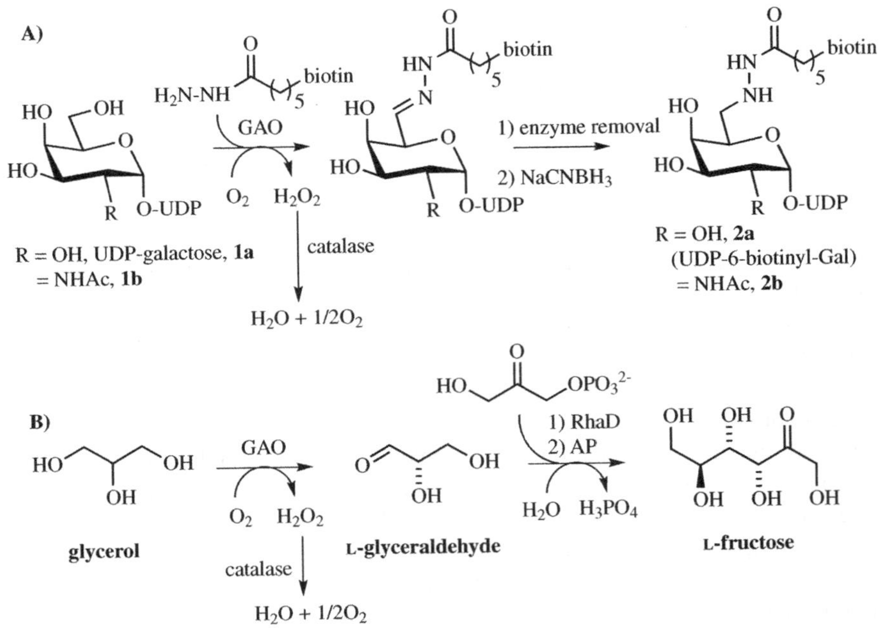

FIGURE 5.3 **A:** Chemoenzymatic synthesis of the biotinylated nucleotide sugars **2a,b**. **B:** One-pot synthesis of L-fructose using a four-enzyme system. RhaD, rhamnulose-1-phosphate aldolase; AP, acid phosphatase.

This strategy for the preparation of sugar-containing polymers complements those using hydrolases (lipases, esterases, and proteases), which give polymers with easily hydrolyzable glycoside or ester/amide linkages, limiting the ruggedness of these materials. Another application for the dialdehydes obtained from C-6 OH oxidation of galactose and lactose, consists of their use as potential protein cross-linkers, as a food-grade alternative to glutaraldehyde [22,23].

Galactose oxidase displays relatively narrow substrate specificity. Even though non-D-galactose-based carbohydrates are not substrates for GAO, other shorter polyol chains, which can be twisted to form a conformation similar to the D-galactose C4–C6 fragment, are oxidized by GAO at different rates [18,24]. Examples are glycerol, salicyl alcohol, xylitol, 3-halogenopropane-1,2-diol, and polyols, which provide the corresponding hydroxylated aldehydes [18,25–27]. An interesting synthetic application of GAO consists of its use in chemoenzymatic one-pot cascade reactions without intermediate recovery steps, such as the preparation of galacturonates [28], 4-deoxy-D-glucosides [29], L-fructose [25], and of biotinylated nucleotide sugars (**2a,b**, Figure 5.3A). These nucleotide derivatives, prepared from UDP-galactose and UDP-*N*-acetylgalactosamine (**1a,b**), are used as labeled substrates for glycosyltransferases (Figure 5.3A) [30]. Likewise, the combination of UDP-GlcNAc-4′-epimerase (from *Campylobacter jejuni*), and GAO for the one-pot chemoenzymatic synthesis of the same biotinylated nucleotide sugars starting from the cheaper UDP-Glucose and from UDP-*N*-acetyl-glucosamine (UDP-GlcNac) was successfully reported [31].

Furthermore, for the one-pot production of L-fructose, Wong and co-workers started from glycerol and dihydroxyacetone using a coupled enzymatic system with GAO, catalase, rhamnulose-1-phosphate aldolase (RhaD) and acid phosphatase (AP). The stereoselective GAO-mediated oxidation of glycerol to L-glyceraldehyde, coupled with the aldolase reaction *in situ*, provides L-fructose in a relatively high yield (Figure 5.3B) [25]. The tandem use of GAO and glycosidases has also been described [32].

From a synthetic standpoint, the GAO-catalyzed production of C-6 oxidized galactose derivatives in aqueous solutions is favorable compared to chemical catalysis, since protective groups are not needed, and only molecular oxygen is required as an oxidant. Moreover, stoichiometric additions of cofactors are not needed, making oxidases more suitable for biocatalytic oxidations than the cofactor-dependent dehydrogenases. On the other hand, the main synthetic limitation of GAO is its reduced substrate tolerance, remaining largely limited to galactose-like substrates. To overcome this drawback, several research groups are applying genetic engineering techniques to design mutants with the desired activity toward unnatural substrates and/or improved performances toward the classical ones [33–35]. In this way, it has been possible to isolate and characterize mutants with improved kinetic performances (mainly lower K_M values) for the natural substrates [34], and with activity toward glucose and glucose derivatives [35]. In this respect, a variant GAO obtained by directed evolution (GAO $M_{3\text{–}5}$) catalyzed the enantioselective oxidation of racemic secondary alcohols, binding hydrophobic substrates. This kinetic resolution was performed in the presence of HRP, to stimulate the GAO activity [17], and afforded after 3 h the (*S*)-isomeric alcohols in excellent enantiomeric excess at a conversion of approximately 50% (Table 5.1) [33]. The same mutant was able to desymmetrize atropisomeric diaryl ethers containing a benzylic hydroxy group by performing an atroposelective oxidation of the *pro*-(*P*) primary alcohol (Figure 5.4) [36]. It has been

TABLE 5.1 GAO-Catalyzed Kinetic Resolution of Secondary Alcohols

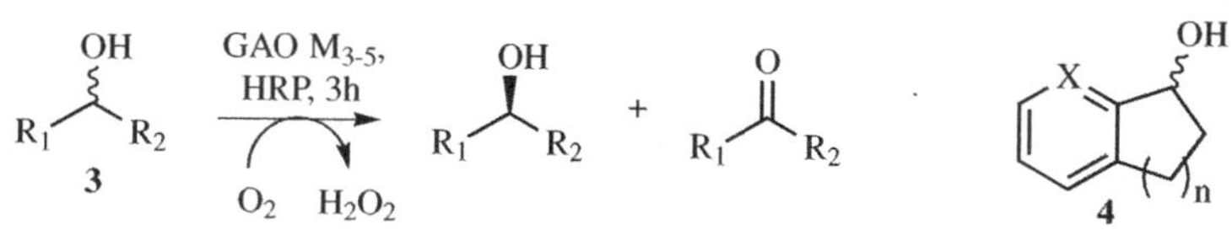

Compound	R_1	R_2	*ee* (%)	Compound	R_1	R_2	*ee* (%)
3a	Ph	Me	99	**3k**	$4\text{-}NO_2\text{-}C_6H_4$	Me	96
3b	$2\text{-}F\text{-}C_6H_4$	Me	61	**3l**	$3\text{-}NO_2\text{-}C_6H_4$	Me	99
3c	$3\text{-}F\text{-} C_6H_4$	Me	99	**3m**	Ph	$CH{=}CH_2$	45
3d	$4\text{-}F\text{-}C_6H_4$	Me	99	**3n**	C_6F_5	Me	65
3e	$4\text{-}Br\text{-}C_6H_4$	Me	99	**3o**	$3,4,5\text{-}MeO\text{-}C_6H_2$	Me	99
3f	$3\text{-}BrC_6H_4$	Me	99	**3p**	Ph	Ethynyl	99
3g	$4\text{-}MeO\text{-}C_6H_4$	Me	99	**3q**	2-pyridyl	Me	67
3h	$3\text{-}MeO\text{-}C_6H_4$	Me	99	**4a**	X=H, $n = 1$		50
3i	$4\text{-}Cl\text{-}C_6H_4$	Me	99	**4b**	X=H, $n = 2$		15
3j	$4\text{-}Me\text{-}C_6H_4$	Me	99	**4c**	X=N, $n = 2$		45

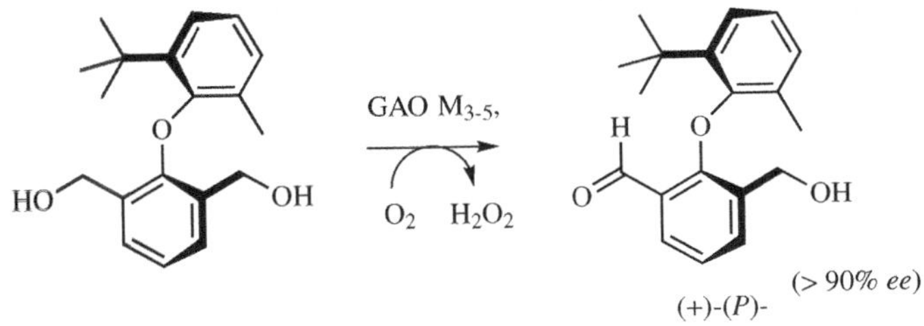

FIGURE 5.4 GAO-mediated desymmetrization of an atropisomeric diaryl ether.

used as a reporter enzyme to rapidly determine the purity of chiral secondary alcohols by using a colorimetric read-out [37].

In the area of analytical applications, the main use of GAO is the determination of galactose or lactose in liquid as well as soft solid matrices, being used in diagnostic kits [38]. Usually the hydrogen peroxide produced in the oxidation is detected colorimetrically (using GAO coimmobilized with peroxidase and a phenolic reagent, which forms a fluorophore by reaction with hydrogen peroxide in the presence of peroxidase) (see Chapter 7). Other monoses, such as glucose, mannose, ribose, xylose, and fructose, do not interfere in the analysis [39].

5.1.1.2 Pyranose Oxidase Pyranose oxidases (PyO, EC 1.1.3.10) are flavoenzymes expressed in many wood-degrading basidiomycete fungi, contributing to the ligninolytic system by producing hydrogen peroxide [40,41]. Unlike glucose oxidase—which acts on the anomeric carbon—they oxidize glucose and a number of common aldopyranoses at their C-2 hydroxyl group in the presence of oxygen to the corresponding 2-keto sugars and hydrogen peroxide. Since its first characterization by Janssen and Ruelius [42], it was suggested that both the presence of a six-membered ring of pyranoid monosaccharides and an equatorially oriented 2-OH group are essential structural requirements for activity. Accordingly, the natural substrates of pyranose oxidase, D-glucose and D-xylose, which are most abundant in wood, are readily oxidized to the corresponding 2-keto derivatives (or 2-dehydrosugars) (Figure 5.5).

Pyranose oxidase displays broader substrate specificity than GAO, and thus ketoses like L-sorbose and aldoses like D-allose and D-galactose, with axial OH groups in positions 3 and 4, are readily oxidized at C-2 (Figure 5.6A) [43]. Deoxygenation is also well tolerated, as in 3-deoxy-D-glucose, which is oxidized at about the same conversion rate as D-glucose. Furthermore, quantitative oxidation of

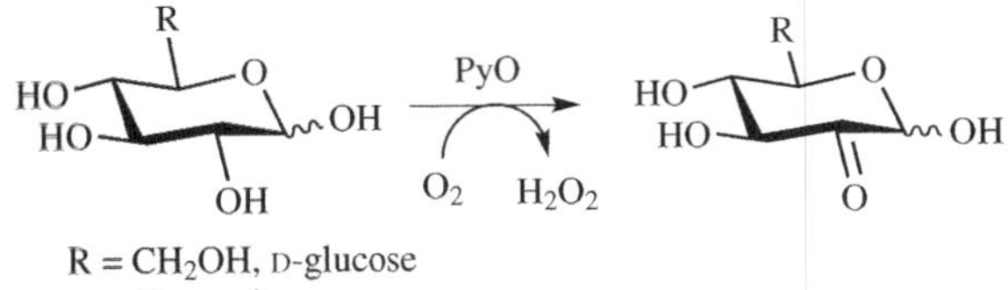

FIGURE 5.5 PyO-catalyzed oxidation of D-glucose and D-xylose.

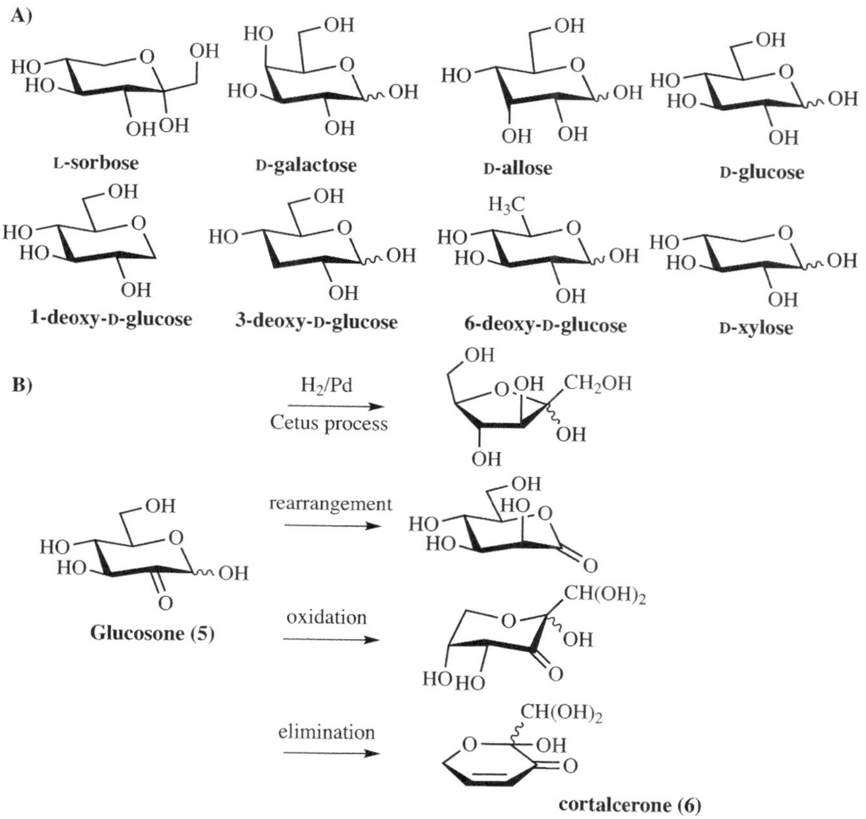

FIGURE 5.6 **A**, representative substrates for PyO; **B**, products formed from glucosone.

1-deoxy- and 6-deoxy-D-glucose is observed at the C-2 OH (Figure 5.6A) [40,44]. Certain compounds, however, are selectively oxidized to their 3-keto derivatives, such as 2-deoxy-D-glucose (converted in good yields), and methyl β-D-gluco- and methyl β-D-galactopyranoside (in lower yields) [44]. The enzyme presumably uses two different binding modes for each type of oxidation, and a model for the minimal structural requirements of the enzyme–substrate interaction was postulated in 1998 [44]. Pyranose oxidase not only oxidizes these substrates very efficiently but also possesses a glycosyl-transfer potential, producing disaccharides from β-glycosides of higher alcohols and phenols like hexyl-, phenyl-, *o*-, and *p*-nitrophenyl [44]. In addition to oxygen, PyO can also use various quinones, complexed metal ions, and radicals as its electron acceptor [45]. Moreover, some of these compounds are even better substrates for the enzyme than oxygen, suggesting that PyO can also play a direct role in the reduction of quinones during ligninolysis [46]. Shortly after its characterization, PyO has received attention as the key biocatalyst in several biotechnological and analytical applications. Some of these are patented, for example, the oxidized 2-keto sugar (osone) obtained in quantitative yield from D-glucose can be

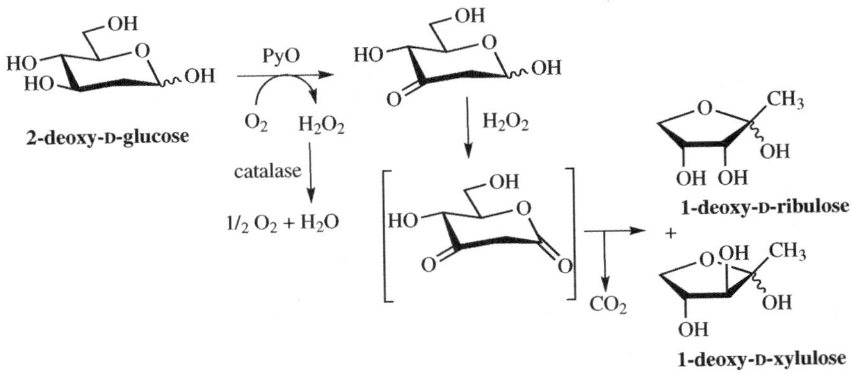

FIGURE 5.7 Formation of 1-deoxy-D-ribulose from PyO-mediated oxidation of 2-deoxy-D-glucose.

reduced catalytically or enzymatically at position C-1 to obtain high yields of virtually pure D-fructose (Figure 5.6B). The selective chemical reduction of the aldehyde group of 2-keto-D-glucose *via* catalytic hydrogenation on Pd/C was patented by the American company Cetus and is known as "Cetus process" [47–49]. Analogously, pure D-tagatose can be obtained by quantitative C-2 oxidation of D-galactose followed by chemical or enzymatic C-1 reduction [50]. The formation of the 2-keto derivative by biotransformation is more favorable than the chemical oxidation both in terms of efficiency and number of steps. In addition, when glucosone **5** is treated under elimination conditions, the β-pyrone antibiotic cortalcerone (**6**) is produced (Figure 5.6B) [40,51].

Other important uses of 2-keto-D-aldoses (osones) include a further oxidation of glucosone at C-1 by GAO to give 2-keto-D-gluconic acid, a precursor for the synthesis of D-isoascorbic acid [52,53]. Also, 2-deoxy-D-glucose is oxidized to the labile 3-keto derivative in 75%, in the presence of catalase to decompose hydrogen peroxide. The 3-keto aldehyde formed is chemically oxidized to 2-deoxy-3-keto-D-gluconic acid, which undergoes subsequent decarboxylation to yield 1-deoxy-D-ribulose together with small amounts of 1-deoxy-D-xylulose (Figure 5.7) [44].

The enolone **7** is used as chemical synthon when benzoylated [54]. It is produced by PyO-mediated oxidation of 1,5-anhydro-D-glucitol, followed by chemical acetylation (Figure 5.8) [44,55].

It has been proposed that the excellent reactivity of PyO with alternative electron acceptors (quinones, radicals) and a range of sugar substrates might be exploited in

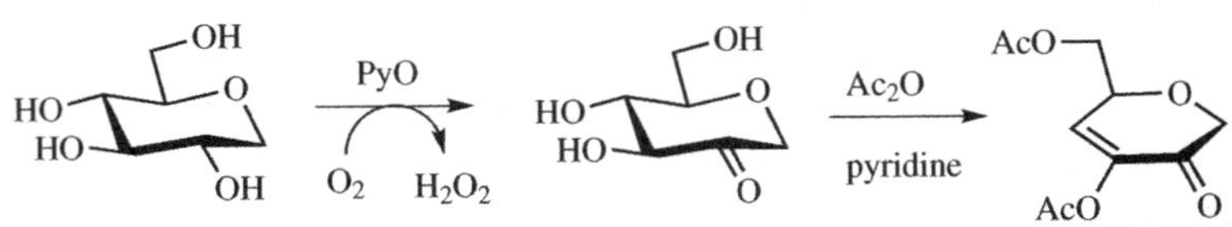

7

FIGURE 5.8 Chemoenzymatic preparation of enolone **7**.

applications such as biocomponent in biofuel cells, and some improvements of the desired features (broader substrate tolerance) has been achieved by enzyme engineering [56]. Enzymatic biofuel cells are a promising future power source with specific applications [57]. Likewise, PyO is currently used for various analytical applications, for example, in clinical chemistry for determination of 1,5-anhydro-D-glucitol—an important marker for glycemic control in diabetic patients [58–61] and for monitoring industrial processes [62–64].

The capacity to use different electron acceptors makes PyO similar to other FAD-containing sugar oxidoreductases that have been classified as "dehydrogenases with other acceptors (EC 1.1.99.-)," such as cellobiose dehydrogenases (CDHs) and pyranose dehydrogenases (PyDHs), probably all having similar biological functions in lignocellulose degradation [65]. Cellobiose dehydrogenases (CDH, EC 1.1.99.18) [66] oxidize the disaccharide cellobiose to cellobionolactone, but also accept other β-1,4-linked disaccharides (like lactose), oligosaccharides, and even cellulose [67]. CDH has been technically used in a few applications, namely in highly selective amperometric biosensors with lower unspecific noise reactions for the measurement of lactose [68], and in the production of lactobionic acid from lactose using continuous enzymatic regeneration of the electron acceptor [69].

As compared to pyranose oxidases, pyranose dehydrogenases (PyDH, EC 1.1.99.29) display a broader substrate specificity and a variable regioselectivity, using substituted benzoquinones and (organo)metallic ions as electron acceptors instead of molecular oxygen [65]. PyDH was first isolated from the basidiomycete fungus *Agaricus bisporus* [70], and subsequently from *Agaricus xanthoderma* [71], *Agaricus meleagris,* [72], and *Macrolepiota rhacodes* [73]. Depending on the structure of the sugar in pyranose form (mono/di/oligosaccharide or glycoside) and the enzyme source, selective monooxidations at C-1, C-2, C-3, or dioxidations at C-2,3 or C-3,4 of the molecule to the corresponding aldonolactones (C-1) or (di)dehydrosugars (aldos(di)uloses) can be performed. Accordingly, new nonnatural sugar derivatives could be produced, isolated, and characterized, together with the more common lactones from C-1 oxidation (Figure 5.9) [65,74].

The sugar oxidases of this family are FAD-dependent enzymes. After the sugar oxidation process, the reduced FADH moiety needs to be reoxidized to start a new catalytic cycle. This is not a problem when molecular oxygen is used as terminal electron acceptor, due to the simplicity of the aerobic reoxidation of the flavin cofactor. On the contrary, in processes employing a flavin dehydrogenase (like PyDH), the electron acceptor (quinones or other molecules) has to be added in stoichiometric amounts and separated from the reaction product, unless *in situ* regeneration systems are applied. Specifically, it has been shown that a wide range of redox mediators such as quinones and complex metal ions can be enzymatically reoxidized employing laccases, a group of oxidases that will be described in Section 5.1.2.

5.1.1.3 Alcohol Oxidase Alcohol oxidases (AOs, EC 1.1.3.13) are flavin dependent, alcohol oxidizing proteins belong to the glucose-methanol-choline (GMC)-oxidoreductase superfamily [75]. Similar to GAO, AO oxidizes a range of (aliphatic) primary alcohols in the presence of oxygen to the corresponding aldehyde with

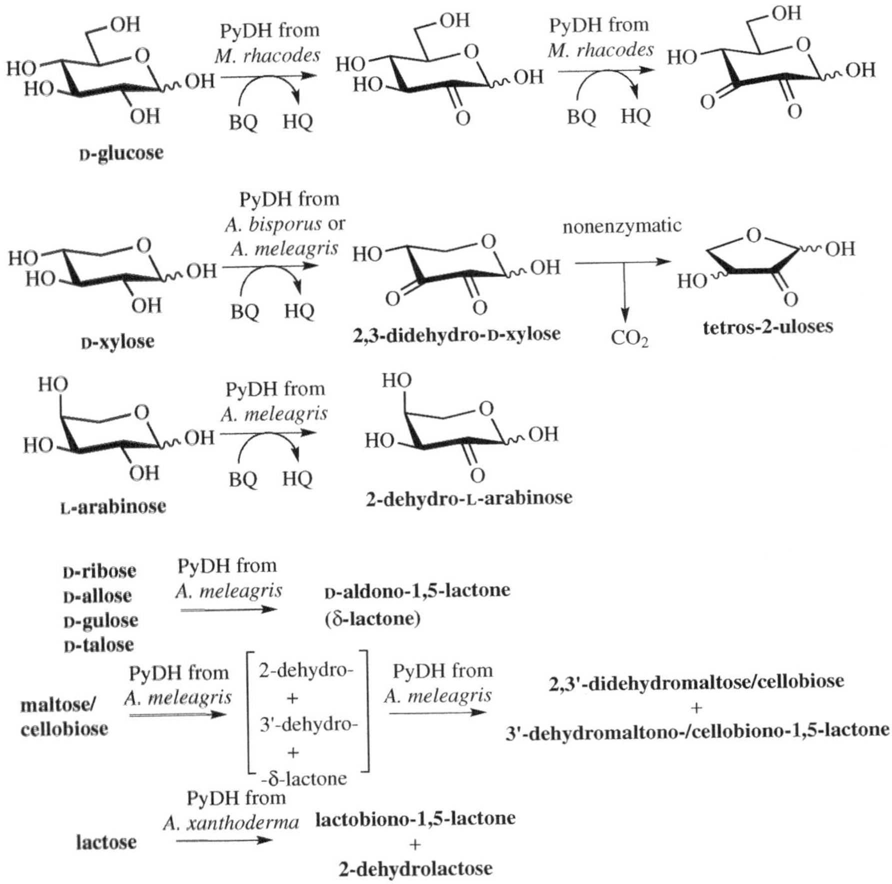

FIGURE 5.9 Oxidation of sugars by different PyDH. BQ: benzoquinone; HQ: hydroquinone.

concomitant production of hydrogen peroxide. However, their substrate profile is complementary, that is, GAO for galactose-derived polyols and AO for unbranched aliphatic short-chain monoalcohols and diols [18,76]. The enzyme effectively oxidizes short-chain saturated and unsaturated primary alcohols, such as methanol and unsaturated linear alcohols of 5 carbon atoms or less. Introduction of further hydroxyl groups in the chain quickly decreases the effectivity of the system. Synthetically, AO can be used for the preparation of functionalized aldehydes under mild aqueous conditions. In addition, it has been used combined with an aldolase (deoxyribose-5-phosphate aldolase (DERA)) in a two-step one-pot reaction, to produce chiral aldol products (see also Chapter 7 for multistep cascade processes). In this way 4-pentenol was oxidized by the AO/catalase system in the presence of acetone and DERA to yield the aldol product **8** in 30% after 5 days in 70% *ee* (Figure 5.10) [18]. Racemic short-chain 2-methyl-1-alkanols could be resolved by enatioselective oxidation with the AO from *Candida boidinii*, *Hansenula* sp., *P. pastoris*, and *Torulopsis methanothermo*, affording reaction mixtures enriched in (*R*)-2-methyl-1-alkanol at incomplete

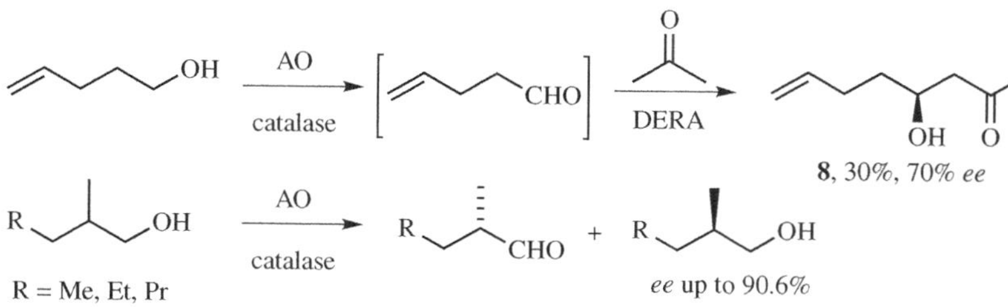

FIGURE 5.10 Cascade reaction involving AO and AO-mediated kinetic resolution of 2-methyl-1-alkanol.

conversions. The highest enantioselectivity was displayed by *C. boidinii* AO, with the greatest enantiomeric excess obtained for (*R*)-2-methyl-1-pentanol (90.6% *ee* at 76% conversion) (Figure 5.10) [77].

5.1.1.4 Glucose Oxidase Glucose oxidase (GOX, EC 1.1.3.4) is an FAD-containing flavoenzyme that catalyzes the oxidation of *β*-D-glucose to D-glucono-*δ*-lactone (which is nonenzymatically hydrolyzed to gluconic acid). In the subsequent oxidative half-reaction the cofactor $FADH_2$ is reoxidized with molecular oxygen producing hydrogen peroxide (Figure 5.11) [78–80].

Since the first description of GOX in 1928 in *Aspergillus niger* extracts [81], it has been found in fungi and insects, being the most common fungal sources *Aspergillus*, *Penicillium*, and *Saccharomyces* species [82]. Wild-type GOX is highly specific for *β*-D-glucose, acting on the anomeric carbon. For instance, using the reaction rate of glucose as reference (100%), only 2-deoxy-D-glucose (20–30%), 4-*O*-methyl-D-glucose (15%), and 6-deoxy-D-glucose (10%) are oxidized by *A. niger* GOX at significant rates [80], thus being its synthetic utility rather limited. On the other hand, it is widely used for bioanalytical and biotechnological applications in chemical, pharmaceutical, food and beverage, clinical chemistry, cosmetics, textile, and other industries [82,83]. GOX has been used in the food industry, as food additive (preservative), for breadmaking, and for the production of dry egg powder, wine, and gluconic acid. Regarding the bioanalytical applications, GOX is commonly used to construct amperometric biosensors for medical use [84–86], and food industry [87]. Moreover, its electrochemical activity makes it potentially useful in fuel cell applications.

5.1.1.5 Glycolate Oxidase Glycolate oxidase (GlyO, (*S*)-2-hydroxyacid oxidase, EC 1.1.3.15) is an FMN-dependent enzyme that catalyzes the oxidation of

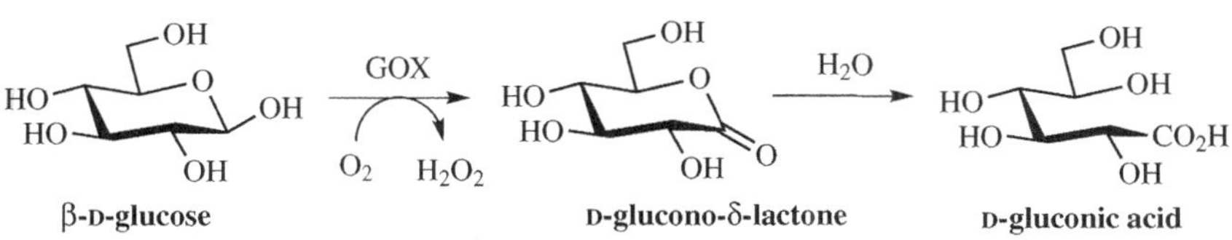

FIGURE 5.11 GOX-catalyzed oxidation of *β*-D-glucose.

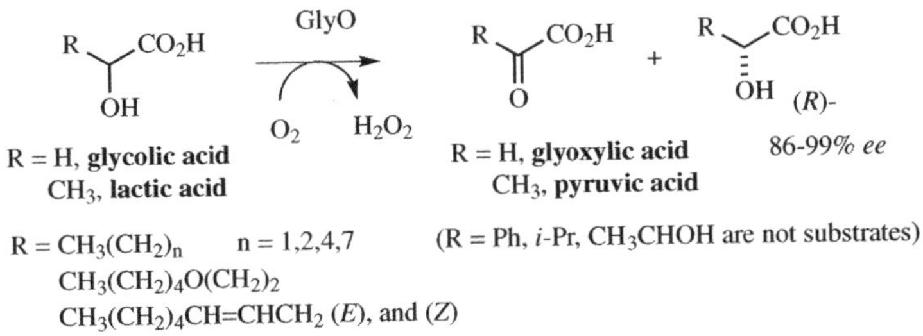

FIGURE 5.12 Glycolate oxidase-mediated oxidation of 2-hydroxy acids.

2-hydroxy acids, such as glycolic or lactic acid, to the corresponding 2-keto acids (glyoxylic or pyruvic acid), while hydrogen peroxide is produced by the reduction of molecular oxygen (Figure 5.12) [88]. The enzyme is located in the peroxisome, with ample distribution in green plants and animals. GlyO possesses a broad substrate tolerance and a strict (*S*)-enantioselectivity. Adam et al. have shown that GlyO from spinach (*Spinacia oleracea*) catalyzes the enantioselective oxidation of a range of racemic 2-hydroxyacids to yield the corresponding 2-ketoacids and enantiomerically pure (*R*)-2-hydroxy acids (Figure 5.12) [89]. Short- and medium-chain 2-hydroxy acids, as well as side chains containing oxygen atoms and alkenes are accepted as substrates, whereas the more sterically demanding mandelic acid (R = Ph), 2-hydroxyisobutyric acid (R = *i*-Pr), and 2,3-dihydroxybutyric acid (R = CH_3CHOH) are not transformed (Figure 5.12).

Apart from the addition of catalase to *in situ* decompose the hydrogen peroxide, enzyme immobilization is also common strategy to improve the operational stability of the system. The coimmobilization of glycolate oxidase and catalase produced a stable catalyst, which afforded glyoxylic acid in 99% yield [90,91]. In a different strategy, Das et al. have coexpressed catalase and GlyO in *P. pastoris* strain NRRL Y-21001 [92]. Permeabilized cells of r*P. pastoris* exhibited activity with a broad range of substrates including 3-phenyllactic acid, 3-indolelactic acid, 3-chlorolactic acid, 2-hydroxybutanoic acid, and 2-hydroxydecanoic acid, resulting in a high degree of (*S*)-selectivity, leaving the (*R*)-isomers intact. The best substrates were 3-chlorolactic acid (110% relative to lactic acid) and 2-hydroxybutanoic acid (120%) (Figure 5.13). Another whole-cell system was designed for the preparation of *N*-(phosphonomethyl)glycine (glyphosate), a broad-spectrum postemergence herbicide (Figure 5.13) [93]. An aqueous mixture of glycolic acid and (aminomethyl)phosphonic acid was oxidized and hydrogenated after separation of the microbial catalyst, to give glyphosate *via* reductive amination.

GlyO from spinach has also been used for deracemization of 2-hydroxy acids, combined with D-lactate dehydrogenase *via* two enantioselective steps. The enzymatic resolution of racemic 2-hydroxy acids by enantioselective glycolate oxidase-catalyzed oxidation and subsequent asymmetric reduction of 2-oxo acids with D-lactate dehydrogenase from *Lactobacillus leichmannii* led to enantiomerically pure (*R*)-2-hydroxy acids in up to 89% yield based on the racemate (Figure 5.14) [94].

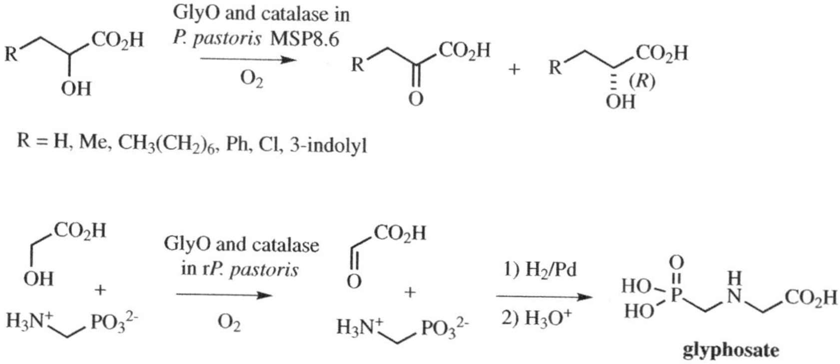

FIGURE 5.13 GlyO-catalyzed oxidations in whole-cell systems.

5.1.2 Laccases and Tyrosinases (Phenol Oxidases)

5.1.2.1 Laccase Laccases (EC 1.10.3.2) are one of the few enzymes that have been studied since the end of nineteenth century, being first described in the exudates of *Rhus vernicifera*, the Japanese lacquer tree [95], and in fungi a few years later [96]. However, they only attracted considerable attention after the beginning of studies of enzymatic degradation of wood by wood-rotting fungi. These enzymes belong to a group of polyphenol oxidases (PPOs) containing copper atoms in the catalytic center, called blue multicopper oxidases (BMCO) [97–104]. They catalyze the four-electron reduction of molecular oxygen to water coupled to four single-electron oxidation of the hydrogen donating (usually aromatic) substrate [105]. The catalytic cycle takes place in different copper centers within the enzyme. In particular, all BMCO are characterized by the presence of at least one type 1 (T_1, blue) copper, together with at least three additional copper ions (T_2 and T_3) organized in a trinuclear cluster (see Section 1.3.3.2); the classification of copper atoms in three types is based on their spectroscopic properties [100,106]. The substrates are oxidized by the T_1 copper atom and the extracted electrons are transferred, probably through a strongly conserved His–Cys–His tripeptide motif, to the trinuclear cluster site, where molecular oxygen is reduced to water [103,107]. The catalytic core of the enzyme is brought back to its

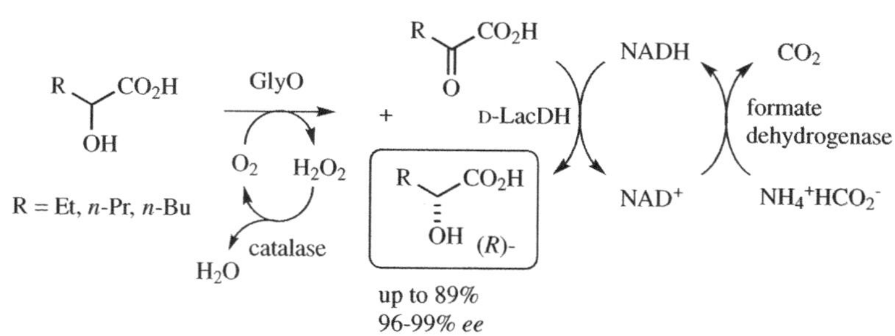

FIGURE 5.14 Preparation of (*R*)-2-hydroxy acids *via* deracemization of racemic 2-hydroxy acids, using GlyO and D-lactate dehydrogenase (D-LacDH).

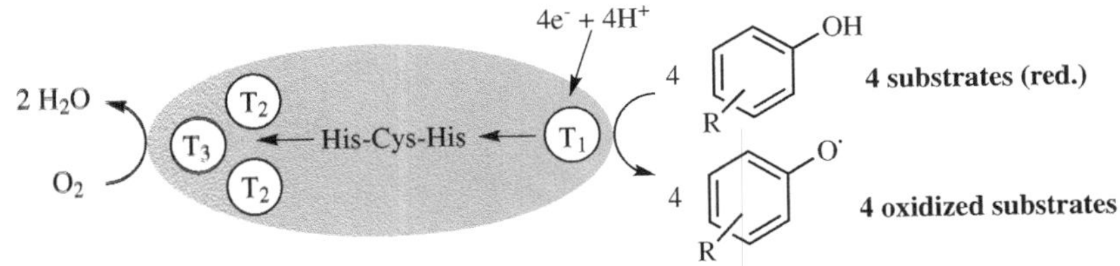

FIGURE 5.15 Catalytic cycle of laccase using phenolic substrates [103].

reduced form by the oxidation of four substrate molecules to produce four radicals, which can subsequently react to form dimers, oligomers, and polymers [106]. The overall outcome of the catalytic cycle is shown in Figure 5.15. The first laccase structures containing the four copper atoms were published in 2002, corresponding to laccases from *Melanocarpus albomyces* and *Trametes versicolor* [108–110].

Laccases have been widely described in plants and fungi (ascomycetes, mainly *Botrytis* sp. and basidiomycetes, mainly *Trametes* sp.), where they are presumably involved in lignin biosynthesis and degradation processes, respectively, as well as in fungal virulence and morphogenesis [105,111,112]. Plant sources of laccase include *R. vernicifera*, *Rhus succedanea* [113], *Acer pseudoplatanus* [114], *Pinus taeda* [115,116], *Populus euramericana* [117], *Liriodendron tulipifera* [118], *Nicotiana tobacum* [119], *Lolium perenne* [120], *Aesculus parviflora* [121], and *Zea mays* [122]. To date, most of the characterized laccases (more than one hundred) have been isolated from fungi [103]. In addition, bacterial laccases, first reported in 1993 [123], seem to be widespread among bacteria, based on homology searches in protein databases and bacterial genomes [124], and experimental data [125]. They are involved in melanin production, spore coat resistance, morphogenesis, and copper homeostasis [107,126]. Laccase-like activity has been found in *E. coli* [127,128], *Streptomycetes* [129,130], *Azospirillum lipoferum* [131], *Marinomonas mediterranea* [132], *Bacillus sphaericus* [133], *Bacillus subtilis* [134–136], *Oceanobacillus iheyensis* [137], *Leptothrix discophora* [138], *Thermus thermophilus* HB27 [139], *Xanthomonas campestris* [140], and several *Pseudomonas* sp. [141–143], among others. Likewise, they have been described in arthropods (mainly insects) and in bovine rumen microflora [144–148]. A database of laccases and related copper oxidases, "Laccase Engineering Database," is available online [149].

For biotechnological applications, the corresponding genes have been isolated and expressed in *A. niger*, *Aspergillus oryzae*, *Aspergillus sojae*, *Aspergillus nidulans*, *Pichia pastoris*, *Pichia methanolica*, *Yarrowia lipolytica*, *Saccharomyces cerevisiae*, *Kluyveromyces lactis*, *Trichoderma reesei*, *Streptomyces lividans*, among other heterologous expression systems [150–161]. The laccase from the ligninolytic fungus *Cyathus bulleri* has been recently expressed in *E. coli* making it the first fungal laccase to be expressed in a bacterial host [162]. Previously, bacterial laccases had also been expressed in *E. coli* [132]. In addition, several commercial preparations of laccases are available [163]. Considering their ample distribution, it is not surprising that laccases from different organisms do vary with respect to their properties such as substrate specificity, optimum pH, stability, and so on. Moreover, several isoenzymes have been detected in many fungal species. In particular, more than one isoenzyme is produced in most white-rot fungi [164,165]. They can therefore be selected for certain specific applications depending on the requirements of the synthetic biocatalyst. Fungal laccases are mostly extracellular, monomeric glycoproteins, with a degree of glycosylation up to 25% and typically three to ten sites where α-linked mannose chains are connected [166]. Plant laccases are more glycosylated, up to 45% [167]. The carbohydrate moiety has been supposed to contribute to the stability in the extracellular hydrophilic medium and to protect the enzyme against proteolysis and free radical attack [105,107]. They show a broad substrate range of, for example, lignin breakdown products such as diversely substituted phenols or hydroquinones and their analogs, aminophenols, polyamines, or various metal ions (like Mn^{2+}, in presence of chelators) [99,107,168]. With respect to other ligninolytic enzymes, laccases can oxidize only phenolic fragments of lignin due to the laccase lower redox potential [169]. The redox potential of fungal laccases usually falls within the range of 0.5–0.8 V [98,170], whereas plant laccases usually display a lower potential. Thus, the complete range of redox potential of laccases spans from 0.4 to 0.8 V [98,170], with typical values of 0.78 V for high potential fungal laccases (from *Trametes hirsuta*) [171], 0.55 V for middle potential laccases (from *Coprinus cinereus*) [172], and 0.43 V for low potential plant laccases (from *R. vernicifera*) [173]. Recently, a new measurement of the redox potential of *T. versicolor* laccase yielded 0.717 V [174], lower than the well-known value of 0.78 V [170,175].

Owing to their high nonspecific oxidation capacity, laccases are useful biocatalysts for organic synthesis and diverse biotechnological applications [98,161,176–184]. Products resulting from laccase-catalyzed oxidation can be loosely classified, according to the type of substrate and mode of oxidation, in products derived from direct oxidation, cross-coupling and domino reactions, or from the use of a laccase-mediator system (LMS).

Laccase-Catalyzed Direct Oxidation of Organic Compounds The laccase-catalyzed direct oxidation of electron rich aromatics produces reactive radical intermediates, which easily undergo self-coupling reactions leading to the formation of C–C and C–O (and C–N) dimers, oligomers, and eventually, polymers. The reactivity of the radical intermediates is often a problem, and thus a compromise between conversion and isolated yield has to be achieved to avoid the extensive polymerization that may take place *via* the radical oxidation mechanism.

Monolignols and derivatives are common substrates for oxidation. For instance, the oxidation of the monolignol coniferyl alcohol and isoeugenol was studied in acetone:water (1:1) mixtures using plant and fungal laccases from *R. vernicifera* and *Pycnoporus coccineus*, respectively. The rate of oxidation depends both on the nature of the substrate and the enzyme (the fungal enzyme being faster). Using isoeugenol and *R. vernicifera* laccase, a mixture of dimeric and tetrameric derivatives has been isolated in approximately 40% yield (Figure 5.16) [185]. Aqueous extracts of the residual compost of the fungus *A. bisporus*, displaying laccase activity, were used for the oxidation of a series of mono- and polyphenolic compounds and anilines (including 2,6-dimethoxyphenol, guaiacol, phenol, and aniline) to form

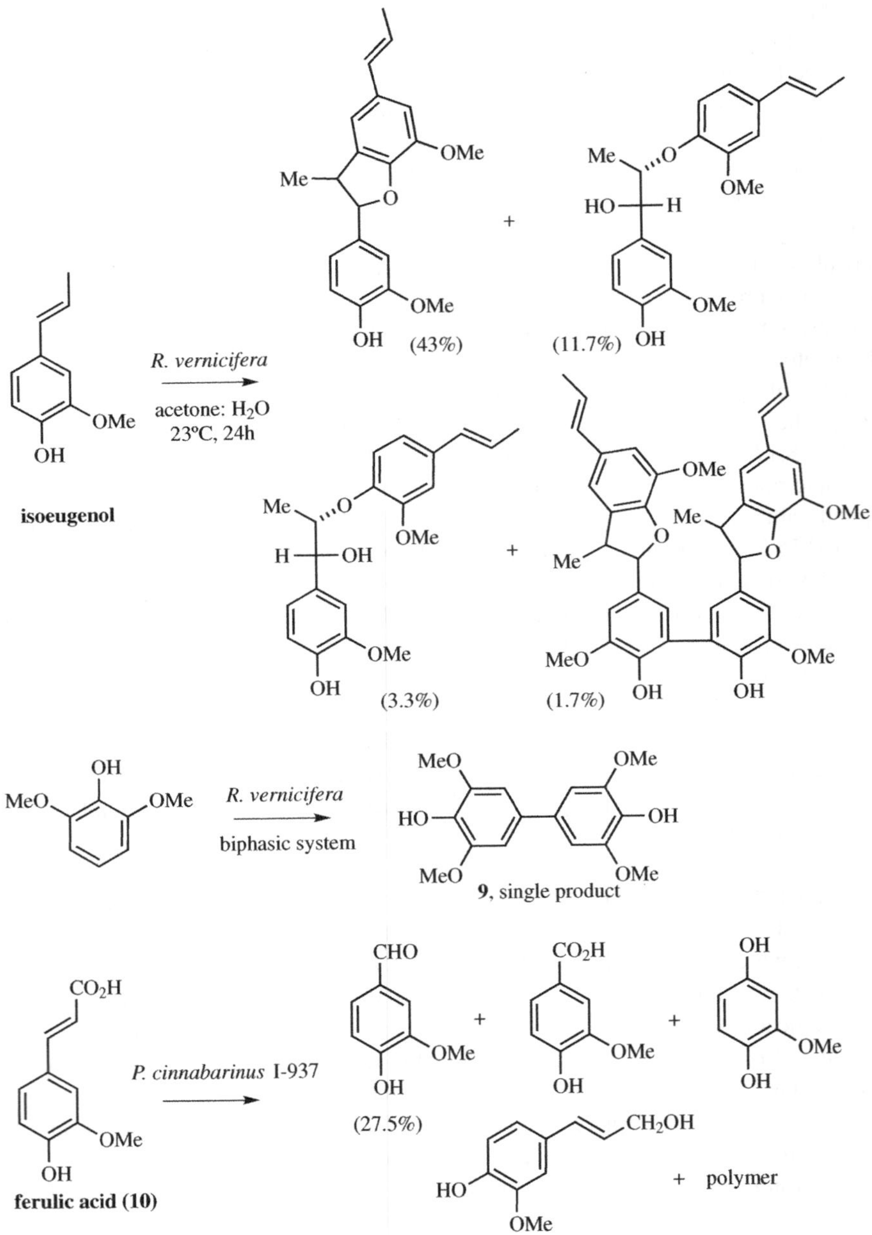

FIGURE 5.16 Laccase-catalyzed oxidation of phenolic compounds.

insoluble (noncharacterized) products after treatment [186]. Recently, the oxidation of 2,6-dimethoxyphenol using laccase from *R. vernicifera* in different water-organic solvent systems was reported by Wan et al. to afford only one product, 3,3′,5,5′-tetramethoxy-1,1′-biphenyl-4,4′-diol (**9**), resulting from *para–para* coupling [187]. In this case, the biphasic system decreased the reaction rate, and consequently the polymerization, enhancing the product selectivity (see below). Furthermore, the lignan ferulic acid (**10**) has been used as a model for the oxidative decomposition of lignin by laccase. In a medium containing glucose and ethanol under aerobic conditions, the white rot fungus *T. versicolor* transformed ferulic acid into coniferyl alcohol, coniferylaldehyde, dihydroconiferyl alcohol, vanillic acid, vanillyl alcohol, 2-methoxyhydroquinone, and 2-methoxyquinone [188]. Using *Pycnoporus cinnabarinus* I-937, the biotransformation of ferulic acid gave mainly vanillin, which was isolated in 27.5%, together with other minor products, such as coniferyl alcohol, vanillic acid, 2-methoxyhydroquinone and dimers and a polymer of coniferyl alcohol (Figure 5.16) [189].

The generation of several dimeric compounds was an advantage in the biomimetic synthesis of natural dimers of the bisabolene-type metabolite (*S*)-curcuphenol **(11)**, isolated from the marine sponge *Didiscus aceratus*. The laccase-mediated oxidation of (*S*)-curcuphenol in a citrate buffer:methanol (MeOH) (3:1) medium led to a mixture of dicurcuphenols resulting from C–C (4,4′ and 4,2′ and 2,2′) and C–O (O–C4′) coupling. It is noteworthy that only the naturally occurring dimers were detected, in a relative yield that parallels that from the sponge source (with the exception of the yield for the 4,4′-dimer, **12a**). Also, the products from 4,2′- and 2,2′-coupling (**12b,c** and **12d,e**, respectively) were obtained as scalemic mixtures of atropisomers (biaryl axial chirality), in agreement to the natural products (Figure 5.17) [190].

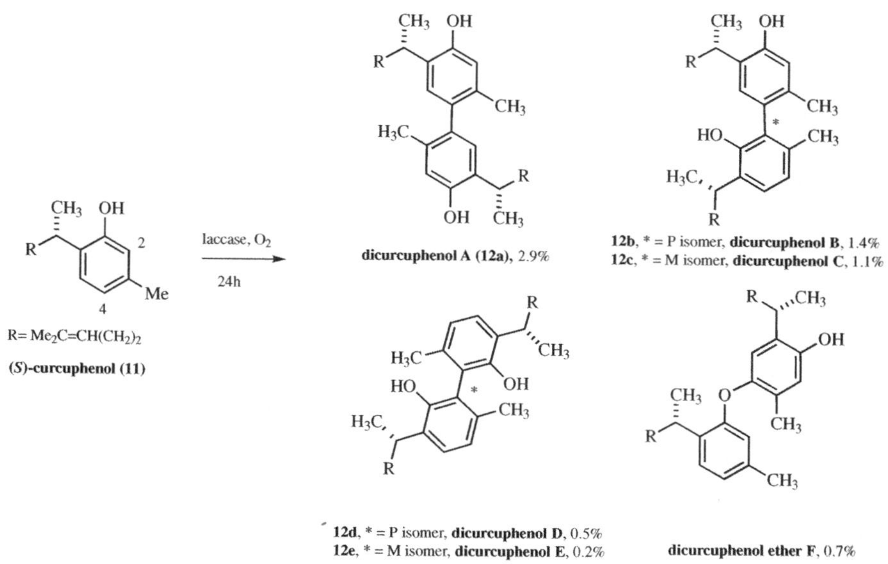

FIGURE 5.17 Synthesis of dicurcuphenols isolated from *D. aceratus*.

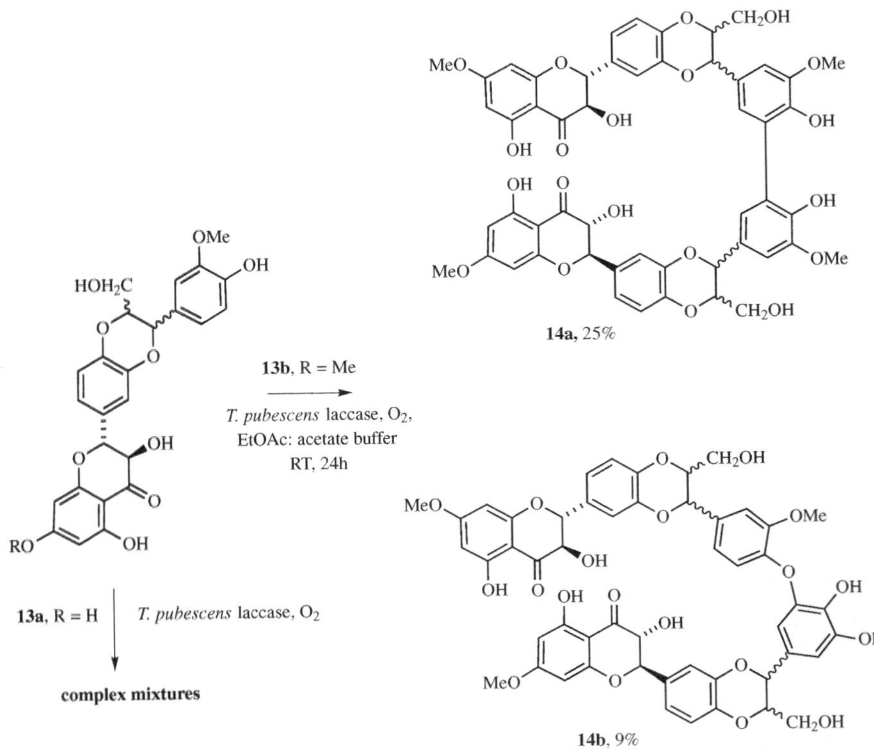

FIGURE 5.18 Laccase-mediated oxidation of 7-*O*-methyl sylibin (**13b**).

In some cases, changes in the reaction conditions or the substrate substitution pattern may alter the polymerization rate or the product distribution. For instance, the reaction of the flavonolignan silybin (**13a**) with *Trametes pubescens* laccase led to complex mixtures of oligomers and polymeric material, whereas the reaction of its 7-*O*-methyl derivative **13b** led to good yields (34%) of only two C–C and C–O dimers, **14a** and **14b**, respectively, in a 2.5:1 ratio (Figure 5.18) [191].

Reaction conditions, including solvent, pH, temperature, and buffer concentration, also influence the product distribution in laccase-mediated couplings. The effect of the solvent in the product distribution was initially reported for the laccase-mediated oxidation of tetrahydronaphthyl derivatives. Accordingly, in the oxidation of 5,6,7,8-tetrahydronaphthalen-2-ol (**15**), the formation of the symmetrical 1,1′ product **16a** was highly favored with respect to the 1,3′ dimer **16b** in the aromatic solvents benzene and toluene (**16a/16b** = 7.2 in benzene vs. 1.3 in methyl *tert*-butyl ether) (Figure 5.19). The reaction was performed in a biphasic system (to confine the enzyme into the aqueous phase) and the dimers were isolated as racemic mixtures of enantiomeric atropisomers in all cases [192]. Another example of the influence of the solvent in the selectivity of the dimerization is the laccase-mediated oxidation of totarol (**17**), a diterpenoid possessing high antibacterial activity. In both biphasic

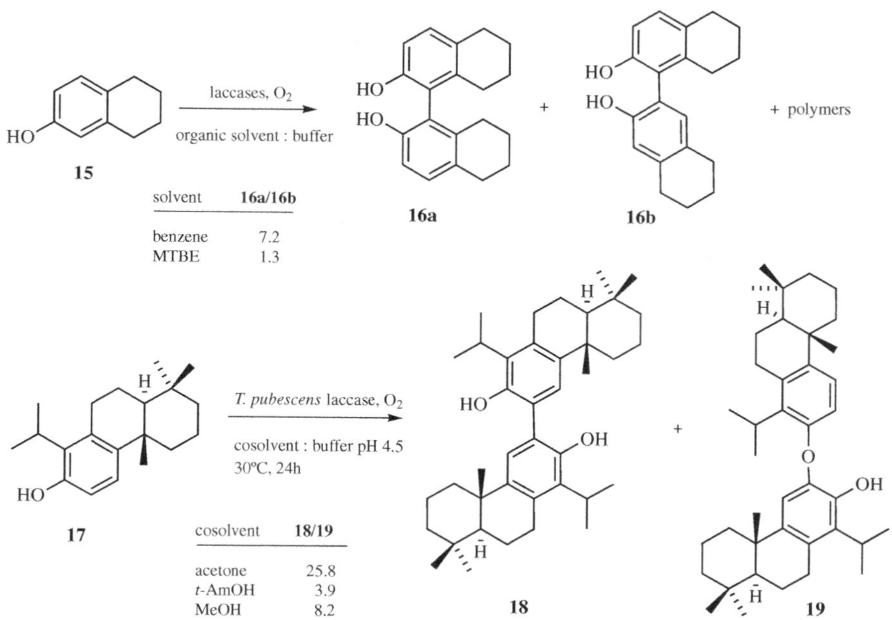

FIGURE 5.19 Solvent effect in laccase-catalyzed dimerizations.

and homogenous aqueous-organic media, the main product was a symmetrical C–C linked dimer **18**, which was obtained in unusually high yields under optimized reaction conditions, for example, in acetone:water (1:1) the dimer **18** was obtained in 85% yield, representing a 96% conversion and a dimer distribution C–C/C–O = 25.8 (Figure 5.19) [193]. As in the previous example, this selectivity is not achieved when using chemical oxidants like iron trichloride or manganese dioxide.

The set-up of a biphasic system improved the selectivity, yield, and the product stability in the oxidation of ferulic (**10**) and sinapinic acid (**20**). Using enzymes from different sources and ethyl acetate as a nonmiscible cosolvent, the racemic *bis*-lactone lignans **21a** and **21b** were obtained in 36% and 97% isolated yield, respectively (Figure 5.20) [194]. Previously, the oxidation of ferulic acid in homogeneous systems (and

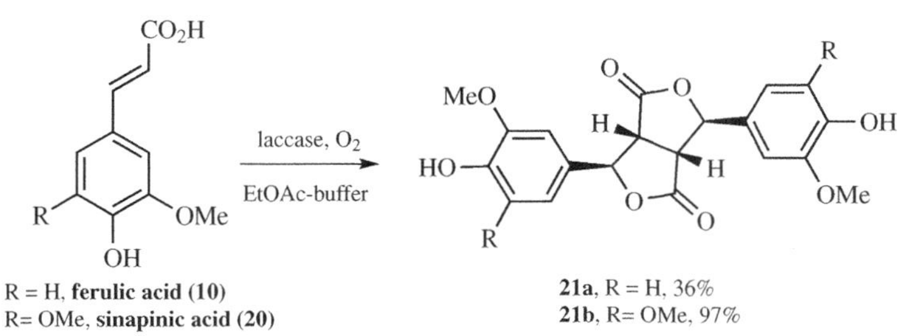

FIGURE 5.20 Improved selectivity of the laccase-mediated oxidation of lignans in a biphasic system.

with different enzyme preparations) was reported to afford mixtures of monomeric and dimeric products [188,189,195]. In addition, the oxidation of ferulic acid by laccase from *Myceliophthora thermophila* in ethyl acetate : phosphate buffer biphasic systems resulted in the production of a yellow phenolic product with potential application as food colorant, whose characterization was not reported [196].

High yields of isolated dimeric products have been obtained for the oxidation of *trans*-resveratrol (**22**), a stilbenic phytoalexin with interesting biological activity. In particular, laccases from *M. thermophila* and from *T. pubescens* gave the dehydrodimer **23** in 31% and 18% isolated yields, respectively (Figure 5.21) [197]. These

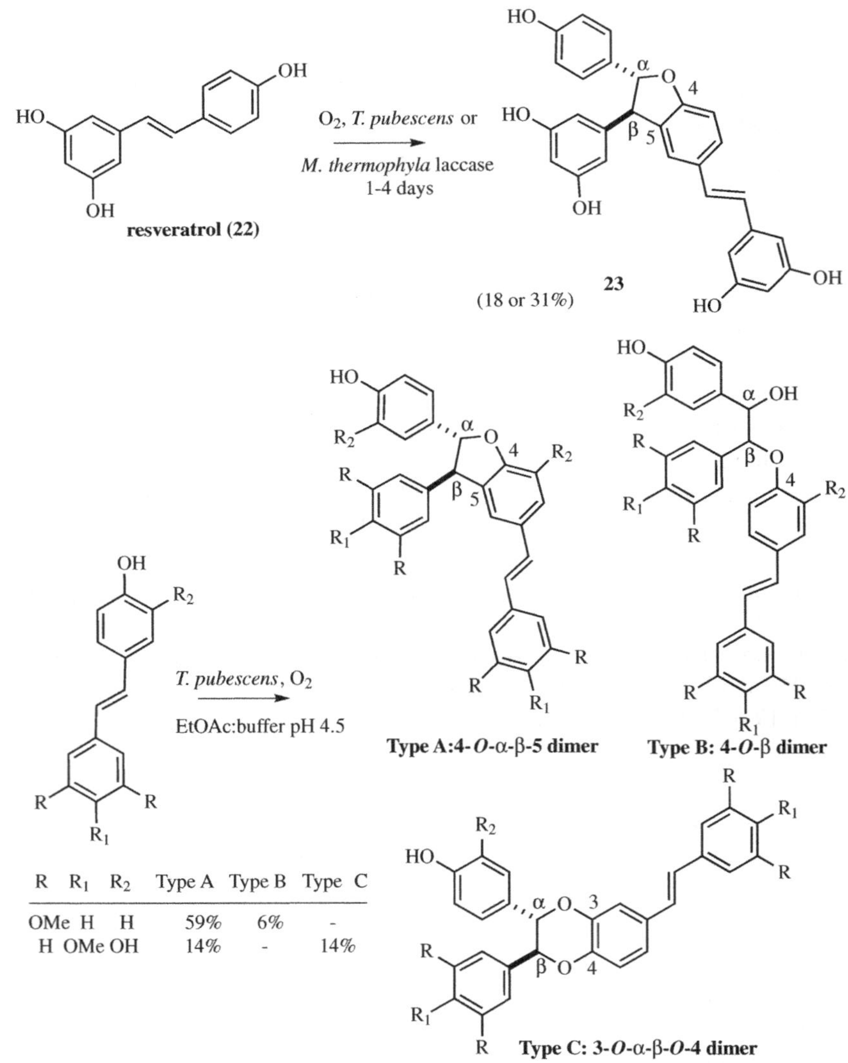

R	R_1	R_2	Type A	Type B	Type C
OMe	H	H	59%	6%	-
H	OMe	OH	14%	-	14%

FIGURE 5.21 Laccase-catalyzed dimerization of resveratrol and related stilbenes.

results compare favorably with the 18% yield reported for the chemically catalyzed dimerization of **22** [198]. The same authors also reported the laccase-mediated oxidation of a series of ten hydroxystilbenes, analogs of resveratrol, using *T. pubescens* laccase in a biphasic system. Three different dimeric products were identified, the main product usually being the 4-*O*-α-β-5 dimer, and a dimerization mechanism based on a radical–radical coupling was proposed (Figure 5.21) [199].

Another type of phenolic derivatives—the *para*-alkyl phenols—leads to reasonable yields of dimers *via* laccase-mediated oxidations as well. In this respect, when *p*-cresol, 3,4-dimethylphenol, tyrosol (*p*-(2-hydroxyethyl)phenol), and 2-*O*-acetyltyrosol were oxidized in a biphasic system using laccase from *M. thermophila* or *T. versicolor*, polycyclic ketones were isolated (approximately 10–20% yields), together with brown–red polymeric materials. These ketones, similar to the "Pumerer's ketone," are formed by initial radical–radical coupling, followed by Michael conjugated addition. In contrast to other dimers, the ketones cannot be further oxidized by the enzyme and thus, accumulate in the reaction medium (Figure 5.22) [200].

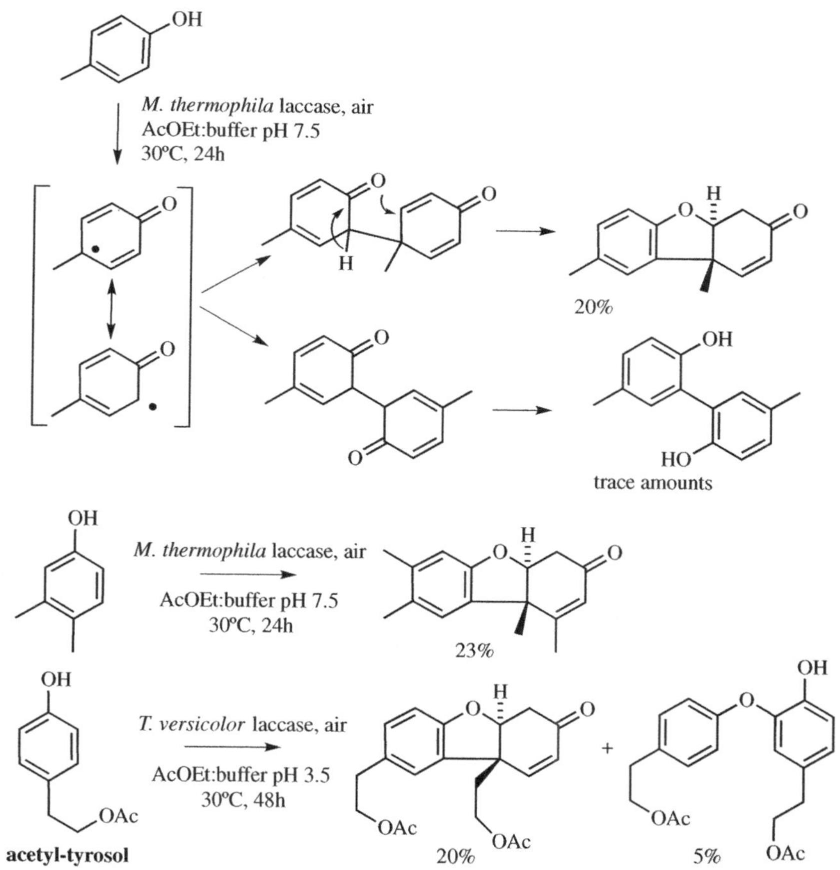

FIGURE 5.22 Laccase-mediated production of dimers from *p*-substituted phenols.

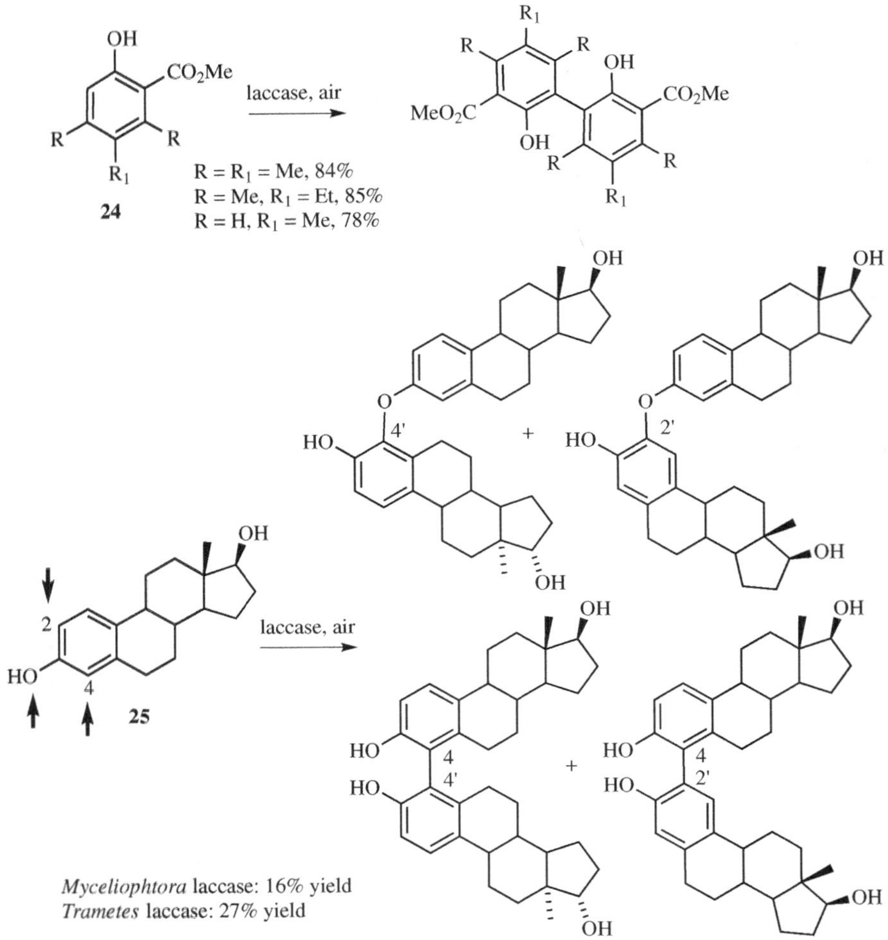

FIGURE 5.23 Laccase-catalyzed oxidation of miscellaneous substrates.

Besides catalyzing the coupling of simple phenols, laccases have been shown to catalyze the oxidation of other natural compounds such as salicylic esters [201], flavonols [202], steroid hormones [203,204], alkaloids [205], catechins [206], procyanidins (condensed tannins) [207], penicillin X [208], and analogs and derivatives of 3-hydroxyanthranilic acid (Figure 5.23) [209–212]. Furthermore, symmetrical biaryls were formed in high yields by oxidative dimerization of a series of methyl and ethyl salicylates of type **24** using *Trametes* laccases in aqueous solutions, leading to best results when a substituent on C5 (*para* to the OH) was present [201]. Flavonols were oxidized mostly in the C-ring upon treatment by *T. versicolor* laccase, in agreement with a predominant dismutation process leading to a pyrilium cation, which is further attacked by oxygen-nucleophiles (Figure 5.23) [202]. Mixtures of dimeric products were obtained from laccase-mediated oxidation of the hormone

17β-estradiol (**25**), along with polymeric material. Thus, four dimers resulting from C–O (O-C2′ and O-C4′) and C–C (4,4′ and 4,2′) coupling were found, in a ratio dependent on the enzyme source used (from *M. thermophila*, immobilized on glass beads, and from *T. pubescens*). Higher yields and selectivities were obtained using *Trametes* laccases in a biphasic ethyl acetate : acetate buffer system, enabling the isolation of the four dimers in 27% overall yield [192,204]. This biotransformation was firstly described in 1973 using a laccase from *Polyporus versicolor*, being actually one of the first examples of the use of enzymes in biphasic systems (Figure 5.23) [213].

The *Vinca* alkaloid vindoline (**26**) is transformed by fungal and plant laccases into a reactive intermediate that subsequently undergoes intramolecular cyclization to an enamine, which ultimately dimerizes (Figure 5.24) [205]. The laccase-catalyzed oxidation of catechin in the presence of gelatin in an aqueous medium forms antioxidant catequin–gelatin conjugates [206]. Laccases also transform procyanidins type-B to type-A, suggesting that in plants this type of conversion might occur by enzymatic means (Figure 5.24) [207]. The phenoxazine chromophore—present in biological pigments and antibiotics—has been prepared *via* laccase-mediated oxidative coupling of *o*-aminophenols. The first step comprises the formation of aminophenoxy radicals, which undergo further coupling and cyclocondensation to give the phenoxazinone system. For instance, actinocin (**28a**), a constituent of the actinomycin antibiotics, was prepared by transformation of 4-metyl-3-hydroxyanthranilic acid (**27a**) using laccase immobilized in polyacrylamide gel in aqueous medium and in 60% acetonitrile [209]. Later on, the preparation of cinnabarinic acid (**28b**), questiomycin A (**28c**) and actinocin (**28a**), was reported using laccase and oxygen in aqueous medium at pH 5 [210,214]. Also, a sulfonate analog of cinnabarinic acid (**28d**) was prepared by laccase-mediated oxidative dimerization of 3-hydroxyorthanilic acid (**27d**), at pH 6 (Figure 5.24) [211]. The use of different sulfonated aminophenol substrates allows for tuneable water solubility of the phenoxazine chromophore as a function of the sulfonyl substituent [212]. Novel phenoxazinones may have potential applications as textile dyes, and several fungal strains were used to test the potential of whole-cells for replacing isolated enzymes as a cheaper industrial-grade biocatalyst for the synthesis of these dyes [215].

Apart from these above-described *in vitro* examples, oxidative couplings are also involved in the biosynthesis of several groups of compounds like lignans, flavonolignans, and alkaloids. However, while the *in vivo* dimerization of propenylphenol derivatives to form lignans generally takes place with high diastero- and enantioselectivity, the *in vitro* reaction displays negligible enantioselectivity [194,197,199]. For instance, the *in vitro* laccase-catalyzed oxidation of coniferyl alcohol (**29**) affords the dimer pinoresinol **30**, as the main product, but in racemic form [216]. Similar results were obtained for the corresponding acid, ferulic acid (see above, Figure 5.20). The rationale for the different *in vivo* and *in vitro* outcomes remains as an open question. Considering that laccases are unable to control regio- or stereoselectivity by themselves, it was speculated whether some other agents might exist. Lewis et al. found a protein in the plant *Forsythia intermedia* which, albeit lacking an oxidative center, was able to *in vitro* effect stereoselective bimolecular radical coupling in the presence of a laccase. In this way, in the presence of the “dirigent” protein from

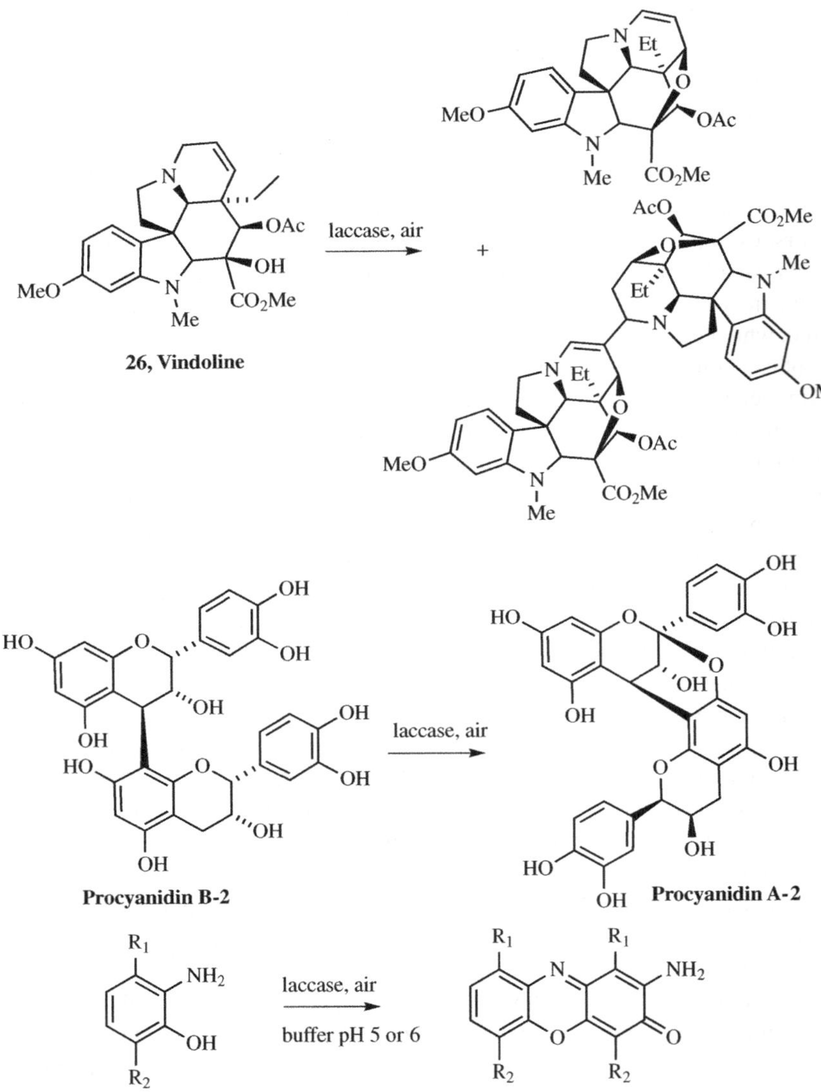

27a, $R_1 = CO_2H$, $R_2 = Me$ — $R_1 = CO_2H$, $R_2 = Me$, **actinocin (28a)**, 53%

27b, $R_1 = CO_2H$, $R_2 = H$ — $R_1 = CO_2H$, $R_2 = H$, **cinnabarinic acid (28b)**, 38%

27c, $R_1 = R_2 = H$ — $R_1 = R_2 = H$, **questiomycin A (28c)**, 29%

27d, $R_1 = SO_3H$, $R_2 = H$ — $R_1 = SO_3H$, $R_2 = H$ **(28d)**

FIGURE 5.24 Laccase-catalyzed oxidation of miscellaneous substrates.

F. intermedia (*Fi*DP), the laccase-mediated oxidative coupling of coniferyl alcohol formed the enantiomerically pure (+)-pinoresinol [217,218]. Moreover, this result has been reproduced by Beifuss et al., isolating an enantiocomplementary dirigent protein from *Arabidopsis thaliana* (*At*DIR6), which allows the preparation of the opposite enantiomer, (-)-pinoresinol, in 78% *ee* (Figure 5.25) [216].

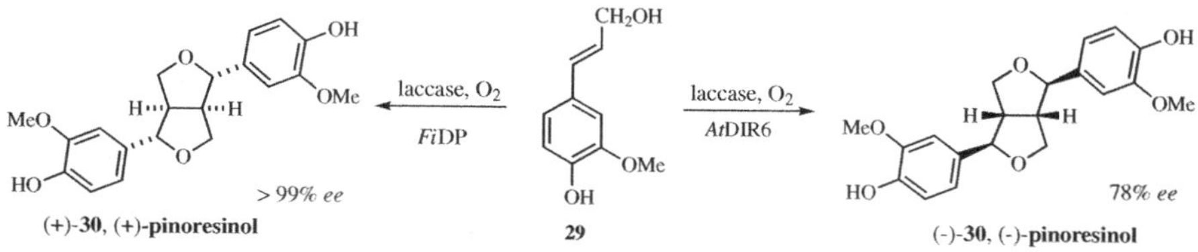

FIGURE 5.25 Enantiocomplementary dirigent proteins for the enantioselective laccase-catalyzed oxidative coupling of phenols.

The capacity of laccases to form polymeric mixtures—which actually constitutes a major drawback of the laccase-mediated preparation of small molecules—can be used as an efficient strategy to build polymers. In addition, laccases can straightforwardly generate polymers that are difficult to prepare by conventional chemical synthesis [219]. Thus, different monomers have been used in laccase-catalyzed polymerizations *via* radical coupling, including catechol and derivatives (coniferyl alcohol, syringic acid), 2,6-dimethylphenol, the flavonoids rutin and catechin, coumarins such as esculin, 1-naphthol, and 4-chloroguaicol, and also acrylamide, 2-hydroxydibenzofuran, phenolic pollutants, bisphenol A, 8-hydroxyquinoline, aniline, and D- and L-3,4-dihydroxyphenylalanine (DOPA) (Table 5.2) [219–238].

It is worth mentioning the preparation of artificial urushi, by Kobayashi et al. [225, 239]. Urushi—the oriental lacquer of Japan and China—is a natural resinous sap from the *R. vernicifera* tree, which is used as a coating agent due to its excellent toughness and long-standing brilliance. It is formed by the cross-linking of urushiol monomer, whose structure is a catechol derivative with unsaturated hydrocarbon chains consisting of mono-, di- and trienes at the 3-, or 4-position of catechol. Kobayashi used urushiol analogs by substituting the hydrocarbon chains by unsaturated fatty acids, and performed the cross-linking under mild conditions without using organic solvents. Moreover, Kobayashi and coworkers prepared poly(rutin) and poly(catechin) by laccase-catalyzed oxidative polymerization of the corresponding flavonoids, and studied their antioxidant and radical scavenging properties [240,241]. The preparation of poly(phenylene oxide), PPhO, can also be achieved using laccase, giving a product consisting of exclusively 1,4-oxyphenylene units. Both 2,6-dimethylphenol and syringic acid (3,5-dimethoxy-4-hydroxybenzoic acid) have been used as monomers to produce substituted PPhO, and a higher molecular weight polymer was obtained using syringic acid, which is decarboxylated during the reaction [234–236].

Laccase-Catalyzed Cross-Coupling and Domino Reactions Another reaction pathway for laccase-generated radicals consists in their cross-coupling reaction with different molecules, in such a way that the reactive intermediates are trapped and do not give rise to the usual set of dimers and oligomeric products. Several examples of the cross-coupling of hydroquinone have been reported. Laccase from *R. vernicifera* catalyzed the formation of two catechin-hydroquinone adducts (linked in the B ring, at C5′ and C2′), where hydroquinone served both as a shuttle oxidant and as a reactant during the reaction (Figure 5.26) [242]. Similarly, the cross-coupling with

TABLE 5.2 Examples of Lacasse-Catalyzed Polymerization *via* Oxidative Coupling Reaction

Substrate	Reaction Conditions	Products	Reference
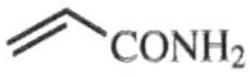	Laccase, H_2O, 65°C, 4 h	Polyacrylamide (MW > 600 × 10^3)	[220]
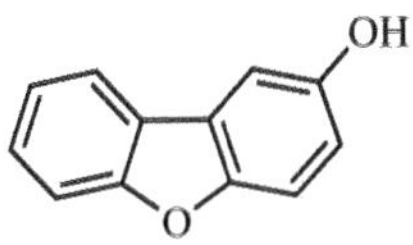	Laccase, pH 5, 30°C, 3 h	C–C and C–O dimers and oligomers	[221]
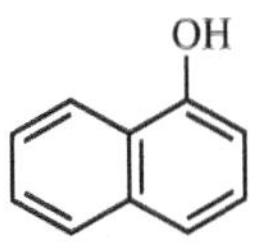	Laccase, buffer pH 5, acetone, 25°C	Orange-colored poly(1-naphthol), average MW = 4920	[222]
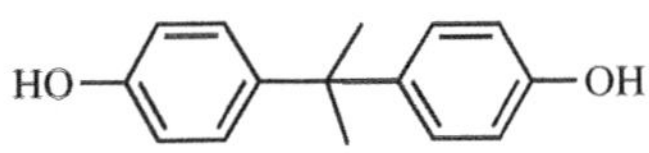	Laccase, pH 6, rt, 4 d	Oligomers linked in 3-3′ (C–C bond)	[223]
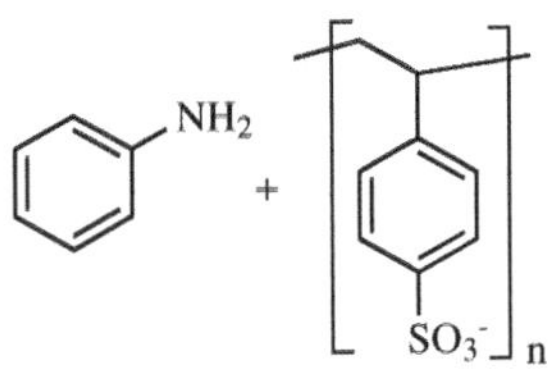	Laccase, pH 3.5-4.4, 20°C	Copolymer	[224]
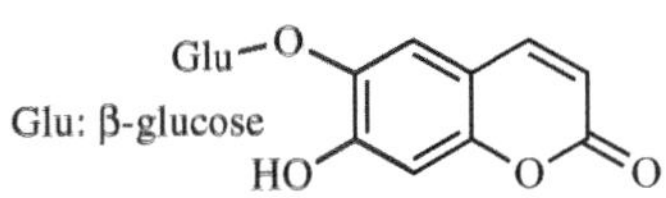 Esculin	*T. versicolor* laccase, MeOH:H_2O, 30:70, pH 7, 20°C, 24 h	Oligomers with C–C and C–O bridges, from both the phenolic and the glucosidic part of the coumarin	[238]
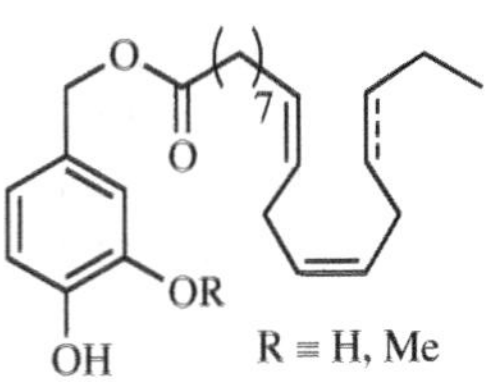	Laccase, no solvent	"Artificial Urushi"	[225]
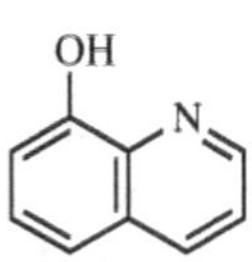 8-Hydroxyquinoline	Laccase, 8% acetone:buffer pH 5, 30°C, 24 h	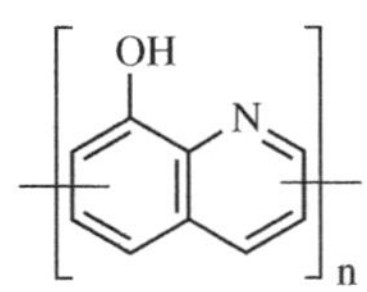 *n* up to 15	[232]

TABLE 5.2 (*Continued*)

Substrate	Reaction Conditions	Products	Reference
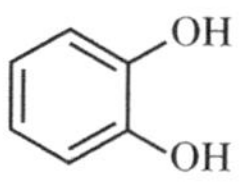	Laccase, acetone: buffer pH 5, 25°C, 48 h	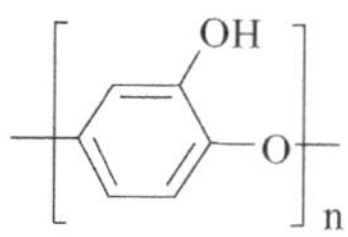 MW=813 ±3	[237]
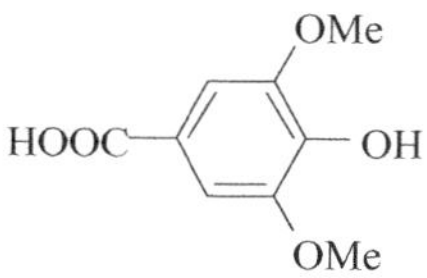 syringic acid	Laccase, acetone:$CHCl_3$: buffer pH 5, rt, 24 h	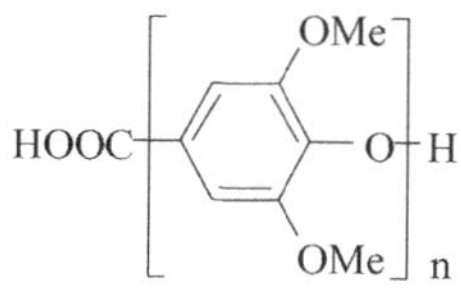 (MW up to 1.8×10^4)	[234,236]
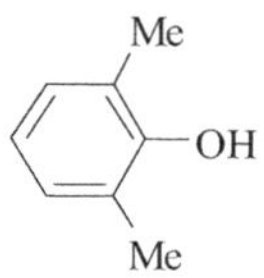	Laccase, dioxane:buffer pH 5, rt, 24 h	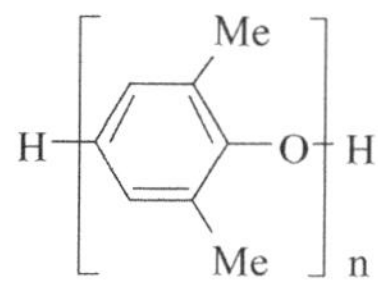 (MW up to 3000)	[235]

the antitumor antibiotic mithramycin afforded an adduct with a covalent link between C2 of hydroquinone and C5 of the polyphenolic core [243]. The efficient *T. versicolor* laccase-catalyzed coupling of the indole alkaloids catharanthine (**31**) and vindoline (**26**) gave the eniminium cationic intermediate **32**, which was reduced *in situ* with $NaBH_4$ to form the synthetically useful dimer anhydrovinblastine (**33**) in 56% isolated yield (Figure 5.26) [244]. The dimer **33** is an advanced precursor of vinblastine, an important anticancer drug mainly used in the treatment of leukemia. The applicability of this bioconversion has been tested through the condensation of catharanthine with the vindoline analog 11-methoxydihydrotabersonine.

The formation of protein-oligosaccharide conjugates—potentially of interest for the food industry—was reported to be initiated by laccases. The formation of hetero-cross-coupling products between the tyrosine side chain of α-casein and phenolic acid of xylan hydrolyzates was catalyzed by *T. hirsuta* laccase [245]. In another application, the *N*-coupling reaction of phenolic substrates with amines in the presence of laccase and oxygen has been carried out. It was proposed that the reaction proceeded *via N*-attack to a cation radical of the phenolic substrate [246]. For instance, laccase from *T. hirsuta* catalyzes the cross-coupling of the methyl propionate derivative **34** to 1*H*-benzotriazole to furnish Tinuvin (**35**), a benzotriazole-base UV-absorber (Figure 5.27) [247]. Conversely, the coupling of *o*-hydroquinones such as dihydrocaffeic acid (**36**), with aliphatic and aromatic primary amines afforded the

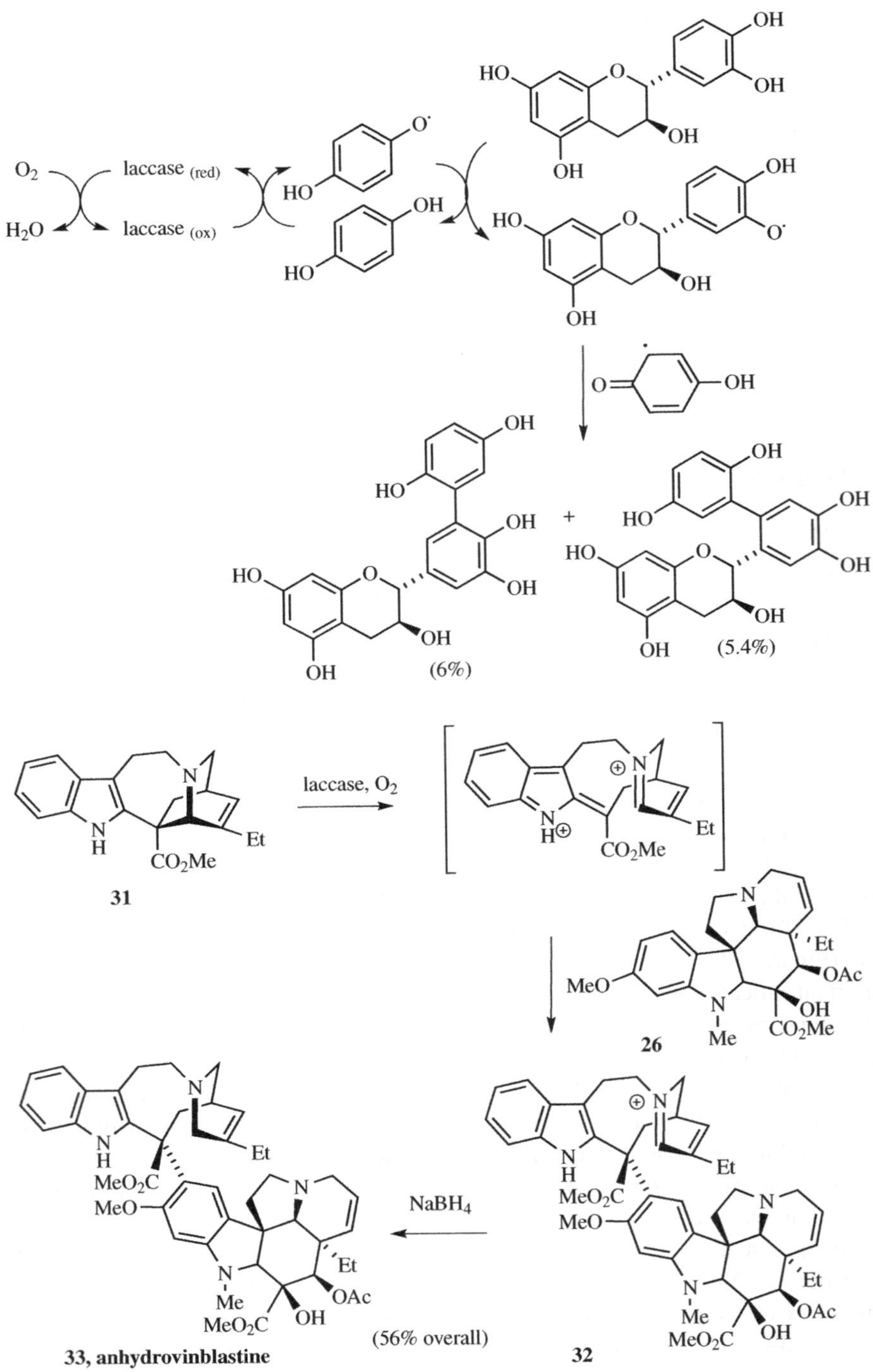

FIGURE 5.26 Laccase-catalyzed cross-coupling reactions.

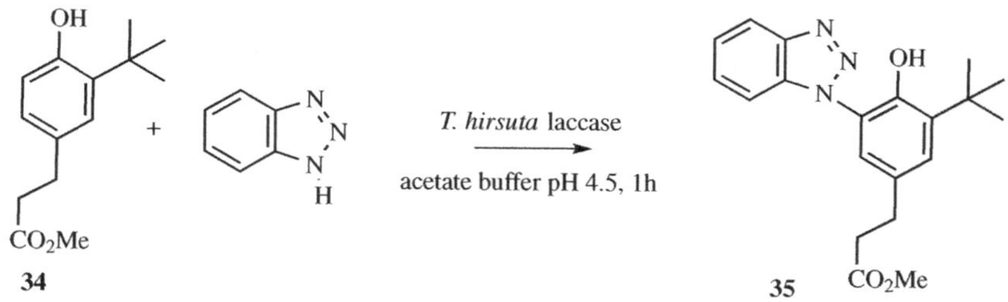

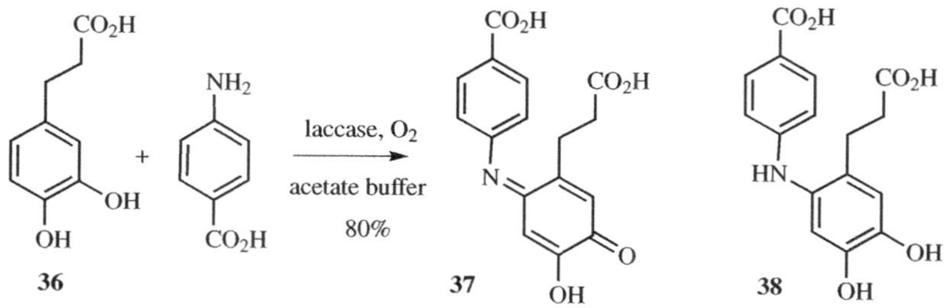

FIGURE 5.27 Laccase-catalyzed *N*-coupling of *N*-containing nucleophiles.

corresponding quinoneimines instead (Figure 5.27). The authors first reported an aminohydroquinone **38** as the product of reaction [246,248], but after a reinvestigation, the corrected structure was shown to be the quinoneimine **37**, derived from an *in situ* generated *o*-quinone [249]. More recently, the proposed structure of the dimers was questioned by the same authors, stating that both hydroquinoid and quinoid systems could be characterized, depending on the spectral method used [250].

An alternative approach within the exploitation of laccase in preparative chemistry consists of oxidizing the phenolic substrates to the respective quinone structures, which may react further with other compounds through different reactions such as aminations, Diels–Alder reactions, and Michael additions (by means of simple and domino reactions) (Figure 5.28).

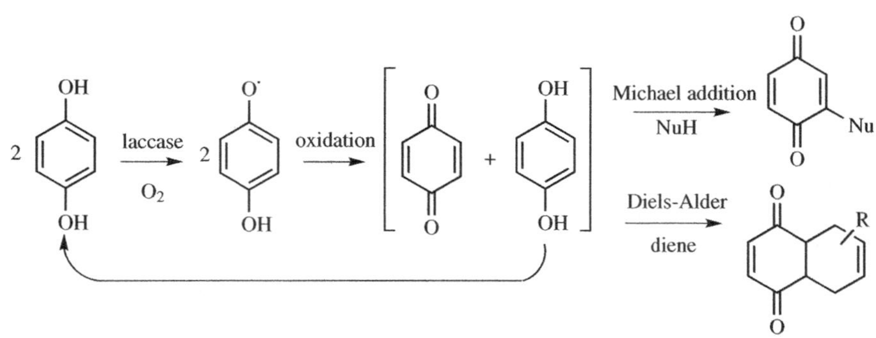

FIGURE 5.28 Formation and use of quinones obtained by laccase-mediated oxidation.

For instance, several laccase-mediated aminations have been reported. Aminoquinones were synthesized by nuclear amination of *p*-hydroquinones with primary aromatic amines in the presence of fungal laccases [249,251,252]. The postulated mechanism involves a Michael addition of the primary amine to the quinone intermediate (Figure 5.29). This strategy was also used to derivatize the unprotected amino acid L-tryptophan [253], *β*-lactam antibiotics like cephalosporins, and novel penicillins [254,255]. The quinone intermediate can also react with solvents such as water, methanol, and other alcohols, to form C–O bond cross-coupling products (Figure 5.29) [256]. Besides *p*-hydroquinones, *o*-hydroquinones have also been used as substrates, yielding the corresponding aminoquinones, as described in the laccase-mediated Michael addition of ^{15}N-sulfapyridine to protocatechuic acid (Figure 5.29) [257].

Diamination can take place as well, depending on the amine concentration, the pH, the reactant solubility, and the type of laccase. Since aminations are pH-dependent, the ratio of mono- to diamination could be controlled in aqueous media using enzymes with different values of optimum pH [249]. Likewise, *p*-hydroquinones produced exclusively diaminobenzoquinones when reacted with aryl and alkyl primary amines in the presence of immobilized commercial laccase (from *M. thermophila*) in aqueous medium with MeOH or dimethylformamide (DMF) as cosolvent, either at room temperature or at 35°C [258]. The *P. cinnabarinus* laccase-catalyzed reaction of 2,5-dihydroxybenzoic acid with 2-aminobenzoic acid at room temperature and pH 5 afforded exclusively the diaminated benzoquinone **39** in 26% yield, resulting from double Michael addition and further decarboxylation (Figure 5.30) [259].

The intermediate quinones can be used in domino reactions (Figure 5.31). For example, the laccase-catalyzed amination of 2,5-dihydroxybenzoic acid derivatives with aromatic and heteroaromatic amines affords cycloheptenes, cyclooctenes, diazaspiro cyclohexenes (**40**), and phenazines (**39**), in a one-pot reaction using either *M. thermophila* laccase at pH 7 or *P. cinnabarinus* laccase at pH 5 [259]. Similarly, 3-substituted-1,2,4-triazolo[4.3-*b*][4.1.2]benzothiadiazine-8-ones (**41**) were prepared in one step using laccase [260]. In addition, *O*-heterocycles derived from benzofurans (**42**–**44**) have been prepared from the reaction of catechols and 1,3-dicarbonyl compounds *via* an oxidation Michael addition domino reaction that is initiated by laccase [261–264]. Different 1,3-dicarbonyls have been used in these reactions, from acyclic to carbocyclic (cyclohexane derivatives) to heterocyclic compounds (pyridinones, quinolinones, and thiocoumarins), with yields of the benzofuran products ranging from 55% to 98%. The reactions were carried out in air at room temperature and in aqueous media, and for the acyclic compounds the addition of a Lewis acid (e.g., scandium triflate) improved the yields by catalyzing the Michael addition [264]. Moreover, the Michael addition can be catalyzed by lipases and thus the benzofuran is obtained through a one-pot, two-enzyme cascade reaction [265]. When crude extracts of *A. bisporus* are used, laccase and tyrosinase present in the extract catalyze the domino reaction between phenol and various cyclic 1,3-dicarbonyls yielding annulated dihydroxybenzofurans of type **42** and **43** in a highly efficient manner [266]. In a new domino process, the *A. bisporus* laccase-catalyzed reaction of *o*-phenylenediamine and benzaldehydes exclusively produces

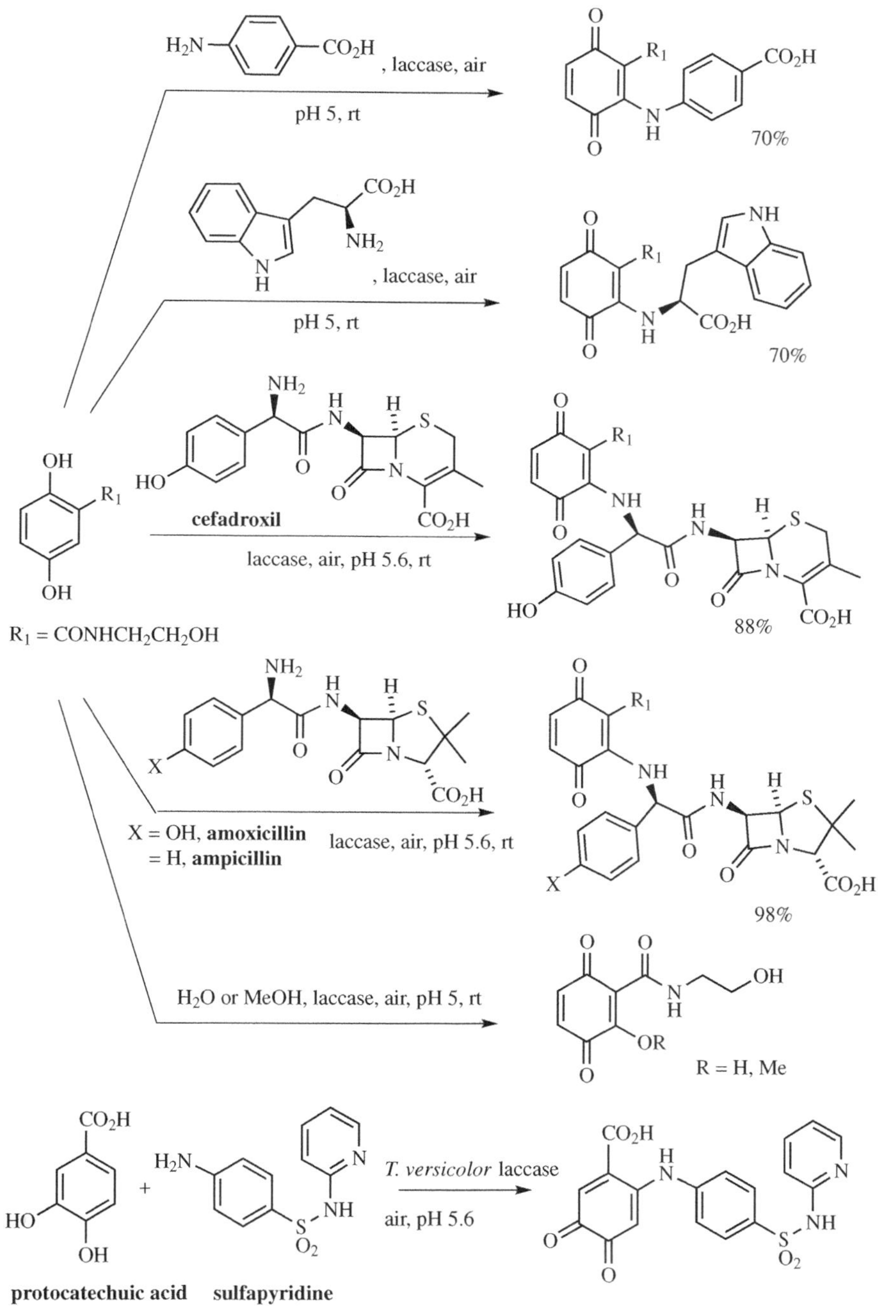

FIGURE 5.29 Laccase-mediated amination reactions.

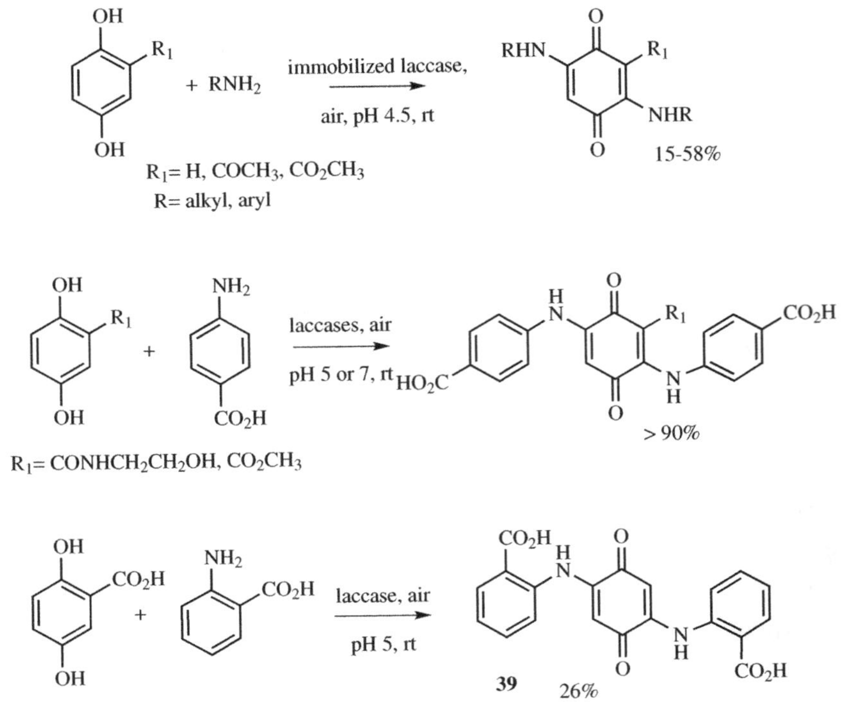

FIGURE 5.30 Laccase-mediated diamination reactions.

2-aryl-1*H*-benzimidazoles (**45**), in moderate-to-high yields (Figure 5.31) [267]. Recently, the preparation of benzopyranocoumarins was reported through a laccase-catalyzed domino reaction of α,β-unsaturated derivatives of coumarins with catechol or 1,4-hydroquinone [268].

Domino reactions have been used for the preparation of naphthoquinones through Diels–Alder reactions of dienes and quinones generated *in situ* by laccase in aqueous medium. The cycloaddition adducts underwent *in situ* aromatization to give the *o*- or *p*-naphthoquinones in good yields (Figure 5.32) [269,270].

Laccase-Mediator System *In vitro*, laccases oxidize electron-rich compounds such as phenols and anilines. Conversely, the redox potential of laccase alone is not high enough to break carbon–hydrogen aliphatic bonds. However, in the presence of a redox cocatalyst or a mediator, oxidation of certain carbon–hydrogen bonds becomes feasible [99,271–273]. The enzyme oxidizes the mediator, which can diffuse away from the enzyme, oxidize a substrate and then the reduced mediator returns to the catalytic cycle. The stoichiometric oxidant is oxygen. The catalytic cycle of the so-called LMS is shown in Figure 5.33 along with a range of common laccase mediators (TEMPO, (2,2,6,6-tetramethylpiperidin-1-yl)oxyl, free radical) and analogs, 2,2′-azino-bis(3-ethylbenzothiazoline-6-sulfonic acid) (ABTS) (2,2′-azino-*bis*(3-ethylbenzothiazoline-6-sulphonic acid), HBT (1-hydroxybenzotriazole), HPI

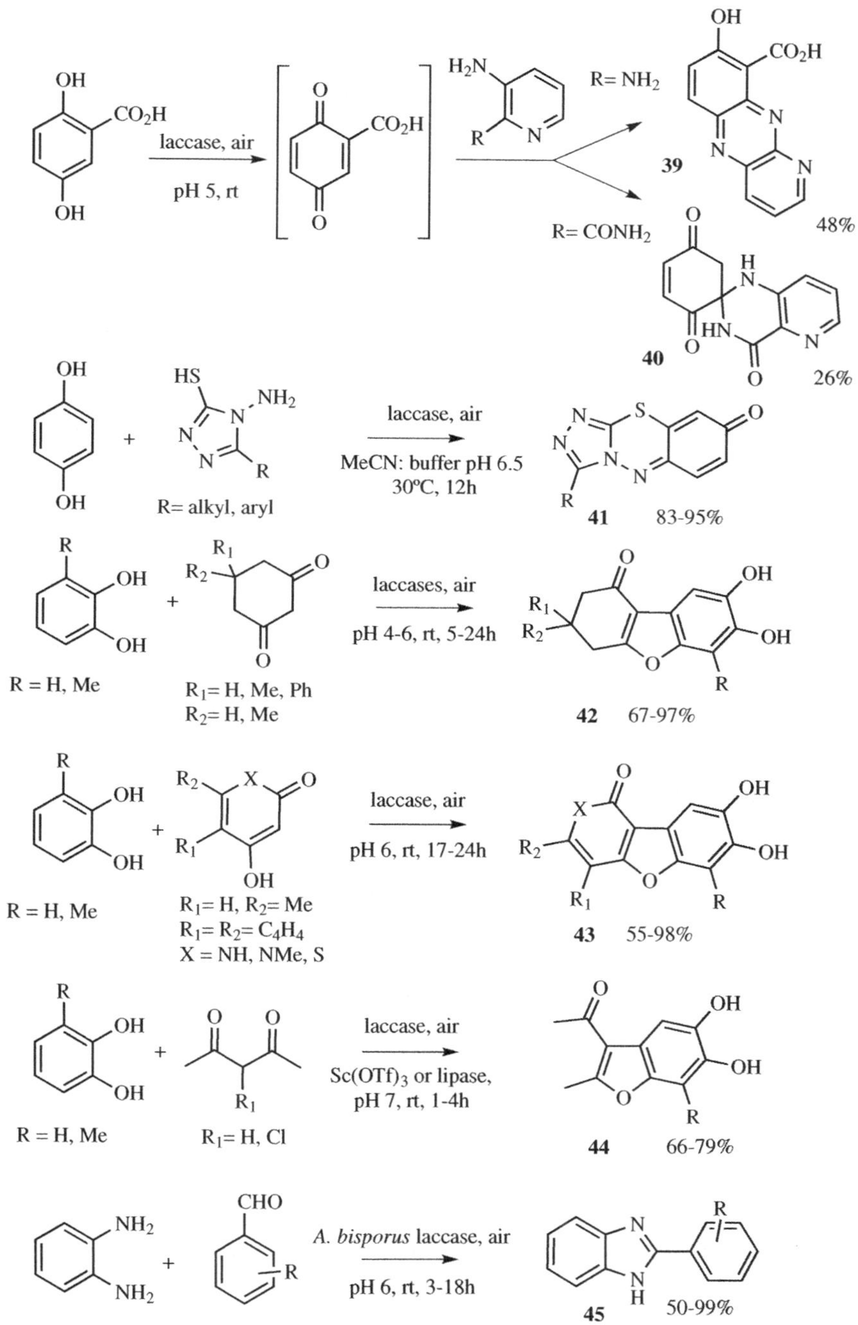

FIGURE 5.31 Laccase-initiated domino reactions of catechols, hydroquinones and diamines.

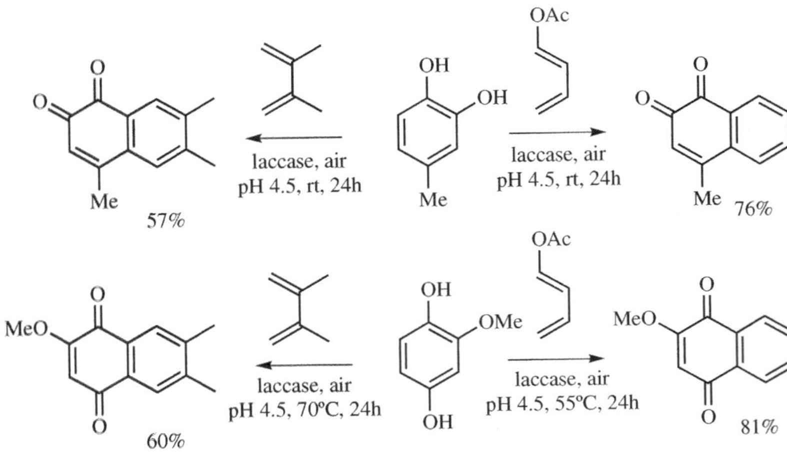

FIGURE 5.32 Laccase-triggered cascade synthesis of naphthoquinones.

(*N*-hydroxyphthalimide), NHA (*N*-hydroxyacetanilide), violuric acid, promazine (PZ), methyl syringate, HAA (*o*-hydroxyanthranilic acid), vanillin, syringaldehyde, ferulic acid, *p*-coumaric acid, sinapic acid) [274–280]. The mediator-substrate oxidation has been reported to occur *via* an ionic mechanism, or two radical mechanisms, namely electron transfer (ET) and hydrogen atom transfer (HAT) [273,275,276,281,282]. Consequently, TEMPO and its analogs favor the ionic pathway whereas mediators such as ABTS follow the electron transfer mechanism and *N*-OH compounds (HBT, HPI, violuric acid) prefer the radical hydrogen atom abstraction pathway [273,275,276,283].

The complete sequence of primary alcohol oxidation to aldehydes (benzyl alcohol to benzaldehyde) in air catalyzed by the laccase-TEMPO system was elucidated recently [284,285]. Laccase oxidizes two molecules of TEMPO to two oxoammonium cations (**46**), and whereas one of them oxidizes the alcohol to aldehyde, being reduced to the corresponding hydroxylamine **47**, the other cation is involved in the regeneration of TEMPO through noncatalyzed comproportionation with the newly formed hydroxylamine (Figure 5.34).

The reasons for the preferred use of LMS over a laccase alone are that mediators can nonenzymatically oxidize compounds with ionization potentials exceeding the redox potentials of laccases. Furthermore, they can also oxidize bulky substrates, which otherwise could hardly enter into the enzyme active site. For example, the redox potential of the enzyme, usually in the range of 0.5–0.8 V for fungal enzymes, can be increased to 1.1 V using the LMS strategy [161]. Consequently these oxidations, which occur in water at mild temperatures (20–45°C) and only require oxygen, producing water as the only by-product, constitute a green and environment friendly oxidation method of xenobiotic compounds [104]. Furthermore, the absence of any complicated cofactor recycling is very attractive as well. These features have led to a widespread use of the LMS in both academic and industrial settings. The industrial and biotechnological (potential) applications of LMS include the detoxification

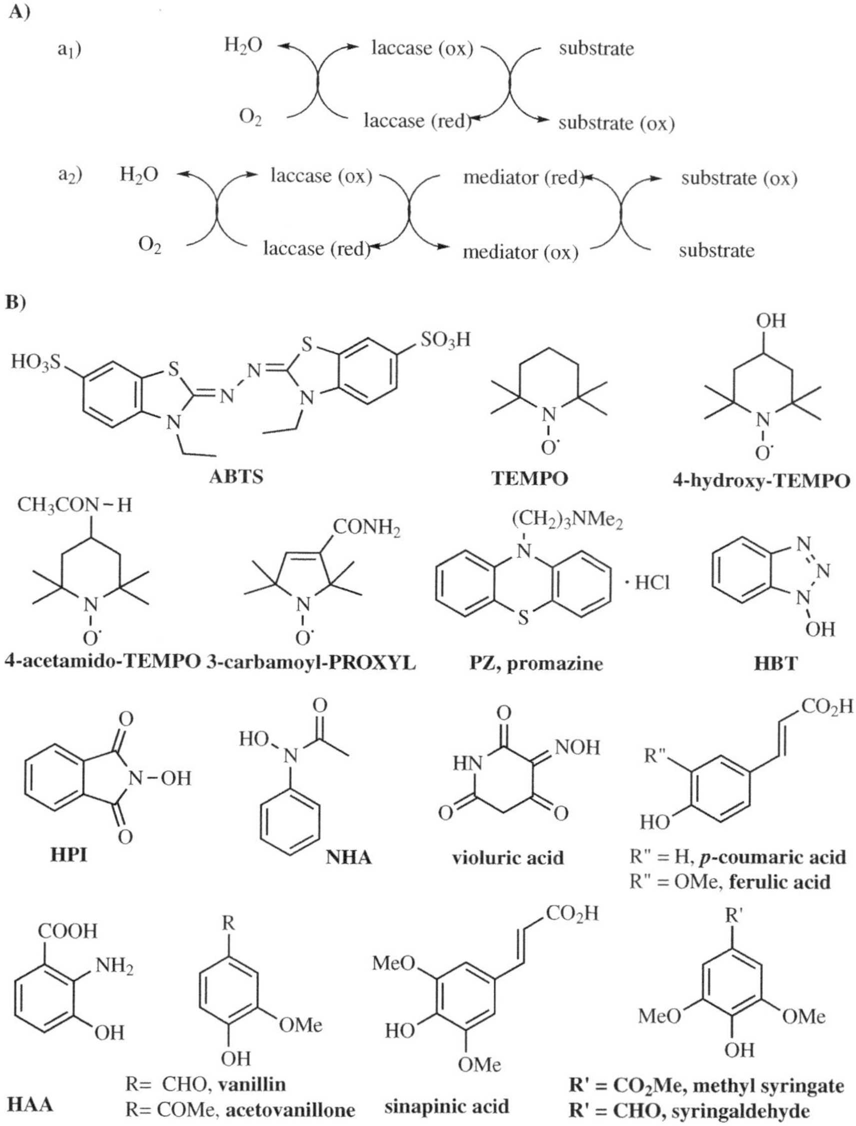

FIGURE 5.33 **A**, laccase-catalyzed substrate oxidation with and without chemical mediators; **B**, common artificial and lignin-derived laccase mediators.

of industrial effluents, mostly from the paper and pulp, textile and petrochemical industries, the use as bioremediation agents to clean up xenobiotics (herbicides, pesticides and certain explosives) in soil, and as cleaning agents for certain water purification systems. The LMS has also been used as catalyst in the pharmaceutical and fine chemistry (cosmetic) industry, and as a tool for medical diagnostics [177]. Representative examples can be found in the delignification and biobleaching of

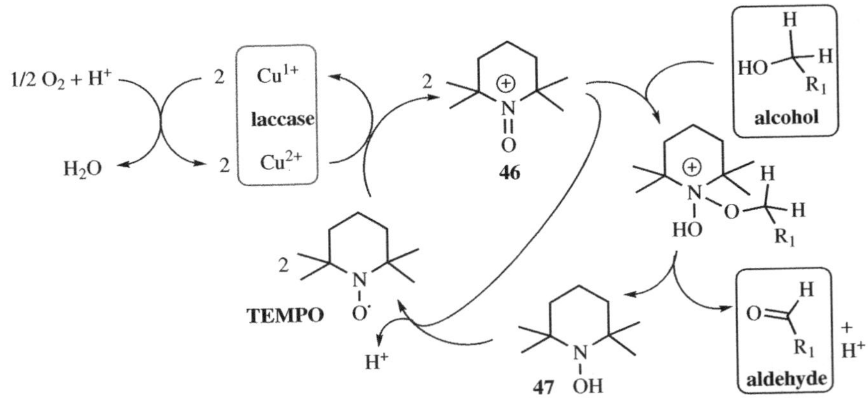

FIGURE 5.34 Postulated mechanism of the laccasse-mediated oxidation using TEMPO as redox mediator [284].

pulp; [271,274,286–292] treatment of wastewater from industrial plants; [293,294] enzymatic modification of fibers and dyes removal processes in the textile, dye, and printing industries; [295–298] enzymatic cross-linking of lignin-based materials to produce medium density fiberboards; [299] detoxification of pollutants and bioremediation; [300–302] detoxification of lignocellulose hydrolysates for ethanol production by yeast; [303,304] selective removal of phenol derivatives to stabilize beverages (fruit juices, wine and beer); [163,305–307] and construction of biosensors and biofuel cells (the use of laccases and other oxidoreductases in biofuel cells is a promising field for nanotechnological applications, which is increasing rapidly) [308–310]. These biotechnological uses are the subject of specialized reviews and will not be covered further in this chapter [177–184,308,311–317]. Applications in synthetic chemistry will be fully discussed in Section Synthetic applications of the Laccase-Mediator System.

SYNTHETIC APPLICATIONS OF THE LACCASE-MEDIATOR SYSTEM The redox potential of laccasse allows only the oxidation of phenolic fragments of lignin, turning them unsuitable for the direct oxidation of functional groups with higher redox potential, such as carbon-hydrogen bonds of alcohols. However, this limitation can be overcome by using the LMS: the target alcohol is oxidized by a higher redox potential mediator, used in catalytic amounts, which is kept in its oxidized form by laccase catalysis (Figure 5.33, a_2) [179,318]. In this way, the LMS allows to oxidize molecules with redox potential (E°) above $+1100$ mV, whereas typical E° values for laccase range from $+475$ to $+790$ mV (at the T_1 site) [319,320]. Besides its redox potential, the mediator acts as a diffusible electron carrier enabling the oxidation of bulky substrates (such as lignin or other high molecular weight polymers), thus circumventing the steric issues related to the access into the enzyme active site. The proper choice of mediator is a key consideration for a given biotransformation, since different mediators may yield different final products starting from the same substrate [161]. Basically, the reason is that substrate oxidation in LMS can take place *via* different

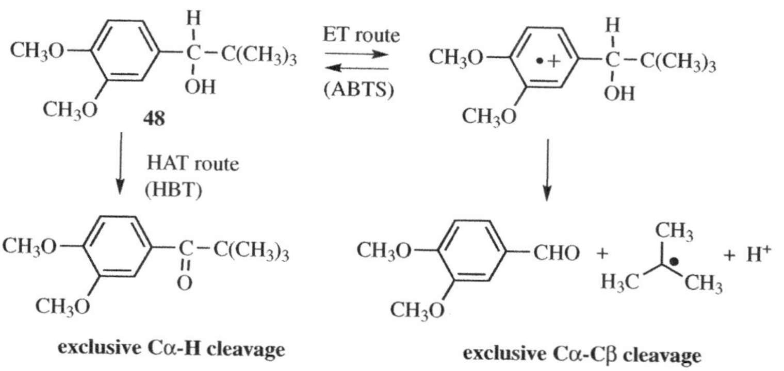

FIGURE 5.35 Oxidation mechanisms followed by ABTS and HBT radicals. ET: electron transfer. HAT: hydrogen atom abstraction.

mechanisms—one ionic and two radical pathways as mentioned before—according to the chemical structure and redox potential of the mediator. For instance, the choice of mediator in the LMS-oxidation of the nonphenolic substrate **48** (resistant to laccase oxidation) determines the product distribution, giving exclusively the aldehyde, product of α,β-cleavage, when using ABTS (*via* the electron transfer route, ET), or the ketone, resulting from hydrogen atom abstraction (HAT route), when using HBT or other *N*-OH mediators (Figure 5.35) [276].

However, although over a hundred redox mediators have been described [161], the most common set of mediators used is displayed in Figure 5.33B. Interestingly, there is a renewed interest in the use of laccase mediators of natural origin, which show an efficiency comparable to the artificial ones and are obtained at lower cost (see below). These so-called ecofriendly mediators are easily available from lignocellulose and are devoid of toxicity [161,321].

The LMS (Figure 5.33, a_2) have found many technological applications [177], as evidenced by the numerous research articles and invention patents published [179,308,322]. Reasons for this interest can be found in the inherently "green" features of these systems, which can work with air at mild temperature and produce water as the only by-product. This is very appealing for the textile and pulp and paper industries, in constant search for efficient and environmentally benign processes (provided that economics can also be matched). The first artificial mediator that was used in the LMS for pulp delignification was ABTS [323], but the most effective mediators for this task are heterocycles bearing *N*-OH groups, such as HBT (*N*-hydroxybenzotriazole; Figure 5.33B). The process has been patented under the trade name "Lignozym® process," and its efficiency has been demonstrated in several pilot plant trials [271]. The study of the laccase-mediated delignification process using lignin model compounds has shown that redox mediators allow these enzymes to oxidize nonphenolic substituents, particularly benzyl alcohol groups [324], as in the efficient oxidation of adlerol to the ketone adlerone using violuric acid as mediator (Figure 5.36) [325]. Also, the oxidation of double bonds conjugated to aromatic rings has been accomplished in lignin model compounds. In the reaction

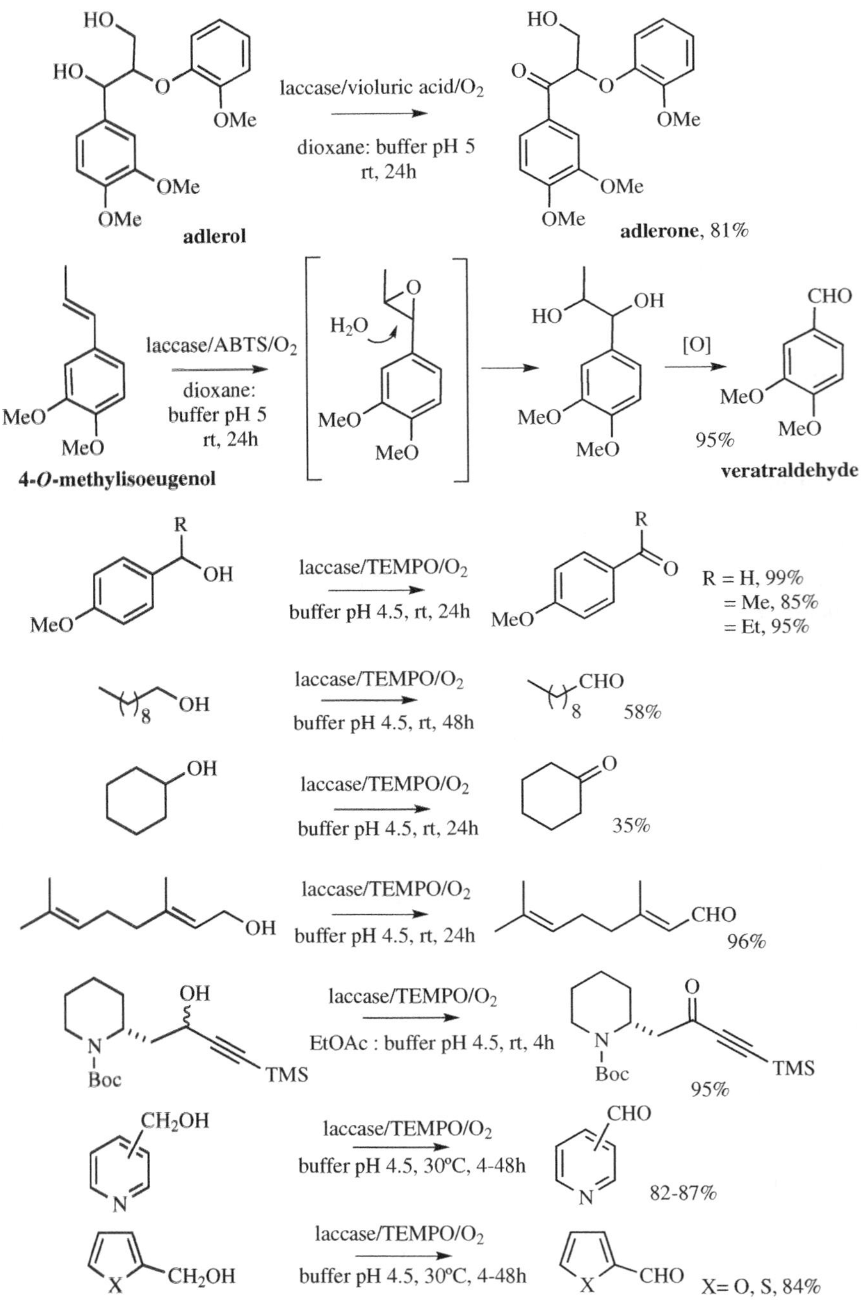

FIGURE 5.36 Oxidation of alcohols by the LMS.

of 4-*O*-methylisoeugenol with the laccase/ABTS system, the corresponding epoxide is formed first, which is then opened to the diol and further oxidatively cleaved to veratraldehyde (Figure 5.36) [326]. Similar results were reported using mediators bearing an *N*-OH functionality (HBT, HPI) [327].

To widen the scope of this oxidation, the biotransformation was applied to the oxidation of benzylic [283,285,318,327–329], allylic [327,328], propargylic [330], and aliphatic alcohols [328,331] to their corresponding aldehydes and ketones (Figure 5.36). The use of cross-linked enzyme aggregates (CLEA) of *T. versicolor* laccase for TEMPO-mediated aerobic oxidation of aliphatic alcohols in aqueous buffer (pH 4) required lower enzyme loadings to achieve similar results than those observed with the free laccase [331]. The evaluation of a dozen different mediators in the oxidation of the model substrate 4-methoxybenzyl alcohol showed that TEMPO was the most effective compound for this kind of reaction, which operates through an ionic mechanism (Figure 5.36) [273,283,318].

The laccase-TEMPO system has been patented for the partial oxidation of cellulose, giving 6-aldehydes and 6-carboxylates [332,333], and has also been used in a more systematic investigation for the oxidation of hydroxyl groups of different sugar derivatives. The LMS (laccase from *T. pubescens* and TEMPO) effectively catalyzed the regioselective oxidation of the primary hydroxyl groups of mono- and disaccharides (e.g., phenyl β-D-glucopyranoside, **49**), and the intermediate aldehydes were oxidized further, *in situ*, to the corresponding glycopyranosiduronates, which were isolated and characterized (Figure 5.37). Both components of the LMS were needed to accomplish the oxidation [334]. This approach, when applied to a water soluble cellulose samples, afforded a low conversion, which could be improved by the previous activation of the cellulose fibers by an alkaline treatment [334]. This LMS has been used for the mild oxidation the glycosylated saponin asiaticoside (**50a**) [335], and a series of natural glycosides, such as amygdalin (**51a**), colchicoside (**52a**), and ginsenoside R_E (**54**) [336,337]. In all cases, the LMS catalyzed the regioselective oxidation of the sugar primary alcohol group to the corresponding carboxylate *via* the intermediate aldehyde (Figure 5.37). In addition, the influence of immobilization and the use of organic cosolvents (acetone, dioxane, DMF, DMSO) on the stability and activity of a laccase from *T. pubescens* were studied using the model compound thiocolchicoside. It is worth noting that the formation of the corresponding sulfoxides (or sulfone) derivatives was never observed [336].

In addition to the oxidation of alcohols, the LMS-catalyzed oxidation of other functional groups has been reported, for example, the oxidation of ethers [338], alkenes [339], amides [340], amines [318], and aromatic methyl groups (Figure 5.38) [327,341]. Both deactivated and activated primary benzyl amines give aldehydes in 40–100% yield after oxidation to the imine and subsequent hydrolysis [318]. With tertiary amines such as substituted *N,N*-diethylbenzylamines, the oxidation can afford the corresponding benzaldehydes or mixtures of *N*-dealkylated products in variable yields [318,328]. A series of primary and secondary hydroxylamines afforded high yields of oximes and nitrones when oxidized using an LMS containing either TEMPO or ABTS as mediators [318]. Activated aromatic aldehydes can be prepared in aqueous acidic medium from the corresponding methyl aromatic compound using an LMS

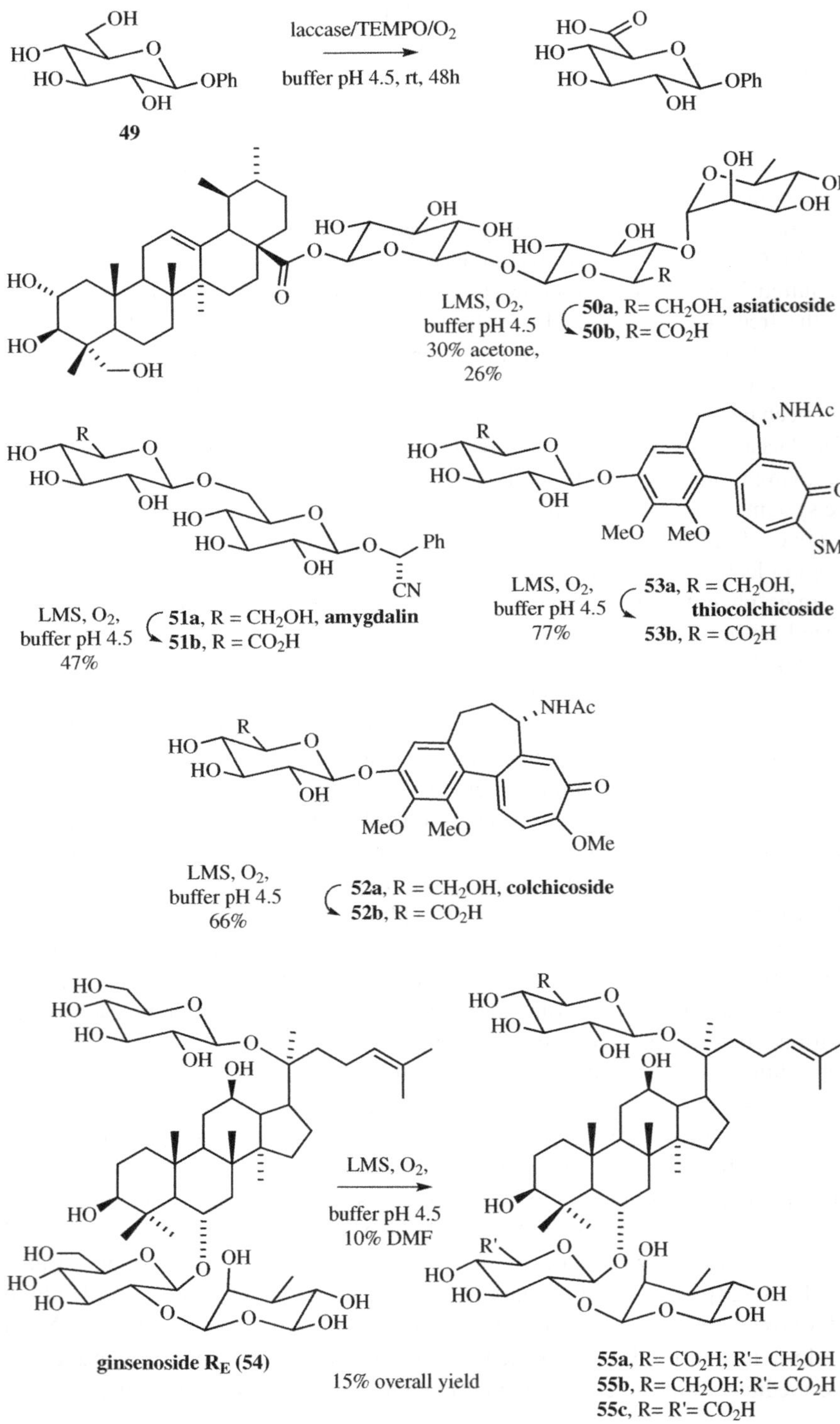

FIGURE 5.37 Laccase/TEMPO-catalyzed regioselective oxidation of glycosides.

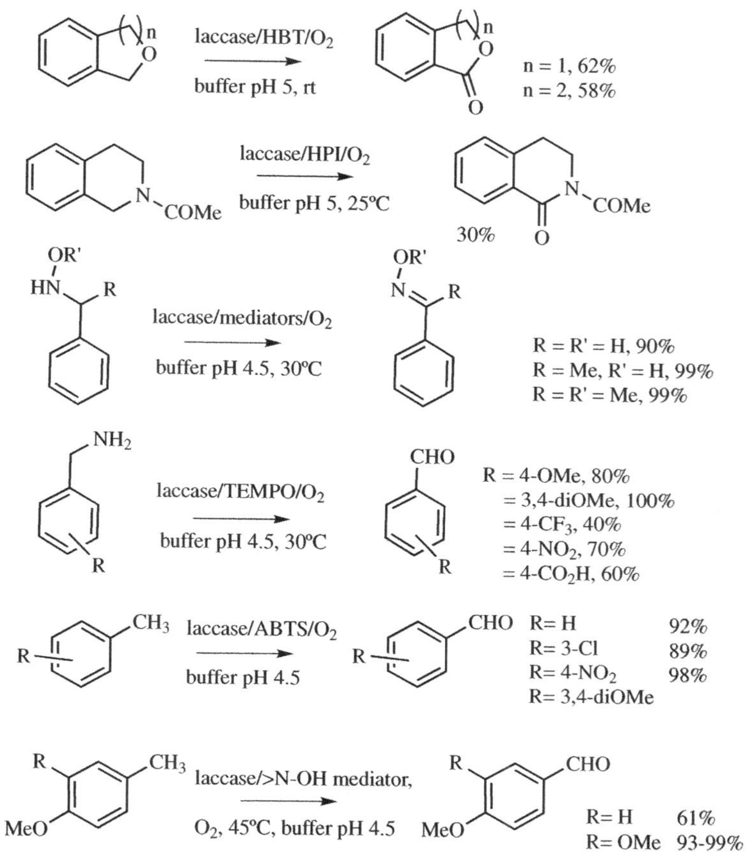

FIGURE 5.38 Laccase/mediator-catalyzed oxidation of ethers, amides and aromatic methyl groups.

containing catalytic amounts of *N*-hydroxy compounds as mediators, such as HBT or 3-amino-*N*-hydroxyphthalimide (3-amino-HPI) (Figure 5.38) [327]. Moreover, high-to-excellent yields of differently substituted aromatic aldehydes have been reported using ABTS as mediator (Figure 5.38) [341].

The oxidation of alkenes was highly dependent on the alkene and mediator used, and the conversions were always under 50%, giving best results with HBT as mediator [339]. Major oxidation products were the corresponding carbonyl compounds, resulting from allylic oxidation (Figure 5.39). Interestingly, with this LMS the oxidation of the polyunsaturated alcohols linalool, geraniol, and nerol, gave the corresponding linalool oxides instead of the expected carbonyl compounds. Recently, the oxidation of unsaturated lipids (fatty acids and sterols) by a laccase/HBT system has been studied [342]. Hydroperoxide- and epoxide-type compounds were the main products from oxidation of unsaturated fatty acids, while the main products detected from the oxidation of sterols were steroid ketones. In particular, the oxidation of

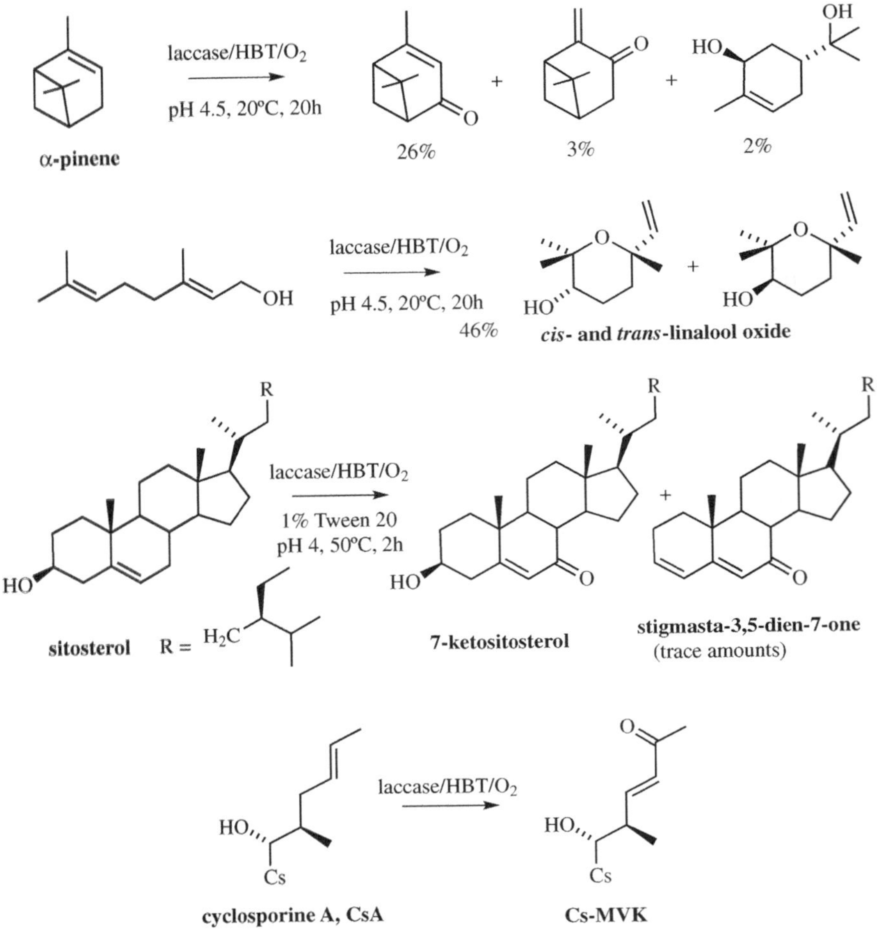

FIGURE 5.39 Laccase/HBT-mediated oxidation of unsaturated compounds.

sitosterol afforded mainly 7-ketositosterol with only traces of the corresponding 7-hydroxysitosterols, and a small amount of stigmasta-3,5-dien-7-one, resulting from dehydration (Figure 5.39). An apparently related allylic oxidation has been recently disclosed on Cyclosporine A, a naturally occurring cyclic nonribosomal peptide endowed with potent immunosuppressive activity. Cyclosporine A (CsA) was converted into cyclosporine methyl vinyl ketone (Cs-MVK) *via* oxidation of its unsaturated side chain by using HBT/laccase-mediated oxidation. The resulting enone is a novel and versatile synthetic intermediate that can be used for the preparation of cyclosporine analogs with therapeutic potential as immunosuppressive agents (Figure 5.39) [343].

Polycyclic aromatic hydrocarbons (PAH) have also been oxidized using LMS [159,278,344]. PAHs are oxidized to quinones using HBT (but not ABTS as mediator), and laccase from *P. cinnabarinus*. Similar results are obtained with "natural"

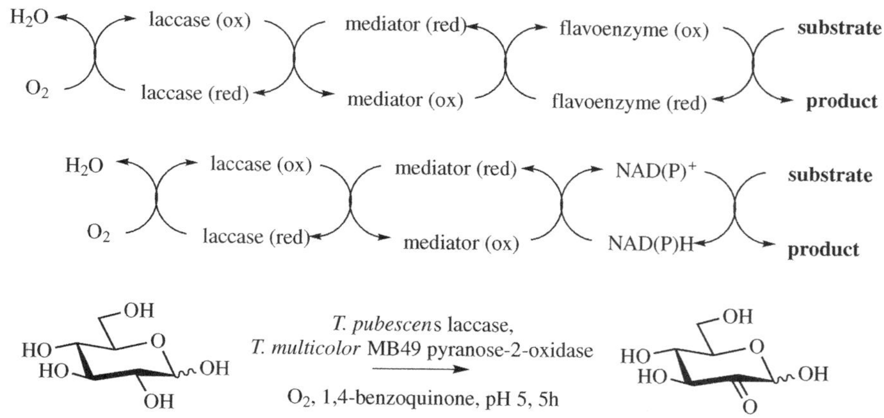

FIGURE 5.40 Laccase-catalyzed regeneration of redox cofactors.

mediators, for example, phenolic compounds derived from lignin degradation, in particular *p*-coumaric acid and *p*-hydroxycinnamic acid [278]. Thus, using fungal laccase and HBT or *p*-coumaric acid, 95% removal of anthracene and benzo[*a*]pyrene and around 50% of pyrene was attained within 24 h [344]. Recently, four laccase isoenzymes from *T. versicolor* have been cloned and expressed in *P. pastoris* [159]. The isoenzymes showed very different behavior with respect to oxidation of PAH. One of them displayed high activity for the oxidation of PAH, and thus, after 72 h more than 80% of fluorene, anthracene, acenaphthene and acenaphthylene were oxidized in the presence of ABTS as mediator. The oxidized products were mainly quinones and diacids [159]. Although laccase alone is able to oxidize phenols, the extent and rate of this transformation was enhanced in the presence of mediators [345,346].

A noteworthy evolution of the LMS is related to its use for the *in situ* regeneration of cofactors, as shown in Figure 5.40 [347]. In this case, the laccase-oxidized mediator in turn oxidizes the cofactor of an oxidase (either flavin or NAD(P)H) in a coupled enzymatic system for the *in situ* regeneration of cofactors, producing H_2O as the only by-product. Using this method, lactose was efficiently oxidized to lactobionic acid in a cascade enzymatic reaction using CDH to oxidize the reducing sugar moiety at C1, whereas the electron acceptor was continuously regenerated by laccase (thus being added in catalytic amounts). Suitable redox mediators were 1,4-benzoquinone, ABTS, and ferricyanide, and best results were obtained with 1,4-benzoquinone and laccase from *T. pubescens*, reaching a specific productivity of up to 32 g $kU^{-1}h^{-1}$ of lactobionic acid [69,348]. More recently, the same LMS system (laccase from *T. pubescens* and 1,4-benzoquinone) has been used for the oxidation of glucose to 2-ketoglucose, catalyzed by pyranose-2-oxidase in a cascade process (*in situ* regeneration of FAD) [349]. Not only flavoenzymes but also oxidized nicotinamide cofactors can be regenerated (Figure 5.40). By combining ABTS-catalyzed oxidation of NAD(P)H with laccase-catalyzed utilization of molecular oxygen as terminal oxidant, a simple chemoenzymatic $NAD(P)^+$ regeneration method is available. This method was tested using an alcohol dehydrogenase from *Thermus* sp. to oxidize secondary alcohols, with promising results [350].

A crucial issue when using an LMS is the fate of the mediators [318]. Theoretically, they are continuously regenerated by the enzymatic system. In practice, however, the situation may be different and some reactions may consume significant amounts of the mediator. Analysis of the reaction products revealed that the predominant pathway to account for the mediator's consumption was deoxygenation, *via* both ionic as well as radical mechanisms. Thus, TEMPO is deoxygenated to afford 2,2,6,6-tetramethylpiperidine, whereas HBT yields benzotriazole. This consumption is important for recalcitrant substrates, which require large amounts of enzyme and mediator. In these conditions, deoxygenation of the mediator may become the predominant process, leading to a rapid deactivation of the laccase [318]. Mediator stability and deleterious interaction of the oxidized mediator with the protein can severely impair the efficiency of the oxidation process. For these reasons is important to select the best combination of enzyme and mediator when using LMS. Further research in this field is needed in order to elucidate the fate and also design improved mediators.

Miscellaneous The capacity of laccase to oxidize phenols and anilines has been used in the detoxification of azo-containing compounds and in the unmasking of amines and carboxylic acids protected with oxidation-sensitive aromatic groups. Being the largest group of industrial colorants, azo dyes are widely found in effluents from the textile industry, and their decomposition is of environmental interest. The laccase-catalyzed decomposition of azo dyes has been studied [351,352], in particular the oxidation of phenolic azo dyes using *Pyricularia oryzae* laccase [353]. The compound is oxidized to a 4-sulfonylhydroperoxide (**56**) and a quinone (**57**), among other products (Figure 5.41). The proposed mechanism involves the formation of a phenoxy radical, which is further oxidized by laccase to products **56** and **57**. The reaction could be accelerated if a combination of laccase and ultrasound treatment were used [351]. In addition to decompose azo dyes, laccase was also reported to catalyze their formation *via* oxidative coupling between *o*-, *m*-, and *p*-methoxyphenols and 3-methyl-2-benzothiazolinone hydrazone [354]. The biotransformation of the anthraquinonic dye Acid Blue 62 by laccases produces colored azo compounds, resulting from coupling and further oxidation of amino-based radicals [355]. Furthermore, the use of laccase (and peroxidase) as selective deprotection agents in peptide synthesis has been reported. In this way, the phenylhydrazide protecting group of both α- and γ-carboxyl groups can be oxidatively removed at pH 4 in aqueous medium under mild conditions, without altering labile amino acids such as methionine and tryptophan in the peptide backbone [356]. Regarding the amino group, Rutjes and coworkers described the oxidative deprotection of *p*-methoxyphenyl (PMP)-protected amines by laccase under mild acidic conditions (Figure 5.41) [357]. The use of ABTS increased the substrate scope and the reaction rate. The transformation of substituted amino imidazoles has been reported to yield newly functionalized imidazoles and acyclic compounds resulting from coupling of two ring-opened molecules of the substrate [358].

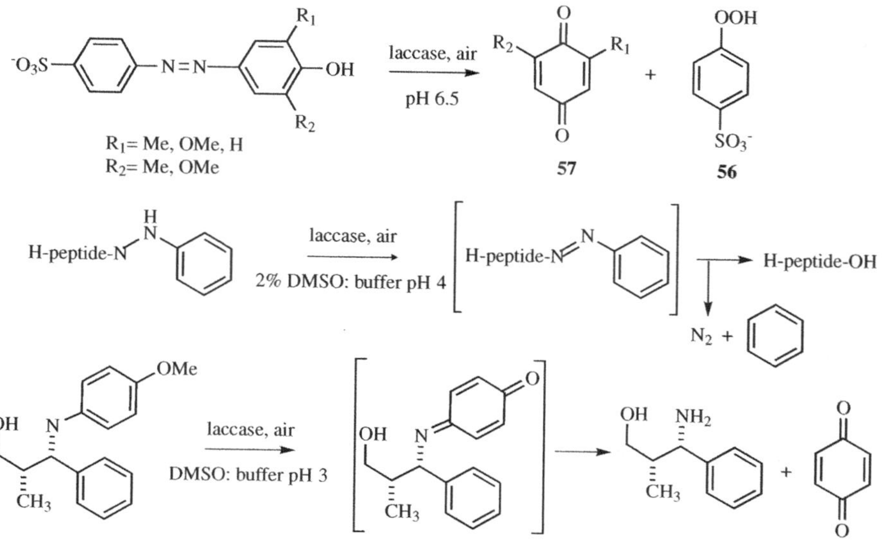

FIGURE 5.41 Laccase-mediated decomposition of azo dyes and carboxyl- and amino-protecting groups.

The low aqueous solubility of many laccase substrates (e.g., lignans, PAH) requires the presence of organic solvents in the reaction medium [151,168,173,302], as exemplified in the number of described synthetic applications using biphasic aqueous-organic systems or significant amounts of cosolvents. The effect of different water miscible cosolvents on laccase activity and stability has been reviewed [359]. Maximum reaction rates remained within the same order of magnitude in aqueous solutions, reverse micelle systems and in the presence of up to 20–30% v/v of several water-miscible solvents [360]. On the other hand, the TEMPO-mediated *T. pubescens* laccase oxidation of glycosides was severely influenced by the presence of water-miscible cosolvents. Using 20% v/v dioxane or acetone, the degree of conversion was significantly lower than in the buffered reaction, and, moreover, with 20% v/v DMSO the reaction did not proceed [336]. The stability of laccases was affected by the cosolvents. For instance, the residual activity of a laccase from *Polyporus pinsitus* decreased substantially after 24 h when the enzyme was dissolved in a 1:1 mixture of buffer with ethanol, DMF, or DMSO, whereas at least 70% of the activity was retained with dioxane, isopropanol, ethylene glycol, and acetonitrile [361]. Immobilized laccases performed successfully in apolar solvents with low water contents [192,362,363]. In some cases, the nature of the solvent had a significant influence on the selectivity of the laccase-mediated oxidation, as described previously for the coupling of tetrahydro-2-naphthol derivatives (Figure 5.19) [192]. The stability and activity of laccase in aqueous solutions of ionic liquids of the imidazolium family has been studied, showing that the stability diminished with the length of the alkyl chain in the ionic liquid [364,365]. In the presence of organic solvents, some laccases are prone to be denaturated or inhibited [366]. In this regard several strategies have been reported for improving the performance of laccases in organic solvents for example, enzyme immobilization [367] and protein engineering [368], which are discussed below.

CATALYST ENGINEERING (IMMOBILIZATION AND MOLECULAR BIOLOGY STRATEGIES) As stated before, the ability of laccases to perform oxidations using air and producing only water as by-product is very appealing for their biotechnological applications. However, this requires the supply of large quantities of stable and active enzymes, unlikely to be realized with native laccases in wild-type strains, due to the low yields usually achieved and the reduced stability in organic solvents. Molecular biology and immobilization strategies have been used to address these drawbacks, in particular, (i) the development of heterologous expression aiming at industrial production; (ii) the enhancement of operational stability and activity; and (iii) the increase of the enzyme redox potential in order to broaden the substrate range.

Heterologous Expression The expression of the high-potential fungal laccases in heterologous systems is an active field of research, given the potential benefits of a dependable production of recombinant laccase over purifying the enzymes directly from wild-type cultures. For instance, recombinant technologies allow expression of a single enzyme in high protein yields, whereas many fungal species secrete several isoenzymes (especially the white-rot fungi) [164,165], lowering the yield and increasing the purification costs. Moreover, the production of laccase by wild-type strains is tightly linked to secondary metabolism and its associated problems (e.g., secretion of proteases that inactivate laccases and uncontrolled fungal growth) [369]. As mentioned, several heterologous expression systems have been used, mainly filamentous fungi (*Aspergillus* and *Trichoderma* sp.) and yeasts (*Saccharomyces, Pichia, Yarrowia* and *Kluyveromyces* sp.) for fungal laccases, and *E. coli* for bacterial laccases [150,161]. Fungal laccases, like other ligninolytic enzymes, are difficult to express in nonfungal systems [155]. Moreover, the high redox-potential laccases, better suited for biocatalysis, are predominantly found in basidiomycetes, which are less amenable to genetic manipulation than ascomycetes. Thus, the enzyme production levels achieved are generally low, in the range of a few hundreds of milligrams per liter [150,161]. Notable exceptions are the laboratory-scale production of 920 mgL^{-1} of *M. albomyces* laccase expressed in *T. reesei* [158], and the use of the basidiomycete *P. cinnabarinus* for the homologous expression of a laccase in a yield of 1200 mgL^{-1}, representing the highest level reported for a recombinant fungal strain [370]. The expression of fungal laccases in bacterial hosts, although highly desirable given their easy cultivation and the large number of expression vectors available, is still a challenge because of the requirement of glycosylation, missing chaperones, and different codon usage, among other issues to be solved. So far, only one example has been disclosed, the C. bulleri laccase expressed in *E. coli* [162]. Recently, a bacterial laccase gene identified through activity-based functional screening of a metagenomic library from mangrove soil was expressed in *E. coli* giving rise to 380 mgL^{-1} of purified enzyme [371]. Finally, addressing the biocatalyst costs, efficient laccase-production bioreactors have been designed using different cultivation techniques such as submerged and solid-state fermentation [369].

The glycosylation pattern of laccases is an important consideration for overexpression and production of these enzymes. Glycosylation is likely to influence native and recombinant enzyme activity, but only recently this topic has been taken into consideration. For instance, the *N*-glycosylation status of *Pycnoporus sanguineus* CelBMD001 laccase was determined and it was found that stability and activity were decreased in the *N*-deglycosylated protein [372]. In addition, *S. cerevisiae* and *P. pastoris* yeast expression systems are known to cause hyperglycosylation of nonnative laccases, which might interfere with protein folding and function [373]. Consequently, the importance of posttranslational glycosylation during heterologous expression of laccases in different yeasts has been assessed. Recently, it was found that a glycan-free *Trametes trogii* laccase produced in a *K. lactis* defective in *O*- and *N*-glycosylation pathways for laccase expression had lower activity than the corresponding recombinant laccase expressed in an unaltered strain [374]. Current data suggest that glycans influence laccase properties significantly and further studies on their role are needed.

Enhancement of Operational Stability and Activity Laccase immobilization on solid supports has been thoroughly investigated as an efficient way to improve enzyme stability and to enable recycling of the biocatalysts. Both chemical and physical methods have been used for immobilization of microbial and plant laccases, resulting usually in increased kinetics, or stability, or both [367]. For instance, immobilization by physical methods has been successful using glass powder, silica gel [375], alumina [295], polyamides and polyacrylamides [209,375], graphite and carbon black [309], alginate [376] polyurethane, gelatin [377], transition metal oxides, vitroceramics [378], molecular sieves [379], among others. Solid supports used for chemical methods include activated sepharose [380–383], DEAE-cellulose [379], glass beads [384–387], activated agarose [388,389], activated carbon [390], chitosan [391–394], epoxide-containing acrylic resins [226,395,396], activated silica and alumina [379,388], and vitroceramics [397]. Immobilization without solid support could be achieved through the preparation of cross-linked enzyme crystals (CLEC) of laccase, which showed very high stability and reuse capability [398]. Similarly, CLEAs of three laccases have been prepared [331]. The immobilization of laccase into cationic polymers (chitosan, polyethyleneimine) cross-linked with epichlorohydrin has promising industrial applications, since these supports are able to promote the coagulation of the insoluble products from laccase-mediated oxidations [395], thus preventing their accumulation at the outlet of the bioreactor (fouling). These reaction products, arising from nonenzymatic coupling of laccase-generated free radicals, tend to accumulate and cause reduction of the flow rate or complete plugging, and constitute an important technical issue [367].

In addition to the more traditional immobilization techniques, the challenge of enhancing the operational stability and activity of laccases has been approached using complementary strategies of enzyme engineering, namely site-directed mutagenesis, directed evolution, or combinations of both. Rational design through site-directed mutagenesis has been used to assess the substrate binding pocket and active site, and for *C*-terminus modification of laccases. A single residue modification in the substrate-binding pocket of recombinant *T. versicolor* Lccβ laccase produced a significant reduction of activity toward the oxidation of aromatics. Thus, when Leu164 was replaced by Val, the resulting single point mutant (Leu164Val) showed a 10-fold decrease in the oxidation of anthracene, and similar results were found when Phe265 was replaced by Ala (Phe265Ala), indicating that these positions contribute to binding and stabilization of bulky hydrophobic aromatic compounds [159]. The study of the binding pocket of *M. albomyces* laccase showed that the carboxylic acid residue Glu235, located at the entrance of the binding site pocket, is crucial in the oxidation of phenolics, presumably by abstraction of a proton from the phenolic group. In a mutant with this anchoring point replaced by Thr (Glu235Thr), the efficiency of oxidation of 2,6-dimethoxyphenol dropped dramatically, while activity toward the nonphenolic ABTS stayed at a similar level [399]. Following an analogous strategy, the substitution of phenylalanine residues for smaller-size alanines in *T. versicolor* laccase, aiming at improved activity toward sterically demanding substrates, was approached with limited success. The double mutant Phe162Ala/Phe332Ala was more efficient than the single ones and the wild strain in the aerobic oxidation of the bulky bisphenol A, but its thermal stability was poor [400]. Likewise, modifications of the *C*-terminus have been studied in laccases from ascomycetes and basidiomycetes. In the ascomycete *M. albomyces*, the final four amino acids reenter the structure to form a plug that ends at a T_3 copper [108]. Deletion of this plug resulted in almost complete loss of activity [401]. Similarly, truncations and/or mutations around the *C*-terminus of the *M. thermophila* laccase, expressed in *S. cerevisiae*, produced widely variable activity [402]. In a different *C*-terminus modification strategy, an ascomycete laccase was linked covalently to small peptides that bind to carotenoids (e.g., lycopene from tomato) [403]. These targeted peptide-laccase fusions displayed enhanced catalytic properties on stained fabrics. The influence of the *C*-terminus on basidiomycete laccases, which do not show an analogous plug, has been less studied. Likewise, directed evolution has been used to improve laccase activity and stability. As mentioned above, most of the work has been performed on the more genetically tractable ascomycetes. A variant with improved expression and approximately 170-fold increase in total activity was obtained after 10 iterative rounds of mutagenesis on a *M. thermophila* laccase expressed in *S. cerevisiae*, together with improved thermostability [402]. Starting also from recombinant *M. thermophila* laccase expressed in *S. cerevisiae* (MtL T2 variant), after five rounds of directed evolution a mutant tolerating a wide array of cosolvents at concentrations as high as 50%, was evolved [301,404–406]. A *S. cerevisiae* strain expressing a *Pleurotus ostreatus* laccase produced variants with greater catalytic activity and stability after two rounds of mutagenesis. The best variant contained only two mutations [407]. In another study, recombinant laccase from *Fome lignosus* expressed in *P. pastoris* was subjected to ethyl methane sulfonate-based random mutagenesis to yield a mutant with fivefold enhancement of total activity, without altering the thermo- and pH stability [408].

Semirational design offers interesting possibilities, combining structure-guided mutagenesis with directed evolution techniques. According to these lines, combinatorial saturation mutagenesis has been used to improve simultaneously functional expression and catalytic activity of a bacterial laccase from *Bacillus licheniformis* [409]. Among fungal laccases, saturation mutagenesis has been applied to two specific residues of a recombinant *M. thermophila* laccase, producing a mutant with threefold higher turnover rates, giving insight into the elements that determine the redox potential of these enzymes [410]. In a further work on this laccase, a variant with two consecutive mutations in the *C*-terminus (Gly614Asp and Glu615Lys) showed approximately threefold better ABTS oxidation kinetics, lower redox potential and increased K_MO_2, suggesting disturbed copper environments at both T_1 and T_2/T_3 sites [411].

The first successful evolution of a basidiomycete laccase was recently described. A high redox potential laccase from the basidiomycete PM1 [412], expressed in *S. cerevisiae*, was subjected to eight rounds of directed evolution in combination with rational approaches to yield a variant displaying very high activity and good stability in terms of temperature, pH range, and organic cosolvents [413]. The semirational mutagenesis of the laccase POXA1b from *P. ostreatus* afforded variants with up to fivefold increased specific activity and higher stability at acidic, neutral and alkaline pH values [414]. Also, a semirational approach was used in CotA laccase from *B. subtilis* to improve the substrate specificity and thus, a variant was found that was 132 times more specific for ABTS, and also displayed increased thermal stability [415,416]. Using a different strategy, laccase chimeras through yeast-mediated homologous recombination of *Trametes* sp. strain C30 laccase cDNA have been constructed. Compared to the parental enzyme, LAC3, a 12-fold increase in the catalytic efficiency of the best-performing hybrid (LAC131) was observed [417].

Increase of the Enzyme Redox Potential The key feature of laccase is the redox potential of the T_1 copper site, which has been determined to vary between 0.4 and 0.8V [98,170]. Theoretically, the oxidative capacity of laccases may be improved, since the redox potential of the T_1 site is still far from the thermodynamic limit of 1.23 V, corresponding to the redox potential for the reduction of oxygen to water. It has been postulated that the distance between the T_1 copper and the axial ligand (usually histidine) modulates the electron-density contribution, being longer in high potential laccases [110]. Working on this hypothesis, laccases from *M. thermophila* and *Rhizoctonia solani* were subjected to mutagenesis by targeting a pentapeptide segment located near the T_1 copper site [418]. Results demonstrated that the replacement of the axial histidine does not significantly affect the redox properties. However, changes in a tripeptide that is likely to be part of the substrate-binding pocket led to a threefold increase of the enzymatic activity [418]. Further studies have been carried out along the same direction [419], but despite some efforts to rationalize the available data [420,421], the large differences among the redox potential values of laccases with not so different environments around the T_1 sites are not fully understood. Clearly, further studies using variants modified in that region are needed. As in

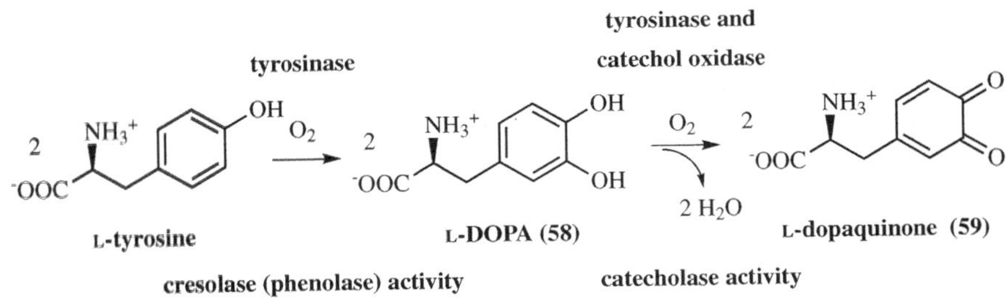

FIGURE 5.42 Enzymatic activity of polyphenol oxidases, using L-tyrosine as substrate.

previous examples, the majority of the protein engineering work has been performed on low redox potential laccases from ascomycetes, aiming at converting them into enzymes of higher redox potential [422]. Regarding the trinuclear copper site where the reduction of dioxygen to water takes place, site-directed mutagenesis studies performed on CotA-laccase from *B. subtilis* suggested that Glu498, located within the dioxygen entrance channel, is the acidic residue responsible of the protonation events involved in the binding of dioxygen and its further reduction [423,424].

5.1.2.2 Tyrosinase and Other Polyphenol Oxidases Similar to laccases, tyrosinases have been first described at the end of nineteenth century, [96,425] and both are multicopper enzymes capable to catalyze the oxidation of aromatic compounds by oxygen. Tyrosinases (EC 1.14.18.1) use molecular oxygen to catalyze two different enzymatic reactions: (i) the *ortho*-hydroxylation of monophenols to *o*-diphenols (referred to as monophenolase/cresolase activity) and (ii) the subsequent two-electron oxidation of these *o*-diphenols to *o*-quinones (EC 1.10.3.1, referred to as diphenolase/catecholase activity). The latter reaction is also catalyzed by the related enzyme catechol oxidase (EC 1.10.3.1), which is unable to mediate the first phenol hydroxylation reaction (Figure 5.42). Thus, phenols and catechols are substrates for tyrosinases, being the *o*-quinones the end products in both cases. Because of this overlapping activity, both tyrosinase and catechol oxidase are collectively designated as PPO, a very common denomination used in the literature. In addition, tyrosinases, catechol oxidases and laccases are often summarized as phenol oxidases, since laccases (EC 1.10.3.2) have no hydroxylase activity but oxidize mono- and diphenols by a radical mechanism (see Section 5.1.2.1). The enzyme classification of PPOs is somewhat ambiguous; enzymes classified as EC 1.10.3.1 show catechol oxidase activity (oxidizing *o*-diphenols to *o*-quinones), whereas those with monophenol oxidase (oxygenase) and catechol oxidase activities are listed as EC 1.14.18.1, and give *o*-quinones if monophenols or *o*-diphenols are available as substrates. In addition, tyrosinases have also been assigned to subclass EC 1.10.3.1, when considering only a part of their enzymatic activity. Moreover, tyrosinases are oxygenases, although they are commonly designated as oxidases. For this reason, they were described as monooxygenases in the overview of oxidoreductase enzymes (Chapter 1) and also included in this chapter according to the common use.

Both tyrosinase and catechol oxidase possess active sites characterized by a pair of coupled copper atoms, CuA and CuB, which are coordinated by three histidine residues each, being referred to as type 3 copper proteins (see Chapter 1 and below in this chapter) [99,426,427]. In addition to these structural features, polyphenoloxidases (tyrosinase and catechol oxidase) can be differentiated from laccase by the inability of PPO to convert *p*-diphenols. Differentiation between laccase and PPO activities can be readily achieved using selective assay methods, such as the phenol oxidase assay that utilizes tyrosine, or the laccase assay using guaiacol, an *o*-substituted phenol not oxidized by PPO [168].

Tyrosinases display a ubiquitous distribution in animals, plants, fungi, and bacteria [428–431]. In most cases, several isoforms are present. Furthermore, whereas bacterial tyrosinases occur as monomers, plant and animal tyrosinases are monomers or dimers, many fungal tyrosinases are tetramers, and those from crustacean are hexamers [428,429,432,433]. The substrate specificity is strongly dependent on the enzyme source, being quite narrow for mammalian enzymes, accepting only L-tyrosine or closely related substrates, and more tolerant for fungal and plant enzymes, where a number of phenols and catechols are oxidized [434,435]. Fungal tyrosinases are the most used and studied, since the enzyme from the edible mushroom *A. bisporus* is commercially available and also easy to obtain in relatively large quantities. Tyrosinases from microorganisms, animals, plants, and insects, exist in both latent (inactive) and active forms [428,436]. In plants and fungi the inactive form predominates; [437] for example, the latent form in *A. bisporus* represented 98–99% of the total tyrosinase activity [438]. Fungal tyrosinases could be activated *in vitro* by acidic shocks [439], surfactants [440], lipids [441], and proteases [442]. *In vivo*, tyrosinase activation was assumed to be the result of ageing [443], exposition to extreme environmental conditions or to pathogens [444,445], and takes place mostly by removal (*via* proteolysis) of the *C*-terminal extension [437]. This mechanism is also operational in a few bacterial tyrosinases that require activation [446,447]. Some bacterial tyrosinases are activated *in vitro* by surfactants, such as the sodium dodecyl sulfate (SDS) activation of tyrosinases from *M. mediterranea* [448].

Regarding the location of these enzymes, fungal tyrosinases are generally cytoplasmic enzymes, whereas bacterial tyrosinases occur as intra- and also extracellular enzymes, which is an advantage for biotechnological applications [430,432]. Tyrosinases are involved in several biological functions, displaying an important role in melanogenesis, for example, the biosynthesis of melanins. These are polyphenolic heteropolymeric pigments distributed in all living organisms. They are formed from the *o*-quinone dopaquinone (**59**), resulting from the tyrosinase activity on tyrosine, through a complex series of chemical steps. In plants and lower organisms, this is the only enzymatically controlled step of melanogenesis, and the pathways then proceeds spontaneously [449–453]. In animals two additional enzymes—closely related to tyrosinase—participate in the pathway [454]. In mammals, tyrosinase-related melanogenesis is responsible for skin, eye and hair pigmentation, which has a fundamental role in the protection of the skin against UV radiation. In fungi, the role of melanin is correlated with the differentiation of reproductive organs and spore formation, virulence of pathogenic fungi, and tissue protection after injury [455,456]. In plants, the activity of tyrosinases results in the characteristic browning reaction that takes place during senescence or at open (cut or injured) surfaces, for instance, in the banana, potato, and grape; however, the function of this reaction is not clear [457]. One possible role is protection of the wound from pathogens or insects [430,458]. In insects, tyrosinase is thought to be involved in wound healing and sclerotization of the cuticle [459]. In bacteria, aside from the obvious photoprotective effect, the function of melanin is not fully understood, but melanin has been suggested to play a role in virulence and also protection against radiation, desiccation and metal ions [460]. Other functions of bacterial tyrosinases include the synthesis of amino acid-based antibiotics like lincomycin [461], and the detoxification of plant defensive phenols by symbiotic bacteria [462]. As shown in Figure 5.42, the cresolase (phenolase) activity of tyrosinase is a hydroxylation reaction, incorporating one atom of molecular oxygen into the substrate, whereas in the following reaction (catecholase activity) the oxygen acts as the electron acceptor. Therefore, tyrosinase first behaves as monooxygenase on phenols and subsequently, acting as oxidase, it transforms the resulting catechols into the corresponding quinones (catecholase activity).

As mentioned before, tyrosinase possesses a coupled binuclear copper active site, each copper atom being coordinated by three histidine residues (type 3 copper center), being the most studied multicopper oxygenase [99]. However, the exact mechanism of reaction of tyrosinases remains partially unclear, giving its complexity. In particular, the presence of two different catalytic activities, tyrosine hydroxylase and catechol oxidase, at the same active site, and the different ratio between both activities in tyrosinases from different biological sources should be explained. Also, an important feature of tyrosinase activity is the existence of a lag period associated with the first hydroxylase activity, which can be shortened by addition of low amounts of *o*-diphenols (such as L-DOPA), and this should be considered in the mechanistic postulates. It is worth mentioning that the first crystal structures of two bacterial tyrosinases from *Streptomyces castaneoglobisporus* and *Bacillus megaterium* have been recently described [463–465]. A model for tyrosinase activity has been proposed, incorporating recent spectroscopic evidence and including both catalytic activities (hydroxylase/cresolase and oxidase/catecholase cycles), the existence of a lag period and suicide inactivation (Figure 5.43) [99,426,427,454,464–467]. Depending on the oxidation state of the copper ions and their linking to oxygen, the active site of tyrosinases can exist in three different states: *deoxy* (Cu^{I}–Cu^{I}), *oxy* (Cu^{II}–O_2–Cu^{II}) and *met* (Cu^{II}–OH–Cu^{II}) [468,469]. In the absence of any substrate, the *met*-state predominates (approximately 85% for mushroom tyrosinase) [99,447]. The *oxy*-state is originated by reversible addition of molecular oxygen to the *deoxy*-state, forming a peroxide between the two copper atoms in a "side-on" conformation. The *oxy*-form is pivotal in the mechanism, being able to coordinate with phenols (hydroxylase cycle, *via* CuA) and *o*-diphenols (oxidase cycle, *via* CuB) [454,464,465]. Both cycles afford an *o*-quinone as they proceed. *In vivo*, most of the resting enzyme is in the *met*-form, which is unable to bind molecular oxygen [470]. In the presence of *o*-diphenols a high affinity complex is formed, resulting in a two-electron transfer to yield the *deoxy*-form and the corresponding *o*-quinone. The *met*-form does not act on monophenols, but has an important affinity for them and binds them through a dead-end pathway, resulting in a lag period [466,471]. The lag period in the cresolase

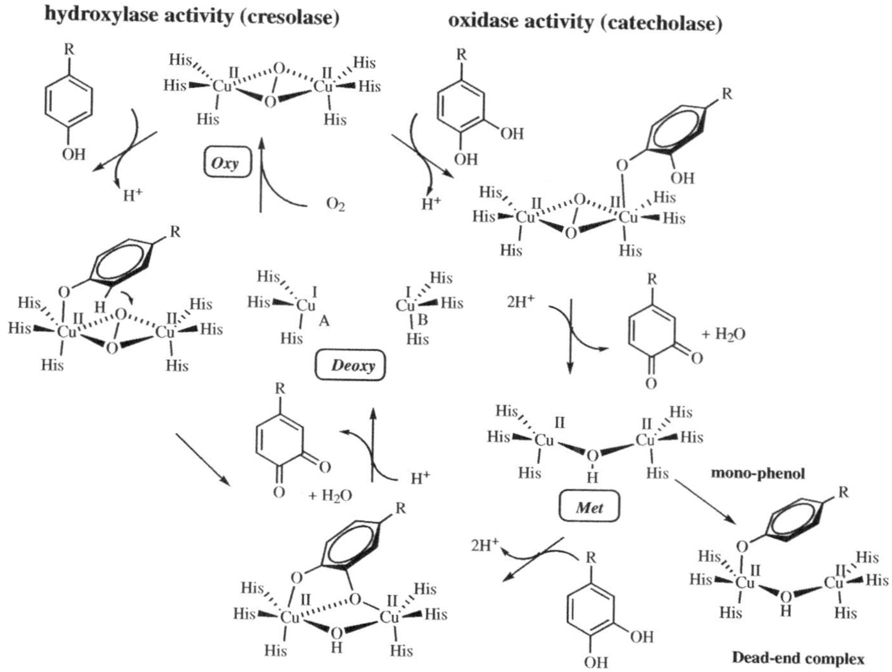

FIGURE 5.43 Mechanism of action for tyrosinase, showing the catalytic cycles for hydroxylase and oxidase activities.

cycle can be abolished by addition of low amounts of *o*-diphenols, which reduce the *met*-form to the cresolase-active *deoxy*-form [434,447,472].

Synthetic Applications of Tyrosinase Since the regiospecific *o*-hydroxylation of phenols is a challenging reaction by means of conventional synthetic methods, the use of biocatalysis constitutes an attractive alternative. However, the final products of the tyrosinase-mediated reaction are quinones, which are quite reactive molecules and, particularly in aqueous medium, generally undergo further nonenzymatic reactions resulting in polymerization, leading to melanins and related polyphenolic pigments. Such side reactions have hampered the use of this enzymatic reaction—oxygenation of phenols to catechols—for applications in synthetic organic chemistry, and thus only a few examples are reported for the tyrosinase-mediated production of *o*-diphenols in aqueous systems (Figure 5.44). In all cases, the formed quinones must be reduced *in situ*. This is usually carried out by adding excess of ascorbic acid, although some controversy has occurred over the effect of ascorbic acid on tyrosinase activity. Some authors report on the inhibitory effect of high concentrations of ascorbic acid [473–475], whereas others mention that it has no effect on the tyrosinase activity [434,476]. In addition, a competitive inhibition of the enzyme is likely to occur when high *o*-diphenol concentrations are reached in the medium [475], thus explaining why high yields of *o*-diphenols are difficult to obtain [474,475]. To circumvent this problem, the combined use of borates and ascorbic acid has recently been introduced

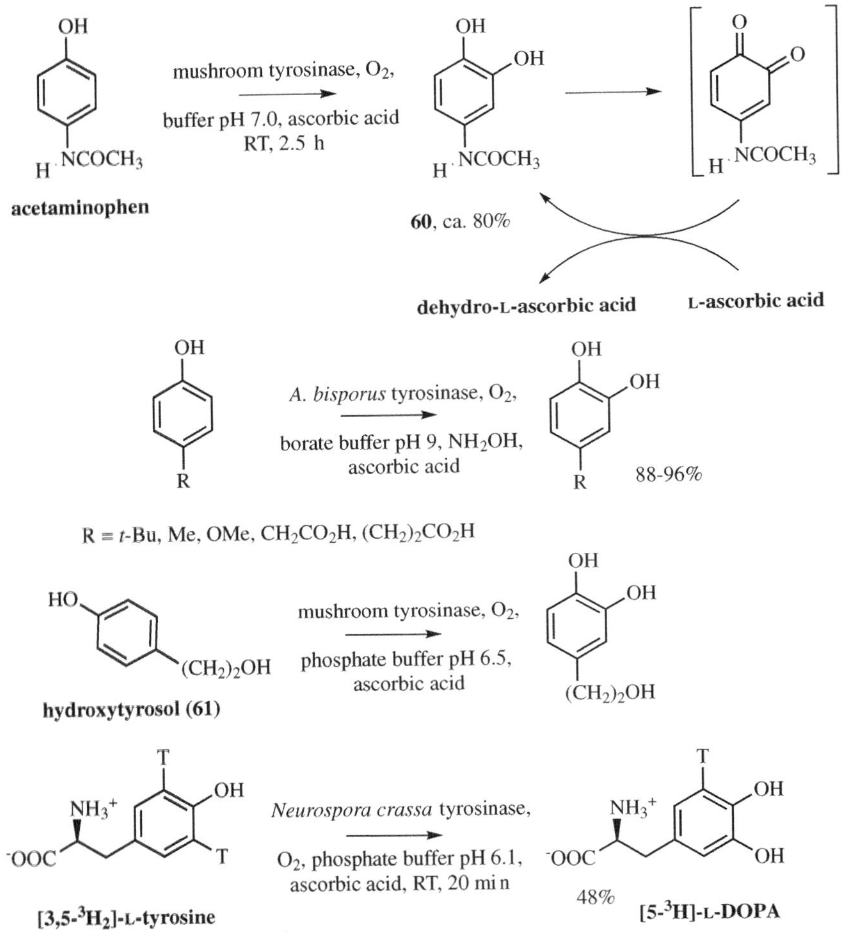

FIGURE 5.44 Tyrosinase-mediated production of *o*-diphenols in aqueous systems.

[477]. In this way, a borate buffer was used to minimize the diphenolase activity of tyrosinase by complexing the *o*-diphenols generated, whereas ascorbic acid reduced the *o*-quinone produced to *o*-diphenol. In addition, hydroxylamine was also present in order to reduce the *met*-form of tyrosinase to the *deoxy*-form, closing the catalytic cycle (Figure 5.43). By using immobilized *A. bisporus* tyrosinase in an aqueous borate buffer with hydroxylamine and ascorbic acid in excess, the highest yields reported so far (> 88%) were obtained for the production of *o*-diphenols from the corresponding *p*-substituted phenols [477]. The first chemoenzymatic synthesis of 3′-hydroxyacetaminophen (**60**), was performed in approximately 80% yield through the tyrosinase-mediated oxidation of acetaminophen in aqueous medium and excess of ascorbic acid [478]. Very good yields of the antioxidant hydroxytyrosol (**61**) have been reported for the tyrosinase-mediated oxidation of commercially available tyrosol in the presence of ascorbic acid and in aqueous phosphate buffer at pH 6.5

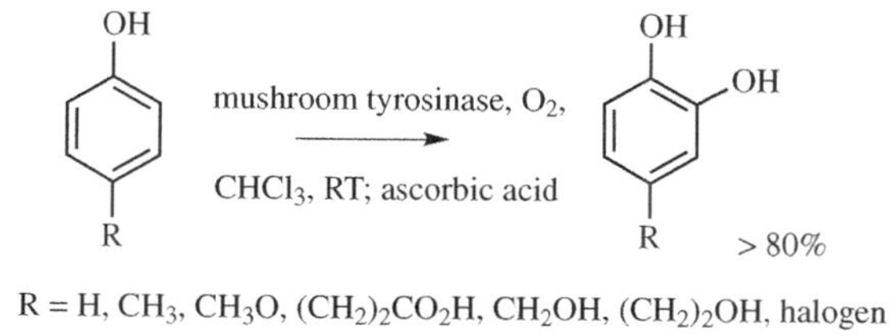

FIGURE 5.45 Tyrosinase-mediated production of *o*-diphenols in chloroform.

[479]. Several isotopomers of L-DOPA have been prepared by means of tyrosinase-catalyzed oxidation of appropriately labeled (with deuterium or tritium) L-tyrosine in the presence of ascorbic acid (Figure 5.44) [480].

The unwanted polymerization of quinones can be reduced by performing the enzymatic reaction in nonaqueous media. In the 1980s, Klibanov showed that mushroom tyrosinase adsorbed on porous glass catalyzed the oxidation of *p*-substituted phenols to quinones in chloroform in nearly quantitative yields [481]. Better reactivity was found for electron donating substituents, whereas bulky phenols (*p*-phenylphenol, 5-indanol, *p-tert*-butylphenol, and 1- and 2-naphthols) and substitution in *ortho* or *meta* positions were not accepted. The quinones produced were then nonenzymatically reduced to catechols by treatment with an aqueous ascorbate solution (Figure 5.45). For other phenols, such as chlorogenic acid and *p*-aminophenol, the specific activity of tyrosinase was found to be much higher in chloroform that in aqueous medium [482]. Klibanov reported on the activation of tyrosinase (and peroxidases) suspended in nonaqueous media when lyophilized in the presence of a number of lyoprotectants, such as phenols, anilines, and polyols [483]. Also, the activation of tyrosinase suspended in chloroform by means of surfactants such as SDS has been described [484]. Within a different approach, the use of mushroom tyrosinase in different micellar systems for the oxidation of several phenols has been well studied, and sometimes is effective to reduce adverse effects such as polymerization and substrate inhibition [359,485,486]. Microemulsions of tyrosinase, as water-restricted systems, have also been investigated, displaying less substrate inhibition and slower inactivation than aqueous systems [487,488]. Recently, bacterial tyrosinases with ample tolerance to water-miscible organic solvents (lower alcohols and DMSO) have been isolated [489].

Envisaging other synthetic applications, the high reactivity of the quinonic products of oxidation is used to trigger selective domino reactions. The first reported combination of enzymatic with nonenzymatic transformations in a domino reaction produced bicyclo[2,2,2]octene-diones (**62**) in good yields through a tyrosinase-initiated hydroxylation-oxidation-Diels–Alder cascade sequence (Figure 5.46) [490]. The reaction was conducted in chloroform:water (99.5:0.5) and best results were obtained when the enzyme was immobilized on glass beads. Moreover, the enzymes (laccase and tyrosinase) from a crude extract of the common mushroom *A. bisporus* catalyzed the domino reaction between phenol and various cyclic 1,3-dicarbonyls to give annulated benzofurans **63** (Figure 5.46) [266]. When the formed quinones bear nucleophilic residues, inter- or intramolecular autocondensation may take place, giving polycyclic compounds, as shown by the oxidation of hydroxytyrosol (**61**)

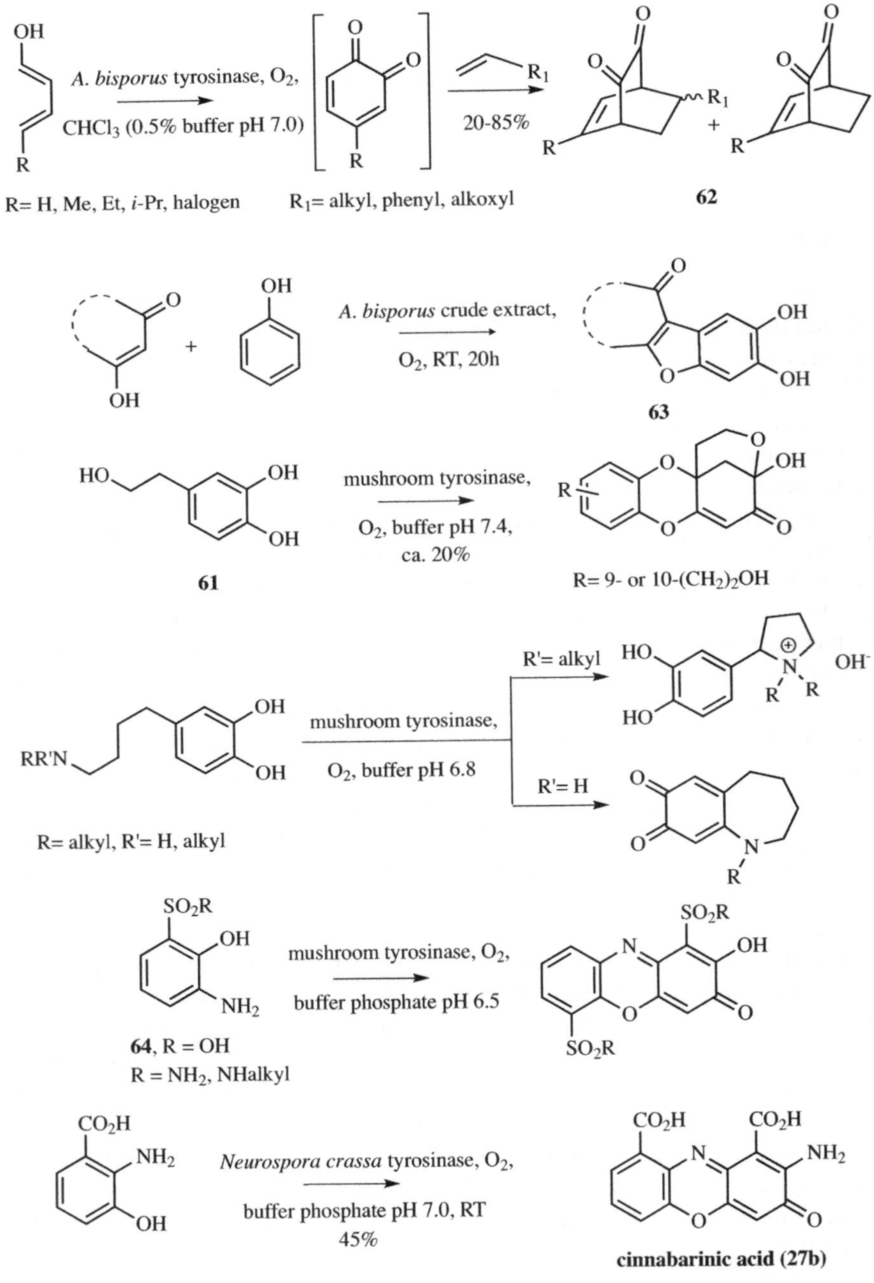

FIGURE 5.46 Tyrosinase-initiated domino (cascade) reactions.

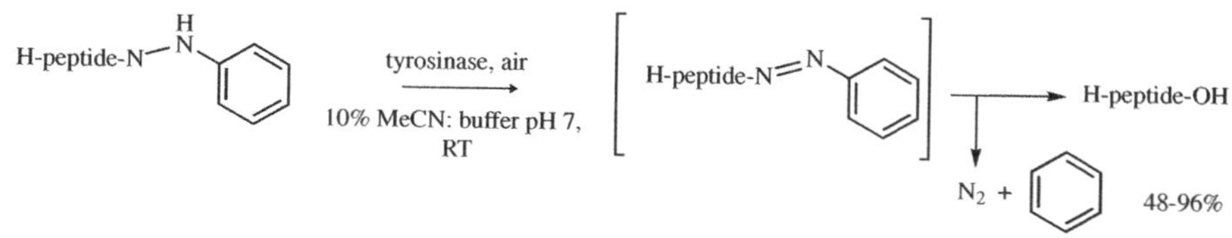

FIGURE 5.47 Tyrosinase-mediated deprotection of phenylhydrazides.

[491]. Similarly, *N,N*-dialkyl-4-aminobutylcatechols give 1,1-dialkylpyrrolidinium salts through a quinone-quinomethane tautomerism, whereas monoalkylated 4-aminobutylcatechols afford bicyclic tetrahydro-1*H*-benzo[*b*]azepine-7,8-diones (Figure 5.46) [492,493]. The oxidation of C–N bonds is also feasible; thus some *o*-aminophenols such as 2-hydroxymetanilic acid (**64**) and derived amides, are converted to the corresponding quinoneimines and in turn to 3-phenoxazinone derivatives according to the reaction depicted in Figure 5.46 [494]. In a similar way, the tyrosinase-mediated oxidation-coupling sequence of 3-hydroxyanthranilic acid produces Cinnabarinic acid (**27b**) in moderate yields [495].

The enzymatic condensation of 4-substituted catechols, such as chlorogenic acid, *N*-methyl-α-methyldopamine and derivatives, with biological nucleophiles like glutathione, proceeds in good yields [496,497]. The tyrosinase-catalyzed oxidation of flavonoids, steroids, and coumarins, do not provide preparatively useful yields [498–502]. Similar to laccase, tyrosinase has been employed for the mild and selective *C*-terminal cleavage of peptides protected as phenylhydrazides. The reaction was performed at room temperature and pH 7, and even sensitive residues, like methionine, were tolerant to these conditions (Figure 5.47) [503,504].

In addition to these synthetic approaches, tyrosinases have been considered for more technologically oriented applications. One important use of PPO takes advantage of the polymerizing properties of the *o*-quinones produced by the enzymatic oxidation and thus, the removal of phenols and related pollutants from wastewaters can be effected by tyrosinase-mediated oxidation and further nonenzymatic polymerization leading to the formation of melanin-like precipitates [316]. In this way, even pentachlorophenol, a recalcitrant xenobiotic used as a pesticide, could be removed [505]. A major advantage of the system results in the utilization of molecular oxygen as the oxidant, which is inexpensive and gives harmless by-products. The methodology could be extended to a two-step procedure, where PPO catalyzed the oxidation of phenols, and the resulting quinone products were adsorbed on chitosan [506,507]. This readily available biopolymer is a waste product of the shellfish industry and stabilized the enzyme as well as adsorbed quinones. Chitosan was found to remove residual color from the aqueous medium after PPO activity had resulted in phenol removal [508]. The *o*-quinones produced by tyrosinase-mediated oxidation may undergo nonenzymatic reaction with nucleophilic sites of polymers leading to cross-linking. It has been shown that tyrosinases can catalyze the formation of covalent bonds between peptides, proteins, and carbohydrates [509–511]. This ability to cross-linking has been applied in the food industry and in material science, for instance, in structure engineering of meat-derived food products [512], grafting of phenolic moieties or silk proteins onto chitosan [511,513–515], formation of hydrogels for skin substitutes [516], high tensile strength adhesives equivalent to that found in nature in mussel glue [517], direct protein cross-linking *via* tyrosine residues [510,518] and in matrices for drug delivery and tissue engineering (mostly based on chitosan) [519], among others [428]. The browning reaction produced by PPO in foods can be either undesirable or beneficial. If undesirable, browning control is achieved by tyrosinase inhibitors such as 4-hexylresorcinol, which is considered as a safe food additive [520]. The melanins produced by quinone polymerization have biotechnological relevance [521], finding application as protective agents against radiation (UV, X-ray, γ-ray), cation exchanger, carrier for drugs, antioxidants, antiviral agents, and immunogens [432,522].

The use of tyrosinase in biosensors is another interesting application that has developed partly as a result of its wide range of phenolic substrates and the importance of detecting phenolic compounds in various circumstances [523]. Biosensors with immobilized tyrosinase have been prepared with solid graphite or composite electrodes, being the solid graphite electrodes more sensitive, enabling the detection of micromolar concentrations of phenols and catechols [453]. A sensor based on the bacterial *Streptomyces antibioticus* tyrosinase was able to detect catechols such as L-DOPA and adrenaline in the low nanomolar range [524]. Using the same enzyme, a novel fluorescent biosensor to detect oxygen was developed [525].

CATALYST ENGINEERING (IMMOBILIZATION AND MOLECULAR BIOLOGY STRATEGIES) Native tyrosinases are generally produced in low quantities. The most used enzyme so far is the fungal *A. bisporus* tyrosinase, mainly because of its long-term commercial availability. However, fungal tyrosinases exhibit low solvent and temperature stability, compared to some bacterial tyrosinases. Commercial products are often contaminated with other enzymes, resulting in preparations of variable quality and activity [526]. Many of these problems can be addressed using molecular biology and immobilization techniques. Tyrosinase immobilization on solid supports, using both chemical and physical methods, is an efficient way to improve operational stability and activity [367,453]. The most common supports include cellulose, collagen membrane, polyacrylamide gel, CH-Sepharose, copper-alginate gel, nylon tubing, hydroxyaluminum–montmorillonite complex, chitosan flakes using glutaraldehyde as cross-linking agent, sodium alginate, glass beads, and gelatin gels. Immobilization protected tyrosinase from inactivation by reaction with quinones, and preserved them from proteolysis [527]. This process also improved the thermal stability [528] and activity yields [529] of fungal tyrosinases, in comparison to soluble enzymes. In addition, the immobilization of tyrosinase on synthetic membranes allows for a more efficient removal of the reaction products from the proximity of the enzyme, thus resulting in decreased product inhibition and hence more effective biocatalysis [530,531].

Another important aspect for biotechnological applications is the development of expression systems for large-scale production, preferably in the extracellular medium. Despite several attempts the overproduction of active fungal tyrosinase has met limited success [532–534]. More recently, the recombinant production of a

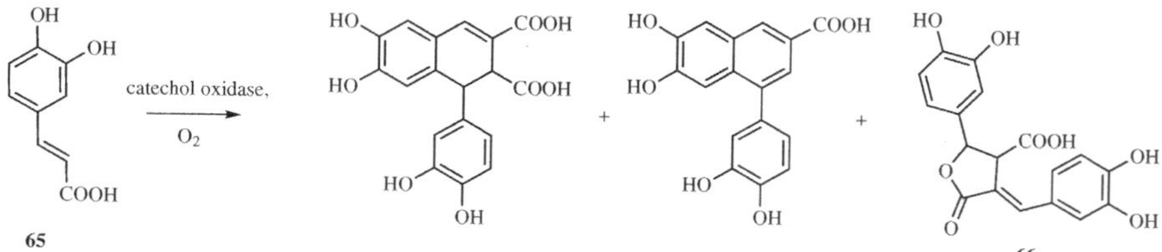

FIGURE 5.48 Catechol oxidase-mediated oxidation of caffeic acid.

P. sanguineus tyrosinase has been performed by heterologous expression in *A. niger*, yielding 20 mgL^{-1} of active enzyme [535]. Contrary to fungal tyrosinases, the bacterial enzymes can be easily expressed in a host like *E. coli* [489,536–538]. Large amounts of protein can be prepared [539], and molecular biology techniques allow for the production of mutants with altered properties. For instance, selective mutations were useful for mechanistic studies, related to the role of the *C*-terminal extension in a bacterial tyrosinase and to the stabilization of the active site [446,540]. Also, directed evolution was used to produce a tyrosinase with higher monophenolase/diphenolase activity ratio [541].

Catechol Oxidases Catechol oxidases (EC 1.10.3.1)—also designated as PPOs—are type 3 copper enzymes that act as catechol oxidases but do not show phenol-hydroxylating activity [542,543]. Most of them were found in plants [544], and that from *Lycopus europaeus* was used in the conversion of caffeic acid (**65**) to neolignans with 2,3-dihydro-1,4-benzodioxin and naphthalene structures, including the novel caffeic acid lactone derivative **66** (Figure 5.48) [545]. Few reports on the application of catechol oxidases for organic synthesis have appeared, since they are devoid of the more synthetically useful oxygenase activity found in tyrosinases.

5.1.3 Oxidases Acting on C–N Bonds

The oxidation of C–N bonds is achieved mainly by amine oxidases, amino acid dehydrogenases, and, less commonly, by laccases, tyrosinases, and peroxidases. When the oxidation is performed by flavin- or metal-dependent amine oxidases, molecular oxygen is the final electron acceptor, generating hydrogen peroxide as the by-product. The flavin-dependent oxidases are the principal enzymes involved, and they are gathered in two structural families, the D-amino acid oxidase family and the monoamine oxidase family [546]. All of them catalyze the oxidation of primary and secondary amines, through the transfer of a hydride equivalent from a C–N bond to the flavin cofactor, which is then reoxidized with molecular oxygen, yielding hydrogen peroxide. In case the presence of hydrogen peroxide caused problems, it is decomposed by addition of catalase.

5.1.3.1 D-Amino Acid Oxidase D-Amino acid oxidase (DAAO, EC 1.4.3.3) was discovered in 1935, but attracted little attention until the 1980s, when its potential biotechnological application produced intense investigations that turned DAAO

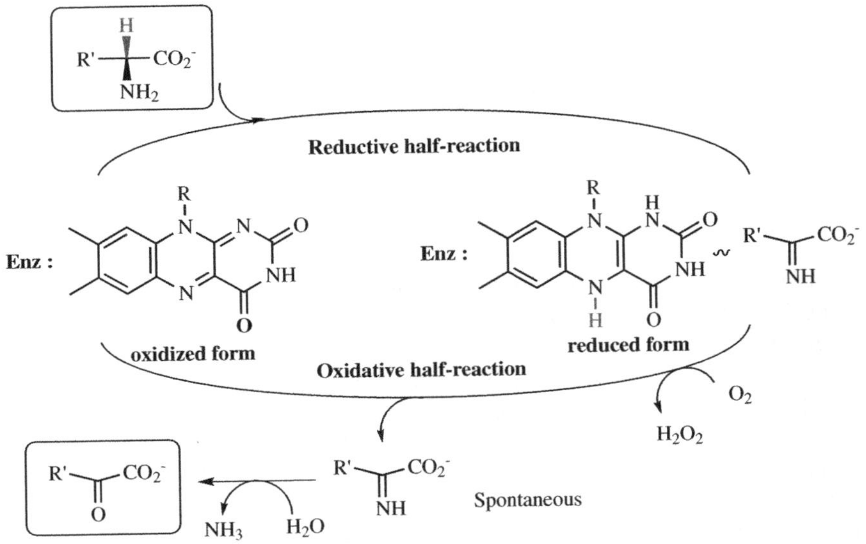

FIGURE 5.49 Catalytic cycle of the reaction catalyzed by D-amino acid oxidase.

into a model for the study of flavoproteins [547–549]. DAAO is an FAD-containing flavoenzyme that catalyzes the oxidative deamination of D-amino acids with strict stereospecificity to give the corresponding imino acid (which hydrolyzes spontaneously to the α-keto acid and ammonia). In the subsequent oxidative half-reaction the colorless $FADH_2$ is reoxidized with molecular oxygen producing hydrogen peroxide (Figure 5.49). In the first, reductive, half-reaction, a direct hydride-ion transfer from the amino acid Cα to the N-5 of the isoalloxazine ring of the flavin occurs without the involvement of amino acid side chains, yielding reduced FAD [546,550]. The release of the imino acid product from the enzymatic complex may take place before o after the reoxidation of $FADH_2$, depending on the enzyme origin and amino acid type [549,551].

DAAO has been found in some bacteria, fungi, and yeast, but it is widespread in mollusk, fish, reptile, amphibia, insects, birds, and mammals (mainly in kidney, lungs, brain) [549,552]. In eucaryotic cells, this enzyme is found in the peroxisomes, thus allowing for the efficient catalase-mediated removal of the cell toxicant hydrogen peroxide, produced in the course of the oxidation [553]. Its biological role is diverse and not completely clarified, ranging from a catabolic function in microorganism (allowing the use of D-amino acids as carbon and nitrogen sources) to a regulatory role in eucaryotic cells (regulating hormone secretion and controlling the levels of the neuromodulator D-serine in the human brain) [552,554].

The most common DAAO enzymes are those from porcine kidney (the first to be commercially available) and from the microorganisms *Trigonopsis variabilis*, *Rhodotorula gracilis*, and *Fusarium solani*. Regarding the biotechnological applications, yeast DAAOs are preferred since they possess higher activity, a stronger binding of the coenzyme FAD, and broader substrate tolerance than mammalian enzymes [555]. In addition, microbial enzymes are more amenable to genetic manipulations and thus DAAO mutants with new or improved properties, such as altered substrate specificity, increased stability, modified cofactor binding properties and different oligomeric states, have been produced [556].

As mentioned above, DAAOs possess a strict stereoselectivity: L-amino acids are not substrates or competitive inhibitors. On the other hand, these enzymes (especially the microbial ones) display a broad substrate tolerance, being able to oxidize several natural and unnatural side chains, preferably nonpolar ones. However, some preferences were found by Tishkov, who divided DAAOs in two groups, according to the preference for amino acids with small apolar side chain (enzymes from *Fusarium oxysporum*, *Candida parapsilosis*, and *C. boidinii*) or with large hydrophobic side chains such as D-Trp, D-Met, D-Val, and D-Phe (enzymes from *A. niger*, *Verticillium luteoalbum*, *R. gracilis*, and *T. variabilis*) [549]. Usually, glycine and the charged (basic and mostly acidic) amino acids are poor substrates.

Amino acid oxidases are used in several biotechnological applications. In addition to synthetic uses, these enzymes find application in analytical determination of D-amino acids and analogs [557], quantitative analysis de D-amino acids [558], development of biosensors [559–561], and therapeutic uses (for tumor treatment, and to prevent ischemia reperfusion injury) [554,562,563].

Synthetic Applications The specificity of DAAO for the D-isomer of amino acids has been exploited mainly for the kinetic resolution and deracemization of racemic substrates (amino acids), as well as for the preparation of α-ketoacids and 7-aminocephalosporanic acid (7-ACA), as shown in the following examples. In the preparation of α-ketoacids, it is a common practice to add catalase in order to decompose the generated hydrogen peroxide, which could decarboxylate the α-keto acid product or deactivate the enzymes [564]. The apoptotic cellular mediator 4-methylthio-2-oxobutyric acid (**67**) was prepared in close to 75% yield from D-methionine using *T. variabilis* DAAO co-immobilized to catalase in an enzymatic stirred tank reactor with *in situ* product removal to avoid decarboxylation. If racemic methionine is used, the unchanged L-methionine can be recovered at the end of the reaction with an optical purity of 99% (Figure 5.50) [565]. The same coimmobilization strategy was used for the oxidation of D-phenylalanine to phenylpyruvate (**68**) [566]. A different approach makes use of permeabilized cells of *R. gracilis* and of *P. pastoris* expressing *T. variabilis* DAAO for the production of phenylpyruvate (**68**) from D-phenylalanine in 97 and 99%, respectively (Figure 5.50) [567,568]. Ionic liquids can be used to increase the substrate solubility and oxygen concentration [569]. Nonnatural aromatic amino acids are also good substrates for the oxidation, as shown by the resolution of racemic mixtures of naphthylalanine and naphthylglycine by *R. gracilis* DAAO [570]. To improve the yield and the scope of the conversion, the enzyme was subjected to rational protein engineering and a single mutation, Met213Gly, led to a reduction in both reaction time and amount of catalyst. Also, the scope of the reaction was increased to include 1- and 2-naphthyl derivatives (Figure 5.50) [570].

The main industrial application of DAAO is in the commercial production of 7-ACA (**71**), from cephalosporin C (**69**), as depicted in Figure 5.50. The enzyme

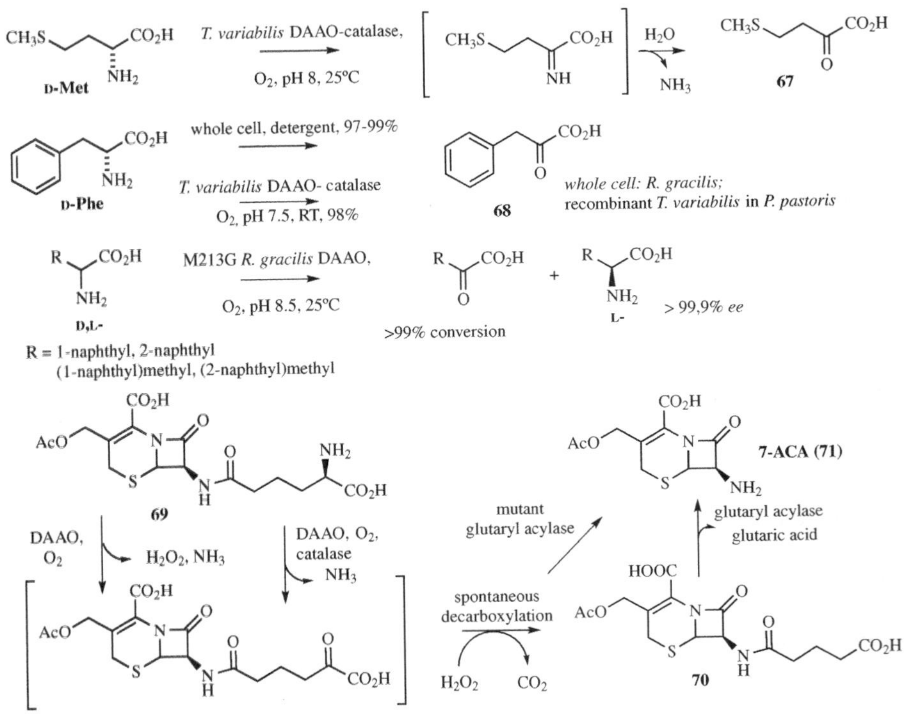

FIGURE 5.50 DAAO-mediated synthesis of α-ketoacids.

catalyzes the initial oxidative deamination of the D-amino acid side-chain in **69**, giving an oxo acid which, in the presence of hydrogen peroxide, spontaneously decarboxylates to glutaryl-7-ACA (**70**) [548]. Subsequently, a glutaryl acylase is used to cleave the modified side-chain to afford 7-ACA. The use of DAAO in this process has been revised [555], and diverse methodologies to improve the stability an efficiency of the catalyst have been used, such as immobilization, design of chimeric enzymes (by fusion of bacterial hemoglobin as an oxygen donor), and use of whole cells systems [548,549]. In addition, the transformation can be carried out in one step, using a multienzymatic system or a fusion protein [571,572]. The main obstacle when using multienzymatic systems is again the production of hydrogen peroxide. To overcome this problem, a three-enzyme system (DAAO, glutaryl acylase, and catalase) was developed [573,574]. In this system, D-amino acid oxidase was coimmobilized with catalase, and thus the product of the oxidation reaction is only the corresponding α-ketoacid, which is subsequently recognized as a substrate by glutaryl acylase and hydrolyzed to produce 7-ACA. The efficiency of this hydrolytic step could be improved with a double mutant of glutaryl acylase from *Pseudomonas* sp. SY-77, affording approximately 80% yield of 7-ACA, when used in the one-pot, three-enzyme system [574]. In a different approach, a new cephalosporin C acylase, produced by a combination of site-directed mutagenesis and directed evolution, was

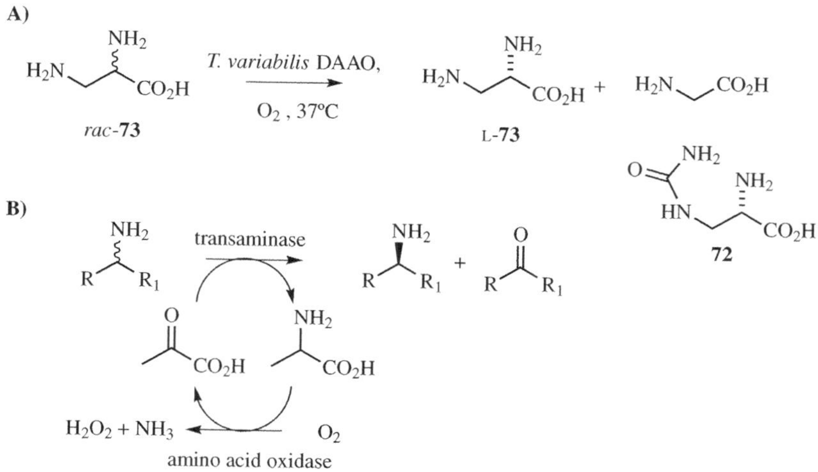

FIGURE 5.51 **A**, DAAO-mediated kinetic resolution of amines; **B**, dual transaminase/amino acid oxidase-mediated kinetic resolution of racemic amines.

recently reported. This mutant is able to convert cephalosporin C (**69**) into 7-ACA, in one step [575].

Several procedures for the kinetic resolution of racemic amines have been developed involving DAAO. During the 1970s, DAAO from porcine kidney was used to make L-pipecolic acid from the racemate *via* oxidation of the D-isomer [576]. In a route to the nonproteinogenic amino acid albizziine (**72**), racemic 2,3-diaminopropionic acid (**73**) was resolved by treatment with resin-bound *T. variabilis* DAAO, yielding the L-isomer with 98% *ee* and the oxidized 3-amino-2-ketopropionic acid, which underwent further transformation to yield glycine as the by-product (Figure 5.51) [577]. An interesting procedure for the kinetic resolution of racemic amines combines the use of amino acid oxidases and transaminases (EC 2.6.1.-). In the first step of the general procedure, the transaminase enantioselectively converts one enantiomer of the racemic amine to the ketone product. Concurrently, the amination of the amine acceptor (pyruvate) takes place, yielding alanine. Then, an amino acid oxidase irreversibly regenerates the amino acceptor using molecular oxygen. Thus, only a catalytic amount of the amine acceptor is used, affecting favorably the reaction by shifting the equilibrium and reducing the inhibition of the transaminases by employing a very low concentration of pyruvate (Figure 5.51). Overall, the amino acid oxidase-catalyzed reduction of molecular oxygen to hydrogen peroxide drives the reaction to completion. The choice of either (*R*)- or (*S*)-selective enzymes allows access to both configurations of chiral amines in excellent *ee* (99%) [578]. A similar principle has been recently used to develop a method for screening transaminase activity on new substrates [579].

Furthermore, oxidases have been used in the deracemization of amino acids and amines, allowing the conversion of a racemate into a nonracemic product in 100% theoretical yield without intermediate separation of materials (see Chapter 3). The most used strategy involving oxidases is a cyclic enantioselective oxidation-nonselective

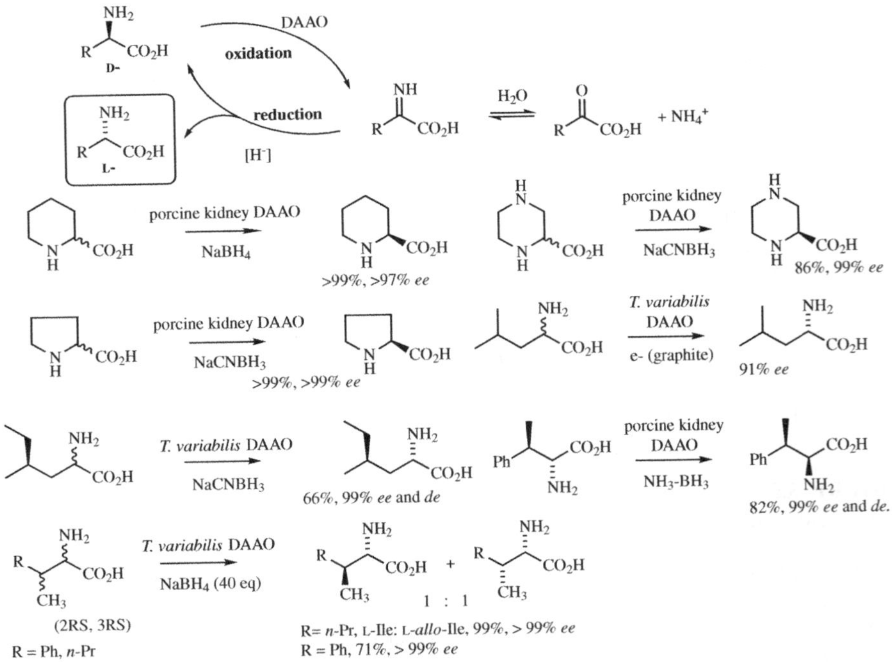

FIGURE 5.52 Preparation of L-amino acids *via* deracemization of racemic amino acids or stereoinversion, using DAAO and a nonselective reducing agent.

reduction sequence. At the onset, one enantiomer is oxidized to the achiral imine intermediate, which is subsequently reduced in a nonselective manner to regenerate the amine of the substrate. The application of this cyclic sequence leads to eventual deracemization of a racemic substrate. Also in these cases, catalase is usually added. This methodology was first realized in 1971 by Hafner and Wellner, who prepared L-amino acids from the D-isomers by the combined use of porcine kidney DAAO and sodium borohydride [580]. Later, Soda et al. developed a one-pot deracemization process for racemic pipecolic acid and proline based on DAAO-catalyzed kinetic resolution of the racemate and nonselective sodium borohydride-mediated reduction of the achiral imine, regenerating the racemic starting material (Figure 5.52) [581,582]. The methodology was further developed by Turner and coworkers, using various amino acid oxidases and reducing agents ($NaBH_4$, $NaCNBH_3$, NH_3–BH_3) (Figure 5.52) [583,584]. The reduction step can also be performed electrochemically, using graphite electrodes [585]. The efficiency of the process is not hampered by the presence of other chiral centers, as shown by Turner in the reactions of γ- and β-substituted α-amino acids [584]. For instance, the four isomers of isoleucine (racemic *allo*- and racemic isoleucine) gave pure L-*allo*- and L-isoleucine *via* interconversion of diastereomers (Figure 5.52).

Other deracemization methodologies use a combination of these enzymes with amino transferases or dehydrogenases of complementary chiral activity, resulting in two enantioselective steps to achieve deracemization. Some examples are shown in

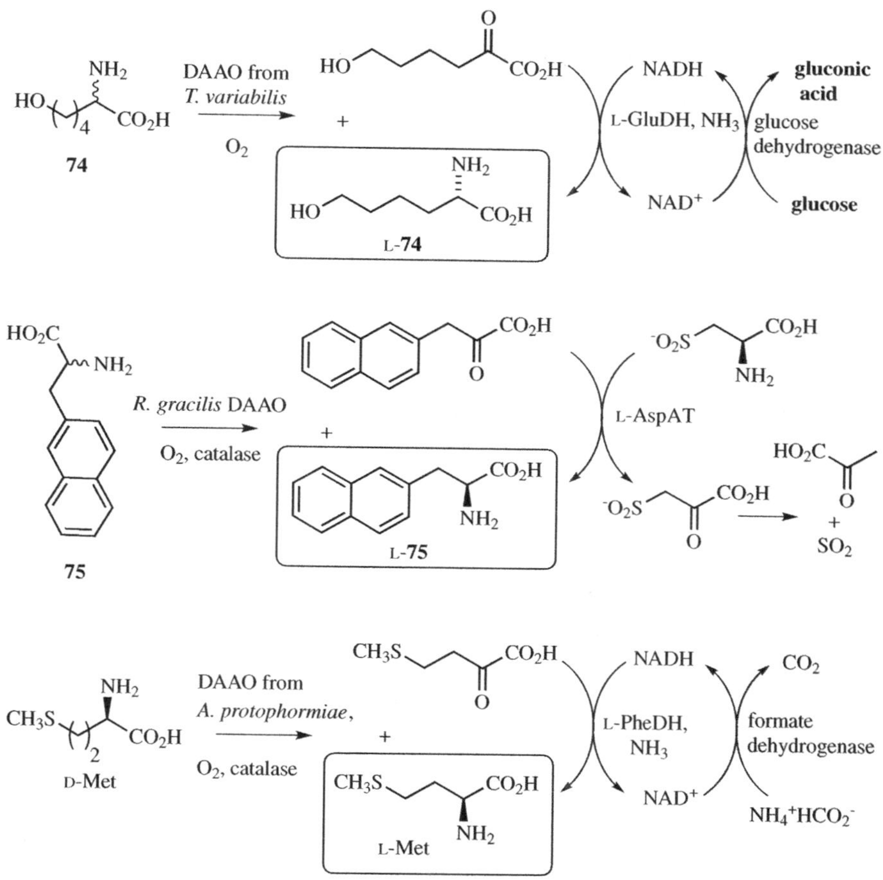

FIGURE 5.53 Preparation of L-amino acids using DAAO in cascade enzymatic reactions. L-GluDH, L-glutamate dehydrogenase; L-AspAT, L-aspartate amino transferase; L-PheDH, L-phenylalanine dehydrogenase.

Figure 5.53 [586]. A procedure involving DAAO and a L-selective dehydrogenase was used to prepare L-6-hydroxynorleucine (L-**74**) from its racemate [587,588]. Thus, racemic 6-hydroxynorleucine was treated with DAAO and catalase (either porcine kidney DAAO and beef liver catalase or *T. variabilis* whole cells) to yield L-6-hydroxynorleucine and the corresponding α-keto acid, which was the substrate of an L-glutamate dehydrogenase to give L-6-hydroxynorleucine, *via* enantioselective reductive amination, in 97% yield and > 98% *ee* (Figure 5.53). The cofactor (NADH) consumed by the dehydrogenase was enzymatically recycled, giving rise to a four-enzyme system. In a related strategy, Servi et al. used an L-amino transferase for the enantioselective reduction of the α-keto acid produced by DAAO-mediated oxidation [589]. In this way, racemic 2-naphthyl alanine (**75**) afforded L-2-naphthyl alanine in almost quantitative yield, using *R. gracilis* DAAO, catalase, *E. coli* L-aspartate amino transferase, and cysteine sulfinic acid as an amino donor (Figure 5.53). DAAO from

Arthrobacter protophormiae was used to produce L-methionine in 100% from the D-isomer in a cascade system of four enzymes (Figure 5.53) [590].

CATALYST ENGINEERING The immobilization of DAAO is a good strategy to increase the stability of the enzyme (e.g., toward hydrogen peroxide), even up to 15,000-fold in certain cases [591]. Moreover, as mentioned above, the coimmobilization of DAAO and catalase in the same support improves the stability and the efficiency of the catalyst [565,566]. However, in biotechnological applications using large bioreactors, the limitations imposed by the immobilization support to the mass transfer have to be considered.

Both rational and directed evolution approaches have been used to produce DAAO variants with new or altered properties [556]. In addition to the above-described examples [570], it is worth mentioning the use of a directed evolution approach to produce *R. gracilis* DAAO with broader substrate tolerance. In particular, two mutants showed a remarkably improved catalytic efficiency on all D-amino acids (including acidic amino acids) and were applied to the analytical determination of the total D-amino acid content in biological samples [592].

5.1.3.2 L-Amino Acid Oxidase L-Amino acid oxidase (LAAO, EC 1.4.3.2) is complementary to DAAO. It is an FAD-containing flavoenzyme that catalyzes the oxidative deamination of L-amino acids with a strict stereospecificity to give the corresponding imino acid, which hydrolyzes spontaneously to the α-keto acid and ammonia; the oxygen consumed is liberated as hydrogen peroxide. Compared to DAAO, the amino acids bind in a mirror image fashion, in agreement with similar mechanisms but opposite stereochemical outcomes [593]. The synthetic applications are mostly deracemizations of racemic amino acids to produce the D-isomers (Figure 5.54). The common sources of LAAO are snake venom and the bacterium *Proteus myxofaciens* [594–596]. The bacterial enzyme has been cloned and overexpressed in *E. coli* K12 and can be used either as a whole cell system or as partially disrupted cells [596]. The nonselective reduction step of the deracemization cycle can be performed with a number of reductants, such as $NaCNBH_3$, NH_3-BH_3, and catalytic transfer hydrogenation using ammonium formate [597]. Excellent yields and optical purities of D-amino acids are obtained with a range of natural and unnatural α-amino acids [584,597].

5.1.3.3 Monoamine Oxidase A number of flavin-dependent amine oxidases are known, of which monoamine oxidase (MAO, EC 1.4.3.4) is the most thoroughly studied. As other flavoproteins, oxidases of this family use molecular oxygen as the oxidant, which is reduced to hydrogen peroxide by a two-electron oxidation of C–N single- to a C–N double bond. A large body of evidence supports hydride transfer as the general mechanism for amine oxidation by members of both the MAO and the DAAO family [546]. Several isoforms of MAO have been extensively studied, in particular, mammal MAO A and B, which catalyze the oxidation of endogenous amine neurotransmitters such as serotonin and dopamine and a wide range of dietary amines and drugs. Consequently, these enzymes have been the subject of intense

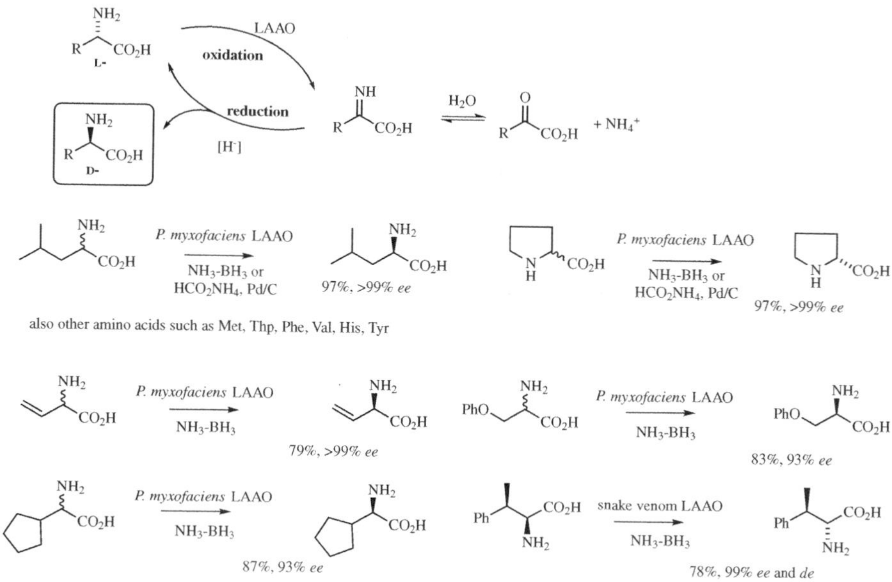

FIGURE 5.54 Preparation of D-amino acids *via* deracemization of racemic amino acids or stereoinversion, using LAAO and a nonselective reducing agent.

study for decades. MAO A and B have the same overall structure [598], but MAO A has a more open active site, producing differences in the substrate specificities of the two isoforms [599]. These oxidases catalyze the oxidation of amines to the corresponding imines, which hydrolyze spontaneously to the carbonyl compound and ammonia (Figure 5.55). The MAO family clearly shows more variety in substrate structure than the amino acid oxidase family.

Interestingly, although these enzymes have received considerable attention with respect to their mechanism and structure, as well as to their substrate specificity, relatively little is known concerning their enantioselectivity toward racemic substrates (since most substrates are primary amines) [600]. In the last years, Turner and coworkers found that MAO from *A. niger* (MAO-N) and its mutants were useful biocatalysts [600]. A flavin-dependent monoamine oxidase from *A. niger* was first described in 1995 (MAO-N, EC 1.4.3.4) [601] and subsequently cloned and overexpressed in *E. coli* [602]. The enzyme catalyzes the oxidation of simple amines such as butylamine, pentylamine, and benzylamine to the corresponding imine (which hydrolyzes to the aldehyde), using molecular oxygen as the oxidant, and has a considerably higher turnover number on many aliphatic and aromatic amines than

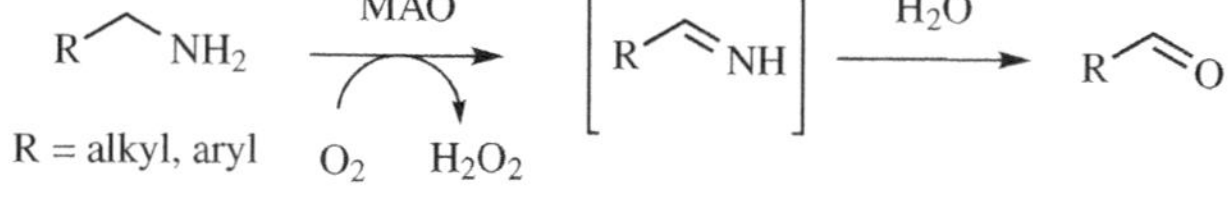

FIGURE 5.55 MAO-mediated oxidation of amines.

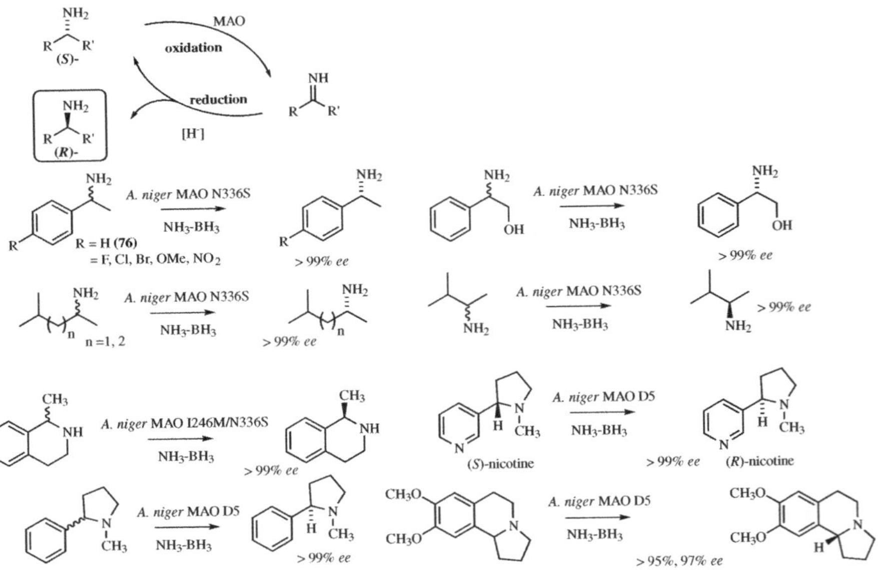

FIGURE 5.56 Preparation of chiral primary, secondary, and tertiary amines *via A. niger* MAO (MAO-N)-catalyzed deracemizations and stereoinversions.

either form of the mammalian enzyme (Figure 5.56). However, the native enzyme displays very low activity toward chiral (primary and secondary) amines such as 1-phenethylamine **76** [602]. To broaden the substrate tolerance, MAO-N was subjected to several rounds of directed evolution, resulting in a number of variants possessing enhanced catalytic activity and considerably broader substrate specificity compared to the wild-type enzyme [603,604]. Thus, only one mutation (Asn336Ser) delivers a highly (*S*)-selective variant, useful for deracemization and stereoinversion of chiral primary amines, particularly 1-alkyl- and 1-arylethylamines (Figure 5.56) [605]. The double mutant Ile246Met/Asn336Ser has improved activity on cyclic secondary amines [606], and the MAO-N-D5 variant, which contains five important mutations (Ile246Met/Asn336Ser/Met348Lys/Thr384Asn/Asp385Ser), shows high activity and (*S*)-enantioselectivity toward a broad range of tertiary amines (Figure 5.56) [607].

These enzymes have also been used in the kinetic resolution of *O*-methyl-*N*-hydroxylamines (also with (*S*)-selectivity, yielding the unreacted (*R*)-isomer and the *E*-oxime) [608], in the preparation of imines in organic solvents [609], chiral amines by reductive amination and *in situ* deracemization [607], as well as in the oxidative desymmetrization of *meso*-pyrrolidines [610] (Figure 5.57). As mentioned previously for other amine oxidases, it is usual to decompose the produced hydrogen peroxide by using catalase (either added or already present in the whole-cell system).

The latter reaction affords optically active 1-pyrrolines, which are useful intermediates in common multicomponent reactions, giving access to alkaloid-like polycyclic compounds **77**, and prolyl peptides found in drugs, such as the protease inhibitors Telaprevir (**78**) and Boceprevir (Figure 5.58) [611–613].

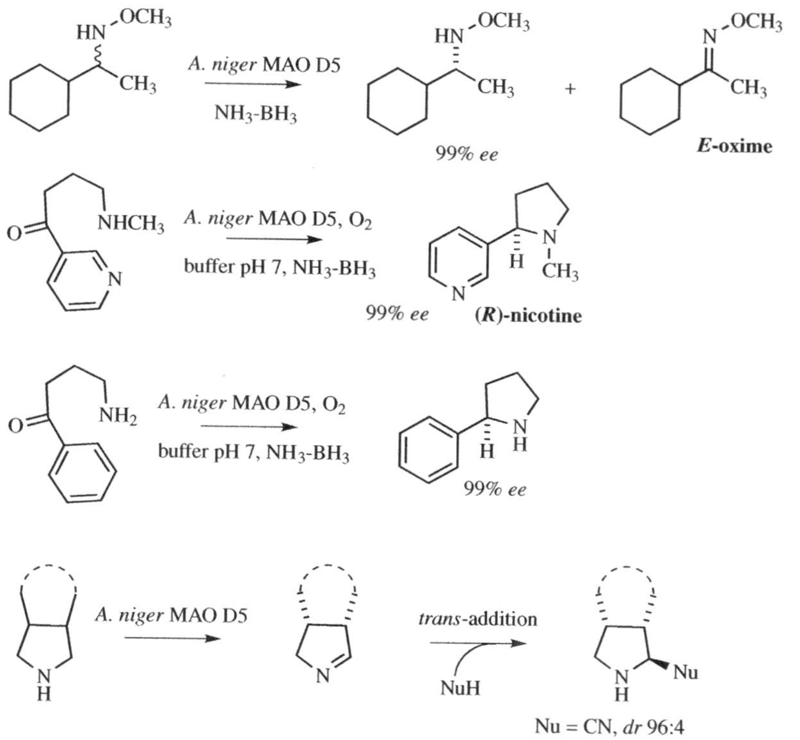

FIGURE 5.57 Miscellaneous MAO-catalyzed reactions. *dr*, diastereomeric ratio.

5.1.3.4 Copper Amine Oxidases The family of the copper amine oxidases is widely distributed among aerobic prokaryotic and eukaryotic organisms. A group of these enzymes (EC 1.4.3.21, primary-amine oxidase, formerly EC 1.4.3.6) catalyzes the oxidative deamination of primary amines to aldehydes (and ammonia) and reduction of molecular oxygen to hydrogen peroxide through a mechanism involving the redox cofactor 2,4,5-trihydroxyphenylalanine quinone (TPQ) and a copper ion, Cu(II) [614]. The cofactor is autocatalytically generated by posttranslation modification of a conserved tyrosine residue and is, therefore, covalently bound to the enzyme [615–617]. The synthetic use of these nonflavin-dependent amine oxidases is scarce. For instance, the kinetic resolution of racemic 1-phenyl-2-aminopropane (amphetamine) by copper amine oxidases from *E. coli* and *Klebsiella oxytoca* was described in 2000, showing a preference for the (*R*)-isomer [618].

5.1.4 Miscellaneous

A number of oxidases have been isolated and characterized. Although less studied, some of them are of preparative (or analytical) value, for instance, cholesterol oxidase, vanillyl alcohol oxidase, and alditol oxidase. For others, such as aldehyde oxidase

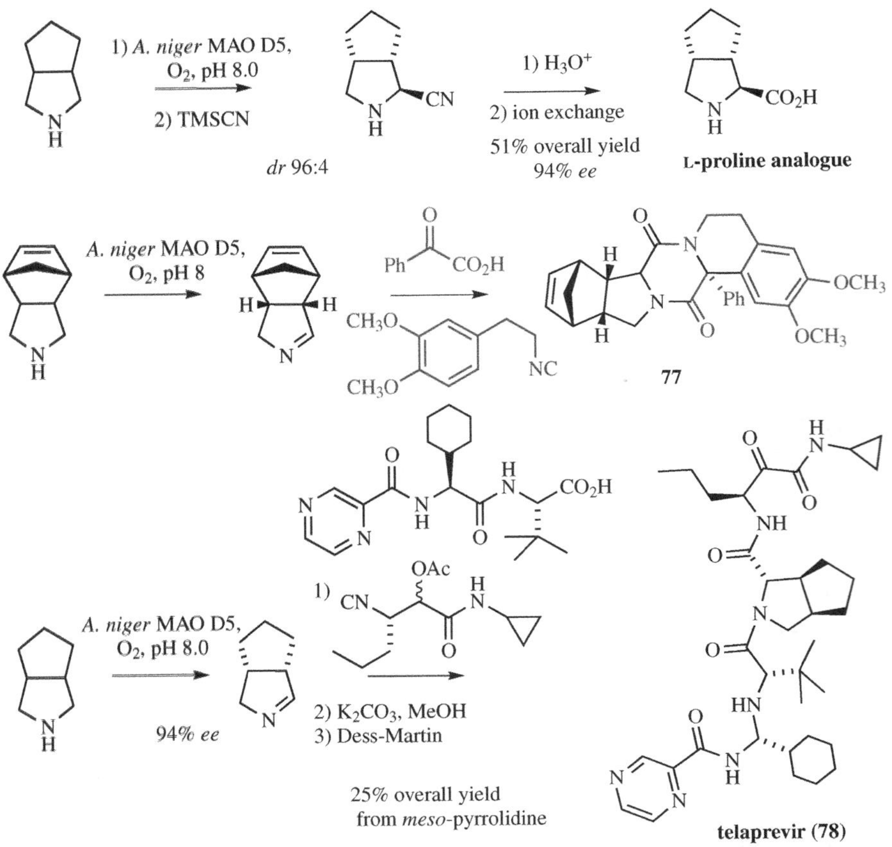

FIGURE 5.58 Preparation of synthetic alkaloids and substituted L-proline derivatives by desymmetrization of *meso*-pyrrolidines and Ugi-type three-component reactions.

(EC 1.2.3.1), lactate oxidase (EC 1.1.3.15), xanthine oxidase (EC 1.17.3.2) or L-lysine α-oxidase (EC 1.4.3.14) [619], their synthetic potential awaits further development.

5.1.4.1 Cholesterol Oxidase Cholesterol oxidase (3β-hydroxysterol oxidase; EC 1.1.3.6) is a bacterial flavin-dependent oxidase that catalyzes the initial step of cholesterol degradation, and other natural sterols that are used as carbon sources for growth of different bacteria. This enzyme catalyzes two reactions: the oxidation of the C_3-OH group of cholesterol (and other sterols) to give cholest-5-en-3-one (or the corresponding Δ^5-3-ketone) and its isomerization to cholest-4-en-3-one (Δ^4-3-ketone) (Figure 5.59) [620]. Cholesterol oxidase has been only found in microorganisms, being produced as a intracellular or extracellular enzyme [621]. The enzyme has several biological roles in the cholesterol metabolism, in the bacterial pathogenesis, and in the biosynthesis of macrolide antifungal antibiotics [622,623].

Regarding the substrate specificity, the studies have been mainly limited to steroids, finding that the hydroxyl group at C-3 has to be equatorial (β), but the double bond

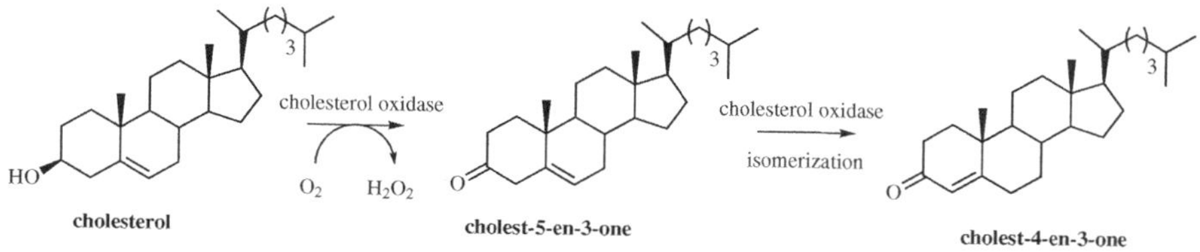

FIGURE 5.59 Cholesterol oxidase-mediated oxidation of cholesterol.

and even the side chain (depending on the enzyme source) are not necessary for the oxidation to proceed. Likewise, various functionalities are tolerated at different positions [624]. Because of its easy reaction with cholesterol, this enzyme has been widely used for the quantification of cholesterol in clinical and food samples and [625], to a less extent, it has also found application as a probe for membrane structure and as an insecticide [626–631]. The preparative use of the enzyme is more limited. However, it has been employed in the preparation of steroid derivatives *via* bioconversion of 3β-hydroxysteroids and in the optical resolution of mono- and bicyclic nonsteroidal allylic alcohols, showing (*S*)-selectivity at the secondary alcohol center [622,624,632–636].

5.1.4.2 Vanillyl Alcohol Oxidase Vanillyl alcohol oxidase (VAO, EC 1.1.3.38) is a covalent flavoenzyme originally isolated from the ascomycete *Penicillium simplicissimum* [637]. It was proposed that its physiological substrate is 4-(methoxymethyl)phenol (**79**), which is demethylated and oxidized to form 4-hydroxybenzaldehyde and methanol, with concomitant reduction of molecular oxygen to hydrogen peroxide (Figure 5.60) [638]. The enzyme displays a broad substrate specificity and high stereoselectivity toward a wide range of phenolic compounds by catalyzing oxidation, deamination, demethylation, dehydrogenation, and hydroxylation reactions [639,640]. For instance, vanillin (**81**), a valuable flavoring compound, is formed by the VAO-mediated conversion of creosol (**80a**), vanillyl alcohol (**80b**), and vanillyl amine (**80c**). This oxidation is promising for the flavor industries as an industrial process to produce vanillin, since the product so obtained can be labeled as natural, starting from creosol or vanillyl amine obtained from natural sources [641]. To improve the process, VAO was subjected to random mutagenesis and single point mutants were generated with up to 40-fold increase in catalytic efficiency with creosol compared to the wild-type enzyme [642]. The 4-allylphenol eugenol (**82**) is stoichiometrically hydroxylated to coniferyl alcohol **28**. By contrast, short-chain 4-alkylphenols are exclusively hydroxylated at the benzylic position and with high stereoselectivity for the (*R*)-isomer [643], whereas dehydrogenation of these substrates is selective for the *cis*- or *trans*-isomer, according to the chain length (Figure 5.60) [638].

5.1.4.3 Alditol Oxidase Alditol oxidase (AldO, EC 1.1.3.41) is a covalent flavoprotein discovered by genome mining in *Streptomyces coelicolor* A3(2), which has been overexpressed in *E. coli* [644,645]. This oxidase performs selective oxidation

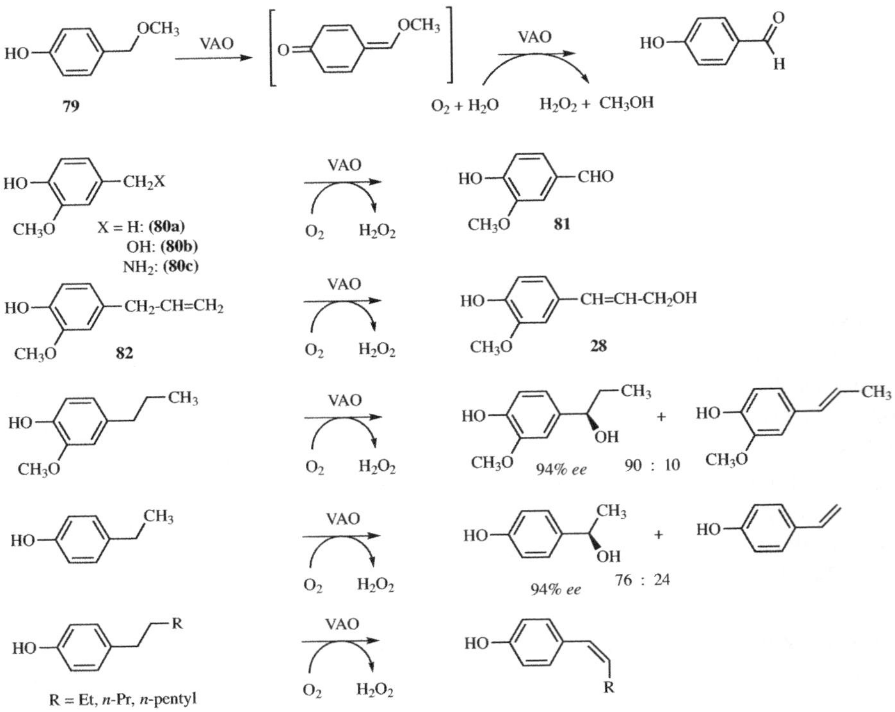

FIGURE 5.60 VAO-mediated oxidation of aromatic substrates.

of the terminal primary hydroxyl group of several alditols, such as xylitol and sorbitol, to afford the aldoses D-xylose and D-glucose respectively, while hydrogen peroxide is produced by the reduction of molecular oxygen (Figure 5.61).

In a recent study, the enzyme was able to oxidize a range of aliphatic and aromatic alcohols [646]. Alcohols containing vicinal hydroxy groups at the C-1 and C-2 positions such as 1,2,4-butanetriol, 1,2-pentanediol, and 1,2-hexanediol were substrates of AldO. The enzyme has shown to be highly enantioselective for the oxidation of 1,2-diols: (*R*)-1-phenyl-1,2-ethanediol was preferentially oxidized with an enantioselectivity (*E* value) of 74. It is important to mention that the products of oxidation of all 1,2-diols examined were the α-hydroxy acids instead of the expected α-hydroxy aldehydes (Figure 5.61). Labeling studies revealed that a second enzymatic oxidation step occurs *via* the initially formed aldehyde hydrate product [646]. Thus, AldO is able to catalyze the oxidation of alcohols into aldehydes and into acids (double oxidation), depending on the specific substrate. The enzyme is an enantioselective biocatalyst for the kinetic resolution of racemic 1,2-diols, giving the corresponding (*R*)-2-hydroxy acids. Considering the industrial relevance of the AldO-mediated oxidation of polyols, a whole cell biocatalyst system was developed, which eliminates the need for enzyme purification [647].

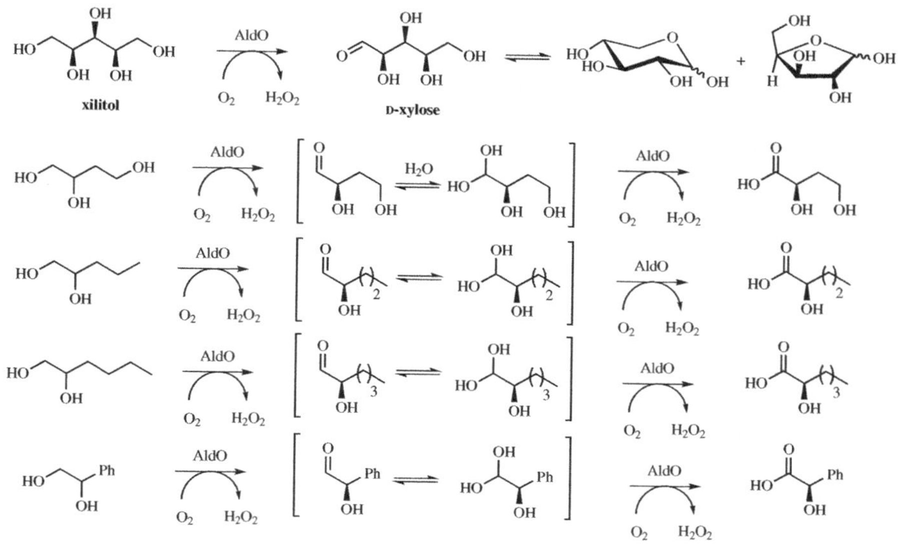

FIGURE 5.61 Alditol oxidase-mediated oxidation of alditols.

5.2 PEROXIDASE-CATALYZED REACTIONS

This group can be differentiated from PPO and laccases, since peroxidases need hydrogen peroxide for activity. Conversely, peroxidases (EC 1.11.1.-) are enzymes that use hydrogen peroxide or some alkyl hydroperoxides as electron acceptors to catalyze a number of oxidations. They are known since the nineteenth century, and their names are derived from their sources, such as horseradish, peanut, and myeloperoxidase, or from their substrates, like lignin, manganese, chloro-, and bromoperoxidase. Details of the individual members of this vast group of enzymes can be found in a freely accessible database, which contains more than 8,000 peroxidase-encoding sequences (access in September, 2011), and in a comprehensive book (with two editions) [648–652]. As widely described in Chapter 1, peroxidases can be divided in two main groups, heme- and nonheme peroxidases. Heme peroxidases are ubiquitous in nature and are classified in two major superfamilies: one mainly found in plants, bacteria and fungi (nonanimal peroxidases, and a second mainly found in animals, fungi and bacteria (animal peroxidases, such as myelo-, lacto-, and eosinophil peroxidase, prostaglandin H synthase, etc.). In addition to these two large superfamilies, heme peroxidases include the group of catalase and haloperoxidases, among others. On the other hand, the nonheme peroxidases are a smaller group, which also includes haloperoxidases, and alkylhydroperoxidases, NADH peroxidases, manganese catalases, and two families of thiol peroxidases (glutathione peroxidase and peroxiredoxins) [648,650]. The active site of the nonheme peroxidases may contain selenium (glutathione peroxidase) [653], vanadium (bromoperoxidase) [654], and flavin (flavoperoxidase) [655]. Because of their widespread distribution, peroxidases possess diverse biological roles. Microbial peroxidases—being mostly extracellular enzymes—are involved in the degradation of aromatic polymers (lignin, humic materials) and in the oxidative detoxification of plant ingredients and organopollutants [422,656,657]. Plant peroxidases are involved in cell-wall biosynthesis (oxidative coupling of phenol compounds during lignification), phytohormone metabolism, pathogen defense, and wound healing [658,659]. Animal/human peroxidases play a defensive role against microbial infections and thyroid peroxidase is essential for the synthesis of the hormone thyroxine.

Heme peroxidases are the most studied, and include the well-known horseradish peroxidase (HRP, EC 1.11.1.7), the haloperoxidases chloroperoxidase (CPO, EC 1.11.1.10), some bromoperoxidases (BPO, EC 1.11.1.-), and iodoperoxidases, which have been investigated for synthetic purposes (there are also nonheme bromo- and iodo peroxidases, mainly isolated from marine organisms). The active site of these enzymes contains iron heme (protoporphyrin IX) as the prosthetic group, linked to the protein by the fifth (proximal) ligand, which is mostly a histidine or, exceptionally, a cysteine. In this latter case the enzymes are known as heme-thiolate proteins [660,661]. Depending on the nature of the fifth ligand and active site residues, the heme can react with either peroxides (in peroxidases) or molecular oxygen (in oxygenases) to form reactive complexes capable of activating C–H bonds [662]. Much effort has been devoted to elucidate the structure of the intermediates and the mechanism of oxidation in a number of heme proteins, and a simplified scheme of the different catalytic cycles involved in peroxidase-mediated reactions is presented in Figure 5.62 [651,658,661,663–665]. Starting from the native (hydro)ferric peroxidase (resting enzyme) [heme(Fe^{III}–H_2O)], binding of hydrogen peroxide affords an extremely short-lived iron-peroxide complex [heme(Fe^{III}–O–OH)] (analogous to Compound 0 of P450s), which is heterolytically cleaved between the oxygen atoms to give water and Compound I, an oxo-ferryl, heme-radical cation complex [heme(Fe^{IV}=O)$^{\bullet +}$]. This intermediate can react along several distinct pathways leading to different types of reaction products. For example, it can perform the classical peroxidase reaction, consisting in an oxidative coupling *via* sequential one-electron oxidations of the substrate (path a), or an oxygen transfer reaction (path b), or halogenation of organic substrates (in the case of haloperoxidases, path c), or even peroxide decomposition (path d) [666,667]. The path a) involves a one-electron transfer from the substrate to give a radical and Compound II. The latter is an oxo-ferryl heme [heme(Fe^{IV}=O)] that reacts with a second substrate molecule resulting in the formation of a second radical and the native ferric enzyme. In summary, within the typical ("classical") heme-peroxidase cycle, two substrate molecules are oxidized by one-electron abstraction (but without oxygen transfer) while one molecule of hydrogen peroxide is consumed and two water molecules are produced. Also, an organic hydroperoxide can function as the electron acceptor, in which case one alcohol molecule and one water molecule are produced. In the path b, the protonated oxo-ferryl species (**83**), is the active intermediate in this two-electron oxygen-transfer reaction of peroxidases [668,669]. Also, this species can deprotonate to form Compound II, and transfer the oxygen to the substrate (for instance, in a sulfoxidation). Within these two-electron oxidations the substrate is monooxygenated, a peroxide serves as oxygen donor and electron acceptor and one molecule of water (or alcohol in the case of using an organic

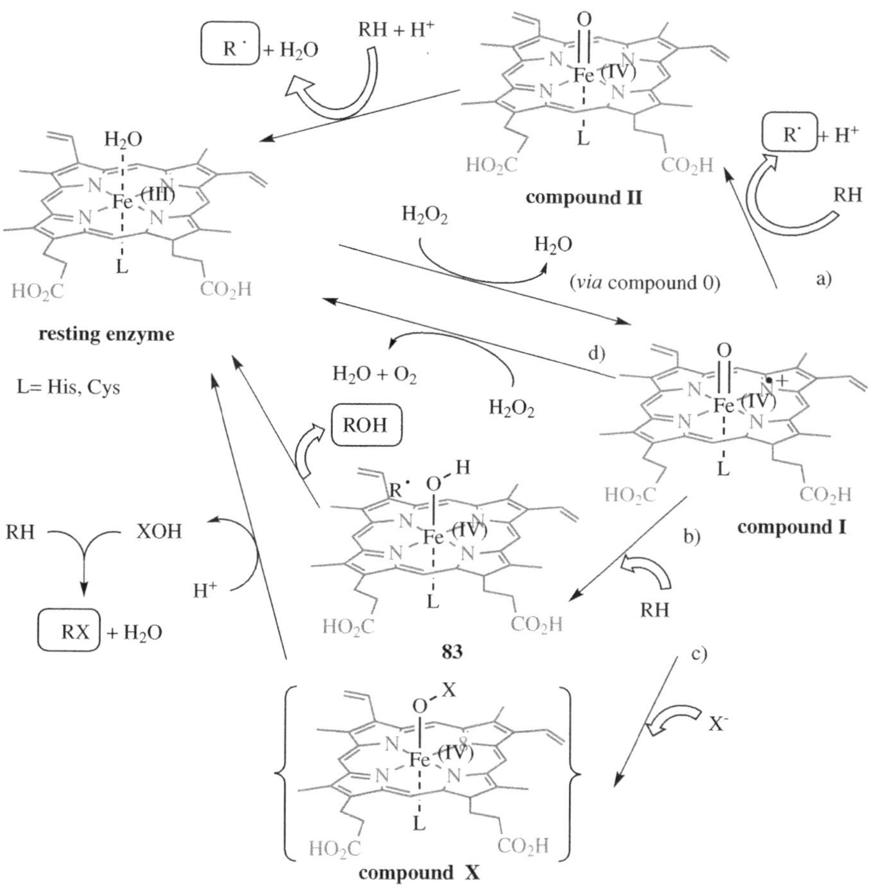

FIGURE 5.62 Catalytic cycles of heme peroxidases [661,665].

peroxide) is produced in the reaction. However, although the oxidation is comparable to those performed by monooxygenases such as cytochrome P450 (resembling the so-called "peroxide shunt"), relatively few examples of this peroxygenase activity of peroxidases have been reported [670,671]. This is due to the steric constraints in the active site and, more importantly, to the presence of histidine as the fifth ligand in most peroxidases, which is useful to activate hydrogen peroxide and favor one-electron oxidation over two-electron oxo transfer from Compound I [663,672]. The fifth ligand is not histidine in the heme-thiolate peroxidases, such as the chloroperoxidase from *Caldariomyces fumago* (CPO). With respect to oxygen transfer reactions, this heme-thiolate peroxidase with iron ligated to cysteine, similar to cytochrome P450, is the most versatile enzyme of the known peroxidases. Another reaction pathway is path c, which takes place in the haloperoxidases. In this path, the intermediate Compound I oxidizes the halide by two electrons, forming hypohalite OX^-, and regenerating the heme resting state [heme(Fe^{III}–H_2O)] through a mechanism still under debate

[673]. Compound X, the intermediate generated immediately upon halide oxidation, a hypothetical ferric hypohalite adduct, has not yet been characterized. It is also under discussion whether OX^- is a ligand of the heme-Fe^{III} at the distal site, whether it is released from the active site and diffuses freely into the bulk solution or whether it is otherwise trapped within the substrate pocket but not coordinated to the heme-Fe^{III}. The hypohalite gives hypohalous acid (HOX) in an acidic aqueous medium (pH < 3). There is evidence pointing at the diffusion of the halogenating species into the bulk solution, in order to halogenate a range of electron-rich substrates (including olefins, aromatics and β-diketones) outside the active site, in a nonselective fashion [661,674,675]. On the contrary, experiments of halogen isotope fractionation indicate that the oxidized intermediate is not in equilibrium with bulk halogenated species, thus suggesting that the oxidized halogen intermediate stays in the active-site pocket and that halogenation occurs within the enzyme active-site channel [676]. Finally, path d involves the heterolytic decomposition of hydrogen peroxide (catalase activity) to form water and molecular oxygen. This activity may be important in the absence of electron-donating substrates, particularly in haloperoxidases [677,678].

According to these main reactivity pathways, the peroxidase-catalyzed reactions are grouped in four categories, as follows.

1. Oxidative dehydrogenation: $2\,RH + H_2O_2 \rightarrow 2\,R^\bullet + 2\,H_2O \rightarrow R\text{-}R$
2. Oxygen-transfer reaction: $RH + H_2O_2 \rightarrow ROH + H_2O$
3. Oxidative halogenation: $RH + H_2O_2 + X^- + H^+ \rightarrow RX + 2\,H_2O$
4. Hydrogen (or organic) peroxide dismutation: $2\,H_2O_2 \rightarrow H_2O + O_2$

The first three transformations are of preparative interest, while hydrogen peroxide dismutation is mostly a side reaction in heme peroxidases, causing a progressive enzyme inactivation [679]. Of particular synthetic relevance is the large variety of oxygen transfer reactions catalyzed by peroxidases, including olefin epoxidation, *S*- and *N*-oxidations, and C–H bond oxidation (allylic, propargylic, and benzylic hydroxylations) [663,665,680,681].

From a synthetic point of view, although peroxidases do not need expensive cofactors, their low operational stability resulting from peroxide-induced deactivation, and relatively low volumetric productivities (space–time yields), constitute a major limitation for preparative applications [667,682]. Attempts to address these issues by using molecular biology strategies, such as random and site-directed mutagenesis, have been partially successful, since the improvements in stability usually implied a decrease in activity [683–686]. A more traditional approach to improve the operational stability of heme peroxidases is to ensure that the concentration of hydrogen peroxide is always kept at a low level. This has been achieved using two methods: feed-on-demand addition de H_2O_2 (for instance, adjusting its delivery rate) and *in situ* generation of hydrogen peroxide from oxygen with a chemical reductant or an enzyme [682,687]. Among the suitable enzymes, GOX is preferred because of its stability, commercial availability and use in glucose sensors that employ the same reaction. It has been used in CPO- and HRP-mediated oxygen transfer reactions, and was coimmobilized to CPO to produce hydrogen peroxide *in situ* from glucose and oxygen [688,689]. In addition to their synthetic use, which will be shown by reaction type below, peroxidases have found applications in the analytical arena—as components of biosensors and immunoassays [659,690,691] in the environmental field—for the removal of pesticides, PAHs, dioxins, and other organic pollutants [692–694], the pulp and paper industry—for Kraft pulp bleaching and wastewater detoxification [695,696], the textile industry for removal of excess dye [297], the food industry—as indicators for food processing [697], and in material science—for the preparation of conducting polymers and biomedical applications [698,699], among others [659].

5.2.1 Peroxidase Mediated Transformations

As already mentioned above, the three types of peroxidase-mediated transformations of synthetic interest are oxidative dehydrogenation, oxidative halogenation, and oxygen-transfer reactions [659,680,681,687].

5.2.1.1 Oxidative Dehydrogenation ($2\ RH + H_2O_2 \rightarrow 2\ R^{\bullet} + 2\ H_2O \rightarrow R\text{-}R$) This type of reaction is called the classical peroxidase reaction since it was the first to be described. It is restricted mainly to heme peroxidases, which catalyze a variety of one-electron oxidations of electron-rich aromatics (phenols and anilines) to give coupling products, from dimers to polymers [680]. The enzyme with the broadest specificity for oxidative dehydrogenation is HRP, a commercially available enzyme occurring as a large family of isoenzymes of which the C isoenzyme is the most abundant (HRP C) [658]. Typical substrates are mono- and diphenols, producing phenoxide radicals that dimerize in a nonenzymatic step to form mostly *o,o'*-biphenyls (and Pummerer's ketone) and, on standing, oligomeric products (Table 5.3, entries 1–3). The oligomerization reactions affect negatively the preparative usefulness of the process. A number of substituted hydroxycinnamic acid and derivatives have been assessed in HRP-mediated oxidation to afford lignan-type coupling adducts and novel spiro compounds (Table 5.3, entries 5 and 6) [700,701]. Usually, both homo-and cross-coupling products are isolated when mixtures of starting materials are used, as in the reaction of methyl sinapate and methyl ferulate, which gives mainly the cross-coupling product (24%) together with the two homodimers (27% combined yield) and oligomers (15%) (Table 5.3, entry 6) [702]. However, useful yields of cross-coupling products have also been reported, for example, arbutin (4'-hydroxyphenyl β-glucoside), which was coupled to sodium gentisate in the presence of hydrogen peroxide and HRP in 50–60% (Table 5.3, entry 7) [703]. The cross-coupling of catharanthine (**31**) and vindoline (**26**) to yield 3',4'-anhydrovinblastine (**33**), intermediate in the synthesis of the anticancer drug vinblastine, was catalyzed by HRP C (Table 5.3, entry 8) [704,705]. Compound **33** was obtained in 40–50% after sodium borohydride-mediated reduction of the eniminium cation resulting from enzymatic coupling, in a similar way to that reported using laccase [244]. The hydrogen peroxide/HRP system was able to oxidize aromatic oximes to afford good yields of the corresponding dimers of iminoxy radicals (Table 5.3, entry 9) [706]. Similar

TABLE 5.3 HRP-Catalyzed Coupling of Phenols and Aromatic Amines

Entry	Substrate	Product	Reference
1	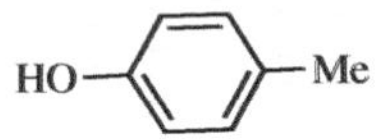	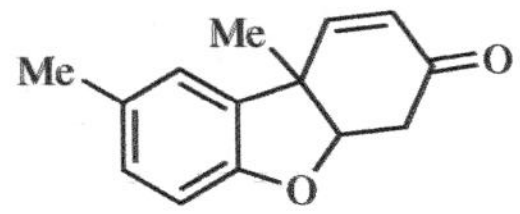 Pummerer's ketone	[707]
2	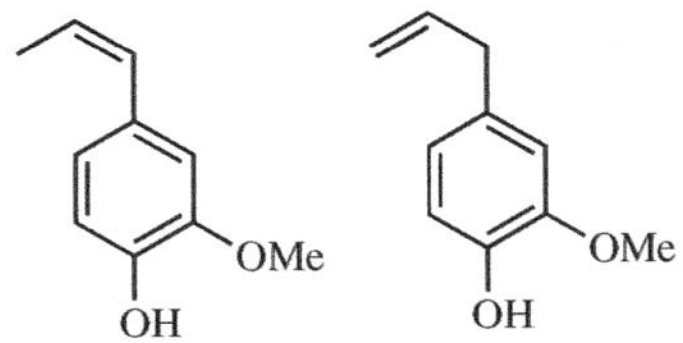	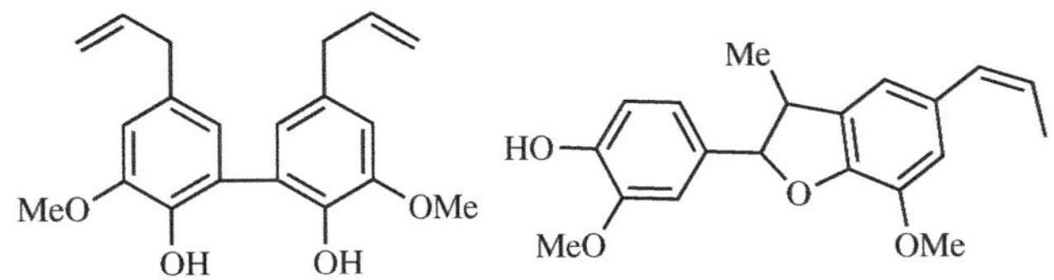	[708]
3	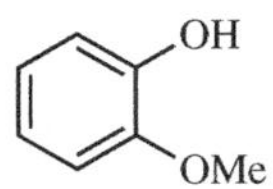 Guaiacol	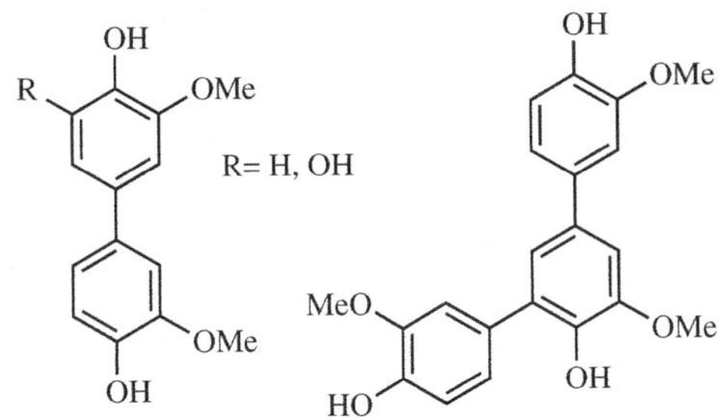 Aapproximately 1%	[709]
4	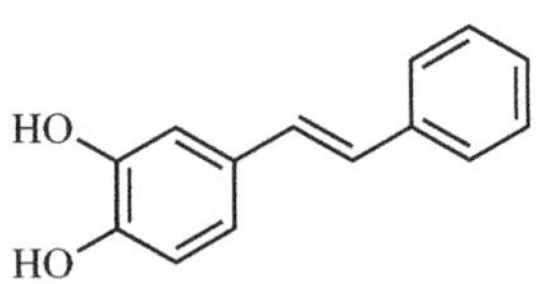	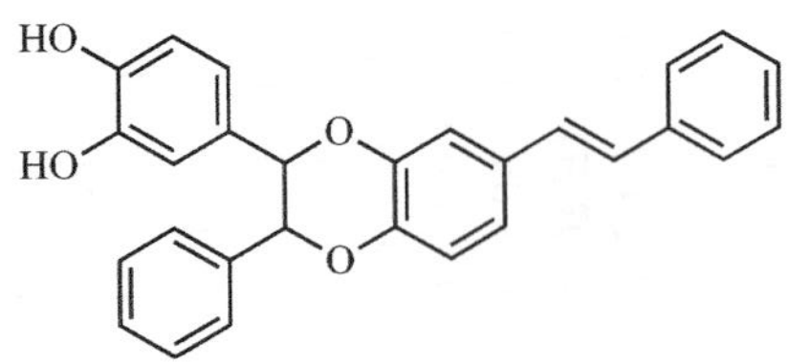	[710]

5	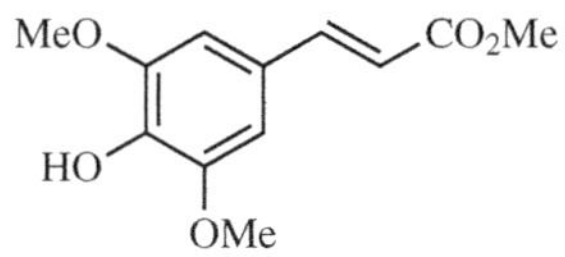 Methyl sinapate	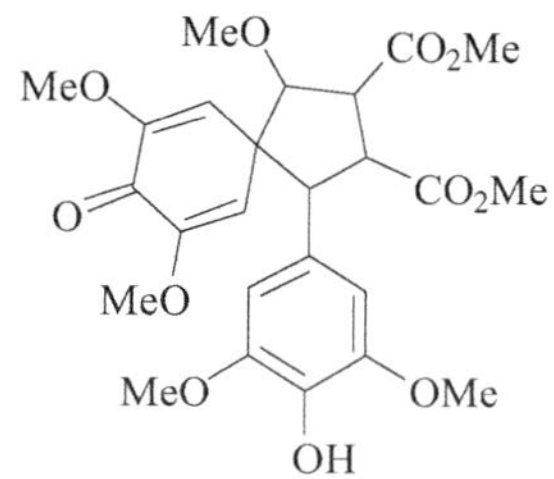	[700]
6	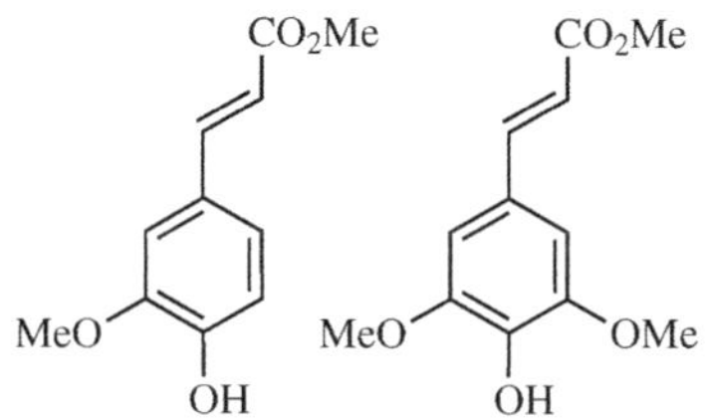 Ferulate: sinapate, 1:1	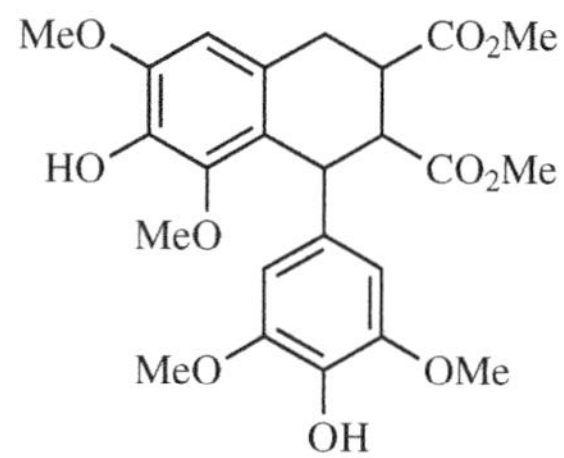 17% 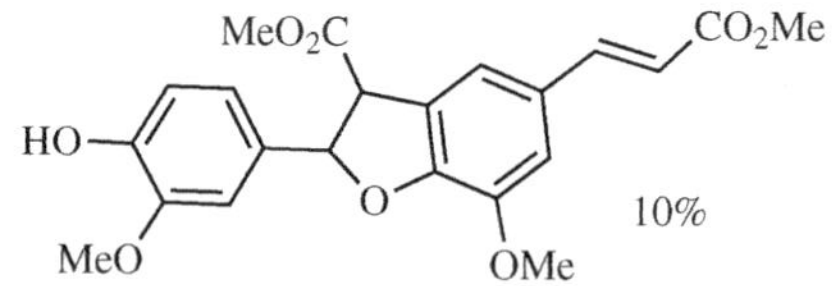 10% 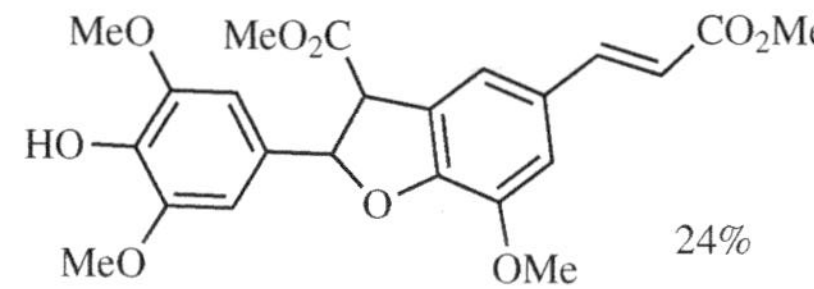 24%	[702]
7	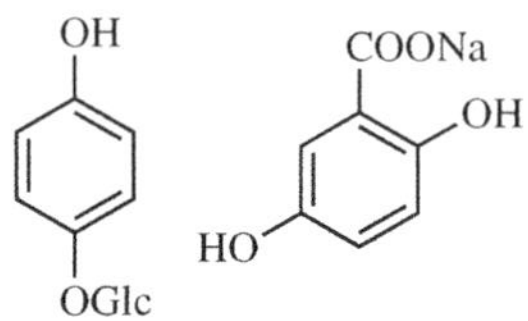 Glc= Glucose	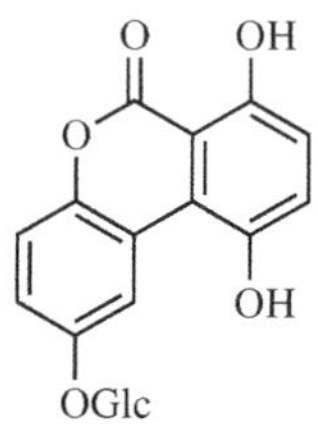 (50-60%)	[703]

(Continued)

TABLE 5.3 (*Continued*)

Entry	Substrate	Product	Reference
8	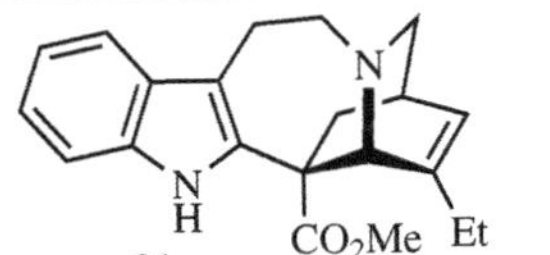 **31** 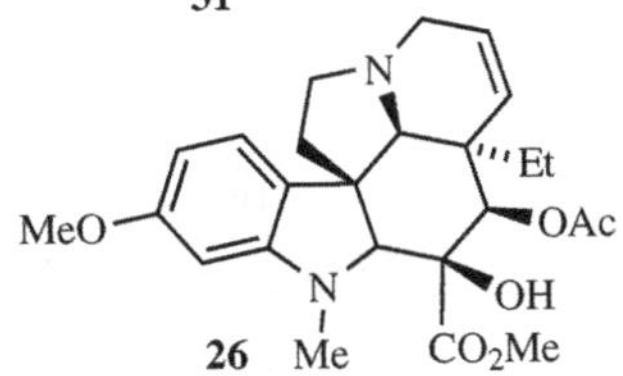 **26**	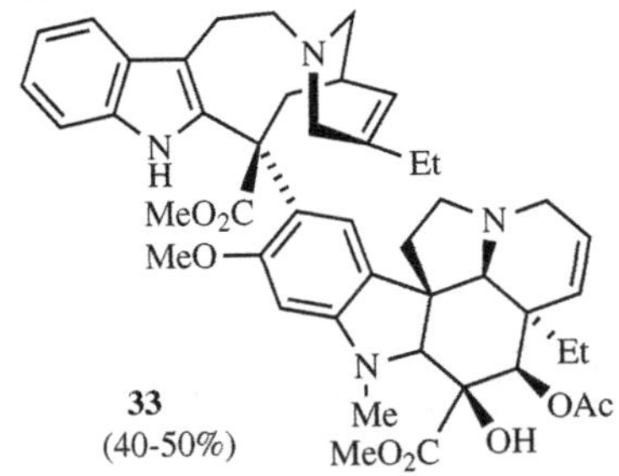 **33** (40-50%)	[704,705]
9	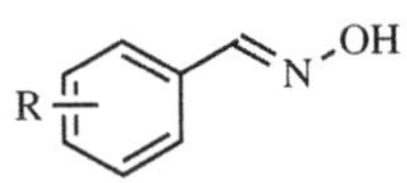 R= H, Me, OH, Cl	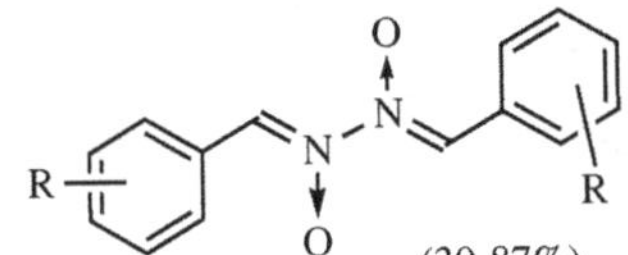 (30-87%)	[706]
10	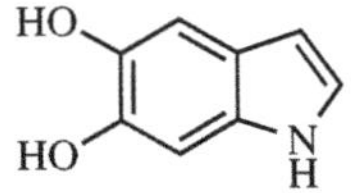	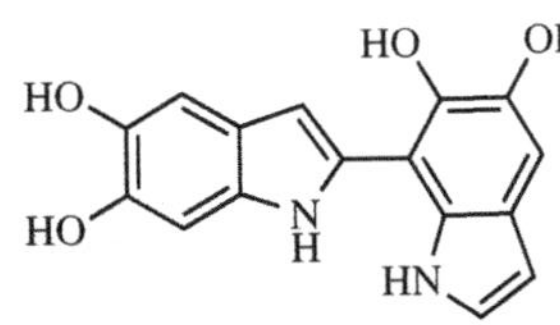 + oligomers	[711]

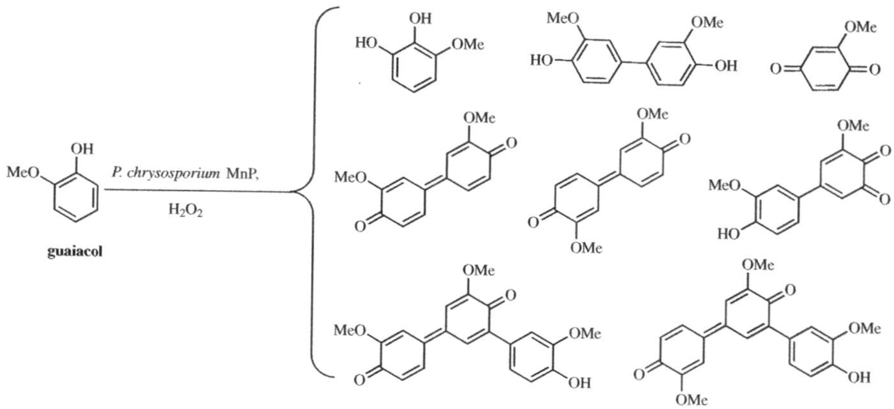

FIGURE 5.63 Products obtained by MnP-catalyzed oxidation of guaiacol.

to laccase, peroxidase has been used as catalyst for mild removal of the phenylhydrazide functionality, used as a carboxy-protecting group in peptide chemistry. The mild conditions used (buffer pH 4 and less than 1mM aqueous H_2O_2) caused no decomposition of the labile amino acids methionine and tryptophan [356].

The oxidation of phenolic compounds mediated by other peroxidases than HRP has been less studied. For instance, the oxidation of guaiacol by the extracellular manganese peroxidase (MnP, EC 1.11.1.13) obtained from the lignin-degrading fungus *Phanerochaete chrysosporium* BKM-F-1767, was recently studied and the structure of the first products was elucidated (Figure 5.63) [712]. The oxidation of guaiacol has been known for a long time and is used as one of the standard assay for peroxidase activity [713].

The reactivity of the radicals formed by the peroxidase-mediated reaction (laccase can also be used) has been employed to prepare polymers of phenols and anilines under mild conditions, constituting interesting alternatives to conventional phenol/formaldehyde resins and even high-performance engineering plastics (Table 5.4) [225,714,715]. For instance, the HRP-mediated polymerization of 2,6-dimethylphenol afforded a very homogeneous PPhO, similar to that produced conventionally [235]. On the other hand, mild polymerization of unsubstituted phenol produced soluble polyphenol having C–C and C–O coupling units with molecular weights ranging from 3 to 6 KDa. Even though 3-ethynylphenol is a good substrate for HRP (Table 5.4) [716], bulkier substituents in the *meta* position, such as isopropyl and *t*-butyl, are not accepted for HRP-mediated polymerization [717]. However, the use of a redox mediator such as *N*-ethyl phenothiazine or phenothiazine-10-propionic acid, triggers the polymerization. This strategy—well known in the laccase field—was used for the first time in the HRP-mediated oxidative polymerization of cardanol, an unsaturated m-C_{15}-phenol [718]. The most used enzymes are HRP and peroxidase from soybean, and the polymerization reactions are performed in aqueous, mixed-, or even totally organic solvent systems [714]. For instance, the polymerization of *p*-(hexyloxy)phenol was conducted in isooctane, using HRP modified by ion-pairing

TABLE 5.4 HRP-Catalyzed Polymerization of Phenols and Aromatic Amines

Substrate	Product	Reference
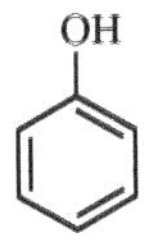	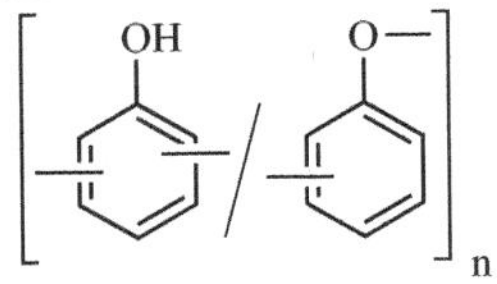	[726]
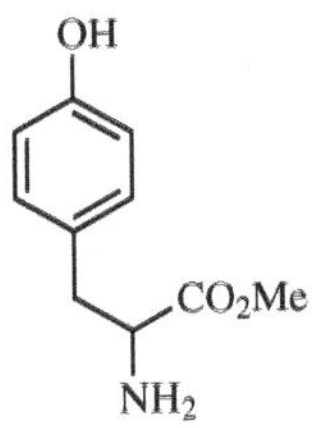	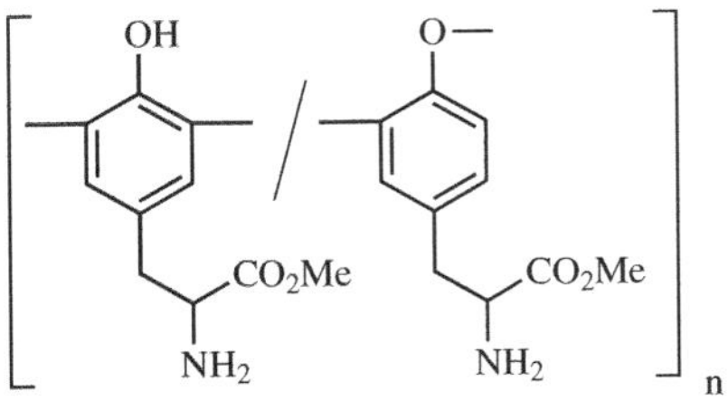 $1500 < n < 4000$	[727]
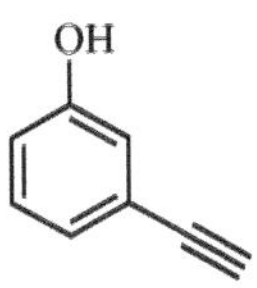	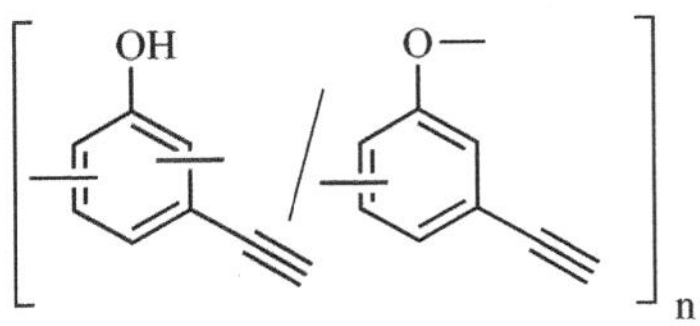	[716]
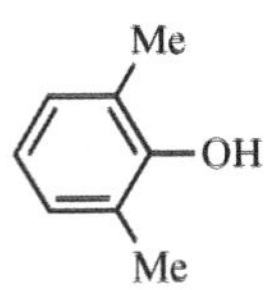	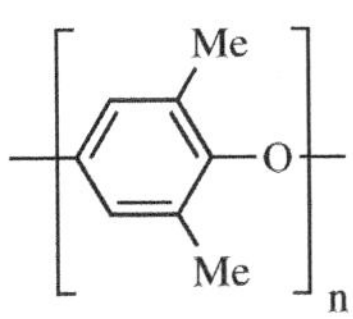	[235]
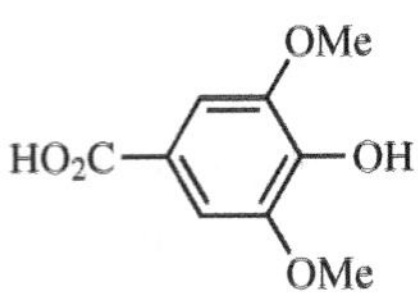 Syringic acid	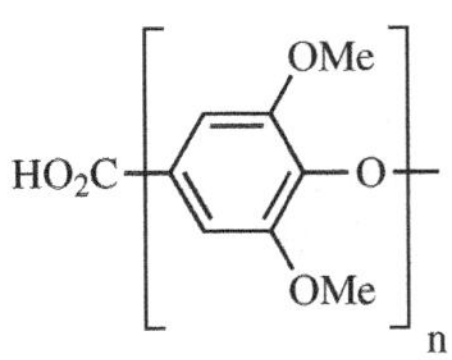 (MW up to 15.000)	[236]
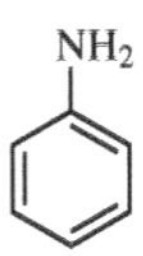	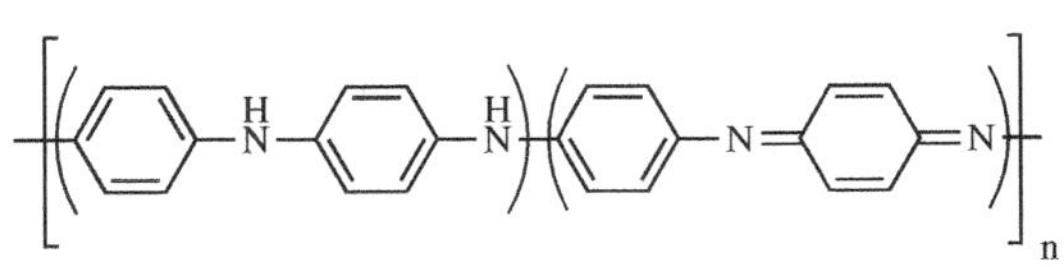	[724,725]
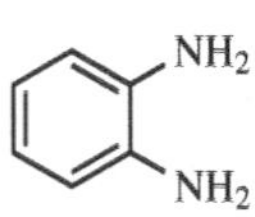	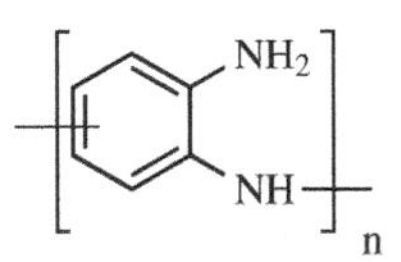 (MW up to 20.000)	[699]

with a surfactant and *t*-butyl hydroperoxide instead of hydrogen peroxide, to furnish a structure where the monomers are linked *via ortho-ortho* C–C coupling [719]. A bienzymatic system (GOX and HRP) was used to polymerize phenol in the presence of glucose without the addition of hydrogen peroxide, which was formed *in situ* by the enzymatic oxidation of glucose [720]. Aniline and derivatives are readily polymerized by HRP, using milder reaction conditions (room temperature and aqueous organic solvents at neutral pH) and simplified purification processes when compared to conventional polymerization methods [721]. Also the polymerization of aniline in the presence of different templates (polymers and metal ions) has been performed [722,723]. For instance, the use of sulfonated polystyrene as template promotes a more linear, *para*-directed polymerization, resulting in a emeraldine salt form of polyaniline, with good water solubility and conducting properties (Table 5.4) [724,725].

The *N*- and *O*-dealkylation of aromatic compounds has been performed in some cases, in particular with *N,N*-dimethylanilines and also methyl arylethers, through an oxidative demethoxylation step [728,729]. The *N*-demethylation of anilines has been accomplished using HRP and also chloroperoxidase [729,730].

5.2.1.2 Oxidative Halogenation ($RH + H_2O_2 + X^- + H^+ \rightarrow RX + 2\ H_2O$)

The halogenation of organic compounds can be performed enzymatically by four groups of enzymes, namely haloperoxidases, nonheme iron halogenases, flavin- and SAM-dependent halogenases, covering electrophilic, radical, and nucleophilic halogenation reactions [673]. Haloperoxidases catalyze electrophilic halogenation reactions, through the oxidative transformation of halides with consumption of one equivalent of hydrogen peroxide *per* halogenated product formed. Traditionally, haloperoxidases are named according to the most electronegative halide that is readily oxidized; thus chloroperoxidases oxidize chloride, bromide and iodide by hydrogen peroxide, BPO react with bromide and iodide, whereas iodoperoxidases are specific for iodide. The first discovered haloperoxidase (described in the 1960s), was a heme-chloroperoxidase [731]. Since then, three classes of haloperoxidases are known, the most important group of the heme-haloperoxidases (found in fungi), the vanadium haloperoxidases (v-HPO) (found in brown and red algae, bacteria, and some fungi), and the less important nonheme, nonmetal haloperoxidases (found only in bacteria) [650,732]. The catalytic mechanism is dependent on the nature of the prosthetic group. The heme haloperoxidases are also known as heme-thiolate haloperoxidases since the fifth ligand of the iron heme is a cysteine, instead of a histidine as in the rest of the heme peroxidases [660,661]. This feature greatly expands the catalytic possibilities of these enzymes, which are the most versatile heme peroxidases. Members of this group are the chloroperoxidase (CPO, EC 1.11.1.10) from the ascomycete *C. fumago*, which for a long time was the only enzyme of this type, and the peroxidase from *Agrocybe aegerita* (APO, EC 1.11.-.-) [733]. The catalytic cycle of the heme peroxidases was shown in Figure 5.62, although it is worth to remark that the specific pathway of halide oxidation is not fully understood yet [673].

The other important group of haloperoxidases is formed by the v-HPO. A simplified scheme of the proposed catalytic cycle for these haloperoxidases is shown

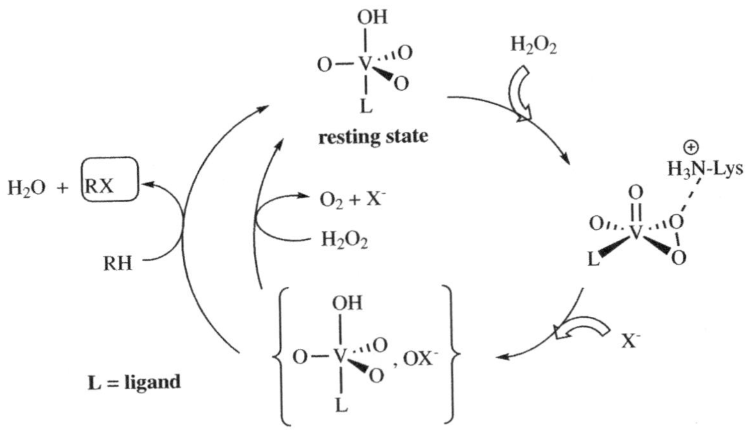

FIGURE 5.64 Proposed catalytic cycle for vanadium haloperoxidases.

in Figure 5.64 [673,734,735]. In the vanadate site, the vanadium is coordinated to the protein by an axial histidine residue. The catalytic reaction is initiated by coordination of one equivalent of hydrogen peroxide to the vanadate site, followed by hydrogen bonding of the peroxo group to a nearby lysine, which is important for halide oxidation because it would increase the oxidation potential. A two-electron oxidation of the halide by the oxo-peroxo-vanadium(V) species affords the formal halonium ion (X^+), probably as hypohalite (OX^-), which reacts with the substrate through an electrophilic halogenation. Alternatively, if no substrate is present, the halogenated species reacts with a second hydrogen peroxide molecule to generate molecular oxygen and the halide (Figure 5.64) [736]. In summary, the v-HPO acts as Lewis acid catalysts of the halide oxidation by hydrogen peroxide.

The most studied halogenating enzyme is the commercially available chloroperoxidase isolated from *C. fumago* (CPO, EC 1.11.1.10), actually the first to be discovered [731,737]. The halogenation reaction is performed at low pH (< 3) and leads to inactivation of the enzyme due to reaction with HOX [738]. The more common substrates are activated aromatic compounds, alkenes and β-dicarbonyl compounds, giving haloaromatics, halohydrins, and mono- and dihalogenated β-dicarbonyl compounds, respectively [739,740]. Representative examples are shown in Table 5.5. The CPO-mediated halogenations usually display low selectivity, but with remarkable exceptions such as the regioselective bromohydration of glycals and the formation of halohydrins from the monoterpene 3-carene [741–743]. The use of CPO immobilized on talc afforded better selectivity than that with the free enzyme for the bromohydroxylation of styrenes, obtaining only the Markovnikov-type halohydrin with no oxidative by-products [744]. In addition to these reactions, in the absence of halide a number of nonhalogenating reactions are catalyzed by CPO (see Section 5.2.1.3).

Another heme-thiolate haloperoxidase is obtained from the agaric mushroom *A. aegerita* (APO, EC 1.11.2.1) [750]. This extracellular enzyme was originally reported as a haloperoxidase but it turned out to be a true peroxygenase, transferring oxygen from peroxides to various organic substrates including aromatic, heterocyclic, and

TABLE 5.5 CPO-Mediated Halogenation Reactions (Reaction Conditions: CPO, H_2O_2, KX, pH 3)

Substrate	Product	Reference
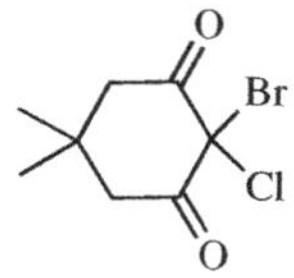	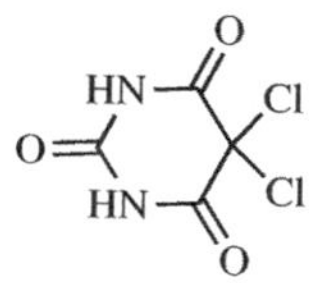	[745]
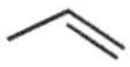	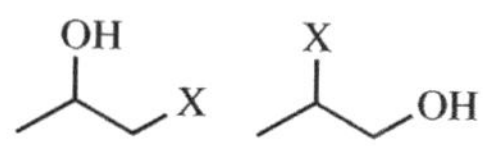 9 : 1 X= Cl, Br, I	[746]
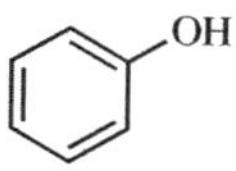	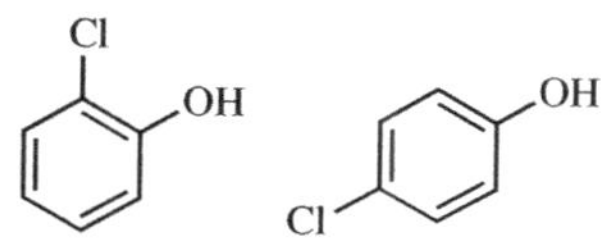	[747]
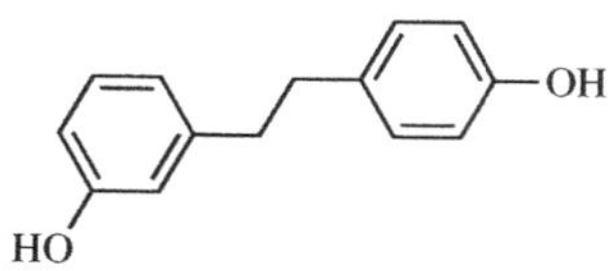 Lunularin	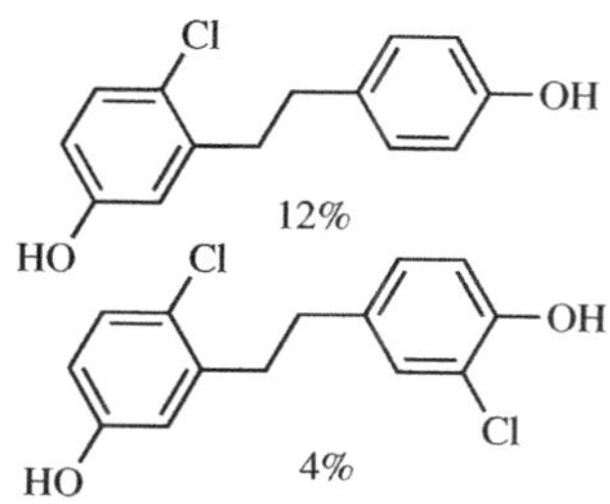 12% 4%	[748]
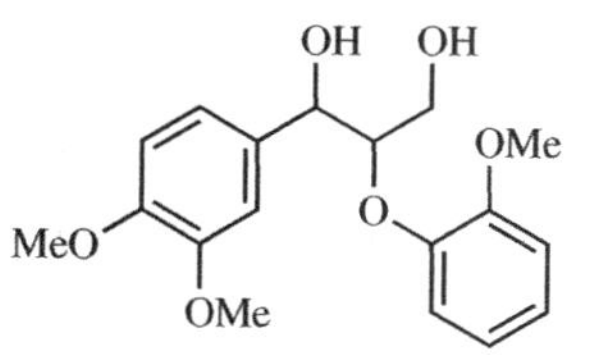	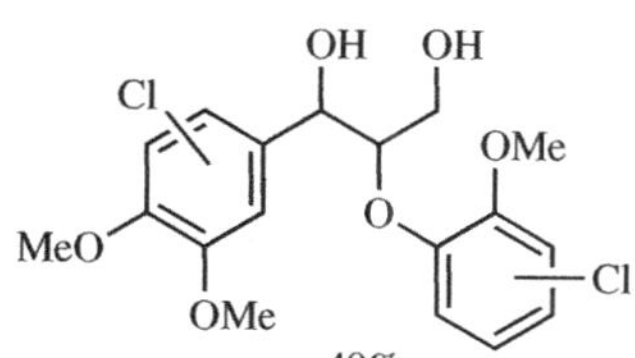 49%	[749]
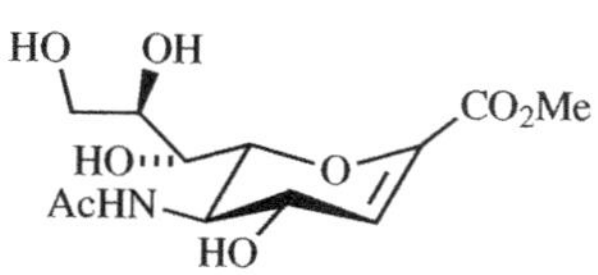	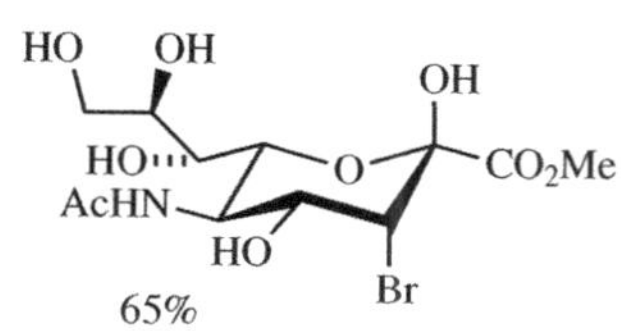 65%	[742]
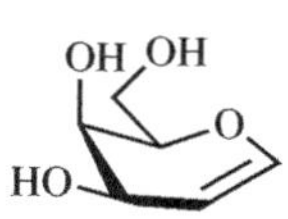	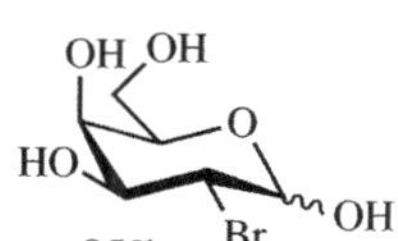 85%	[741]

(*Continued*)

TABLE 5.5 ***(Continued)***

Substrate	Product	Reference
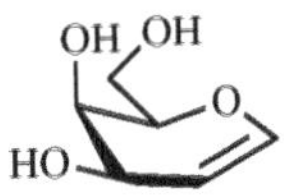	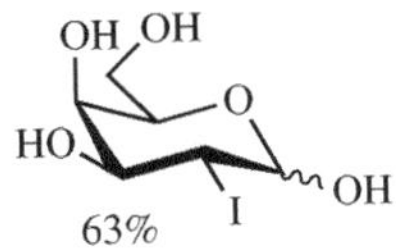 63%	[741]
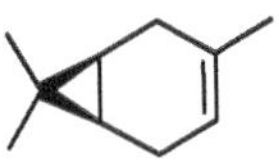 3-carene	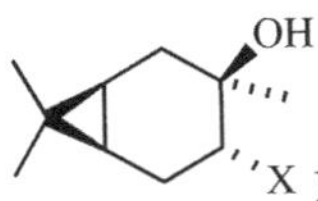 X=Cl, Br, I X= Cl, Br, I	[743]
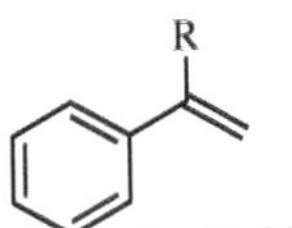 R = H, Me	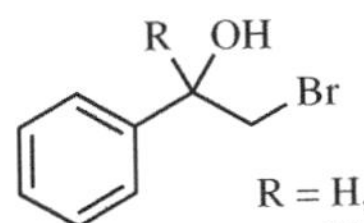 R = H, 95% = Me, 92%	[744]
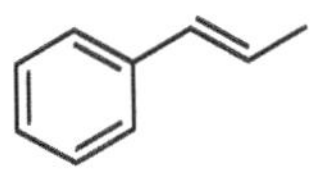	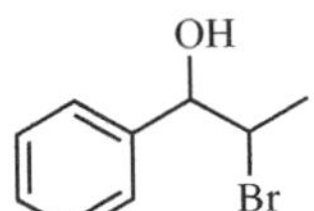 90%	[744]

aliphatic compounds. Hofrichter and coworkers proposed the designation of aromatic peroxygenase (APO) for the enzymes in this group and suggested to group them in a new subsubclass of peroxide-dependent enzymes (EC 1.11.2.-), which was accepted in 2011 [733]. APO displays strong brominating and weak chlorinating and iodinating activities; in the presence of potassium bromide (KBr), phenol is brominated in *o*- and *p*-positions in a 1:4 ratio [750]. The most important activity of APO is related to oxygen transfer reactions, in particular the hydroxylation of aromatic rings (Section 5.2.1.3).

The group of the vanadium peroxidases offers interesting examples of halogenating biocatalysts, since they do not show the inherent instability of the heme group toward oxidative degradation [751]. In addition, they are much more resistant toward heat, detergent, and solvent denaturation [752]. However, deactivation occurs at low pH (pH 4–5) and high H_2O_2 concentrations (100 mM) [753]. The most important sources of this group of haloperoxidases are marine organisms, particularly red, brown, and some green algae [732]. The group includes vanadium chloroperoxidase (v-CPO) from the terrestrial fungus *Curvularia inaequalis* [754], and BPO from the algae *Corallina officinalis*, *Ascophyllum nodosum*, and *Corallina pilulifera*, among others [755,756].

The bromoperoxidase from *C. pilulifera* was able to brominate a number of aromatic and heteroaromatic compounds, and substituted alkenes, such as anisole, 1-methoxynaphthalene, thiophene, styrene, cyclohexene, *trans*-cinnamic acid, *trans*-cinnamyl alcohol, and *cis*-propenylphosphonic acid, but always obtaining mixtures

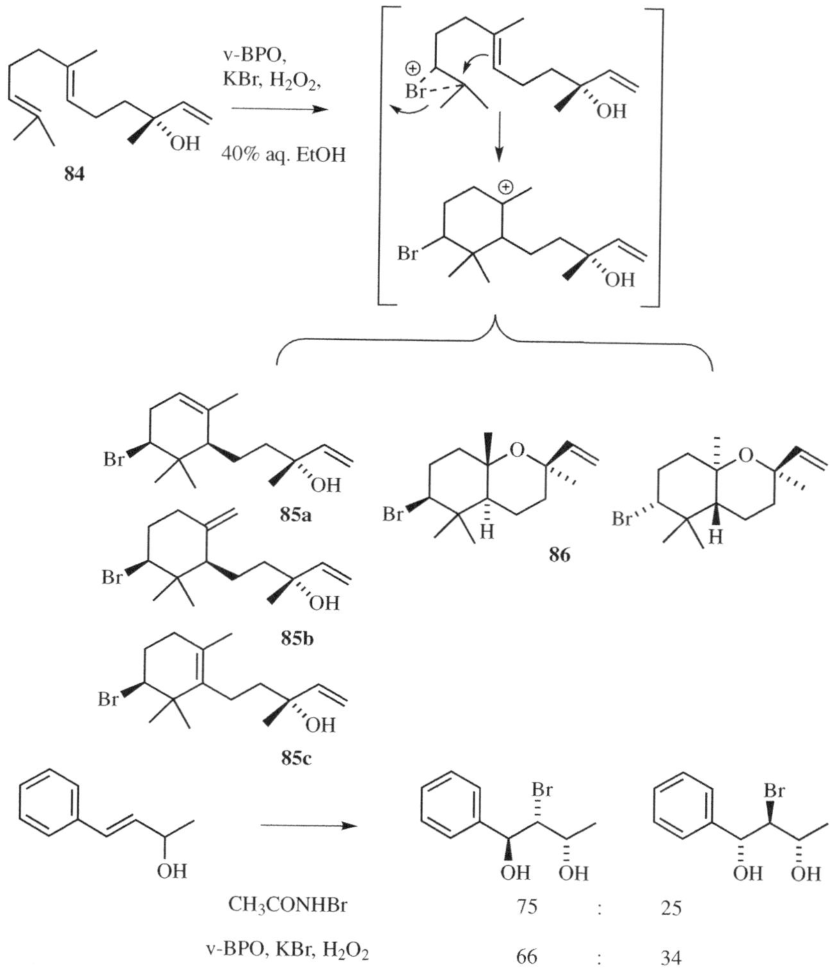

FIGURE 5.65 Selectivity in the v-BPO-catalyzed bromination of (*E*)-(+)-nerolidol and (*E*)-4-phenyl-3-buten-2-ol. v-BPO-catalyzed formation of natural products from (*E*)-(+)-nerolidol.

of regio- and stereoisomers [757]. This lack of selectivity was originally explained assuming a free diffusion of the oxidized bromine species, once formed in the active site. However, more recently it was described the bromination and cyclization of the terpene (*E*)-(+)-nerolidol (**84**), mediated by vanadium bromoperoxidase (v-BPO) isolated from marine red algae, producing the marine natural products α-, β-, and γ-snyderol as single diastereomers (**85a–c**), and a diastereomeric mixture of (+)-3β-bromo-8-epicaparrapi oxide (**86**) (Figure 5.65) [758]. The observed diastereoselectivity is the first example of v-BPO-catalyzed enantioselective bromination and cyclization of sesquiterpenes, and suggests that the terpene docks within the active site in a specific orientation in order to react with the brominating species over one face of the olefin. Moreover, a direct brominating action of the enzyme is suggested for the bromination of indoles [759], and during the bromohydrin formation of (*E*)-4-phenyl-3-buten-2-ol mediated by v-BPO from *C. officinalis*. A mixture of two diastereomers of 2-bromo-1-phenylbutane-1,3-diol was obtained in a ratio (66/34) different from that obtained chemically with *N*-bromoacetamide (75/25) (Figure 5.65) [760].

Vanadium haloperoxidases are promising biocatalysts, albeit their major drawback is the mild acidic pH range for maximum activity, whereas for many applications the required pH values are above 7. To improve the activity of v-HPO at mild alkaline pH for industrial applications, directed evolution on the vanadium chloroperoxidase from *C. inaequalis* was performed, resulting in a triple mutant (Pro395Asp/Leu241Val/Thr343Ala) showing a 100-fold increase in activity at pH 8, and also sixfold better brominating activity at pH 5 [761].

In principle, if the halide is replaced by nitrite, the peroxidase-mediated oxidation would lead to a nitrating species and thus, to an enzymatic nitration reaction. This is the case and HRP, CPO, and other peroxidases are able to mediate the nitration of phenolic compounds, as shown in Table 5.6, for the reaction mediated by soybean peroxidase [762].

5.2.1.3 Oxygen-Transfer Reactions ($RH + H_2O_2 \rightarrow ROH + H_2O$)

Remarkably, oxygen-transfer reactions are the most interesting processes catalyzed by peroxidases, instead of the "classical" oxidative dehydrogenation reaction [663,681,763]. Most of the synthetic applications of these enzymes are based on oxygen-transfer reactions, which allow the enantioselective introduction of an oxygen atom into organic substrates in a mild and controlled fashion, using low cost and environmentally benign oxidants. As just one oxygen atom is transferred from hydrogen peroxide to the substrate, the enzymes act as monooxygenases, catalyzing heteroatom (sulfur and nitrogen) oxidation, epoxidation, and C–H oxidation (benzylic, allylic, indole, benzofuran and alcohol oxidation). Haloperoxidases perform these reactions when there is no halogen to be oxidized by the enzyme [5].

Heteroatom Oxidation Sulfoxidation is the most important of these oxidations. The enantioselective preparation of sulfoxides is of relevance in the field of asymmetric synthesis (either as building blocks or stereodirecting groups) and in the pharmaceutical industry (a number of drugs are sulfoxides). Several types of enzymes have been used for this reaction (mono- and dioxygenases, peroxidases, proteases), and peroxidases play an important role. A number of these enzymes mediate the clean oxidation of organic sulfides to sulfoxides, without any further reaction to the sulfone. The reaction is often performed to demonstrate the oxygen-transfer capabilities of peroxidases, and hydrogen peroxide is the most used oxidant, although *t*-butyl- and some other organic hydroperoxides have been occasionally used. When performing these oxidations an issue to be solved is the slow, noncatalyzed background oxidation that erodes the enantioselectivity. To keep this to a minimum, the concentration of hydrogen peroxide must be maintained as low as possible. As mentioned previously, feed-on-demand and *in situ* generation of hydrogen peroxide are the common strategies used, for example, slow peroxide addition *via* syringe pumps, the use of a

TABLE 5.6 Nitration of Phenols Mediated by Soybean Peroxidase (SBPO) [762]

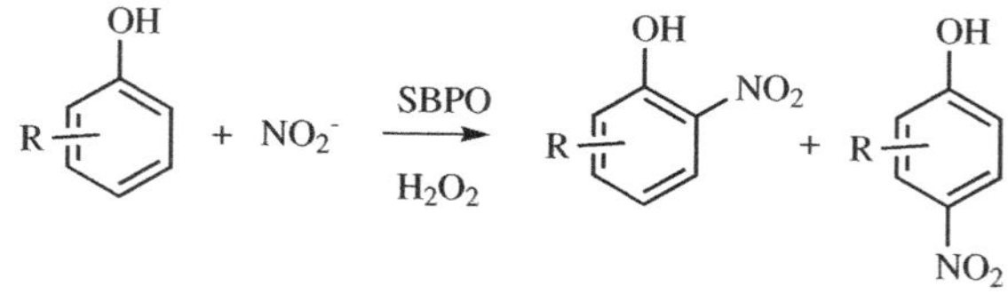

Substrate	Product	
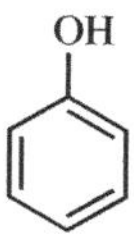	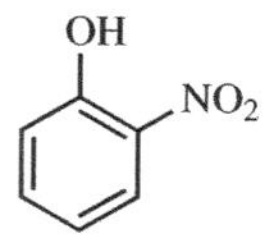 12%	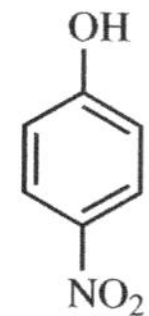 14%
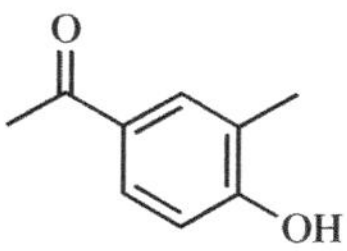	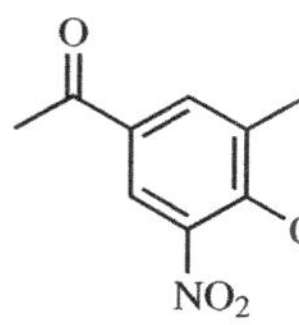 58%	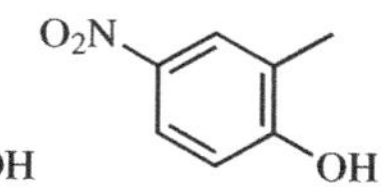 27%
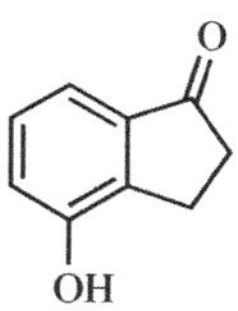	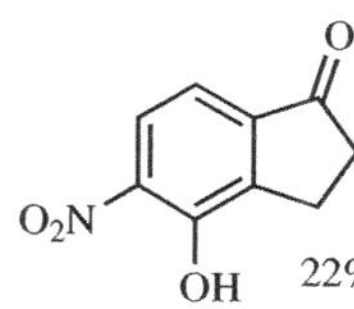 22%	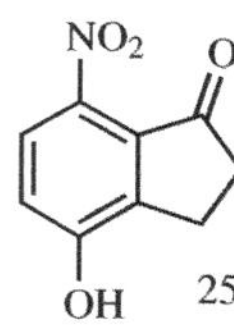 25%
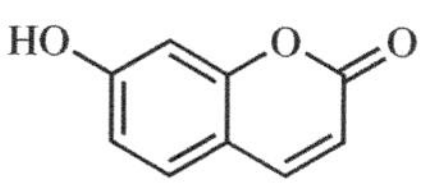	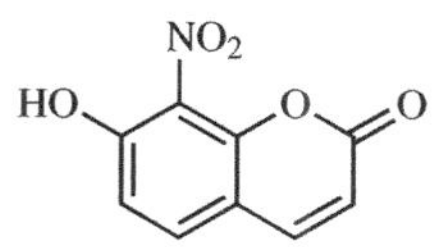 41%	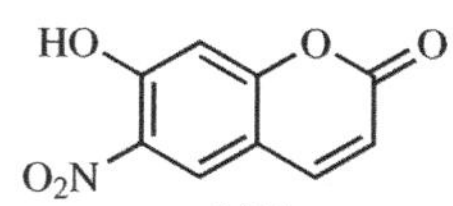 20%
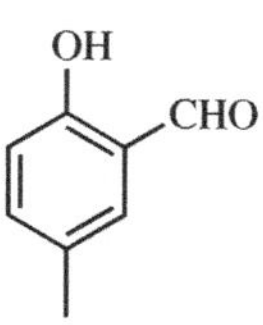	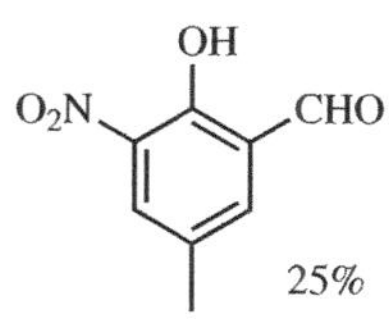 25%	

peroxide-stat, *in situ* production of hydrogen peroxide using an oxidase/peroxidase bienzymatic system, an electrochemical or photochemical method, or else a chemoenzymatic system with chemical generation of hydrogen peroxide through Pd-mediated catalysis (Figure 5.66) [682,764–769]. A comparison of the various methods used to generate hydrogen peroxide for the CPO-mediated oxidation of thioanisole is available [768]. Several oxidases have functioned as partners in the bienzymatic system,

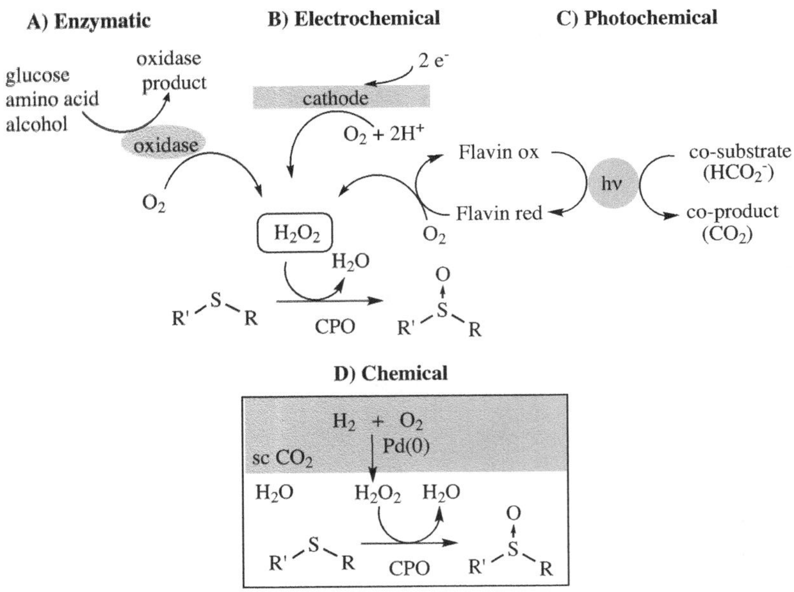

FIGURE 5.66 Methods for *in situ* generation of hydrogen peroxide in the CPO-mediated oxidation of sulfides. $scCO_2$: supercritical CO_2.

such as GOX (EC 1.1.3.4), D-amino acid oxidase (EC 1.4.3.3), and AO (EC 1.1.3.13) [688,770–772].

Many sulfides have been tested as substrates of several peroxidases, giving (*R*)-sulfoxides with moderate-to-excellent enantiomeric excess (up to > 99% *ee*) [680,773]. The most efficient and selective peroxidase in sulfoxidation reactions is CPO from *C. fumago*, which was also the first to be used in an enantioselective sulfoxidation, as reported by Kobayashi in 1987 [774]. A selection of the sulfides used in the CPO-mediated sulfoxidation is presented in Table 5.7, showing that electronic and, particularly, steric factors affect the outcome of the reaction. In some cases, the use of cosolvents, such as 1:1 *t*-butyl alcohol:water mixtures improved the enantiomeric excess by suppressing the uncatalyzed reaction [775].

CPO oxidizes alkyl aryl, dialkyl, and alkyl vinyl sulfides and some cyclic sulfides with up to 99% *ee* for the (*R*)-isomer, being the biocatalyst of choice to test a sulfoxidation reaction. Some more demanding substrates, such as methionine derivatives, were also good substrates for CPO, affording the (*R*)-sulfoxide in good yield and up to 82% *ee*, depending on the protecting group used [781]. Other peroxidases have been used to perform sulfoxidations, yet all of them displayed lower activity and much less selectivity, such as HRP, lignin, and manganese peroxidase, myeloperoxidase, lactoperoxidase, and vanadium-containing peroxidases [773,782–784]. For instance the enantioselectivity of the HRP-mediated sulfoxidation is moderate (<75% *ee*) (Table 5.8). This could be greatly improved by using molecular engineering [785]. The replacement of Phe41 with leucine (HRP Phe41Leu) led to a mutant capable of catalyzing sulfoxidations with greater than 94% *ee* (Table 5.8). This has been

TABLE 5.7 Selected Sulfides Used in the CPO-Mediated Sulfoxidation Reaction

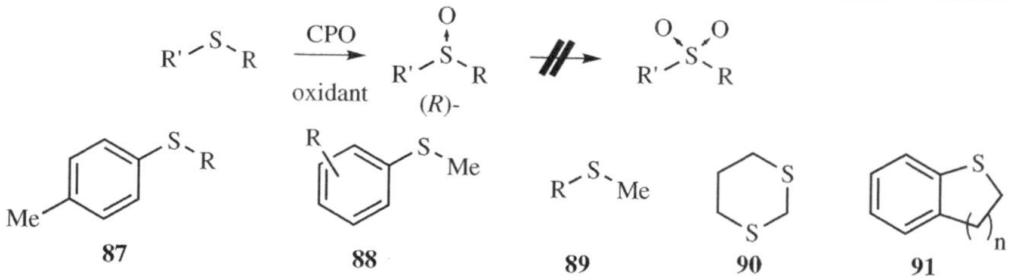

Substrate	R	Oxidant	Yield (%)	*ee* (%)	Reference
87a	Me	H_2O_2	92	99	[742,776]
87a	Me	*t*-BuOOH	80	70	[742,776]
87b	Et	H_2O_2	50	68	[776]
87b	Et	*t*-BuOOH	50	68	[776]
87c	Pr	H_2O_2	53	5	[776]
87c	Pr	*t*-BuOOH	30	5	[776]
88a	H	H_2O_2	90–100	99	[742,775]
88a	H	*t*-BuOOH	90	80	[776]
88b	*p*-Cl	H_2O_2	77	90	[776]
88b	*p*-Cl	*t*-BuOOH	44	85	[777,778]
88c	*p*-NO_2	H_2O_2	10	80	[776]
88c	*p*-NO_2	*t*-BuOOH	16	80	[776]
88d	*p*-F	H_2O_2	100	97	[776]
88d	*p*-F	*t*-BuOOH	90	70	[776]
88e	*p*-MeCONH	H_2O_2	86	67	[776]
88e	*p*-MeCONH	*t*-BuOOH	86	70	[776]
88f	*p*-MeO	H_2O_2	66	100	[742]
88f	*p*-MeO	*t*-BuOOH	71	92	[777]
88g	*m*-MeO	H_2O_2	24	27	[776]
88g	*m*-MeO	*t*-BuOOH	30	37	[776]
89a	2-pyridyl	H_2O_2	100	99	[776]
89a	2-pyridyl	*t*-BuOOH	30	37	[776]
89b	Vinyl	H_2O_2	>98	>98	[779]
89c	Cyclopentyl	H_2O_2	>98	>98	[779]
89d	Cyclohexyl	H_2O_2	85	85	[779]
89e	*n*-Bu	*t*-BuOOH	54	38	[777]
89f	*n*-Pentyl	H_2O_2	75	>98	[779]
89g	*i*-Pr	H_2O_2	>98	>98	[779]
89h	*t*-Bu	H_2O_2	30	35	[779]
89i	CH_2SCH_3	H_2O_2	75	>98	[779]
90		*t*-BuOOH	63	20	[777]
91a	$n = 1$	H_2O_2	99	99	[780]
91b	$n = 2$	H_2O_2	80	96	[780]

TABLE 5.8 Improvement in the Selectivity of HRP-Mediated Sulfoxidations by Using HRP Phe41Leu

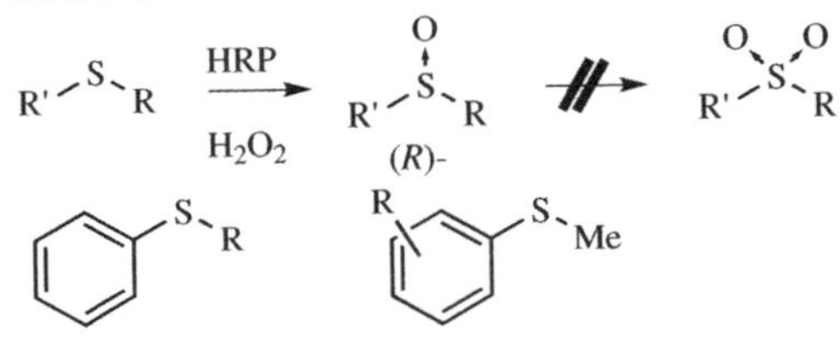

Substrate	R	Enzyme	Yield (%)	*ee* (%)	Reference
92a	Me	HRP	95	46	[786]
92a	Me	HRP-F41L	–	97	[787]
92b	Et	HRP	–	35	[785]
92b	Et	HRP-F41L	–	94	[785]
92c	Pr	HRP	–	12	[787]
92c	Pr	HRP-F41L	–	94	[787]
92d	Cyclohexyl	HRP	–	7	[787]
92d	Cyclohexyl	HRP-F41L	–	94	[787]
88h	Me	HRP	47	68	[786]
88h	Me	HRP-F41L	–	97	[785]
88i	CN	HRP	–	64	[776]
88i	CN	HRP-F41L	–	90	[776]
88e	*p*-MeCONH	HRP	–	78	[785]
88e	*p*-MeCONH	HRP-F41L	–	94	[785]
88f	*p*-MeO	HRP	84	30	[786]
88f	*p*-MeO	HRP-F41L	–	97	[785]
88j	2-naphthyl	HRP	–	69	[785]
88j	2-naphthyl	HRP-F41L	–	99	[785]

rationalized invoking a better access to the ferryl oxygen in the active site, which is responsible for the monooxygenase activity.

Within the vanadium peroxidase group, the BPO from *C. officinalis* and *A. nodosum* gave good (*R*)-enantioselectivity in the sulfoxidation of substituted aryl methyl sulfides [755,756,788,789]. Plant peroxidase from *C. cinereus* (CiP) mediated the oxidation of aryl methyl sulfides to give (*S*)-oxidized products with reasonable yields and *ee* [771,772]. The oxidation of sulfides has also been tried using whole cells, although in this case several enzymes can effect the sulfoxidation. White-rot fungi, that promote lignin degradation using extracellular laccases and peroxidases, were used in the oxidation of alkyl phenyl sulfides, affording good conversions and 12 up to 99% *ee* of the (*S*)-sulfoxide after 48 h [790]. Also, the filamentous fungi *Botrytis cinerea*, *Trichoderma viride*, and *Eutypa lata*, oxidized thioanisole and butyl phenyl sulfide to give the (*S*)-sulfoxide with *B. cinerea*, and the (*R*)-product with the other fungi [791]. The decrease of sulfoxide yield along reaction time, together with an increase of the enantiomeric excess (up to > 95%) suggests that the stereoselectivity is in part due the overoxidation of the minor sulfoxide to sulfone.

In the sulfoxidation reaction, hydrogen peroxide is the most used oxidant, but also organic peroxides are accepted by the peroxidases. When the organic peroxide contains stereogenic centers, the possibility of chiral discrimination arises, originating a kinetic resolution of the oxidant hydroperoxide. Interestingly, racemic hydroperoxides oxidize sulfides to sulfoxides enantioselectively under CPO catalysis [742]. In this reaction, not only the sulfoxides but also the hydroperoxide and the corresponding alcohol were produced in optically active form by enzyme-catalyzed kinetic resolution. The reaction is not suitable to obtain preparative amounts of chiral hydroperoxides since conversions significantly higher than 50% are needed to afford the hydroperoxide in good enantiomeric excess [680]. Methods for improving the catalytic performance of peroxidases in sulfoxidations are usually related to the control of the hydrogen peroxide concentration to prevent enzyme inactivation and the uncatalyzed oxidation of the substrates, to the increase of the solubility of the reactants and to the improvement of the operational stability of the catalyst. Addition of short-chain polyethylene glycols to increase the substrate solubility in aqueous media was effective to achieve good conversions with complete enantioselectivity in the CPO-mediated oxidation of thioanisole [792]. Hydrophilic ionic liquids and alcohols have been used as cosolvents for these reactions, with mixed effects in selectivity and reaction rates [793,794]. By means of the *in situ* enzymatic generation of hydrogen peroxide, good results were obtained by coimmobilization of CPO and GOX into polyurethane foams [689]. By generating the oxidant inside the peroxidase-containing matrix, beneficial effects were detected, such as the suppression of the uncatalyzed background oxidation and better oxidation efficiency, reaching 100% yield of thioanisole sulfoxide with 99% *ee* [682,689]. Other immobilization procedures were used, such as encapsulation and preparation of CPO-coated magnetic nanoparticles, affording satisfactory results [795–797].

Peroxidases are able to perform *N*-oxidations as well. This reaction has received less attention and only a few examples are described in the literature. The CPO-mediated oxidation of arylamines produces the corresponding nitroso compounds (Figure 5.67) [798–800]. Another peroxidase used in *N*-oxidations of aromatic amines is HRP. Aliphatic amines are not substrates of HRP, with the exception of the tertiary amine of certain indole alkaloids and some *N*-(3-indoleethyl) cyclic amines, in which

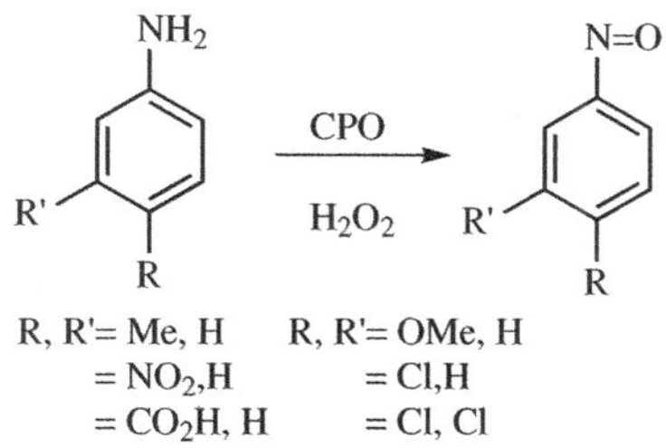

FIGURE 5.67 CPO-mediated oxidation of aromatic amines.

the indole ring facilitates the electronic transfer to give the corresponding iminium species [801,802].

Epoxidation Peroxidases, mostly heme-thiolate peroxidases, form epoxides from olefins by a direct oxygen-transfer reaction or, indirectly, *via* halohydrin formation and further chemical or enzymatic ring closure. Epoxides are of special interest in asymmetric synthesis since they can give bifunctional compounds with up to two new chiral centers. In general, native HRP does not perform epoxidations, whereas various mutants can produce optically active styrene oxide and derivatives. However, this reaction has limited synthetic value because of the large amount of rearranged aldehydes formed as by-products [787]. On the other hand, the use of CPO to catalyze epoxidations, first described in 1983, is a high yielding an enantioselective reaction [803]. The enantioselective version of the reaction was reported 10 years later [804,805]. For aliphatic alkenes, the size of the substituents is a major limitation. Thus, *cis*-alkenes and *gem*-substituted alkenes not too distant from the chain terminus are efficiently epoxidized by CPO, leading to products with high enantiomeric excess (Table 5.9). If the size of the R_2 group increases the reaction becomes sluggish and allylic hydroxylation may predominate, as seen with 3-alkenes such as **94e**, where allylic alcohols are the major products. The *trans*-isomers are highly unreactive toward epoxidation, giving mostly products of allylic oxidation, accompanied by further oxidation to the aldehyde. Terminal alkenes are suicide inhibitors, leading to alkylation of the heme group and subsequent enzyme deactivation [806]. Directed evolution was applied to CPO in order to improve the stability of the enzyme, resulting in a fourth-generation mutant that was completely resistant to heme alkylation and, moreover, displayed enhanced epoxidation activity [807].

The CPO-mediated epoxidation is a slow process and the oxidative deactivation of the catalyst may be comparable to the rate of epoxidation. Thus, it is important to maintain low levels of hydrogen peroxide throughout the reaction, for example, by slow addition or *in situ* generation of hydrogen peroxide or, much simpler, by using a less aggressive oxidant, such as *t*-butylhydroperoxide (TBHP). With styrene derivatives this is the oxidant of choice, instead of hydrogen peroxide, and better yields of epoxides are obtained while the enantioselectivity remains almost unchanged [804]. The epoxidation of phenyl olefins is presented in Table 5.10. In some cases the products are labile and the ring opened products are seen, such as the diols **98** and **99**. Indene was oxidized in anhydrous glycerol to avoid the hydrolysis of the epoxide that spontaneously takes place in aqueous medium, constituting the first report on the use of CPO in a nonaqueous medium [811].

More complex substrates have been tested in CPO-mediated epoxidations. For instance, some dienes were selectively epoxidized by using CPO to furnish alkene epoxides useful as bifunctional monomers for polymerization (Figure 5.68) [813]. Conjugated dienic esters gave mixtures of epoxides and products of allylic oxidation, depending on the substitution pattern of the dienic system. The simpler methyl 2,4-pentadienoate afforded the terminal monoepoxide and a cyclodimer (Figure 5.68) [814]. The oxidation of conjugated cyclic dienes using the CPO/TBHP

TABLE 5.9 CPO-Mediated Epoxidation of Olefins

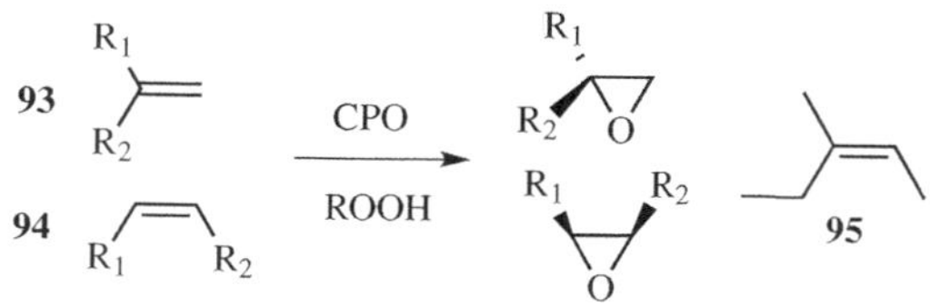

Compound	R_1	R_2	Oxidant	Yield (%)	*ee* (%)	Reference
93a	$PhOCH_2$	Me	H_2O_2	22	89	[808]
93b	$EtCO_2CH_2$	Me	H_2O_2	34	94	[808]
93c	$CH_3(CH_2)_3CH_2$	Me	H_2O_2	23	95	[808]
93d	$BrCH_2$	Me	*t*-BuOOH	61	62	[809]
93e	$Br(CH_2)_2$	Me	*t*-BuOOH	88	93	[809]
93f	$Br(CH_2)_3$	Me	*t*-BuOOH	89	95	[809]
93g	$Br(CH_2)_4$	Me	*t*-BuOOH	33	87	[809]
93h	$Br(CH_2)_5$	Me	*t*-BuOOH	42	50	[809]
93i	Cyclohexyl	Me	H_2O_2	10	74	[805]
95			H_2O_2	50	81	[805]
94a	$CH_3(CH_2)_3$	Me	H_2O_2	78/100	96/95	[805]/[810]
94b	$CH_3(CH_2)_4$	Me	H_2O_2	82	92	[805]
94c	$CH_3CH_2CH(CH_3)$	Me	H_2O_2	33	94	[805]
94d	$(CH_3)_2CH$	Me	H_2O_2	28	66	[805]
94e	$CH_3(CH_2)_2$	Et	H_2O_2	40[a]	95	[810]
94f	$CH_3(CH_2)_3$	Et	H_2O_2	20	95	[810]
Monosubstituted alkenes						
94g	$CH_3(CH_2)_4$	H	H_2O_2	2/2	0/10	[810]/[808]
94h	$PhOCH_2$	H	H_2O_2	1	46	[808]
trans-Alkenes						
(*E*)- **94a**	$CH_3(CH_2)_3$	Me	H_2O_2	3[b]	–	[810]

[a]Plus 50% of products of allylic oxidation.
[b]Plus 35% of the aldehyde 2-heptenal.

system gave monoepoxides in high yields, or products of allylic oxidation depending on the ring size [815].

One of the first applications of the CPO-mediated epoxidation in total synthesis was the concise preparation of (*R*)-mevalonolactone (**100**). Enantioselective epoxidation of ethyl 3-methyl-3-butenoate yielded the corresponding epoxide in 67% yield and 93% *ee*. Nucleophilic opening of the epoxide and simple manipulation of functional groups gave the natural lactone through an efficient chemoenzymatic route (Figure 5.69) [764].

C–H Oxidation (Benzylic/Allylic, Indole, Benzofuran, and Alcohol Oxidation)

Selective hydroxylation of hydrocarbons is of paramount importance in synthetic

TABLE 5.10 CPO-Mediated Epoxidation of Phenyl Olefins

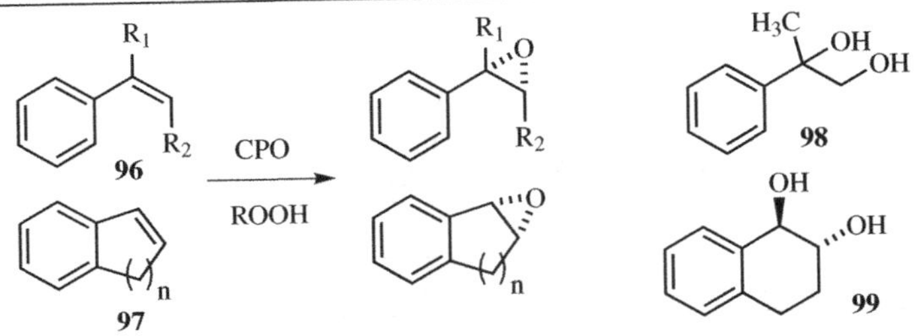

Compound	R_1	R_2	Conversion (%)	*ee* (%)	Reference
96a	H	H	89	49	[805]/[808]
96b	H	Me	67	96	[805]
96c	Me	H	55 (**98**)	89	[808]
97a	$n = 1$		90	30	[811]
97b	$n = 2$		85 (**99**)	97	[805]
97b	$n = 2$		48 (**99**)	85	[812]

chemistry and the search for methods to accomplish this transformation is always intense. The most common biocatalysts for this reaction are heme-containing proteins, such as monooxygenases, but peroxidases are rarely used. Exceptions are the heme-thiolate haloperoxidases CPO and APO, which can mediate oxygen-transfer reactions such as benzylic/allylic hydroxylation, indole, benzofuran, and alcohol oxidation.

BENZYLIC/ALLYLIC HYDROXYLATION After the first report on the CPO-mediated allylic oxidation of cyclohexene to afford cyclohex-2-en-1-ol, the scope of the reaction was extended to straight-chain aliphatic internal olefins and to the benzylic hydroxylation of short-chain alkyl benzenes (Figure 5.70) [803,810,816]. It is noteworthy the reversal in the enantioselectivity of the benzylic hydroxylation when changing

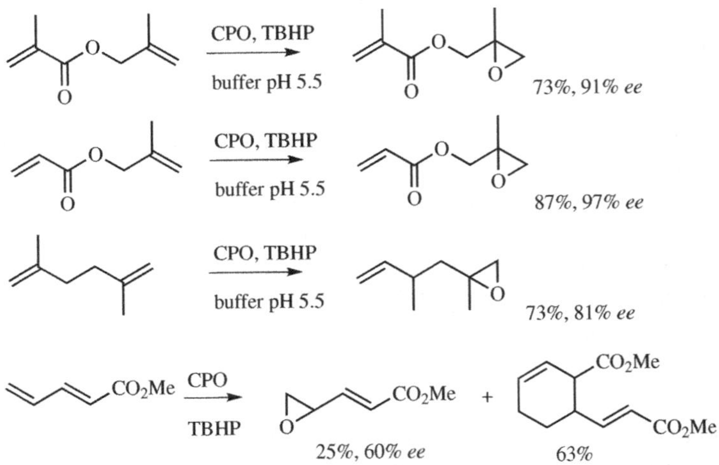

FIGURE 5.68 CPO-mediated oxidation of dienic systems. TBHP, *t*-butylhydroperoxide.

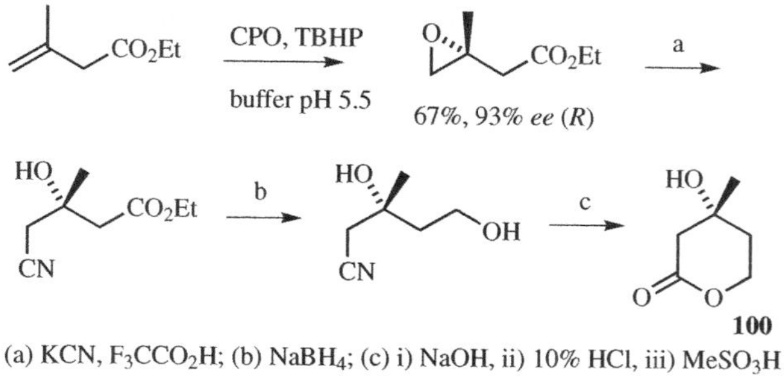

FIGURE 5.69 CPO-mediated epoxidation as a key step in the preparation of (*R*)-mevalonolactone.

from ethyl ((*R*)- selective) to propyl group ((*S*)-selective) as side chain. In the CPO-mediated oxidation of *p*-xylene, only one of the two methyl groups reacts, giving mainly the corresponding benzylic alcohol, together with *p*-methylbenzaldehyde and some *p*-toluic acid [817]. By optimizing the pH of the reaction, the rate of inactivation was lowered and the yield of benzylic oxidation improved [818].

A related reaction is the enantioselective α-hydroxylation of alkynes to give secondary propargylic alcohols (Table 5.11) [819]. Primary propargylic alcohols are further oxidized to aldehydes in good yields [820].

Other heme-thiolate peroxidase active in this type of oxygen-transfer reactions is the aromatic peroxygenase obtained from the agaric mushroom *A. aegerita* (APO,

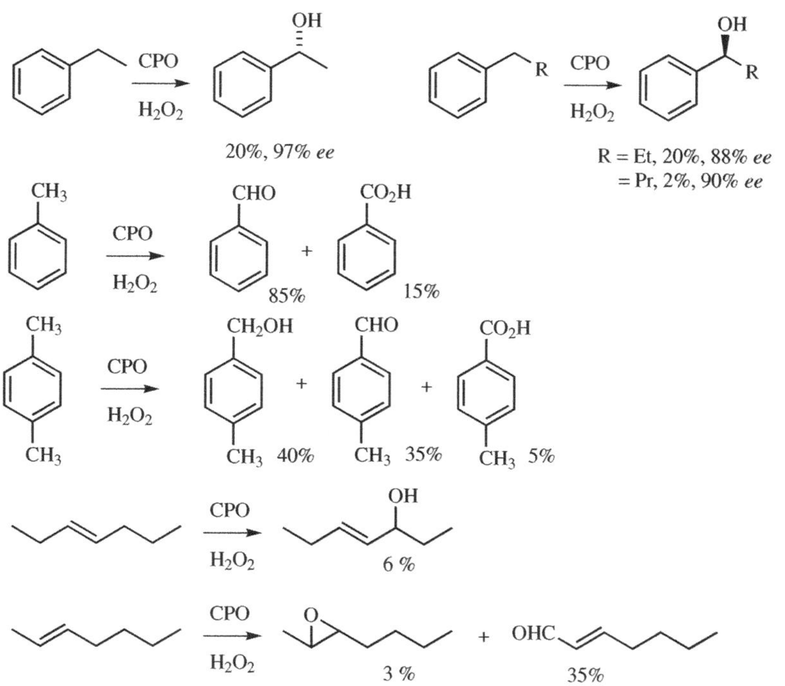

FIGURE 5.70 CPO-mediated allylic/benzylic oxidations.

TABLE 5.11 CPO-Mediated Propargylic Hydroxylation of Internal Alkynes

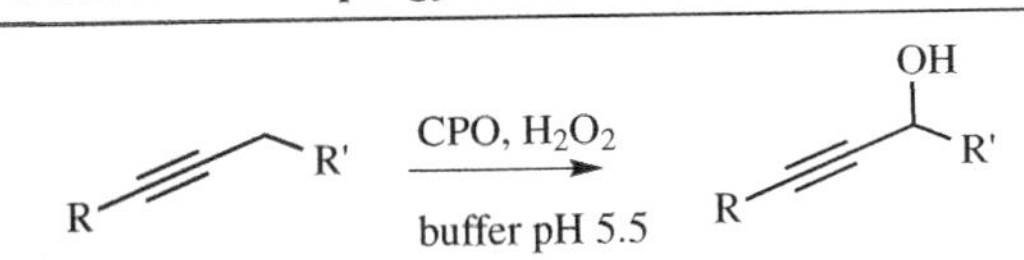

R	R′	*ee* (%)	Yield (%)	Absolute Configuration
CH_3	CH_3	57	7	*R*
CH_3CH_2	CH_3	91	26	*R*
$CH_3(CH_2)_2$	CH_3	87	30	*R*
$CH_3(CH_2)_3$	CH_3	78	8	*R*
Ph	CH_3	86	15	*R*
$AcOCH_2$	CH_3	95	52	*R*
$BrCH_2$	CH_3	94	65	*R*
$AcOCH_2CH_2$	CH_3	83	26	*R*
$BrCH_2CH_2$	CH_3	94	25	*R*
$AcOCH_2$	CH_2CH_3	87	8	n.d.
$COCH_3$	CH_3	No reaction	–	–

n.d, not determined.

EC 1.11.2.1), originally reported as a haloperoxidase. The most important activity of APO is related to oxygen transfer reactions, in particular the hydroxylation of aromatic rings. In this regard, the APO-mediated oxidation of toluene gives, in addition to the benzylic oxidation of the side chain, *p*- and *o*-cresol as well as methyl-*p*-benzoquinone. Naphthalene hydroxylation proceeds regioselectively and leads to the formation of 1-naphthol and traces of 2-naphthol (Figure 5.71) [661,750,821].

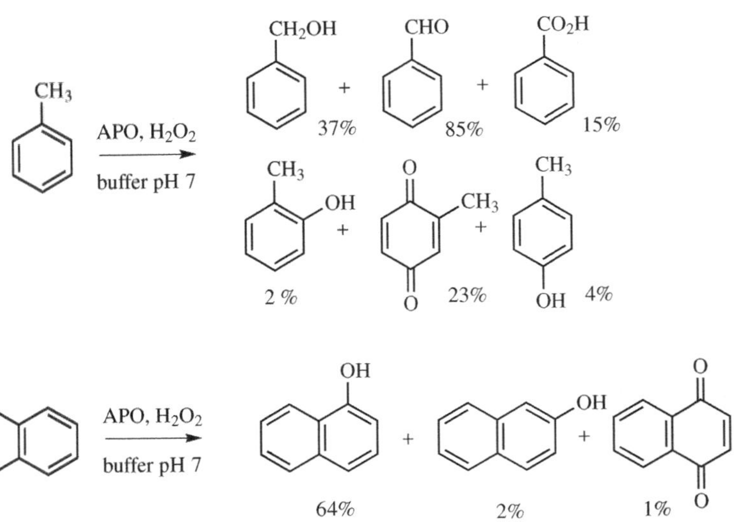

FIGURE 5.71 APO-mediated benzylic and aromatic hydroxylations.

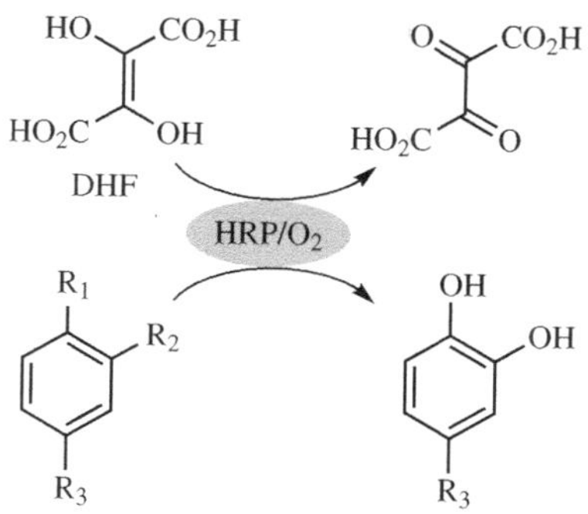

R_1 = OH, R_2 = R_3 = H
R_1 = OH, R_2 = H, R_3 = $CH_2CH(NH_2)CO_2H$, L-tyrosine
R_1 = OH, R_2 = H, R_3 = $CH(NH_2)CO_2H$
R_1 = H, R_2 = OH, R_3 = $CH(OH)CH_2NHCH_3$, L-phenylephrine

FIGURE 5.72 HRP-catalyzed hydroxylation of phenols.

HRP catalyzes the hydroxylation of activated aromatic compounds by molecular oxygen using dihydroxyfumaric acid (DHF) as a hydrogen donor [822]. Later, Klibanov et al. increased the yield up to 70%, turning the reaction into a synthetically attractive process [823,824]. For example, catechol was produced from phenol, L-DOPA from L-tyrosine, D-(-)-3,4-dihydroxyphenylglycine from D-(-)-*p*-hydroxyglycine, and L-epinephrine (adrenaline) from L-(-)-phenylephrine (Figure 5.72). The last three compounds are drugs of industrial relevance.

ALCOHOL OXIDATION Again, in contrast to other peroxidases, CPO is capable to oxidize primary alcohols to aldehydes, with a preference for allylic, propargylic, and benzylic alcohols. Hydrogen peroxide or TBHP can be used as oxidants [820,825,826]. The reaction is selective and no acids are produced, except with reactive aldehydes, such as 5-hydroxymethylfurfural, which partially affords the corresponding acid *via* direct oxygen transfer (confirmed using labeled hydrogen peroxide) (Figure 5.73) [827].

Several cyclopropylcarbinols have been subjected to CPO-mediated oxidation with TBHP to give the corresponding cyclopropylaldehydes, and the enzyme showed some recognition of the ring sterocenters, enabling kinetic resolutions. The oxidation of *cis*-cyclopropylcarbinols shows much higher enantioselectivity than with the *trans*-isomers, as shown in Table 5.12 [828].

The aromatic peroxygenase APO oxidizes aromatic alcohols into the corresponding aldehydes and further into benzoic acids (Figure 5.74) [750]. Thus, this enzyme

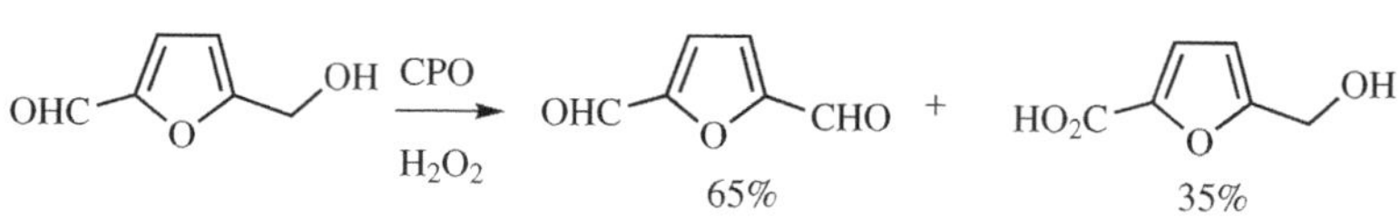

FIGURE 5.73 CPO-mediated oxidation of 5-hydroxymethylfurfural.

TABLE 5.12 CPO-Mediated Enantioselective Oxidation of Cyclopropylcarbinols

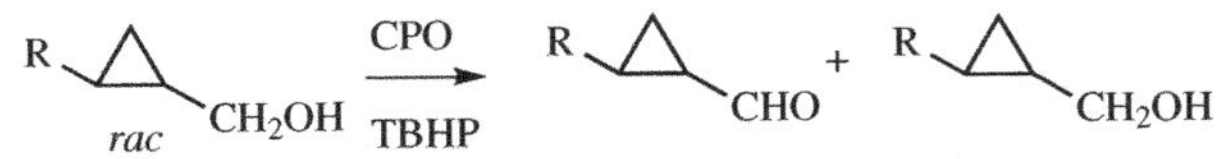

	Aldehyde		Alcohol	
R	Yield (%)	*ee* (%)	Yield (%)	*ee* (%)
Ph	34	3	60	15
$AcOCH_2$	39	–	57	20
$Br(CH_2)_2$	48	18	31	21
$Br(CH_2)_3$	33	31	46	22

	Aldehyde		Alcohol	
R	Yield (%)	*ee* (%)	Yield (%)	*ee* (%)
Ph	7	65	>90	18
$PhCH_2$	32	66	45	95
CH_3	37	90	60	37
CH_3CH_2	30	89	57	57
$CH_3(CH_2)_2$	44	82	42	93
$AcOCH_2$	40	92	57	57
$Br(CH_2)_2$	35	91	50	83

catalyzes—in addition to hydroxylations and halogenations—the oxidation of aromatic alcohols and aldehydes, *O*-dealkylation, phenol oxidation, and the catalase-like destruction of hydrogen peroxide [661,733]. Considering the range of reactions catalyzed by APO and the need of only peroxide to act, it is likey that this enzyme could become a useful biocatalyst.

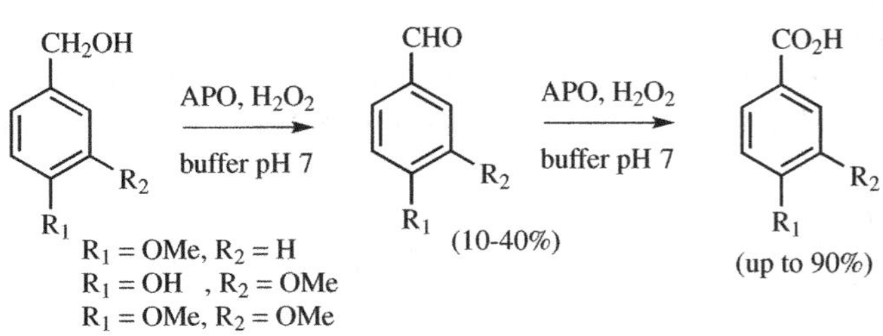

FIGURE 5.74 APO-mediated oxidation of aromatic alcohols.

INDOLE AND BENZOFURAN OXIDATION The CPO-mediated oxidation of indoles readily produces 2-oxindoles in nearly quantitative yields (Figure 5.75) [829].

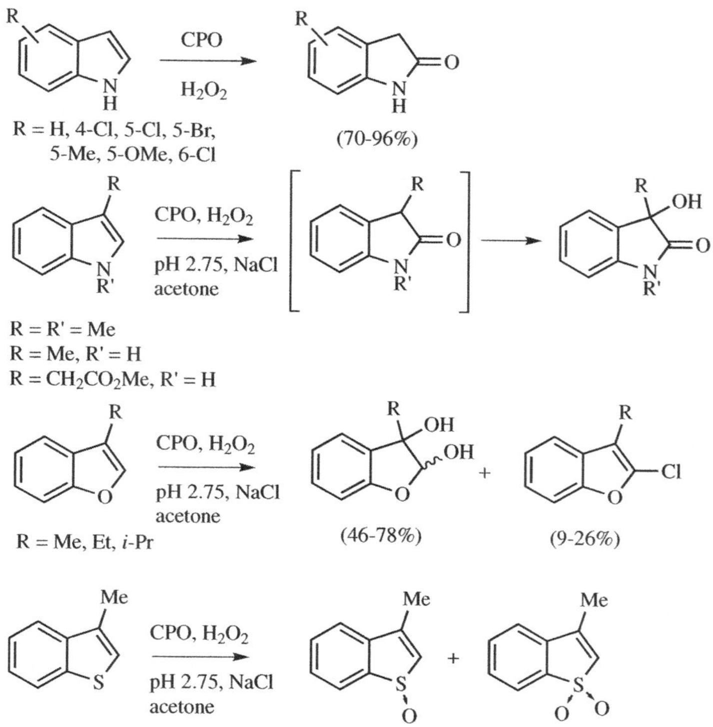

FIGURE 5.75 CPO-mediated oxidation of indoles, benzofurans and benzothiophenes.

Coimmobilized CPO with GOX on molecular sieves has been successfully used for *in situ* generation of hydrogen peroxide in this reaction [830]. Indoles bearing an alkyl substituent on C-3 gave the corresponding α-hydroxy lactams upon long reaction times [831,832]. In the presence of chloride ion (to improve the yields), 3-alkyl benzofurans gave the corresponding 2,3-diols (predominantly *trans*) as main products, with variable amounts of the chlorinated benzofuran. The labile diols, obtained as racemic mixtures, could undergo acid catalyzed dehydration to the corresponding lactones, depending on the reaction conditions. Finally, benzothiophenes gave the corresponding sulfoxides and sulfones under the same oxidation conditions [831].

5.3 CONCLUDING REMARKS

This chapter has provided a comprehensive overview on the performances that oxidases and peroxidases can provide, comprising a wide number of biocatalytic applications in many areas like organic synthesis, analytical chemistry, biotechnology, natural products, and so on. It is worth mentioning the wide range of environmental applications found for oxidases, in particular lacasses, which are able to oxidize an ample array of substrates utilizing molecular oxygen and producing water as the only by-product. Moreover, the use of redox mediators expands the substrate spectrum, constituting a robust and efficient system with industrial applications, for example, pulp bleaching, and wastewater treatment. Considering the state-of-the-art given in Chapter 3 (dehydrogenases), Chapter 4 (oxygenases) and Chapter 5 (oxidases and peroxidases), the ample diversity of oxidative enzymes that can be found in nature is completed. Overall, a multidisciplinary field has emerged, combining areas like biochemistry, molecular biology, organic synthesis, and applications.

REFERENCES

1. Punniyamurthy T, Velusamy S, Iqbal J. Chem Rev 2005;105:2329–2364.
2. Wong OA, Shi Y. Chem Rev 2008;108:3958–3987.
3. Sheldon RA, Arends IWCE. Adv Synth Catal 2004;346:1051–1071.
4. Matsuda T, Yamanaka R, Nakamura K. Tetrahedron: Asymmetry 2009;20:513–557.
5. Burton SG. Trends Biotechnol 2003;21:543–549.
6. Hollmann F, Arends IWCE, Buehler K, Schallmey A, Buhler B. Green Chem 2011;13:226–265.
7. Monti D, Ottolina G, Carrea G, Riva S. Chem Rev 2011;111:4111–4140.
8. International Union of Biochemistry and Molecular Biology (NC-IUBMB), http://www.chem.qmul.ac.uk/iubmb/.
9. Cooper JAD, Smith W, Bacila M, Medina H. J Biol Chem 1959;234:445–448.
10. Ogel ZB, Brayford D, McPherson MJ. Mycol Res 1994;98:474–480.
11. Pereira AM, Kemmelmeier C. Braz J Microbiol 2000;31:129–134.
12. Baron AJ, Stevens C, Wilmot C, Seneviratne KD, Blakeley V, Dooley DM, Phillips SEV, Knowles PF, McPherson MJ. J Biol Chem 1994;269:25095–25105.
13. Hartmans S, De Vries HT, Beijer P, Brady RL, Hofbauer M, Haandrikman AJ. In: Gatenholm P, Tenkanen M, editors. *Hemicelluloses: Science and Technology*. Washington, DC: American Chemical Society;2004. pp. 360–371.
14. Sun L, Petrounia IP, Yagasaki M, Bandara G, Arnold FH. Protein Eng 2001;14:699–704.
15. Whittaker JW. Arch Biochem Biophys 2005;433:227–239.
16. Whittaker JW. Chem Rev 2003;103:2347–2363.
17. Tressel P, Kosman DJ. Biochem Biophys Res Commun 1980;92:781–786.
18. Siebum A, van Wijk A, Schoevaart R, Kieboom T. J Mol Catal B: Enzym 2006;41:141–145.
19. Baig MM, Cerda JJ. Anal Biochem 1979;98:429–432.
20. Parikka K, Leppanen A-S, Pitkanen L, Reunanen M, Willfor S, Tenkanen M. J Agric Food Chem 2010;58:262–271.
21. Liu X-C, Dordick JS. J Am Chem Soc 1999;121:466–467.
22. Schoevaart R, Kieboom T. Carbohydr Res 2002;337:899–904.
23. van Wijk A, Siebum A, Schoevaart R, Kieboom T. Carbohydr Res 2006;341:2921–2926.
24. Xu F, Golightly EJ, Schneider P, Berka RM, Brown KM, Johnstone JA, Baker DH, Fuglsang CC, Brown SH, Svendsen A, Klotz AV. Appl Biochem Biotechnol 2000;88: 23–32.

25. Franke D, Machajewski T, Hsu C-C, Wong C-H. J Org Chem 2003;68:6828–6831.
26. Klibanov AM, Alberti BN, Marletta MA. Biochem Biophys Res Commun 1982;108:804–808.
27. Root RL, Durrwachter JR, Wong CH. J Am Chem Soc 1985;107:2997–2999.
28. Schoevaart R, Kieboom T. Top Catal 2004;27:3–9.
29. Schoevaart R, Kieboom T. Tetrahedron Lett 2002;43:3399–3400.
30. Bulter T, Schumacher T, Namdjou DJ, Gallego RG, Clausen H, Elling K. Chem Bio Chem 2001;2:884–894.
31. Namdjou DJ, Sauerzapfe B, Schmiedel J, Drager G, Bernatchez S, Bernatchez S, Wakarchuk WW, Elling L. Adv Synth Catal 2007;349:314–318.
32. Fialova P, Namdjou DJ, Ettrich R, Prykrilova V, Rauvolfova J, Krenec K, Kuzma M, Elling L, Bezouska K, Kren V. Adv Synth Catal 2005;347:997–1006.
33. Escalettes F, Turner NJ. Chem Bio Chem 2008;9:857–860.
34. Wilkinson D, Akumanyi N, Hurtado-Guerrero R, Dawkes H, Knowles RP, Phillips SEV, McPherson MJ. Protein Eng Des Sel 2004;17:141–148.
35. Sun W, Bulter T, Alcalde M, Petrounia IP, Arnold FH. Chem Bio Chem 2002;3:781–783.
36. Yuan B, Page A, Worrall CP, Escalettes F, Willies SC, McDouall JJW, Turner NJ, Clayden J. Angew Chem Int Ed 2010;49:7010–7013.
37. Truppo MD, Escalettes F, Turner NJ. Angew Chem Int Ed 2008;47:2639–2641.
38. Crueger A, Crueger W. In: Fogarty WM, Kelly CT, editors. *Microbial Enzymes and Biotechnology*. 2nd ed. London: Elsevier; 1990. pp. 177–226.
39. Olsson B, Lundbäck H, Johansson G. Anal Chim Acta 1985;167:123–136.
40. Giffhorn F. Appl Microbiol Biotechnol 2000;54:727–740.
41. Daniel G, Volc J, Kubatova E. Appl Environ Microbiol 1994;60:2524–2532.
42. Janssen FW, Ruelius HW. Biochim Biophys Acta 1968;167:501–510.
43. Volc J, Sedmera P, Musilek V. Collect Czech Chem Commun 1980;45:950–955.
44. Freimund S, Huwig A, Giffhorn F, Köpper S. Chem Eur J 1998;4:2442–2455.
45. Leitner C, Volc J, Haltrich D. Appl Env Microbiol 2001;67:3636–3644.
46. Ander P, Marzullo L. J Biotechnol 1997;53:115–131.
47. Neidleman S, Amon WF, Geigert J. Cetus Corporation, Patent number 4,246,347, United States, 1981.
48. Greigert J, Neidleman SL, Hirano DS. Carbohydr Res 1983;113:159–162.
49. Leitner C, Neuhauser W, Volc J, Kulbe KD, Nidetzky B, Haltrich D. Biocatal Biotransform 1998;16:365–382.
50. Freimund S, Huwig A, Giffhorn F, Köpper S. J Carbohydr Chem 1996;15:115–120.
51. Gabriel J, Volt J, Kubitova E, Palkovi Z, Pospisek M. Carbohydr Res 1994;252:297–301.
52. Greigert J, Neidleman SL, Hirano DS, Wolf B, Panschar B. Carbohydr Res 1983;113:163–165.
53. Roper H. In: Lichtentahler FW, editor. *Carbohydrates as Organic Raw Materials*. Weinheim: Verlag Chemie; 1991. pp. 267–288.
54. Lichtentahler FW, Cuny E, Martin D, Ronninger S, Weber T. In: Lichtentahler FW, editor. *Carbohydrates as Organic Raw Materials*. Weinheim: Verlag Chemie; 1991. pp. 207–246.

55. Freimund S, Kopper S. Carbohydr Res 1998;308:195–200.
56. Spadiut O, Pisanelli I, Maischberger T, Peterbauer C, Gorton L, Chaiyen P, Haltrich D. J Biotechnol 2009;139:250–257.
57. Tasca F, Timur S, Ludwig R, Haltrich D, Volc J, Antiochia R, Gorton L. Electroanal 2007;19:294–302.
58. Namba N, Watanabe F, Tokuda M, Mino M, Furuya E. Diabetes Res Clin Pract 1994;24:55–61.
59. Chusney GD, Phillipa M, Pickup JC, Clin Chim Acta 1995;235:91–99.
60. Fukumura Y, Tajima S, Oshitani S, Ushijima Y, Kobayashi I, Hara F, Yamamoto S, Yabuuchi M. Clin Chem 1994;40:2013–2016.
61. Tanabe T, Umegae Y, Koyashiki Y, Kato Y, Fukahori K, Tajima S, Yabuuchi M. Clin Chem 1994;40:2006–2012.
62. Jurgens H, Kabuss R, Plumbaum T, Weigl B, Kretzmer G, Schugerl K, Ignatzek AK, Giffhorn F. Anal Chim Acta 1994;298:141–149.
63. Olsson L, Mandenius CF, Volc J. Anal Chem 1990;62:2688–2691.
64. Weigl B, Hitzmann B, Kretzmer G, Schugerl K, Huwig A, Giffhorn F. J Biotechnol 1996;50:93–106.
65. Peterbauer CK, Volc J. Appl Microbiol Biotechnol 2010;85:837–848.
66. Henriksson G, Johansson G, Pettersson G. J Biotechnol 2000;78:93–113.
67. Pricelius S, Ludwig R, Lant N, Haltrich D, Guebitz GM. Appl Microb Biotechnol 2009;85:75–83.
68. Larsson T, Elmgren M, Lindquist SE, Gorton L, Tessema M, Henriksson G. Anal Chim Acta 1996;331:207–215.
69. Ludwig R, Ozga M, Zamocky M, Peterbauer C, Kulbe KD, Haltrich D. Biocatal Biotransform 2004;22:97–104.
70. Volc J, Kubatova E, Wood DA, Daniel G. Arch Microbiol 1997;167:119–125.
71. Kujawa M, Volc J, Halada P, Sedmera P, Divne C, Sygmund C, Leitner C, Peterbauer C, Haltrich D. FEBS J 2007;274:879–894.
72. Sygmund C, Kittl R, Volc J, Halada P, Kubatova E, Haltrich D, Peterbauer CK. J Biotechnol 2008;133:334–342.
73. Volc J, Kubatova E, Daniel G, Sedmera P, Haltrich D. Arch Microbiol 2001;176:178–186.
74. Sedmera P, Halada P, Kubatova E, Haltrich D, Prikrylova V, Volc J. J Mol Catal B: Enzym 2006;41:32–42.
75. Cavener DR. J Mol Biol 1992;223:811–814.
76. Menon V, Hsieh C-T, Fitzpatrick PF. Bioorg Chem 1995;23:42–53.
77. Clark DS, Geresh S, DiCosimo R. Bioorg Med Chem Lett 1995;5:1383–1388.
78. Witt S, Wohlfahrt G, Schomburg D, Hecht H-J, Kalisz HM. Biochem. J 2000;347:553–559.
79. Hatzinikolaou DG, Hansen OC, Macris BJ, Tingey A, Kekos D, Goodenough P, Stougaard P. Appl Microbiol Biotechnol 1996;46:371–381.
80. Leskovac V, Trivic S, Wohlfahrt G, Kandrac J, Pericin D, Int J Biochem Cell Biol 2005;37:731–750.
81. Muller D. Biochem Z 1928;199:136–170.

82. Bankar SB, Bule MV, Singhal RS, Ananthanarayan L. Biotechnol Adv 2009;27:489–501.
83. Wong CM, Wong KH, Chen XD. Appl Microbiol Biotechnol 2008;78:927–938.
84. Wilkins E, Atanasov P. Med Eng Phys 1996;18:273–288.
85. Santoni T, Santianni D, Manzoni A, Zanardi S, Mascini M. Talanta 1997;44:1573–1580.
86. Newman JD, Turner APF. Biosens Bioelectron 2005;20:2435–2453.
87. Mello LD, Kubota LT. Food Chem 2002;77:237–256.
88. Adam W, Heckel F, Saha-Möller CR, Schreier P. J Organomet Chem 2002;661:17–29.
89. Adam W, Lazarus M, Boss B, Saha-Moller CR, Humpf H-U, Schreier P. J Org Chem 1997;62:7841–7843.
90. Seip JE, Fager SK, Gavagan JE, Gosser LW, Anton DL, DiCosimo R. J Org Chem 1993;58:2253–2259.
91. Seip JE, Fager SK, Gavagan JE, Anton DL, DiCosimo R. Bioorg Med Chem 1994;2:371–378.
92. Das S, Glenn JHI, Subramanian M. Biotechnol Progr 2010;26:607–615.
93. Gavagan JE, Fager SK, Seip JE, Clark DS, Payne MS, Anton DL, DiCosimo R. J Org Chem 1997;62:5419–5427.
94. Adam W, Lazarus M, Saha-Moller CR, Schreier P. Tetrahedron: Asymmetry 1998;9:351–355.
95. Yoshida H. J Chem Soc, Perkin Trans 1883;43:472–486.
96. Bertrand G. Compt Rend Acad Sci Paris 1896;123:463–465.
97. Hoegger PJ, Kilaru S, James TY, Thacker JR, Kues U. FEBS J 2006;273:2308–2326.
98. Alcalde M. In: Polaina J, MacCabe AP, editors. *Industrial Enzymes: Structure, Function and Applications. Volume. 26*, Netherlands-Springer; 2007. pp. 461–476.
99. Solomon EI, Sundaram UM, Macho TE. Chem Rev 1996;96:2563–2605.
100. Solomon EI, Chen P, Metz M, Lee SK, Palmer AE. Angew Chem Int Ed 2001;40:4570–4590.
101. Rodgers CJ, Blanford CF, Giddens SR, Skamnioti P, Armstrong FA, Gurr SJ. Trends Biotechnol 2010;28:63–72.
102. Giardina P, Faraco V, Pezzella C, Piscitelli A, Vanhulle S, Sannia G. Cell Mol Life Sci 2010;67:369–385.
103. Baldrian P. FEMS Microbiol Rev 2006;30:215–242.
104. Riva S. Trends Biotechnol 2006;24:219–226.
105. Thurston CF. Microbiology 1994;140:19–26.
106. Quintanar L, Stoj C, Taylor AB, Hart PJ, Kosman DJ, Solomon EI. Acc Chem Res 2007;40:445–452.
107. Claus H. Micron 2004;35:93–96.
108. Hakulinen N, Kiiskinen LL, Kruus K, Saloheimo M, Paananen A, Koivula A, Rouvinen J. Nat Struct Mol Biol 2002;9:601–605.
109. Bertrand T, Jolivalt C, Briozzo P, Caminade E, Joly N, Madzak C, Mougin C. Biochemistry 2002;41:7325–7333.
110. Piontek K, Antorini M, Choinowski T. J Biol Chem 2002;277:37663–37669.
111. Mayer AM, Staples RC. Phytochemistry 2002;60:551–565.

112. Ranocha P, Chabannes M, Chamayou S, Danoun S, Jauneau A, Boudet A-M, Goffner D. Plant Physiol 2002;129:145–155.
113. Keilin D, Mann T. Nature 1939;143:23–24.
114. R. Sterjiades, Dean JFD, Eriksson K-EL. Plant Physiol 1992;99:1162–1168.
115. Bao W, O'Malley DM, Whetten R, Sederoff RR. Science 1993;260:672 - 674.
116. Sato Y, Wuli B, Sederoff RR, Whetten R. J Plant Res 2001;114:147–155.
117. Ranocha P, McDougall G, Hawkins S, Sterjiades R, Borderies G, Stewart D, Cabanes-Macheteau M, Boudet AM, Goffner D. Eur J Biochem 1999;259:485–495.
118. LaFayette PR, Eriksson K-EL, Dean JFD. Plant Mol Biol 1999;40:23–35.
119. Kiefer-Meyer M-C, Gomord V, O'Connell A, Halpin C, Faye L. Gene 1996;178:205–207.
120. Gavnholt B, Larsen K, Rasmussen SK. Plant Sci 2002;162:873–885.
121. Wosilait W, Nason A, Terrell AN. J Biol Chem 1954;206:271–282.
122. Caparros-Ruiz D, Fornale S, Civardi L, Puigdomenech P, Rigau J. Plant Sci 2006;171:217–225.
123. Givaudan A, Effosse A, Faure D, Potier P, Bouillant ML, Bally R. FEMS Microbiol Lett 1993;108:205–210.
124. Alexandre G, Zhulin IB. Trends Biotechnol 2000;18:41–42.
125. Sharma P, Goel R, Capalash N. World J Microbiol Biotechnol 2007;23:823–832.
126. Martins LO, Soares CM, Pereira MM, Teixeira M, Costa T, Jones GH, Henriques AO. J Biol Chem 2002;277:18849–18859.
127. Grass G, Rensing C. Biochem Biophys Res Commun 2001;286:902–908.
128. Kim C, Lorenz WW, Hoopes JT, Dean JFD. J Bacteriol 2001;183:4866–4875.
129. Endo K, Hosono K, Beppu T, Ueda K. Microbiology 2002;148:1767–1776.
130. Suzuki T, Endo K, Ito M, Tsujibo H, Miyamoto K, Inamori Y. Biosci Biotechnol Biochem 2003;67:2167–2175.
131. Diamantidis G, Effosse A, Potier P, Bally R. Soil Biol Biochem 2000;32:919–927.
132. Sanchez-Amat A, Lucas-Elio P, Fernandez E, García-Borrón JC, Solano F. Biochim Biophys Acta 2001;1547:104–116.
133. Claus H, Filip Z. Microbiol Res 1997;152:209–216.
134. Hullo MF, Moszer I, Danchin A, Martin-Verstraete I. J Bacteriol 2001;183:5426–5430.
135. Martins LO, Soares CM, Pereira MM, Teixeira M, Costa T, Jones GH, Henriques AO. J Biol Chem 2002;277:18849–18859.
136. Wang C, Zhao M, Lu L, Wei X, Li T. African J Biotechnol 2011;10:2186–2192.
137. Takami H, Takaki Y, Chee G, Nucleic Acids Res. 2002;30:3927–3935.
138. Lee FA, William CG. J Bacteriol 1987;169:1279–1285.
139. Miyazaki K. Extremophiles 2005;9:415–425.
140. Lee Y, Hendson M, Panopoulos NJ, Schroth MN. J Bacteriol 1994;176:173–188.
141. Francis CA, Tebo BM. Appl Environ Microbiol 2001;67:4272–4278.
142. Okazaki M, Sugita T, Shimizu M, Ohode Y, Iwamoto K, de Vrind-deJong EW, de Vrind JPM, Corstjens PLAM. Appl Environ Microbiol 1997;63:4793–4799.
143. Brouwers GJ, de Vrind JPM, Corstjens PLAM, Cornelis P, Baysse C, de Vrind-deJong EW. Appl Environ Microbiol 1999;65:1762–1768.

144. Dittmer NT, Kanost MR. Insect Biochem Mol Biol 2010;40:179–188.
145. Coy MR, Salem TZ, Denton JS, Kovaleva ES, Liu Z, Barber DS, Campbell JH, Davis DC, Buchman GW, Boucias DG, Scharf ME. Insect Biochem Mol Biol 2010;40:723–732.
146. Thomas BR, Yonekura M, Morgan TD, Czapla TH, Hopkins TL, Kramer KJ. Insect Biochem 1989;19:611–622.
147. Kramer KJ, Kanost MR, Hopkins TL, Jiang H, Zhu YC, Xu R, Kerwin JL, Turecek F. Tetrahedron 2001;57:385–392.
148. Beloqui A, Pita M, Polaina J, Martınez-Arias A, Golyshina OV, Zumarraga M, Yakimov MM, Garcıa-Arellano H, Alcalde M, Fernandez VM, Elborough K, Andreu JM, Ballesteros A, Plou FJ, Timmis KN, Ferrer M, Golyshin PN. J Biol Chem 2006;281:22933–22942.
149. Sirim D, Wagner F, Wang L, Schmid RD, Pleiss J. Database 2011, DOI: 10.1093/database/bar1006.
150. Piscitelli A, Pezzella C, Giardina P, Faraco V, Sannia G. Bioeng Bugs 2010;1:252–262.
151. Gianfreda L, Xu F, Bollag J-M. Biorem J 1999;3:1–26.
152. Yaver DS, Xu F, Golightly EJ, Brown KM, Brown SH, Rey MW, Schneider P, Halkier T, Mondorf K, Dalboge H. Appl Environ Microbiol 1996;62:834–841.
153. Otterbein L, Record E, Longhi S, Asther M, Moukha S. Eur J Biochem 2000;267:1619–1625.
154. Guo M, Lu F, Pu J, Bai D, Du L. Appl Microbiol Biotechnol 2005;69:178–183.
155. Jönsson LJ, Saloheimo M, Penttila M. Current Genetics 1997;32:425–430.
156. Cassland P, Jönsson LJ. Appl Environ Microbiol 1999;52:393–400.
157. Jolivalt C, Madzak C, Brault A, Caminade E, Malosse C, Mougin C. Appl Microbiol Biotechnol 2005;66:450–456.
158. Kiiskinen LL, Kruus K, Bailey M, Ylosmaki E, Siika-Aho M, Saloheimo M. Microbiology 2004;150:3065–3074.
159. Koschorreck K, Richter SM, Swierczek A, Beifuss U, Schmid RD, Urlacher VB. Arch Biochem Biophys 2008;474:213–219.
160. Koschorreck K, Richter SM, Ene AB, Roduner E, Schmid RD, Urlacher VB. Appl Microbiol Biotechnol 2008;79:217–224.
161. Kunamneni A, Camarero S, García Burgos C, Plou FJ, Ballesteros A, Alcalde M. Microbial Cell Factories 2008;7:32, DOI:10.1186/1475–2859.
162. Salony GN, Baranwal R, Chhabra M, Mishra S, Chaudhuri TK, Bisaria VS. Biochim Biophys Acta 2008;1784:259–268.
163. Xu F. Ind Biotechnol 2005;1:38–50.
164. Blaich R, Esser K. Arch Microbiol 1975;103:271–277.
165. Hoshida H, Nakao M, Kanazawa H, Kubo K, Hakukawa T, Morimasa K, Akada R, Nishizawa Y. J Biosci Bioeng 2001;92:372–380.
166. Ryden LG, Hunt LT. J Mol Evol 1993;36:41–66.
167. Morozova OV, Shumakovich GP, Gorbacheva MA, Shleev SV, Yaropolov AI. Biochemistry (Moscow) 2007;72:1136–1150.
168. Burton SG. Curr Org Chem 2003;7:1317–1331.
169. Evans C, Hedger JN. In: Gadd GM, editor. *Fungi in Bioremediation*. UK: British Mycological Society, Cambridge University Press; 2001. pp. 1–26.

170. Shleev S, Tkac J, Christenson A, Ruzgas T, Yaropolov AI, Whittakerd JW, Gorton L. Biosens Bioelectron 2005;20:2517–2554.
171. Shleev SV, Christenson A, Serezhenkov V, Burbaev D, Yaropolov AI, Gorton L, Ruzgas T. Biochem J 2005;385:745–754.
172. Schneider P, Caspersen MB, Mondorf K, Halkier T, Skov LK, Ostergaard PR, Brown KM, Brown SH, Xu F. Enzyme Microb Technol 1999;25:502–508.
173. Yaropolov AI, Skorobogatko OV, Vartanov SS, Varfolomeyev SD. Appl Biochem Biotechnol 1994;49:257–280.
174. Matijosyte I, Arends IWCE, Sheldon RA, Vries Sd. Inorg Chim Acta 2008;361:1202–1206.
175. Reinhammar BRM. Biochim Biophys Acta 1972;275:245–259.
176. Witayakran S, Ragauskas AJ. Adv Synth Catal 2009;351:1187–1209.
177. Rodríguez Couto S, Toca Herrera JL. Biotechnol Adv 2006;24:500–513.
178. Kunamneni A, Ballesteros A, Plou FJ, Alcalde M. In: Mendez-Vilas A, editor. *Communicating Current Research and Educational Topics and Trends in Applied Microbiology*. Formatex, Badajoz, Spain, 2007. pp. 233–245.
179. Morozova OV, Shumakovich GP, Shleev SV, Yaropolov YI. Appl Biochem Microbiol 2007;43:523–535.
180. Widsten P, Kandelbauer A. Enzyme Microb Technol 2008;42:293–307.
181. Harms H, Schlosser D, Wick LY. Nat Rev Microbiol 2011;9:177–192.
182. Smith M, Thurston CF, Wood DA. In: Messerschmidt A, editor. *Multi-copper Oxidases*. Singapore: World Scientific; 1997. pp. 201–224.
183. Alcalde M, Ferrer M, Plou FJ, Ballesteros A. Trends Biotechnol 2006;24:281–287.
184. Gianfreda L, Rao MA. Enzyme Microb Technol 2004;35:339–354.
185. Shiba T, Xiao L, Miyakoshi T, Chen C-L. J Mol Catal B: Enzym 2000;10:605–615.
186. Trejo-Hernandez MR, Lopez-Munguia A, Ramirez RQ. Process Biochem 2001;36:635–639.
187. Wan Y, Du Y, Miyakoshi T. Science in China Series B: Chemistry 2008;51:669–676.
188. Nishida A, Fukuzumi T. Phytochemistry 1978;17:417–419.
189. Falconnier B, Lapierre C, Lesage-Meessen L, Yonnet G, Brunerie P, Colonna-Ceccaldi B, Corrieu G, Asther M. J Biotechnol 1994;37:123–132.
190. Cichewicz RH, Clifford LJ, Lassen PR, Cao X, Freedman TB, Nafie LA, Deschamps JD, Kenyon VA, Flanary JR, Holman TR, Crews P. Bioorg Med Chem 2005;13:5600–5612.
191. Gazak R, Sedmera P, Marzorati M, Riva S, Kren V. J Mol Catal B: Enzym 2008;50:87–92.
192. Intra A, Nicotra S, Riva S, Danieli B. Adv Synth Catal 2005;347:973–977.
193. Ncanana S, Baratto L, Roncaglia L, Riva S, Burton SG. Adv Synth Catal 2007;349:1507–1513.
194. Tranchimand S, Tron T, Gaudin C, Iacazio G. J Mol Catal B: Enzym 2006;42:27–31.
195. Carunchio F, Crescenzi C, Girelli AM, Messina A, Tarola AM. Talanta 2001;55:189–200.
196. Mustafa R, Muniglia L, Rovel B, Girardin M. Food Res Int 2005;38:995–1000.
197. Nicotra S, Cramarossa MR, Mucci A, Pagnoni UM, Riva S, Forti L. Tetrahedron 2004;60:595–600.

198. Wang MF, Jin Y, Ho C-T. J Agric Food Chem 1999;47:3974–3977.

199. Ponzoni C, Beneventi E, Cramarossa MR, Raimondi S, Trevisi G, Pagnoni UM, Riva S, Forti L. Adv Synth Catal 2007;349:1497–1506.

200. Navarra C, Goodwin C, Burton S, Danieli B, Riva S. J Mol Catal B: Enzym 2010;65:52–57.

201. Ciecholewski S, Hammer E, Manda K, Bose G, Nguyen vTH, Langer P, Schauer F. Tetrahedron 2005;61:4615–4619.

202. Ghidouche S, Es-Safi N-E, Ducrot P-H. Tetrahedron Lett 2008;49:619–623.

203. Auriol M, Filali-Meknassi Y, Tyagi RD, Adams CD. Water Res 2007;41:3281–3288.

204. Nicotra S, Intra A, Ottolina G, Riva S, Danieli B. Tetrahedron: Asymmetry 2004;15:2927–2931.

205. Eckenrode F, Peczynska-Czoch W, Rosazza JP. J Pharm Sci 1982;71:1246–1250.

206. Chung JE, Kurisawa M, Uyama H, Kobayashi S. Biotechnol Lett 2003;25:1993–1997.

207. Osman AM, Wong KKY. Tetrahedron Lett 2007;48:1163–1167.

208. Agematu H, Tsuchida T, Kominato K, Shibamoto N, Yoshioka T, Nishida H, Okamoto R, Shin T, Murao S. J Antibiotics 1993;46:141–148.

209. Osiadacz J, Al-Adhami AJH, Bajraszewska D, Fischer P, Peczynska-Czoch W. J Biotechnol 1999;72:141–149.

210. Giurg M, Piekielska K, Gebala M, Ditkowski B, Wolanski M, Peczynska-Czoch W, Mlochowski J. Synth Commun 2007;37:1779–1789.

211. Bruyneel F, Enaud E, Billottet L, Vanhulle S, Marchand-BrynaertJ. Eur J Org Chem 2008;72–79.

212. Bruyneel F, Payen O, Rescigno A, Tinant B, Marchand-Brynaert J. Chem Eur J 2009;15:8283–8295.

213. Lugaro G, Carrea G, Cremonesi P, Casellato MM, Antonini E. Arch Biochem Biophys 1973;159:1–6.

214. Eggert C, Temp U, Dean J, Eriksson K. FEBS Lett 1995;376:202–206.

215. Polak J, Jarosz-Wilkolazka A. Microbial Cell Factories 2010;9:51–62.

216. Pickel B, Constantin M-A, Pfannstiel J, Conrad J, Beifuss U, Schaller A. Angew Chem Int Ed 2010;49:202–204.

217. Davin LB, Wang H-B, Crowell AL, Bedgar DL, Martin DM, Sarkanen S, Lewis NG. Science 1997;275:362–365.

218. Halls SC, Davin LB, Kramer DM, Lewis NG. Biochemistry 2004;43:2587–2595.

219. Aktas N, Tanyolaç A. Bioresour Technol 2003;87:209–214.

220. Ikeda R, Tanaka H, Uyama H, Kobayashi S. Macromol Rapid Commun 1998;19:423–425.

221. Jonas U, Hammer E, Schauer F, Bollag J-M. Biodegradation 1998;8:321–327.

222. Aktas N, Çiçek H, Unal AT, Kibarer G, Kolankaya N, Tanyolaç A. Bioresour Technol 2001;80:29–36.

223. Uchida H, Fukuda T, Miyamoto H, Kawabata T, Suzuki M, Uwajima T. Biochem Biophys Res Commun 2001;287:355–358.

224. Karamyshev AV, Shleev SV, Koroleva OV, Yaropolov AI, Sakharov IY. Enzyme Microb Technol 2003;33:556–564.

225. Uyama H, Kobayashi S. Curr Org Chem 2003;7:1387–1397.

226. Hublik G, Schinner F. Enzyme Microb Technol 2000;27:330–336.
227. Ceylan H, Kubilay S, Aktas N, Sahiner N. Bioresour Technol 2008;99:2025–2031.
228. Aktas N, Tanyolaç A. J Mol Catal B: Enzym 2003;22:61–69.
229. Tanaka T, Yamamoto M, Takahashi M, Fujii T, Taniguchi M. J Chem Eng Jpn 2003;36:1101–1106.
230. Streltsov AV, Shumakovich GP, Morozova OV, Gorbacheva MA, Yaropolov AI. Appl Biochem Microbiol 2008;44:264–270.
231. Vasileva IS, Morozova OV, Shumakovich GP, Shleev SV, Sakharov IY, Yaropolov AI. Synth Met 2007;157:684–689.
232. Ncanana S, Burton S. J Mol Catal B: Enzym 2007;44:66–71.
233. Eisenman HC, Mues M, Weber SE, Frases S, Chaskes S, Gerfen G, Casadevall A. Microbiology 2007;153:3954–3962.
234. Ikeda R, Uyama H, Kobayashi S. Macromolecules 1996;29:3053–3054.
235. Ikeda R, Sugihara J, Uyama H, Kobayashi S. Macromolecules 1996;29:8702–8705.
236. Ikeda R, Sugihara J, Uyama H, Kobayashi S. Polym Int 1998;47:295–301.
237. Aktas N, Sahiner N, Kantoglu O, Salih B, Tanyolac A. J Polym Environ 2003;11:123–128.
238. Anthoni J, Humeau C, Maia ER, Chebil L, Engasser J-M, Ghoul M. Eur Food Res Technol 2010;231:571–579.
239. Kobayashi S, Uyama H, Ikeda R. Chem Eur J 2001;7:4754–4760.
240. Kurisawa M, Chung JE, Uyama H, Kobayashi S. Biomacromolecules 2003;4:1394–1399.
241. Kurisawa M, Chung JE, Uyama H, Kobayashi S. Macromol Biosci 2003;3:758–764.
242. Hosny M, Rosazza JPN. J Agric Food Chem 2002;50:5539–5545.
243. Anyanwutaku I, Petroski RJ, Rosazza JPN. Bioorg Med Chem 1994;2:543–551.
244. Sagui F, Chirivì C, Fontana G, Nicotra S, Passarella D, Riva S, Danieli B. Tetrahedron 2009;65:312–317.
245. Selinheimo E, Lampila P, Mattinen M-L, Buchert J. J Agric Food Chem 2008;56:3118–3128.
246. Mikolasch A, Hammer E, Jonas U, Popowski K, Stielow A, Schauer F. Tetrahedron 2002;58:7589–7593.
247. Schroeder M, Pereira L, Couto SR, Erlacher A, Schoening K-U, Cavaco-Paulo A, Guebitz GM. Enzyme Microb Technol 2007;40:1748–1752.
248. Pilz R, Hammer E, Schauer F, Kragl U. Appl Microbiol Biotechnol 2003;60:708–712.
249. Niedermeyer THJ, Mikolasch A, Lalk M. J Org Chem 2005;70:2002–2008.
250. Mikolasch A, Hahn V, Manda K, Pump J, Illas N, Gordes D, Lalk M, Gesell-Salazar M, Hammer E, Julich W-D, Rawer S, Thurow K, Lindequist U, Schauer F. Amino Acids 2010;39:671–683.
251. Manda K, Hammer E, Mikolasch A, Niedermeyer T, Dec J, Jones AD, Benesi AJ, Schauer F, Bollag J-M. J Mol Catal B: Enzym 2005;35:86–92.
252. Niedermeyer T, Lalk M. J Mol Catal B: Enzym 2007;45:113–117.
253. Manda K, Hammer E, Mikolasch A, Gordes D, Thurow K, Schauer F. Amino Acids 2006;31:409–419.
254. Mikolasch A, Niedermeyer THJ, Lalk M, Witt S, Seefeldt S, Hammer E, Schauer F, Gesell M, Hessel S, Julich W-D, Lindequist U. Chem Pharm Bull 2006;54:632–638.

255. Mikolasch A, Niedermeyer THJ, Lalk M, Witt S, Seefeldt S, Hammer E, Schauer F, Gesell M, Salazar S, Hessel S, Julich W-D, Lindequist U. Chem Pharm Bull 2007;55: 412–416.
256. Manda K, Gordes D, Mikolasch A, Hammer E, Schmidt E, Thurow K, Schauer F. Appl Microbiol Biotechnol 2007;76:407–416.
257. Bialk HM, Hedman C, Castillo A, Pedersen JA. Environ Sci Technol 2007;41:3593–3600.
258. Wellington KW, Steenkamp P, Brady D. Bioorg Med Chem 2010;18:1406–1414.
259. Hahn V, Davids T, Lalk M, Schauer F, Mikolasch A. Green Chem 2010;12:879–887.
260. Bhalerao UT, Muralikrishna C, Rani BR. Tetrahedron 1994;50:4019–4024.
261. Leutbecher H, Conrad J, Klaiber I, Beifuss U. Synlett 2005;3126–3130.
262. Hajdok S, Leutbecher H, Greiner G, Conrad J, Beifuss U. Tetrahedron Lett 2007;48:5073–5076.
263. Hajdok S, Conrad J, Leutbecher H, Strobel S, Schleid T, Beifuss U. J Org Chem 2009;74:7230–7237.
264. Witayakran S, Gelbaum L, Ragauskas AJ. Tetrahedron 2007;63:10958–10962.
265. Witayakran S, Ragauskas AJ. Eur J Org Chem 2009;358–363.
266. Leutbecher H, Hajdok S, Braunberger C, Neumann M, Mika S, Conrad J, Beifuss U. Green Chem 2009;11:676–679.
267. Leutbecher H, Constantin M-A, Mika S, Conrad J, Beifuss U. Tetrahedron Lett 2011;52:604–607.
268. Kidwai M, Poddar R, Diwaniyan S, Kuhad RC. Synth Commun 2011;41:695–706.
269. Witayakran S, Zettili A, Ragauskas AJ. Tetrahedron Lett 2007;48:2983–2987.
270. Witayakran S, Ragauskas AJ. Green Chem 2007;9:475–480.
271. Call HP, Mucke I. J Biotechnol 1997;53:163–202.
272. ten Have R, Teunissen PJM. Chem Rev 2001;101:3397–3413.
273. Fabbrini M, Galli C, Gentili P. J Mol Catal B: Enzym 2002;16:231–240.
274. Bourbonnais R, Paice MG, Freiermuth B, Bodie E, Borneman S. Appl Env Microbiol 1997;63:4627–4632.
275. Baiocco P, Barreca AM, Fabbrini M, Galli C, Gentili P. Org Biomol Chem 2003;1:191–197.
276. Galli C, Gentili P. J Phys Org Chem 2004;17:973–977.
277. Astolfi P, Brandi P, Galli C, Gentili P, Gerini MF, Greci L, Lanzalunga O. New J Chem 2005;29:1308–1317.
278. Camarero S, Cañas AI, Nousiainen P, Record E, Lomascolo A, Martinez MJ, Martinez AT. Environ Sci Technol 2008;42:6703–6709.
279. Camarero S, Ibarra D, Martınez MJ, Martınez AT. Appl Env Microbiol 2005;71:1775–1784.
280. Díaz-González M, Vidal T, Tzanov T. Appl Microbiol Biotechnol 2011;89:1693–1700.
281. Minisci F, Recupero F, Cecchetto A, Gambarotti C, Punta C, Faletti R, Paganelli R, Pedulli GF. Eur J Org Chem 2004;109–119.
282. Fabbrini M, Galli C, Gentili P. J Mol Catal B: Enzym 2002;18:169–171.
283. D'Acunzo F, Baiocco P, Fabbrini M, Galli C, Gentili P. Eur J Org Chem 2002;24:4195–4201.

284. Tromp SA, Matijosite I, Sheldon RA, Arends IWCE, Mul G, Kreutzer MT, Moulijn JA, de Vries S. ChemCatChem 2010;2:827–833.
285. Arends IWCE, Li Y-X, Ausan R, Sheldon RA. Tetrahedron 2006;62:6659–6665.
286. Arias ME, Arenas M, Rodrıguez J, Soliveri J, Ball SA, Hernandez M. Appl Environ Microbiol 2003;69:1953–1958.
287. Camarero S, García O, Vidal T, Colom J, del Río JC, Gutiérrez A, Gras JM, Monje R, Martínez MJ, Martínez AT. Enzyme Microb Technol 2004;35:113–120.
288. Ibarra D, Camarero S, Romero J, Martínez MJ, Martínez AT. J Chem Technol Biotechnol 2006;81:1159–1165.
289. Bajpai P. Biotechnol Progr 1999;15:147–157.
290. Rochefort D, Leech D, Bourbonnais R. Green Chem 2004;6:14–24.
291. Camarero S, Ibarra D, Martinez MJ, Romero J, Gutierrez A, Del Rio JC. Enzyme Microb Technol 2007;40:2171.
292. Kumar A, Vanamala A, Kumar R. FEBS J 2005;272:6–8.
293. Bergbauer M, Eggert C, Kraepelin G. Appl Microbiol Biotechnol 1991;35:105–109.
294. Berrio J, Plou FJ, Ballesteros A, Martinez AT, Martinez MJ. Biocatal Biotransform 2007;25:130–134.
295. Abadulla E, Tzanov T, Costa S, Robra KH, Cavaco-Paulo A, Gübitz GM. Appl Environ Microbiol 2000;66:3357–3362.
296. Kunamneni A, Ghazi I, Camarero S, Ballesteros A, Plou FJ, Alcalde M. Process Biochem 2008;43:169–178.
297. Galante YM, Formantici C. Curr Org Chem 2003;7:1399–1422.
298. Claus H, Faber G, König H. Appl Microbiol Biotechnol 2002;59:672–678.
299. Widsten P, Tuominen S, Qvintus-Leino P, Laine JE. Wood Sci Technol 2004;38:521–528.
300. Bollag JM, Chu HL, Rao MA, Gianfreda L. J Environ Qual 2003;32:63–69.
301. Zumarraga M, Plou FJ, Garcia-Arellano H, Ballesteros A, Alcalde M. Biocatal Biotransform 2007;25:219–228.
302. Torres E, Bustos-Jaimes I, Le Borgne S. Appl Catal B: Environ 2003;46:1–15.
303. Jonsson LJ, Palmqvist E, Nilvebrant NO, Hahn HB. Appl Microbiol Biotechnol 1998;49:691–697.
304. Larsson S, Reimann A, Nilvebrant NO, Jonsson LJ. Appl Biochem Biotechnol 1999;77–79:91–103.
305. Cantarelli C, Brenna O, Giovanelli G, Rossi M. Food Biotechnol 1989;3:203–214.
306. Servili M, De Stefano G, Piacquadio P, Sciancalepore V. Am J Enol Viticul 2000;51:357–361.
307. Minussi RC, Pastore GM, Durán N. Trends Food Sci Technol 2002;13:205–216.
308. Kunamneni A, Plou FJ, Ballesteros A, Alcalde M. Recent Patent Biotechnol 2007;2:10–24.
309. Ghindilis A. Biochem Soc Trans 2000;28:84–89.
310. Ghindilis AL, Gavrilova VP, Yaropolov AI. Biosens Bioelectron 1992;7:127–131.
311. Majeau J-A, Brar SK, Tyagi RD. Bioresour Technol 2010;101:2331–2350.
312. Maciel MJ, Silva AC, Ribeiro HCT. Electron J Biotechnol 2010, 13:1–12, http://dx.DOI.org/10.2225/vol2213-issue2226-fulltext-2222.

313. Brijwani K, Rigdon A, Vadlani PV. Enzyme Research 2010, DOI:10.4061/2010/149748.

314. Desai SS, Nityanand C. Asian J Biotechnol 2011;3:98–124.

315. Shraddha S, Shekher R, Sehgal S, Kamthania M, Kumar A. Enzyme Research 2011, DOI:10.4061/2011/217861.

316. Durán N, Esposito E. Appl Catal, B 2000;28:83–99.

317. Osma JF, Toca-Herrera JL, Rodriguez-Couto S. Enzyme Research 2010, DOI: 10.4061/2010/918761.

318. Wells A, Teria M, Eve T. Biochem Soc Trans 2006;34:304–308.

319. Fernández-Sánchez C, Tzanov T, Gübitz GM, Cavaco-Paulo A. Bioelectrochemistry 2002;58:149–156.

320. Johannes C, Majcherczyk A. J Biotechnol 2000;78:193–199.

321. Cañas A, Camarero S. Biotechnol Adv 2010;28:694–705.

322. Rodgers CJ, Blanford CF, Giddens SR, Skamnioti P, Armstrong FA, Gurr SJ. Trends Biotechnol 2010;28:63–72.

323. Bourbonnais R, Paice MG. FEBS Lett 1990;267:99–102.

324. H. Xu, Lai Y-Z, Slomczynski D, Nakas JP, Tanenbaum SW. Biotechnol Lett 1997;19:957–960.

325. Barreca AM, Fabbrini M, Galli C, Gentili P, Ljunggren S. J Mol Catal B: Enzym 2003;26:105–110.

326. Chen C-L, Potthast A, Rosenau T, Gratzl JS, Kirkman AG, Nagai D, Miyakoshi T. J Mol Catal B: Enzym 2000;8:213–219.

327. Fritz-Langhals E, Kunath B. Tetrahedron Lett 1998;39:5955–5956.

328. Fabbrini M, Galli C, Gentili P, Macchitella D. Tetrahedron Lett 2001;42:7551–7553.

329. Potthast A, Rosenau T, Chen CL, Gratzl JS. J Mol Catal A: Chem 1996;108:5–9.

330. Barilli A, Belinghieri F, Passarella D, Lesma G, Riva S, Silvani A, Danieli B. Tetrahedron: Asymmetry 2004;15:2921–2925.

331. Matijosyte I, Arends IWCE, Vries Sd, Sheldon RA. J Mol Catal B: Enzym 2010;62:142–148.

332. Jetten JM, Van Den Dool R, Hartingsveldt W, Besemer AC. in WO 00/50463, 2000.

333. Viikari L, Kruus K, Buchert J. in WO 99/23117, 1999.

334. Marzorati M, Danieli B, Haltrich D, Riva S. Green Chem 2005;7:310–315.

335. Monti D, Candido A, Silva MMC, Kren V, Riva S, Danieli B. Adv Synth Catal 2005;347:1168–1174.

336. Baratto L, Candido A, Marzorati M, Sagui F, Riva S, Danieli B. J Mol Catal B: Enzym 2006;39:3–8.

337. Chirivi C, Sagui F, Riva S. In: Whittall J, Sutton PW, editors. *Practical Methods for Biocatalysis and Biotransformations*. Chippenham: John Wiley & Sons; 2010. pp. 240–244.

338. d'Acunzo F, Baiocco P, Galli C. New J Chem 2003;27:329–332.

339. Niku-Paavola M-L, Viikari L. J Mol Catal B: Enzym 2000;10:435–444.

340. Coniglio A, Galli C, Gentili P, Vadala R. J Mol Catal B: Enzym 2008;50:40–49.

341. Potthast A, Rosenau T, Chen C-L, Gratzl JS. J Org Chem 1995;60:4320–4321.

342. Molina S, Rencoret J, del Rio JC, Lomascolo A, Record E, Martinez AT, Gutierrez A. Appl Microbiol Biotechnol 2008;80:211–222.

343. Yang Z, Pattamana K, Molino BF, Haydar SN, Cao Y, Bois F, Maeng J, Hemenway MS, Rich JO, Khmelnitsky YL, Friedrich T, Peace D, Michels P. Synlett 2009;18:2935–2938.

344. Cañas AI, Alcalde M, Plou F, Martinez MJ, Martinez AT, Camarero S. Environ Sci Technol 2007;41:2964–2971.

345. D'Acunzo F, Galli C, Masci B. Eur J Biochem 2002;269:5330–5335.

346. Kurniawati S, Nicell JA. Enzyme MicrobTechnol 2007;41:353–361.

347. Hollmann F, Arends IWCE, Buehler K. ChemCatChem 2010;2:762–782.

348. van Hecke W, Bhagwat A, Ludwig R, Dewulf J, Haltrich D, van Langenhove H. Biotechnol Bioeng 2009;102:1475–1482.

349. Van Hecke W, Salaheddin C, Ludwig R, Dewulf J, Haltrich D, Langenhove HV. Bioresour Technol 2009;100:5566–5573.

350. Aksu S, Arends IWCE, Hollmann F. Adv Synth Catal 2009;351:1211–1216.

351. Tauber MM, Gubitz GM, Rehorek A. Bioresour Technol 2008;99:4213–4220.

352. Zille A, Gornacka B, Rehorek A, Cavaco-Paulo A. Appl Environ Microbiol 2005;71:6711–6718.

353. Chivukula M, Renganathan V. Appl Environ Microbiol 1995;61:4374–4377.

354. Setti L, Giuliani S, Spinozzi G, Pifferi PG. Enzyme Microb Technol 1999;25:285-289.

355. Pereira L, Coelho AV, Viegas CA, Ganachaud C, Iacazio G, Tron T, Robalo MP, Martins LO. Adv Synth Catal 2009;351:1857–1865.

356. Semenov AN, Lomonsova IV, Berezin VI, Titov MI. Biotechnol Bioeng 1993;42:1137—1141.

357. Verkade JMM, Hemert LJCv, Quaedflieg PJLM, Schoemaker HE, Schurmann M, Delft FLv, Rutjes FPJT. Adv Synth Catal 2007;349:1332–1336.

358. Schafer A, Specht M, Hetzheim A, Francke W, Schauer F. Tetrahedron 2001;57:7693–7699.

359. Rodakiewicz-Novak J. Top Catal 2000;11/12:419–434.

360. Mozhaev VV, Khmelnitsky YL, Sergeeva MV, Belova AB, Klyachko NL, Levashov AV, Martinek K. Eur J Biochem 1989;184:597–602.

361. D'Acunzo F, Barreca AM, Galli C. J Mol Catal B: Enzym 2004;31:25–30.

362. Vacurov AV, Gladilin AK, Levashov AV, Khmelnitsky YL. Biotechnol Lett 1994;16:175–178.

363. van Erp SHM, Kamenskaya EO, Khmelnitsky YL. Eur. J. Biochem. 1991;202:379–384.

364. Domínguez A, Rodríguez O, Tavares APM, Macedo EA, Longo MA, Sanromán MÁ. Bioresour Technol 2011;102:7494–7499.

365. Rodríguez O, Cristóvão RO, Tavares APM, Macedo EA. Appl Biochem Biotechnol 2011;164:524–533.

366. Rodakiewicz-Nowak J, Kasture SM, Dudek B, Haber J. J Mol Catal B: Enzym 2000;11:1–11.

367. Durán N, Rosa MA, D'Annibale A, Gianfreda L. Enzyme Microb Technol 2002;31:907–931.

368. Strohmeier GA, Pichler H, May O, Gruber-Khadjawi M. Chem Rev 2011;111:4141–4164.

369. Rodriguez-Couto S, Toca-Herrera JL. Biotechnol Adv 2007;25:558–569.
370. Alves AMCR, Record E, Lomascolo A, Scholtmeijer K, Asther M, Wessels JGH, Wosten HAB. Appl Environ Microbiol 2004;70:6379–6384.
371. Ye M, Li G, Liang WQ, Liu YH. Appl Microbiol Biotechnol 2010;87:1023–1031.
372. Vite-Vallejo O, Palomares LA, Dantan-Gonzalez E, Ayala-Castro HG, Martinez-Anaya C, Valderrama B. J. Folch-Mallol. Enzyme Microb Technol 2009;45:233—239.
373. Bohlin C, Jonsson LJ, Roth RL, Van Zyl WH. Appl Biochem Biotechnol 2006;129:195–214.
374. Ranieri D, Colao MC, Ruzzi M, Romagnoli G, Bianchi MM. FEMS Yeast Res 2009;9:892–902.
375. Ruiz AI, Malave AJ, Felby C, Griebenow K. Biotechnol Lett 2000;22:229–233.
376. Davis S, Burns RG. Appl Microbiol Biotechnol 1990;32:721–726.
377. Crecchio C, Ruggiero P, Pizzigallo MDR. Biotechnol Bioeng 1995;48:585–591.
378. Cordi L, Minussi RC, Freire RS, Durán N. Afr J Biotechnol 2007;6:1255–1259.
379. Zamorani A, Spettoli P, Lante A, Crapisi A, Passini G. Ital J. Food Sci 1993;5:409–414.
380. De Stefano G, Piacquadio P, Sciancalepore V. Biotechnol Tech. 1996;10:857–860.
381. Piacquadio P, DeStefano G, Sammartino M, Sciancalepore V. Biotechnol. Tech. 1997;11:515–517.
382. Sammartino M, Piacquadio P, DeStefano G, Sciancalepore V. Industrie delle Bevande 1998;27:367–369.
383. Piacquadio P, DeStefano G, Sammartino M, Sciancalepore V. Industrie delle Bevande 1998;27:378–383.
384. Rogalski J, Wojtas-Wasilewska M, Apalovic R, Leonowicz A. Biotechnol Bioeng 1991;37:770–777.
385. Rogalski J, Jozwik E, Hatakka A, Leonowicz A. J Mol Catal A: Chem 1995;95:99–108.
386. Satoh I, Sakurai I. Ann N Y Acad Sci 1998;864:493–496.
387. Luterek J, Gianfreda L, Wojtas-Wasilewska M, Cho NS, Rogalski J, Jaszek H, Malarczyk E, Staszczak M, Fink-Boots M, Leonowicz A. Holzforschung 1998;52:589–595.
388. Brenna O, Bianchi E. Biotechnol Lett 1994;16:35–40.
389. Reyes P, Pickard MA, Vazquez-Duhalt R, Biotechnol Lett 1999;21:875–880.
390. Davis S, Burns RG. Appl Microbiol Biotechnol 1992;37:474–479.
391. D'Annibale A, Stazi SR, Vinciguerra V, Di Mattia E, Sermanni GG. Process Biochem 1999;34:697–706.
392. Zhang J, Xu Z, Chen H, Zong Y. Biochem Eng J 2009;45:54–59.
393. Vasquez-Duhalt R, Tinoco R, D'Antonio P, Topoleski LDT, Payne GF. Bioconjugate Chem 2001;12:301–306.
394. Delanoy G, Li Q, Yu J. Int J Biol Macromol 2005;35:89–95.
395. D'Annibale A, Stazi SR, Vinciguerra V, Sermanni GG. J Biotechnol 2000;77:265–273.
396. Boller T, Meier C, Menzler S. Org Proc Res Dev. 2002;6:509–519.
397. Peralta-Zamora P, Pereira CM, Tiburtius ERL, Moraes SG, Rosa MA, Minussi RC, Durán N. Appl Catal B: Environ 2003;42:131–144.
398. Roy JJ, Abraham TE. J Chem Technol Biotechnol 2006;81:1836–1839.

399. Kallio JP, Auer S, Jänis J, Andberg M, Kruus K, Rouvinen J, Koivula A, Hakulinen N. J Mol Biol 2009;392:895–909.
400. Galli C, Gentili P, Jolivalt C, Madzak C, Vadalà R. Appl Microbiol Biotechnol 2011;91:123–131.
401. Andberg M, Hakulinen N, Auer S, Saloheimo M, Koivula A, Rouvinen J, Kruus K. FEBS J 2009;276:6285–6300.
402. Bulter T, Alcalde M, Sieber V, Meinhold P, Schlachtbauer C, Arnold FH. Appl Environ Microbiol 2003;69:987–995.
403. Janssen GG, Baldwin TM, Winetzky DS, Tierney LM, Wang H, Murray CJ. J Peptide Res 2004;64:10–24.
404. Alcalde M, Zumárraga M, Polaina J, Ballesteros A, Plou FJ. Comb Chem High Throughput Screen 2006;9:719–727.
405. Alcalde M, Bulter T, Zumárraga M, García-Arellano H, Mencía M, Plou FJ, Ballesteros A. J Biomol Screen 2005;10:624–631.
406. Zumarraga M, Bulter T, Shleev S, Polaina J, Martınez-Arias A, Plou FJ, Ballesteros A, Alcalde M. Chem Biol 2007;14:1052–1064.
407. Festa G, Autore F, Fraternali F, Giardina P, Sannia G. Proteins 2008;72:25–34.
408. Hu M-R, Chao, Y-P., Zhang G-Q, Yang X-Q, Xue Z-Q, Qian S-J. Biomol Eng 2007;24:619–624.
409. Koschorreck K, Schmid RD, Urlacher VB. BMC Biotechnol 2009, 9, DOI: 10.1186/1472–6750.
410. Zumarraga M, Vaz Domínguez C, Camarero S, Shleev S, Polaina J, Martínez-Arias A, Ferrer M, De Lacey AL, Fernández VM, Ballesteros A, Plou FJ, Alcalde M. Comb Chem High Throughput Screen 2008;11:807–816.
411. Zumárraga M, Camarero S, Shleev S, Martínez-Arias A, Ballesteros A, Plou FJ, Alcalde M. Proteins 2008;71:250–260.
412. Coll PM, Fernandez-Abalos JM, Villanueva JR, Santamaria R, Perez P. Appl Environ Microbiol 1993;59:2607–2613.
413. Mate D, Garcıa-Burgos C, Garcıa-Ruiz E, Ballesteros A, Camarero S, Alcalde M. Chem Biol 2010;17:1030–1041.
414. Miele A, Giardina P, Notomista E, Piscitelli A, Sannia G, Faraco V. Mol Biotechnol 2010;46:149–156.
415. Gupta N, Lee FS, Farinas ET. J Mol Cat B: Enzymatic 2010;62:230–234.
416. Gupta N, Farinas ET. Protein Eng Des Sel 2010;23:679–682.
417. Cusano AM, Mekmouche Y, Meglecz E, Tron T. FEBS J 2009;276:5471–5480.
418. Xu F, Berka RM, Wahleithner JA, Nelson BA, Shuster JR, Brown SH, Palmer AE, Solomon EI. Biochem J 1998;334:63–70.
419. Palmer AE, Szilagyi RK, Cherry JR, Jones A, Xu F, Solomon EI. Inorg Chem 2003;42:4006–4017.
420. Solomon EI, Szilagyi RK, George SD-B, Basumallick L. Chem Rev 2004;104:419–458.
421. Li H, Webb SP, Ivanic J, Jensen JH. J Am Chem Soc 2004;126:8010–8019.
422. Martınez AT, Ruiz-Dueñas FJ, Martınez MJ, del Rıo J, Gutierrez A. Curr Opin Biotechnol 2009;20:348–357.
423. Bento I, Silva CS, Chen Z, Martins LO, Lindley PF, Soares CM. BMC Struct Biol 2010;10:28.

424. Chen Z, Durao P, Silva CS, Pereira MM, Todorovic S, Hildebrandt P, Bento I, Lindley PF, Martins LO. Dalton Trans 2010;39:2875–2882.

425. Bertrand G. Compt Rend Acad Sci Paris 1896;122:1215–1217.

426. Itoh S, Fukuzumi S. Acc Chem Res 2007;40:592–600.

427. Rolff M, Schottenheim J, Decker H, Tuczek F. Chem Soc Rev 2011;40:4077–4098.

428. Halaouli S, Asther M, Sigoillot J-C, Hamdi M, Lomascolo A. J Appl Microbiol 2006;100:219–232.

429. van Gelder CW, Flurkey WH, Wichers HJ. Phytochemistry 1997;45:1309–1323.

430. Mayer AM. Phytochemistry 2006;67:2318–2331.

431. Marusek CM, Trobough NM, Flurkey WH, Inlow JK. J Inorg Chem 2006;100:108–123.

432. Claus H, Decker H. Syst Appl Microbiol 2006;29:3–14.

433. Brack A, Hellmann N, Decker H. Photochem Photobiol 2008;84:692–699.

434. Burton S. Catal Today 1994;22:459–487.

435. Espın JC, Varon R, Fenoll L, Gilabert MG, Garcıa-Ruız PA, Tudela J, Garcıa-Canovas F. Eur J Biochem 2000;267:1270–1279.

436. Whitaker JR. In: Wong DWS, editor. *Food Enzymes: Structure and Mechanism.* New York: Chapman Hall; 1995. pp. 271–307.

437. Flurkey WH, Inlow JK. J Inorg Biochem 2008;102:2160–2170.

438. van Leeuwen J, Wichers HJ. Mycol Res 1999;103:413–418.

439. Ichishima E, Maeba H, Amikura T, Sakata H. Biochim Biophys Acta 1984;786:25–31.

440. Espin JC, Wichers HJ. J Agric Food Chem 1999;47:3518–3525.

441. Lee SY, Cho MY, Hyun JH, Lee KM, Homma KI, Natori S, Kawabata SI, Iwanaga S, Lee BL. Eur J Biochem 1998;257:615–621.

442. Espin JC, van Leeuwen J, Wichers HJ. J Agric Food Chem 1999;47:3509–3517.

443. Burton KS, Love ME, Smith JF. Enzyme Microb Technol 1993;15:736–741.

444. Soler-Rivas C, Arpin N, Olivier JM, Wichers HJ. Mycol Res 1997;101:375–382.

445. Jolivet S, Arpin M, Wichers HJ, Pellon G. Mycol Res 1998;102:1459–1483.

446. Fairhead M, Thony-Meyer L. FEBS J 2010;277:2083–2095.

447. Fairhead M, Thony-Meyer L. New Biotechnol 2012, DOI:10.1016/j.nbt.2011.1005.1007.

448. Lopez-Serrano D, Sanchez-Amat A, Solano F. Pigment Cell Res 2002;15:104–111.

449. Raper HS. Physiol Rev 1928;8:245–282.

450. Mason HS. J Biol Chem 1948;172:83–99.

451. Kobayashi T, Vieira WD, Potterf B, Sakai C, Imokawa G, Hearing VJ. J Cell Sci 1995;108:2301–2309.

452. Olivares C, Jimenez-Cervantes C, Lozano JA, Solano F, Garcia-Borron JC. Biochem J 2001;354:131–139.

453. Seo S-Y, Sharma VK, Sharma N. J Agric Food Chem 2003;51:2837–2853.

454. Olivares C, Solano F. Pigment Cell Melanoma Res 2009;22:750–760.

455. Bell AA, Wheeler MH. Annu Rev Phytopathol 1986;24:411–451.

456. Jacobson ES. Clin Microbiol Rev 2000;13:708–717.

457. Martinez MV, Whitaker JR. Trends Food Sci Technol 1995;6:195–200.

458. Walker JR, Ferrar PH. Biotechnol Genet Eng Rev 1998;15:457–498.

459. Sugumaran M. Pigment Cell Res 2002;15:2–9.
460. Plonka PM, Grabacka M. Acta Biochim Pol 2006;53:429–443.
461. Michalik J, Emilianowicz-Czerska W, Switalski L, Raczynska-Bojanowska K. Antimicrob Agents Chemother 1975;8:526–531.
462. Nikitina VE, Vetchinkina EP, Ponomareva EG, Gogoleva YV. Microbiology 2010;79:327–333.
463. Matoba Y, Kumagai T, Yamamoto A, Yoshitsu H, Sugiyama M. J Biol Chem 2006;281:8981–8990.
464. Decker H, Schweikardt T, Tuczek F. Angew Chem Int Ed 2006;45:4546–4550.
465. Sendovski M, Kanteev M, Ben-Yosef VS, Adir N, Fishman A. J Mol Biol 2011;405:227–237.
466. Cabanes J, Chazarra S, Garcia-Carmona F. J Theo Biol 2002;214:321–328.
467. Munoz-Munoz JL, Garcia-Molina F, Varon R, Garcia-Ruız P, Tudela J, Garcia-Canovas F, Rodrıguez-Lopez JN. IUBMB Life 2010;62:539–547.
468. Decker H, Tuczek F. Trends Biochem Sci 2000;25:392–397.
469. Sanchez-Ferrer A, Rodriguez-Lopez JN, Garcia-Canovas F, Garcia-Carmona F. Biochim Biophys Acta 1995;1247:1–11.
470. Lerch K. In: Sigel H, editor. *Metal ions in Biological Systems.* Volume 13, New York: Marcel Dekker; 1981. pp. 143–186.
471. García Molina F., Muñoz JL, Varón R, Rodríguez-López JN, García-Cánovas F, Tudela J. Int J Biochem Cell Biol 2007;39:238–252.
472. Espin JC, Wichers HJ. Biochim Biophys Acta 2001;1544:289–300.
473. Andrawis A, Varda K. J Food Biochem 1990;14:103–115.
474. de Faria RO, Rotuno-Moure V, Lopes de Almeida MA, Krieger N, Mitchell DA. Food Technol Biotechnol 2007;45:287–294.
475. Pialis P, Saville BA. Enzyme Microb Technol 1998;22:261–268.
476. Ros JR, Rodriguez-Lopez JN, Garcia-Canovas F. Biochem J 1993;295:309–312.
477. Marín-Zamora ME, Rojas-Melgarejo F, García-Cánovas F, García-Ruiz PA. J Biotechnol 2009;139:163–168.
478. Valero E, Lozano MI, Varon R, García-Carmona F. Biotechnol Progr 2003;19:1632–1638.
479. Espın JC, Soler-Rivas C, Cantos E, Tomas-Barberan F, Wichers HJ. J Agric Food Chem 2001;49:1187–1193.
480. Kozlowska M, Kalnski R, Kanska M. J Labelled Compd Radiopharm 2005;48:235–240.
481. Kazandjian RZ, Klibanov AM. J Am Chem Soc 1985;107:5448–5450.
482. Kermasha S, Tse M. J Chem Technol Biotechnol 2000;75:475–483.
483. Dai L, Klibanov AM. Proc Natl Acad Sci U S A 1999;96:9475–9478.
484. Burton SG, Duncan JR. Biotechnol Lett 1995;17:627–630.
485. Rojo M, Gomez M, Isorna P, Estrada P. J Mol Catal B: Enzym 2001;11:857–865.
486. Yang Z, Robb DA. Biocatal Biotransform 2005;23:423–430.
487. Vulfson EM, Ahmed G, Gill I, Kozlov IA, Goodenough P, Law B. Biotechnol Lett 1991;13:91–96.
488. Rodakiewicz-Nowak J, Ito M. Adv Colloid Interface Sci 2005;284:674–679.

489. Shuster V, Fishman A. J Mol Microbiol Biotechnol 2009;17:188–200.
490. Muller GH, Lang A, Seithel DR, Waldmann H. Chem Eur J 1998;4:2513–2522.
491. Vogna D, Pezzella A, Panzella L, Napolitano A, d'Ischia M. Tetrahedron Lett 2003;44:8289–8292.
492. Land EJ, Ramsden CA, Riley PA, Yoganathan G. Org Biomol Chem 2003;1:3120.
493. Land EJ, Ramsden CA, Riley PA. Acc Chem Res 2003;36:300–308.
494. Rescigno A, Bruyneel F, Padiglia A, Sollai F, Salis A, Marchand-Brynaert J, Sanjust E. Biochim Biophys Acta 2011;1810;799–807.
495. Toussaint O, Lerch K. Biochemistry 1987;26:8567–8571.
496. Panzella L, Napolitano A, d'Ischia M. Bioorg Med Chem 2003;11:4797–4805.
497. Pizarro N, Torre Rdl, Joglar J, Okumura N, Perfetti X, Lau SS, Monks TJ. Chem Res Toxicol 2008;21:2272–2279.
498. Kubo I, Nihei K-i, Shimizu K. Bioorg Med Chem 2004;12:5343–5347.
499. Pezzella A, Lista L, Napolitano A, d'Ischia M. Chem Res Toxicol 2005;18:1413.
500. Guerneve CL, Sanoner P, Drilleau J-F, Guyot S. Tetrahedron Lett 2004;45:6673–6677.
501. Bhalerao UT, Muralikrishna C, Pandey G. Synth Commun 1989;19:1303–1307.
502. Janse van Rensburg W, Ferreira D, Malan E, Steenkamp JA. Phytochemistry 2000;53:285–292.
503. Volkert M, Koul S, Muller GH, Lehnig M, Waldmann H. J Org Chem 2002;67:6902–6910.
504. Muller GH, Waldmann H. Tetrahedron Lett 1999;40:3549–3552.
505. Montiel AM, Fernandez FJ, Marcial J, Soriano J, Barrios-Gonzales J, Tomasini A. Biotechnol Lett 2004;26:1353–1357.
506. Sun W-Q, Payne GF, Moas MSGL, Chu JH, Wallace KK. Biotechnol Progr 1992;8:179–186.
507. El-Shora HM, Metwally M. Biotechnology 2008;7:305–310.
508. Wada S, Ichikawa H, Tatsumi K. Biotechnol Bioeng 1993;42:854–858.
509. Aberg CM, Chen T, Olumide A, Raghavan SR, Payne GF. J Agric Food Chem 2004;52:788–793.
510. Halaouli S, Asther M, Kruus K, Guo L, Hamdi M, Sigoillot J-C, Asther M, Lomascolo A. J Appl Microbiol 2005;98:332–343.
511. Freddi G, Anghileri A, Sampaio S, Buchert J, Monti P, Taddei P. J Biotechnol 2006;125:281–294.
512. Lantto R, Puolanne E, Kruus K, Buchert J, Autio K. J Agric Food Chem 2007;55:1248–1255.
513. Wu LQ, Embree HD, Balgley BM, Smith PJ, Payne GF. Environ Sci Technol 2002;36:3446–3454.
514. Anhgileri A, Lantto R, Kruus K, Arosio C, Freddi G. J Biotechnol 2007;127:508–519.
515. Lenhart JL, Chaubal MV, Payne GF, Barbari TA. Gross RA, Kaplan DL, Swift G, editors. In: Enzymatic modification of chitosan by tyrosinase. *Enzymes in Polymer Synthesis*. ACS Symposium Series 684., Washington DC, 1998. pp 188–198.
516. Kane JB, Tompkins RG, Yarmush ML, Burke JF. In: Ratner BD, Hoffman AS, Schoen FJ, Lemons JE, editors. Chapter 7.10: Burn Dressings. *Biomaterials Science: An Introduction to Materials in Medicine*. San Diego: Academic Press; 1996. pp. 360–370.

517. Yamamoto H, Kuno S, Nagai A, Nishida A, Yamauchi S, Ikeda K. Int J Biol Macromol 1990;12:305–310.
518. Thalman CR, Lotzbeyer T, Eur. Food Res. Technol. 2002;214:276–281.
519. Lee KY, Mooney DJ. Chem Rev 2001;101:1869–1879.
520. Kim Y-J, Uyama H. Cell Mol Life Sci 2005;62:1707–1723.
521. Meredith P, Sarna T. Pigment Cell Res 2006;19:572–594.
522. Nosanchuk JD, Casadevall A. Cell Microbiol 2003;5:203–223.
523. Marko-Varga JG, Emneus J, Gorton L, Ruzgas T. Trends Anal Chem 1995;14:319–328.
524. Streffer K, Vijgenboom E, Tepper AWJW, Makower A, Scheller FW, Canters GW, Wollenberger U. Anal Chim Acta 2001;427:201–210.
525. Zauner G, Lonardi E, Bubacco L, Aartsma TJ, Canters GW, Tepper AW. Chemistry 2007;13:7085–7090.
526. Flurkey A, Cooksey J, Reddy A, Spoonmore K, Rescigno A, Inlow J, Flurkey WH. J Agric Food Chem 2008;56:4760–4768.
527. Wada S, Ichikawa H, Tatsumi K. Water Sci Technol 1992;26:2057–2059.
528. Munjal N, Sawhney SK. Enzyme Microb Technol 2002;30:613–619.
529. Hamann MCJ, Saville BA. Food Bioprod Process 1996;74:47–52.
530. Edwards W, Leukes WD, Rose PD, Burton SG. Enzyme Microb Technol 1999;25:769–773.
531. Boshoff A, Edwards W, Leukes WD, Rose PD, Burton SG. Desalination 1998;115:307–312.
532. Wichers HJ, Recourt K, Hendriks M, Ebbelaar CEM, Biancone G, Hoeberichts FA, Mooibroek H, Soler-Rivas C. Appl Microbiol Biotechnol 2003;61:336–341.
533. Kupper U, Linden M, Cao K, Lerch K. Current Genetics 1990;18:331–335.
534. Fujita Y, Uraga Y, Ichisima E. Biochim Biophys Acta 1995;1261:151–154.
535. Halaouli S, Record E, Casalot L, Hamdi M, Sigoillot J-C, Asther M, Lomascolo A. Appl Microbiol Biotechnol 2006;70:580–589.
536. Ruan L, He W, He J, Sun M, Yu Z. Antonie Van Leeuwenhoek 2005;87:283–288.
537. Cabrera-Valladares N, Martinez A, Pinero S, Lagunas-Munoz VH, Tinoco R, de Anda R, Vazquez-Duhalt R, Bolivar F, Gosset G. Enzyme Microb Technol 2006;38:772–779.
538. Della-Cioppa G, Garger SJ, Sverlow GG, Turpen TH, Grill LK. Nat Biotechnol 1990;8:634–638.
539. Ito M, Inouye K. J Biochem 2005;138:355–362.
540. Jackman MP, Huber M, Hajnal A, Lerch K. Biochem J 1992;282(Pt.3):915–918.
541. Ben-Yosef VS, Sendovski M, Fishman A. Enzyme Microb Technol 2010;47:372–376.
542. Gerdemann C, Eicken C, Krebs B. Acc Chem Res 2002;35:183–191.
543. Koval IA, Gamez P, Belle C, Selmeczi K, Reedijk J. Chem Soc Rev 2006;35:814–840.
544. Eicken C, Krebs B, Sacchettini JC. Curr Opin Struct Biol 1999;9:677–683.
545. Rompel A, Fischer H, Meiwes D, Büldt-Karentzopoulos K, Magrini A, Eicken C, Gerdemann C, Krebs B. FEBS Lett 1999;445:103–110.
546. Fitzpatrick PF. Arch Biochem Biophys 2010;493:13–25.
547. Krebs HA. Biochem J 1935;29:1620–1644.

548. Pollegioni L, Molla G, Sacchi S, Rosini E, Verga R, Pilone MS. Appl Microbiol Biotechnol 2008;78:1–16.

549. Tishkov VI, Khoronenkova SV. Biochemistry (Moscow) 2005;70:40–54.

550. Porter DJ, Voet JG, Bright HJ. J Biol Chem. 1977;252:4464–4473.

551. Pilone MS. Cell Mol Life Sci 2000;57:1732–1747.

552. Pollegioni L, Piubelli L, Sacchi S, Pilone MS, Molla G. Cell Mol Life Sci 2007;64:1373–1394.

553. Momoi K, Fukui K, Watanabe F, Miyake Y. FEBS Lett 1988;238:180–184.

554. Khoronenkova SV, Tishkov VI. Biochemistry (Moscow) 2008;73:1511–1518.

555. Pilone MS, Pollegioni L. Biocatal Biotransform 2002;20:145–159.

556. Pollegioni L, Sacchi S, Caldinelli L, Boselli A, Pilone MS, Piubelli L, Molla G. Curr Protein Pept Sci 2007;8:600–618.

557. Sacchi S, Pollegioni L, Pilone MS, Rossetti C. Biotechnol Tech 1998;12:149–153.

558. Caldinelli L, Motteran L, Sacchi S, Piubelli L, Boselli A, Mothet JP, Pollegioni L, Pilone MS. In: Konno R, Brückner H, D'Aniello A, Fisher G, Fujii N, Homma. H, editors. *D-amino Acids: A New Frontier in Amino Acid and Protein Research—practical Methods and Protocols*. New York: Nova Biomedical Books; 2006. pp. 135–168.

559. van Staden JF, Stefan RI, Aboul-Enein HY. Fresenius J Anal Chem 2000;367:178–180.

560. Wu X, van Wie BJ, Kidwell DA. Biosens Bioelectron 2004;20:879–886.

561. Obzansky DM, Rabin BR, Simons DM, Tseng SY, Severino DM, Eggelte H, Fisher M, Harbron S, Stout RW, Di Paolo MJ. Clin Chem 1991;37:1513–1518.

562. Stegman LD, Zheng H, Neal ER, Ben-Yoseph O, Pollegioni L, Pilone MS, Ross BD. Hum Gene Ther 1998;9:185–193.

563. Katsuki H, Nonaka M, Shirakawa H, Kume T, Akaike A. J Pharmacol Exp Ther 2004;311:836–844.

564. Buto S, Pollegioni L, D'Angiuro L, Pilone MS. Biotechnol Bioeng 1994;44:1288–1294.

565. Garcia-Garcia M, Martinez-Martinez I, Sanchez-Ferrer A, Garcia-Carmona F. Biotechnol Progr 2008;24:187–191.

566. Fernandez-Lafuente R, Rodriguez V, Guisan JM. Enzyme Microb Technol 1998;23:28–33.

567. Upadhya R, Nagajyothi S, Bhat SG. Process Biochem 1999;35:7–13.

568. Tan Q, Song Q, Zhang Y, Wei D. Appl Biochem Biotechnol 2007;136:279–289.

569. Lutz-Wahl S, Trost E-M, Wagner B, Manns A, Fischer L. J Biotechnol 2006;124:163–171.

570. Caligiuri A, D'Arrigo P, Rosini E, Tessaro D, Molla G, Servi S, Pollegioni L. Adv Synth Catal 2006;348:2183–2190.

571. Luo H, Yu H, Li Q, Shen Z. Enzyme Microb Technol 2004;35:514–518.

572. Luo H, Li Q, Yu H, Shen Z. Biotechnol Lett 2004;26:939–945.

573. Lopez-Gallego F, Batencor L, Hidalgo A, Mateo C, Fernandez-Lafuente R, Guisán JM. Adv Synth Catal 2005;347:1804–1810.

574. Lopez-Gallego F, Bentancor L, Sio CF, Reis CR, Jimenez PN, Guisan JM, Quax WJ, Fernandez-Lafuente R. Adv Synth Catal 2008;350:343.

575. Pollegioni L, Lorenzi S, Rosini E, Marcone GL, Molla G, Verga R, Cabri W, Pilone MS. Protein Sci 2005;14:3064–3076.

576. Rodwell WV. Methods Enzymol 1971;17B, 174–188.
577. Dobrovinskaya NA, Archer I, Hulme AN. Synlett 2008;513–516.
578. Truppo MD, Turner NJ, Rozzell JD. Chem Commun 2009;45:2127–2129.
579. Hopwood J, Truppo MD, Turner NJ, Lloyd RC. Chem Commun 2011;47:773–775.
580. Hafner EW, Wellner D. Proc Natl Acad Sci USA 1971;68:987–991.
581. Huh JW, Yokoigawa K, Esaki N, Soda K. Biosci Biotechnol Biochem 1992;56:2081–2082.
582. Huh JW, Yokoigawa K, Esaki N, Soda K. J Ferment Bioeng 1992;74:189–190.
583. Beard TM, Turner NJ. Chem Commun 2002;246–247.
584. Enright A, Alexandre F-R, Roff G, Fotheringham IG, Dawson MJ, Turner NJ. Chem Commun 2003;2636–2637.
585. Markle W, Lutz S. Electrochim Acta 2008;53:3175–3180.
586. Servi S, Tessaro D, Pedrocchi-Fantoni G. Coord Chem Rev 2008;252:715–726.
587. Hanson RL, Schwinden MD, Banerjee A, Brzozowski DB, Chen BC, Patel BP, McNamee CG, Kodersha GA, Kronenthal DR, Patel RN, Szarka LJ. Bioorg Med Chem 1999;7:2247–2252.
588. Patel RN. Biomol Eng 2001;17:167–182.
589. Caliguri A, D'Arrigo P, Gefflaut T, Molla G, Pollegioni L, Rosini E, Rossi C, Servi S. Biocatal Biotransform 2006;24:409–413.
590. Findrik Z, Vasic-Racki D. Biotechnol Bioeng 2007;98:956–967.
591. Betancor L, Hidalgo A, Fernández-Lorente G, Mateo C, Rodríguez V, Fuentes M, López-Gallego F, Fernández-Lorente R, Guisan JM. Biotechnol Progr 2003;19:784–788.
592. Sacchi S, Rosini E, Molla G, Pilone MS, Pollegioni L. Protein Eng Des Sel 2004;17:517–525.
593. Pawelek PD, Cheah J, Coulombe R, Macheroux P, Ghisla S, Vrielink A. EMBO J 2000;19:4204–4215.
594. Zuliani JP, Kayano AM, Zaqueo KD, Neto AC, Sampaio SV, Soares AM, Stabeli RG. Protein Pept Lett 2009;16:908–912.
595. Du XY, Clemetson KJ. Toxicon 2002;40:659–665.
596. Pantaleone DP, Geller AM, Taylor PP. J Mol Catal B: Enzym 2001;11:795–803.
597. Alexandre F-R, Pantaleone DP, Taylor PP, Fotheringham IG, Ager DJ, Turner NJ. Tetrahedron Lett 2002;43:707–710.
598. De Colibus L, Li M, Binda C, Lustig A, Edmondson DE, Mattevi A. Proc Natl Acad Sci USA 2005;102:12684–12689.
599. Edmondson DE, Binda C, Mattevi A. Arch Biochem Biophys 2007;464:269–276.
600. Turner NJ. Chem Rev 2011;111:4073–4087.
601. Schilling B, Lerch K. Biochim Biophys Acta 1995;1243:529—537.
602. Sablin SO, Yankovskaya V, Bernard S, Cronin CN, Singer TP. Eur J Biochem 1998;253:270–279.
603. Alexeeva M, Enright A, Dawson MJ, Mahmoudian M, Turner NJ. Angew Chem Int Ed 2002;41:3177.
604. Alexeeva M, Carr R, Turner NJ. Org Biomol Chem 2003;1:4133–4137.
605. Carr R, Alexeeva M, Enright A, Eve TSC, Dawson MJ, Turner NJ. Angew Chem, Int Ed 2003;42:4807–4810.

606. Carr R, Alexeeva M, Dawson MJ, Gotor-Fernandez V, Humphrey CE, Turner NJ. ChemBioChem 2005;6:637–639.
607. Dunsmore CJ, Carr R, Fleming T, Turner NJ. J Am Chem Soc 2006;128:2224–2225.
608. Eve TSC, Wells A, Turner NJ. Chem Commun 2007;1530–1531.
609. Woo JCG, Wang XQ, Silverman RB. J Org Chem 1995;60:6235–6236.
610. Kohler V, Bailey KR, Znabet A, Raftery J, Helliwell M, Turner NJ. Angew Chem Int Ed 2010;49:2182–2184.
611. Znabet A, Zonneveld J, Janssen E, De Kanter FJJ, Helliwell M, Turner NJ, Ruijter E, Orru RVA. Chem Commun 2010;46:7706–7708.
612. Znabet A, Ruijter E, de Kanter FJJ, Koehler V, Helliwell M, Turner NJ, Orru RVA. Angew Chem Int Ed 2010;49:5289–5292.
613. Znabet A, Polak MM, Janssen E, de Kanter FJJ, Turner NJ, Orru RVA, Ruijter E. Chem Commun 2010;46:7918–7920.
614. Qiao C, Ling K-Q, Shepard EM, Dooley DM, Sayre LM. J Am Chem Soc 2006;128:6206–6219.
615. Kim M, Okajima T, Kishishita S, Yoshimura M, Kawamori A, Tanizawa K, Yamaguchi H. Nat Struct Biol 2002;9:591–596.
616. Cai D, Klinman JP. J Biol Chem 1994;269:32039–32042.
617. Chen Z-w, Datta S, DuBois JL, Klinman JP, Mathews FS. Biochemistry 2010;49:7393–7402.
618. Hacisalihoglu A, Jongejan A, Jongejan JA, Duine JA. J Mol Catal B: Enzym 2000;11:81–88.
619. Lukasheva EV, Berezov TT. Biochemistry (Moscow) 2002;67:1152–1158.
620. Vrielink A, Ghisla S. FEBS J 2009;276:6826–6843.
621. MacLachlan J, Wotherspoon ATL, Ansell RO, Brooks CJW. J Steroid Biochem Mol Biol 2000;72:169–195.
622. Doukyu N. Appl Microbiol Biotechnol 2009;83:825–837.
623. Kreit J, Sampson NS. FEBS J 2009;276:6844–6856.
624. Biellmann JF. Chirality 2001;13:34–39.
625. Aparicio JF, Martın JF. Mol BioSyst 2008;4:804–809.
626. Pollegioni L, Piubelli L, Molla G. FEBS J 2009;276:6857–6870.
627. Rouquette-Jazdanian AK, Pelassy C, Breittmayer JP, Aussel C. Cell Signalling 2006;18:105–122.
628. Xu X, Bittman R, Duportail G, Heissler D, Vilcheze C, London E. J Biol Chem 2001;276:33540–33546.
629. Lange Y, Ye J, Steck TL. J Biol Chem 2005;280:36126–36131.
630. Purcell JP, Greenplate JT, Jennings MG. Biochem Biophys Res Commun 1993;196:1406–1413.
631. Corbin DR, Greenplate JT, Wong EY, Purcell JP. Appl Environ Microbiol 1994;60:4239–4244.
632. Alexander DL, Fisher JF. Steroids 1995;60:290–294.
633. Guo L-W, Wilson WK, Pang J, Shackleton CHL. Steroids 2003;68:31–42.
634. Dieth S, Tritsch D, Biellmann JF. Tetrahedron Lett 1995;36:2243–2246.
635. Labaree D, Hoyte RM, Hochberg RB. Steroids 1997;62:482–486.

636. Lee K, Biellmann J. Tetrahedron 1988;44:1135–1139.
637. de Jong E, van Berkel WJH, van der Zwan RP, de Bont JAM. Eur J Biochem 1992;208:651–657.
638. van den Heuvel RHH, Fraaije MW, Mattevi A, Laane C, van Berkel WJH. J Mol Catal B: Enzym 2001;11:185–188.
639. Fraaije MW, Veeger C, Berkel WJHv. Eur J Biochem 1995;234:271–277.
640. van den Heuvel RHH, Fraaije MW, Laane C, van Berkel WJH. J Bacteriol 1998;180:5646–5651.
641. Dubal SA, Tilkari YP, Momin SA, Borkar IV. Advanced Biotech 2008;3:20–31.
642. van den Heuvel RHH, van den Berg WAM, Rovida S, van Berkel WJH. J Biol Chem 2004;279:33492–33500.
643. Drijfhout FP, Fraaije MW, Jongejan H, Berkel WJHv, Franssen MCR. Biotechnol Bioeng 1998;59:171–177.
644. Heuts DPHM, Hellemond EWv, Janssen DB, Fraaije MW. J Biol Chem 2007;282:20283–20291.
645. Forneris F, Heuts DPHM, Delvecchio M, Rovida S, Fraaije MW, Mattevi A. Biochemistry 2008;47:978–985.
646. van Hellemond EW, Vermote L, Koolen W, Sonke T, Zandvoort E, Heuts DP, Janssen DB, Fraaije MW. Adv Synth Catal 2009;351:1523–1530.
647. van Bloois E, Winter RT, Janssen DB, Fraaije MW. Appl Microbiol Biotechnol 2009;83:679–687.
648. Passardi F, Theiler G, Zamocky M, Cosio C, Nicolas Rouhier, Teixera F, Margis-Pinheiro M, Ioannidis V, Penel C, Falquet L, Dunand C. Phytochemistry 2007;68:1605–1611.
649. Koua D, Cerutti L, Falquet L, Sigrist CJ, Theiler G, Hulo N, Dunand C. Nucleic Acids Res 2009;37:D261–D266.
650. PeroxiBase-The peroxidase database. http://peroxibase.toulouse.inra.fr.
651. Dunford HB. *Heme Peroxidases*. New York: Wiley-VCH; 1999.
652. Dunford HB. *Peroxidases and Catalases: Biochemistry, Biophysics, Biotechnology, and Physiology*. 2nd ed. Hoboken, NJ: John Wiley & Sons; 2010.
653. Epp O, Ladenstein R, Wendel A. Eur J Biochem 1983;133:51–69.
654. Deboer E, van Kooyk Y., Tromp MGM, Plat H, Wever R. Biochim Biophys Acta 1986;869:48–53.
655. Dolin MI. J Biol Chem 1957;225:557.
656. Martinez AT, Speranza M, Ruiz-Duenas FJ, Ferreira P, Camarero S, Guillen F, Martinez MJ, Gutierrez A, del Rio JC. Int Microbiol 2005;8:195–204.
657. Hammel KE, Cullen D. Curr Opin Plant Biol 2008;11:349–355.
658. Veitch NC. Phytochemistry 2004;65:249–259.
659. Azevedo AM, Martins VC, Prazeres DM, Vojinović V, Cabral JM, Fonseca LP. Biotechnol Annu Rev 2003;9:199–247.
660. Omura T. Biochem Biophys Res Commun 2005;338:404–409.
661. Hofrichter M, Ullrich R. Appl Microbiol Biotechnol 2006;71:276–288.
662. Green MT, Curr. Opin. Chem. Biol. 2009;13:84–88.
663. van Rantwijk F, Sheldon RA. Curr Opin Biotechnol 2000;11:554–564.

664. Makris TM, Denisov IG, Sligar SG. Biochem Soc Trans 2003;31:516–519.
665. Ullrich R, Hofrichter M. Cell Mol Life Sci 2007;64:271–293.
666. Blank LM, Ebert BE, Buehler K, Buehler B. Antioxid Redox Signaling 2010;13:349–394.
667. Valderrama B, Ayala M, Vazquez-Duhalt R. Chem Biol 2002;9:555–565.
668. Terner J, Palaniappan V, Gold A, Weiss R, Fitzgerald MM, Sullivan AM, Hosten CM. J Inorg Biochem 2006;100:480–501.
669. Green MT, Dawson JH, Gray HB. Science 2004;304:1653–1656.
670. Matsunaga I, Sumimoto T, Ayata M, Ogura H. FEBS Lett 2002;528:90–94.
671. Matsunaga I, Shiro Y. Curr Op Chem Biol 2004;8:127–132.
672. Hersleth HP, Ryde U, Rydberg P, Gorbitz CH, Andersson KK. J Inorg Biochem 2006;100:460–476.
673. Butler A, Sandy M. Nature 2009;460:848–854.
674. Vaillancourt FH, Yeh E, Vosburg DA, Garneau-Tsodikova S, Walsh CT. Chem Rev 2006;106:3364–3378.
675. Wagner C, Omari ME, Konig GM. J Nat Prod 2009;72:540–553.
676. Reddy CM, Xu L, Drenzek NJ, Sturchio NC, Heraty LJ, Kimblin C, Butler A. J Am Chem Soc 2002;124:14526–14527.
677. Sun W, Kadima TA, Pickard MA, Dunford HB. Biochem Cell Biol 1994;72:321–331.
678. Hernandez-Ruiz J. Arnao MB, Hiner ANP, Garcia-Canovas F, Acosta M. Biochem J 2001;354:107–114.
679. Vlasits J, Jakopitsch C, Bernroitner M, Zamocky M, Furtmuller PG, Obinger C. Arch Biochem Biophys 2010;500:74–81.
680. Adam W, Lazarus M, Saha-Möller CR, Weichold O, Hoch U, Häring D, Schreier P. Adv Biochem Eng/Biotechnol 1999;63:73–108.
681. van Deurzen MPJ, van Rantwijk F, Sheldon RA. Tetrahedron 1997;53:13183–13220.
682. van de Velde F, van Rantwijk F, Sheldon RA. Trends Biotechnol 2001;19:73–80.
683. Cherry JR, Lamsa MH, Schneider P, Vind J, Svendsen A, Jones A, Pedersen AH. Nat Biotechnol 1999;17:379–384.
684. Miyazaki-Imamura C, Oohira K, Kitagawa R, Nakano H, Yamane T, Takahashi H. Protein Eng 2003;16:423–428.
685. Miyazaki C, Takahashi H. FEBS Lett 2001;509:111–114.
686. Valderrama B, Garcia-Arellano H, Giansanti S, Baratto MC, Pogni R, Vazquez-Duhalt R. FASEB J 2005;20:1233–1235.
687. Colonna S, Gaggero N, Richelmi C, Pasta P. Trends Biotechnol 1999;17:163–168.
688. Okrasa K, Guibé-Jampel E, Therisod M. J Chem Soc, Perkin Trans 1 2000;1077–1079.
689. Van de Velde F, Lourenço ND, Bakker M, van Rantwijk F, Sheldon RA. Biotechnol Bioeng 2000;69:286–291.
690. Ramos MC, Torijas MC, Diaz AN. Sens Actuators, B 2001;73:71–75.
691. Gaspar S, Popescu IC, Gazaryan IG, Bautista AG, Sakharov IY, Mattiasson B, Csoregi E. Electrochim Acta 2000;46:255–264.
692. Field JA, Sierra-Alvarez R. Chemosphere 2008;71:1005–1018.
693. Haritash AK, Kaushik CP. J Hazard Mat 2009;169:1–15.

694. Aitken MD. Chem Eng J 1993;52:B49–B58.
695. Ferrer I, Dezotti M, Duran N. Biotechnol Lett 1991;13:577–582.
696. Wagner M, Nicell JA. Water Res 2001;35:485–495.
697. Burnette FS. J Food Sci 2006;42:1–6.
698. Folkes LK, Greco O, Dachs GU, Stratford MRL, Wardman P. Biochem Pharmacol 2002;63:265–272.
699. Kobayashi S, Kaneko I, Uyama H. Chem Lett 1992;393–394.
700. Setälä H, Pajunen A, Kilpeläinen I, Brunow G. J Chem Soc, Perkin Trans 1 1994;1163–1165.
701. H. Setälä, Pajunen A, Rummakko P, J. Sipila, Brunow G, J. Chem. Soc., Perkin Trans. 1 1999;461–464.
702. Saliu F, Tolppa E-L, Zoia L, Orlandi M. Tetrahedron Lett 2011;52:3856–3860.
703. Kiso T, Shizuma M, Murakami H, Kiryu T, Hozono K, Terai T, Nakano H. J Mol Catal B: Enzym 2007;45:50–56.
704. Goodbody AE, Endo T, Vukovic J, Kutney JP, Choi LSL, Misawa M. Planta Med 1988;54:136–140.
705. Sottomayor M, DiCosmo F, Barceló AR. Enzyme Microb Technol 1997;21:543–549.
706. Fukunishi K, Kitada K. Naito I, Synthesis 1991;237–238.
707. Hewson DW, Dunford HB. J Biol Chem 1976;251:6043–6052.
708. Krawczyk AR, Lipkowska E, Wrobel JT. Collect Czech Chem Commun 1991;56:1147–1150.
709. Kobayashi A, Koguchi Y, Kanzaki H, Kajiyama S-i. Biosci Biotechnol Biochem 1994;58:133–134.
710. Donelly DMX, Murphy FG, Polonski J, Prangé T. J Chem Soc, Perkin Trans 1 1987;2719–2722.
711. d'Ischia M, Napolitano A, Tsiakas K, Prota G. Tetrahedron 1990;46:5789–5796.
712. Hwang S, Lee CH, Ahn IS. J Ind Eng Chem 2008;14:487–492.
713. Gallager PH. Biochem J 1923;17:515–529.
714. Kobayashi S, Makino A. Chem Rev 2009;109:5288–5353.
715. Gross RA, Kumar A, Kalra B. Chem Rev 2001;101:2097–2124.
716. Tonami H, Uyama H, Kobayashi S, Fujita T, Taguchi Y, Osada K. Biomacromolecules 2000;1:149–151.
717. Tonami H, Uyama H, Kobayashi S, Kubota M. Macromol Chem Phys 1999;200:2365–2371.
718. Won K, Kim Y, An E, Lee Y, Song B. Biomacromolecules 2004;5:1–4.
719. Angerer PS, Studer A, Witholt B, Li Z. Macromolecules 2005;38:6248–6250.
720. Uyama H, Kurioka H, Kobayashi S. Polym J 1997;29:190–192.
721. Xu P, Singh A, Kaplan DL. Adv Polym Sci 2006;194:69–94.
722. Thiyagarajan M, Samuelson LA, Kumar J, Cholli AL. J Am Chem Soc 2003;125:11502–11503.
723. Cui A, Singh A, Kaplan DL. Biomacromolecules 2002;3:1353–1358.
724. Samuelson LA, Anagnostopoulos A, Alva KS, Kumar J, Tripathy SK. Macromolecules 1998;31:4376–4378.

725. Liu W, Anagnostopoulos A, Bruno FF, Senecal K, Kumar J, Tripathy S, Samuelson L. Synth Met 1999;101:738–741.

726. Reihmann M, Ritter H. Adv Polym Sci 2006;194:1–49.

727. Fukuoka T, Tachibana Y, Tonami H, Uyama H, Kobayashi S. Biomacromolecules 2002;3:768–774.

728. Meunier G, Meunier B. J Biol Chem 1985;260:10576–10582.

729. Kedderis GL, Hollenberg PF. J Biol Chem 1984;259:3663–3668.

730. Okazaki O, Guengerich FP. J Biol Chem 1993;268:1546–1552.

731. Morris DR, Hager LP. J Biol Chem 1966;241:1763–1768.

732. Butler A, Walker JV. Chem Rev 1993;93:1937–1944.

733. Hofrichter M, Ullrich R, Pecyna M, Liers C, Lundell T. Appl Microbiol Biotechnol 2010;87:871–897.

734. Littlechild J, Rodriguez EG, Isupov M. J Inorg Biochem 2009;103:617–621.

735. Hemrika W, Renirie R, Macedo-Ribeiro S, Messerschmidt A, Wever R. J Biol Chem 1999;274:23820–23827.

736. Everett RR, Soedjak HS, Butler A. J Biol Chem 1990;265:15671–15679.

737. Libby RD, Beachy TM, Phipps AK. J Biol Chem 1996;271:21820.

738. Wagenknecht HA, Woggon WD. Chem Biol 1997;4:367–372.

739. Yaipakdee P, Robertson LW. Phytochemistry 2001;57:341–347.

740. Niedan V, Pavasars I, Oberg G. Chemosphere 2000;41:779–785.

741. Liu KKC, Wong C-H. J Org Chem 1992;57:3748–3750.

742. Fu H, Kondo H, Ichikawa Y, Look GC, Wong C-H. J Org Chem 1992;57:7265–7270.

743. Kaup BA, Piantini U, Wust M, Schrader J. Appl Microbiol Biotechnol 2007;73:1087–1096.

744. Aoun S, Baboulene M. J Mol Catal B: Enzym 1998;4:101–109.

745. Franssen MCR, van der Plas HC. Bioorg Chem 1987;15:59–70.

746. Geigert J, Neidleman SL, Dalietos DJ, DeWitt SK. Appl Environ Microbiol 1983;45:366–374.

747. Wannstedt C, Rotella D, Siuda JF. Bull Environ Contam Toxicol 1990;44:282–287.

748. Speicher A. J Prakt Chem 2000;342:162–168.

749. Ortiz-Bermudez P, Srebotnik E, Hammel KE. Appl Environ Microbiol 2003;69:5015–5018.

750. Ullrich R, Nüske J, Scheibner K, Spantzel J, Hofrichter M. Appl Environ Microbiol 2004;70:4575–4581.

751. Renirie R, Pierlot C, Aubry J-M, Hartog AF, Schoemaker HE, Alsters PL, Wever R. Adv Synth Catal 2003;345:849–858.

752. Deboer E, Plat H, Tromp MGM, Wever R, Franssen MCR, Van der Plas HC, Meijer EM, Schoemaker HE. Biotechnol Bioeng 1987;30:607–610.

753. Soedjak HS, Walker JV, Butler A. Biochemistry 1995;34:12689–12696.

754. Van Schijndel JWPM, Vollenbroek EGM, Wever R. Biochim Biophys Acta 1993;1161:249–256.

755. Andersson M, Willetts A, Allenmark S. J Org Chem 1997;62:8455–8458.

756. ten Brink HB, Tuynman A, Dekker HL, Hemrika W, Izumi Y, Oshiro T, Schoemaker HE, Wever R. Inorg Chem 1998;37:6780–6784.
757. Itoh N, Hasan A, Izumi Y, Yamada H. Eur J Biochem 1988;172:477–484.
758. Carter-Franklin JN, Butler A. J Am Chem Soc 2004;126:15060–15066.
759. Tschirret-Guth RA, Butler A. J Am Chem Soc 1994;116:411–412.
760. Coughlin P, Roberts S, Rush C, Willets A. Biotechnol Lett 1993;15:907–912.
761. Hasan Z, Renirie R, Kerkman R, Ruijssenaars HJ, Hartog AF, Wever R. J Biol Chem 2006;281:9738–9744.
762. Budde CL, Beyer A, Munir IZ, Dordick JS, Khmelnitsky YL. J Mol Catal B: Enzym 2001;15:55–64.
763. Dembitsky VM. Tetrahedron 2003;59:4701–4720.
764. Lakner FJ, Hager LP. J Org Chem 1996;61:3923–3925.
765. Pasta P, Carrea G, Monzani E, Gaggero N, Colonna S. Biotechnol Bioeng 1999;62:489–493.
766. Lutz S, Vuorilehto K, Liese A. Biotechnol Bioeng 2007;98:525.
767. Kohlmann C, Lutz S. Eng Life Sci 2006;6:170–174.
768. Perez DI, Grau MM, Arends IW, Hollmann F. Chem Commun 2009;6848–6850.
769. Karmee SK, Roosen C, Kohlmann C, Lutz S, Greiner L, Leitner W. Green Chem 2009;11:1052–1055.
770. Pezzotti F, Okrasa K, Therisod M. Tetrahedron: Asymmetry 2005;16:2681–2683.
771. Pezzotti F, Therisod M. Tetrahedron: Asymmetry 2007;18:701–704.
772. Okrasa K, Falcimaigne A, Guibe-Jampel E, Therisod M. Tetrahedron: Asymmetry 2002;13:519–522.
773. Wojaczynska E, Wojaczynski J. Chem Rev 2010;110:4303–4356.
774. Kobayashi S, Nakano M, Kimura T, Schaap AP. Biochemistry 1987;26:5019–5022.
775. van Deurzen MPJ, Remkes IJ, van Rantwijk F, Sheldon RA. J Mol Catal A: Chem 1997;17:329–337.
776. Colonna S, Gaggero N, Casella L, Carrea G, Pasta P. Tetrahedron: Asymmetry 1992;3:95–106.
777. Colonna S, Gaggero N, Manfredi A, Casella L, Gullotti M, Carrea G, Pasta P. Biochemistry 1990;29:10465–10468.
778. Colonna S, Gaggero N, Manfredi A, Casella L, Gullotti M. J Chem Soc, Chem Commun 1988;1451–1452.
779. Colonna S, Gaggero N, Carrea G, Pasta P. J Chem Soc Chem Commun 1997;439–440.
780. Allenmark SG, Andersson MA. Chirality 1998;10:246–252.
781. Holland HL, Brown FM, Lozada D, Mayne BW, Szerminski R, van Vliet AJ. Can J Chem 2002;80:633–639.
782. Baciocchi E, Gerini MF, Harvey PJ, Lanzalunga O, Mancinelli S. Eur J Biochem 2000;267:2705.
783. Tuynman A, Schoemaker HE, Wever R. Chemical Monthly 2000;131:687–695.
784. Tuynman A, Vink MKS, Dekker HL, Schoemaker HE, Wever R. Eur J Biochem 1998;258:906–913.
785. Ozaki S, Ortiz de Montellano PR. J Am Chem Soc 1994;116:4487–4488.

786. Colonna S, Gaggero N, Carrea G, Pasta P. J Chem Soc, Chem Commun 1992;357–358.
787. Ozaki S, Ortiz de Montellano PR. J Am Chem Soc 1995;117:7056–7064.
788. Andersson MA, Allenmark SG. Tetrahedron 1998;54:15293–15304.
789. ten Brink HB, Schoemaker HE, Wever R. Eur J Biochem 2001;268:132–138.
790. Ricci LC, Comasseto JV, Andrade LH, Capelari M, Cass QB, Porto ALM. Enzyme Microb Technol 2005;36:937–946.
791. Pinedo-Rivilla C, Aleu J, Collado IG. J Mol Catal B: Enzym 2007;49:18–23.
792. Spreti N, Germani R, Incani A, Savelli G. Biotechnol Progr 2004;20:96–101.
793. Dzyuba SV, Klibanov AM. Biotechnol. Lett. 2003;25:1961–1965.
794. Chiappe C, Neri L, Pieraccini D. Tetrahedron Lett 2006;47:5089–5093.
795. Trevisan V, Signoretto M, Colonna S, Pironti V, Strukul G. Angew Chem Int Ed 2004;43:4097–4099.
796. de Hoog HM, Nallani M, Cornelissen J, Rowan AE, Nolte RJM, Arends IWCE. Org Biomol Chem 2009;7:4604–4610.
797. Wang W, Xu Y, Wang DIC, Li Z. J Am Chem Soc 2009;131:12892–12893.
798. Corbett MD, Chipko BR, Baden DG. Biochem J 1978;175:353–360.
799. Doerge DR, Corbett MD. Chem Res Toxicol 1991;4:556–560.
800. Yadav P, Sharma JK, Singh VK, Yadav KDS. Biocatal Biotransform 2010;28:222–226.
801. Ling KQ, Li WS, Sayre LM. J Am Chem Soc 2008;130:933–944.
802. Elmarakby SA, Duffel MW, Goswami A, Sariaslani FS, Rosazza JPN. J Med Chem 1989;32:674–679.
803. McCarthy M-B, White RE. J Biol Chem 1983;258:9153–9158.
804. Colonna S, Gaggero N, Casella L, Carrea G, Pasta P. Tetrahedron: Asymmetry 1993;4:1325–1330.
805. Allain EJ, Hager LP. J Am Chem Soc 1993;115:4415–4416.
806. Dexter AF, Hager LP. J Am Chem Soc 1995;117:817–818.
807. Rai GP, Zong Q, Hager LP. Israel J Chem 2000;40:63–70.
808. Dexter AF, Lakner FJ, Campbell RA, Hager LP. J Am Chem Soc 1995;117:6412–6413.
809. Lakner FJ, Cain KP, Hager LP. J Am Chem Soc 1997;1997;443–444.
810. Zaks A, Dodds DR. J Am Chem Soc 1995;117:10419–10424.
811. Manoj KM, Lakner FJ, Hager LP. J Mol Catal B: Enzym 2000;9:107–111.
812. Sanfilippo C, D'Antona N, Nicolosi G. Biotechnol Lett 2004;26:1815–1819.
813. Hu S, Gupta P, Prasad AK, Gross RA, Parmar VS. Tetrahedron Lett 2002;43:6763–6766.
814. Bougioukou DJ, Smonou I. J Mol Catal B: Enzym 2002;17:257–259.
815. Sanfilippo C, Nicolosi G. Tetrahedron: Asymmetry 2002;13:1889–1892.
816. Miller VP, Tschirret-Guth RA, Ortiz de Montellano PR. Arch Biochem Biophys 1995;319:333–340.
817. Morgan JA, Lu Z, Clark DS. J Mol Catal B: Enzym 2002;18:147–154.
818. Park JB, Clark DS. Biotechnol Bioeng 2006;93:1190–1195.
819. Hu S, Hager LP. J Am Chem Soc 1999;121:872–873.
820. Hu S, Hager LP. Biochem Biophys Res Commun 1998;253:544–546.
821. Ullrich R, Hofrichter M. FEBS Lett 2005;579:6247–6250.

822. Buhler DR, Mason HS. Arch Biochem Biophys 1961;92:424–437.
823. Dordick JS, Klibanov AM, Marletta MA. Biochemistry 1986;25:2946–2951.
824. Klibanov AM, Berman Z, Alberti BN. J Am Chem Soc 1981;103:6263–6264.
825. Geigert J, Dalietos DJ, Neidleman SL, Lee TD, Wadsworth J. Biochem Biophys Res Commun 1983;114:1104–1108.
826. Kiljunen E, Kanerva LT. J Mol Catal B: Enzym 2000;9:163–172.
827. van Deurzen MPJ, van Rantwijk F, Sheldon RA. J Carbohydr Chem 1997;16:299–309.
828. S. Hu, Dordick JS. J Org Chem 2002;67:314–317.
829. van Deurzen MPJ, van Rantwijk F, Sheldon RA. J Mol Catal B: Enzym 1996;2:33–42.
830. Jung D, Streb C, Hartmann M. Microporous Mesoporous Mater 2008;113:523–529.
831. Alvarez RG, Hunter IS, Suckling CJ, Thomas M, Vitinius U. Tetrahedron 2001;57:8581–8587.
832. Kren V, Kawulokova L, Sedmera P, Polasek M, Lindhorst TK, van Pee KH. Liebigs Ann. 1997;2379–2383.

CHAPTER 6

Hydrolase-Mediated Oxidations

6.1 HYDROLASE PROMISCUITY AND *IN SITU* PERACID FORMATION. PERHYDROLASES VS. HYDROLASES. OTHER PROMISCUOUS HYDROLASE-MEDIATED OXIDATIONS

Previous chapters of this book have shown that nature provides a useful and ample toolbox of redox enzymes able to catalyze many different reactions, and often with broad substrate spectrum as well. Such diversity provides a useful framework for using these biocatalysts for practical purposes, either as isolated enzymes or as whole-cells. In fact, there are already a number of industrial processes operating with redox enzymes leading to *on-spec* yields and enantioselectivities (see Chapters 7–8 for further details on industrial biocatalysis applied to redox processes).

Interestingly, based on their "biocatalytic promiscuity," other nonredox enzymes can catalyze (or mediate) oxidative processes under certain reaction conditions, leading to promising practical approaches as well. In a broad sense, the term "biocatalytic promiscuity" defines the ability of a given enzyme to catalyze several chemical transformations. Such catalytic performances may be more or less far from their natural role. Obviously, this feature greatly enhances the practical possibilities that biocatalysis may offer, since an ample number of nonnatural reactions can be efficiently tackled by means of one biocatalyst. Furthermore, since molecular basis for such biocatalytic promiscuity are currently being elucidated for many enzymatic systems, it is expected that a more rational genetic evolution of enzymes will be made in the coming years, thus developing novel biocatalysts with enhanced promiscuous activities [1–9].

The different types of biocatalytic promiscuity can be categorized in several levels, depending on the performance:

i. Use of analog substrates for the same enzyme. This is probably the most common case of biocatalytic promiscuity observed in biotransformations. In fact, many enzymes are able to accept a broad substrate spectrum, thus enhancing the practical possibilities in organic synthesis.

Redox Biocatalysis: Fundamentals and Applications, First Edition.
Daniela Gamenara, Gustavo A. Seoane, Patricia Saenz-Méndez, and Pablo Domínguez de María.

ii. Reactions derived from the imperfect control of the nonnatural reactants in the active site. Other "not expected" chemical interactions between amino acidic residues of the active site and nonnatural substrates may trigger new chemical transformations.
iii. Totally different catalytic performances, using the same amino acidic residues of the enzymatic catalytic machinery.
iv. New reactions derived from different amino acidic residues within the active site.

Within enzymes, hydrolases are probably the most important group of catalysts when biocatalytic applications are envisaged. Although the natural role of hydrolases is the cleavage of ester and amide bonds of oils, short chain esters, or proteins, this group of enzymes often displays an outstanding capability for catalyzing nonnatural reactions, such as C–C bond formations (e.g., aldol and Michael additions, Henry and Knoevenagel reactions, etc.), C–heteroatom and heteroatom–heteroatom bond formations, or oxidative processes, to cite some relevant examples [3–6,9–11]. Since hydrolases are very stable enzymes, able to catalyze processes in a wide number of nonconventional media, with broad substrate spectra and no need of external cofactors, such an extra capability of catalyzing different reactions is clearly an asset for practical uses. Moreover, given that the development of green and efficient oxidative protocols are presently considered as one of the key-areas for green chemistry developments in pharmaceutical industries (see Chapters 3–5) [12], the use of hydrolases for promiscuous oxidative-mediated processes represents a promising synergy that may be explored, and it is certainly being so, for practical problem-solving strategies. This chapter will deal with this topic in detail, ranging from the rational basis of such oxidative performances mediated by hydrolases, to applications in several important fields.

One of these promiscuous activities is the ability of certain hydrolases to form peracids using carboxylic acids—or analogous activated substrates, for example, esters—as acyl donors, and H_2O_2 as nucleophile (Figure 6.1) [13–15].

Actually, this is an interesting reaction for practical purposes, since *in situ* produced peracids are very useful for further oxidative processes. The often-imposed limitations in peracid storage and transportation, due to safety reasons, makes that approach very promising for many applications, since only the needed low peracid concentrations would be formed *in situ* in the actual place of the application (see examples reported in Sections 6.2–6.4).

The molecular basis for perhydrolytic activity of some hydrolases has been recently studied by several research groups, providing some interesting rationale hints behind this promiscuous activity. The starting point was the discovery that nonheme haloperoxidases, enzymes also able to use H_2O_2 as substrate, harbor a catalytic triad, Ser–Asp–His in the active site, analogous to that reported for the rest of serine-hydrolases [5,16–23]. Likewise, several groups reported that not only some hydrolases were able to catalyze perhydrolysis (and even act as nonheme haloperoxidases), but also that some nonheme haloperoxidases were able to catalyze promiscuous

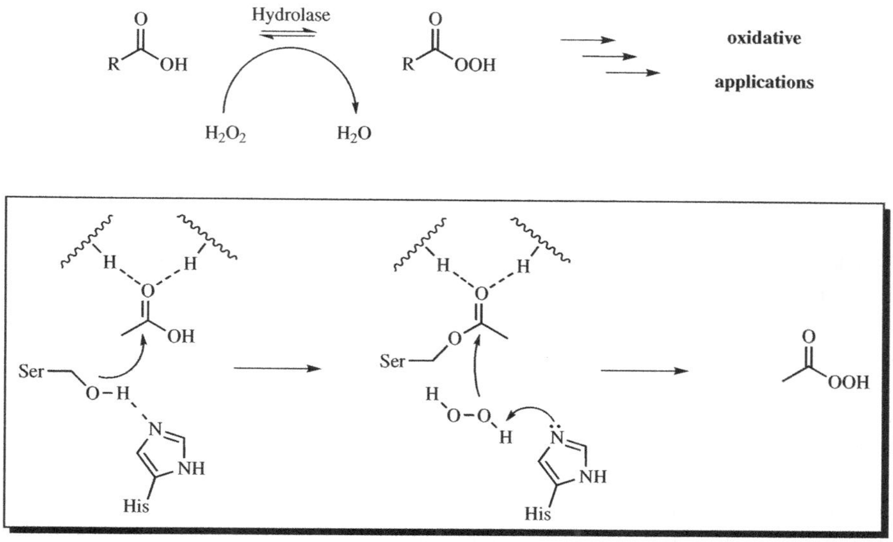

FIGURE 6.1 Hydrolase-catalyzed formation of peracids using H_2O_2 as nucleophile and carboxylic acids as acyl donors, and postulated mechanism [13–15].

hydrolytic reactions to a (more or less) low extent [5,24–27]. It was concluded that both groups share the same catalytic mechanism—enzymes *just* enable the formation of peracid in a first step; in a subsequent step, the peracid is responsible for the oxidation in strictly chemical fashion. This fact is also supported by the complete absence of enantioselectivity that is observed in these oxidative steps, clearly suggesting that no chiral microenvironments (e.g., an active site) are actually involved in the oxidation [5,16–27]. Thus, this biocatalytic approach is often regarded as an "enzyme-mediated" oxidation, since hydrolases are not actually oxidizing the substrate, and not as an "enzyme-catalyzed" oxidation.

With regard to the intrinsic mechanism of (per)hydrolases, several recent studies have shed some light on the chemical causes of the reaction [27–30]. Experiments performed using mutants of subtilisin Carlsberg, as well as of *Pseudomonas fluorescens* esterase, have shown that a single amino acidic variation within the active site is enough to enhance the perhydrolytic activity, compared to the hydrolytic one. Such single mutation allows the formation of a hydrogen bond between hydrogen peroxide and enzyme in the transition state, which stabilizes the intermediate and thus triggers the formation of peroxycarboxylic acids. In the case of *P. fluorescens* esterase, the single substitution in the active site, Leu29Pro, increased the specific constant of the enzyme for peracetic acid (PAA) formation more than 100-fold (Figure 6.2) [27–29].

In this area, another interesting hydrolase-mediated reaction is the direct epoxidation of α,β-unsaturated carbonylic compounds with H_2O_2 as nucleophile, catalyzed by *Candida antarctica* lipase B (CAL-B). The serine residue present in the active site of all hydrolases is not involved in the catalysis, since mutants lacking that amino

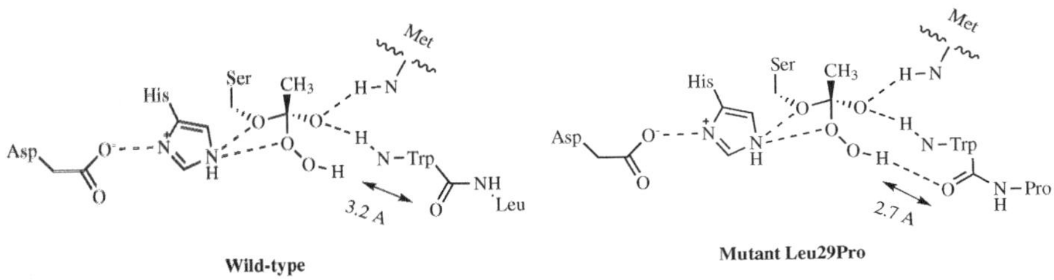

FIGURE 6.2 Proposed molecular basis of perhydrolase activity in *P. fluorescens* esterase, pinpointing the importance of one single mutation in the distance for enabling a hydrogen bond to stabilize the hydrogen peroxide in the active site [27–29].

acid are also active for such epoxidation. In this case, the promiscuous reaction proceeds with the participation of His224 as a base, together with the activation of the carbonylic group within the active site of the enzyme (Figure 6.3) [31].

Other lipase-based promiscuous activities for alcohol oxidations through a synergy of the enzyme and ionic liquids have also been recently reported [32]. In fact, combination of nonnatural enzymatic performances together with medium engineering may also open new research avenues for practical biocatalysis. In this regard, ionic liquids may offer many diverse possibilities to adapt applications, due to the well-known versatility of ionic liquids by choosing the proper cations and anions.

6.2 HYDROLASE-MEDIATED *BULK* OXIDATIONS IN AQUEOUS MEDIA (E.G., BLEACHING, DISINFECTION, ETC.)

The use of peracids in aqueous media provides interesting oxidative applications. For instance, PAA has been studied in sanitization of surfaces, medical-instrument or food-processing equipment sterilization, laundry care applications like disinfection, bleaching, destaining, deodorizing, and so on. In many of these applications, organic peracids are more effective than hydrogen peroxide, thus needing lower concentrations for a proper oxidative performance. Moreover, PAA decomposes into nontoxic

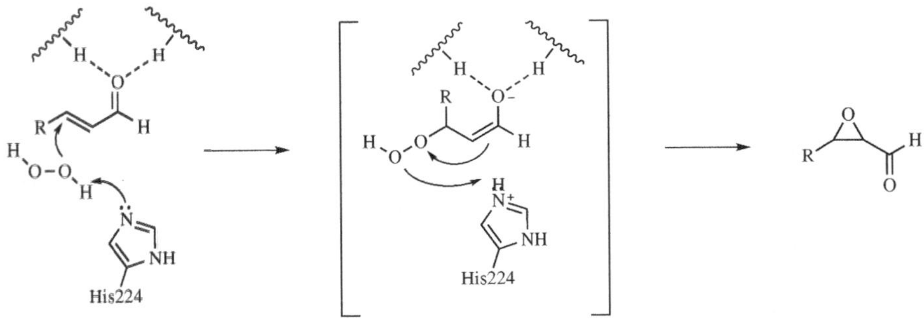

FIGURE 6.3 Proposed mechanism for the direct epoxidation of α,β-unsaturated carbonylic compounds catalyzed by CAL-B. Numbering (His224) is for CAL-B [31].

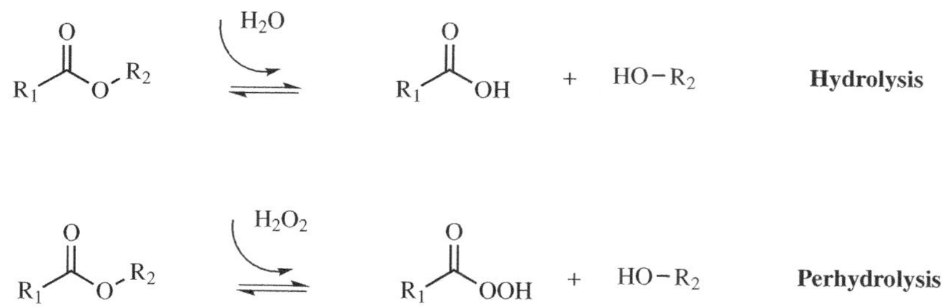

FIGURE 6.4 Nucleophilic competition between H_2O/H_2O_2 when hydrolase-mediated oxidations are intended [5,36].

oxygen, water, and acetic acid. In virtue of these promising properties, PAA has been approved by the US Environmental Protection Agency (US EPA) as a pesticide and by the Food and Drug Administration (FDA) for direct food contact and food contact surfaces, or for disinfecting medical devices and supplies [33]. Yet, as stated before, peracids have problems of storage, corrosiveness, stability, and so on. Their syntheses usually involve acid catalysts (typically H_2SO_4) at high temperatures, which also do not improve the ecological footprint, or the safety handling of the overall process. Thus, enzymatic *in situ* production of peracids may represent a promising alternative to make use of their inherent advantages, but without compromising the other related aspects of safety and ecology. In this respect, the use of hydrolases (lipases, proteases) for these aqueous, oxidative, *bulk* applications (e.g., laundry or bleaching) was assessed some decades ago [5,15,34–38].

Although there are some examples of hydrolase-catalyzed peracid formation for these *bulk* applications that are performed in nonaqueous solvents [39], the usual approach is conducted in aqueous environments (the actual place for the oxidative action of the peracid). Hence, by considering the mechanism of hydrolases, a nucleophilic competition between the H_2O/H_2O_2 will be created for the acyl–enzyme complex, leading either to ester hydrolysis or to perhydrolysis (Figure 6.4) [5,36].

To provide efficient catalysts for the *in situ* peracid formation, the ratio perhydrolysis-rate to hydrolysis-rate (P/H) in the reaction mixture is an important parameter to be considered. Lipases often display low P/H values (<0.25) in aqueous conditions, and therefore they do not meet the *on-spec* standards for an industrial application in oxidative *bulk* uses. Pioneering works in the field reported, some decades ago, on the genetic improvement of *Pseudomonas mendocina* lipase, starting from a P/H value of 0.25 (actually a rather good value for lipases) and reaching P/H of 1 by two single mutations close to the active site [5,36]. These old data can be intuitively linked to recent studies (Figure 6.2), where molecular basis for perhydrolase activity were determined [27–30].

In recent years, the field has experienced a tremendous development, derived from the discovery of several perhydrolases with high P/H values (wild-type values of ~ 1 or higher) that furthermore could be enhanced by directed evolution means. In this respect, one outstanding case is the perhydrolase from *Mycobacterium smegmatis* (or *M. smegmatis* with acyltransferase activity, MsAcT), for which values of

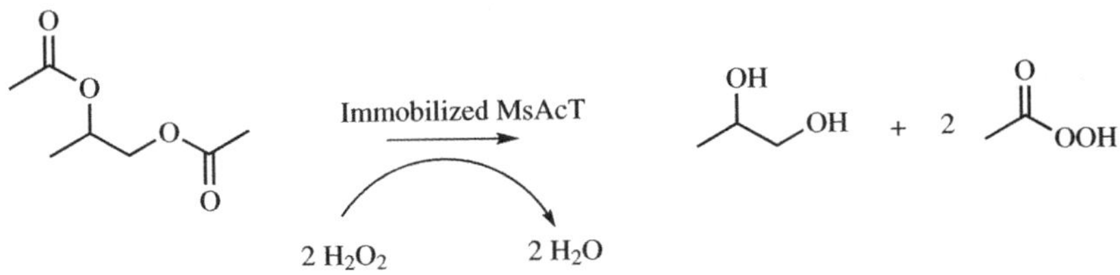

FIGURE 6.5 Decontamination strategy by *in situ* production of peracetic acid with immobilized perhydrolase [33,53].

P/H > 5 could be achieved [40]. In fact, this is a very interesting hydrolase, also able to catalyze esterification reactions in aqueous conditions [41–44]. This can be achieved by the singular octameric structure of the protein, which creates a great restriction in the shape of the active site, thus limiting solvent accessibility and therefore enhancing the hydrophobicity, allowing synthetic reactions in bulk aqueous conditions [41–44]. Actually, this enzyme has shown useful applications in the *in situ* oxidative discoloration of dyes [45], decontamination [46], tooth whitening (e.g., using a buffer system with 50 mM H_2O_2, 25 mM of propylene glycol diacetate and approximately 4 ppm of enzyme) [47], bleaching, textile color modification, and disinfection of surfaces [48–52]. In this area, a recent approach was described by immobilizing MsAcT in carbon nanotubes, and further incorporating the immobilized derivatives into polymers and latex-based paints. These bioactive composites were able to generate 11 mM of PAA in 20 min to kill > 99% spores charged. Propylene glycol diacetate was used as the acyl donor (Figure 6.5), and derivatives were used during several days and in several cycles, thus providing the proof-of-concept of a rather promising strategy for decontamination, since through immobilization enzyme costs may be efficiently tackled by means of an adequate recovery and reusability [33,53].

In the same area, other promising enzymes with high perhydrolytic activities in aqueous media are the group of the cephalosporin acetyl hydrolases, and the group of carbohydrate esterases (e.g., acetyl xylan esterases). Herein, from P/H values of ~ 3 in wild-type enzymes, which are already excellent values as such, especially when compared with lipases, enhancements up to P/H values of 6.5–6.6 have been reported. Applications of formed peracids are analogous to the above-mentioned examples [54–63]. Other recent aqueous, *bulk* applications of peracids make use of (mixtures of) different hydrolases [64,65], for instance, use of different proteases in the prevention and control of biofilms [66].

Subtilisin Carlsberg has also been matter of study for perhydrolytic reactions in *bulk* aqueous media. Whereas wild-type enzyme does not display any significant activity for perhydrolysis, some mutants are actually active. Higher activities were observed in those mutants in which the formation of the hydrogen bond between the hydrogen peroxide and the enzyme was enabled, in alignment with results observed for other (per)hydrolases (Figure 6.2) [5,27,30]. Herein, a double mutant of subtilisin Carlsberg, Thr59Ala/Leu217Trp was reported to have an increased perhydrolytic

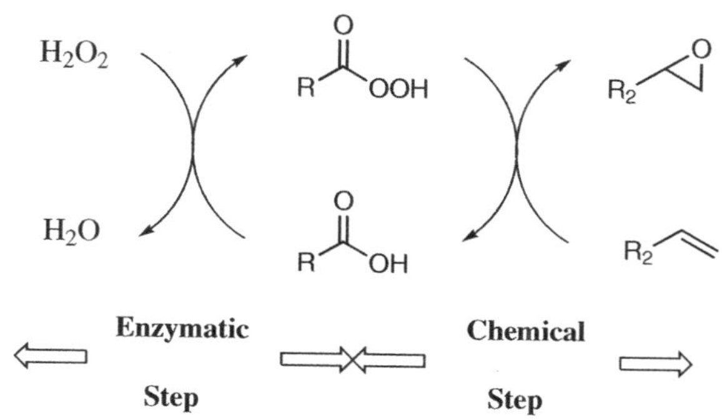

FIGURE 6.6 Proof-of-concept for the lipase-mediated epoxidation of alkenes, using hydrogen peroxide as promiscuous nucleophile [13–15].

activity [30,67]. Different aqueous *bulk* applications of these mutants have been reported, analogous to those described above for the other enzymes (e.g., bleaching, disinfection, etc.) [67–69].

6.3 LIPASE-MEDIATED OXIDATIONS: PRILESHAJEV EPOXIDATIONS AND BAEYER–VILLIGER REACTIONS

One asset of hydrolases is their ability for catalyzing reactions in organic media, often in an efficient, selective, and environment friendly fashion. Since water is not present, other nucleophiles (e.g., alcohols, amines, oximes, etc.) can be used to perform enantioselective reactions, as reported in the seminal work of Zacks and Klibanov in the 1980s [70]. Triggered by these findings, some years later some research papers reported the proof-of-concept on the lipase-mediated epoxidation of alkenes catalyzed by the *in situ* formed peracid (the so-called Prileshajev epoxidation) (Figure 6.6) [13–15].

Immobilized CAL-B displays often the best results, both in toluene, or in water:hexane biphasic system. Other nonconventional solvents like ionic liquids have been reported for this reaction as well [71,72]. Several aliphatic carboxylic acids (from octanoic to hexadecanoic acid) were successfully tested in these seminal works, affording epoxidation of a number of alkenes, such as terminal nonbranched aliphatic and cyclic alkenes. Further studies revealed a broad acceptance of carboxylic acids as acyl donors for the epoxidation reactions [73]. Likewise, better results were obtained when hydrogen peroxide was added stepwise, to avoid enzyme deactivation. In this respect, enzyme immobilization usually leads to much higher stability for the process [5,33,74–77]. Another relevant aspect of these reaction is that carboxylic acids may be added in catalytic amounts, since the peracid concentration will be always low, with the subsequent formation of carboxylic acid again (after epoxidation) (Figure 6.6). Several kinetic studies of the reactions have been reported [78–81], from which a ping–pong Bi–Bi mechanism appears to fit well with observed experimental data [79].

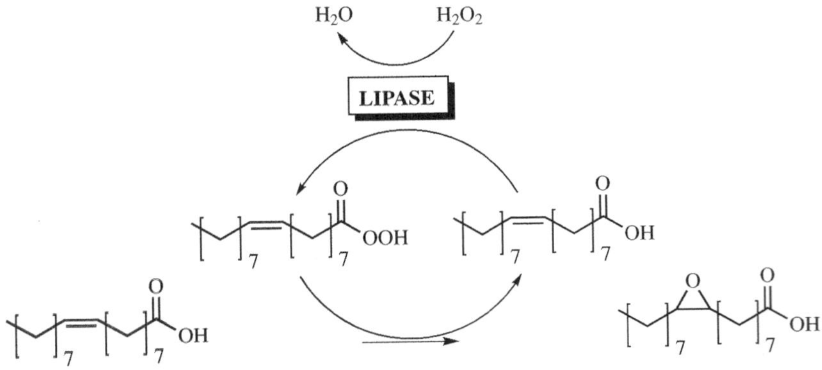

FIGURE 6.7 Schematic self-epoxidation of fatty acids catalyzed by lipases, taking oleic acid as model substrate [82–87].

The herein discussed biocatalytic-mediated epoxidation has been studied in a wide number of practical applications, reported by many research groups [5,9,10]. One relevant area is the self-epoxidation of fatty acids, to yield the corresponding epoxidized derivatives. Fatty acids act as acyl donors for the lipase, and at the same time are self-epoxidized by the *in situ* formed fatty peracids (Figure 6.7) [82–87].

Reactions are often performed either in toluene [88,89] or in other organic solvents [90]. Furthermore, solvent-free approaches have recently been introduced as well [82–84]. In one of these examples [82], a simultaneous esterification and epoxidation of oleic acid was shown, affording alkylepoxyestearates. Such compounds are relevant building blocks for a number of industrial applications as plasticizer for polyvinyl chloride (PVC), as well as other plastic materials. This interesting concept has also been extended to unsaturated plant oils (e.g., soybean oils), from which analogous useful raw materials can be produced. In this case, epoxidation can occur both by the addition of catalytic amounts of a carboxylic acid, as acyl donor for the lipase, or directly with some transesterification from the oil to afford peracids (Figure 6.8) [81,91–94].

In some cases, the technology was further used for the production of novel derivatives (e.g., polyesters, or lignin-derived materials) entirely made from renewable resources [93,94]. This is certainly an area of research that will develop in the coming decades.

The above-described technologies operate under very mild reaction conditions. Apart from obvious advantages (e.g., less by-product formation, energy savings, etc.), these mild approaches also allow the set-up of one-pot-multistep reactions that may lead to versatile platforms, yielding one or another product depending on the reaction conditions. As an outstanding example of this, lipase-mediated processes for the deprotection and modification of unsaturated trimethylsilyl ethers were reported. Depending on the reaction conditions, and specifically depending on the acyl donor used (ethyl acetate (EtOAc), diethyl carbonate, or fatty acids), several outcomes were achieved (Figure 6.9) [95].

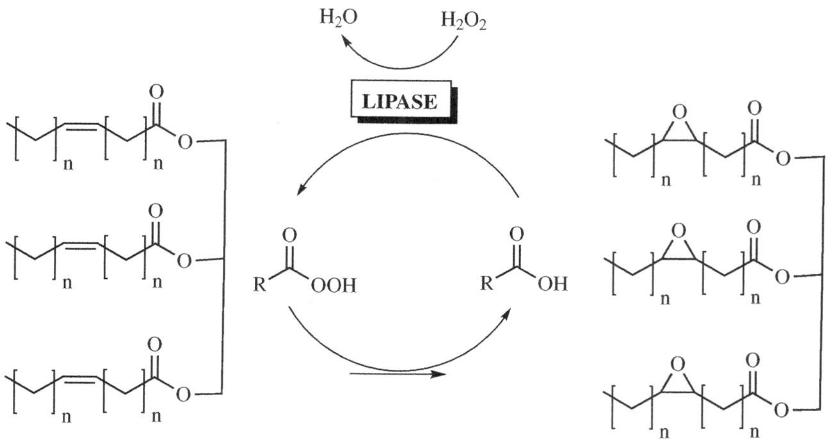

FIGURE 6.8 Lipase-mediated epoxidation of natural oils [81,91–94].

In this elegant and useful example, EtOAc forms PAA, catalyzed by the lipase, thus epoxidizing and yielding acetic acid, which is subsequently used by the enzyme to esterify it with the (deprotected) fatty alcohol, since acetic acid is strong enough for the deprotection of the trialkylsilyl group. On the other hand, fatty acids are not strong enough to deprotect such group, and therefore only epoxidations of protected alcohols are achieved when fatty acids are used as acyl donors. Finally, diethyl carbonate yields rather unstable acids after peracid reduction *via* epoxidation. Thus, after epoxidation

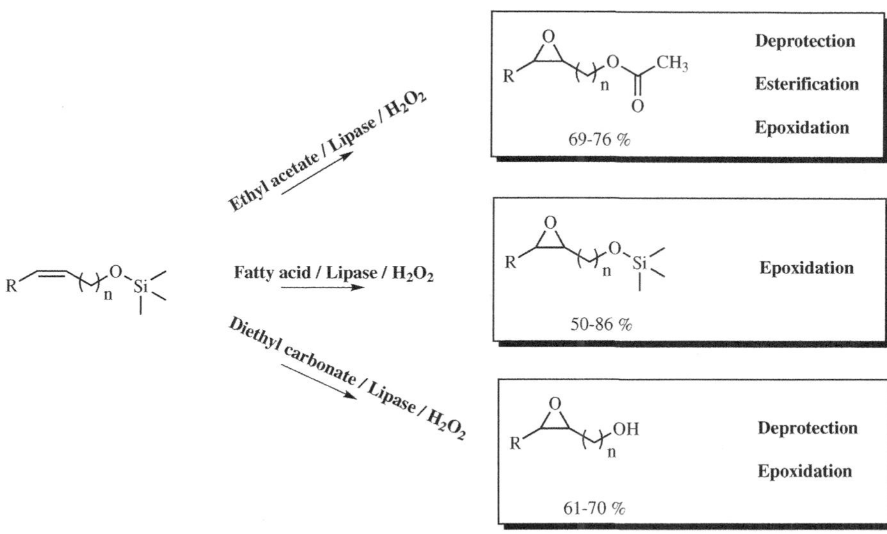

FIGURE 6.9 Reaction pathways for the lipase-catalyzed (and mediated) conversions of unsaturated trimethylsilyl ethers. Variable outcomes are achieved depending on the acyl donor employed [95].

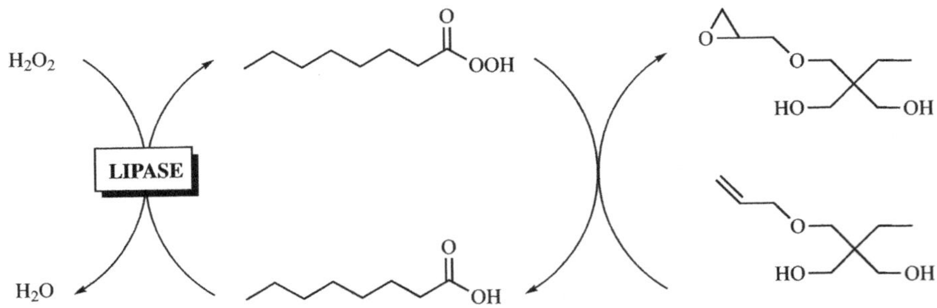

FIGURE 6.10 Production of glycidyl ethers by chemo-enzymatic epoxidation of alkene precursors [98].

and deprotection, esterifications cannot proceed further (see also Figure 6.11 for dimethyl carbonate-based processes). Thus a selective epoxidation and deprotection to afford epoxidized fatty alcohols is achieved with this latter approach [95].

Analogous epoxidations of unsaturated polymers have been reported for polybutadienes and other synthetic polymers, once again catalyzed by CAL-B under very mild reaction conditions. Interestingly, some regioselectivities in the alkene epoxidation were reported [96,97]. Likewise, the production of glycidyl ethers by chemoenzymatic epoxidation of their allyl precursors was recently reported. These compounds are very useful for many industrial applications. A recent example shows how, by applying the enzymatic approach, the use of carcinogenic epichlorohydrin could be circumvented. Caprylic acid was used as acyl donor, and a full optimization of reaction parameters (temperature, amount of hydrogen peroxide, way of dosing, etc.) was conducted (Figure 6.10) [98].

Likewise, the use of alkyl-carbonates (e.g., diethyl- or dimethylcarbonate) as solvents and acyl donors for the lipase-mediated epoxidations has been studied [95,99]. The formed carbonic acid monoalkyl ester is unstable, and therefore decomposes leading to CO_2 and the corresponding alkanol (either methanol or ethanol). Thus, an "acid-free" epoxidation can be achieved. This can be useful to control the reaction outcome (see Figure 6.9) [95], or to perform epoxidations in solvent-free conditions (Figure 6.11) [99]. Other approaches using lactones as acyl-donors and solvents have also been described, proving to be useful for the epoxidation of an ample number of alkenes (e.g., styrene-based substrates) [100].

The direct use of aqueous hydrogen peroxide for lipase-mediated oxidative reactions is the common way of adding the nucleophile, especially in aqueous applications, since hydrogen peroxide can safely be handled in diluted form (see Section 6.1) [5,9,10]. For applications in nonconventional media, the addition of water may represent a drawback that needs to be addressed. Thus, recently a new approach using urea–hydrogen peroxide complex (UHP), a safer oxidant compared to the direct use of hydrogen peroxide, was introduced [101,102]. UHP is an anhydrous form of hydrogen peroxide, and has the potential of releasing hydrogen peroxide slowly to the reaction mixture. Thus, the reaction proceeds with fewer amounts of water, as well as

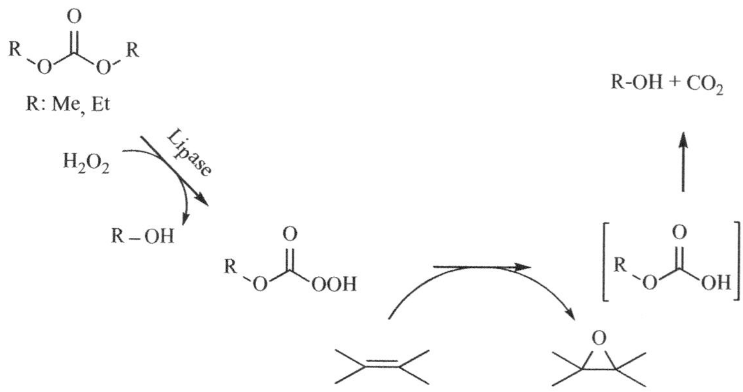

FIGURE 6.11 "Acid-free" approach for the lipase-mediated epoxidation, using dialkyl carbonates as solvents and acyl donors [95,99].

of hydrogen peroxide, leading to higher enzyme stability. The strategy was successfully applied for the synthesis of *N*-alkyloxaziridines by oxidizing *N*-alkylamines under very mild reaction conditions (Figure 6.12) [102].

As stated previously, the lipase-mediated epoxidation proceeds out of the active site of the enzyme. Hydrolases catalyze only the formation of the peracid, which is the actual oxidant. Thus, produced epoxides are always racemic. However, several works have reported either the maintenance of the optical purity during epoxidation [103], or even the induction of enantioselectivity by using chiral acids as acyl donors. In the latter, chiral peracids were formed, enabling the formation of asymmetric epoxides by transmitting the chiral information during the epoxidation step. Enantiomeric excesses were moderate (40–80%), by using an immobilized *Pseudomonas* G6 lipase, and a variety of structurally different carboxylic acids (Figure 6.13) [104–106].

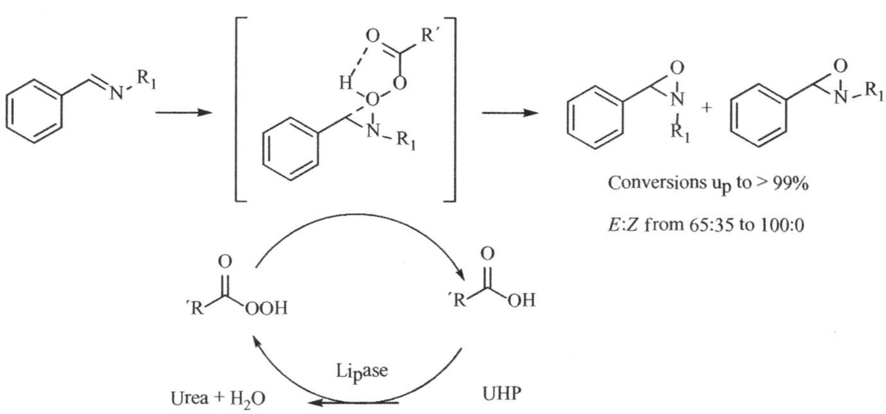

FIGURE 6.12 Lipase-mediated oxidation of *N*-alkylamines to afford *N*-alkyloxaziridines using UHP complex as oxidant [102].

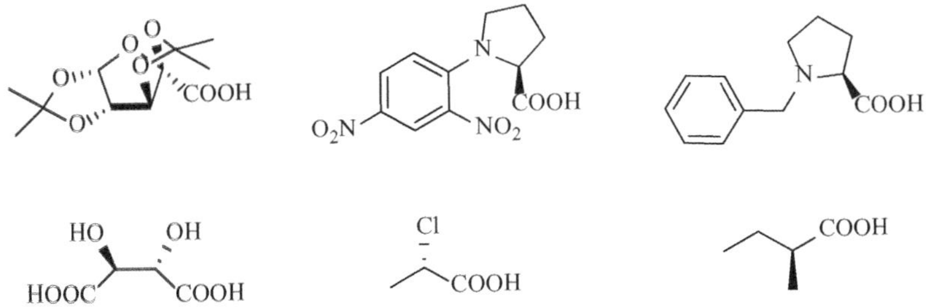

FIGURE 6.13 Chiral carboxylic acids employed in the enantioselective lipase-mediated epoxidation of alkenes [104–106].

Another approach to generate value out of the racemic epoxides produced by the herein described lipase-mediated approaches would be the subsequent enantioselective epoxide opening, to afford chiral diols. In this respect, the one-pot combination of CAL-B-mediated epoxidation with whole cells of *Sphingomonas* sp. HXN-200 containing epoxide hydrolases represents a promising approach for efficient enantioselective syntheses with environment-friendly biocatalysts and diminished waste production (one-pot process). In several examples, high yields and enantioselectivities were reported (Figure 6.14) [107].

It is worth mentioning that some economic studies based on reaction and reactor design have been performed by several authors [74,75,108–110]. To assure an economic attractiveness to these approaches, enzyme stability must still be enhanced. In fact, in the production of (racemic) styrene oxide by means of CAL-B, approximately 90–95% of associated costs to the process are related to the immobilized enzyme [108]. Considering the improvements observed in the understanding of perhydrolytic performance (see Sections 6.1 and 6.2), this challenge might be surely tackled in coming years, by developing more active enzymes that may be produced at lower costs.

Apart from Prileshajev epoxidations, as well as some other oxidative processes [111], peracids, formed *in situ* by lipases, can also perform the Baeyer–Villiger

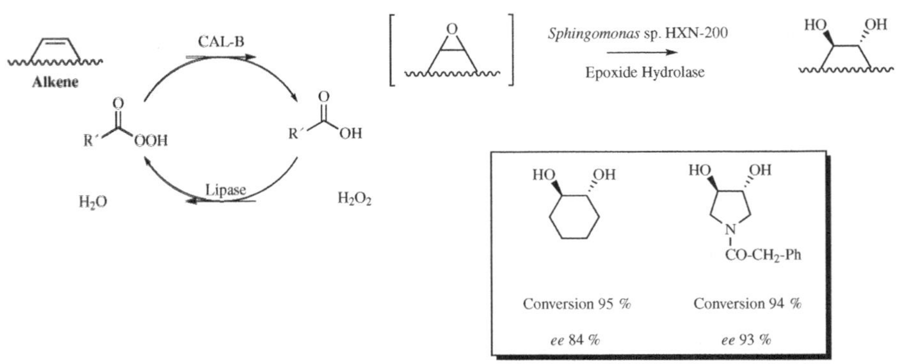

FIGURE 6.14 One-pot approach involving two enzymes to afford chiral diols from alkenes [107].

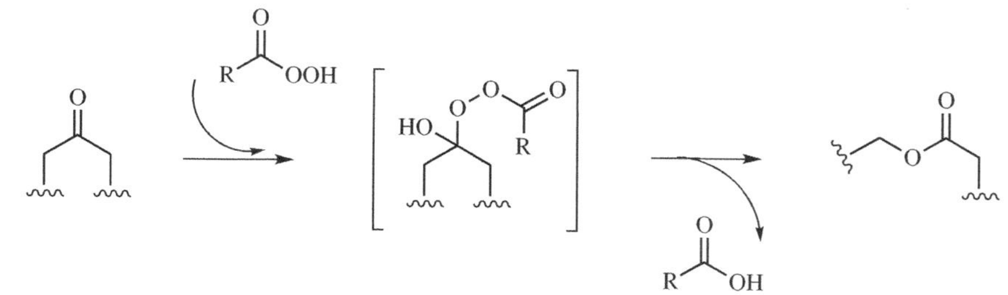

FIGURE 6.15 Baeyer–Villiger reaction with peracids.

oxidation of ketones to afford esters (Figure 6.15), albeit often at low reaction rates (due to the low reactivity of the peracid at room temperature to conduct the oxidation of ketones) [112–114].

Despite Baeyer–Villiger reaction proceeds outside of the active site of the enzyme—in the same fashion as for epoxidations—in some cases some chiral induction was observed in these lipase-mediated reactions (Figure 6.16) [112]. It has been suggested that lipase forms the peracid more rapidly with one enantiomer of the racemic keto acid, which performs intramolecular Baeyer–Villiger reaction, and thus some enantioenrichment can be observed.

In order to enhance the peracid-mediated Baeyer–Villiger oxidation, the incorporation of solvents like hydrogen-bond-donating ionic liquids has been proposed [115]. The possibility of designing ionic liquids that might be compatible with CAL-B, and at the same time could accelerate the second step (peracid oxidation), is a further example of the broad possibilities that these neoteric solvents may bring to chemistry in general, and to biocatalysis in particular [72]. High yields were observed in some cases for the different Baeyer–Villiger reactions (Table 6.1).

6.4 HYDROLASE-MEDIATED OXIDATION AND PROCESSING OF LIGNOCELLULOSIC MATERIALS

There is currently an enormous interest in processing lignocellulosic biomass to afford the raw materials, fuels, and commodities that will be industrially used in the future, both for energy and chemical processing [116]. Yet, lignocellulose has a recalcitrant and complex structure—mainly composed by lignin, hemicellulose, and

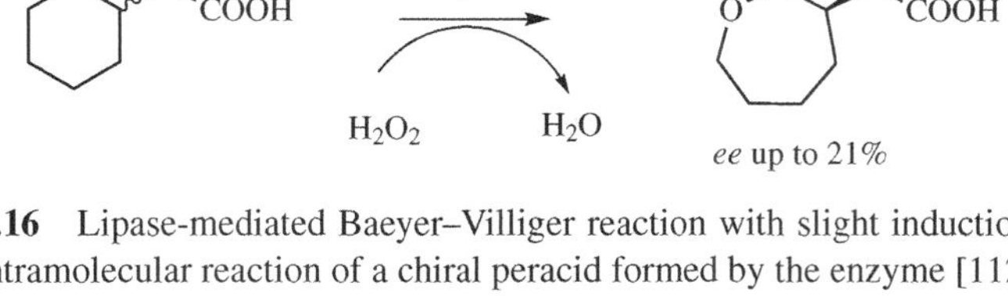

FIGURE 6.16 Lipase-mediated Baeyer–Villiger reaction with slight induction of chirality due to the intramolecular reaction of a chiral peracid formed by the enzyme [112].

TABLE 6.1 Lipase-Mediated Baeyer–Villiger Reactions Performed in Ionic Liquids, in 5 h Reaction Time [115]

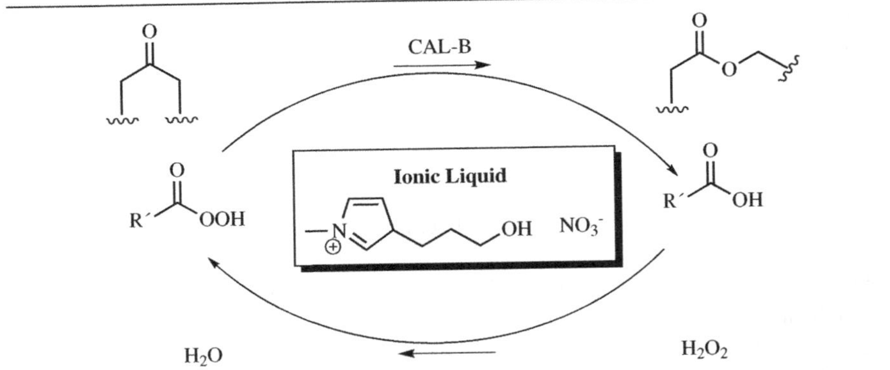

Substrate	Product	Yield (%)
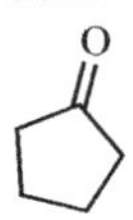	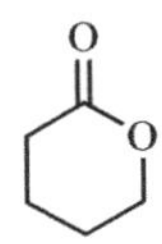	99
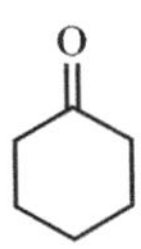	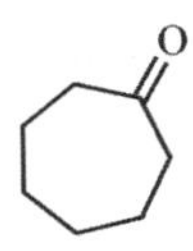	45
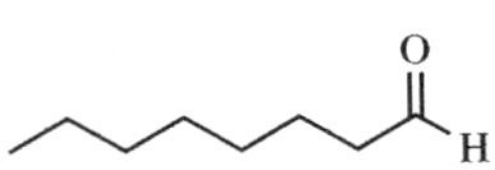	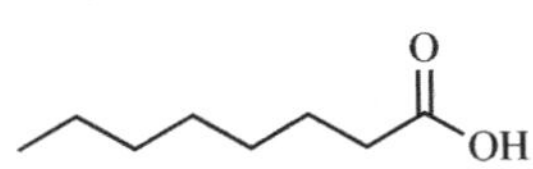	98
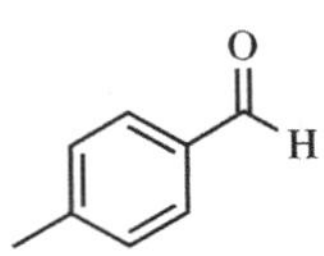	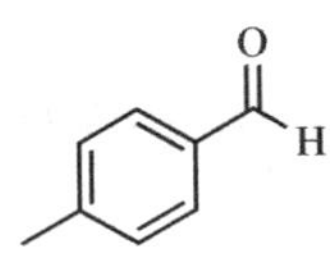	14

cellulose—and the provision of efficient and sustainable technologies for its processing is not trivial at all. In this area, one important bottleneck is the production of sugars from polysaccharides, that may be further fermented by microorganisms to produce a wide number of industrially sound products [116–118]. To this end, polysaccharides from wood must be accessible to enzymes (e.g., cellulases), to provide an attractive hydrolytic rate for industrial purposes. Among the number of methods proposed for such aim, several groups have studied the use of PAA to *in situ* degrade lignin, thus enhancing the subsequent enzymatic activity toward polysaccharides. PAA oxidizes the hydroxyl groups in lignin side chains to carbonyl groups, thus cleaving β-aryl bonds, reducing lignin molecular weight. Likewise, PAA hydroxylates the phenolic rings in lignin to form hydroquinones. Such removal of lignin increases the accessibility

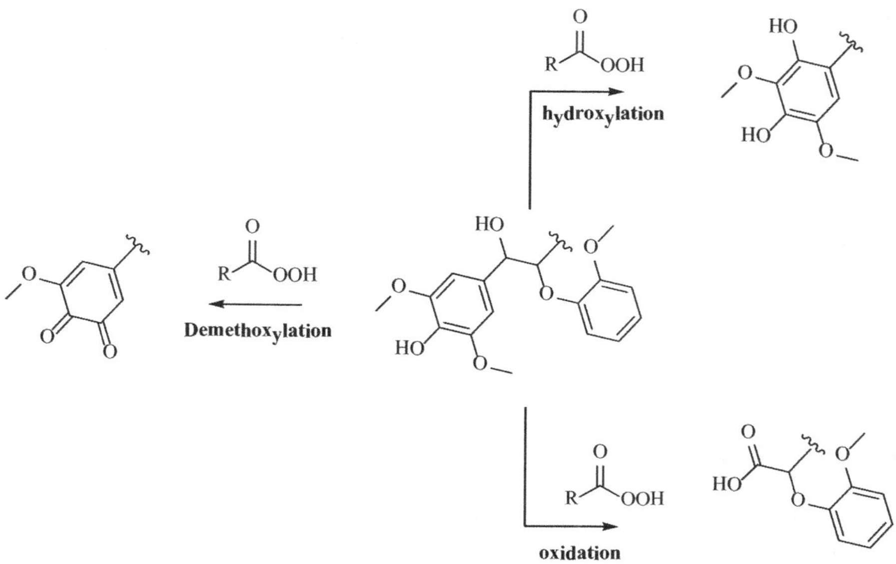

FIGURE 6.17 Some proposed pathways for the peracid-based oxidation and degradation of the lignin structure [119].

of hemicellulose and cellulose for subsequent enzymatic hydrolysis (Figure 6.17) [119–121].

Yet, as stated in Section 6.2 (aqueous *bulk* applications), PAA cannot be transported, nor stored due to safety and corrosion issues. Thus, it must be generated *in situ* and obviously under mild reaction conditions that may not degrade lignocellulosic

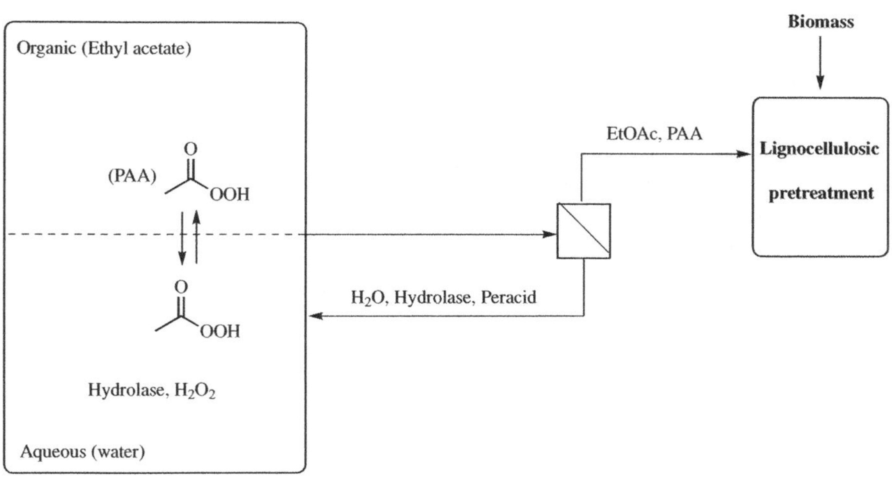

FIGURE 6.18 Conceptual approach for the hydrolase-mediated delignification via *in situ* formation of peracetic acid.

components at the same time. Based on the developments reported in this chapter, an alternative would be the enzymatic generation of PAA. Recent studies [119,120] focus on this aspect by using the native and mutants of *P. fluorescens* esterase [27–29], with promising productions of PAA in aqueous media, by using EtOAc as acyl donor. Under these conditions, significant improvements in cellulose hydrolysis by cellulases were observed, using aspen wood as model substrate, and after several cycles of peracid formation and processing (Figure 6.18).

Furthermore, treatment of lignocellulose with PAA would eventually sterilize the raw material, thus preparing them for the subsequent fermentative steps. Although costs and operational windows of this approach are still far from being a practical robust application, it is obvious that the development of this strategy may represent a useful research line for the coming decades.

6.5 CONCLUDING REMARKS

This chapter has addressed a topic that, in a broad sense, provides practical and academic insights. Starting some decades ago from the discovery of a promiscuous activity in lipases, the peracid formation using hydrogen peroxide as nucleophile, the field has rapidly developed by the discovery of perhydrolases, as well as the identification of some amino acidic motifs that enable a higher peracid formation. From these hints, this chapter has focused on application in aqueous *bulk* media, as well as in synthetic processes in different organic, nonconventional media to perform epoxidations or Baeyer–Villiger reactions, using different acyl donors and reaction set-ups. Finally, the first proof-of-concept on using the strategy for biomass processing has also been addressed. As topics for the future, the stability of enzyme in the usually harsh peracidic media needs to be addressed, together with reduction of costs in enzyme formation, and/or immobilization and recovery of the catalysts.

REFERENCES

1. O'Brien PJ, Herschlag D. Chem Biol 1999;6:R91–R105.
2. Copley SD. Curr Opin Chem Biol 2003;7:265–272.
3. Bornscheuer UT, Kazlauskas RJ. Angew Chem Int Ed 2004;43:6032–6040.
4. Bugg TDH. Bioorg Chem 2004;32:367–375.
5. Carboni-Oerlemans C, Domínguez de María P, Tuin B, Bargeman G, van der Meer A, van Gemert R. J Biotechnol 2006;126:140–151.
6. Hult K, Berglung P. Trends Biotechnol 2007;25:231–238.
7. Nobeli I, Favia AD, Thornton JM. Nat Biotechnol 2009;27:157–167.
8. Khersonsky O, Tawfik DS. Annu Rev Biochem 2010;79:471–505.
9. Busto E, Gotor-Fernández V, Gotor V. Chem Soc Rev 2010;39:4504–4523.
10. Bornscheuer UT, Kazlauskas RJ. Hydrolases. In: *Organic Chemistry*. 2nd ed. Weinheim: VCH-Wiley; 2005.

11. Hari Krishna S, Karanth NG. Catal Rev Sci Eng 2002;44:499–591.
12. Constable DJ, Dunn PJ, Hayler JD, Humphrey GR, Leazer JL, Lindemann RJ, Lorenz K, Manley J, Pearlman BA, Wells A, Zaks A, Zhang TY. Green Chem 2007;9:411–420.
13. Björkling F, Godtfredsen SE, Kirk O. J Chem Soc Chem Commun 1990;1301–1303.
14. Björkling F, Frykman H, Godtfredsen SE, Kirk O. Tetrahedron 1992;48:4587–4592.
15. Kirk O, Christensen MW, Dambus T, Godtfredsen SE. Biocatalysis 1994;11:65–77.
16. van Pee KH, Sury G, Lingens F. Biol Chem Hoppe-Seyler 1987;368:1225–1232.
17. Wiessner W, van Pee KH, Lingens F. J Mol Chem 1988;263:13725–137832.
18. Pelletier I, Altenbuchner J, Mattes R. Biochim Biophys Acta 1995;1250:149–157.
19. Pelletier I, Altenbuchner J. Microbiology 1995;141:459–468.
20. Hofmann B, Toelzer S, Pelletier I, Altenbuchner J, van Pee KH, Hecht HJ. J Mol Biol 1998;279:889–900.
21. Kataoka M, Honda K. Eur J Biochem 2000;267:3–10.
22. Picard M, Gross J, Luebbert E, Toelzer S, Krauss S, van Pee KH, Berkessel A. Angew Chem Int Ed 1997;36:1196–1199.
23. Song JK, Ahn HJ, Kim HS, Song BK. Biotechnol Lett 2006;28:849–856.
24. Hecht HJ, Sobek H, Haag T, Pfeiffer O, van Pee KH. Proc Natl Acad Sci USA 1994;1:532–537.
25. Kirk O, Conrad LS. Angew Chem Int Ed 1999;38:977–979.
26. Cheeseman JD, Tocilj A, Park S, Schrag JD, Kazlauskas RJ. Acta Crystallogr, Sect D: Biol Crystallogr 2004;D60:1237–1243.
27. Bernhardt P, Hult K, Kazlauskas RJ. Angew Chem Int Ed 2005;44:2742–2746.
28. Kazlauskas RJ, Bernhardt P, WO2006/133079. 2006.
29. Yin DL, Bernhardt P, Morley KL, Jiang Y, Cheeseman JD, Purpero V, Schrag JD, Kazlauskas RJ. Biochemistry 2010;49:1931–1942.
30. Lee W, Vojcic L, Despotovic D, Prodanovic R, Maurer KH, Schwaneberg U, Zacharias M. Theor Chem Acc 2010;125:375–386.
31. Svedendahl M, Carlqvist P, Branneby C, Allner O, Frise A, Hult K, Berglund P, Brinck T. ChemBioChem 2008;9:2443–2451.
32. Sharma UK, Sharma N, Kumar R, Kumar R, Sinha AK. Org Lett 2009;11:4846–4848.
33. Dinu CZ, Zhu G, Bale SS, Anand G, Reeder PJ, Sanford K, Whited G, Kane RS, Dordick JS. Adv Funct Mater 2010;20:392–398.
34. Wiersema RJ, Stanislowski AG, Gray GL, Poulouse AJ, Power SD. US5030240. 1991.
35. Poulouse AJ, Anderson SA. US5108457, 1992.
36. Estell DA. J Biotechnol 1993;28:25–30.
37. Poulouse AJ. US5352594, 1994.
38. Stanislowski AG, Heitfeld FA, Wiersema RJ, Poulouse AJ, US5364554, 1994.
39. gen. Klaas MR, Steffens K, Patett N. J Mol Catal B: Enzym 2002;19–20:499–505.
40. Amin N, Boston MG, Bott RR, Cervin MA, Concar EM, Gustwiller ME, Jones B, Liebeton K, Miracle GS, Oh H, Poulouse AJ, Ramer SW, Scheibel JJ, Weyler W, Whited G. WO05/056782. 2005.
41. Pettrone FA, Whited GM, Goodhue CT, US5114850. 1992.

42. Mathews I, Soltis M, Saldajeno M, Ganshaw G, Sala R, Weyler W, Cervin MA, Whited G, Bott R. Biochemistry 2007;46:8969–8979.
43. Bott RR, Cervin MA, Yao J. WO2008/019069. 2008.
44. Bott RR, Cervin MA. WO2009/002480. 2009.
45. Barnett CC, Sala RF. WO2010/101867. 2010.
46. Cervin MA, Whited G. US2009/0311395. 2009.
47. Concar EM, Poulouse AJ. WO2007/103050. 2007.
48. Harding F. WO2007/067473. 2007.
49. Concar EM, Estell D, Oh H. WO2007/106293. 2007.
50. Kumar M, Mcdonald HC, Poulouse AJ, Rand T, Wang H. WO2008/051491. 2008.
51. Barnett C. WO2008/140988. 2008.
52. Vermeersch L, Redling E, Ashton W, Barnett CC, Krouwer AJJ, Pericu P, Sala RF. WO2010/139601. 2010.
53. Dinu CZ, Dordick JS, Kane RS, Sanford K, Whited GM, Zhu G. WO2010/123534. 2010.
54. Dicosimo R, Cahill WR, Dipietro DG, Hann EC, Reynolds RA. WO2010/039961. 2010; WO2010/039956. 2010.
55. Dicosimo R, Gavagan JE, Payne MS, Cooling FB. US2010/0041752. 2010.
56. Dicosimo R, Ben-Bassat A, Cahill WR, Didietro DG, Hann EC, Payne MS, Reynolds RA, Zolandz RR. WO2010/039953. 2010.
57. Dicosimo R, Payne MS, Yin T, WO2010/039958. 2010.
58. Dicosimo R, Payne MS, Hann EC. WO2010/019544. 2010.
59. Dicosimo R, Gavagan JE, Payne MS, Cooling FB. US2009/0005590. 2009.
60. Dicosimo R, Gavagan JE, Payne MS, Cooling FB. WO2009/067279. 2009.
61. Dicosimo R, Gavagan JE, Payne MS, Cooling FB, US 2008/0176299. 2008.
62. Dicosimo R, Wagner L. WO2007/044666. 2007.
63. Dicosimo R, Payne MS, Hann EC, Croud VB, Gavagan JE, Wagner LW. WO2006/119060. 2006.
64. Righetto Z. WO2009/044098. 2009.
65. Dubreucq E, Weiss A, Moulin G. US2007/0244021. 2007.
66. Barnett C, Kumar M, Whited GM, WO2009/085743. 2009.
67. Wieland S, Polanyi-Bald L, Prueser I, Stehr R, Maurer KH. US2005/021773. 2005.
68. Stehr R, Bessler C, Maurer KH, Wieland S, Hoven N, Prüser I. WO2005/124012. 2005.
69. O'Connell T, Maurer KH, Weber T. WO2009/015951. 2009.
70. Zacks A, Klibanov AM. Proc Natl Acad Sci USA 1985;82:3192–3196.
71. MadeiraLau R, van Rantwijk F, Seddon KR, Sheldon RA. Org Lett 2000;2:4189–4191.
72. Domínguez de María P. (Ed). *Ionic Liquids in Biotransformations and Organocatalysis: Solvents and Beyond*. Hoboken, NJ: John Wiley & Sons; 2012.
73. gen. Klaas MR, Warwel S. J Mol Catal A: Chem 1997;117:311–319.
74. Cuperus FP, Bouwer ST, Kramer GFH, Derksen JTP. Biocatalysis 1994;9:89–96.
75. Kramer GFH, Bouwer ST, van Gemert RW, Derksen JTP, Cuperus FP. Catal Today 1994;22:537–547.
76. Miura Y, Yamane T. Biotechnol Lett 1997;19:611–613.

77. Tzialla AA, Pavlidis IV, Felicissimo MP, Rudolf P, Gournis D, Stamatis H. Bioresour Technol 2010;101:1587–1594.
78. Yadav GD, Devi KM. J Am Oil Chem Soc 2001;78:347–351.
79. Yadav GD, Devi KM. Biochem Eng J 2002;10:93–101.
80. Yadav GD, Borkar IV. AIChe J 2006;52:1235–1247.
81. Hilker I, Bothe D, Prüss J, Warnecke H.-J. Chem Eng Sci 2001;56:427–432.
82. Orellana-Coca C, Billakanti JM, Mattiasson B, R. Hatti-Kaul. J Mol Catal B: Enzym 2007;44:133–137.
83. Törnvall U, Orellana-Coca C, Hatti-Kaul R, Adlercreutz D. Enzyme Microb Technol 2007;40:447–451.
84. Orellana-Coca C, Törnvall U, Adlercreutz D, Mattiasson B, Hatti-Kaul R. Biocatal Biotransform 2005;23:431–437.
85. gen. Klaas MR, Warwel S. Synth Commun 1998;28:251–260.
86. Warwel S, gen.Klaas MR. J Mol Catal B: Enzym 1995;1:29–35.
87. gen. Klaas MR, Warwel S. J Am Oil Chem Soc 1996;73:1453–1457.
88. Skouridou V, Stamatis H, Kolisis FN. Biocatal Biotransform 2003;21:285–290.
89. Skouridou V, Stamatis H, Kolisis FN. J Mol Catal B: Enzym 2003;21:67–69.
90. Moreira MA, Bitencourt TB, da Graça Nascimento M. Synth Commun 2005;35:2107–2114.
91. gen. Klaas MR, Warwel S. Ind Crops Prod 1999;9:125–132.
92. Abdul Rahman MB, Udaiyappan AF, Salleh AB, Raja Abdul Rahman NZ. WO2010/098651. 2010.
93. Kudanga T, Prasetyo EN, Sipilä J, Nyanhongo GS, Guebitz GM. Process Biochem 2010;45:1557–1562.
94. Uyama H, Kuwabara M, Tsujimoto T, Kobayashi S. Biomacromolecules 2003;4:211–215.
95. gen. Klaas MR, Kunz M, Warwel S. J Mol Catal B: Enzym 1999;7:283–289.
96. Jarvie AWP, Overton N, St. Pourçain CB. Chem Commun 1998;177–178.
97. Jarvie AWP, Overton N, St. Pourçain CB. J Chem Soc Perkin Trans 1 1999;2:171–2176.
98. Tufvesson P, Adlercreutz D, Lundmark S, Manea M, Hatti-Kaul R. J Mol Catal B: Enzym 2008;54:1–6.
99. gen. Klaas MR, Warwel S. Org Lett 1999;1:1025–1026.
100. Xu Y, Bo NR, Khaw J, Li Z. Green Chem 2009;11:2047–2051.
101. Ankudey EG, Olivo HF, Peeples TL. Green Chem 2006;8:923–926.
102. Bitencourt TB, da Graça Nascimento M. Green Chem 2009;11:209–214.
103. Moreira MA, da Graça Nascimento MG. Catal Commun 2007;8:2043–2047.
104. Sarma K, Bhati N, Borthakur N, Goswami A. Tetrahedron 2007;63:8735–8741.
105. Sarma K, Borthakur N, Goswami A, Tetrahedron Lett2007;48:6776–6778.
106. Sarma K, Goswami A, Goswami BC. Tetrahedron: Asymmetry 2009;20:1295–1300.
107. Xu Y, Li A, Jia X, Li Z. Green Chem 2011;13:2452–2458.
108. Borole AP, Davison BH. Appl Biochem Biotechnol. 2007;136–140:437–449.
109. Wiles C, Hammond MJ, Watts P. Beilstein J Org Chem 2009;5:27.

110. Hangström AEV, Törnvall U, Nordblad M, Hatti-Kaul R, Woodley JM. Biotechnol Prog 2011;27:67–76.
111. de Zoete MC, van Rantwijk F, Maat L, Sheldon RA. Recl Trav Chim Pays-Bas 1993;112:462–463.
112. Lemoult SC, Richardson PF, Roberts SM, J Chem Soc Perkin Trans 1 1995;89–91.
113. Rios MY, Salazar E, Olivo HF. Green Chem 2007;9:459–462.
114. Pchelka BK, Gelo-Pujic M, Guibe-Jampel E. J Chem Soc Perkin Trans 1 1998;2625–2628.
115. Kotlewska AJ, Van Rantwijk F, Sheldon RA, Arends IWCE. Green Chem 2011;13:2154–2160.
116. Kamm B, Gruber PR, Kamm M. (Eds). *Biorefineries – Industrial Processes and Products Status Quo and Future Directions*. Weinheim: Wiley-VCH; 2010.
117. Domínguez de María P. ChemSusChem 2011;4:327–329.
118. vom Stein T, Grande PM, Kayser H, Sibilla F, Leitner W, Domínguez de María P. Green Chem 2011;13:1772–1777.
119. Brasileiro LB, Colodette JL, Pilo-Veloso D. Quim Nova 2001;24:819–829.
120. Duncan S, Jing Q, Katona A, Kazlauskas RJ, Schilling J, Tschirner U, Aldajani WW. Appl Biochem Biotechnol 2010;160:1637–1652.
121. Yin DT, Jing Q, AlDajani WW, Duncan S, Tschirner U, Schilling J, Kazlauskas RJ. Bioresour Technol 2011;102:5183–5192.

CHAPTER 7

Bridging Gaps: From Enzyme Discovery to Bioprocesses

7.1 CONTEXT

Former chapters of this book have focused on different aspects related to redox enzymes, covering some fundamentals of redox biocatalysis (see Chapters 1–2), as well as the biochemical characterization of these enzymes, together with the biocatalytic strategies that conform a wide palette of applications that these biocatalysts may offer to chemists of a broad range of disciplines (see Chapters 3–6). Once that information has been provided—state-of-the-art, what can be done with redox enzymes, and so on—Chapters 7 and 8 of the book deal with important aspects of industrial biocatalysis in general, and for redox enzymes in particular. Thus, emphasis will be put in using enzymes in such a way that economics and technical feasibility of a process can be aligned, with a clear focus on direct practical uses of these enzymes. To separate these features, in Chapter 7 relevant parameters and issues of a practical biocatalytic reaction will be discussed, whereas in Chapter 8 a number of examples of industrially applied redox biocatalysis will be described in depth. Rather than providing a comprehensive compilation of all the reported technical approaches, focus will be put on providing reasons of why biocatalysis is implemented in such a way at practical scale (and not in another one), and which would be the alternatives to assure an economic and technical success when dealing with redox enzymes.

In general, along the large development that often proceeds from identification of a novel biocatalyst until an industrial application can be finally provided, there are a number of crucial steps that must be successfully tackled and accomplished. Obviously, the first step is the identification of an enzyme that may perform a desired reaction, and preferably displaying some activity toward the desired substrate already as well (albeit this is not absolutely mandatory, as some recent examples have shown, see Chapter 8). Starting from increasing knowledge in biocatalysis (Chapters 3–6), together with the development of genome data banks, and other techniques like metagenomics, there are already a wide number of alternatives to start identifying a biocatalyst that may be potentially useful for a desired reaction. Subsequently, once

Redox Biocatalysis: Fundamentals and Applications, First Edition.
Daniela Gamenara, Gustavo A. Seoane, Patricia Saenz-Méndez, and Pablo Domínguez de María.

successful enzyme identification is realized, genetic tools allow the cloning of such enzyme and its overexpression in an external, heterologous host. This allows the production of the biocatalyst in large quantities, obviously needed when an industrial application is envisaged. Finally, a significant improvement of the enzyme (e.g., directed evolution) is often needed to deliver a biocatalyst able to compete at industrial level in terms of efficiency, kinetics, enantio- or regioselectivity, stability, and so on. In this regard, to enable the evolution of the enzyme by means of such directed evolution strategies, there is a need for high-throughput-screening methods that may enable the selective, sensitive, and straightforward discrimination of those mutants that are improving the targeted parameter, while rejecting the worse ones and avoid, to a large extent, the existence of false positives or other background responses [1–5].

Addressing all these steps and challenges in providing a biocatalyst into the industrial arena, Section 7.2 will discuss, from a chemical viewpoint, which high-throughput-screening methods are already available for the rapid and sensitive identification of novel or improved (mutants) redox enzymes, providing successful examples in each case. Subsequently, Section 7.3 will deal with a paradigmatic case, the baker's yeast redox enzymes, to show how from a wild-type complex mixture of redox enzymes present in *Saccharomyces cerevisiae*—what often leads to opposite enantioselectivities, thus invalidating them for practical applications—the separate cloning, characterization, and overexpression of each of these enzymes may enable the provision of robust and trustable biocatalysts, useful for different substrates in an exquisite manner and with reproducible results. Later on, in Section 7.4, aspects dealing with the medium engineering, and advantages and disadvantages on the use of either free enzymes or whole-cells will be briefly discussed, thus bridging the gap between having already a robust mutated (improved) biocatalyst at hand, and further steps to reach the final and desired industrial application (bioprocess). The chapter will finish with a glimpse on the vision of future (practical) biocatalysis. This will cover aspects regarding the use of multistep reactions, or domino reactions, comprising different enzymes or enzymes and chemocatalysts, to avoid waste generation during downstream processing, as well as simplifying the process from a practical viewpoint. Likewise, last sections will also discuss other approaches for making redox biocatalysis more competitive, like for instance the performance of redox biocatalysis in water-free, neat substrates.

7.2 ENZYME DIRECTED EVOLUTION AND HIGH-THROUGHPUT-SCREENING OF BIOCATALYSTS

When a certain wild-type enzyme is identified and assessed for a given practical application, namely a biocatalytic reaction, it usually appears that the catalytic properties of such wild-type enzymes are not sufficient for a practical application. With regard to these properties, this can be easily exemplified in enzymes providing low stabilities or poor solvent tolerance under real reaction conditions, low selectivities, or activities for a certain substrate, insufficient thermostability, among other common issues [6–9]. Therefore, it is necessary to improve such wild-type enzymes to provide

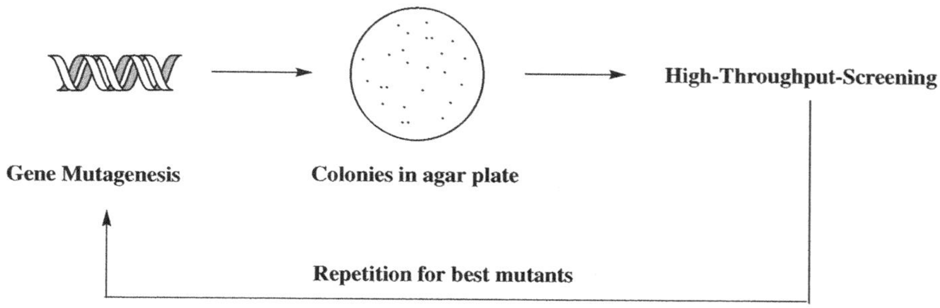

FIGURE 7.1 General concept for the directed evolution [6–9].

features that can be compatible with reaction systems, and with a practical application. In this respect, to overcome these limitations of wild-type enzymes, during the last decades the genetic technique of directed evolution (and related approaches) has been largely studied and developed, thus providing nowadays the possibility of addressing all these bottlenecks, aiming at delivering powerful biocatalysts able to perform reactions in the challenging industrial conditions.

By definition, in a nutshell, directed evolution pursues to imitate natural evolution by performing repeated cycles of gene mutagenesis, expression, and (high-throughput-)screening of the mutants until a desired degree of improvement is achieved in the focused property (Figure 7.1) [6–9]. Thus, after generation of a wide number of new enzymes, namely mutants, it would be expected that some of them would be better suited toward the screened property (*you get what you screen for*). With an adequate and sensitive high-throughput-screening method, such mutants should be identified in a straightforward manner. Once identified, these mutants would be subjected to another round of directed evolution to afford even better biocatalysts. In many examples, after several cycles of evolution, impressive results are achieved. The development of directed evolution strategies has come along with the incorporation of rational and computer-aided tools and methods, thus opening the possibility of predicting which parts of the enzymes (namely "hot spots") might be more prone to provide the best mutants for the improvement of a certain enzymatic property [6–9].

The first attempts for directed evolution used chemicals, light, or mutator strains to generate the library. Nowadays, the most commonly used approach is that of error prone polymerase chain reaction (epPCR). This strategy is based on how the fidelity of the PCR process is perturbed with the addition of different concentrations of $MgCl_2$, $MnCl_2$, and nucleotides. Interestingly, the mutation rate can be easily controlled by these additions, thus providing a very efficient and easily adapted technique for many proteins, with the possibility of deciding how many mutations per protein molecule should be produced, what in fact allows to adjust the technique to all enzymes (e.g., based on enzyme size, calculate one mutation per enzyme, etc). Likewise, other strategies for enzyme mutation and evolution are DNA shuffling [10,11] and site-directed mutagenesis, the latter approach comprising several subclasses and options, in some cases with really sophisticated approaches [6–9].

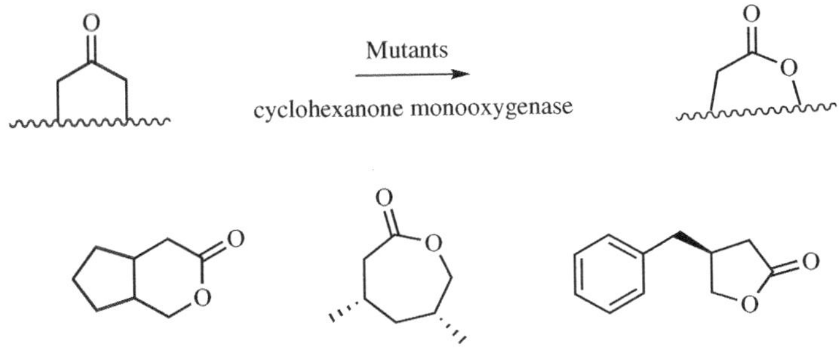

FIGURE 7.2 Performance of mutants of cyclohexanone monooxygenase from *Acinetobacter* sp. Enantiomeric excess of mutants ranged from 75% to 99%, whereas wild-type performances were around 50–60% *ee*. In some cases, inversion of the enantioselectivity was observed as well [6,12].

Recent literature shows several outstanding examples on the use of directed evolution for improving the performance of redox enzymes [6]. For instance, mutants of cyclohexanone monooxygenase from *Acinetobacter* sp. led to improved enantioselectivities in Baeyer–Villiger reactions, when compared to wild-type performances that, in this particular case, often led to very poor sterobias. Several rounds of epPCR were carried out for the proper enzyme improvement (Figure 7.2) [12].

By looking at other enzyme properties, the same biocatalyst was also mutated for the efficient desymmetrization of 4-hydroxy-cyclohexanone, to afford mutants able to deliver both enantiomers (depending on the mutant) with high enantiomeric excesses. This approach would provide access to both important building blocks. The formed lactones rearranged spontaneously without loss of the enantioselectivity (Figure 7.3) [13]. Likewise, the same enzyme was also mutated to produce biocatalysts enabling the oxidation of prochiral thioethers [14]. All these examples reveal the diversity that can be achieved by directed evolution techniques. By changing the aim of the approach (e.g., substrate type), different enzymes adapted to the goal can be

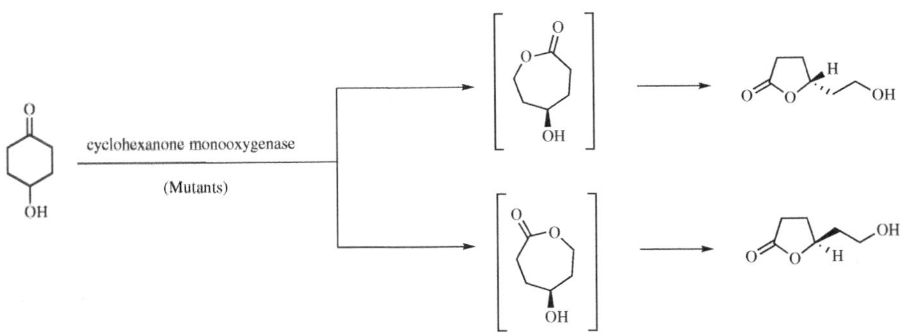

FIGURE 7.3 Desymmetrization of 4-hydroxy-cyclohexanone catalyzed by mutants of cyclohexanone monooxygenases produced by epPCR, providing access to both enantiomers depending on the mutant [13].

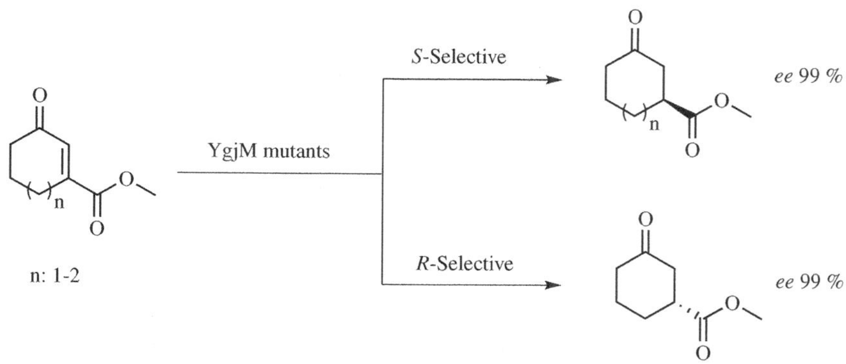

FIGURE 7.4 Biocatalytic reduction of unsaturated keto esters catalyzed by mutants of enoate reductase YgjM from *B. subtilis* [6,15].

produced. As stated before, the classic adage for directed evolution is *you get what you screen for*.

In this area, another outstanding example of directed evolution has been reported for enoate reductases YgjM from *Bacillus subtilis*. This enzyme is actually a homolog of the well-known old yellow enzyme (OYE), enzymes catalyzing the reduction of activated olefinic bonds of several unsaturated keto esters in an enantioselective fashion. Once again, different mutants of this enzyme were generated, enabling the provision of both enantiomers with high enantiomeric excesses in both cases (Figure 7.4) [6,15].

The directed evolution approach is very useful at academic level to understand and predict enzymatic mechanisms. Remarkably, as previously stated in this chapter, it is not common that a wild-type enzyme could be directly used for industrial applications, because the properties of the enzyme often need to be genetically improved to be able to catalyze the desired reaction under *real* conditions. Therefore, in virtue of its importance directed evolution techniques have also found wide use in industry, becoming routine nowadays in many cases for which an improved enzyme is needed for a practical approach. Herein some examples will be given (see Chapter 8 for further cases of industrial applications, and previous chapters of this book as well).

In this area, a recent important case is represented by the Codexis route to produce cholesterol-lowering drugs (e.g., Lipitor®) using ketoreductases combined with other enzymes and chemocatalysts. The biocatalytic reduction displayed excellent yields (96%) and enantiomeric excesses (>99.5%), enabling an attractive process both from an economic and environmental viewpoint, with diminished waste production for the overall process. For this set-up, glucose dehydrogenase (GDH) was used, with addition of glucose equivalents, to afford the required and efficient cofactor regeneration of the reduction cycle (Figure 7.5) [16]. To obtain the improved mutants for this process, a DNA shuffling approach was used. By this means GDH was finally improved by a factor of 13 and ketoreductase activity by a factor of 7, after several cycles of genetic shuffling. To illustrate the improvement of the process, with the tailored enzymes the reaction was completed in 8 h with a substrate loading of

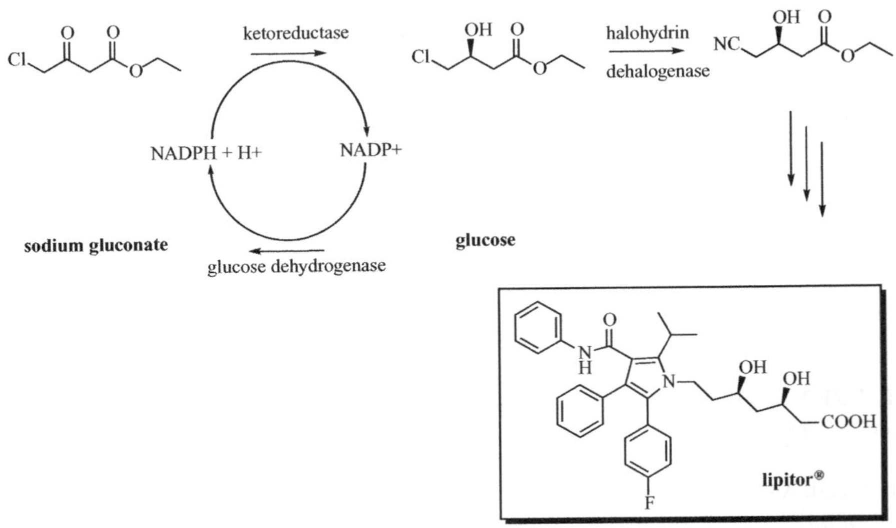

FIGURE 7.5 Biocatalytic steps—including a redox process—involved in the synthesis of the cholesterol lowering drug Lipitor (see also Figure 7.8 in Chapter 8 and further discussion there) [16].

160 gL^{-1} and with diminished amounts of enzymes added, what certainly allows economic and environmental assets for a modern industrial application [16].

Likewise, another enzyme, *Lactobacillus kefir* alcohol dehydrogenase (LKADH) served as template for further genetic improvement, involving up to eight rounds of directed evolution strategies, to finally develop a mutant able to catalyze the enantioselective reduction of tetrahydrothiophene-3-one to afford a useful building block for the pharmaceutical synthesis of sulopenem, a potent antibacterial agent (Figure 7.6) [17]. The usefulness of that enzyme is noteworthy for this substrate, which is synthetically problematic in virtue of the nearly spatially symmetrical (β-CH_2 vs. β-S) ketone used as starting material. By means of chemocatalysts, the accessibility of such molecule would be in fact much more challenging. Therefore, it was necessary to evolve an oxidoreductase to enable a biocatalytic platform for such synthesis. In this process *L. kefir* alcohol dehydrogenase was also coupled with GDH, again for cofactor regeneration issues, with concomitant addition of equivalents of glucose to provide an electron source for the redox process. The optimized reaction was successfully scaled up to 100 kg of product [17].

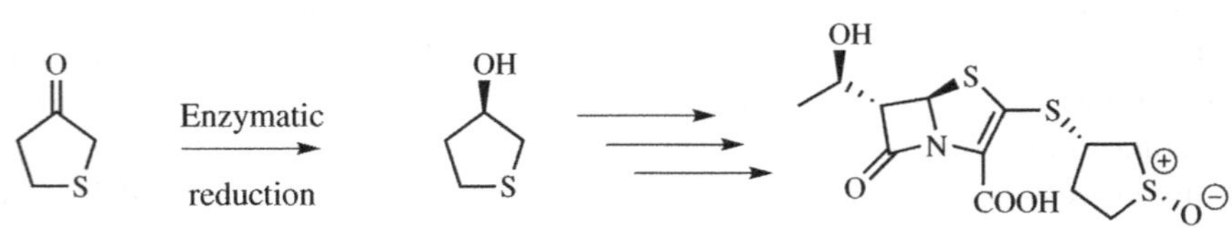

sulopenem

FIGURE 7.6 Biocatalytic step in the synthesis of sulopenem (see also Figure 7.5 in Chapter 8 and further discussion there) [17].

As observed above in all these examples, directed evolution approaches to develop more robust and active biocatalysts are nowadays almost routine in any biocatalytic laboratory, especially at industrial level, where economic and productivity constraints are necessarily more challenging and pressuring. Regardless of the evolution approach that may be undertaken (e.g., epPCR, DNA shuffling, etc.), in all of these processes involving directed evolution a crucial step is the setup of a suitable and sensitive screening method that may enable the discrimination of active (improved) mutants, while providing a diminished number of false positives. In other words, it has to be realized that normally it is necessary to produce thousands of mutants of an enzyme to identify one biocatalyst that may perform a certain reaction better than the wild-type enzyme. Thus, it is necessary to screen many mutants in a reduced period to assure the successfulness of the project.

Yet, these high-throughput-screening methods are often not easy to develop or to implement, for example, due to similarities between substrate and product, which leads to mixed responses in the applied method. Despite these difficulties, there are also a number of successful general cases that sometimes may be used for many examples [1–5]. In the following part of this section, screening methods successfully reported for redox enzymes will be briefly described and discussed.

As starting point, the easiest way for screening many oxidoreductases (and many other enzymes by making use of a coupled system, see below), lies in the possibility of monitoring the NAD(P)H/NAD(P)$^+$ variation at 340 nm, based on the absorbance of the reduced cofactor at that wavelength. This feature can often be used in a straightforward manner not only for screening oxidoreductases but also to assess mutants of other enzymes by smartly coupling the enzymatic performance with an oxidoreduction process involving a NAD(P)-dependent enzyme. Once the reaction conditions are carefully set, by measuring at 340 nm kinetics can be easily monitored. In this area, for instance, the method was successfully used to identify mutants of aldolases and hydroxynitrile lyases by combining such enzymes with a coupling system comprising NADH-dependent benzaldehyde dehydrogenase (BALD) (Figure 7.7). Thus, aldolases and hydroxynitrile lyases would generate aldehydes, which would be subsequently oxidized *in situ* by BALD, and monitored at 340 nm [18].

The above-described method, coupling of a certain enzyme with an oxidoreductase to measure at 340 nm, can also be employed for the identification of nitrilases and nitrile hydratases, by incorporating some modifications to the above-described path. In this case, nitrilases and nitrile hydratases (plus amidases) would form ammonia as one of the products of the reaction. Thus, for this approach a glutamate dehydrogenase (GluDH) was used, forming glutamate out of such ammonia, upon external addition of 2-ketoglutarate. This second reaction leads to the subsequent formation of NAD$^+$, which can be easily monitored at 340 nm (Figure 7.8) [18].

Apart from the above-described enzyme types that can be screened in a high-throughput manner by coupling them with oxidoreductases, other creative ways, for example, for measuring the activity of esterases have been reported, also based on analogous concepts [18]. In this respect, Bornscheuer et al. developed a sensitive high-throughput-screening method for the identification of active and enantioselective esterases, based on the formation of acetic acid upon esterase-catalyzed

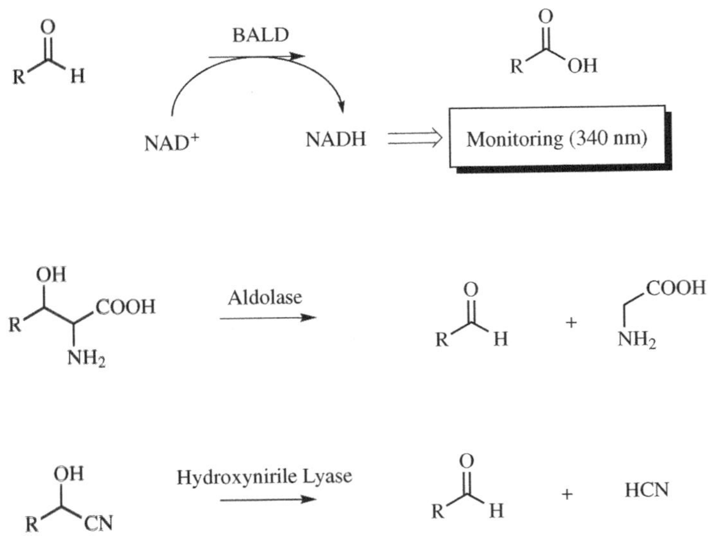

FIGURE 7.7 Screening of different mutants of aldolases and hydroxynitrile lyases by combining them with NADH-dependent benzaldehyde dehydrogenase (BALD) [18].

hydrolysis (by using acetate substrates). Such acetic acid undergoes an *in situ* multistep enzymatic process that enables the final formation of NADH, which once again can be easily monitored at 340 nm (Figure 7.9) [19]. The method was successfully used for screening esterases (native and mutants) toward different acetate substrates.

An analogous approach was used to put forward a high-throughput-screening method for measuring the enantiomeric excess of chiral alcohols. In this case, two oxidoreductases were used, with one of the enzymes recognizing only one of the enantiomers, whereas the other enzyme recognized the opposite isomer. Interestingly, the method could be generalized for many chemical approaches in which chiral alcohols are expected to be formed (Figure 7.10). Again the changes in the $NAD(P)^+/NAD(P)H$, measured at 340 nm, were monitored [20].

In another analogous approach, the kinetic resolution of epoxides was coupled with two enzymes to measure the enantiomeric excess of different catalysts performing the epoxide ring opening. To this end, LKADH, and horse-liver alcohol dehydrogenase (HLADH) were used, and again the formation of NAD(P)H was monitored at 340 nm

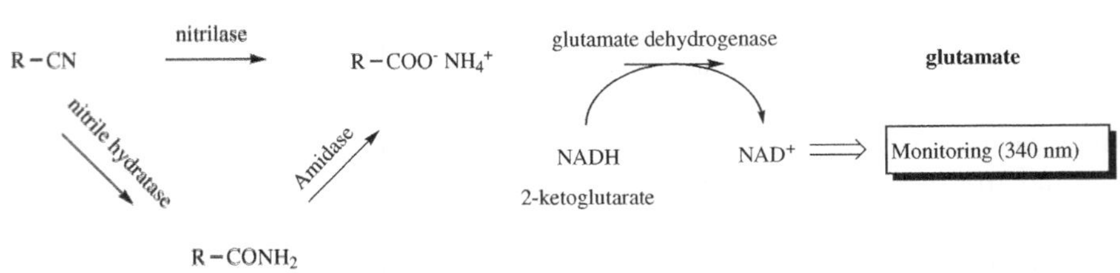

FIGURE 7.8 Identification of mutants with nitrilase or nitrile hydratase activity using the $NADH/NAD^+$ measurement at 340 nm [18].

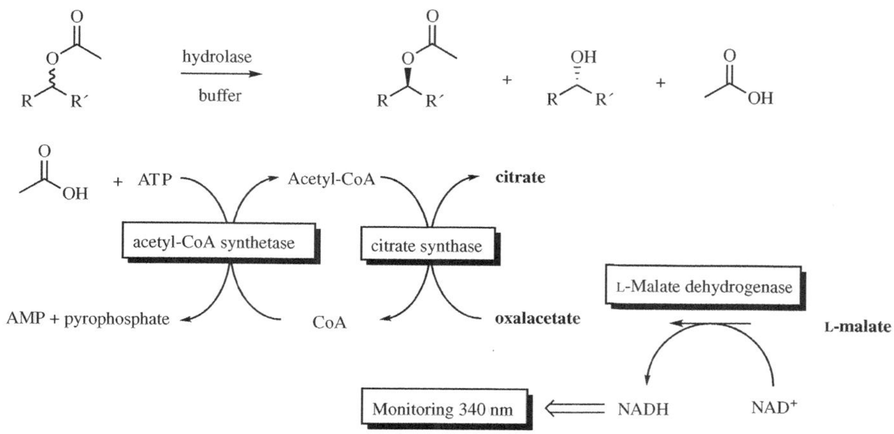

FIGURE 7.9 High-throughput-screening to identify active and enantioselective hydrolases using oxidoreductases, developed by Bornscheuer et al. [19].

(Figure 7.11) [21]. In this case, the approach makes use of the different cofactor selectivity that both enzymes display, being LKADH selective for NADPH, whereas HLADH is so for NADH. Thus, in this way it is straightforward to measure which enantiomers were formed and in which proportion. Importantly, the screening method is decoupled with the catalytic formation of chiral diols, which allows the use of the method for a rather ample palette of catalysts.

Apart from using these strategies of cofactor monitoring at 340 nm for the identification of novel different enzymes and mutants, the system is also valid for direct screening of novel or mutated oxidoreductases [1–4,18]. Thus, very recently it was

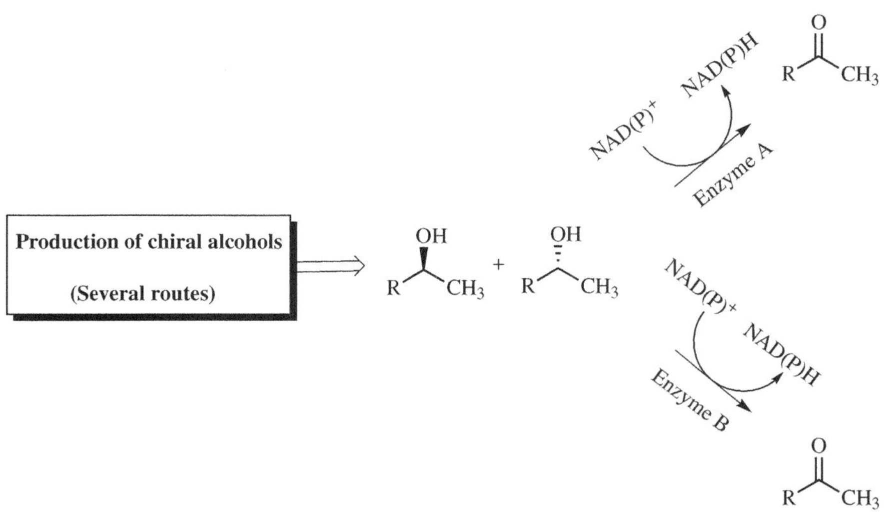

FIGURE 7.10 High-throughput-screening method for measuring the enantioselectivity of alcohols, based on two oxidoreductases displaying opposite enantioselectivity [20].

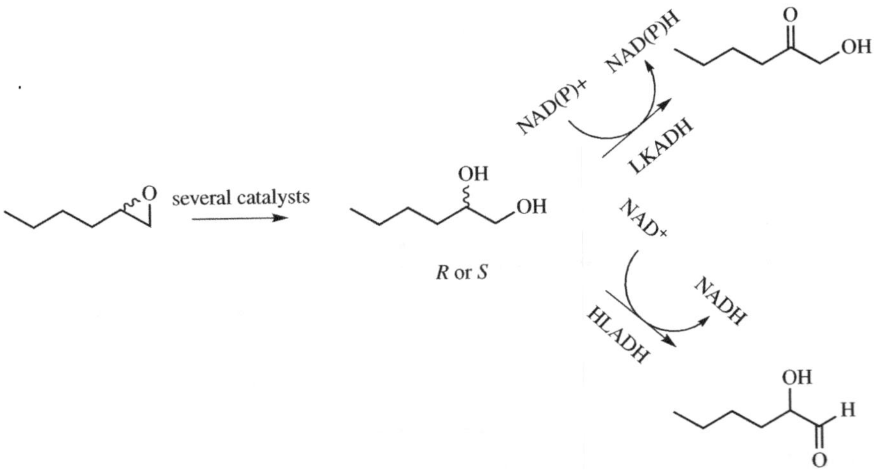

FIGURE 7.11 Measuring the enantiomeric excess produced by catalytic epoxide hydrolysis [21].

used for the characterization and mutant development of pentaerythritol tetranitrate reductases, *via* site-saturated libraries, making use of an automated anaerobic robotic platform. Interestingly, once optimized, the method proved to be very versatile toward many substrates of these enzymes, leading to the identification of a number of useful mutants (Figure 7.12). Once again, directed evolution is useful for a broad

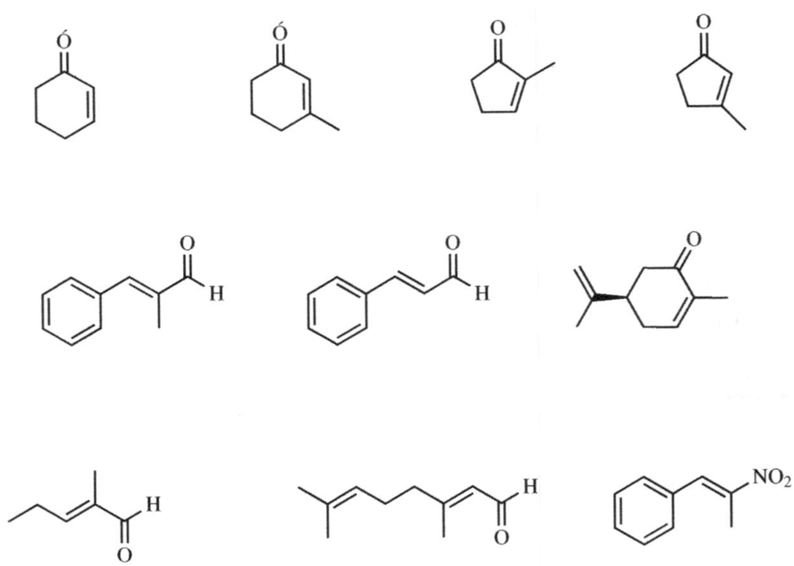

FIGURE 7.12 Diversity of substrates screened toward site-saturated libraries of pentaerythritol tetranitrate reductases by measuring the change of NADPH at 340 nm, using an automated anaerobic kinetic robotic platform [22].

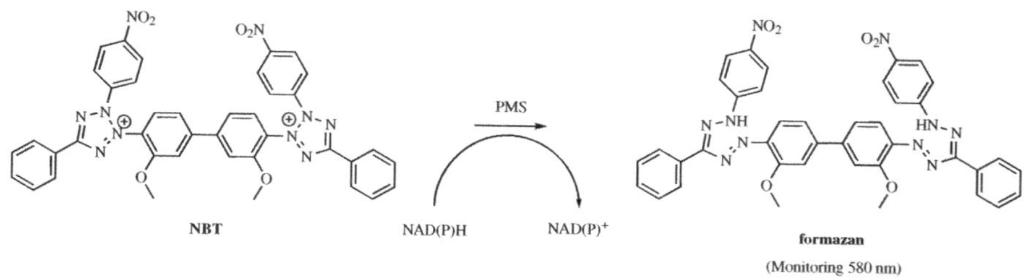

FIGURE 7.13 Production of insoluble blue–purple formazan by NAD(P)H in the presence of phenazine methosulfate (PMS) as mediator.

range of applications and can tackle many enzyme properties, provided that a high-throughput-screening method is available [22].

Despite the simplicity of the above-described methods—several approaches and enzymes, but at the very end *just* measuring absorbance of $NAD(P)^+$/NAD(P)H at 340 nm—in some cases the use of crude cell extracts containing other enzymes may create a number of interferences and erroneous responses when screening for mutants. Due to the enormous amount of mutants that are expected to be assessed in the quest of an active one, the introduction of further purification steps may make the approach totally unsuitable from a practical viewpoint in the laboratory. Thus, apart from monitoring NAD^+ changes at 340 nm, other *ad hoc* approaches have been developed, when needed, for different types of redox enzymes.

One of these options is to induce a color formation during the reaction, by means of different reagents that are concomitantly added to the screening reaction system. For instance, the use of nitro-blue tetrazolium chloride (NBT) to produce the insoluble blue–purple formazan, which can be measured by absorbance at 580 nm, in the presence of mediator phenazine methosulfate (PMS), is a useful and simple approach that has found use in the area of oxidoreductases and high-throughput-screening (Figure 7.13).

This strategy has been successfully used for the directed evolution of a thermostable phosphite dehydrogenase aiming at a more efficient NAD(P)H regeneration [23], as well as for measuring kinetics of oxidoreductases in crude cell lysates, where impurities do not allow the simple monitoring of NAD^+ changes at 340 nm in a sensitive manner [24]. Furthermore, the same reagent was used for a colorimetric discrimination of chiral alcohols in an analogous trend as reported before (Figure 7.11) [21]. In that latter case, the colorimetric assay was successfully used for the inversion of enantioselectivity of P450 monooxygenase by means of directed evolution. To this end, two oxidoreductases were identified, cloned, and overexpressed. One was BRD from *Micrococcus luteus*, and the other one RDR from *Devosia riboflavina*, which were enantiospecific for both enantiomers of the intended target (Figure 7.14). Thus, by playing with the addition of different types of cofactor, either NADH or NADPH, it was possible to easily discriminate and measure the enantioselectivities of the delivered mutants [25].

Other analogous approaches based on color formation have been used for evolving the thermostability of lignolytic oxidoreductases, namely peroxidases and laccases,

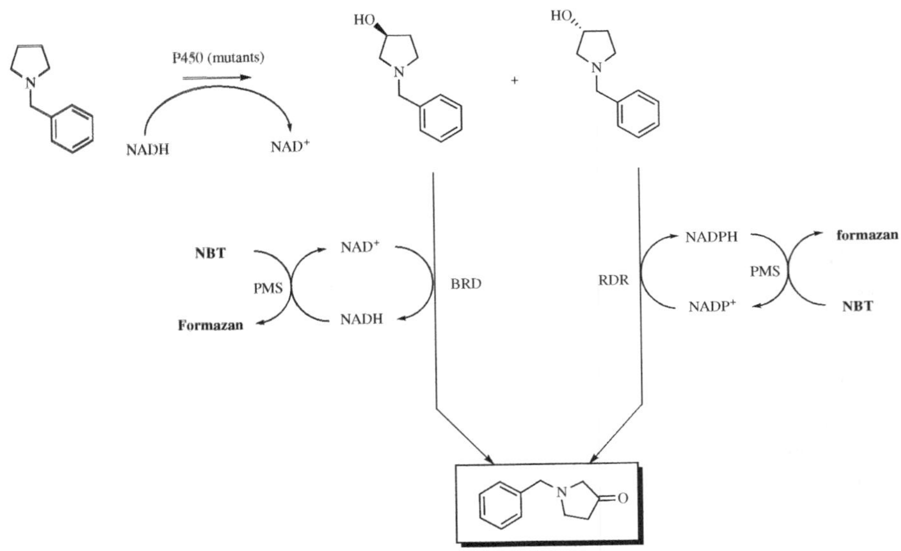

FIGURE 7.14 High-throughput-screening assay for the assessment of mutants of P450 monooxygenase [25].

using ABTS (2,2′-azino-bis(3-ethylbenzothiazoline-6-sulfonic-acid) as the colorimetric reagent [26]. Also, Turner and coworkers used a peroxidase-coupled assay for the assessment of mutant libraries of amine oxidases [27,28]. These enzymes produce hydrogen peroxide as a by-product of their reaction. Thus, this feature was used for the combination of H_2O_2 with the addition of 3,3′-diaminobenzidine (DAB) in the presence of a peroxidase. This leads to the formation of a dark pink insoluble product, enabling the identification of active colonies in Petri dishes, and significantly simplifying the practical work when dealing with mutants (Figure 7.15) [27,28].

Apart from these reported methods, based on *in situ* color formation, other strategies focus on fluorescence-based reactions, since fluorescence provides highly sensitive responses, making it especially useful in high-throughput-screening methods aiming at the development of mutants [1–5]. For instance, the alcohol dehydrogenase (ADH)-catalyzed oxidation of different aromatic alcohols led to formation of

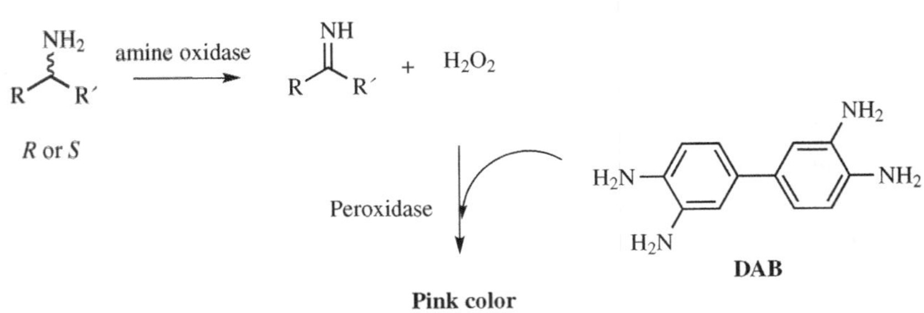

FIGURE 7.15 Coupling amine oxidases with peroxidases for screening [27,28].

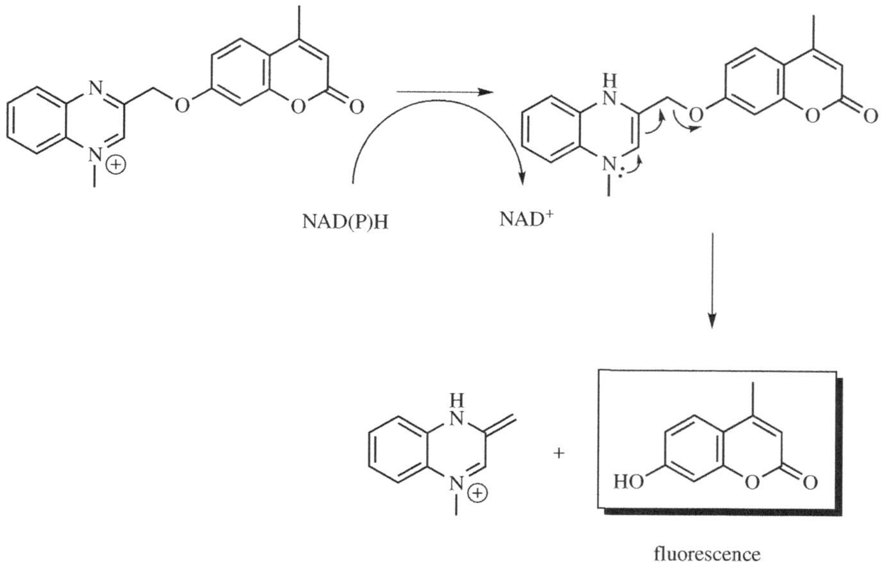

FIGURE 7.16 Fluorescence formation for the detection of NAD(P)H [30].

fluorescent naphthaldehydes and quinoline aldehydes. The approach enabled the selective measure of the activity of different classes (isoenzymes) of ADH in human sera and other body fluids and tissues. Thus, this approach might provide tools in the diagnosis of alcohol abuse, or alcohol tolerance, especially in real crude samples, where interferences of other enzymes or other components may represent a tough hurdle for a sensitive analysis [29].

Furthermore, another approach widely used in redox enzymes, with regard to fluorescence, is the development of a pathway aiming to release *in situ* umbelliferone-like chemicals. These compounds are fluorescent (λ_{ex} 340-360 nm; λ_{em} 450–460 nm), sensitive enough, and useful for screening purposes. For instance, highly sensitive fluorescent methods for NAD(P)H detection (e.g., for cancer therapy) were developed in that way (Figure 7.16). Also in this case, the application of fluorescence-based methods allowed the screening in crude samples containing impurities that could clearly hamper other detection screening methods [30].

An analogous approach based on the development of umbelliferone-like molecules has been put forward for the screening of libraries in the quest of novel enantioselective ADHs [31]. In this particular case, enzymes catalyze the oxidation of substrates, usually a ketone, leading to unstable components that spontaneously decompose to form umbelliferone. For that second step, bovine serum albumin (BSA) was used to catalyze the process, based on the instability of the enzymatic products. Several oxidoreductases were successfully screened by using this approach (Figure 7.17). As a drawback, however, the screening by substrate discriminate only those biocatalysts that are able to convert such substrate into the desired product, and therefore cannot be easily generalized [31].

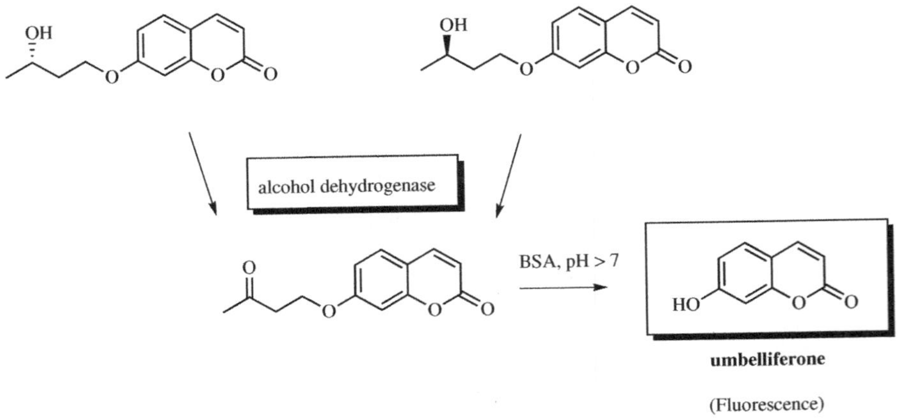

FIGURE 7.17 Fluorometric assay reported for alcohol dehydrogenases. BSA, Bovine serum albumin [31].

Apart from dehydrogenases, strategies relying on the formation of umbelliferone-based fluorescence have found utility for the identification of Baeyer–Villiger monooxygenases. To this end, several approaches were followed. One uses analogous substrates as reported above for dehydrogenases (Figure 7.17). In addition, other substrates for Baeyer–Villiger monooxygenases (e.g., 2-coumaryloxy ketones) were successfully employed as well (Figure 7.18) [32,33].

In the same area (fluorescence-based methods for high-throughput screening), Arnold and coworkers reported a different approach for the high-throughput-screening of cytochrome P450 monooxygenases that are able to oxidize naphthalene-based substrates, by cloning and overexpressing two enzymes in the same *Escherichia coli* host [34,35]. Those enzymes were mutants of cytochrome P450 monooxygenase, and the second one was horseradish peroxidase (HRP), which is able to couple the phenolic compounds formed by the mutant monooxygenase. The coupled compound had a λ_{ex} of 350 nm, and a λ_{em} of 460 nm. The method also worked by adding free HRP to the reaction system (Figure 7.19) [34,35].

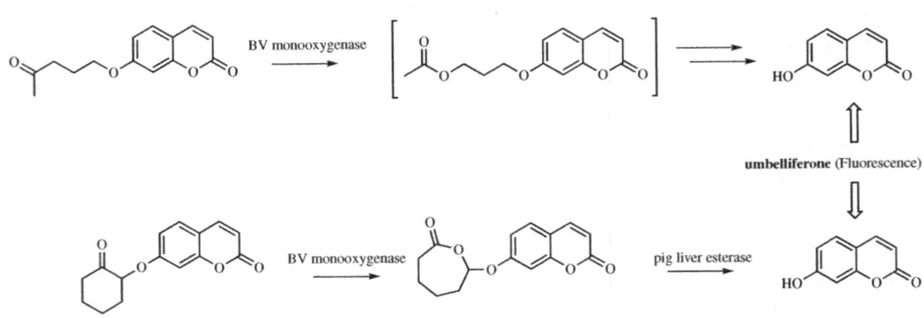

FIGURE 7.18 Umbelliferone-based fluorescence assays for Baeyer–Villiger monooxygenases [32,33].

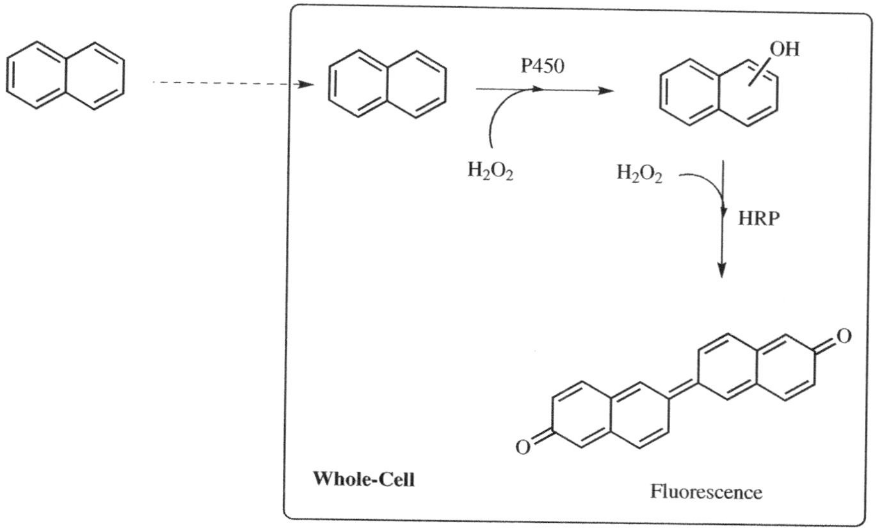

FIGURE 7.19 High-throughput-screening of cytochrome P450 monooxygenases by coupling the oxidation with horseradish peroxidase (HRP) [34,35].

To sum up this section, there are already an ample number of approaches for the efficient high-throughput-screening of novel redox enzymes. Methods are continuously being developed, once a demand for such an approach is needed.

7.3 SUCCESSFUL CASE: BAKER'S YEAST REDOX ENZYMES, THEIR CLONING, AND SEPARATE OVEREXPRESSION

As it has been clearly stated in the present book (Chapters 3–5), the applications of redox enzymes are currently a mature field, with a broad number of outstanding examples, even at industrial scale (see Chapter 8) [36,37]. In this area, one predominant and paradigmatic case is the use of *S. cerevisiae* (the so-called baker's yeast) as the biocatalyst for enantioselective reduction of an ample number of ketones, both aliphatic and aromatic ones. Actually, the substrate diversity of baker's yeast oxidoreductases is extremely broad, and thus many compounds in different reaction conditions, often providing high selectivities and enantioselectivities, have been reported as substrates so far [38,39]. Well-known for a long time, baker's yeast was domesticated approximately 4000 years ago, aiming at producing bread and beverages by performing alcoholic fermentation of different natural resources (e.g., starches, potatoes, wheat mill, etc.). Later on, at the beginning of the twentieth century, Neuberg and Nord reported for the first time on the oxidation of *n*-pentanal catalyzed by yeast cells, as well as on the enantioselective reduction of prochiral ketones, even at gram scale. Those are probably some of the first examples of biotransformations applied to asymmetric organic synthesis [40,41]. Subsequently, along the twentieth century the biocatalytic application of these baker's yeast cells started to be reported in a much broader extent, thus showing the enormous substrate acceptance of baker yeast oxidoreductases [39]. Despite this volume of research and effort using baker's yeasts as biocatalyst, even nowadays novel applications with excellent enantioselectivities are often described with wild-type baker's yeast. For instance, the reduction of keto-alcohols to the corresponding chiral diols can be performed by using nonhazardous and biodegradable baker's yeasts cells [42]. In fact, it represents a fantastic way to be initiated in biocatalysis, since nowadays baker's yeast are commercially available in many supermarkets at really accessible prices.

Yet, despite the above-described ample and useful approach of baker's yeast as biocatalyst to perform simple and smart synthetic organic reactions, it was also observed that enantioselectivities and activities toward many substrates were often poor and not reproducible at all. Thus, it was a common fact that depending on the origin of the baker's yeast (or research groups using baker's yeast from the same supplier but from different lots), totally opposite results in terms of enantioselectivities were reported for a certain substrate. The clarification of these findings became evident when the genome of baker's yeast was totally elucidated. An in-depth evaluation of such genome revealed the existence of an ample number of different redox enzymes, responsible for different metabolic pathways and/or different substrates. In addition, it was observed that these redox enzymes displayed often overlapping activities and even differing enantioselectivities when nonnatural substrates were assessed, under nonnatural reaction conditions. Conclusively, it turned out clear that it was worth to look at the complex enzymatic redox machinery present in that baker's yeast, and actually several oxidoreductases started to be identified, separated, and biochemically characterized [43]. By following this line along years of research, further biochemical studies revealed that up to approximately 50 different redox enzymes exist in baker's yeast, gathered in different biochemical families and performing different natural roles within the cell. In Table 7.1 such diversity is briefly summarized [44].

Thus, once the enormous diversity of redox enzymes present in baker's yeast was realized, considerable attempts were paid trying to diminish the *negative* effects of some of these enzymes toward a certain substrate (e.g., undesired enantioselectivity), while keeping the activity of the desired biocatalysts, or at least those displaying similar stereobias, in a higher extent. To tackle that issue, at the very beginning several *medium engineering* strategies were applied to enhance the enantioselectivity of whole-cells of baker's yeast in asymmetric reductions, with more or less moderate successful results (see Chapter 3) [45–47]. For instance, approaches comprised the modification of the substrate structure [48,49], careful control of substrate concentration [48,50,51], change of growth condition in baker's yeast [52,53], use of organic solvents and biphasic systems [50,5 4–59], or the inclusion of enzyme inhibitors in the culture medium [60–65], among other examples reported in the literature. Since only moderately successful results were obtained and, furthermore, due to the fact that these results were often empirical and no systematic rules could be drawn out of them, the subsequent step was to metabolically engineer baker's yeast cells, to knock out some of the redox enzymes, while overexpressing other oxidoreductases that are more active and selective for the desired substrate target. With this latter approach, several successful results were reported [66–70]. Yet, the system of redox

TABLE 7.1 Diversity of Baker Yeast Redox Enzymes [44]

Protein	Family	ORF	Mature Mw (Da)	Predicted pI
L-Lactate dehydrogenase	Lactate dehydrogenases	YML054C	56.58	6.36
D-Lactate dehydrogenase	Lactate dehydrogenases	YDL174C	62.16	5.90
D-Lactate dehydrogenase	Lactate dehydrogenases	YDL178W	59.26	5.91
D-Lactate dehydrogenase	Lactate dehydrogenases	YEL071W	55.09	6.66
Long-chain alcohol dehydrogenase	Alcohol dehydrogenases	YDL168W	40.93	6.41
Alcohol dehydrogenase I	Alcohol dehydrogenases	YOL086C	36.74	6.30
Alcohol dehydrogenase II	Alcohol dehydrogenases	YMR303C	36.62	6.37
Alcohol dehydrogenase III	Alcohol dehydrogenases	YMR083W	37.16	7.37
Alcohol dehydrogenase IV	Alcohol dehydrogenases	YGL256W	50.62	8.71
Alcohol dehydrogenase V	Alcohol dehydrogenases	YBR145W	37.51	6.15
Sorbitol dehydrogenase	Medium-chain alcohol dehydrogenases	YJR159W	38.06	6.52
Putative oxidoreductase	Medium-chain alcohol dehydrogenases	YLR070C	38.46	5.93
Putative oxidoreductase	Medium-chain alcohol dehydrogenases	YAL061W	46.08	5.88
Putative oxidoreductase	Medium-chain alcohol dehydrogenases	YAL060W	41.53	6.43
Putative oxidoreductase	Medium-chain alcohol dehydrogenases	YMR318C	39.51	6.38
Putative oxidoreductase	Medium-chain alcohol dehydrogenases	YCR105W	39.35	7.18
ζ-crystalline homolog	Medium-chain alcohol dehydrogenases	YBR046C	37.02	8.89
α-acetoxyketone reductase	Short-chain alcohol dehydrogenases	YOL151W	38.06	5.82
Putative oxidoreductase	Short-chain alcohol dehydrogenases	YDR541C	38.47	6.09
Putative oxidoreductase	Short-chain alcohol dehydrogenases	YGL157W	37.94	5.70
Putative oxidoreductase	Short-chain alcohol dehydrogenases	YGL039W	38.12	5.79
Putative oxidoreductase	Short-chain alcohol dehydrogenases	YLR426W	36.72	9.24
Putative oxidoreductase	Short-chain alcohol dehydrogenases	YBR159W	38.60	9.65
Putative oxidoreductase	Short-chain alcohol dehydrogenases	YDL114W	34.17	9.64
Putative oxidoreductase	Short-chain alcohol dehydrogenases	YMR226C	29.05	6.20
Putative oxidoreductase	Short-chain alcohol dehydrogenases	YIR035C	27.38	5.83
Putative oxidoreductase	Short-chain alcohol dehydrogenases	YIL124W	32.72	9.34

(Continued)

TABLE 7.1 ***(Continued)***

Protein	Family	ORF	Mature Mw (Da)	Predicted pI
Putative oxidoreductase	Short-chain alcohol dehydrogenases	YKL107W	34.52	6.37
3-oxoacyl-(acyl-carrier-protein) reductase	Short-chain alcohol dehydrogenases	YKL055C	31.20	9.17
Aryl alcohol dehydrogenase	Short-chain alcohol dehydrogenases	YNL331C	41.86	6.51
Putative aryl alcohol dehydrogenase	Short-chain alcohol dehydrogenases	YCR107W	40.78	6.58
Putative aryl alcohol dehydrogenase	Short-chain alcohol dehydrogenases	YDL243C	36.88	6.40
Putative aryl alcohol dehydrogenase	Short-chain alcohol dehydrogenases	YJR155W	32.47	9.01
Putative oxidoreductase	Short-chain alcohol dehydrogenases	YFL057C	16.49	7.87
Putative aryl alcohol dehydrogenase	Short-chain alcohol dehydrogenases	YOL165C	15.69	5.78
Aldo-keto reductase	Aldose reductase	YDR368W	34.62	6.88
Putative oxidoreductase	Aldose reductase	YOR120W	34.95	7.98
D-arabinose dehydrogenase	Aldose reductase	YBR149W	38.41	5.63
Putative oxidoreductase	Aldose reductase	YJR096W	32.21	6.71
Putative oxidoreductase	Aldose reductase	YHR104W	37.01	6.70
Putative oxidoreductase	Aldose reductase	YDL124W	35.45	5.76
Putative oxidoreductase	D-hydroxyacid dehydrogenase	YPL113C	44.88	7.03
Putative oxidoreductase	D-hydroxyacid dehydrogenase	YGL185C	42.85	8.27
Putative oxidoreductase	D-hydroxyacid dehydrogenase	YNL274C	38.72	5.98
Putative oxidoreductase	D-hydroxyacid dehydrogenase	YOR388C	41.61	6.11
Putative oxidoreductase	D-hydroxyacid dehydrogenase	YPL275W	26.36	9.41
Fatty acid synthase	Multifunctional	YKL182W	228.68	5.71
		YPL231W	206.89	5.27

ORF, open reading frame.

enzymes in baker's yeast turned out to be much more complex than it was initially expected (see Table 7.1), and these approaches could not be generalized for all enzymes from baker's yeast and for all organic substrates [44]. Thus, it became quickly understood that only by performing the separate cloning and overexpression of each redox enzyme in a heterologous host, such enzymes might provide selective biotransformations in a reproducible way. Preliminary studies showed positive results, even at gram scale [71,72]. Consequently, more comprehensive studies were performed, assessing each enzyme on its own toward an ample number of substrates comprising different chemical structures. For some of these substrates (e.g., some α-alkyl-β-hydroxyesters, or α-keto-β-lactams), in fact, it was observed that even each separate enzyme was not completely enantio- or diastereoselective, thus leading to mixed products that were then characterized for each biocatalyst [73–75]. Despite this low enantio-, or diastereoselectivity, it is worth to mention that once these enzymes have been cloned and overexpressed, they might be then easily subjected to directed evolution to afford biocatalysts that may operate under better bias toward such substrates. To this end, approaches reported in the first section of this chapter (directed evolution and high-throughput-screening) may be directly used.

Despite these not so inspiring results, for some other substrates the assessment of separate enzymes led to very successful results, providing very useful biocatalysts for an ample number of substrates (Figure 7.20), which were converted to chiral building blocks by several of these enzymes with high enantio- and diastereoselectivities [76–78]. The examples reported in Figure 7.20, as such, represent an outstanding case on how modern approaches of molecular biology—in this case, gene identification, cloning, and overexpression—can assist organic chemists in identifying novel diastereoselective biocatalysts for many applications. As it can be rapidly envisaged, approaches like directed evolution can still provide novel insights in the field, giving access to other not so successful substrates in the future.

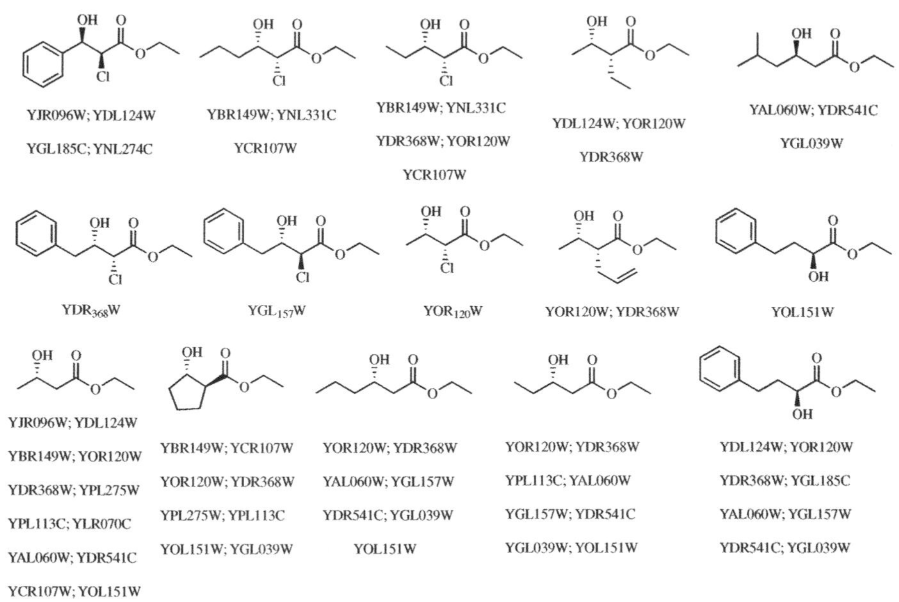

FIGURE 7.20 Selected examples of products enantio- and diastereoselectively obtained by separate redox enzymes from baker's yeast, cloned and overexpressed in *E. coli*. Only those examples providing exquisite enantio- and diastereoselectivity have been depicted [76–78].

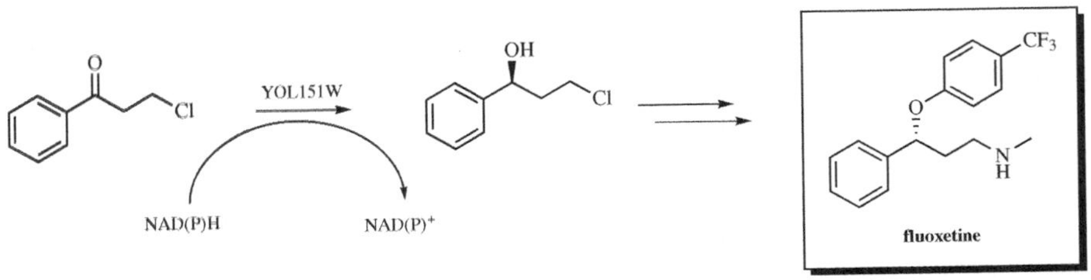

FIGURE 7.21 Synthesis of optically active building blocks (*ee* >99%) for the fluoxetine production, by means of separate recombinant YOL151W from baker yeast [79].

This enormous potential that redox enzymes of baker's yeast possess, when conveniently cloned and separately overexpressed in heterologous host, has started to attract the attention of several research groups interested in producing optically active building blocks for the synthesis of different pharmaceuticals. As a recent example, the use of recombinant baker's yeast YOL151W (Table 7.1) to produce (*S*)-3-chloro-1-phenyl-1-propanol was reported. This compound is a chiral intermediate in the synthesis of antidepressant drugs, for example, fluoxetine, a drug that belongs to the selective serotonin reuptake inhibitor (Figure 7.21) [79]. Interestingly, this particular enzyme was active both with NADPH and NADH as cofactor, thus enhancing the synthetic possibilities of the catalyst for practical applications, since cofactor regeneration may be then tackled in a more efficient way within the whole-cell biocatalyst, by means of both of them.

Likewise, another recent important example is the production of ethyl-(*S*)-4-chloro-3-hydroxy butanoate, a useful building block for the synthesis of cholesterol lowering drugs (e.g., Lipitor, see also Figure 7.5, and Chapter 8). Actually, when screened separately, several baker's yeast oxidoreductases exhibited high activities and enantioselectivities toward the ketone substrate. Interestingly, several of these enzymes afforded the (*S*) enantiomer, whereas another one provided the (*R*) isomer (Figure 7.22). Thus, access to different compounds may be reached by

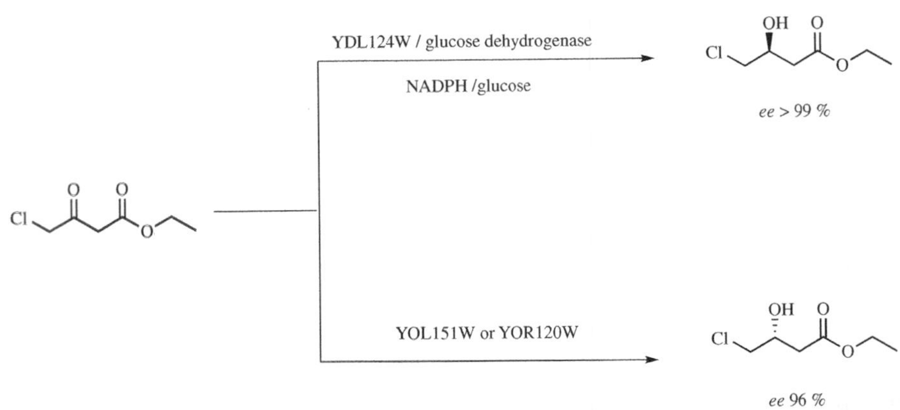

FIGURE 7.22 Different redox enzymes from bakers yeast displaying different enantioselectivity to afford, on demand, both enantiomers of ethyl-4-chloro-3-hydroxy butanoate [80].

means of these separate enzymes, without problems of reproducibility or erratic enantioselectivities [80].

To summarize the state-of-the-art, this section has been addressing the importance of cloning and overexpressing enzymes for biocatalytic purposes, exemplifying the importance that this could have for a practical application. As case study, the complexity that wild-type baker's yeast has in terms of different oxidoreductases has been shown. Such diversity often leads to nonreproducible results in terms of activity and enantioselectivity, depending on the activity of the enzymatic mixture toward that substrate. Conversely, by cloning and overexpressing the enzymes in a separate way, reproducible and robust biocatalysts can be achieved. Furthermore, if directed evolution strategies are also incorporated, it appears clear that modern biocatalysis has a powerful palette of molecular biology techniques to reach industrial targets.

7.4 WHOLE-CELLS VS. ISOLATED ENZYMES: MEDIUM ENGINEERING

Once an active redox biocatalyst has been identified, cloned, overexpressed, and genetically improved toward a desired substrate (Sections 7.2 and 7.3), the next step is the integration of the biocatalyst in a process scheme. Within the field of redox enzymes, usually one of the questions that are addressed is that of using either free, isolated enzymes (with an appropriate cofactor regeneration system) (see Chapter 2 for fundamentals, and Chapters 3–5 for a wide number of examples of application), or whole-cells, in which cofactor is automatically regenerated within the biocatalysts. Before discussing in this section the advantages and disadvantages of using either enzymes or cells, the key message that must be given is that there are no unique solutions, and that often the decision on making use of one or the other technology depends on specific aspects of each research group or industry (e.g., tradition to use free cells, availability of a laboratory to operate with recombinant microorganisms, etc.). As it will be provided in Chapter 8, there are a number of outstanding examples at industrial scale both using free enzymes and whole-cells, and therefore economic terms are not always discriminating one approach over the other one.

The use of isolated enzymes often provides a better reaction control, together with the absence of side reactions that other enzymes of the cell might bring. In addition, many isolated enzymes tolerate a high substrate and product concentration, which makes the process attractive from an industrial viewpoint. However, the use of isolated enzymes also provides some drawbacks that have necessarily to be addressed. One is the common low stability of the biocatalysts as isolated enzymes, what may certainly hamper the economics of the process, and may oblige to use huge amounts of biocatalysts for the production step. Besides this, another problem that necessary needs to be tackled is that of cofactor regeneration, either by adding an ancillary substrate that can also be converted by the same enzyme, or by adding a second enzyme that may be coupled to the catalytic cycle. Last but not least, albeit the situation is rapidly changing, nowadays there are still not many commercially available redox enzymes that may be applied as isolated enzymes. Therefore, in many occasions the expertise from molecular biologists is needed in research groups to provide the necessary knowledge for the production of biocatalysts at laboratory scale to develop whole-cells, and later on to purify the enzymes.

Following the latter argumentation, the application of whole-cells is definitely simpler and cheaper, once the cells overexpressing the desired enzymes have been developed, than the use of isolated enzymes. In fact, there is no need of purifying enzymes or isolating them, and bottlenecks related to cofactor regeneration issues can be easily circumvented. Likewise, usually enzymes are much more stable in whole-cells than in a free state, thus providing a promising framework for many applications at competitive costs [81]. Yet, whole-cell use may also bring some disadvantages. For instance, the existence of unwanted side reactions cannot be excluded, and must be experimentally assessed case by case, under the desired reaction conditions. Moreover, some mass-transfer limitations may occur, since the contact between the enzymes and the substrates (or during the release of products) is not as straightforward as when isolated enzymes are employed. Moreover in many countries, recombinant cells can be used only under especial regulations and safety standards. Albeit recombinant microorganisms typically used in biocatalysis are nonpathogenic, the need of investment and the existing regulations may hamper researchers in the use of whole-cells, and push them for the use of commercially available isolated enzymes.

Furthermore, as it has become clear in this section, and also in other parts of this book, one of the most important issues to be tackled with redox enzymes is how to recycle enzyme cofactors in an efficient way, in order to drive thermodynamic equilibria to higher conversions. To this end, approaches like *in situ* product removal (ISPR) techniques can be used to remove the most volatile compound by an air flow, thus resulting in a shift of the equilibrium toward the formation of higher product outcomes. The strategy was successfully implemented by using lyophilized whole-cells of *E. coli* with overexpressed ADH-‘A’ from *Rhodococcus ruber* DSM 44541 in the enantioselective reduction of different prochiral ketones. To recycle the cofactor, isopropanol was added as ancillary substrate, and the formed acetone was stripped by the ISPR approach. Thus, impressive conversions of >99% with >99% *ee* were achieved in short reaction times (<5 h) (Figure 7.23) [82].

Apart from cofactor regeneration issues, another important challenge to be solved, when practical biotransformations are considered, is the poor solubility of many organic compounds in aqueous systems. Thus, several efforts have been invested to provide creative solutions for these issues, since reaction conditions need to be operated under acceptable substrate concentration conditions. Some options would be the provision of either cosolvents miscible with buffer systems, or the set-up of biphasic systems (see also Chapter 8 for examples of these two approaches at industrial scale). In general, biphasic systems are very useful because they allow a straightforward downstream processing, high substrate loadings, and at the same time less inhibitory problems created by high concentrations of products or substrates in the proximity of the enzyme. In virtue of this, many examples have been reported (Chapter 8), and novel approaches are being continuously developed. One alternative to these systems is the use of ionic liquids as extractive phase. For instance, Weuster-Botz and coworkers have reported on the highly efficient reduction of 2-octanone

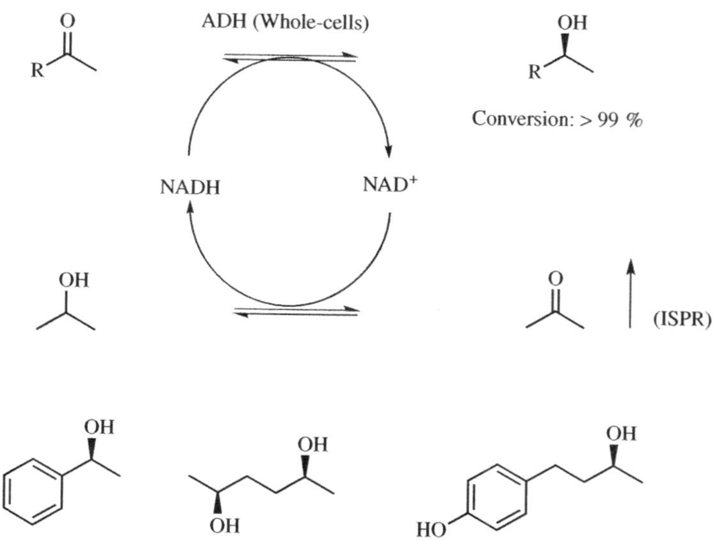

FIGURE 7.23 Process with ISPR to tackle thermodynamic hurdles [82].

to afford (*R*)-2-octanol in high yields and enantioselectivities by setting a biphasic medium buffer-ionic liquid. From the different ionic liquids (ILs) assessed, 1-hexyl-1-methylpyrrolidinium bis(trifluoromethylsulfonyl)imide, [HMPL][NTF], turned out to be the most suitable one. As biocatalysts, recombinant *E. coli* cells overexpressing both *Lactobacillus brevis* ADH (LBADH), and formate dehydrogenase (FDH) from *Candida boidinii* for cofactor regeneration, were used. Under optimized conditions, a productivity of $\sim$ 180 $gL^{-1}d^{-1}$, with a yield of 95% and enantiomeric excess of 99.7% was achieved (Figure 7.24) [83].

Other novel strategies aiming at solving the substrate solubility issues focus on the elimination of organic solvents to enhance the substrate concentration, and to significantly reduce the waste production of the overall process. For instance, Greiner et al. have reported on a biphasic medium composed of the substrate (solvent-free) and buffer, to perform the enantioselective reduction of hardly water-soluble substrates. To this end, ADH from *L. brevis* was used (isolated enzymes in this case), and GDH from *Bacillus* sp. was also added to recycle the cofactor upon addition of glucose. With this concept, several long-chain aliphatic ketones were successfully reduced by the enzyme (Figure 7.25) [84].

Furthermore, very recently lyophilized *E. coli* whole-cells with overexpressed oxidoreductases were directly used in neat substrates, thus leading to so-called "solvent-free" and "water-free" systems [81]. In this approach, by carefully controlling the water activity of the lyophilized cells, a great stability of the biocatalysts can be achieved, together with efficient performances in enantioselective reductions. This concept may be very useful for practical applications at high substrate loadings, and especially in substrates that are prone to be unstable in water, such as 3-butyn-2-one. Interestingly, the produced alcohol is an excellent building block for the production

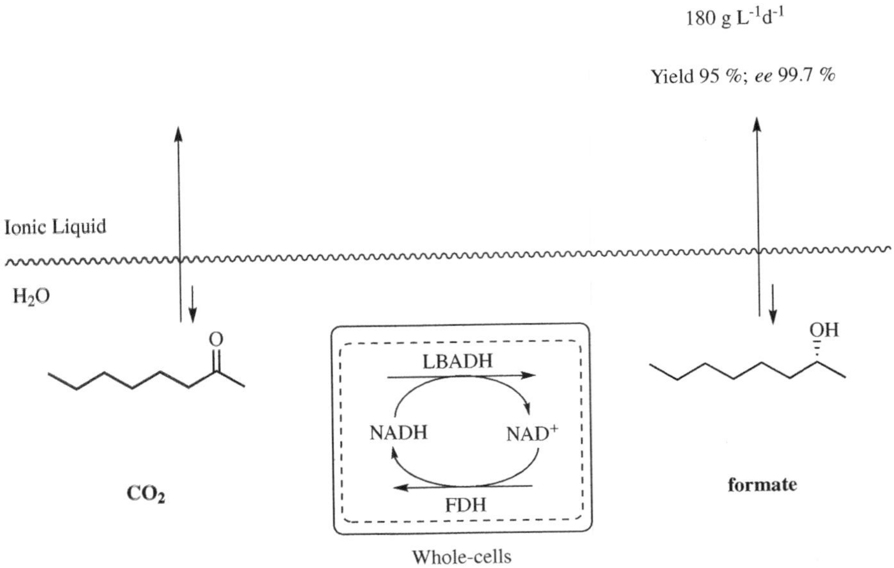

FIGURE 7.24 Set-up of a biphasic medium composed of ionic liquid and buffer to afford the enantioselective reduction of 2-octanone catalyzed by recombinant whole-cells [83].

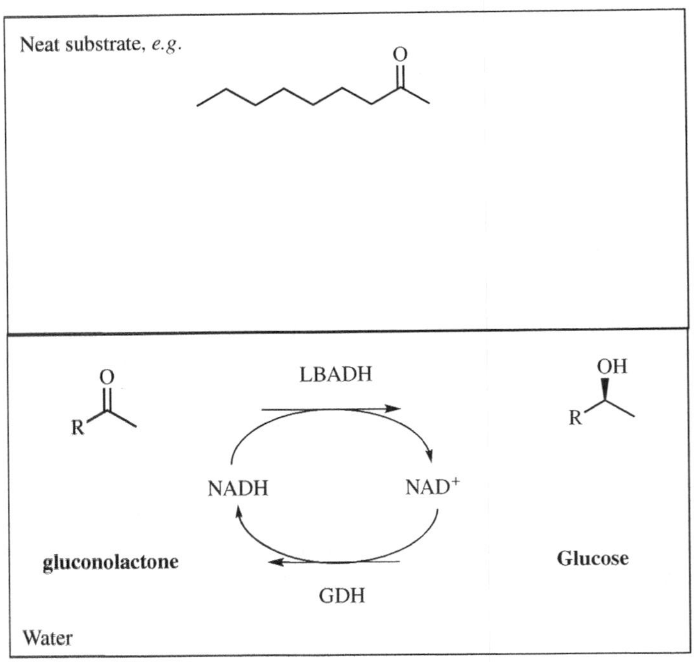

FIGURE 7.25 Conceptual solvent-free biphasic approach for the enzymatic reduction of hardly-water-soluble ketones [83].

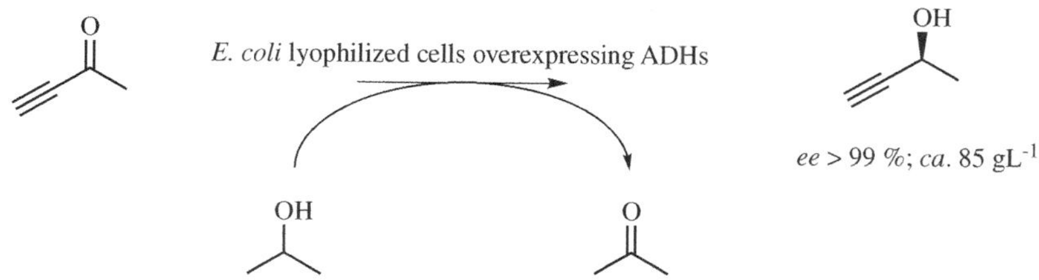

FIGURE 7.26 Enzymatic reduction using lyophilized whole cells with overexpressed oxidoreductases performed in neat substrates [81].

of pharmaceuticals like antiasthma drugs, β2-adrenergic stimulants, or protease inhibitors. By means of this strategy, up to $\sim 85\ gL^{-1}$ of enantiomerically pure substrate were achieved under nonoptimized conditions, clearly showing the potential that the concept may have in the coming years (Figure 7.26) [81].

As standard protocol, the gene of the desired enzyme (in this case, *Candida parapsilosis* carbonyl reductase, CPCR) must be cloned and overexpressed in *E. coli*. Once this step is successfully performed, cells are equilibrated with different values of water activity, by means of different inorganic salts. To carry out the biotransformation, substrate and cosubstrate (in this case, isopropanol for cofactor regeneration issues) are mixed and cells are added at room temperature. For the work-up, suspended cells are filtrated, and cosubstrates (remnant of isopropanol and formed acetone) are evaporated, thus affording the desired product with high enantioselectivity. A further integration of, for instance, *in situ* acetone removal would be highly beneficial to even enhance the productivities and reduce the reaction times.

Another important topic regarding redox enzymes and process set-up is the implementation of enzymatic membrane reactors to perform continuous processes with high productivities, while keeping enantioselectivities in the high standard levels [85,86]. The rationale behind is rather simple—the use of a ceramic membrane that are able to retain enzymes or whole-cells where a flow of (co)products is passing [87]. For instance, the reduction of methyl acetoacetate was performed in continuous operation with recombinant *E. coli* whole-cells expressing ADH from *L. brevis*, together with FDH from *C. boidinii* for cofactor regeneration. The system remained stable for over 7 weeks, providing a productivity of $\sim 58\ gL^{-1}d^{-1}$, with a maximal conversion of >99% and enantiomeric excess of >99% (Figure 7.27) [88].

7.5 BEYOND: MULTISTEP DOMINO BIOCATALYTIC PROCESSES

As reported in previous sections, there are a number of tools and strategies to identify, characterize, improve, and implement enzymatic redox reactions in a reactor setup. Some of these setups have also reached the commercial scale (Chapter 8). Being the implementation of such technologies based on redox enzymes a rather consolidated field, even at industrial scale, academia has started to pave the way

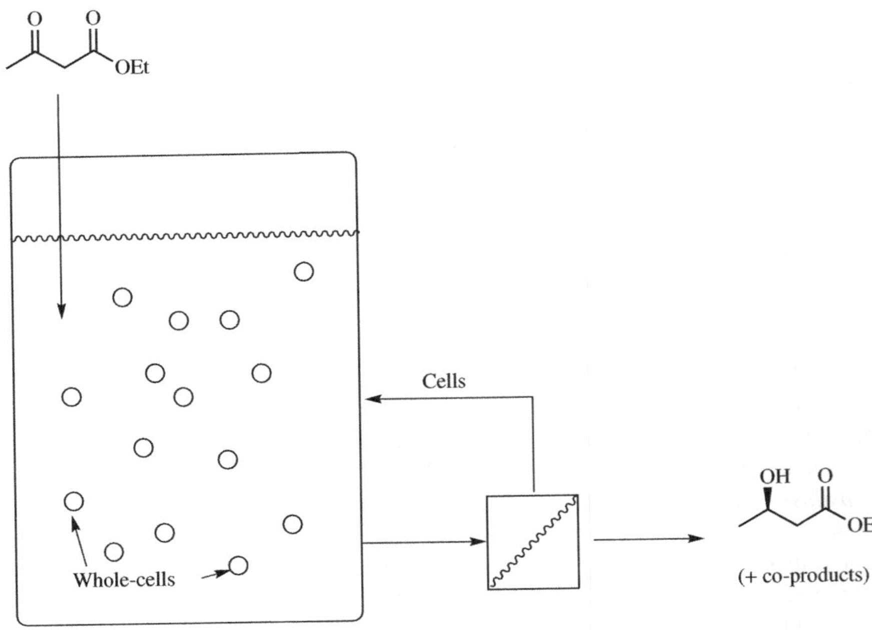

FIGURE 7.27 Set-up of a continuous process with ultrafiltration membrane of molecular weight cut-off of 10 kDa [88].

for the next generation of biocatalytic systems. One of these new approaches is the development of multistep processes (or domino reactions). In a nutshell, the rationale behind such strategy consists on mimicking what a living cell is doing—catalyzing consecutive reactions—thus establishing a pathway of synthetic reactions in a reactor, by combining different catalysts or reagents. In this way processes can be significantly simplified, and furthermore a number of downstream operational units can be overcome, since processes are conducted in one pot, either in one or in several steps. The concept has started to be developed, already with some interesting examples, and it is expected that will be a major field of research in the next decades, probably ending up with several processes at industrial scale as well [89,90]. Actually, using isolated enzymes with a coupled biocatalyst for cofactor regeneration can be strictly considered a multistep, cascade process as such. Thus, in these conditions two or more enzymes are combined in the same reactor system. Some other examples, already reported in this chapter, comprise the combination of oxidoreductases with halohydrin dehalogenases to afford building blocks for the cholesterol lowering drug Lipitor, providing an improved environmental profile when compared with a process involving separate reaction steps (Figure 7.5, see also Figure 7.8 in Chapter 8) [16]. Following these strategies, in the present section some other modern approaches for cascade processes involving redox enzymes will be briefly described.

An elegant approach comprising several oxidoreductases in the same reactor system is the deracemization of alcohols leading to stereoinversions. Basically, one of the oxidoreductases catalyzes the oxidation of one of the enantiomers to the ketone. The

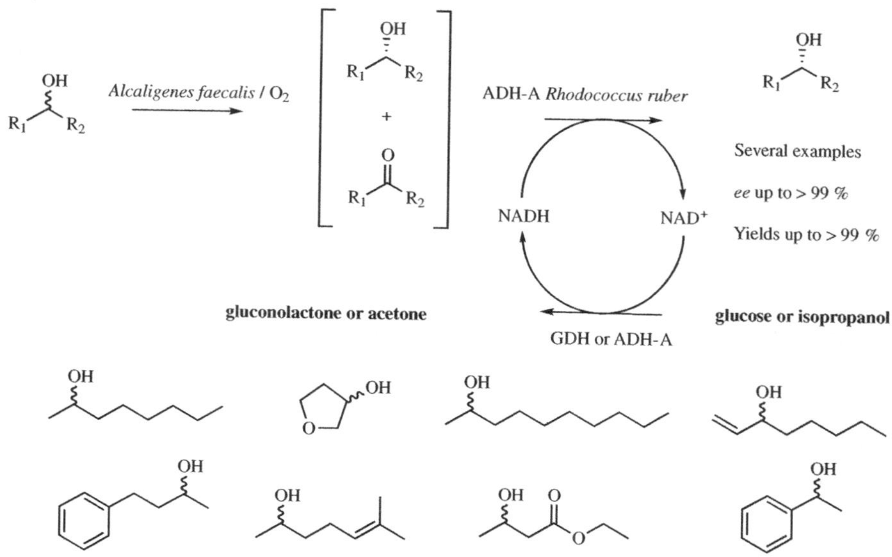

FIGURE 7.28 Deracemization of secondary alcohols through a biocatalytic selective oxidation and reduction [91].

second oxidoreductase catalyzes the reduction of the ketone to the alcohol enantiomer with the opposite configuration (see Chapter 3) [90]. In one of these approaches, the deracemization of different secondary alcohols was performed by coupling fresh cells of *Alcaligenes faecalis* (with O_2) together with ADH-'A' from *R. ruber*. Under optimized reaction conditions, the intermediate ketone was not observed, and high enantiomeric excesses were achieved for a broad spectrum of substrate alcohols. Remarkably, the concept was successfully applied by using two ways of regenerating the cofactor, either by means of the same enzyme (ADH-'A') and isopropanol addition, or by using glucose and GDH system (Figure 7.28) [91].

An analogous and very elegant approach, aiming at the same deracemization, used two oxidoreductases with different cofactor bias (either NADH or NADPH). Thus, by "decoupling" the cofactor regeneration cycles it was possible to show how the first ADH would actually perform the oxidation of a racemic alcohol, whereas the second ADH would perform the reduction of the formed ketone. In this case, the first cycle (oxidation) involved the (*R*)-selective ADH from *L. kefir* (LKADH) and the FMN-dependent YcnD from *B. subtilis*, whereas in the second cycle again ADH-'A' from *R. ruber*, together with FDH from *C. boidinii* were used (Figure 7.29) [92].

Other reported examples have focused on the combination of metal catalysis with oxidoreductases, for example, enzymatic sulfoxidation catalyzed by chloroperoxidase from *Caldariomyces fumago*, in which the required hydrogen peroxide was *in situ* generated *via* Pd(0)-catalyzed formation from hydrogen and oxygen in supercritical CO_2 [93]. Another recent example discloses the synthesis of homoallylic alcohols by means of a chemoenzymatic one-pot oxidation–allylation cascade, thus providing elongated homoallylic alcohols in water and in excellent yields. To this end, galactose oxidase from *Fusarium* NRRL 2903 was used in combination with

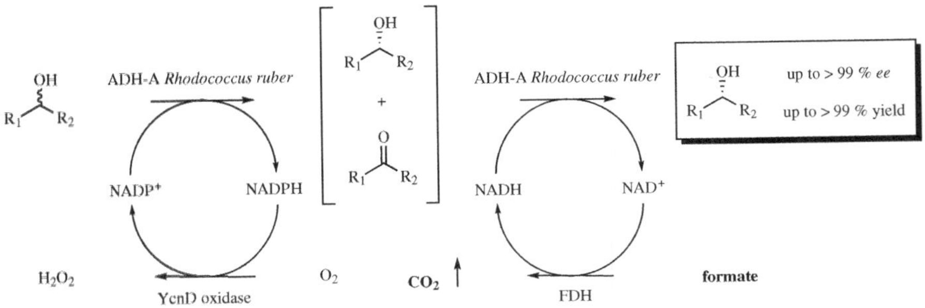

FIGURE 7.29 Deracemization *via* stereoinversion of secondary alcohols by concurrent oxidation and reduction cycles [92].

HRP and ABTS (2,2′-azino-bis(3-ethylbenzothiazoline-6-sulfonic-acid) to remove the hydrogen peroxide formed during the alcohol oxidation (Figure 7.30) [94].

Another successful example is represented by a palladium-catalyzed Suzuki coupling combined with oxidoreductases to afford different chiral aromatic diols by means of enantio- and diastereoselective ketone reductions (Figure 7.31) [95,96]. Interestingly, after finding conditions that were compatible with both reactions, by simply changing the oxidoreductase, access to all enantiomers or diastereomers was possible. Overall, it represents a further example on the versatility of biocatalysis in general, and redox enzymes in particular, for the production of high added value products.

An analogous approach was developed aiming at combining the well-known Wittig reaction with, again, oxidoreductases. In this case, α,β-unsaturated ketones were

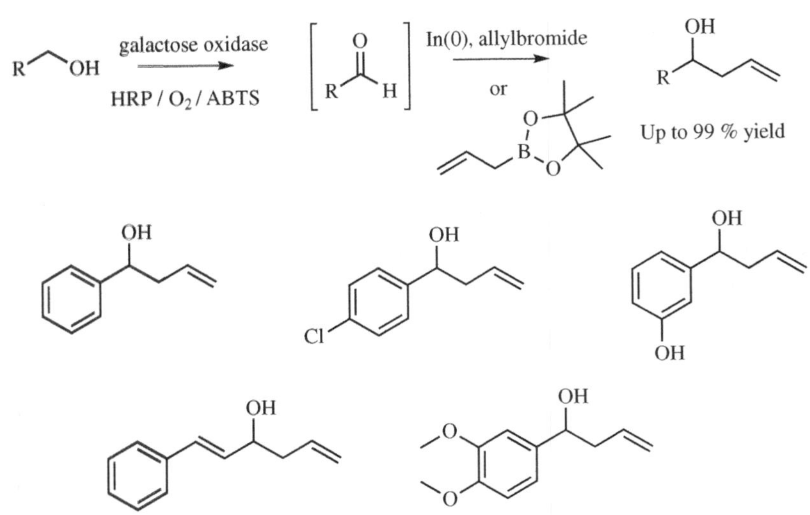

FIGURE 7.30 Concept for the chemoenzymatic synthesis of homoallylic alcohols *via* a chemoenzymatic one-pot oxidation–allylation cascade, and selected examples [94].

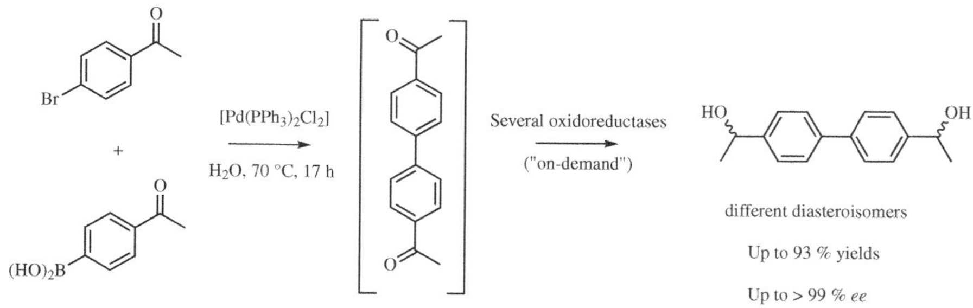

FIGURE 7.31 Chemoenzymatic one-pot combination of palladium-catalyzed cross coupling together with enzymatic reduction [95,96].

formed *in situ* and subsequently enantioselectively reduced by different oxidoreductases with different enantioselectivities, namely the (*R*)-selective ADH from *L. kefir*, or the (*S*)-selective ADH from *Rhodococcus* sp. Isopropanol was used for cofactor regeneration. Conversions ranging 60–90% and excellent enantiomeric excesses (typically >99%) were achieved (Figure 7.32) under optimized reaction conditions [97].

In addition to the reported examples, combinations of organo- and biocatalytic approaches have started to appear as well [98]. Being organocatalysts and enzymes typically applied under very mild and environment friendly conditions, the successful combination of them to provide powerful synthetic synergies was somehow expected. In this case, as proof-of-concept a highly versatile and diastereoselective synthesis

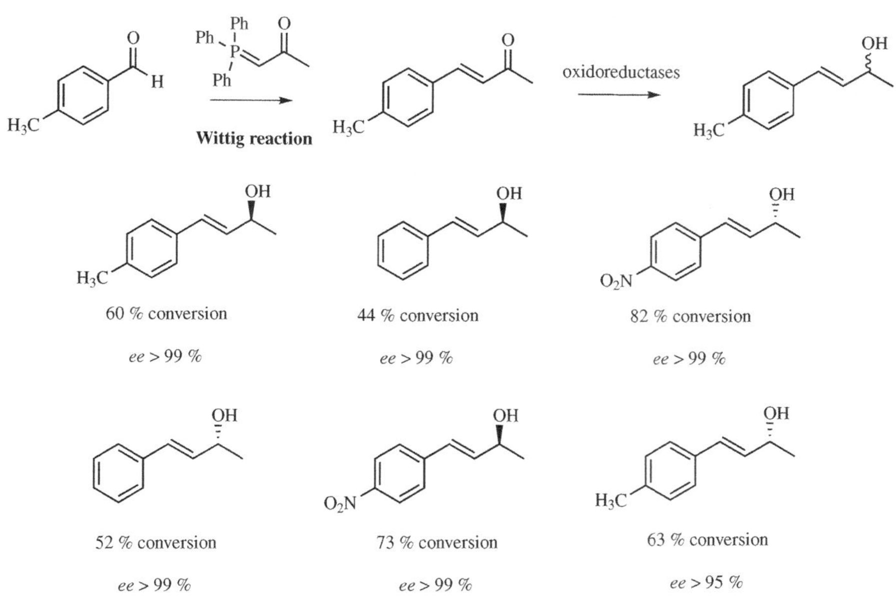

FIGURE 7.32 One-pot two-step synthesis of hydrophobic allylic alcohols combining a Wittig reaction and enzymatic ketone reduction [97].

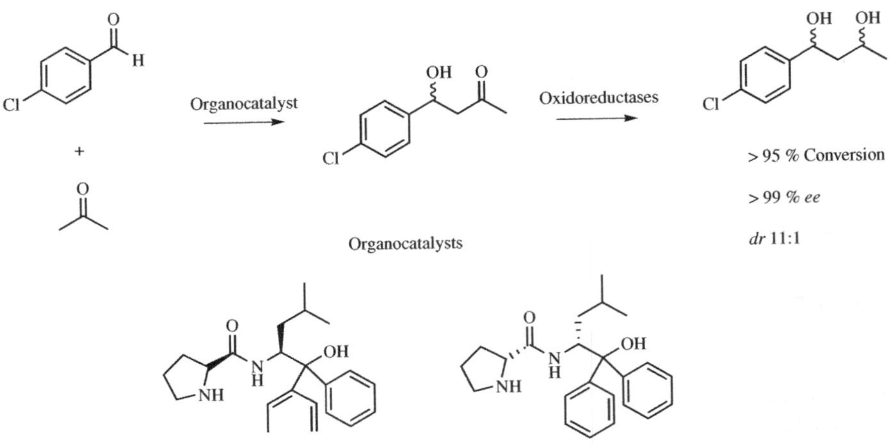

FIGURE 7.33 Combination of organocatalysis and biocatalysis to afford chiral 1,3-diols [98].

of diols was reported, by combining proline-like organocatalysis together with different oxidoreductases with complementary enantioselectivity. Outstanding results in terms of conversion, enantiomeric excess, and diastereoselectivities were achieved (Figure 7.33) [98].

7.6 CONCLUDING REMARKS

Chapters 1–5 of this book have addressed the diversity of redox enzymes, providing a comprehensive state-of-the-art of oxidoreductases, biochemical features, and so on. Subsequently, Chapter 7 has dealt with the necessary techniques and tools needed to convert knowledge—enzymes catalyzing a certain reaction—in a robust bioprocess with the chance to be scaled-up and transferred to commercial level. Thus, biocatalysis starts (ideally) with the identification of an enzyme able to catalyze a biotransformation with the desired substrate. To this end, sources of biocatalysts (see Chapter 1) must be screened in the quest for activities. Other approaches like metagenomics and screening of genome data banks have also gained importance in the last years. Once that biocatalyst screening is performed, identified (wild-type) hits often need to be genetically improved to enable a competitive biotransformation. For such purpose, directed evolution techniques have been broadly developed in the last decade, providing a palette of tools for the enzymatic design. Some outstanding examples are already available (see Chapters 7 and 8). With the goal of the genetic improvement in mind, the development of efficient and sensitive high-throughput-screening methods are crucial. Thus, in this chapter the most common methods for the screening of redox enzymes have been discussed, ranging from *simple* monitoring of NAD(P)H changes at 340 nm to more sophisticated fluorescence-based approaches. Depending on the complexity of the system, and on the possibility of background reactions, the selection of a screening method is rather important for the success of the future biotransformation.

Once a promising mutant is created, suitable reaction conditions must be set as well. In this respect, most of the organic substrates are hardly soluble in aqueous systems (usual reaction media for enzymes). To overcome this limitation, either water miscible cosolvents are applied, or the set-up of biphasic systems is envisaged. The latest developments also suggest that some oxidoreductases may be active in water-free systems, what certainly would open new research avenues in applied biocatalysis. In addition, the set-up of enzyme membrane reactors has proven to be very useful for redox-based enzymatic processes. Finally, another important issue that must be tackled is the provision of an efficient cofactor regeneration system. In this regard, both the use of isolated enzymes—with addition of two enzymes or complementary substrates—or the set-up of whole-cells systems have reached the level of established technology. Depending on factors that are usually decided on a case-by-case scenario, both systems are presently being used at industrial level (see also Chapter 8 for examples and Chapter 2 for a more in-depth study of cofactor regeneration systems).

Finally, in this chapter, future aspects of biocatalytic integration in general, and involving redox enzymes in particular, have been discussed. More in detail, multi-step cascade reactions are currently being developed, and it is expected that they will be routinely applied at industrial scale within the next decades. In this area, several outstanding examples of combinations of enzyme–enzyme, enzyme–metal, or enzyme–organocatalyst have already been reported at the proof-of-principle level, and discussed herein. In general, such approaches provide a more simplified operational unit, with diminished waste formation during downstream processing, and avoiding the use of large amounts of hazardous chemicals, since many toxic substrates may be generated *in situ* in the reactor. The demonstration that redox enzymes can be efficiently combined in such integrated processes, once reaction conditions have been complemented and set, represents a promising research line for the next years.

REFERENCES

1. Goddard JP, Reymond JL. Trends Biotechnol 2004;22:363–370.
2. Goddard JP, Reymond JL. Curr Opin Biotechnol 2004;15:314–322.
3. Reymond JL. Food Technol Biotechnol 2004;42:265–269.
4. Reymond JL, Babiak P. Adv Biochem Eng/Biotechnol 2007;105:31–58.
5. Medici R, Domínguez de María P, Otten L, Straathof AJJ. Adv Synth Catal 2011;353:2369–2376.
6. Reetz MT. Angew Chem Int Ed 2011;50:138–174.
7. Johannes TW, Zhao H. Curr Opin Microbiol 2006;9:261–267.
8. Turner NJ. Nat Chem Biol 2009;5:567–573.
9. Jäckel C, Kast P, Hilvert D. Annu Rev Biophys Biomol Struct 2008;37:153–173.
10. Stemmer WPC. Nature 1994;370:389–391.

11. Powell KA, Ramer SW, del Cardayré SB, Stemmer WPC, Tobin MB, Longchamp PF, Huisman GW. Angew Chem Int Ed 2001;40:3948–3959.
12. Mihovilovic MD, Rudroff F, Winninger A, Schneider T, Schulz F, Reetz MT. Org Lett 2006;8:1221–1224.
13. Reetz MT, Brunner B, Schneider T, Schulz F, Clouthier CM, Kayser MM. Angew Chem Int Ed 2004;43:4075–4078.
14. Reetz MT, Daligault F, Brunner B, Hinrichs H, Deege A. Angew Chem Int Ed 2004;43:4078–4081.
15. Bougioukou DJ, Kille S, Taglieber A, Reetz MT. Adv Synth Catal 2009;351:3287–305.
16. Ma SK, Gruber J, Davis C, Newman L, Gray D, Wang A, Grate J, Huisman GW, Sheldon RA. Green Chem 2010;12:81–86.
17. Liang J, Mundorff E, Voladri R, Jenne S, Gilson L, Conway A, Krebber A, Wong J, Huisman G, Truesdell S, Lalonde J. Org Proc Res Dev 2010;14:188–192.
18. Reisinger C, van Assema F, Schrümann M, Hussain Z, Remler P, Schwab H. J Mol Catal B: Enzym 2006;39:149–155.
19. Baumann M, Stürmer R, Bornscheuer UT. Angew Chem Int Ed 2001;40:4201–4204.
20. Li Z, Bütikofer L, Witholt B. Angew Chem Int Ed 2004;43:1698–1702.
21. Dey S, Powell DR, Hu C, Berkowitz DB. Angew Chem Int Ed 2007;46:7010–7014.
22. Hulley ME, Toogood HS, Fryszkowska A, Mansell D, Stephens GM, Gardiner JM, Scrutton NS. ChemBioChem 2010;11:2433–2447.
23. Johannes TW, Woodyer RD, Zhao H. Appl Environ Microbiol 2005;71:5728–5734.
24. Mayer KM, Arnold FH. J Biomol Screening 2002;7:135–130.
25. Tang WL, Li Z, Zhao H. Chem Commun 2010;46:5461–5463.
26. García-Ruiz E, Maté D, Ballesteros A, Martínez AT, Alcalde M. Microb Cell Fact 2010;9:17.
27. Alexeeva M, Enright A, Dawson MJ, Mahmoudian M, Turner NJ. Angew Chem Int Ed 2002;41:3177–3180.
28. Carr R, Alexeeva M, Enright A, Eve TSC, Dawson MJ, Turner NJ. Angew Chem Int Ed 2003;42:4807–4810.
29. Vallee BL, US 5162203. 1992.
30. Roeschlaub CA, Maidwell NL, Reza Rezai M, Sammes PG. Chem Commun 1999;1637–1638.
31. Klein G, Reymond JL. Bioorg Med Chem Lett 1998;8:1113–1116.
32. Gutiérrez MC, Sleegers A, Simpson HD, Alphand V, Furstoss R. Org Biomol Chem 2003;1:3500–3506.
33. Sicari R, Chen LS, Marzaioli AJ, Reymond JL. Adv Synth Catal 2005;347:1041–1050.
34. Joo H, Arisawa A, Lin Z, Arnold FH. Chem Biol 1999;6:699–706.
35. Joo H, Lin Z, Arnold FH. Nature 1999;399:670–673.
36. Matsuda T, Yamanaka R, Nakamura K. Tetrahedron: Asymmetry 2009;20:513–557.
37. Moore JC, Pollard DJ, Kosiek B, Devine PN. Acc Chem Res 2007;40:1412–1419.
38. Servi S, Synthesis 1990, 1–25.
39. Sybesma WFH, Straathof AJJ, Jongejan JA, Pronck JT, Heinen JJ. Biocatal Biotransform 1998;16:95–134.

40. Neuberg C. Biochem Z 1914;62:482–488.
41. Neuberg C, Lewite A. Biochem Z 1918;91:257–256.
42. Acetti D, Brenna E, Fuganti C, Gatti FG, Serra S. Eur J Org Chem 2010;1:142–151.
43. Kuhn A, van Zyl C, van Tonder A, Prior BA. Appl Environ Microbiol 1995;61:1580–1585.
44. Stewart JD, Rodríguez S, Kayser MM. In: *Enzyme Technology for Pharmaceutical and Biotechnological Applications*. New York: Marcel Dekker, Inc.; 2001. pp. 175–208.
45. Shieh WR, Sih CJ. Tetrahedron: Asymmetry 1993;4:1259–1269.
46. Nakamura K, Kawai Y, Nakajima N, Ohno A. J Org Chem 1991;56:4778–4783.
47. Shieh WR, Gopalan AS, Sih CJ. J Am Chem Soc 1985;107:2993–2994.
48. Zhou BN, Gopalin AS, Van Middlesworth F, Shieh WR, Sih CJ. J Am Chem Soc 1983;105:5925–5926.
49. Nakamura K, Miyai T, Nozaki K, Ushio K, Oka S, Ohno A. Tetrahedron Lett 1986; 27:3155–3156.
50. D'Arrigo P, Pedrocchi Fantoni G, Servi S, Strini A. Tetrahedron: Asymmetry 1997; 8:2375–2379.
51. Wipf B, Kupfer E, Bertazzi R, Leuenberger HGW. Helv Chim Acta 1983;66:485–488.
52. Ehrler J, Giovannini F, Lamatsch B, Seebach D. Chimia 1986;40:172–173.
53. Kometani T, Yoshii H, Kitatsuji E, Nishimura H, Matsuno R. J Ferment Bioeng 1993;76:33–37.
54. Nakamura K, Kondo SI, Kawai Y, Ohno A, Tetrahedron Lett 1991, 7075–7078.
55. Nakamura K, Kondo SI, Kawai Y, Ohno A. Bull Chem Soc Jpn 1993;66:2738–2743.
56. Jayasinghe LY, Smallridge AJ, Trewhella MA. Tetrahedron Lett 1993;34:3949–3950.
57. North M. Tetrahedron Lett 1996;37:1699–1702.
58. Medson C, Smallridge AJ, Trewhella MA. Tetrahedron: Asymmetry 1997;8:1049–1054.
59. Cui JN, Teraoka R, Ema T, Sakai T, Utaka M. Tetrahedron Lett 1997;38(17):3021–3024.
60. Nakamura K, Inoue K, Oshio K, Oka S, Ohno A. Chem Lett. 1987;16:679–682.
61. Guo ZW, Sih CJ. J Am Chem Soc 1989;111:6836–6841.
62. Nakamura K, Kawai Y, Oka S, Ohno A. Bull Chem Soc Jpn 1989;62:875–879.
63. Nakamura K, Kawai Y, Miyai T, Ohno A. Tetrahedron Lett 1990;31:3631–3632.
64. Nakamura K, Kawai Y, Ohno A. Tetrahedron Lett 1990;31:267–270.
65. Kawai Y, Kondo SI, Tsujimoto M, Nakamura K, Ohno A. Bull Chem Soc Jpn 1994;67:2244–2347.
66. Rodríguez S, Kayser MM, Stewart JD. J Am Chem Soc 2001;123:1547–1555.
67. Rodríguez S, Kayser M, Stewart JD. Org Lett 1999;1:1153–1155.
68. Stewart JD, Reed KW, Martínez CA, Zhu J, Chen G, Kayser MM. J Am Chem Soc 1998;120:3541–3548.
69. Katz M, Hahn-Hägerdal B, Gorwa-Grauslund MF. Enzyme Microb Technol 2003;33:163–172.
70. Katz M, Frejd T, Hahn-Hägerdal B, Gorwa-Grauslund MF. Biotechnol Bioeng 2003;84:573–582.
71. Kaluzna I, Andrei AA, Bonilla M, Martzen MR, Stewart JD. J Mol Catal B: Enzym 2002;17:101–105.

72. Walton AZ, Stewart JD. Biotechnol Progr 2004;20:403–411.
73. Rodríguez S, Schroeder KT, Kayser MM, Stewart JD. J Org Chem 2000;65:2586–2587.
74. Yang Y, Kayser MM, Rochon FD, Rodríguez S, Stewart JD. J Mol Catal B: Enzym 2005;32:167–174.
75. Kayser MM, Drolet M, Stewart JD. Tetrahedron: Asymmetry 2005;16:4004–4009.
76. Kaluzna IA, Feske BD, Wittayanan W, Ghiviriga I, Stewart JD. J Org Chem 2005;70:342–345.
77. Padhi SK, Kaluzna IA, Buisson D, Azerad R, Stewart JD. Tetrahedron: Asymmetry 2007;18:2133–2138.
78. Kaluzna IA, Matsuda T, Sewell AK, Stewart JD. J Am Chem Soc 2004;126:12827–12832.
79. Choi YH, Choi HJ, Kim D, Uhm KN, Kim HK. Appl Microbiol Biotechnol 2010;87:185–193.
80. Jung J, Park HJ, Uhm KN, Kim D, Kim HK. Biochim Biophys Acta 2010;1804:1841–1849.
81. Jakoblinnert A, Mladenov R, Paul A, Sibilla F, Schwaneberg U, Ansorge-Schumacher MB. Domínguez de María P. Chem Commun 2011;47:12230–12232.
82. Goldberg K, Edegger K, Kroutil W, Liese A. Biotechnol Bioeng 2006;95:192–198.
83. Bräutigam S, Dennewald D, Schürmann M, Lutje-Spelberg J, Pitner WR, Weuster-Botz D. Enzyme Microb Technol 2009;45:310–316.
84. Kohlmann C, Robertz N, Leuchs S, Greiner L, Na'amnieh S. Green Chem 2011;13:3093–3095.
85. Goldberg K, Schroer K, Lütz S, Liese A. Appl Microbiol Biotechnol 2007;76:249–255.
86. Goldberg K, Schroer K, Lütz S, Liese A. Appl Microbiol Biotechnol 2007;76:237–248.
87. Lütz S, Rao NN, Wandrey C. Chem Eng Technol 2006;29(12)1404–1415.
88. Schroer K, Mackfeld U, Wei Tan I, Wandrey C, Heuser F, Bringer-Meyer S, Weckbecker A, Hummel W, Daußmann T, Pfaller R, Liese A, Lütz S. J Biotechnol 2007;132:438–444.
89. Bruggink A, Schoevaart R, Kieboom T. Org Proc Res Dev 2003;7:622–640.
90. Ricca E, Brucher B, Schrittwieser JH. Adv Synth Catal 2011;353:2239–2262.
91. Voss CV, Gruber CG, Kroutil W. Angew Chem Int Ed 2008;47:741–745.
92. Voss CV, Gruber CG, Faber K, Knaus T, Macheroux P, Kroutil W. J Am Chem Soc 2008;130:13969–13972.
93. Karmee SK, Roosen C, Kohlmann C, Lutz S, Greiner L, Leitner W. Green Chem 2009;11:1052–1055.
94. Fuchs M, Schober M, Pfeffer J, Kroutil W, Birner-Gruenberger R, Faber K. Adv Synth Catal 2011;353:2354–2358.
95. Burda E, Bauer W, Hummel W, Gröger H. ChemCatChem 2010;2:67–72.
96. Burda E, Hummel W, Gröger H. Angew Chem Int Ed 2008;47:9551–9554.
97. Krausser M, Hummel W, Gröger H. Eur J Org Chem 2007;5175–5179.
98. Baer K, Krausser M, Burda E, Hummel W, Berkessel A, Gröger H. Angew Chem Int Ed 2009;48:9355–9358.

CHAPTER 8

Industrial Applications of Biocatalytic Redox Reactions: From Academic Curiosities to Robust Processes

8.1 MOTIVATION: DRIVERS FOR INDUSTRIAL BIOCATALYTIC PROCESSES

The ample range of redox biocatalytic reactions, usually with broad substrate spectra, described in previous chapters of this book, has attracted the interest of industries to assess whether such biotransformations would be able to meet industrial goals. This has successfully triggered the set-up of several biocatalytic processes at industrial scale in different chemical segments, from pharmaceuticals or fine chemicals to a range of low-added-value products produced in bulk [1–6]. To fully reach the requirements of a practical industrial process, it is obvious that biocatalysts must be technically and economically integrated in process pipelines. Moreover, continuous improvement and innovation are being brought along the entire process, to assure that costs can be tackled in an optimum way, and that emerging breakthroughs are rapidly incorporated in the set-ups as well [7–9]. Besides, some recent examples, for example, terminal alkene bioproduction (e.g., isobutylene, propylene) *via* metabolic engineered whole-cells [10], or highly efficient transaminase-catalyzed processes [11], are further cases that clearly indicate that the implementation of biocatalytic strategies at industrial level will become much more established in the coming decades. Furthermore, the discovery of new groups of oxidoreductases, for example, nitrile reductases, may nicely open new and important research avenues with potential industrial applications, in virtue of the importance that nitrile chemistry has in many industrial steps [12].

As largely stated in previous chapters of this book, there are several key drivers that can be pinpointed as important assets of enzymes when industrial biotransformations are envisaged. These drivers are the main motivation for industries to look at enzymes as promising and efficient biocatalysts:

- Selectivity. Enzymes have been evolved in nature during millions of years to display an exquisite selectivity toward their natural substrates, under their natural

Redox Biocatalysis: Fundamentals and Applications, First Edition.
Daniela Gamenara, Gustavo A. Seoane, Patricia Saenz-Méndez, and Pablo Domínguez de María.

reaction conditions. Interestingly, when nonnatural reactions and/or nonnatural substrates are involved, some enzymes provide often high enantio-, regio-, and/or chemoselectivity. In many cases, the displayed selectivity is higher than that observed for chemocatalysts, thus becoming an attractive alternative to replace other less efficient catalysts for a certain reaction or substrate. This is actually the most well-known and valued property of enzymes, when a practical application is envisaged. Thus, a single catalyst can reach many products (from different substrates or reactions) with further industrial interest, market, or use.

- Waste reduction. Processes involving enzymes are usually set up with high atom economies and efficiencies. Likewise, albeit not always, in many cases milder solvents for the milieu are used in biocatalysis also. Therefore, the set-up of enzymatic processes may reduce the ecological footprint that production of chemicals may bring. This is certainly an aspect that will be increasingly important in the coming future, due to environmental concerns, as well as to the envisaged establishment of more strict regulations in terms of residues allowed from chemical industries. By using enzymes there is the potential to set up efficient processes that at the same time may lead to less waste formation, while keeping *on-spec* selectivities for product quality.
- Energy savings. Since reaction conditions for enzymatic processes are usually very mild (namely ambient pressure and moderate temperatures, $<100°C$), the application of enzymatic processes can contribute to tackle problems derived from consumption of fossil fuels and thus environmental problems associated to that use. Again, envisaged future regulations may trigger the search for enzymatic processes, which are expected to account for a significant amount of (new) catalytic processes that will be implemented.
- Less by-products formation. In virtue of the aforementioned reaction conditions, the use of enzymes often leads to less formation of unexpected compounds, thus simplifying the work-up procedures, and enhancing the overall yields of a given process. Likewise, other undesired properties that may often be found in commercial products, for example, color formation, may be circumvented due to the mild reaction conditions applied with enzymes.
- Access to new molecules/markets. Enzymes are biodegradable catalysts. This may confer some advantages when products for certain markets (e.g., food industry) are envisaged. The (industrial) anticipation in developing a novel bioprocess for a certain market gap may provide excellent opportunities for building a strong IP position (intellectual property) in a given field. Herein, a smart combination of nonhazardous solvents with mild reaction conditions, together with biodegradable and nontoxic biocatalysts may provide nice synergies. Industries have already started to look at them as a future strategy (e.g., aroma industry).

8.2 KEY ASPECTS IN INDUSTRIAL BIOCATALYTIC PROCESSES

Obviously, to convert mere "academic curiosities," for example, enzymes catalyzing reactions in nonnatural media and with nonnatural substrates, into robust industrial biocatalytic processes—thus benefiting from the above-mentioned enzyme added values—many aspects must be considered, integrated, and improved within a certain biotransformation. To put it in simple terms, enzymatic processes must be able to compete in economic figures with their chemical counterparts, to assure that there will be a chance for their industrial implementation and eventually to replace a current process by a new one. In this respect, several key aspects of modern biocatalysis can be pinpointed as crucial for such practical implementation (see previous chapters for deeper discussions on these topics):

- Cloning and overexpression of biocatalysts. Production (biosynthesis) of enzymes performed by wild-type microorganisms is seldom high enough to assure an "industrial" provision of the enzyme, though there are some exceptions in which wild-type microorganisms can be directly used as biocatalysts. However, molecular biology techniques have spectacularly developed in the last decades, thus providing the framework to clone and overexpress a certain enzyme within a heterologous host (often bacteria or yeasts). This strategy permits the "on-demand" production and accessibility of biocatalysts, which are produced by *simple* fermentative processes. Likewise, applying this technology allows the sustainable broad industrial use of biocatalysts from sensitive sources (e.g., enzymes from plants or animals, see Chapter 1), from which a "wild-type" extraction would never reach sustainable targets, given the enormous amount of resources that would be needed to provide large-scale quantities of the enzyme. Some outstanding examples of that approach may be the use of hydroxynitrile lyases from plants, or the recently cloning and overexpression of pig liver esterase (PLE) in several heterologous hosts, leading to the provision of not only one, but up to six PLE isoenzymes with different enantioselectivities and bias toward substrates, that can now be applied separately in pharmaceutical industry. In addition, in the case of PLE the microbial origin of the cloned ones avoids problems related to virus transmission from animal sources (pigs), as well as social barriers to the use of them (e.g., religious beliefs) [13–15]. Last but not least, the implementation of molecular biology techniques (cloning and overexpression) allows the industrial production of mutant enzymes as well.
- Genetic improvement of enzymes. Furthermore, when industrial processes are envisaged, it is important to note that catalytic performances usually displayed by wild-type enzymes are not enough for the strong requirements of an industrial application using nonnatural substrates. It is important to mention that these nonnatural substrates are often structurally very different from natural ones, thus allowing enzymatic activities but at rates that may be several orders of magnitude lower than those observed for natural substrates. Furthermore, industrial reactions may not be operated under analogous natural conditions for the enzyme, again leading to low stabilities and poor biocatalytic activities for the desired process. To overcome this drawback, directed evolution techniques allow the genetic improvement of the biocatalysts to adapt them to a *real* application, either in terms of enantioselectivity, activity, or stability of the enzyme

toward, for example, an organic solvent. For all of these aspects there are outstanding examples of improved enzymes, even by several orders of magnitude, to reach the required industrial standards of competitiveness (see Chapter 7 for further detailed information, and further in this chapter, Sections 8.3 and 8.4 for several selected industrial examples of genetically improved enzymes).

- Metagenomics. It appears obvious that not all the enzymes are valid for industrial uses in nonnatural conditions, or with nonnatural substrates. In fact many enzymes will display a restricted substrate spectrum to only their natural products. Others will not show activity at all, or will have low stability under nonnatural conditions, for example, in the presence of organic solvents. However many other enzymes are actually very useful for practical applications, since they display activity toward different substrates and under a diverse range of reaction conditions. Therefore, the identification of novel enzymes that may serve as promising future industrial biocatalysts is a key aspect in industrial biotransformations, albeit usually this can become a tough challenge. To add to this challenge, the proportion of biodiversity that is accessible by conventional fermentative approaches stands is almost negligible compared to the whole biodiversity existent in the earth. To overcome this issue, metagenomic approaches offer the possibility of cloning and overexpression of enzymes of nonculturable microorganisms, thus clearly enhancing the opportunities that enzyme-catalysis can bring for practical applications. This field is being developed very fast, with already some successful outstanding cases of important enzymes found *via* these approaches [16,17].
- Biocatalytic promiscuity. As widely discussed in Chapter 6, many enzymes can catalyze not only their natural reaction with their natural substrate, but conversely provide a rather broad palette of reactions, often with ample substrate spectra. Obviously, these features are very important assets when industrial applications with enzymes are envisaged, since by means of just one biocatalyst several reactivities, and consequently different products, can be reached. The biochemical basis for enzymatic promiscuity are important academic tasks, since a correct and rational understanding of them may indeed facilitate the development of novel processes based on the same biocatalyst [18–25].

With regard to redox biocatalysis, the option of using enzymes for industrial processes has moved pretty quickly from the mere "lab curiosity" to become, in many cases, the "first choice" of consideration in industries, when novel redox reactions or novel substrates are foreseen [26]. In this specific area, a crucial issue to be tackled is the provision of an efficient cofactor regeneration system that may afford competitive prices for a given biotransformation. As reported in this book, cofactors for redox enzymes are very expensive, and therefore their direct use in stoichiometric amounts is largely prohibited due to economic reasons. Thus, cofactors must be added in catalytic amounts (or not added at all, if whole-cells are used), to assure an economic competitiveness of the entire cycle. Herein, from the wide number of alternatives for cofactor regeneration methods that have been put forward in the literature (see

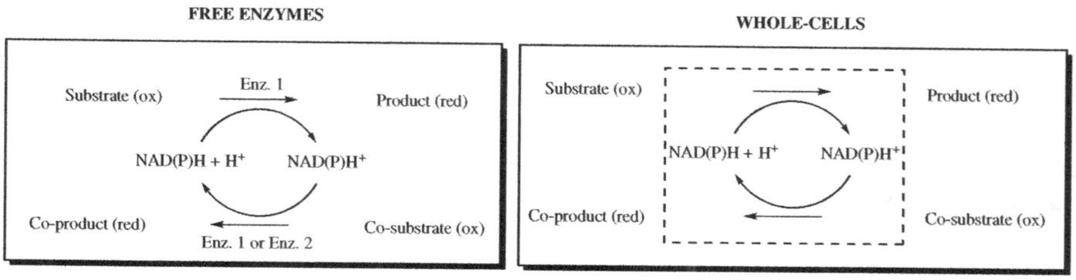

FIGURE 8.1 Cofactor regeneration strategies that are being used in industrial applications [26–28].

also Chapter 2 for a further and detailed discussion on the topic), only some of these approaches have found industrial relevance so far (Figure 8.1) [26–28].

As observed (Figure 8.1), industrial cofactor regeneration systems may involve the use of free enzymes or whole-cells. When the free enzyme alternative is preferred, coupled enzymes—either one enzyme for both (co-)substrates or two different ones—are used and catalytic amounts of cofactor need to be added to the reactor system. In addition to this approach, another alternative involves the direct use of whole-cells as biocatalyst. In this case, neither free enzymes nor cofactors need to be added, since all of them occur within the cell and cofactor is *in situ* regenerated by the enzymatic machinery of the whole-cell. Both approaches are possible for industrial applications, and it is rather arguable to put one approach before the other one. Advantages and disadvantages of both approaches were further discussed in Chapter 7 of this book.

Until now, a number of redox biocatalytic processes have been successfully implemented at industrial scale by different industries, with the goal of tackling different synthetic problems. As stated before, depending on the application, enzyme, or industry, either whole-cells or free enzymes are used. In Figure 8.2 some selected examples are depicted [29–31].

In the coming sections, several selected examples of industrial applications of redox enzymes, involving both coupled enzymes and whole-cells, will be discussed.

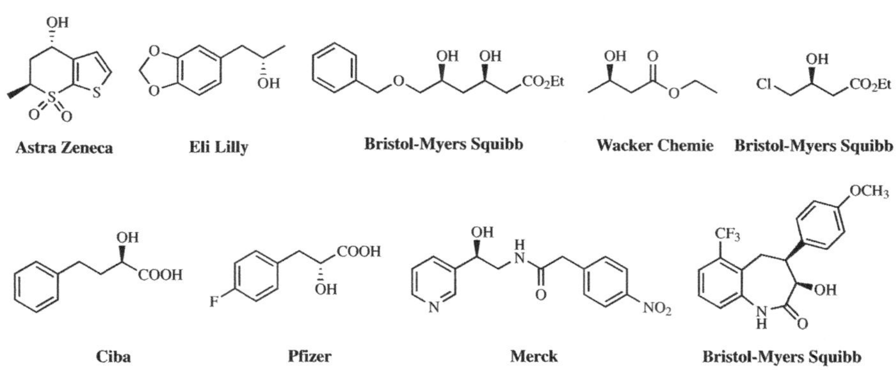

FIGURE 8.2 Selected examples of compounds produced involving enzyme-catalyzed redox processes [29–31].

Likewise, emerging approaches like metabolic engineering or the discovery of novel enzymes to add to the biocatalytic portfolio will be briefly mentioned as well.

8.3 INDUSTRIAL BIOCATALYTIC REDOX PROCESSES: FREE ENZYMES

Several examples of industrial redox enzymatic processes using free enzymes as biocatalysts have been reported so far. In general, industrial emphasis is given to the development of an efficient cofactor regeneration system, together with high substrate loadings despite the expected poor solubility of organic substrates in aqueous media (e.g., medium engineering approaches). Due to associated costs, the efficiency in cofactor regeneration is crucial to afford a competitive biocatalytic redox application at practical scale. On the other hand, virtually all the reported industrial processes dealing with redox enzymes, and specifically with free enzymes, are devoted to asymmetric synthesis, for example, enantioselective reduction–oxidations, to afford chiral compounds. Clearly, since market prices of these chemicals are rather high, the use of enzymes (oxidoreductases) may become economically feasible. Moreover, the provision of chemocatalytic enantioselective redox processes is often not trivial, thus enhancing the added value that enzymes may bring. Conversely, a much more challenging approach would be to produce low-added-value (nonchiral) products with oxidoreductases. In these cases, when no chiral induction is envisaged, the motivation for biocatalysis can be found in the waste reduction achieved by using enzymes, or in the significantly lower by-product formation, due to the mild reaction conditions applied in biocatalytic approaches. Conclusively, although asymmetric processes are presently core technologies of many industries, the latter nonchiral approaches are believed to become very important in the coming years in biocatalysis. Herein, a remaining challenge will be to reduce costs associated to this technology enabling market competition, which could be reached by using whole-cells and immobilizing them to afford stable and highly active biocatalysts [26].

With respect to chiral synthesis and the use of oxidoreductases as free enzymes, several dehydrogenases displaying outstanding performances have been reported. For instance, alcohol dehydrogenase from *Rhodococcus erythropolis* has been assessed by several industries, since it provides (as wild-type enzyme) a broad substrate spectrum ranging from aromatic to aliphatic ketones leading in most of the cases to high conversions (>95%) and enantioselectivities (>99% *ee*) (Figure 8.3) [32].

To assure high volumetric loading of organic substrates in typical aqueous solutions used in enzyme catalysis, the set-up of biphasic systems is common in industrial biotransformations. In this respect, the same alcohol dehydrogenase from *R. erythropolis* has been used in biphasic systems buffer:heptane (4:1) for the efficient industrial production of enantiomeric alcohols *via* ketone reduction with relatively high substrate loadings (~150 mM) in low reaction times (< 24 h). In these cases, formate dehydrogenase (FDH) from *Candida boidinii* was used to close the cofactor regeneration cycle by oxidizing formate to carbon dioxide, being the reaction equilibrium efficiently shifted toward the product formation. Interestingly, it is important

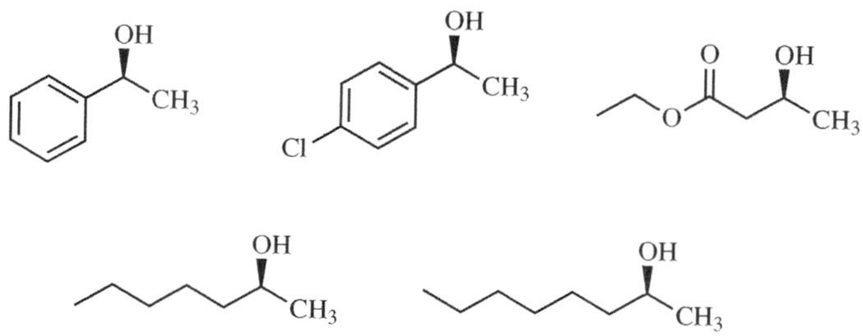

FIGURE 8.3 Products obtained from the ample substrate spectrum of alcohol dehydrogenase from *R. erythropolis* [32].

to mention that FDH is very sensitive to most organic solvents. In this case, however, high stabilities were found with alkanes like hexane or heptane (Figure 8.4). Thus, albeit the incorporation of these solvents may negatively influence the ecological footprint of the process, being both oil-based solvents, the desired high productivities demanded in an industrial approach can be reached [32–34].

Apart from setting-up biphasic systems, the same alcohol dehydrogenase from *R. erythropolis* was used for the effective synthesis of (*S*)-3,5-*bis*trifluoromethyl-phenylethanol, which is a useful building block for the synthesis of NK-1 receptor antagonists. In this case, substrate (ketone) was added directly in aqueous solutions, with an adequate stirring speed to assure full homogenization of the emulsion. When FDH was used as second enzyme for cofactor regeneration, productivities of ~100–110 $gL^{-1}d^{-1}$ were reached. Notably, the reaction concept was largely improved by using glucose dehydrogenase (GDH) as the second enzyme, since that enzyme is 10-fold more active than FDH with regard to the cofactor regeneration. In this latter case, glucose was added to assure the electron source for the redox process. Conclusively, the equilibrium is efficiently shifted to the production side,

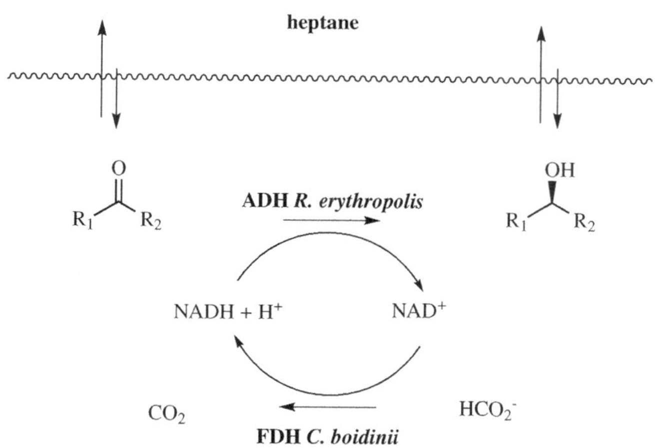

FIGURE 8.4 Concept for the biphasic system for industrial redox biocatalysis [32–34].

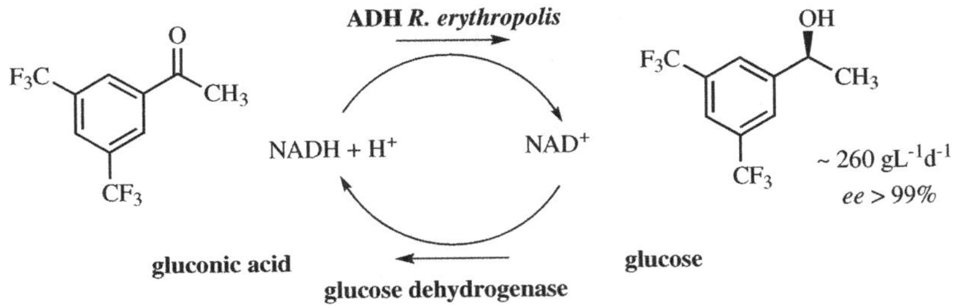

FIGURE 8.5 Enzyme-catalyzed production of (*S*)-3,5-*bis*trifluoromethylphenylethanol by using GDH as coupled enzyme for cofactor regeneration [35].

and therefore under optimized reaction conditions excellent *on-spec* productivities of ~260 $gL^{-1}d^{-1}$ were achieved (Figure 8.5) [35].

Another interesting enzyme that has shown promising use as "free enzyme" in industrial redox biocatalytic processes is the alcohol dehydrogenase from *Lactobacillus kefir* alcohol dehydrogenase (LKADH) [36,37]. In this case, however, activity and selectivity of the wild-type enzyme was not sufficient to assure *on-spec* production of chiral alcohols, as often observed for wild-type enzymes. Thus, for all the herein discussed processes, several rounds of genetic improvement of the enzyme *via* directed evolution techniques were needed (up to five point mutations, compared to wild-type enzyme), representing an excellent example on how modern molecular biology techniques can assist industrial biocatalysis to reach productivity targets. In this way, an efficient process for the production of (*R*)-2-methylpentanol was reported by means of kinetic resolution of the aldehyde (*via* reduction). In this case, the same enzyme was able to use isopropanol as ancillary cosubstrate to yield acetone, and thus to regenerate the cofactor cycle. Under optimized reaction conditions, high yields (46%, for a kinetic resolution with a maximum theoretical yield of 50%) from substrate loadings of ~220 gL^{-1} were achieved in less than 24 h reaction time, with high enantiomeric excesses (>98%) (Figure 8.6) [36].

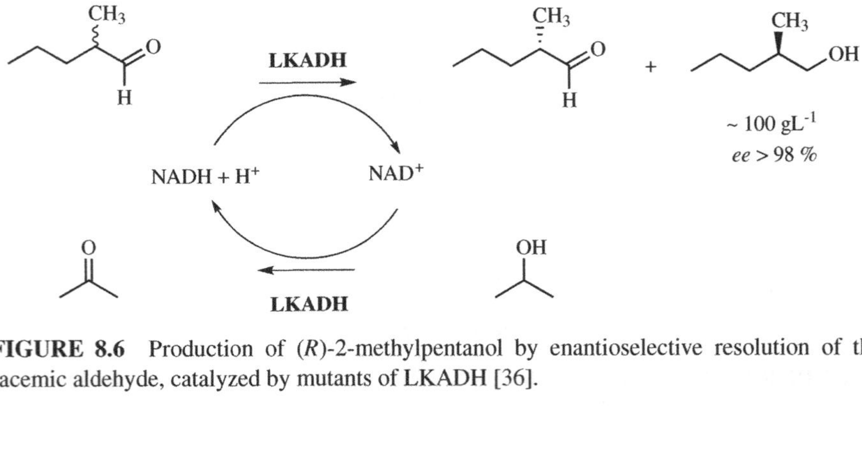

FIGURE 8.6 Production of (*R*)-2-methylpentanol by enantioselective resolution of the racemic aldehyde, catalyzed by mutants of LKADH [36].

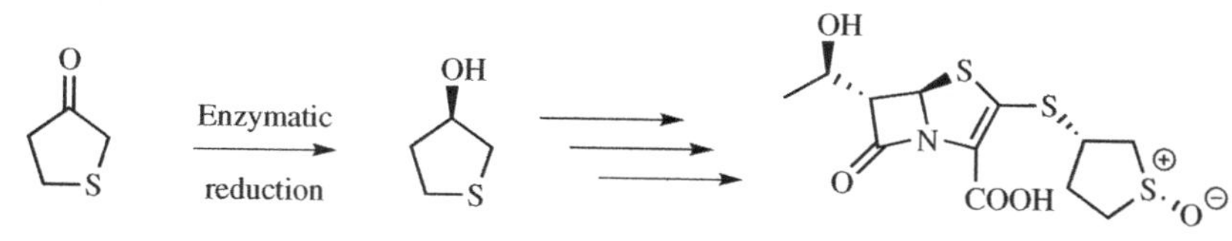

FIGURE 8.7 Biocatalytic step in the synthesis of sulopenem [37].

Moreover, the same LKADH has also served as template for further genetic improvements (up to eight rounds of directed evolution) to develop a mutant able to catalyze the enantioselective reduction of tetrahydrothiophene-3-one to afford a useful building block for the pharmaceutical synthesis of sulopenem, a potent antibacterial (Figure 8.7) [37]. The usefulness of that enzyme is noteworthy for this substrate, which is synthetically very challenging in virtue of the nearly spatially symmetrical (β-CH_2 vs. β-S) ketone used as starting material. By means of chemocatalysts, the accessibility of such molecule would be in fact much more challenging. In this process LKADH was coupled with GDH for cofactor regeneration, in the same way as depicted in Figure 8.5, that is, with addition of equivalents of glucose to provide an electron source for the redox process. The optimized reaction was successfully scaled up to 100 kg product [37].

Another relevant industrial redox process is represented by the synthesis of Montelukast, a leukotriene receptor antagonist that is used to control the symptoms of asthma and allergies [38]. For this substrate, the major challenge was the highly hydrophobic nature of the substrate, with a logP of ~7. It appears obvious, therefore, that a significant amount of organic solvent should be used to provide efficient substrate dissolution in the reaction media. To that end, a ketoreductase (from *Codexis* code CDX-026, its origin was not reported) was evolved toward high volumes of organic (co-)solvents by means of directed evolution strategies. Thus, under optimized conditions the activity of the biocatalyst was improved by approximately 3000-fold compared to the wild-type performance. For the reaction set-up, the best conditions resulted in a 1:5:3 mixture of toluene–isopropanol–buffer system (v/v/v). Under such processing conditions, a robust process that is able to compete with other chemical approaches, in terms of selectivity and productivity, was set-up (Table 8.1). Overall, the strategy is a nice example on how biocatalysis must (and is able to) be adapted to process set-up to afford competitive conditions for the production of a certain compound, depending on the market needs [38].

Likewise, ketoreductases have also been employed in a key step for the synthesis of cholesterol lowering drugs (e.g., Lipitor®), combined with other enzymes and chemocatalysts. The biocatalytic reduction displayed excellent yields (96%) and enantiomeric excesses (>99.5%), enabling an attractive process both from an economic and environmental viewpoint (Figure 8.8). For this case, again GDH was used, with addition of glucose equivalents, to afford efficient cofactor regeneration [39].

Furthermore, the use of several amino acid dehydrogenases for the reductive amination of α-keto-acids to afford (nonnatural) amino acids has also been extensively

TABLE 8.1 Synthesis of Montelukast. Comparison of Chemo-Catalyzed Route and the New Developed Enzymatic Approach [38]

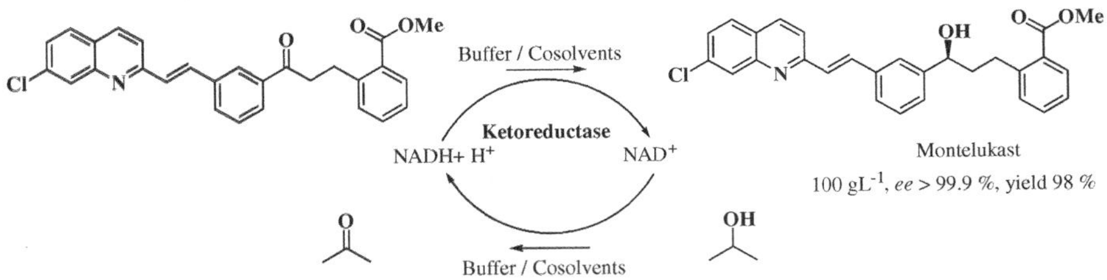

	Chemical Route[a]	Biocatalytic Route
Ketone loading	100 gL^{-1}	100 gL^{-1}
Reaction conditions	Moisture sensitive	Aqueous
Catalyst loading	At least 1.8 equiv	3–5 wt%
Optical purity	>99% (after crystallization)	>99.9% (no crystallization needed)
Yield	85–90%	90–98%

[a]Chemical catalyst used was (-)-B-chlorodiisopinocampheylborane.

used at industrial scale, even for decades now. In this area an outstanding example is represented by the synthesis of L-*tert*-leucine in combination with FDH from *C. boidinii* as coupled enzyme for cofactor regeneration (formate to carbon dioxide to shift the equilibrium). Herein, using free enzymes, yields of >70% and superb productivities of >600 $gL^{-1}d^{-1}$ were already reported in the 1990s [26]. Apart from this free-enzyme approach, these processes have also been implemented by using whole cells as biocatalysts (see Section 8.4).

In the same area, phenylalanine dehydrogenase from *Thermoactinomyces intermedius* has been used for the production of an intermediate for the pharmaceutical synthesis of Saxagliptin (BMS-477118), a dipeptidyl peptidase IV inhibitor for the

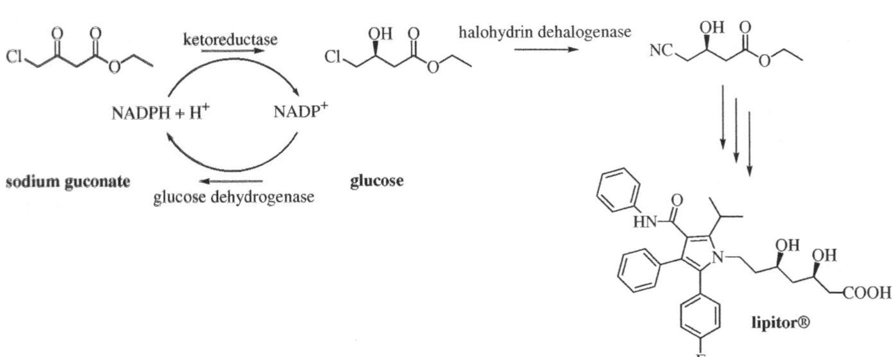

FIGURE 8.8 Biocatalytic steps—including a redox process—involved in the synthesis of the cholesterol lowering drug Lipitor [39].

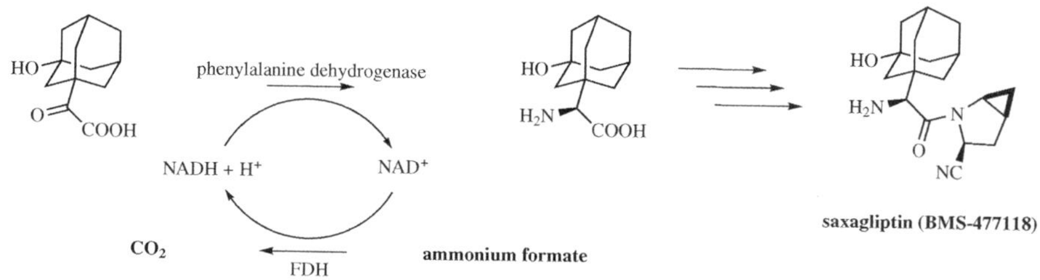

FIGURE 8.9 Synthesis of saxagliptin involving a biocatalytic oxidative step [40].

treatment of type 2 diabetes mellitus. Also in this case, FDH was used for cofactor regeneration, using ammonium formate both as electron and nitrogen donor for the amino acid synthesis (Figure 8.9) [40].

Notably, the same enzyme (*T. intermedius* phenylalanine dehydrogenase) is also able to accept other substrates with high efficiency and selectivity. Thus, it has been used for the synthesis of allysine ethylene acetal, a building block that is used in the production of Vanlev, a vasopeptidase inhibitor (Figure 8.10). As stated above in this chapter, this is an excellent example on how some enzymes can accept more than one substrate, thus enhancing the value that they may have for industrial practical purposes [41].

Other examples of amino acid dehydrogenases used at industrial scale have been reported as well, with several outstanding examples even at kg scale. In Figure 8.11 some other representative examples of this useful technology are depicted [29,42–45].

Using amino acid dehydrogenases, another very elegant strategy, recently reported, is to perform the kinetic resolution of amino acids by means of (*R*)-amino acid oxidase of *Trigonopsis variabilis* to afford the α-keto-acid and the untouched enantiomer of the amino acid and to couple that reaction with a second enzyme, a (*S*)-amino acid dehydrogenase, to yield the desired compound in theoretical yields of 100% (Figure 8.12) [46]. In other words, a multistep enzymatic process that affords important products with high efficiency and diminished waste formation, since less downstream steps need to be applied, as a novel trend that will be widely used in the coming decades (see Chapter 7 for further details). In this specific case, several enzymes acted "in concert" for an efficient production pattern.

Related to the above-described amino acid decarboxylases, ω-transaminases have emerged in the last years as powerful and promising biocatalysts for many industrial purposes [29,47]. Herein, transaminases have also developed from being considered

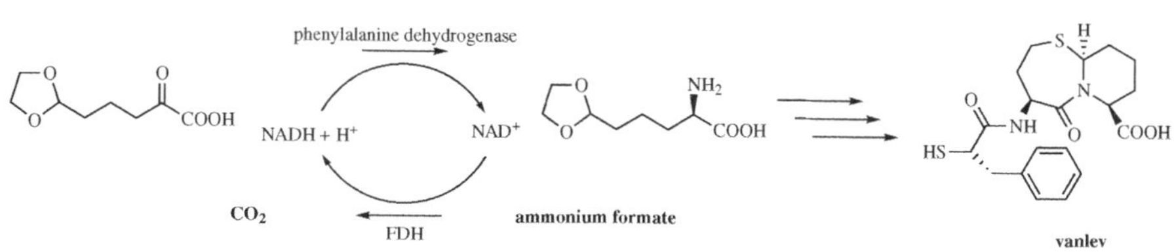

FIGURE 8.10 Synthesis of vanlev, involving a biocatalytic oxidative step [51].

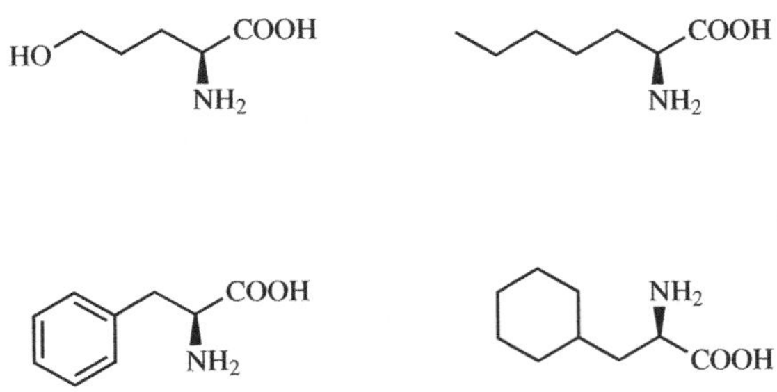

FIGURE 8.11 Further examples of compounds produced *via* reductive amination of α-ketoacids [29].

as mere "academic curiosities" to become important key player in the arena of industrial biocatalysis. Albeit these amino transferase enzymes are formally not regarded as redox enzymes, they actually catalyze the net transfer of an amino group to a carbonyl group, thus representing a reductive amination of the substrate (acceptor), coupled with the oxidation of the cosubstrate (donor). Transaminases use pyridoxal-5′-phosphate as organic cofactor (Figure 8.13), and require an amino donor and an amino acceptor. Pyridoxyl-5′-phosphate is able to form a Schiff base with the amino group, thus triggering the amino transfer from one organic molecule to the other.

Traditionally it was considered that alanine was the best amino donor for transaminases, producing pyruvic acid as by-product. However, it is important to mention that in the last years transaminases that are able to accept cheap and readily available isopropylamine have been reported or evolved [47]. In this chapter, in virtue of the increasing importance of these biocatalysts in the industrial biocatalysis arena, some insights and relevant examples on transaminases for industrial processes will be briefly reported. Until now, a broad number of different structures have been studied for transamination reactions, leading in most of the cases to outstanding enantioselectivities, though in many cases with still low substrate loadings and/or productivities, together with a number of inhibition problems. In Figure 8.14, some representative examples are depicted.

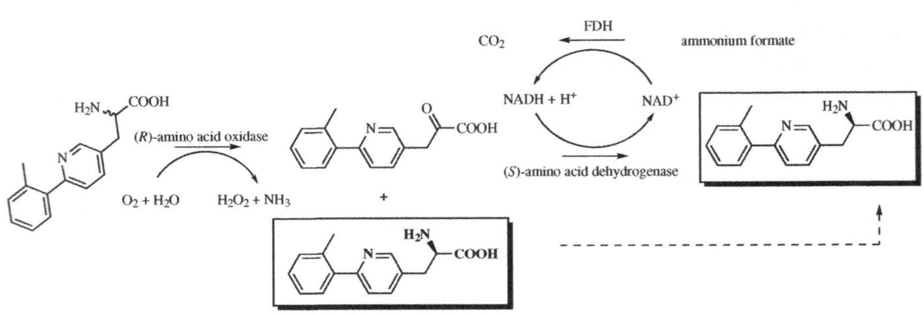

FIGURE 8.12 Multistep enzymatic process for the production of nonnatural amino acids [46].

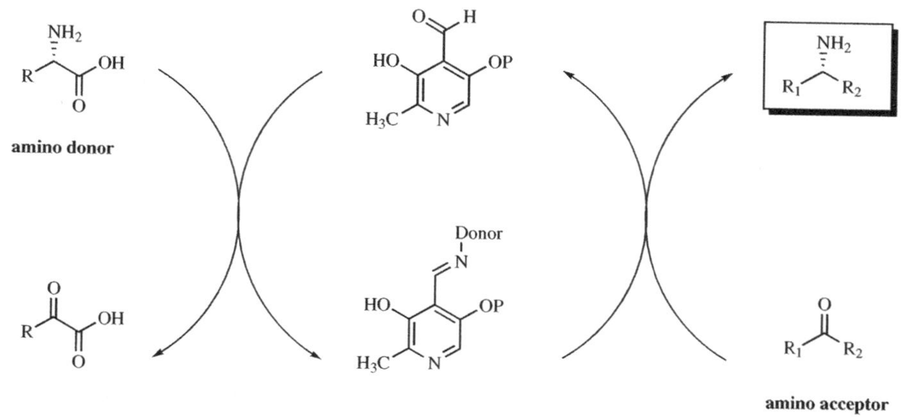

FIGURE 8.13 Catalytic cyclic of transaminases, involving an amino donor and an amino acceptor.

A recent outstanding example of transaminases applied to industrial biocatalysis is the biocatalytic manufacture of sitagliptin, which actually will replace the current chemocatalytic route at industrial level [11,48]. Herein, the starting point was a wild-type transaminase that was not able to convert the desired substrate at all, since it was too bulky to be accepted within the enzyme active site. As stated in Section 8.2, by means of protein engineering and directed evolution approaches, a mutant transaminase was evolved, which was able to convert the bulky substrate efficiently into the desired chiral amine. Under optimized process conditions, almost full conversion (>90%), representing >200 gL^{-1}, with high enantiomeric excess (>99.5%) were reached (Figure 8.15), and leading to a diminished waste formation when compared with the chemical process. Therefore, this is another outstanding example on how modern industrial biocatalysis can be adapted to a certain process by smartly modifying the biocatalyst toward the desired process. The development of an engineered enzyme that may compete under such industrial conditions is challenging. However, it is important to stress that once such a mutant is developed and efficiently cloned and overexpressed, it can be produced "on-demand" at large scale

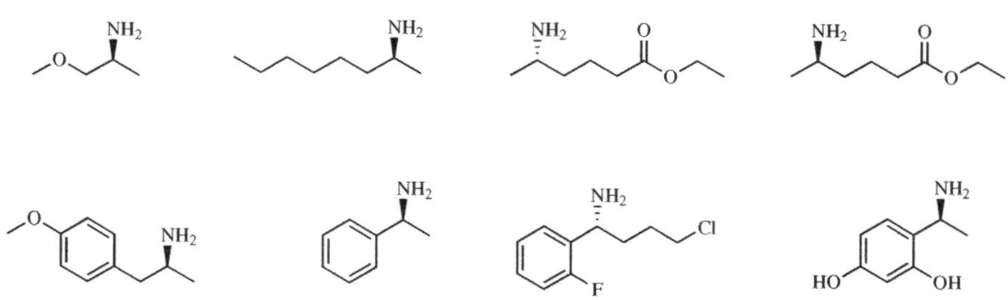

FIGURE 8.14 Representative examples of chiral amines produced by transaminase-catalyzed reaction of the corresponding carbonyl compounds [29,47].

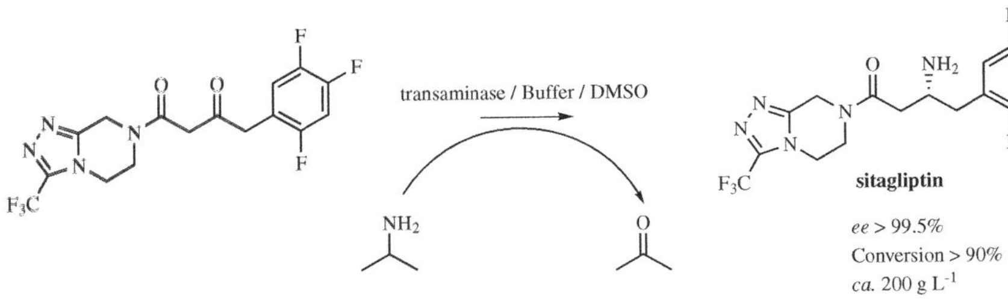

FIGURE 8.15 Transaminase-catalyzed process for the sitagliptin manufacture [11,48].

by conventional fermentative processes, thus assuring the large availability of the biocatalyst for market needs.

Apart from dehydrogenases (or transaminases), other oxidative enzymes like lipoxygenases have been assessed for industrial processes as well, with several interesting cases of applied biocatalysis [49]. As a recent example, soybean lipoxygenase was used for the synthesis of several 10 g batches of (5*Z*,8*Z*,11*Z*,13*E*)-(15*S*)-15-hydroxyeicosan-5,8,11,13-tetraenoic acid (15*S*)-HETE), to be used in clinical trials for the treatment of dry-eye syndrome (Figure 8.16). In this case the process was a chemoselective approach, where lipoxygenase created the asymmetric peroxide that was subsequently reduced to the unsaturated hydroxy carboxylic acid while keeping the enantiomeric excess in a high *on-spec* level [50].

To finish this section of industrial uses of free enzymes, it is important to mention that in addition to the above-described examples, other applied redox processes involving free biocatalysts make use of the biocatalytic promiscuity that some enzymes may display. For instance, several hydrolases are able to accept hydrogen peroxide as nucleophile to afford peracids, which can be used for *in situ* oxidations and epoxidations, either in aqueous media or in nonaqueous media [21]. While oxidations in aqueous media were assessed for cleaning, disinfecting and bleaching strategies, the oxidation in organic media afforded several efficient epoxidations, oxidations and Baeyer–Villiger reactions [21]. Since this is an emerging topic, the ample number of possible applications of these approaches were widely discussed separately in Chapter 6 of the present book.

8.4 INDUSTRIAL BIOCATALYTIC REDOX PROCESSES—WHOLE-CELLS: THE "DESIGNER BUG" CONCEPT AND BEYOND (METABOLIC ENGINEERING)

A central strategy to perform industrial redox biocatalytic processes is the direct use of whole-cells as biocatalysts [26]. Despite the use of whole-cells must be examined

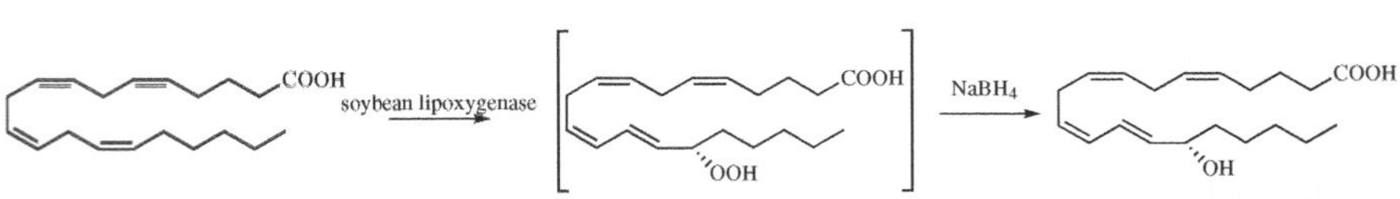

FIGURE 8.16 Chemo-enzymatic synthesis of 15*S*-HETE [50].

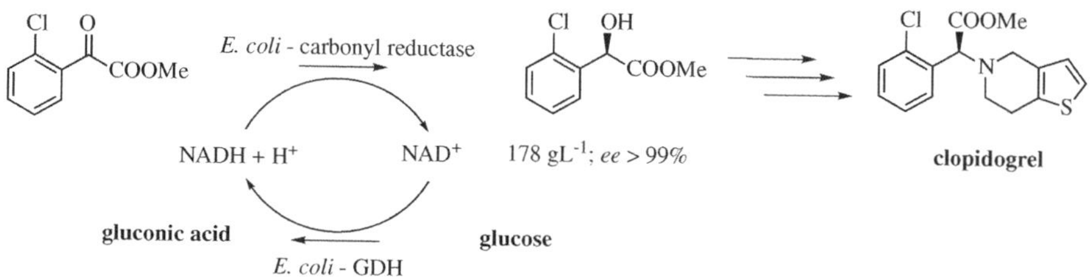

FIGURE 8.17 Enzymatic synthesis of a precursor of Clopidogrel by using whole-cells overexpressing two oxidoreductases [51].

in a case by case scenario to avoid problems related to by-product formation, side-reactions, and so on; when such problems can be avoided, the use of whole-cells may significantly simplify the cofactor regeneration issues related to redox biocatalysis. Likewise, the use of whole-cells avoids the need of purifying enzymes from broken cells or fermentative broths, thus contributing to reduce biocatalyst costs significantly. In Chapter 7 the choice, pros and cons of whole-cells and free enzymes were already discussed.

Within the whole-cell approach, in order to reach high productivities in short reaction times, the cloning and overexpression of the oxidoreductases is often crucial (albeit not always necessary) to assure that a huge proportion of the required enzyme(s) is present in the whole-cell. Thus, as an example of this the highly enantioselective synthesis of methyl (*R*)-*o*-chloromandelate with whole-cells of recombinant *Escherichia coli* containing an overexpressed carbonyl-reductase and a GDH was recently reported. The afforded compound is a key building block for the production of clopidogrel, a platelet aggregation inhibitor administered to atherosclerotic patients with risk of heart attack. By using whole-cells as biocatalysts, excellent productivities of ~178 $gL^{-1}d^{-1}$ with outstanding enantioselectivities (*ee* >99%) were achieved under optimized reaction conditions (Figure 8.17) [41].

As observed in the above-described example, instead of overexpressing *just one* enzyme, in the field of whole-cells-based industrial biocatalysis, a crucial concept is that of the "designer bug" [52,53]. Basically, the approach implies the cloning and overexpression of several enzymes in the same whole-cell, by carefully controlling the activities of each one (toward a certain reaction), and the overexpression level of each of the enzymes. By means of this strategy, high levels of conversions, selectivities, and efficiency can be achieved, since the biocatalyst, namely the whole-cell, is specifically adapted for a certain reaction and toward a certain substrate (Figure 8.18) [52,53].

As a relevant example of this technology, designer bugs were constructed by cloning GDH from *Bacillus subtilis* together with either (*R*)-enantioselective alcohol dehydrogenase from *L. kefir* (also used as free enzyme in several processes, see Section 8.3 for details), or the (*S*)-enantioselective alcohol dehydrogenase from *R. erythropolis* (again see Section 8.3), depending on the desired configuration of the resulting alcohol. With that approach, a wide range of useful chiral alcohols were obtained in high enantioselectivities and productivities (Figure 8.19) [52,53]. It is thus important to stress that depending on market demands, either (*R*)- or (*S*)- enantiomers

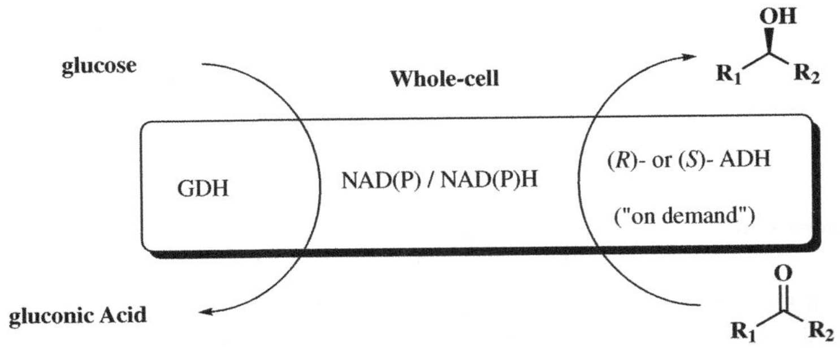

FIGURE 8.18 Concept for the "designer bug" approach [52,53].

can be produced, simply by choosing the right designer bug, producing it by growing the cells, and using it directly in reactor set-ups under very mild reaction conditions.

Playing with the versatility of the designer-bug system, an analogous concept was built using (*R*)-alcohol dehydrogenase from *Lactobacillus brevis* alcohol dehydrogenase (LBADH) together with GDH from *Thermoplasma acidophilum* [54]. In this case, the approach was envisioned for the efficient synthesis of enantiopure aliphatic terminal (*S*)-epoxides. Herein, a first biocatalytic step to introduce the chiral center within the molecule, was combined with a subsequent chemical epoxide cyclization to form the chiral epoxide (Figure 8.20) [54]. Due to the high inherent chemical reactivity of epoxides, they are very valuable building blocks for further synthetic procedures. In this case, by means of the designer bug, the chiral center is already incorporated in the oxirane ring.

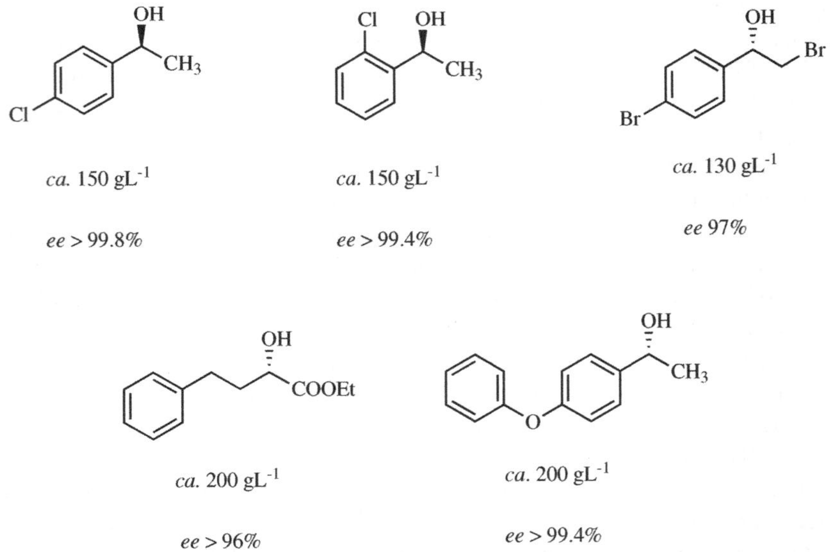

FIGURE 8.19 Examples of compounds and productivities achieved by building whole cell biocatalysts as designer bugs [52,53].

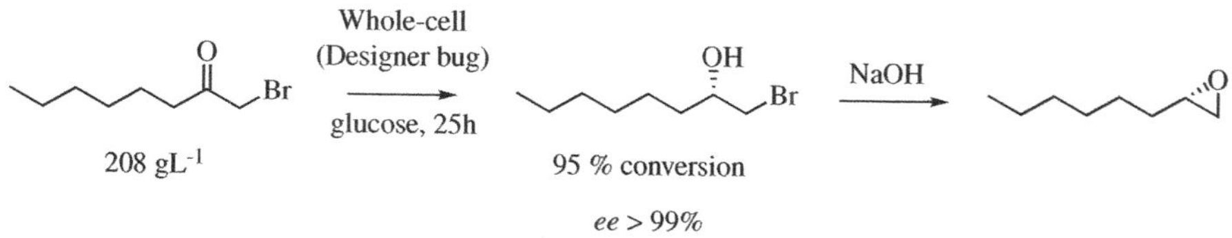

FIGURE 8.20 Designer bug approach for the highly efficient production of terminal (*S*)-epoxides [54].

Interestingly, an analogous approach was assessed by the same research group to evaluate the biocatalytic aldehyde reduction, from cinnamyl aldehyde to the aroma chemical cinnamyl alcohol. The latter compound is actually a very versatile building block in aroma industries, and in this case only natural sources (substrate and enzyme) were used for its manufacture. Remarkably, high substrate inputs in cinnamyl aldehyde (~166 gL^{-1}) with high yields (77%) were achieved (Figure 8.21) [55]. It is important to note the difficulty that such reaction may have when conventional chemocatalysts are used, due to the high reactivity of α,β-unsaturated aldehydes, which are markedly prone to suffer condensation, by-product formation, and so on. As stated in the first section of this chapter (Section 8.1), another advantage of biocatalysis is that by using enzymes, a mild and selective approach is provided, thus diminishing the risks associated to such secondary reactions. Moreover, once process parameters are optimized to build a robust process, high productivities are achieved as well.

Apart from alcohol dehydrogenases, the same designer-bug concept has been applied by combining amino acid dehydrogenases (see Section 8.3 for insights in the use of these biocatalysts as free enzymes) with FDH from *C. boidinii*, to set up the synthesis of nonnatural amino acids. In this regard, efficient production of L-*tert*-leucine [56], as well as of L-neopentylglycine [57], have been reported by means of these whole-cells specifically designed for this process (Figure 8.22).

Following this designer-bug concept, recently a novel biocatalyst was developed, leading to an excellent production of 620 gL^{-1} of a chiral compound in <24 h of reaction time [58]. To this end, a β-ketoacyl-ACP reductase (FabG) from *Bacillus* sp. ECU0013 was overexpressed in *E. coli*, together with GDH for cofactor regeneration (upon addition of equivalents of cheap and readily available glucose). Again in this example, once reaction conditions were optimized, the process proceeded smoothly at room temperature and ambient pressure, virtually without need of external cofactor addition (Figure 8.23) [58].

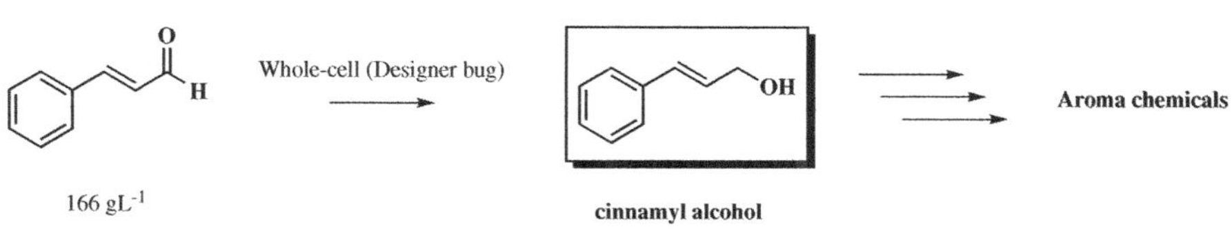

FIGURE 8.21 Biocatalytic aldehyde reduction with interest in the aroma chemical industry, by means of whole-cells as designer bugs [55].

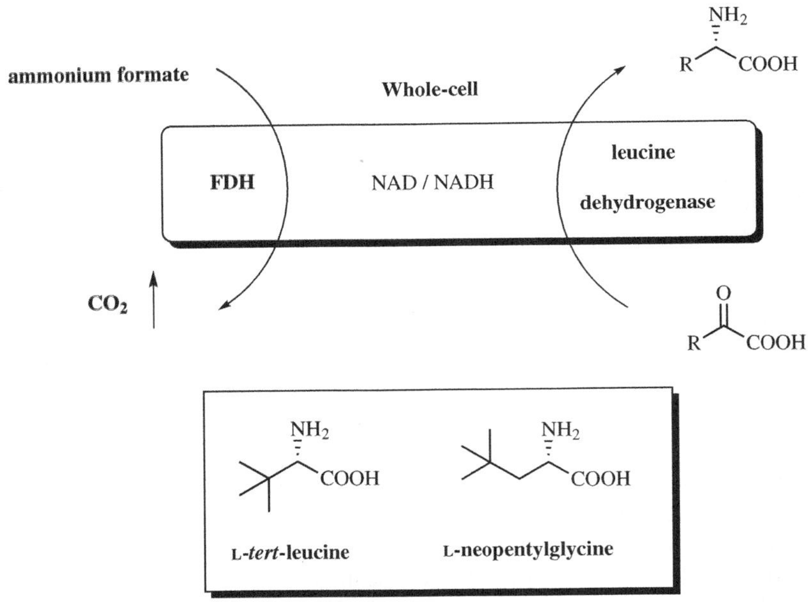

FIGURE 8.22 Designer bugs composed of FDH from *C. boidinii* and leucine dehydrogenase from *Bacillus cereus* cloned and overexpressed in *E. coli* [56,57].

The designer bug concept has been reported as well for other oxidoreductases. For instance, an *E. coli* whole-cell overexpressing old yellow enzyme (OYE) from *Candida macedoniensis* (an enoate reductase), coupled with GDH, was able to perform the conversion of ketoisophorone to (6*R*)-levodione with high productivities (~25 $gL^{-1}h^{-1}$) on 100 gL^{-1} scale (Figure 8.24) [29,59,60].

Furthermore, there are also examples of whole-cells overexpressing transaminases and lyases at industrial scale. As stated in the previous section, although these enzymes are formally not classified as oxidoreductases, they actually catalyze a reductive process over carbonyl groups to form amines (transaminases) or α-hydroxy ketones (lyases) [47,61,62]. Therefore, they will be briefly explained in this section as well. With regard to transaminases, the (*S*)-transaminase from *Bacillus megaterium* overexpressed in *E. coli* was used for the kinetic resolution of racemic amines, affording approximately 60 gL^{-1} of chiral amines in 24 h. Notably, the use of a whole-cell

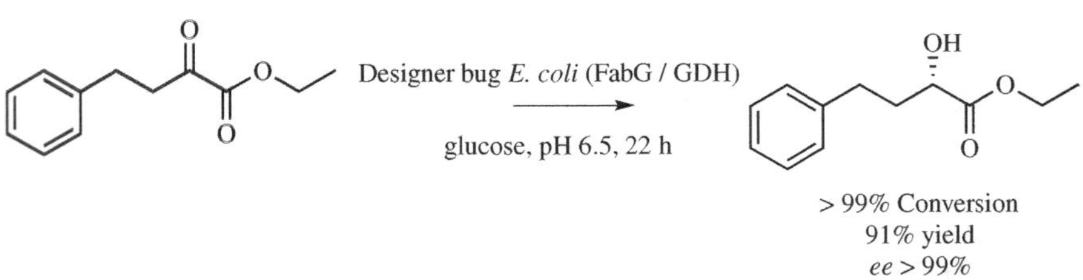

FIGURE 8.23 Novel designer bug affording up to 620 gL^{-1} of a building block [58].

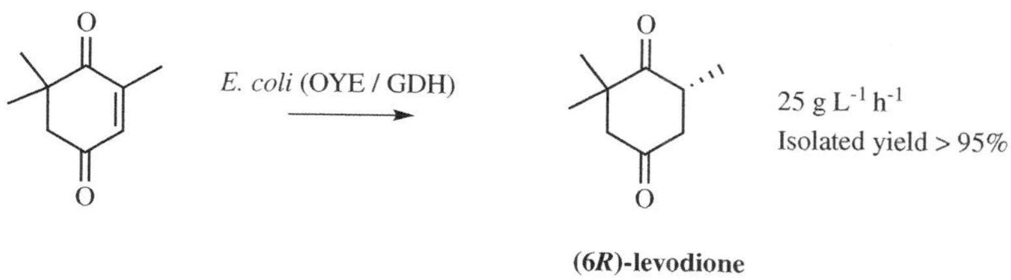

FIGURE 8.24 Designer bug containing OYE and GDH [29,59,60].

made the addition of organic cofactor, pyridoxal-5′-phosphate, completely unnecessary (Figure 8.25), representing a further added-value that whole-cells may bring for industrial biocatalysis. In this case, pyruvic acid was added as amino acceptor, thus producing alanine along the course of the biocatalytic process [63].

Regarding lyases, an analogous approach has been proposed using a thiamine diphosphate-dependent enzyme (ThDP-enzyme), benzaldehyde lyase (BAL) overexpressed in *E. coli* in a biphasic system MTBE-buffer (MTBE: methyl-*tert*-butyl-ether). High productivities of different chiral α-hydroxy ketones ($\sim$80–100 gL^{-1}, 24 h) with high enantiomeric excesses (*ee* >98%) were reported (Figure 8.26) [64]. Also in this case, the use of a whole-cell made the addition of organic cofactor (ThDP in this case) unnecessary to afford efficient biocatalytic processes. Remarkably, BAL was even able to catalyze the cross condensation between two different aldehydes in a selective fashion, thus providing access to very useful building blocks under very mild reaction conditions. In fact, lyases are also another diverse group of enzymes from which many practical applications can be envisaged [61].

As previously stated in this chapter (Section 8.2), one of the key features that make enzymes important for industrial applications is the biocatalytic promiscuity that some of these biocatalysts may display [18–22]. In this respect, with regard to lyases and their potential industrial use in the production of hydroxyketones, a relevant case is that of pyruvate decarboxylase (PDC) to produce (*R*)-phenylacetylcarbinol (*R*)-(PAC), an intermediate for the synthesis of (-)-ephedrine [20,62]. In this particular case, the carboligation between benzaldehyde and acetaldehyde is a promiscuous activity of PDC, which is actually a ThDP-enzyme, and not the *natural* reaction

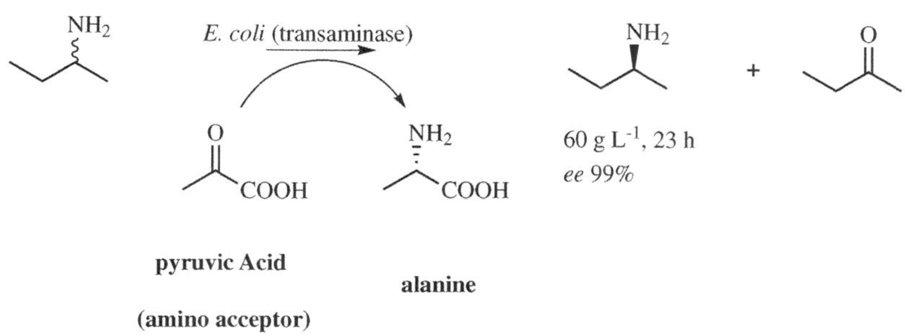

FIGURE 8.25 Kinetic resolution of amines catalyzed by a transaminase overexpressed in *E. coli* whole-cells [63].

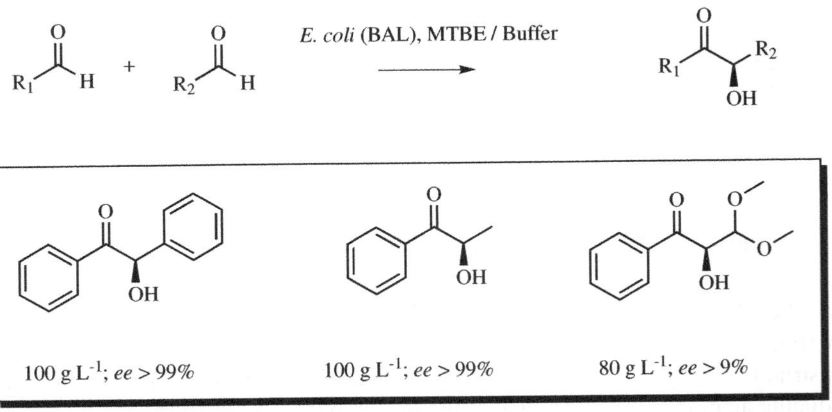

FIGURE 8.26 Examples of enzymatic C-C bond formation using *E. coli* whole-cells over-expressing benzaldehyde lyase (BAL) [64].

that PDC catalyzes, that is, decarboxylation of pyruvic acid. The process was set up industrially as early as in the 1920s, using *Saccharomyces cerevisiae* as whole-cell containing PDC. During the last century further optimizations were performed, either at the level of process development (e.g., by the introduction of supercritical CO_2 as solvent), or at the level of the enzyme, for example, by using other PDC sources like *Candida utilis*. Productivities in (*R*)-PAC of ~100 $gL^{-1}d^{-1}$ with enantiomeric excesses of 98%, are presently achieved at industrial level (Figure 8.27) [64].

Within industrial enzymatic redox strategies, another interesting field is represented by the oxidation of different alcohols to carboxylic acids, some of them with sound applications like the production of acetic acid performed by acetic acid bacteria [65]. Two different dehydrogenases are involved in the process, alcohol dehydrogenase and acetaldehyde dehydrogenase, the first one catalyzing the oxidation to aldehyde, and the second one oxidizing such aldehyde to carboxylic acid [65–69]. Several microorganisms are able to oxidize other alcohols different from ethanol. For instance, *Gluconobacter oxydans* is able to form propionic acid from *n*-propanol when growing with glycerol as the carbon source. In this case, up to 37 gL^{-1} of calcium propionate within 30 h and up to 56 gL^{-1} of sodium propionate within 70 h

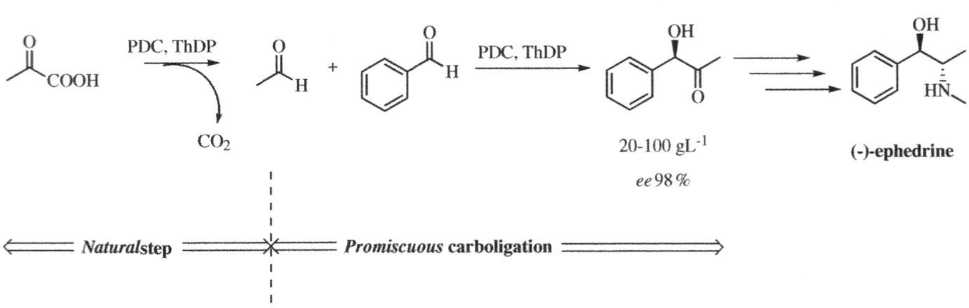

FIGURE 8.27 Chemo-enzymatic industrial production of (-)-ephedrine, catalyzed partly by PDC involving a promiscuous step [64].

were reported, when fed-batch processes were set up [65]. In the same area, Molinari et al. showed that *Acetobacter pasteurianus* was able to oxidize different alcohols like *n*-propanol, *n*-butanol, 2-methylbutanol, or 3-methylbutanol to the corresponding carboxylic acids, yielding 45–60 gL^{-1}, albeit in several days (long reaction times) of fermentative (fed-batch) approaches [66]. Cells have been also subjected to immobilization in the case of *Acetobacter aceti* for the production of either phenylacetaldehyde or phenylacetic acid, starting from 2-phenylethanol, depending on the reaction conditions applied (one or two- phase systems for the *in situ* extraction of the produced aldehyde) [67]. In general, it has to be mentioned that in these reported examples most of the produced carboxylic acids are nonchiral and simple structures that may be used as commodity chemicals in different industrial segments. Therefore, market prices for such carboxylic acids need to be low, and therefore the volumetric productivities achieved in these examples, in most of the cases below 70 gL^{-1}, certainly complicate the downstream processing in an economic way. In this respect, it might be that metabolic engineering approaches will enable, in the coming future, the provision of these and other molecules at competitive prices, since in this latter case cheap sugar residues (e.g., from cellulosic materials and agricultural wastes) will be used for the fermentation, biomass, and substrate production, and finally for the acid biosynthesis.

Apart from the production of nonchiral carboxylic acids, in the open literature there are also some examples of production of chiral carboxylic acids *via* that oxidative route [68,69]. For instance, Molinari et al. [68] assessed the enantioselective oxidation of prochiral 2-methyl-1,3-propanediol performed by *A. pasteurianus*, to form (*R*)-3-hydroxy-2-methylpropionic acid with 97% enantiomeric excess and 25 gL^{-1}. The afforded compound is a very useful building block for the synthesis of captopril, a widely used antihypertensive drug (Figure 8.28) [68].

Likewise, the deracemization of racemic mandelic acid carried out by *Pseudomonas polycolor* and *Micrococcus freudenreichii* has been reported, with a production of (*R*)-mandelic acid of 31 gL^{-1} *via* oxidation–reduction of one of the

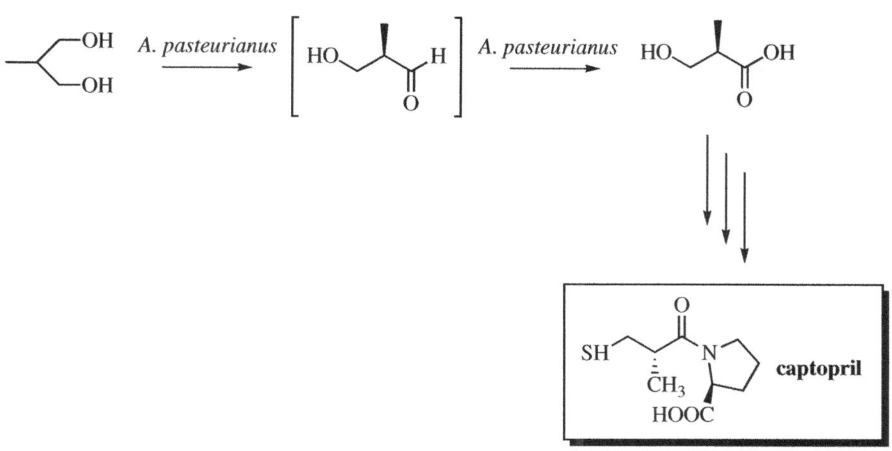

FIGURE 8.28 Production of (*R*)-3-hydroxy-2-methyl-propionic acid performed by *A. pasteurianus* [68].

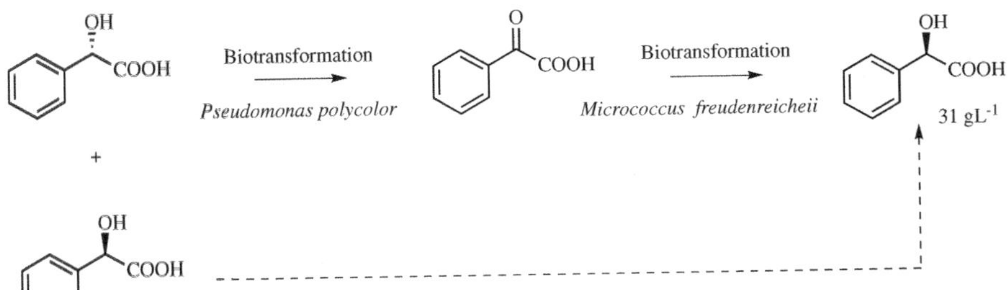

FIGURE 8.29 Deracemization of mandelic acid [69].

enantiomers, whereas the other one remained unaltered during the whole biotransformation (Figure 8.29) [69].

The use of different oxygenases for industrial purposes has also been assessed in detail, with some applications, albeit these enzymes often possess some drawbacks like low stabilities in the reaction conditions, low catalytic rates for nonnatural substrates, and so on [70]. Despite these drawbacks, some industrially sound applications have been reported. For instance, a process for the selective monohydroxylation of aromatic carboxylic acids was reported [71]. In this case, growing cells of *Beauveria bassiana* were used, with a rather broad substrate spectrum of aromatic carboxylic acids (Figure 8.30) [71]. As stated previously in this chapter, such broad acceptance of different substrates is an important asset when biocatalyst with potential industrial use are considered, since by using just one biocatalyst, access to a broad number of products can be achieved.

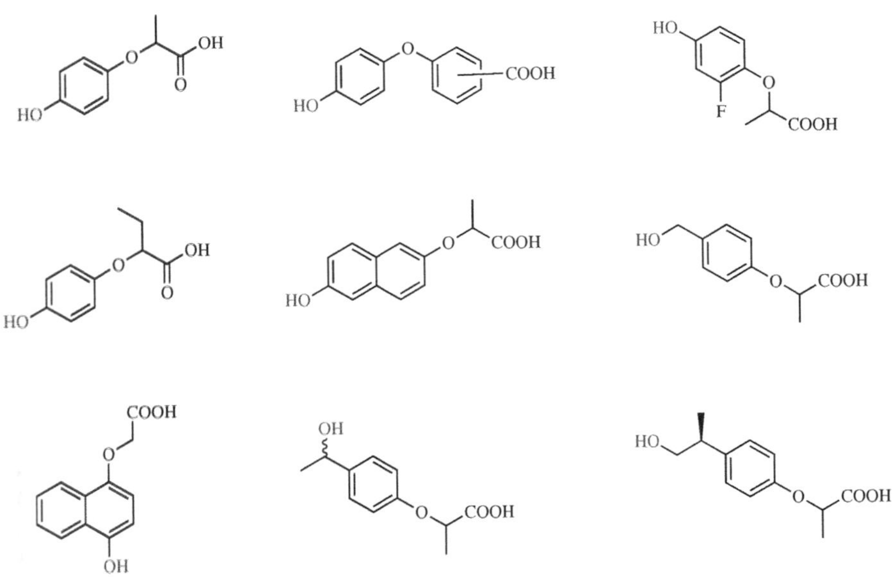

FIGURE 8.30 Different aromatic carboxylic acids monohydroxylated (in grey) by *B. bassiana* whole-cells [71].

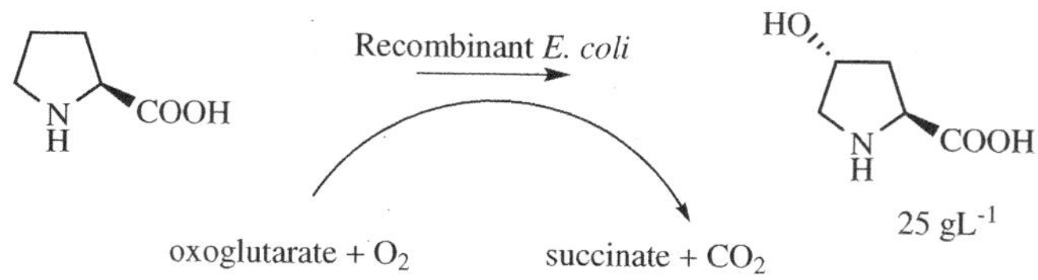

FIGURE 8.31 Production of *trans*-4-hydroxy-L-proline catalyzed by a recombinant *E. coli* [72].

In other cases recombinant whole-cells were also built. For instance, the production of *trans*-4-hydroxy-L-proline was reported by means of a recombinant *E. coli* strain in which a proline-4-hydroxylase gene was introduced and overexpressed, leading to a production of 25 gL^{-1} of the hydroxylated amino acid in 96 h (Figure 8.31). Such component is useful for several pharmaceutical syntheses, and is often obtained by collagen hydrolysis, with the subsequent waste formation. In this case however, still further optimization of the process should be accomplished to afford a robust industrial application [72].

Likewise, biotransformations were assessed for the industrial oxidation of 2-methylquinoxaline to 2-quinoxalinecarboxylic acid [73]. A 14.000 L run producing 20.5 kg of acid was reported by using *Absidia repens* whole-cells. Yet, here again productivity was not enough to warrant an industrial use, and therefore a second microbial process involving *Pseudomonas putida* was reported. In this case productivities of ~10 gL^{-1} of acid (86% yield) were achieved. For industrial purposes, the optimization of the enzyme(s) involved in the process by means of directed evolution techniques would probably afford more *on-spec* outcomes (Figure 8.32). Importantly, the set-up of this biosynthetic approach avoided the use of hazardous chemicals in the chemical synthesis of the carboxylic acid [73].

Moreover, the synthesis of pravastatin, a selective inhibitor of 3-hydroxy-3-methylglutaryl CoA reductase, a key enzyme for cholesterol biosynthesis, was produced by means of cytochrome P450 monooxygenase from *Streptomyces carbophilus*, leading to a highly selective hydroxylation process (Figure 8.33) [74].

With regard to whole cells and cloning and overexpression of enzymes, another field that must be briefly commented is that of metabolic engineering [10]. Such approach involves the holistic optimization of the metabolism of a certain microorganism to afford a desired compound. Once the strain is engineered and optimized, a *simple* fermentative process starting from cheap raw materials (e.g., glucose from lignocellulosic materials, agricultural wastes, etc.) leads to product formation. Apart

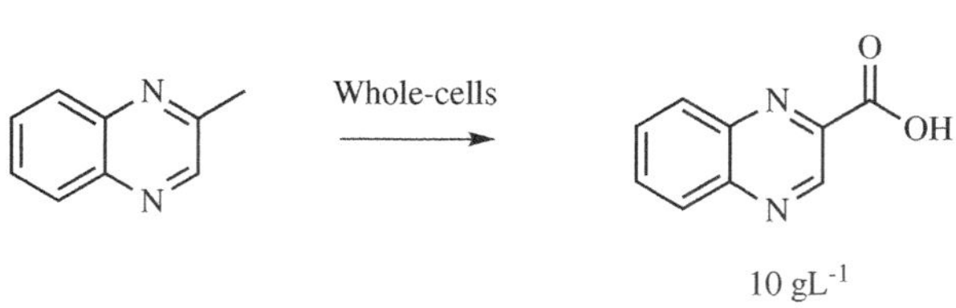

FIGURE 8.32 Biotransformation to produce 2-quinoxalinecarboxylic acid [73].

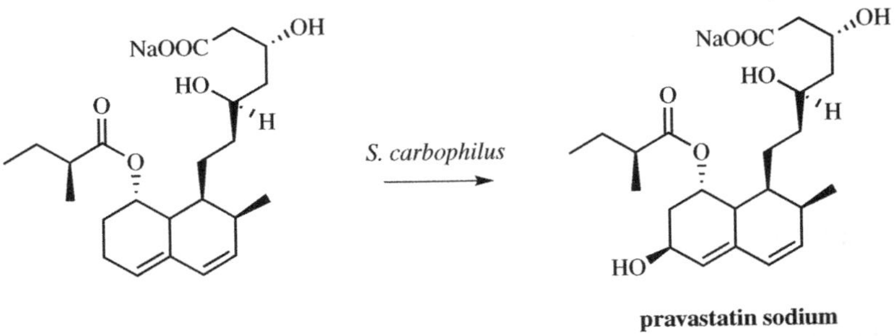

FIGURE 8.33 Production of pravastatin by means of *S. carbophilus* cells [74].

from the use of nonfood-competing and cheap feedstock, another important and costly aspect with this approach is the downstream processing, since a more or less diluted product (typically $< 50\ gL^{-1}$) will be accumulated in the fermentation broth. Furthermore, external enzymes or even complete biochemical pathways can be incorporated to the host, thus enhancing enormously the possibilities that biotechnology may provide. A recent outstanding example is the production of terminal alkenes (isobutylene, propylene) by means of *E. coli* cells and starting from simple glucose as the carbon source. In that specific case the downstream processing was straightforward, since these terminal alkenes are gases that are rapidly stripped out from the fermentative broth [10]. In other cases more cumbersome and expensive procedures need to be applied to provide a highly pure and *on-spec* product.

With respect to redox enzymes, several examples of metabolic engineering approaches have already been reported [75,76]. For instance, a recombinant *E. coli* in which the tryptophan biochemical pathway was modified to enhance the indol production, and furthermore *P. putida* genes encoding naphthalene dioxygenase were cloned and overexpressed. As a result of these modifications, the strain resulted into a powerful indigo producer biocatalyst (Figure 8.34) [75].

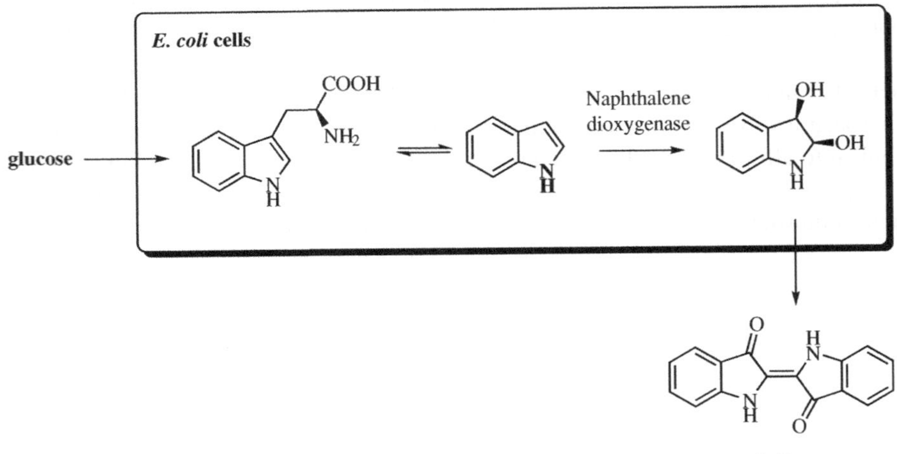

FIGURE 8.34 Production of indigo by recombinant *E. coli* cells [75].

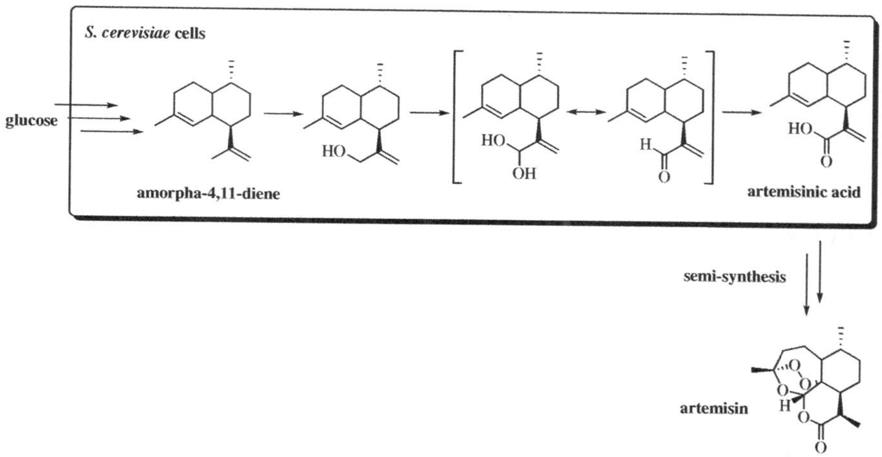

FIGURE 8.35 Engineered yeasts for the semisynthesis of artemisin [76].

Likewise, artemisinic acid, a precursor of the antimalarial drug artemisin, can now be produced by means of engineered yeasts up to 100 mgL^{-1} [76]. Later on, artemisinic acid can be converted in artemisin by semisynthetic approaches. To this end, *S. cerevisiae* was engineered with a mevalonate pathway, amorphadiene synthase, and a novel cytochrome P450 monooxygenase, to carry out the three-step oxidation of amorpha-4,11-diene to artemisinic acid. Herein, it appears obvious that more improvements in productivity of the engineered yeast are needed to reach *on-spec* industrial requirements (Figure 8.35), albeit in this case, and given the high market price of artemisin, probably a lower production in gL^{-1}, when compared to other cases, will be needed [76].

8.5 CONCLUDING REMARKS AND FUTURE PERSPECTIVES

By looking at the examples reported, it appears clear that in almost all cases redox biocatalysis is focused on the development of high-added-value chiral compounds, which always provide high market prices that certainly contribute to cover costs associated to biocatalyst development and production. It is expected, however, that redox biocatalysis will take over more and more areas for the synthesis of medium- and low-added-value products (bulk chemistry), in the quest of milder and more environment friendly reaction conditions with less by-product formation and energy savings as main drivers. To reach these goals, it is important to develop very robust biocatalytic processes that may economically compete with such low market prices of these molecules. To this end, whole-cells will be surely the chosen approach, and immobilization of them for reusing during several months will be almost mandatory to compete in terms of costs. There are already several examples of biocatalysts operating at industrial level in that direction, for example, for hydrolysis of nitriles to yield acrylamide or glycolic acid, to cite some relevant examples [77,78]. An example of this quest for redox enzymes is represented by a recently reported recombinant whole-cell that is able to oxidize hydroxymethylfurfural (HMF), a platform

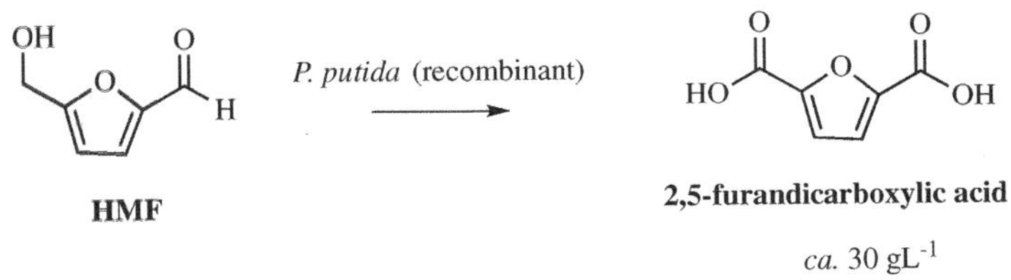

FIGURE 8.36 Oxidation of HMF to 2,5-furandicarboxylic acid catalyzed by recombinant whole-cells of *P. putida* [79].

chemical that is obtained from glucose and fructose dehydration, that is, coming from renewable sources [79]. In this case, engineered cells of *P. putida* overexpressing HMF/furfural oxidoreductase from *Cupriavidus basilensis* were able to produce 2,5-furandicarboxylic acid (Figure 8.36), a promising bio-based platform chemical that might be useful as a greener substitute of terephthalate in the manufacture of polyesters [79]. Under optimized conditions, a production of $\sim$30 gL^{-1} of oxidized acid was reached, by applying fed-batch experiments. Further optimization of these values, together with a proper process development (e.g., cell immobilization), might lead to a future bio-based process for platform chemicals production.

In addition to exploiting the already available and well-known field of oxidoreductases, from an industrial viewpoint there is a clear need to identify novel redox biocatalytic processes that may add to the current portfolio. Last years have witnessed the discovery and characterization of novel enzymatic systems that might bring industrially sound applications in the coming future. Remarkably, in some cases the "proof-of-concept" was reported decades ago [80]. As an example, it was observed that mutant strains of *Alcaligenes eutrophus* (nowadays renamed to *Ralstonia eutropha*) growing in the presence of benzoic acid were able to accumulate the *cis*-1,2-dihydroxylation product (Chapter 4, Section "Regioselectivity of the dihydroxylation of arene *cis*-dihydrodiols"). Later on, other mutants of *P. putida* were shown to display an analogous behavior, suggesting a metabolic defect in the pathway producing catechol from benzoate [81,82]. More recently it was reported that the process could be optimized to a production of >250 g product with 95% *ee* within 18 h biotransformation at 30°C (Figure 8.37) [83].

This finding triggered other groups to undertake considerable further research by means of classic organic synthesis starting from that building block. Remarkably, many sugar-related compounds so produced are very relevant for biological and medicinal purposes [84–87]. Likewise, other substituted benzoic derivatives have been assessed as substrates for these microorganisms, leading to novel and promising

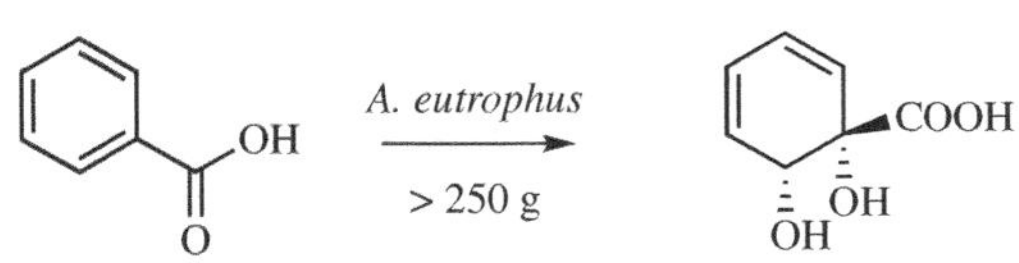

FIGURE 8.37 Production of (1*S*,2*R*)-1,2-dihydroxycyclohexa-3,5-diene-1-carboxylic acid by biotransformation [83].

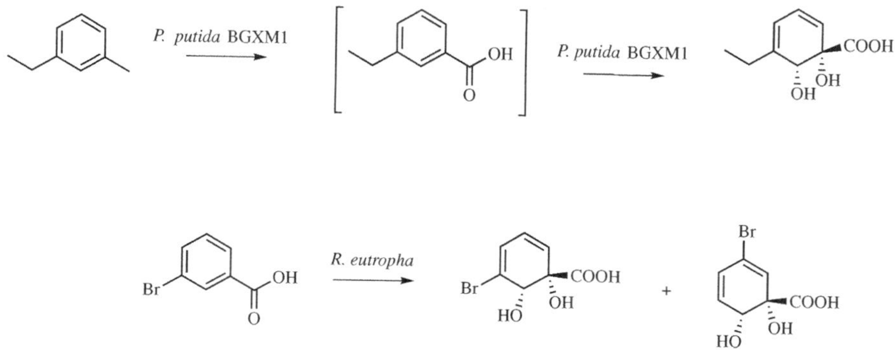

FIGURE 8.38 Exploring the substrate spectrum of aromatic biodihydroxylations [88,89].

building blocks (Figure 8.38) [88,89]. It is clearly expected that this field will boost in the coming years, surely with some industrial applications as well.

Another emerging area with regard to novel redox biocatalytic process is that of nitrile reductases [12]. Actually, the reduction of nitriles to afford primary amines is an industrially sound reaction to which there has not been, so far, a biocatalytic counterpart reported. Remarkably, nitrile reductases, namely 7-cyano-7-deazaguanine reductase (QueF), are involved in the biosynthesis of hypermodified nucleosides present in transfer RNA (tRNA). tRNAs undergo extensive posttranscriptional processing resulting in a broad variety of modified nucleosides. Herein, QueF converts intermediate 7-cyano-7-deazaguanine (preQ$_o$) to 7-aminomethyl-7-deazaguanine (preQ$_1$), which is subsequently modified to render queuosine, a guanosine derivative found in tRNA, with the presumable role of improving the fidelity of translation (Figure 8.39) [90–97].

The proposed mechanism of nitrile reductases suggests a fascinating performance involving two molecules of cofactor (NADPH) in each catalytic turnover for the provision of the four electrons together with several key amino acidic residues highly conserved within the active site of QueFs (e.g., Cys, His, Asp). The enzymatic reaction starts with the nucleophilic attack of the thiol group of a cysteine to the nitrile, leading to the formation of an imidothioate, which is subsequently reduced by the first NADPH by hydride transfer, affording a thiohemiaminal. Later on, another NADPH transfers the second hydride provoking the breakdown of the structure and liberating the cysteine. Remarkably, the imidothioate acts as a "protecting group" of the imine, which otherwise would be prone to hydrolysis by nucleophilic attack of water. Thus, the initial proton is provided by the nucleophilic attack of the cysteine; cofactors

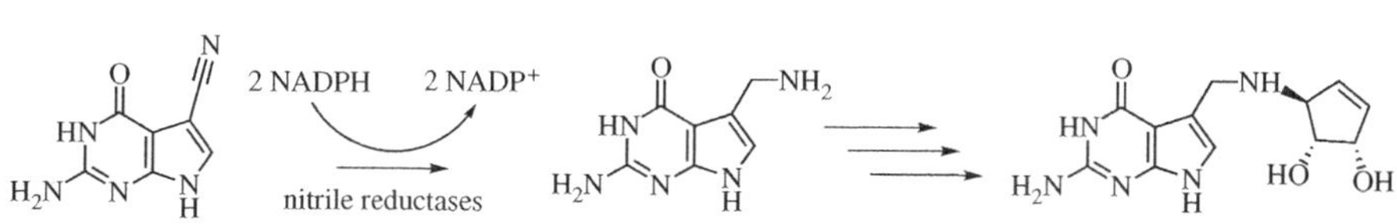

preQ$_o$ **preQ$_1$** **queuosine**

FIGURE 8.39 Natural biochemical pathway of nitrile reductases [12,90–97].

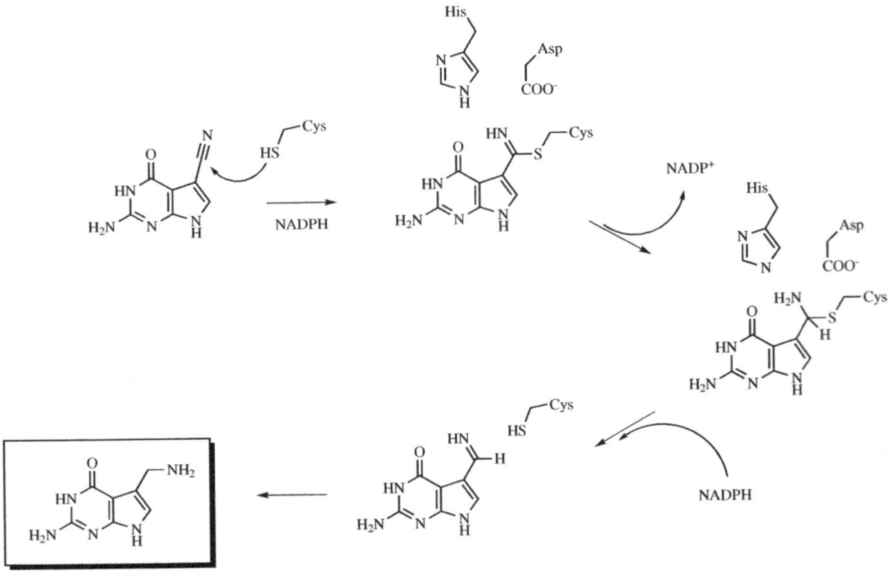

FIGURE 8.40 Postulated mechanism of nitrile reductases [12,90–97].

(2 NADPH) transfer two hydride (four electrons) and two additional protons; and the last proton is provided by the histidine residue (Figure 8.40) [12,90–97].

Furthermore, very recently several oxidoreductases that are able to reduce different imines were reported. Again the reduction of imines has presently almost no precedents in biocatalysis, and given the potential importance of such reaction at industrial level; it would be of utmost interest to identify biocatalysts able to catalyze such reduction [29,47,98]. Actually, whole-cells of *Candida parapsilosis* and *Streptomyces* sp. GF3587 and GF3546 have been reported as able to catalyze such imine reduction (Figure 8.41) [29,99,100]. The inherent instability of most of these imines in aqueous conditions, being prone to hydrolysis, may explain why the assessment of such biocatalytic reactions has been, to date, rather scarce.

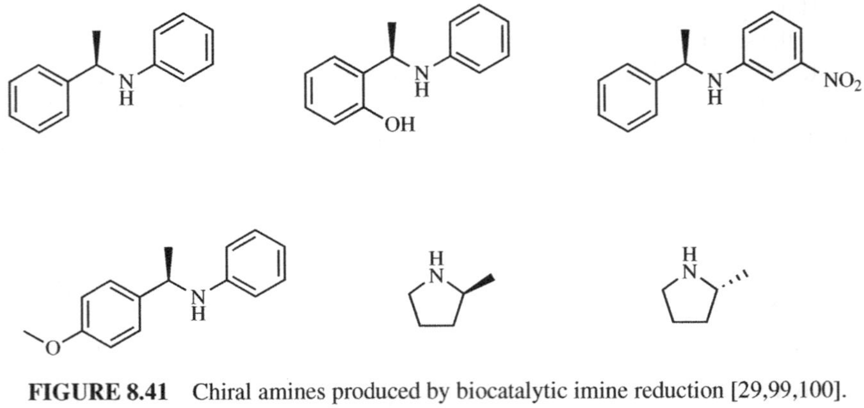

FIGURE 8.41 Chiral amines produced by biocatalytic imine reduction [29,99,100].

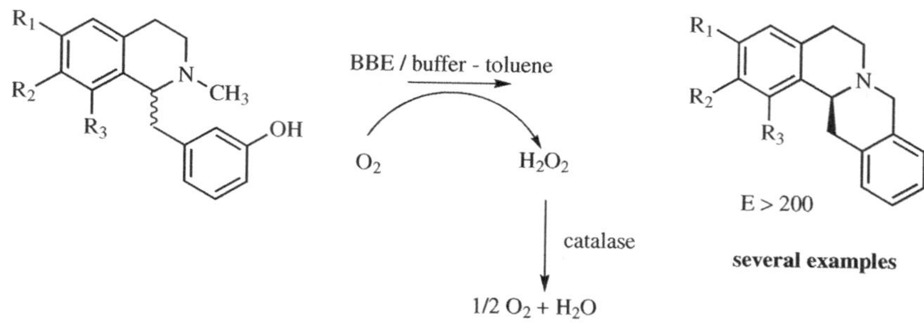

FIGURE 8.42 Enzymatic kinetic resolutions by means of the oxidative C-C intramolecular coupling catalyzed by berberine bridge enzyme (BBE) of *E. californica* (California poppy). Catalase is normally added to eliminate the formed hydrogen peroxide, which leads to enzyme denaturation [101].

Likewise, very recently Kroutil and coworkers characterized the berberine bridge enzyme (BBE) of *Escholzia californica* (California poppy), cloned and overexpressed in *Pichia pastoris*. This enzyme is able to catalyze an intramolecular oxidative C–C coupling by activating a methyl group attached to tertiary nitrogen, at the cost of molecular oxygen. Interestingly, the substrate spectrum of BBE is broader than the natural compound, displaying in all studied substrates high enantioselectivities (E >200) when kinetic resolutions were conducted. Likewise, BBE showed an outstanding tolerance to many organic solvents, both miscible and immiscible with water, for example, in water:toluene mixtures of up to >90% v/v toluene. Thus, this type of novel enzymes/reactions may become very relevant for organic chemists, since novel applications of enzyme catalysis may be easily envisaged (Figure 8.42) [101].

This chapter has focused on the enormous possibilities that redox enzymes can provide to industrially sound applications. By discussing an ample number of current real applications, it has been shown how biocatalytic redox processes can no longer be regarded as mere "academic curiosities," but rather as serious increasingly important alternatives to many chemical and pharmaceutical segments, in which the need of efficient and more environment friendly (bio)catalysts is crucial. Thus, when an adequate biocatalyst design is combined with a proper and robust process development, redox enzymes are actually able to compete in the industrial arena, usually providing much better frameworks than those given by other more conventional chemocatalysts. As stated at the beginning of this chapter, reasons for using enzymes in industry must be initially pointed out as drivers—the urgent environmental need to provide better reaction conditions, the need of higher selectivities with fewer by-products and lower energy consumption, and/or the possibilities of reaching new markets with "green labels" or with the synthesis of novel molecules. Importantly, of these drivers one has to also provide an economic attractiveness for using enzymes in industrial environments. This can be achieved by making use of widely developed techniques like cloning and overexpressing—to provide large amounts of biocatalysts—genetic enzyme improvement or evolution of an enzyme toward a certain substrate or toward certain reaction conditions, and finally, the link to a robust process development (e.g., high substrate loadings, smart use of either a monophasic or a biphasic system, efficient work-up, etc.) that may lead, altogether, to a competitive industrial biotransformation. Remarkably, apart from these requisites, in the case of redox enzymes, the problem of setting an economic cofactor regeneration system adds other challenges to the practical implementation. As discussed in this chapter, this can be overcome by using either free enzymes or whole-cells, displaying both systems inherent advantages and disadvantages, but being both approaches useful for industrial biotransformations. As a main take-home message of this chapter, biocatalysis, in general, and the field of redox biocatalysis in particular, is currently gaining an important place in many industries, and it is obviously expected to play an increasingly crucial role in the coming decades.

REFERENCES

1. Adams JP, Collis AJ, Henderson RK, Sutton PW. In: Whittall J, Sutton P, editors. *Practical Methods for Biocatalysis and Biotransformations*. Chapter 1. Chichester UK: John Wiley & Sons; 2009.
2. Patel RN. Coord Chem Rev 2008;252:659–701.
3. Ran N, Zhao L, Chen Z, Tao J. Green Chem 2008;10:361–372.
4. Woodley JM. Trends Biotechnol 2008;26:321–327.
5. Akoh CC, Chang SW, Lee GC, Shaw JF. J Agric Food Chem 2008;56:10445–10451.
6. Schoemaker HE, Mink D, Wubbolts M. Science 2003;299:1694–1697.
7. Santacoloma PA, Sin G, Gernaey KV, Woodley JM. Org Proc Res Dev 2011;15:203–212.
8. Tufvesson P, Lima-Ramos J, Nordblad M, Woodley JM. Org Proc Res Dev 2011;15:266–274.
9. Jiménez-González C, Poechlauer P, Broxterman QB, Yang BS, Ende D, Baird J, Bertsch C, Hannah RE, Dell'Orco P, Noorman H, Yee S, Reintjens R, Wells A, Massonneau V, Manley J. Org Proc Res Dev 2011;15:900–911.
10. Domínguez de María P. ChemSusChem 2011;4:327–329.
11. Savile C, Janey JM, Mundorff EC, Moore JC, Tam S, Jarvis WR, Colbeck JC, Krebber A, Fleitz FJ, Brands J, Devine PN, Huisman GW, Hughes GJ. Science 2010;329:305–309.
12. Domínguez de María P. ChemCatChem 2011;3:1683–1685.
13. Domínguez de María P, García-Burgos CA, Bargeman G, van Gemert RW. Synthesis 2007;10:1439–1452.
14. Böttcher D, Brüsehaber E, Doderer K, Bornscheuer UT. Appl Microbiol Biotechnol 2007;73:1282–1289.
15. Hermann M, Kietzmann MU, Ivanci M, Zenzmaier C, Luiten RGM, Skranc W, Wubbolts M, Winkler M, Birner-Gruenberger R, Pichler H, Schwab H. J Biotechnol 2008;133:301–310.
16. Beloqui A, Domínguez de María P, Golyshin P, Ferrer M. Curr Opin Microbiol 2008;11:240–248.
17. Gabor E, Liebeton F, Niehaus J, Eck J. Biotechnol J 2007;2:201–206.
18. O'Brien PJ, Herschlag D. Chem Biol. 1999;6:R91–R105.

19. Copley SD. Curr Opin Chem Biol 2003;7:265–272.
20. Bornscheuer UT, Kazlauskas RJ. Angew Chem Int Ed 2004;43:6032–6040.
21. Carboni-Oerlemans C, Domínguez de María P, Tuin B, Bargeman G, van der Meer A, van Gemert R. J Biotechnol 2006;126:140–151.
22. Hult K, Berglung P. Trends Biotechnol 2007;25:231–238.
23. Nobeli I, Favia AD, Thornton JM. Nat Biotechnol 2009;27:157–167.
24. Khersonsky O, Tawfik DS. Annu Rev Biochem 2010;79:471–505.
25. Busto E, Gotor-Fernández V, Gotor V. Chem Soc Rev 2010;39:4504–4523.
26. de Wildeman SA, Sonke T, Schoemaker HE, May O. Acc Chem Res 2007;40:1260–1266.
27. Goldberg K, Schroer K, Lütz S, Liese A. Appl Microbiol Biotechnol 2007;76:237–248.
28. Goldberg K, Schroer K, Lütz S, Liese A. Appl Microbiol Biotechnol 2007;76:249–255.
29. Hollmann F, Arends IWCE, Holtmann D. Green Chem 2011;13:2285–2314.
30. Hollmann F, Arends IWCE, Bühler K, Schallmey A, Bühler B. Green Chem 2011;13:226–265.
31. Liese A, Seebach K, Wandrey C. *Industrial Biotransformations*. Weinheim: Wiley-VCH; 2006.
32. Hummel W, Abokitse K, Drauz KH, Rollmann C, Gröger H. Adv Synth Catal 2003;345:143–159.
33. Gröger H, Hummel W, Rollmann C, Chamouleau F, Hüsken H, Werner H, Wunderlich C, Abokitse K, Drauz KH, Buchholz S. Tetrahedron 2004;60:633–640.
34. Gröger H, Hummel W, Buchholz S, Drauz KH, van Nguyen T, Rollmann C, Hüsken H, Abokitse K. Org Lett 2003;5:173–176.
35. Pollard D, Truppo M, Pollard J, Chen CY, Moore J. Tetrahedron: Asymmetry 2006;17:554–559.
36. Gooding OW, Voladri R, Bautista A, Hopkins T, Huisman G, Jenne S, Ma S, Mundorff EC, Savile MM. Org Proc Res Dev 2010;14:119–126.
37. Liang J, Mundorff E, Voladri R, Jenne S, Gilson L, Conway A, Krebber A, Wong J, Huisman G, Truesdell S, Lalonde J. Org Proc Res Dev 2010;14:188–192.
38. Liang J, Lalonde J, Borup B, Mitchell V, Mundorff E, Trinh N, Kochrekar DA, Cherat RN, Pai GG. Org Proc Res Dev 2010;14:193–198.
39. Ma SK, Gruber J, Davis C, Newman L, Gray D, Wang A, Grate J, Huisman GW, Sheldon RA. Green Chem 2010;12:81–86.
40. Hanson RL, Goldberg SL, Brzozowski DB, Tully TP, Cazzulino D, Parker WL, Lyngberg OK, Vu TC, Wong MK, Patel RN. Adv Synth Catal 2007;349:1369–1378.
41. Hanson RL, Howell JM, LaPorte TL, Donovan MJ, Cazzulino DL, Zannella V, Montana MA, Nanduri VB, Schwarz SR, Eiring RF, Durand SC, Wasylyk JM, Parker WL, Liu MS, Okuniewicz FJ, Chen BC, Harris JC, Natalie KJ, Ramig K, Swaminathan S, Rosso VW, Pack SK, Lotz BT, Bernot PJ, Rusowicz A, Lust DA, Tee KS, Venit JJ, Szarka LJ, Patel RN. Enzyme Microb Technol 2000;26:348–358.
42. Galkin A, Kulakova L, Yoshimura T, Soda K, Esaki N, Appl Environ Microbiol 1997;63:4651–4656.
43. Hanson RL, Singh J, Kissick TP, Patel RN, Szarka LJ, Mueller RH. Bioorg Chem 1990;18:116–130.

44. Hanson RL, Schwinden MD, Banerjee A, Brzozowski DB, Chen B.-C, Patel BP, McNamee CG, Kodersha GA, Kronenthal DR, Patel RN, Szarka LJ. Bioorg Med Chem 1999;7:2247–2252.
45. Krix G, Bommarius AS, Drauz K, Kottenhahn M, Schwarm M, Kula MR. J Biotechnol 1997;53:29–39.
46. Chen Y, Goldberg SL, Hanson RL, Parker WL, Gill I, Tully TP, Montana MA, Goswami A, Patel RN. Org Proc Res Dev 2011;15:241–248.
47. Höhne M, Bornscheuer UT. ChemCatChem 2009;1:42–51.
48. Mutti FG, Sattler J, Tauber K, Kroutil W. ChemCatChem 2011;3:109–111.
49. Nanda S, Yadav JS. J Mol Catal B: Enzym 2003;26:3–28.
50. Conrow RE, Harrison P, Jackson M, Jones S, Kronig C, Lennon IC, Simmonds S. Org Proc Res Dev 2011;15:301–304.
51. Ema T, Okita N, Ide S, Sakai T. Org Biomol Chem 2007;5:1175–1176.
52. Gröger H, Chamouleau F, Orologas N, Rollmann C, Drauz KH, Hummel W, Weckbecker A, May O. Angew Chem Int Ed 2006;45:5677–5680.
53. Gröger H, Rollmann C, Chamouleau F, Sebastien I, May O, Wienand W, Drauz KH. Adv Synth Catal 2007;349:709–712.
54. Berkessel A, Rollmann C, Chamouleau F, Labs S, May O, Gröger H. Adv Synth Catal 2007;349:2697–2704.
55. Chamouleau F, Hagedorn C, May O, Gröger H. Flavour Fragance J 2007;22:169–172.
56. Gröger H, May O, Werner H, Menzel A, Altenbuchner J. Org Proc Res Dev 2006;10:666–669.
57. Menzel A, Werner H, Altenbchner J, Gröger H. Eng Life Sci 2004;4:573–576.
58. Ni Y, Li CX, Zhang J, Shen ND, Bornscheuer UT, Xu JH. Adv Synth Catal 2011;353:1213–1217.
59. Kataoka M, Kotaka A, Thiwthong R, Wada M, Nakamori S, Shimizu S. J Biotechnol 2004;114:1–9.
60. Kataoka M, Kotaka A, Hasegawa A, Wada M, Yoshizumi A, Nakamori S, Shimizu S. Biosci Biotechnol Biochem 2002;66:2651–2657.
61. Brovetto M, Gamenara D, Saenz Méndez P, Seoane GA. Chem Rev 2011;111:4346–4403.
62. Hoyos P, Sinisterra JV, Molinari F, Alcántara AR, Domínguez de María P. Acc Chem Res 2010;43:288–299.
63. Hanson RL, Davis BL, Chen Y, Goldberg SL, Parker WL, Tully TP, Montana MA, Patel RN. Adv Synth Catal 2008;350:1367–1375.
64. Domínguez de María P, Stillger T, Pohl M, Kiesel M, Liese A, Gröger H, Trauthwein H. Adv Synth Catal 2008;350:165–173.
65. Svitel J, Sturdik E. Enzyme Microb Technol 1995;17:546–550.
66. Molinari F, Villa R, Aragozzini F, Cabella P, Barbeni M, Squarcia F. J Chem Technol Biotechnol 1997;70:294–298.
67. Gandolfi R, Cavenago K, Gualandris R, Sinisterra JV, Molinari F. Process Biochem 2004;39:747–751.
68. Molinari F, Gandolfi R, Villa R, Urban E, Kiener A. Tetrahedron: Asymmetry 2003;14:2041–2043.

69. Takahashi E, Nakamichi K, Furui M. J Ferment Bioeng 1995;80:247–250.
70. van Beilen JB, Duetz WA, Schmid A, Witholt B. Trends Biotechnol 2003;21: 170–177.
71. Dingler C, Ladner W, Krei GA, Cooper B, Hauer B. Pesticide Science 1996;46:33–35.
72. Shibasaki T, Hashimoto S, Mori H, Ozakf A. J Biosci Bioeng 2000;90:522–525.
73. Wong JW, Watson HA, Bouressa JF, Burns MP, Cawley JJ, Doro AE, Guzek DB, Hintz MA, McCormick EL, Scully DA, Siderewicz JM, Taylor WJ, Truesdell SJ, Wax RG. Org Proc Res Dev 2002;6:477–481.
74. Serizawa N, Matsuoka T. Biochim Biophys Acta 1991;1084:35–40.
75. Berry A, Dodge TC, Pepsin M, Weyler W. J Ind Microbiol Biotechnol 2002;28: 127–133.
76. Ro DK, Paradise EM, Quellet M, Fischer KJ, Newman KL, Ndungu JM, Ho KA, Eachus RA, Ham TS, Kirby J, Chang MCY, Withers ST, Shiba Y, Sarpong R, Keasling JD. Nature 2006;440:940–943.
77. Yamada H, Kobayashi M. Biosci Biotechnol Biochem 1996;60:1391–1400.
78. Panova A, Mersinger L, Liu Q, Foo T, Roe D, Spillan W, Sigmund A, Ben-Bassat A, Wagner L, O'Keefe D, Wu S, Petrillo K, Payne M, Breske S, Gallagher F, DiCossimo R. Adv Synth Catal 2007;349:1462–1474.
79. Koopman F, Wierckx N, de Winde JH, Ruijssenaars HJ. Bioresour Technol 2010;101:6291–6296.
80. Reiner AM, Hegeman GD. Biochemistry 1971;10:2530–2536.
81. Jenkins GN, Ribbons DW, Widdowson DA, Slawin AMZ, Williams DJ, J Chem Soc, Perkin Trans 1 1995, 2647-2655.
82. Engesser K-H, Schmidt E, Knackmuss H-J. Appl Environ Microbiol 1980;39:68–73.
83. Myers AG, Siegel DR, Buzard DJ, Charest MG. Org Lett 2001;3:2923–2926.
84. Pilgrim S, Kociok-Köhm G, Lloyd MD, Lewis SE. Chem Commun 2011;47:4799–4801.
85. Parker MH, Maryanoff BE, Reitz AB. Synlett 2004;12:2095–2098.
86. Palframan MJ, Kociok-Köhm G, Lewis SE. Org Lett 2011;13:3150–3153.
87. Mihovilovic MD, Leisch HG, Mereiter K. Tetrahedron Lett 2004;45:7087–7090.
88. Banwell MG, Edwards AJ, Lupton DW, Whited G. Aust J Chem 2005;58:14–17.
89. Griffen JA, Le Coz AM, Kociok-Köhn G, Khan MA, Stewart AJW, Lewis SE. Org Biomol Chem 2011;9:3920–3928.
90. Van Lanen SG, Reader JS, Swairjo MA, de Crecy-Lagard V, Lee B, Iwata-Reuyl D. Proc Natl Acad Sci. USA 2005;102:4264–4269.
91. Iwata-Reuyl D, Crecy-Lagard V, Van Lanen SG. US7364882. 2008.
92. Reader S, Metzgar D, Schimmel P, Crécy-Lagard V. J Biol Chem 2004;279:6280–6285.
93. Iwata-Reuyl D. Bioorg Chem 2003;31:24–43.
94. Iwata-Reuyl D. Curr Opin Chem Biol 2008;12:126–133.
95. Kim Y, Zhou M, Moy S, Morales J, Cunningham MA, Joachimiak A. J Mol Biol 2010;404:127–137.
96. Lee BWK, Van Lanen SG, Iwata-Reuyl D. Biochemistry, 2007;46:12844–12854.
97. Swairjo MA, Reddy RR, Lee B, Van Lanen SG, Brown S, de Crécy-Lagard V, Iwata-Reuyl D, Schimmel P, Acta Cryst. 2005;F61:945–948.

98. Gamenara D, Domínguez de María P. Biotechnol Adv 2009;27:278–285.
99. Vaijayanthi T, Chadka A. Tetrahedron: Asymmetry 2008;19:93–96.
100. Mitsukura K, Suzuki M, Tada K, Yoshida T, Nagasawa T. Org Biomol Chem 2010;8:4533–4535.
101. Schrittwieser JH, Resch V, Sattler JH, Lienhart W.-D, Durchschein K, Winkler A, Gruber K, Macheroux P, Kroutil W. Angew Chem Int Ed 2011;50:1068–1071.

INDEX

Note: Page numbers followed by f and t indicate figures and tables, respectively.

Redox Biocatalysis: Fundamentals and Applications, First Edition.
Daniela Gamenara, Gustavo A. Seoane, Patricia Saenz-Méndez, and Pablo Domínguez de María.

Printed and bound by CPI Group (UK) Ltd, Croydon, CR0 4YY
18/09/2026
14965010-0001